Trigonometry

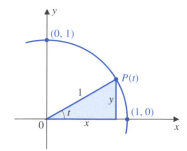

$$\sin t = y \qquad \cos t = x$$

$$\tan t = \frac{\sin t}{\cos t} \qquad \cot t = \frac{\cos t}{\sin t}$$

$$\sec t = \frac{1}{\cos t} \qquad \csc t = \frac{1}{\sin t}$$

$$(\sin t)^2 + (\cos t)^2 = 1$$

$$\sin t_1 \sin t_2 = \frac{1}{2}[\cos(t_1 - t_2) - \cos(t_1 + t_2)]$$

$$\sin(t_1 \pm t_2) = \sin t_1 \cos t_2 \pm \cos t_1 \sin t_2$$

$$\cos t_1 \cos t_2 = \frac{1}{2}[\cos(t_1 - t_2) + \cos(t_1 + t_2)]$$

$$\cos(t_1 \pm t_2) = \cos t_1 \cos t_2 \mp \sin t_1 \sin t_2$$

$$\sin t_1 \cos t_2 = \frac{1}{2}[\sin(t_1 - t_2) + \sin(t_1 + t_2)]$$

Law of Sines: $\quad \dfrac{\sin \alpha}{\alpha} = \dfrac{\sin \beta}{\beta} = \dfrac{\sin \gamma}{\gamma}$

Law of Cosines: $\quad c^2 = a^2 + b^2 - 2ab \cos \gamma$

Common Series

$$\sin t = \sum_{n=0}^{\infty} \frac{(-1)^n t^{2n+1}}{(2n+1)!} = t - \frac{t^3}{3!} + \frac{t^5}{5!} - \cdots \qquad e^t = \sum_{n=0}^{\infty} \frac{t^n}{n!} = 1 + t + \frac{t^2}{2!} + \frac{t^3}{3!} + \cdots$$

$$\cos t = \sum_{n=0}^{\infty} \frac{(-1)^n t^{2n}}{(2n)!} = 1 - \frac{t^2}{2!} + \frac{t^4}{4!} - \cdots \qquad \frac{1}{1-t} = \sum_{n=0}^{\infty} t^n = 1 + t + t^2 + \cdots, \qquad |t| < 1$$

The Greek Alphabet

Alpha	A	α	Eta	H	η	Nu	N	ν	Tau	T	τ
Beta	B	β	Theta	Θ	θ	Xi	Ξ	ξ	Upsilon	Υ	υ
Gamma	Γ	γ	Iota	I	ι	Omicron	O	o	Phi	Φ	ϕ
Delta	Δ	δ	Kappa	K	κ	Pi	Π	π	Chi	X	χ
Epsilon	E	ϵ	Lambda	Λ	λ	Rho	P	ρ	Psi	Ψ	ψ
Zeta	Z	ζ	Mu	M	μ	Sigma	Σ	σ	Omega	Ω	ω

Numerical Methods

Numerical Methods

FOURTH EDITION

J. Douglas Faires
Youngstown State University

Richard Burden
Youngstown State University

BROOKS/COLE
CENGAGE Learning™

Australia • Brazil • Japan • Korea • Mexico • Singapore • Spain • United Kingdom • United States

BROOKS/COLE
CENGAGE Learning™

Numerical Methods,
Fourth Edition
J. Douglas Faires and Richard Burden

Vice President, Editorial Director: *P.J. Boardman*

Publisher: *Richard Stratton*

Senior Sponsoring Editor: *Molly Taylor*

Assistant Editor: *Shaylin Walsh Hogan*

Editorial Assistant: *Alexander Gontar*

Associate Media Editor: *Andrew Coppola*

Senior Marketing Manager: *Jennifer Pursley Jones*

Marketing Coordinator: *Michael Ledesma*

Marketing Communications Manager: *Mary Anne Payumo*

Content Project Manager: *Jill Quinn*

Art Director: *Linda May*

Manufacturing Planner: *Doug Bertke*

Rights Acquisition Specialist: *Shalice Shah-Caldwell*

Production Service: *Cenveo Publisher Services*

Cover Designer: *Wing Ngan*

Cover Image: *AKIRA INOUE/Getty Images*

Compositor: *Cenveo Publisher Services*

For product information and technology assistance, contact us at
Cengage Learning Customer & Sales Support,
1-800-354-9706

For permission to use material from this text or product,
submit all requests online at
www.cengage.com/permissions.
Further permissions questions can be emailed to
permissionrequest@cengage.com.

Library of Congress Control Number: 2012935435

ISBN-13: 978-0-495-11476-5
ISBN-10: 0-495-11476-6

Brooks/Cole
20 Channel Center Street
Boston, MA 02210
US

Cengage Learning is a leading provider of customized learning solutions with office locations around the globe, including Singapore, the United Kingdom, Australia, Mexico, Brazil and Japan. Locate your local office at **international.cengage.com/region**

Cengage Learning products are represented in Canada by Nelson Education, Ltd.

For your course and learning solutions, visit
www.cengage.com.
Purchase any of our products at your local college store or at our preferred online store **www.cengagebrain.com**.

Instructors: Please visit **login.cengage.com** and log in to access instructor-specific resources.

Printed in the United States of America
1 2 3 4 5 6 7 16 15 14 13 12

Contents

Preface

About the Text

The teaching of numerical approximation techniques to undergraduates is done in a variety of ways. The traditional Numerical Analysis course discusses approximation methods, and provides mathematical justification for those methods. A Numerical Methods course emphasizes the choice and application of techniques to solve problems in engineering and the physical sciences over the derivation of the methods.

The books used in Numerical Methods courses differ widely in both intent and content. Sometimes a book written for Numerical Analysis is adapted for a Numerical Methods course by deleting the more theoretical topics and derivations. The advantage of this approach is that the leading Numerical Analysis books are mature; they have been through a number of editions, and they have a wealth of proven examples and exercises. They are also written for a full year's coverage of the subject, so they have methods that can be used for reference even when there is not sufficient time for discussing them in the course. The weakness of using a Numerical Analysis book for a Numerical Methods course is that material will need to be omitted, and students can have difficulty distinguishing what is important from what is tangential.

The second type of book used for a Numerical Methods course is one that is specifically written for a service course. These books follow the established line of service-oriented mathematics books, similar to the technical calculus books written for students in business and the life sciences, and the statistics books designed for students in economics, psychology, and business. However, the engineering and science students for whom the Numerical Methods course is designed have a much stronger mathematical background than students in other disciplines. They are quite capable of mastering the material in a Numerical Analysis course, but they do not have the time for—nor the interest in—the theoretical aspects of such a course. What they need is a sophisticated introduction to the approximation techniques used to solve the problems that arise in science and engineering. They also need to know why the methods work, what type of errors to expect, and when a method might lead to difficulties. Finally, they need information, with recommendations, regarding the availability of high-quality software for numerical approximation routines. In such a course the mathematical analysis is reduced due to a lack of time, not because of the mathematical abilities of the students.

The emphasis in this edition of *Numerical Methods* is on the intelligent application of approximation techniques to the type of problems that commonly occur in engineering and the physical sciences. The book is designed for a one-semester course, but contains at least 50% more material than is likely to be covered, so instructors have flexibility in topic coverage, and students have a reference for future work. The techniques covered are essentially the same as those included in our book designed for the Numerical Analysis course (See *Numerical Analysis, 9e*). However, the emphasis in the two books is quite different. In *Numerical Analysis*, a book with more than 800 text pages, each technique is given a mathematical justification before the implementation of the method is discussed. If some portion of the justification is beyond the mathematical level of the book, then it is referenced, but the book is, for the most part, mathematically self-contained.

In *Numerical Methods*, each technique is motivated and described from an implementation standpoint. The aim of the motivation is to convince the student that the method is reasonable both mathematically and computationally. A full mathematical justification is included only if it is concise and adds to the understanding of the method.

A number of software packages are available to produce symbolic and numerical computations. Predominant among the items for sale are Maple®, *Mathematica*®, and MATLAB®. In addition, Sage, a free open-source mathematical system licensed under the GNU Public License, can be very useful for a student of numerical techniques. Sage connects either locally to your own Sage installation or to a Sage server on the network. Information about this system can be found at http://www.sagemath.org.

There are several versions of the software packages for most common computer systems, and student versions are generally available. Although the packages differ in philosophy, packaging, and price, they all can be used to obtain accurate numerical approximations. So, having a package available can be very useful in the study of approximation techniques. The results in most of our examples and exercises have been generated using problems for which exact values can be determined because this permits the performance of the approximation method to be monitored. Exact solutions can often be obtained quite easily using the packages that perform symbolic computation.

In past editions we have used Maple as our standard package. In this edition we have changed to MATLAB because this is the software most frequently used by schools of engineering, where the course is now frequently being taught. We have added MATLAB examples and exercises, complete with M-files, whenever we felt that this system would be beneficial, and have discussed the approximation methods that MATLAB provides for applying a numerical technique.

Software is included with and is an integral part of this edition of *Numerical Methods*. Our website includes programs for each method discussed in C, FORTRAN, and Pascal, and a worksheet in Maple, *Mathematica*, and MATLAB. There are also Java applets for each of the programs. Previous exposure to one of these systems is valuable but not essential.

The programs permit students to generate all the results that are included in the examples and to modify the programs to generate solutions to problems of their choice. The intent of the software is to provide students with programs that will solve most of the problems that they are likely to encounter in their studies.

Occasionally, exercises in the text contain problems for which the programs do not give satisfactory solutions. These are included to illustrate the difficulties that can arise in the application of approximation techniques and to show the need for the flexibility provided by the standard general purpose software packages that are available for scientific computation. Information about the standard general purpose software packages is discussed in the text. Included are those in packages distributed by netlib, the International Mathematical and Statistical Library (IMSL), the National Algorithms Group (NAG), and the specialized techniques in EISPACK and LINPACK.

New for this Edition

We have substantially rewritten the fourth edition due to our decision to use MATLAB as our basic system for generating results. MATLAB is a collection of professional programs that can be used to solve many problems, including most problems requiring numerical techniques. In fact, MATLAB is the software package that most engineers and scientists will use in their professional careers. However, we do not find it as convenient to use for a teaching tool as Maple and *Mathematica*. In past editions, and in our *Numerical Analysis* book, we have used Maple to illustrate the steps in our numerical techniques because this system generally follows our algorithm structure very closely. Abandoning this system meant that we had to expand our discussion in many instances to ensure that students would follow all the required steps of the techniques we discuss.

In summary, this edition introduces the student to the techniques required for numerical approximation, describes how the professional software available in MATLAB approaches the solution to problems, and gives expanded details in the Examples and Illustrations that accompany the methods. MATLAB code is illustrated as it appears in that system wherever it is relevant, and the output that MATLAB provides is clearly documented in a condensed MATLAB style. Students who have read the material have had no difficulty implementing the procedures and generating our results.

In addition to the incorporation of MATLAB material, some of the most noticeable changes for the fourth edition are:

- Our treatment of Numerical Linear Algebra has been extensively expanded. We have added a section on the singular value decomposition at the end of Chapter 9. This required a complete rewrite and considerable expansion of the early part of this chapter to include more material on symmetric and orthogonal matrices. The chapter is approximately 40% longer than in the previous edition, and contains many new examples and exercises.

- All the Examples in the book have been rewritten to better emphasize the problem being solved before the solution is given. Additions have been made to the Examples to include the computations required for the first steps of iteration processes so that students can better follow the details of the techniques.

- New Illustrations have been added where appropriate to discuss a specific application of a method that is not suitable for the problem statement-solution format that the Examples now assume.

- A number of sections have been expanded, and some divided, to make it easier for instructors to assign problems immediately after the material is presented. This is particularly true in Chapter 9.

- Numerous new historical notes have been added, primarily in the margins where they can be considered independent of the text material. Much of the current material used in Numerical Methods was developed in middle of the 20th century. Students should be aware of this, and realize that this is an area of current interest.

- The bibliographic material has been updated to reflect new editions of books that we reference. New sources have been added that were not previously available.

As always with our revisions, every sentence was examined to determine if it was phrased in a manner that best relates what we are trying to describe.

We have also updated all the programming code to the latest releases that were available for each of the programming systems, and we will post updated versions of the Maple, *Mathematica*, and MATLAB at the book's website:

http://www.math.ysu.edu/~faires/Numerical-Methods

Supplements

Student Study Guide

A *Student Study Guide* is available with this edition and contains worked-out solutions to many of the problems. The first two chapters of this *Guide* are available on the website for the book in PDF format so that prospective users can tell if they find it sufficiently useful to justify the purchase of the *Guide*. The authors do not have the remaining *Guide* material for the remaining chapters available in this format, however. These can only be obtained from the publisher at www.cengagebrain.com.

Instructor's Manual

The publisher can provide instructors with an *Instructor's Manual* that provides answers and solutions to all the exercises in the book. Computation results in the *Instructor's Manual* were regenerated for this edition using the programs on the website to ensure compatibility among the various programming systems.

SolutionBuilder

This online instructor database offers complete solutions to all exercises in the text, allowing you to create customized, secure solutions printouts (in PDF format) matched exactly to the problems you assign in class. Sign up for access at www.cengage.com/SolutionBuilder.

Presentation Material

We are particularly excited about a set of classroom lecture slides prepared by Professor John Carroll of Dublin City University, which are designed to accompany the presentations in the book. These slides present examples, hints, and step-by-step animations of important techniques in Numerical Methods. They are available on the website for the book:

http://www.math.ysu.edu/~faires/Numerical-Methods

The slides were created using the Beamer package of LaTeX, and are in PDF format.

Possible Course Suggestions

Numerical Methods is designed to allow instructors flexibility in the choice of topics, as well as in the level of theoretical rigor and in the emphasis on applications. In line with these aims, we provide references for many of the results that are not demonstrated in the text and for the applications that are used to indicate the practical importance of the methods. The text references cited are those most likely to be available in college libraries and have been updated to reflect recent editions. All referenced material has been indexed to the appropriate locations in the text, and Library of Congress call information for reference material has been included to permit easy location if searching for library material.

The following flowchart indicates chapter prerequisites. Most of the possible sequences that can be generated from this chart have been taught by the authors at Youngstown State University.

The additional material in this edition should permit instructors to prepare an under-graduate course in Numerical Linear Algebra for students who have not previously studied Numerical Methods or Numerical Analysis. This could be done by covering Chapters 1, 6, 7, and 9.

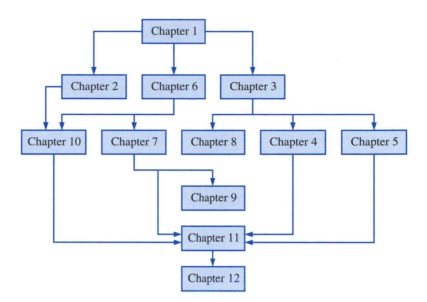

Acknowledgments

We have been fortunate to have had many of our students and colleagues give us their impressions of earlier editions of this book, and those of our other book, *Numerical Analysis*. We very much appreciate this effort and take all of these comments and suggestions very seriously. We have tried to include all the suggestions that complement the philosophy of the book, and are extremely grateful to all those who have taken the time to contact us about ways to improve subsequent versions.

We would particularly like to thank the following, whose suggestions we have used in this and previous editions.

- John Carroll–Dublin City University

- Willian Duncan–Louisiana State University

- Saroj Kumar Sahani–Birla Institute of Techonology & Science

- Misha Shvartsman–St. Thomas University

- Dale Smith–Bridgewater State University

- Dennis Smolarski–Santa Clara University

- Emel Yavuz–Istanbul Kultur University

In addition, we would like to thank the faculty of the Department of Mathematics at Youngstown State University for being so supportive of our work over the years. Even though we have now been retired for the better part of a decade, we are still treated as regular colleagues, albeit without the onus of committee work. Finally, we would like to thank our two student assistants, Jena Baun and Ashley Bowers, who did excellent work with much of the tedious details of manuscript presentation. They admirably followed in the footsteps of so many excellent students we have had the pleasure to work with over the years.

Mathematical Preliminaries and Error Analysis

1.1 Introduction

This book examines problems that can be solved by methods of approximation, techniques called *numerical methods*. We begin by considering some of the mathematical and computational topics that arise when approximating a solution to a problem. Nearly all the problems whose solutions can be approximated involve continuous functions, so calculus is the principal tool to use for deriving numerical methods and verifying that they solve the problems. The calculus definitions and results included in the next section provide a handy reference when these concepts are needed later in the book.

There are two things to consider when applying a numerical technique. The first and most obvious is to obtain the approximation. The equally important second objective is to determine a safety factor for the approximation: some assurance, or at least a sense, of the accuracy of the approximation. Sections 1.3 and 1.4 deal with a standard difficulty that occurs when applying techniques to approximate the solution to a problem:

- Where and why is computational error produced and how can it be controlled?

The final section in this chapter describes various types and sources of mathematical software for implementing numerical methods.

1.2 Review of Calculus

Limits and Continuity

The limit of a function at a specific number tells, in essence, what the function values approach as the numbers in the domain approach the specific number. The limit concept is basic to calculus, and the major developments of calculus were discovered in the latter part of the seventeenth century, primarily by Isaac Newton and Gottfried Leibnitz. However, it was not until 200 years later that Augustus Cauchy, based on work of Karl Weierstrass, first expressed the limit concept in the form we now use.

We say that a function f defined on a set X of real numbers has the **limit** L at x_0, written $\lim_{x \to x_0} f(x) = L$, if, given any real number $\varepsilon > 0$, there exists a real number $\delta > 0$ such that $|f(x) - L| < \varepsilon$ whenever $0 < |x - x_0| < \delta$. This definition ensures that values of the function will be close to L whenever x is sufficiently close to x_0. (See Figure 1.1.)

Figure 1.1

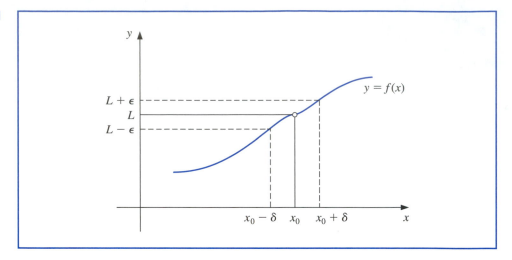

A function is said to be continuous at a number in its domain when the limit at the number agrees with the value of the function at the number. So a function f is **continuous** at x_0 if $\lim_{x \to x_0} f(x) = f(x_0)$.

A function f is **continuous on the set** X if it is continuous at each number in X. We use $C(X)$ to denote the set of all functions that are continuous on X. When X is an interval of the real line, the parentheses in this notation are omitted. For example, the set of all functions that are continuous on the closed interval $[a, b]$ is denoted $C[a, b]$.

The limit of a sequence of real or complex numbers is defined in a similar manner. An infinite sequence $\{x_n\}_{n=1}^{\infty}$ **converges** to a number x if, given any $\varepsilon > 0$, there exists a positive integer $N(\varepsilon)$ such that $|x_n - x| < \varepsilon$ whenever $n > N(\varepsilon)$. The notation $\lim_{n \to \infty} x_n = x$, or $x_n \to x$ as $n \to \infty$, means that the sequence $\{x_n\}_{n=1}^{\infty}$ converges to x.

Continuity and Sequence Convergence

If f is a function defined on a set X of real numbers and $x_0 \in X$, then the following are equivalent:

a. f is continuous at x_0.

b. If $\{x_n\}_{n=1}^{\infty}$ is any sequence in X converging to x_0, then

$$\lim_{n \to \infty} f(x_n) = f(x_0).$$

All the functions we consider when discussing numerical methods are continuous because this is a minimal requirement for predictable behavior. Functions that are not continuous can skip over points of interest, which can cause difficulties when we attempt to approximate a solution to a problem.

More sophisticated assumptions about a function generally lead to better approximation results. For example, a function with a smooth graph would normally behave more predictably than would one with numerous jagged features. Smoothness relies on the concept of the derivative.

Differentiability

If f is a function defined in an open interval containing x_0, then f is **differentiable** at x_0 when

$$f'(x_0) = \lim_{x \to x_0} \frac{f(x) - f(x_0)}{x - x_0}$$

exists. The number $f'(x_0)$ is called the **derivative** of f at x_0. The derivative of f at x_0 is the slope of the tangent line to the graph of f at $(x_0, f(x_0))$, as shown in Figure 1.2.

Figure 1.2

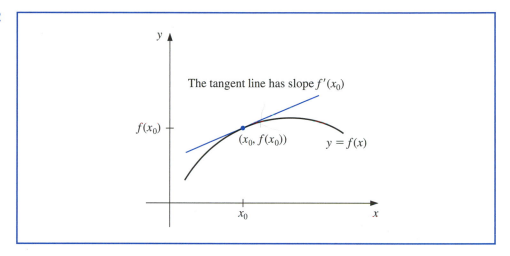

A function that has a derivative at each number in a set X is **differentiable** on X. Differentiability is a stronger condition on a function than continuity in the following sense.

Differentiability Implies Continuity

If the function f is differentiable at x_0, then f is continuous at x_0.

The set of all functions that have n continuous derivatives on X is denoted $C^n(X)$, and the set of functions that have derivatives of all orders on X is denoted $C^\infty(X)$. Polynomial, rational, trigonometric, exponential, and logarithmic functions are in $C^\infty(X)$, where X consists of all numbers at which the function is defined.

The next results are of fundamental importance in deriving methods for error estimation. The proofs of most of these can be found in any standard calculus text.

Mean Value Theorem

If $f \in C[a, b]$ and f is differentiable on (a, b), then a number c in (a, b) exists such that (see Figure 1.3)

$$f'(c) = \frac{f(b) - f(a)}{b - a}.$$

Figure 1.3

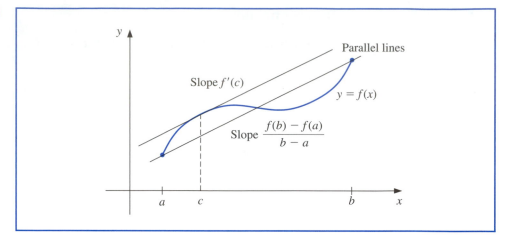

The following result is frequently used to determine bounds for error formulas.

Extreme Value Theorem

If $f \in C[a, b]$, then c_1 and c_2 in $[a, b]$ exist with $f(c_1) \leq f(x) \leq f(c_2)$ for all x in $[a, b]$. If, in addition, f is differentiable on (a, b), then the numbers c_1 and c_2 occur either at endpoints of $[a, b]$ or where f' is zero.

The values where a continuous function has its derivative 0 or where the derivative does not exist are called *critical points* of the function. So the Extreme Value Theorem states that a maximum or minimum value of a continuously differentiable function on a closed interval can occur only at the critical points or the endpoints.

Our first example gives some illustrations of applications of the Extreme Value Theorem and MATLAB.

Example 1   Use MATLAB to find the absolute minimum and absolute maximum values of

$$f(x) = 5\cos 2x - 2x \sin 2x$$

on the intervals **(a)** [1, 2], and **(b)** [0.5, 1].

Solution   The solution to this problem is one that is commonly needed in calculus. It provides a good example for illustrating some commonly used commands in MATLAB and the response to the commands that MATLAB gives. In our presentations of MATLAB material, input statements appear left-justified using a `typewriter-like` font. To add emphasis to the responses from MATLAB, these appear centered and in cyan type.

For better readability, we will delete the ≫ symbols needed for input statements as well as the blank lines from MATLAB responses. Other than these changes, the statements will agree with that of MATLAB.

The following command defines $f(x) = 5\cos 2x - 2x \sin 2x$ as a function of x.

```
f = inline('5*cos(2*x)-2*x*sin(2*x)','x')
```

and MATLAB responds with (actually, the response is on two separate lines, but we will compress the MATLAB responses, here and throughout)

Inline function: $f(x) = 5 * cos(2 * x) - 2 * x * sin(2 * x)$

We have now defined our base function $f(x)$. The x in the command indicates that x is the argument of the function f.

To find the absolute minimum and maximum values of $f(x)$ on the given intervals, we also need its derivative $f'(x)$, which is

$$f'(x) = -12 \sin 2x - 4x \cos 2x.$$

Then we define the function $fp(x) \equiv f'(x)$ in MATLAB to represent the derivative with the inline command

```
fp = inline('-12*sin(2*x)-4*x*cos(2*x)','x')
```

By default, MATLAB displays only a five-digit result, as illustrated by the following command which computes $f(0.5)$:

```
f(0.5)
```

The result from MATLAB is

$ans = 1.8600$

We can increase the number of digits of display with the command

```
format long
```

Then the command

```
f(0.5)
```

produces

$ans = 1.860040544532602$

We will use this extended precision version of MATLAB output in the remainder of the text.

(a) The absolute minimum and maximum of the continuously differentiable function f occur only at the endpoints of the interval $[1, 2]$ or at a critical point within this interval. We obtain the values at the endpoints with

```
f(1),f(2)
```

and MATLAB responds with

$ans = -3.899329036387075, \quad ans = -0.241008123086347$

To determine critical points of the function f, we need to find zeros of $f'(x)$. For this we use the `fzero` command in MATLAB:

```
p =fzero(fp,[1,2])
```

and MATLAB responds with

$$p = 1.358229873843064$$

Evaluating f at this single critical point with

```
f(p)
```

gives

$$ans = -5.675301337592883$$

In summary, the absolute minimum and absolute maximum values of $f(x)$ on the interval $[1, 2]$ are approximately

$$f(1.358229873843064) = -5.675301337592883 \quad \text{and} \quad f(2) = -0.241008123086347.$$

(b) When the interval is $[0.5, 1]$ we have the values at the endpoints given by

$$f(0.5) = 5\cos 1 - 1\sin 1 = 1.860040544532602 \quad \text{and}$$
$$f(1) = 5\cos 2 - 2\sin 2 = -3.899329036387075.$$

However, when we attempt to determine critical points in the interval $[0.5, 1]$ with the command

```
p1 = fzero(fp,[0.5 1])
```

MATLAB returns the response

??? Error using ==> fzero at 293

This indicates that MATLAB could not find a solution to this equation, which is the correct response because f is strictly decreasing on $[0.5, 1]$ and no solution exists. Hence the approximate absolute minimum and absolute maximum values on the interval $[0.5, 1]$ are

$$f(1) = -3.899329036387075 \quad \text{and} \quad f(0.5) = 1.860040544532602. \qquad ■$$

The following five commands plot the function on the interval $[0.5, 2]$ with titles for the graph and axes on a grid.

```
fplot(f,[0.5 2])
title('Plot of f(x)')
xlabel('Values of x')
ylabel('Values of f(x)')
grid
```

Figure 1.4 shows the screen that results from these commands. They confirm the results we obtained in Example 1. The graph is displayed in a window that can be saved in a variety of forms for use in technical presentations.

Figure 1.4

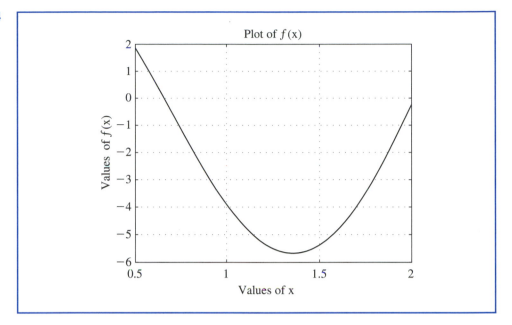

The next result is the Intermediate Value Theorem. Although its statement is not diffi-cult, the proof is beyond the scope of the usual calculus course.

Intermediate Value Theorem

If $f \in C[a, b]$ and K is any number between $f(a)$ and $f(b)$, then there exists a number c in (a, b) for which $f(c) = K$. (Figure 1.5 shows one of the three possibilities for this function and interval.)

Figure 1.5

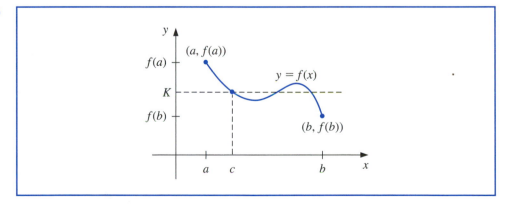

Example 2   Show that $x^5 - 2x^3 + 3x^2 - 1 = 0$ has a solution in the interval $[0, 1]$.

Solution   Consider the function defined by $f(x) = x^5 - 2x^3 + 3x^2 - 1$. The function f is continuous on $[0, 1]$. In addition,

$$f(0) = -1 < 0 \qquad \text{and} \qquad 0 < 1 = f(1).$$

The Intermediate Value Theorem implies that a number x exists in $(0, 1)$ with $x^5 - 2x^3 + 3x^2 - 1 = 0$.   ■

As seen in Example 2, the Intermediate Value Theorem is used to help determine when solutions to certain problems exist. It does not, however, give an efficient means for finding these solutions. This topic is considered in Chapter 2.

Integration

The integral is the other basic concept of calculus. The **Riemann integral** of the function f on the interval $[a, b]$ is the following limit, provided it exists:

$$\int_a^b f(x)\, dx = \lim_{\max \Delta x_i \to 0} \sum_{i=1}^n f(z_i)\, \Delta x_i,$$

where the numbers $x_0, x_1, \ldots, x_n$ satisfy $a = x_0 < x_1 < \cdots < x_n = b$ and where $\Delta x_i = x_i - x_{i-1}$, for each $i = 1, 2, \ldots, n$, and z_i is arbitrarily chosen in the interval $[x_{i-1}, x_i]$.

A function f that is continuous on an interval $[a, b]$ is also Riemann integrable on $[a, b]$. This permits us to choose, for computational convenience, the points x_i to be equally spaced in $[a, b]$ and for each $i = 1, 2, \ldots, n$, to choose $z_i = x_i$. In this case,

$$\int_a^b f(x)\, dx = \lim_{n \to \infty} \frac{b - a}{n} \sum_{i=1}^n f(x_i),$$

where the numbers shown in Figure 1.6 as x_i are $x_i = a + (i(b - a)/n)$.

Figure 1.6

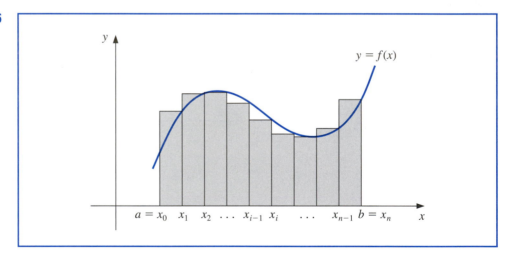

Two more basic results are needed in our study of numerical methods. The first is a generalization of the usual Mean Value Theorem for Integrals.

Mean Value Theorem for Integrals

If $f \in C[a, b]$, g is integrable on $[a, b]$, and $g(x)$ does not change sign on $[a, b]$, then there exists a number c in (a, b) with

$$\int_a^b f(x)g(x)\, dx = f(c) \int_a^b g(x)\, dx.$$

When $g(x) \equiv 1$, this result reduces to the usual Mean Value Theorem for Integrals. It gives the **average value** of the function f over the interval $[a, b]$ as

$$f(c) = \frac{1}{b - a} \int_a^b f(x) \, dx.$$

(See Figure 1.7.)

Figure 1.7

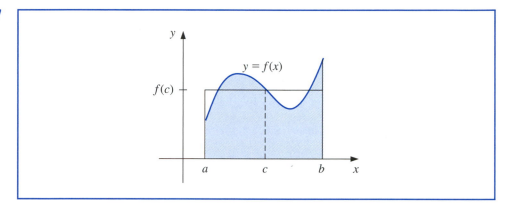

Taylor Polynomials and Series

The final result in this review from calculus describes the development of the Taylor polynomials. The importance of the Taylor polynomials to the study of numerical analysis cannot be overemphasized, and the following result is used repeatedly.

Taylor's Theorem

Suppose $f \in C^n[a, b]$ and $f^{(n+1)}$ exists on $[a, b]$. Let x_0 be a number in $[a, b]$. For every x in $[a, b]$, there exists a number $\xi(x)$ between x_0 and x with

$$f(x) = P_n(x) + R_n(x),$$

where

$$P_n(x) = f(x_0) + f'(x_0)(x - x_0) + \frac{f''(x_0)}{2!}(x - x_0)^2 + \cdots + \frac{f^{(n)}(x_0)}{n!}(x - x_0)^n$$

$$= \sum_{k=0}^{n} \frac{f^{(k)}(x_0)}{k!}(x - x_0)^k$$

and

$$R_n(x) = \frac{f^{(n+1)}(\xi(x))}{(n + 1)!}(x - x_0)^{n+1}.$$

Here $P_n(x)$ is called the **nth Taylor polynomial** for f about x_0, and $R_n(x)$ is called the **truncation error** (or *remainder term*) associated with $P_n(x)$. The number $\xi(x)$ in the truncation error $R_n(x)$ depends on the value of x at which the polynomial $P_n(x)$ is being evaluated, so it is actually a function of the variable x. However, we should not expect to

Brook Taylor (1685–1731) described this series in 1715 in the paper *Methodus incrementorum directa et inversa.* Special cases of the result, and likely the result itself, had been previously known to Isaac Newton, James Gregory, and others.

be able to explicitly determine the function $\xi(x)$. Taylor's Theorem simply ensures that such a function exists, and that its value lies between x and x_0. In fact, one of the common problems in numerical methods is to try to determine a realistic bound for the value of $f^{(n+1)}(\xi(x))$ for values of x within some specified interval.

The infinite series obtained by taking the limit of $P_n(x)$ as $n \to \infty$ is called the *Taylor series* for f about x_0. The term *truncation error* in the Taylor polynomial refers to the error involved in using a truncated (that is, finite) summation to approximate the sum of an infinite series.

In the case $x_0 = 0$, the Taylor polynomial is often called a **Maclaurin polynomial**, and the Taylor series is called a *Maclaurin series*.

Example 3   Let $f(x) = \cos x$ and $x_0 = 0$. Determine

(a)   the second Taylor polynomial for f about x_0; and

(b)   the third Taylor polynomial for f about x_0.

Solution   Since $f \in C^\infty(\mathbb{R})$, Taylor's Theorem can be applied for any $n \geq 0$. Also,

$$f'(x) = -\sin x, \ \ f''(x) = -\cos x, \ \ f'''(x) = \sin x, \quad \text{and} \quad f^{(4)}(x) = \cos x,$$

so

$$f(0) = 1, \ \ f'(0) = 0, \ \ f''(0) = -1, \quad \text{and} \quad f'''(0) = 0.$$

(a) For $n = 2$ and $x_0 = 0$, we have

$$\cos x = f(0) + f'(0)x + \frac{f''(0)}{2!}x^2 + \frac{f'''(\xi(x))}{3!}x^3$$

$$= 1 - \frac{1}{2}x^2 + \frac{1}{6}x^3 \sin \xi(x),$$

where $\xi(x)$ is some (generally unknown) number between 0 and x. (See Figure 1.8.)

Figure 1.8

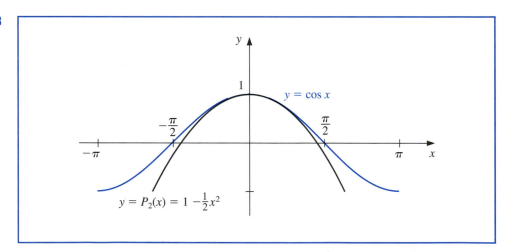

y

1

$y = \cos x$

$-\frac{\pi}{2}$ $\frac{\pi}{2}$

$-\pi$ π x

$y = P_2(x) = 1 - \frac{1}{2}x^2$

When $x = 0.01$, this becomes

$$\cos 0.01 = 1 - \frac{1}{2}(0.01)^2 + \frac{1}{6}(0.01)^3 \sin \xi(0.01) = 0.99995 + \frac{10^{-6}}{6} \sin \xi(0.01).$$

The approximation to $\cos 0.01$ given by the Taylor polynomial is therefore 0.99995. The truncation error, or remainder term, associated with this approximation is

$$\frac{10^{-6}}{6} \sin \xi(0.01) = 0.1\overline{6} \times 10^{-6} \sin \xi(0.01),$$

where the bar over the 6 in $0.1\overline{6}$ is used to indicate that this digit repeats indefinitely. Although we have no way of determining $\sin \xi(0.01)$, we know that all values of the sine lie in the interval $[-1, 1]$, so a bound for the error occurring if we use the approximation 0.99995 for the value of $\cos 0.01$ is

$$|\cos(0.01) - 0.99995| = 0.1\overline{6} \times 10^{-6} |\sin \xi(0.01)| \leq 0.1\overline{6} \times 10^{-6}.$$

Hence the approximation 0.99995 matches at least the first five digits of $\cos 0.01$, and

$$0.9999483 < 0.99995 - 1.\overline{6} \times 10^{-6} \leq \cos 0.01$$

$$\leq 0.99995 + 1.\overline{6} \times 10^{-6} < 0.9999517.$$

The error bound is much larger than the actual error. This is due in part to the poor bound we used for $|\sin \xi(x)|$. It is shown in Exercise 16 that for all values of x, we have $|\sin x| \leq |x|$. Since $0 \leq \xi < 0.01$, we could have used the fact that $|\sin \xi(x)| \leq 0.01$ in the error formula, producing the bound $0.1\overline{6} \times 10^{-8}$.

(b) Since $f'''(0) = 0$, the third Taylor polynomial with remainder term about $x_0 = 0$ has no x^3 term. It is

$$\cos x = 1 - \frac{1}{2}x^2 + \frac{1}{24}x^4 \cos \tilde{\xi}(x),$$

where $0 < \tilde{\xi}(x) < 0.01$. The approximating polynomial remains the same, and the approximation is still 0.99995, but we now have much better accuracy assurance. Since $|\cos \tilde{\xi}(x)| \leq 1$ for all x, we have

$$\left| \frac{1}{24}x^4 \cos \tilde{\xi}(x) \right| \leq \frac{1}{24}(0.01)^4(1) \approx 4.2 \times 10^{-10}.$$

So

$$|\cos 0.01 - 0.99995| \leq 4.2 \times 10^{-10},$$

and

$$0.99994999958 = 0.99995 - 4.2 \times 10^{-10}$$

$$\leq \cos 0.01 \leq 0.99995 + 4.2 \times 10^{-10} = 0.99995000042. \qquad \blacksquare$$

Example 3 illustrates the two basic objectives of numerical methods:

- Find an approximation to the solution of a given problem.

- Determine a bound for the accuracy of the approximation.

The second and third Taylor polynomials gave the same result for the first objective, but the third Taylor polynomial gave a much better result for the second objective.

Illustration   We can also use the third Taylor polynomial and its remainder term found in Example 3 to approximate $\int_0^{0.1} \cos x \, dx$. We have

$$\int_0^{0.1} \cos x \, dx = \int_0^{0.1} \left(1 - \frac{1}{2} x^2 \right) dx + \frac{1}{24} \int_0^{0.1} x^4 \cos \tilde{\xi}(x) \, dx$$

$$= \left[x - \frac{1}{6} x^3 \right]_0^{0.1} + \frac{1}{24} \int_0^{0.1} x^4 \cos \tilde{\xi}(x) \, dx$$

$$= 0.1 - \frac{1}{6} (0.1)^3 + \frac{1}{24} \int_0^{0.1} x^4 \cos \tilde{\xi}(x) \, dx.$$

Therefore

$$\int_0^{0.1} \cos x \, dx \approx 0.1 - \frac{1}{6} (0.1)^3 = 0.0998\overline{3}.$$

A bound for the error in this approximation is determined from the integral of the Taylor remainder term and the fact that $|\cos \tilde{\xi}(x)| \leq 1$ for all x:

$$\frac{1}{24} \left| \int_0^{0.1} x^4 \cos \tilde{\xi}(x) \, dx \right| \leq \frac{1}{24} \int_0^{0.1} x^4 |\cos \tilde{\xi}(x)| \, dx$$

$$\leq \frac{1}{24} \int_0^{0.1} x^4 \, dx = \frac{(0.1)^5}{120} = 8.\overline{3} \times 10^{-8}.$$

The true value of this integral is

$$\int_0^{0.1} \cos x \, dx = \sin x \bigg]_0^{0.1} = \sin 0.1 \approx 0.099833416647,$$

so the actual error for this approximation is 8.3314×10^{-8}, which is within the error bound.   □

    MATLAB can be used to obtain these results by first defining $f(x) = \cos x$ and the second Taylor polynomial $T2(x) \equiv T_2(x) = 1 - \frac{1}{2} x^2$ with

```
f = inline('cos(x)','x')
T2 = inline('1-0.5.*x.^ 2','x')
```

The next commands evaluate f at 0.01, $T2$ at 0.01 and compute the error in approximating $\cos(0.01)$ with the $T_2(0.01)$.

```
y1 = f(0.01), y2 = T2(0.01), err1 = abs(y1-y2)
```

giving

$$y1 = 0.999950000416665, \quad y2 = 0.999950000000000,$$

$$err1 = 4.166652578518892e - 010$$

To obtain a graph similar to Figure 1.8 requires creating an M-file which can load more than one command at a time. We need this in order to define both $f(x)$ and $T2(x)$ if we want to plot them both on the same graph.

An M-file is created by selecting File on the MATLAB toolbar. Then select New and Script. The three statements are entered and the result is saved as a file named ourfunction1. The M-file consists of the following commands:

```
function Y = ourfunction1(x)
Y(:,1)=cos(x(:));
Y(:,2)=1-0.5.*x(:).^2;
```

From the worksheet, we need to enter the following command to create a reference to the function M-file:

```
fh = @ourfunction1
```

The response from MATLAB is

$$fh = @ourfunction1$$

The graph shown in Figure 1.9 can then be created with the commands

```
fplot(fh,[-pi pi])
grid
```

Figure 1.9

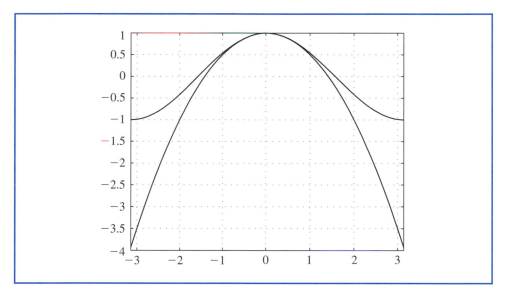

We can also compute the integrals of $f(x)$ and the Taylor polynomial $T_2(x)$ on the interval $[0, 0.1]$ using MATLAB. We use the commands

```
q1 = quad('cos(x)',0,0.1)
q2 = quad('1-0.5*x.^2',0,0.1)
```

and MATLAB produces

$$q1 = 0.099833416646828, \quad q2 = 0.099833333333333$$

EXERCISE SET 1.2

1.   Show that the following equations have at least one solution in the given intervals.
   a. $x \cos x - 2x^2 + 3x - 1 = 0$,   [0.2, 0.3] and [1.2, 1.3]
   b. $(x - 2)^2 - \ln x = 0$,   [1, 2] and [e, 4]
   c. $2x \cos(2x) - (x - 2)^2 = 0$,   [2, 3] and [3, 4]
   d. $x - (\ln x)^x = 0$,   [4, 5]

2.   Find intervals containing solutions to the following equations.
   a. $x - 3^{-x} = 0$
   b. $4x^2 - e^x = 0$
   c. $x^3 - 2x^2 - 4x + 3 = 0$
   d. $x^3 + 4.001x^2 + 4.002x + 1.101 = 0$

3.   Show that the first derivatives of the following functions are zero at least once in the given intervals.
   a. $f(x) = 1 - e^x + (e - 1) \sin((\pi/2)x)$,   [0, 1]
   b. $f(x) = (x - 1) \tan x + x \sin \pi x$,   [0, 1]
   c. $f(x) = x \sin \pi x - (x - 2) \ln x$,   [1, 2]
   d. $f(x) = (x - 2) \sin x \ln(x + 2)$,   [−1, 3]

4.   Find $\max_{a \le x \le b} |f(x)|$ for the following functions and intervals.
   a. $f(x) = (2 - e^x + 2x)/3$,   [0, 1]
   b. $f(x) = (4x - 3)/(x^2 - 2x)$,   [0.5, 1]
   c. $f(x) = 2x \cos(2x) - (x - 2)^2$,   [2, 4]
   d. $f(x) = 1 + e^{-\cos(x-1)}$,   [1, 2]

5.   Let $f(x) = x^3$.
   a.   Find the second Taylor polynomial $P_2(x)$ about $x_0 = 0$.
   b.   Find $R_2(0.5)$ and the actual error when using $P_2(0.5)$ to approximate $f(0.5)$.
   c.   Repeat (a) with $x_0 = 1$.
   d.   Repeat (b) for the polynomial found in (c).

6.   Let $f(x) = \sqrt{x + 1}$.
   a.   Find the third Taylor polynomial $P_3(x)$ about $x_0 = 0$.
   b.   Use $P_3(x)$ to approximate $\sqrt{0.5}$, $\sqrt{0.75}$, $\sqrt{1.25}$, and $\sqrt{1.5}$.
   c.   Determine the actual error of the approximations in (b).

7.   Find the second Taylor polynomial $P_2(x)$ for the function $f(x) = e^x \cos x$ about $x_0 = 0$.
   a.   Use $P_2(0.5)$ to approximate $f(0.5)$. Find an upper bound for error $|f(0.5) - P_2(0.5)|$ using the error formula, and compare it to the actual error.
   b.   Find a bound for the error $|f(x) - P_2(x)|$ in using $P_2(x)$ to approximate $f(x)$ on the interval [0, 1].
   c.   Approximate $\int_0^1 f(x)\, dx$ using $\int_0^1 P_2(x)\, dx$.
   d.   Find an upper bound for the error in (c) using $\int_0^1 |R_2(x)\, dx|$, and compare the bound to the actual error.

8.   Find the third Taylor polynomial $P_3(x)$ for the function $f(x) = (x - 1) \ln x$ about $x_0 = 1$.
   a.   Use $P_3(0.5)$ to approximate $f(0.5)$. Find an upper bound for error $|f(0.5) - P_3(0.5)|$ using the error formula, and compare it to the actual error.
   b.   Find a bound for the error $|f(x) - P_3(x)|$ in using $P_3(x)$ to approximate $f(x)$ on the interval [0.5, 1.5].
   c.   Approximate $\int_{0.5}^{1.5} f(x)\, dx$ using $\int_{0.5}^{1.5} P_3(x)\, dx$.
   d.   Find an upper bound for the error in (c) using $\int_{0.5}^{1.5} |R_3(x)\, dx|$, and compare the bound to the actual error.

9.   Use the error term of a Taylor polynomial to estimate the error involved in using $\sin x \approx x$ to approximate $\sin 1°$.

10. Use a Taylor polynomial about $\pi/4$ to approximate $\cos 42°$ to an accuracy of 10^{-6}.

11. Let $f(x) = e^{x/2} \sin(x/3)$. Use MATLAB to determine the following.

   a. The third Maclaurin polynomial $P_3(x)$.

   b. A bound for the error $|f(x) - P_3(x)|$ on $[0, 1]$.

12. Let $f(x) = \ln(x^2 + 2)$. Use MATLAB to determine the following.

   a. The Taylor polynomial $P_3(x)$ for f expanded about $x_0 = 1$.

   b. The maximum error $|f(x) - P_3(x)|$ for $0 \leq x \leq 1$.

   c. The Maclaurin polynomial $\tilde{P}_3(x)$ for f.

   d. The maximum error $|f(x) - \tilde{P}_3(x)|$ for $0 \leq x \leq 1$.

   e. Does $P_3(0)$ approximate $f(0)$ better than $\tilde{P}_3(1)$ approximates $f(1)$?

13. The polynomial $P_2(x) = 1 - \frac{1}{2}x^2$ is to be used to approximate $f(x) = \cos x$ in $[-\frac{1}{2}, \frac{1}{2}]$. Find a bound for the maximum error.

14. The nth Taylor polynomial for a function f at x_0 is sometimes referred to as the polynomial of degree at most n that "best" approximates f near x_0.

   a. Explain why this description is accurate.

   b. Find the quadratic polynomial that best approximates a function f near $x_0 = 1$ if the tangent line at $x_0 = 1$ has equation $y = 4x - 1$, and if $f''(1) = 6$.

15. The *error function* defined by

$$\text{erf}(x) = \frac{2}{\sqrt{\pi}} \int_0^x e^{-t^2} \, dt$$

gives the probability that any one of a series of trials will lie within x units of the mean, assuming that the trials have a normal distribution with mean 0 and standard deviation $\sqrt{2}/2$. This integral cannot be evaluated in terms of elementary functions, so an approximating technique must be used.

   a. Integrate the Maclaurin series for e^{-t^2} to show that

$$\text{erf}(x) = \frac{2}{\sqrt{\pi}} \sum_{k=0}^{\infty} \frac{(-1)^k x^{2k+1}}{(2k+1)k!}.$$

   b. The error function can also be expressed in the form

$$\text{erf}(x) = \frac{2}{\sqrt{\pi}} e^{-x^2} \sum_{k=0}^{\infty} \frac{2^k x^{2k+1}}{1 \cdot 3 \cdot 5 \cdots (2k+1)}.$$

   Verify that the two series agree for $k = 1, 2, 3$, and 4. [*Hint:* Use the Maclaurin series for e^{-x^2}.]

   c. Use the series in (a) to approximate $\text{erf}(1)$ to within 10^{-7}.

   d. Use the same number of terms used in (c) to approximate $\text{erf}(1)$ with the series in (b).

   e. Explain why difficulties occur using the series in (b) to approximate $\text{erf}(x)$.

16. In Example 3 it is stated that x we have $|\sin x| \leq |x|$. Use the following to verify this statement.

   a. Show that for all $x \geq 0$ we have $f(x) = x - \sin x$ is non-decreasing, which implies that $\sin x \leq x$ with equality only when $x = 0$.

   b. Reach the conclusion by using the fact that for all values of x, $\sin(-x) = -\sin x$.

1.3 Round-Off Error and Computer Arithmetic

The arithmetic performed by a calculator or computer is different from the arithmetic that we use in our algebra and calculus courses. From your past experience, you might expect that we always have as true statements such things as $2 + 2 = 4$, $4 \cdot 8 = 32$, and $(\sqrt{3})^2 = 3$.

In standard *computational* arithmetic we expect exact results for $2 + 2 = 4$ and $4 \cdot 8 = 32$, but we will not have precisely $(\sqrt{3})^2 = 3$. To understand why this is true we must explore the world of finite-digit arithmetic.

In our traditional mathematical world we permit numbers with an infinite number of digits. The arithmetic we use in this world *defines* $\sqrt{3}$ as that unique positive number which when multiplied by itself produces the integer 3. In the computational world, however, each representable number has only a fixed and finite number of digits. This means, for example, that only rational numbers—and not even all these—can be represented exactly. Since $\sqrt{3}$ is not rational, it is given an approximate representation within the machine, a representation whose square will not be precisely 3, although it will likely be sufficiently close to 3 to be acceptable in most situations. In most cases, then, this machine representation and arithmetic is satisfactory and passes without notice or concern, but at times problems arise because of this discrepancy.

Error due to rounding should be expected whenever computations are performed using numbers that are not powers of 2. Keeping this error under control is extremely important when the number of calculations is large.

The error that is produced when a calculator or computer is used to perform real-number calculations is called **round-off error**. It occurs because the arithmetic performed in a machine involves numbers with only a finite number of digits, with the result that calculations are performed with only approximate representations of the actual numbers. In a typical computer, only a relatively small subset of the real number system is used for the representation of all the real numbers. This subset contains only rational numbers, both positive and negative, and stores the fractional part, together with an exponential part.

Binary Machine Numbers

In 1985, the IEEE (Institute for Electrical and Electronic Engineers) published a report called *Binary Floating Point Arithmetic Standard 754–1985*. An updated version was published in 2008 as *IEEE 754-2008*. This provides standards for binary and decimal floating point numbers, formats for data interchange, algorithms for rounding arithmetic operations, and for the handling of exceptions. Formats are specified for single, double, and extended precisions, and these standards are generally followed by all microcomputer manufacturers using floating-point hardware.

For example, double precision real numbers require a 64-bit (binary digit) representation. The first bit is a sign indicator, denoted s. This is followed by an 11-bit exponent, c, called the **characteristic**, and a 52-bit binary fraction, f, called the **mantissa**. The base for the exponent is 2.

The normalized form for the nonzero double precision numbers has $0 < c < 2^{11} - 1 = 2047$. Since c is positive, a bias of 1023 is subtracted from c to give an actual exponent in the interval $(-1023, 1024)$. This permits adequate representation of numbers with both large and small magnitude. The first bit of the fractional part of a number is assumed to be 1 and is not stored in order to give one additional bit of precision to the representation, Since 53 binary digits correspond to between 15 and 16 decimal digits, we can assume that a number represented using this system has at least 15 decimal digits of precision. Thus, numbers represented in normalized double precision have the form

$$(-1)^s 2^{c-1023}(1 + f).$$

Illustration   Consider the machine number

0 10000000011 1011100100010000000000000000000000000000000000000000.

The leftmost bit is $s = 0$, which indicates that the number is positive. The next 11 bits, 10000000011, give the characteristic and are equivalent to the decimal number

$$c = 1 \cdot 2^{10} + 0 \cdot 2^9 + \cdots + 0 \cdot 2^2 + 1 \cdot 2^1 + 1 \cdot 2^0 = 1024 + 2 + 1 = 1027.$$

The exponential part of the number is, therefore, $2^{1027-1023} = 2^4$. The final 52 bits specify that the mantissa is

$$f = 1 \cdot \left(\frac{1}{2}\right)^1 + 1 \cdot \left(\frac{1}{2}\right)^3 + 1 \cdot \left(\frac{1}{2}\right)^4 + 1 \cdot \left(\frac{1}{2}\right)^5 + 1 \cdot \left(\frac{1}{2}\right)^8 + 1 \cdot \left(\frac{1}{2}\right)^{12}.$$

As a consequence, this machine number precisely represents the decimal number

$$(-1)^s 2^{c-1023}(1+f) = (-1)^0 \cdot 2^{1027-1023}\left(1 + \left(\frac{1}{2} + \frac{1}{8} + \frac{1}{16} + \frac{1}{32} + \frac{1}{256} + \frac{1}{4096}\right)\right)$$

$$= 27.56640625.$$

However, the next smallest machine number is

$$0 \ 10000000011 \ 1011100100001111111111111111111111111111111111111111,$$

and the next largest machine number is

$$0 \ 10000000011 \ 1011100100010000000000000000000000000000000000000001.$$

This means that our original machine number represents not only 27.56640625, but also half of the real numbers that are between 27.56640625 and the next smallest machine number, as well as half the numbers between 27.56640625 and the next largest machine number. To be precise, it represents any real number in the interval

$$[27.5664062499999982236431605997495353221893310546875,$$
$$27.5664062500000017763568394002504646778106689453125).$$ □

The smallest normalized positive number that can be represented has $s = 0$, $c = 1$, and $f = 0$, and is equivalent to the decimal number

$$2^{-1022} \cdot (1 + 0) \approx 0.225 \times 10^{-307}.$$

The largest normalized positive number that can be represented has $s = 0$, $c = 2046$, and $f = 1 - 2^{-52}$, and is equivalent to the decimal number

$$2^{1023} \cdot (1 + (1 - 2^{-52})) \approx 0.17977 \times 10^{309}.$$

Numbers occurring in calculations that have too small a magnitude to be represented result in **underflow**, and are generally set to 0 with computations continuing. However, numbers occurring in calculations that have too large a magnitude to be represented result in **overflow** and typically cause the computations to stop. Note that there are two representations for the number zero; a positive 0 when $s = 0$, $c = 0$, and $f = 0$ and a negative 0 when $s = 1$, $c = 0$, and $f = 0$.

Decimal Machine Numbers

The use of binary digits tends to complicate the computational problems that occur when a finite collection of machine numbers is used to represent all the real numbers. To examine these problems, we now assume, for simplicity, that machine numbers are represented in the normalized *decimal* form

$$\pm 0.d_1 d_2 \ldots d_k \times 10^n, \quad 1 \le d_1 \le 9, \quad 0 \le d_i \le 9$$

for each $i = 2, \ldots, k$. We call numbers of this form *k-digit decimal machine numbers*.

Any positive real number within numerical range of the machine can be normalized to achieve the form

$$y = 0.d_1 d_2 \ldots d_k d_{k+1} d_{k+2} \ldots \times 10^n.$$

The error that results from replacing a number with its floating-point form is called *round-off error* regardless of whether the rounding or chopping method is used.

The **floating-point form** of y, denoted by $fl(y)$, is obtained by terminating the mantissa of y at k decimal digits. There are two ways of performing the termination. One method, called **chopping**, is to simply chop off the digits $d_{k+1}d_{k+2}\ldots$ to obtain

$$fl(y) = 0.d_1 d_2 \ldots d_k \times 10^n.$$

The other method of terminating the mantissa of y at k decimal points is called **rounding**. If the $k+1$st digit is smaller than 5, then the result is the same as chopping. If the $k+1$st digit is 5 or greater, then 1 is added to the kth digit and the resulting number is chopped. As a consequence, rounding can be accomplished by simply adding $5 \times 10^{n-(k+1)}$ to y and then chopping the result to obtain $fl(y)$. Note that when rounding up, the exponent n could increase by 1. In summary, when rounding we add one to d_k to obtain $fl(y)$ whenever $d_{k+1} \ge 5$, that is, we round up; when $d_{k+1} < 5$, we chop off all but the first k digits, so we round down.

The next examples illustrate floating-point arithmetic when the number of digits being retained is quite small. Although the floating-point arithmetic that is performed on a calculator or computer will retain many more digits, the problems this arithmetic can cause are essentially the same regardless of the number of digits. Retaining more digits simply postpones the awareness of the situation.

Example 1 Determine the five-digit **(a)** chopping and **(b)** rounding values of the irrational number π.

Solution The number π has an infinite decimal expansion of the form $\pi = 3.14159265\ldots$. Written in normalized decimal form, we have

$$\pi = 0.314159265\ldots \times 10^1.$$

(a) The floating-point form of π using five-digit chopping is

$$fl(\pi) = 0.31415 \times 10^1 = 3.1415.$$

(b) The sixth digit of the decimal expansion of π is a 9, so the floating-point form of π using five-digit rounding is

$$fl(\pi) = (0.31415 + 0.00001) \times 10^1 = 3.1416. \qquad ▪$$

The relative error is generally a better measure of accuracy than the absolute error because it takes into consideration the size of the number being approximated.

There are two common methods for measuring approximation errors.

The approximation p^* to p has **absolute error** $|p - p^*|$ and **relative error** $|p - p^*|/|p|$, provided that $p \ne 0$.

Example 2 Determine the absolute and relative errors when approximating p by p^* when

(a) $p = 3.000$ and $p^* = 3.100$;

(b) $p = 0.003000$ and $p^* = 0.003100$;

(c) $p = 3000$ and $p^* = 3100$.

Solution First we need to write these numbers in standard floating-point form:

(a) For $p = 0.3000 \times 10^1$ and $p^* = 0.3100 \times 10^1$ the absolute error is 0.1, and the relative error is $0.333\overline{3} \times 10^{-1}$.

(b) For $p = 0.3000 \times 10^{-3}$ and $p^* = 0.3100 \times 10^{-3}$ the absolute error is 0.1×10^{-4}, and the relative error is $0.333\overline{3} \times 10^{-1}$.

(c) For $p = 0.3000 \times 10^4$ and $p^* = 0.3100 \times 10^4$, the absolute error is 0.1×10^3, and the relative error is again $0.333\overline{3} \times 10^{-1}$.

We often cannot find an accurate value for the true error in an approximation. Instead we find a bound for the error, which gives us a "worst-case" error.

This example shows that the same relative error, $0.333\overline{3} \times 10^{-1}$, occurs for widely varying absolute errors. As a measure of accuracy, the absolute error can be misleading and the relative error more meaningful, because the relative error takes into consideration the size of the value. ∎

Finite-Digit Arithmetic

The arithmetic operations of addition, subtraction, multiplication, and division performed by a computer on floating-point numbers also introduce error. These arithmetic operations involve manipulating binary digits by various shifting and logical operations, but the actual mechanics of the arithmetic are not pertinent to our discussion. To illustrate the problems that can occur, we simulate this *finite-digit arithmetic* by first performing, at each stage in a calculation, the appropriate operation using exact arithmetic on the floating-point representations of the numbers. We then convert the result to decimal machine-number representation. The most common round-off error producing arithmetic operation involves the subtraction of nearly equal numbers.

Example 3   Use four-digit decimal chopping arithmetic to simulate the problem of performing the computer operation $\pi - \frac{22}{7}$.

Solution   The floating-point representations of these numbers are

$$fl(\pi) = 0.3141 \times 10^1 \qquad \text{and} \qquad fl\left(\frac{22}{7}\right) = 0.3142 \times 10^1.$$

Performing the exact arithmetic on the floating-point numbers gives

$$fl(\pi) - fl\left(\frac{22}{7}\right) = -0.0001 \times 10^1,$$

which converts to the floating-point approximation of this calculation:

$$p^* = fl\left(fl(\pi) - fl\left(\frac{22}{7}\right)\right) = -0.1000 \times 10^{-2}.$$

Although the relative errors using the floating-point representations for π and $\frac{22}{7}$ are small,

$$\left|\frac{\pi - fl(\pi)}{\pi}\right| \le 0.0002 \quad \text{and} \quad \left|\frac{\frac{22}{7} - fl\left(\frac{22}{7}\right)}{\frac{22}{7}}\right| \le 0.0003,$$

the relative error produced by subtracting the nearly equal numbers π and $\frac{22}{7}$ is about 700 times as large:

$$\left|\frac{\left(\pi - \frac{22}{7}\right) - p^*}{\left(\pi - \frac{22}{7}\right)}\right| \approx 0.2092.$$

∎

Rounding arithmetic in MATLAB involves the use of the function `round`. This function rounds to the nearest whole number and the function `fix` chops off the fractional part. Suppose we define the numbers $x1$, $x2$, $x3$, and $x4$ with the MATLAB command

```
x1=-123.4,x2=-123.5,x3=123.4,x4=123.5
```

MATLAB responds with,

$$x1 = -1.234000000000000e + 002$$

$$x2 = -1.235000000000000e + 002$$

$$x3 = 1.234000000000000e + 002$$

$$x4 = 1.235000000000000e + 002$$

Invoking the commands `round` and `fix` gives, for `round`,

$$ans = -123, \quad ans = -124, \quad ans = 123, \quad ans = 124$$

and, for `fix`,

$$ans = -123, \quad ans = -123, \quad ans = 123, \quad ans = 123$$

Exercise 12 illustrates how these functions can be used to perform rounding and chopping arithmetic to a specific number of digits.

EXERCISE SET 1.3

1. Compute the absolute error and relative error in approximations of p by p^*.

   a. $p = \pi, p^* = \frac{22}{7}$ b. $p = \pi, p^* = 3.1416$

   c. $p = e, p^* = 2.718$ d. $p = \sqrt{2}, p^* = 1.414$

   e. $p = e^{10}, p^* = 22000$ f. $p = 10^\pi, p^* = 1400$

   g. $p = 8!, p^* = 39900$ h. $p = 9!, p^* = \sqrt{18\pi} \, (9/e)^9$

2. Perform the following computations (i) exactly, (ii) using three-digit chopping arithmetic, and (iii) using three-digit rounding arithmetic. (iv) Compute the relative errors in (ii) and (iii).

   a. $\dfrac{4}{5} + \dfrac{1}{3}$ b. $\dfrac{4}{5} \cdot \dfrac{1}{3}$

   c. $\left(\dfrac{1}{3} - \dfrac{3}{11}\right) + \dfrac{3}{20}$ d. $\left(\dfrac{1}{3} + \dfrac{3}{11}\right) - \dfrac{3}{20}$

3. Use three-digit rounding arithmetic to perform the following calculations. Compute the absolute error and relative error with the exact value determined to at least five digits.

   a. $133 + 0.921$ b. $133 - 0.499$

   c. $(121 - 0.327) - 119$ d. $(121 - 119) - 0.327$

   e. $\dfrac{\frac{13}{14} - \frac{6}{7}}{2e - 5.4}$ f. $-10\pi + 6e - \dfrac{3}{62}$

   g. $\left(\dfrac{2}{9}\right) \cdot \left(\dfrac{9}{7}\right)$ h. $\dfrac{\pi - \frac{22}{7}}{\frac{1}{17}}$

4.   Repeat Exercise 3 using three-digit chopping arithmetic.

5.   Repeat Exercise 3 using four-digit rounding arithmetic.

6.   Repeat Exercise 3 using four-digit chopping arithmetic.

7.   The first three nonzero terms of the Maclaurin series for the arctan x are $x - \frac{1}{3}x^3 + \frac{1}{5}x^5$. Compute the absolute error and relative error in the following approximations of π using the polynomial in place of the arctan x:

   a. $4\left[\arctan\left(\frac{1}{2}\right) + \arctan\left(\frac{1}{3}\right)\right]$ **b.** $16\arctan\left(\frac{1}{5}\right) - 4\arctan\left(\frac{1}{239}\right)$

8.   The two-by-two linear system

$$ax + by = e,$$
$$cx + dy = f,$$

where a, b, c, d, e, f are given, can be solved for x and y as follows:

$$\text{set } m = \frac{c}{a}, \quad \text{provided } a \neq 0;$$

$$d_1 = d - mb;$$

$$f_1 = f - me;$$

$$y = \frac{f_1}{d_1};$$

$$x = \frac{(e - by)}{a}.$$

Solve the following linear systems using four-digit rounding arithmetic.

   a. $1.130x - 6.990y = 14.20$ **b.** $1.013x - 6.099y = 14.22$
        $8.110x + 12.20y = -0.1370$ $-18.11x + 112.2y = -0.1376$

9.   Suppose the points (x_0, y_0) and (x_1, y_1) are on a straight line with $y_1 \neq y_0$. Two formulas are available to find the x-intercept of the line:

$$x = \frac{x_0 y_1 - x_1 y_0}{y_1 - y_0} \quad \text{and} \quad x = x_0 - \frac{(x_1 - x_0)y_0}{y_1 - y_0}.$$

   a.   Show that both formulas are algebraically correct.

   b.   Use the data $(x_0, y_0) = (1.31, 3.24)$ and $(x_1, y_1) = (1.93, 4.76)$ and three-digit rounding arithmetic to compute the x-intercept both ways. Which method is better, and why?

10.   The Taylor polynomial of degree n for $f(x) = e^x$ is $\sum_{i=0}^{n} x^i / i!$. Use the Taylor polynomial of degree nine and three-digit chopping arithmetic to find an approximation to e^{-5} by each of the following methods.

   a. $e^{-5} \approx \sum_{i=0}^{9} \frac{(-5)^i}{i!} = \sum_{i=0}^{9} \frac{(-1)^i 5^i}{i!}$ **b.** $e^{-5} = \frac{1}{e^5} \approx \frac{1}{\sum_{i=0}^{9} 5^i / i!}$

An approximate value of e^{-5} correct to three digits is 6.74×10^{-3}. Which formula, (a) or (b), gives the most accuracy, and why?

11.   A rectangular parallelepiped has sides 3 cm, 4 cm, and 5 cm, measured to the nearest centimeter.

   a.   What are the best upper and lower bounds for the volume of this parallelepiped?

   b.   What are the best upper and lower bounds for the surface area?

12.   The following MATLAB M-file rounds or chops a number x to t digits where rnd = 1 for rounding and rnd = 0 for chopping.

```
function [res] = CHIP(rnd,t,x)
% This program is used to round or chop a number x to a specific number t
% of digits.
if x == 0
  w = 0;
else
  ee = fix(log10(abs(x)));
  if abs(x) > 1
     ee = ee + 1;
  end;
  if rnd == 1
     w = round(x*10^(t-ee))*10^(ee-t);
  else
     w = fix(x*10^(t-ee))*10^(ee-t);
  end;
end;
res = w ;
```

Verify that the procedure works for the following values.

a. $x = 124.031, t = 5$ **b.** $x = 124.036, t = 5$

c. $x = -0.00653, t = 2$ **d.** $x = -0.00656, t = 2$

13.   The binomial coefficient

$$\binom{m}{k} = \frac{m!}{k!\,(m-k)!}$$

describes the number of ways of choosing a subset of k objects from a set of m elements.

a.   Suppose decimal machine numbers are of the form

$$\pm 0.d_1 d_2 d_3 d_4 \times 10^n, \quad \text{with } 1 \le d_1 \le 9,\ 0 \le d_i \le 9, \quad \text{if } i = 2, 3, 4 \quad \text{and} \quad |n| \le 15.$$

What is the largest value of m for which the binomial coefficient $\binom{m}{k}$ can be computed for all k by the definition without causing overflow?

b.   Show that $\binom{m}{k}$ can also be computed by

$$\binom{m}{k} = \left(\frac{m}{k}\right)\left(\frac{m-1}{k-1}\right)\cdots\left(\frac{m-k+1}{1}\right).$$

c.   What is the largest value of m for which the binomial coefficient $\binom{m}{3}$ can be computed by the formula in (b) without causing overflow?

d.   Use the equation in (b) and four-digit chopping arithmetic to compute the number of possible 5-card hands in a 52-card deck. Compute the actual and relative errors.

1.4   Errors in Scientific Computation

In the previous section we saw how computational devices represent and manipulate numbers using finite-digit arithmetic. We now examine how the problems with this arithmetic can compound and look at ways to arrange arithmetic calculations to reduce this inaccuracy.

The loss of accuracy due to round-off error can often be avoided by a careful sequencing of operations or a reformulation of the problem. This is most easily described by considering a common computational problem.

Illustration   The quadratic formula states that the roots of $ax^2 + bx + c = 0$, when $a \neq 0$, are

$$x_1 = \frac{-b + \sqrt{b^2 - 4ac}}{2a} \quad \text{and} \quad x_2 = \frac{-b - \sqrt{b^2 - 4ac}}{2a}. \tag{1.1}$$

Consider this formula applied to the equation $x^2 + 62.10x + 1 = 0$, whose roots are approximately

$$x_1 = -0.01610723 \quad \text{and} \quad x_2 = -62.08390.$$

We will again use four-digit rounding arithmetic in the calculations to determine the root. In this equation, b^2 is much larger than $4ac$, so the numerator in the calculation for x_1 involves the *subtraction* of nearly equal numbers. Because

$$\sqrt{b^2 - 4ac} = \sqrt{(62.10)^2 - (4.000)(1.000)(1.000)}$$

$$= \sqrt{3856. - 4.000} = \sqrt{3852.} = 62.06,$$

we have

$$fl(x_1) = \frac{-62.10 + 62.06}{2.000} = \frac{-0.04000}{2.000} = -0.02000,$$

a poor approximation to $x_1 = -0.01611$, with the large relative error

$$\frac{|-0.01611 + 0.02000|}{|-0.01611|} \approx 2.4 \times 10^{-1}.$$

On the other hand, the calculation for x_2 involves the *addition* of the nearly equal numbers $-b$ and $-\sqrt{b^2 - 4ac}$. This presents no problem because

$$fl(x_2) = \frac{-62.10 - 62.06}{2.000} = \frac{-124.2}{2.000} = -62.10$$

has the small relative error

$$\frac{|-62.08 + 62.10|}{|-62.08|} \approx 3.2 \times 10^{-4}.$$

To obtain a more accurate four-digit rounding approximation for x_1, we change the form of the quadratic formula by *rationalizing the numerator*:

$$x_1 = \frac{-b + \sqrt{b^2 - 4ac}}{2a} \left(\frac{-b - \sqrt{b^2 - 4ac}}{-b - \sqrt{b^2 - 4ac}} \right) = \frac{b^2 - (b^2 - 4ac)}{2a(-b - \sqrt{b^2 - 4ac})},$$

The roots x_1 and x_2 of a general quadratic equation are related to the coefficients by the fact that

$$x_1 + x_2 = -\frac{b}{a} \quad \text{and} \quad x_1 x_2 = \frac{c}{a}.$$

This is a special case of Vièta's formula for the coefficients of polynomials.

which simplifies to an alternate quadratic formula

$$x_1 = \frac{-2c}{b + \sqrt{b^2 - 4ac}}. \tag{1.2}$$

Using Eq. (1.2) gives

$$fl(x_1) = \frac{-2.000}{62.10 + 62.06} = \frac{-2.000}{124.2} = -0.01610,$$

which has the small relative error 6.2×10^{-4}.

The rationalization technique can also be applied to give the following alternative quadratic formula for x_2:

$$x_2 = \frac{-2c}{b - \sqrt{b^2 - 4ac}}. \tag{1.3}$$

This is the form to use if b is a negative number. In the Illustration, however, the mistaken use of this formula for x_2 would result in not only the subtraction of nearly equal numbers, but also the division by the small result of this subtraction. The inaccuracy that this combination produces,

$$fl(x_2) = \frac{-2c}{b - \sqrt{b^2 - 4ac}} = \frac{-2.000}{62.10 - 62.06} = \frac{-2.000}{0.04000} = -50.00,$$

has the large relative error 1.9×10^{-1}.   □

Nested Arithmetic

Example 1   Evaluate $f(x) = x^3 - 6.1x^2 + 3.2x + 1.5$ at $x = 4.71$ using three-digit arithmetic.

Solution   Table 1.1 gives the intermediate results in the calculations.

Table 1.1

	x	x^2	x^3	$6.1x^2$	$3.2x$
Exact	4.71	22.1841	104.487111	135.32301	15.072
Three-digit (chopping)	4.71	22.1	104.	134.	15.0
Three-digit (rounding)	4.71	22.2	105.	135.	15.1

Remember that chopping (or rounding) is performed after each calculation.

To illustrate the calculations, let us first look at those involved with finding x^3 using three-digit rounding arithmetic.

First we find

$$x^2 = 4.71^2 = 22.1841, \quad \text{which rounds to } 22.2.$$

Then we use this value of x^2 to find

$$x^3 = x^2 \cdot x = 22.2 \cdot 4.71 = 104.562, \quad \text{which rounds to } 105.$$

Also,

$$6.1x^2 = 6.1(22.2) = 135.42, \quad \text{which rounds to } 135,$$

and

$$3.2x = 3.2(4.71) = 15.072, \quad \text{which rounds to } 15.1.$$

Using finite-digit arithmetic, the way in which we add the results can affect the final result. Suppose that we add left to right. Then for chopping arithmetic we have

Three-digit (chopping): $f(4.71) = ((104. - 134.) + 15.0) + 1.5 = -13.5,$

and for rounding arithmetic we have

Three-digit (rounding): $f(4.71) = ((105. - 135.) + 15.1) + 1.5 = -13.4.$

(You should carefully verify these results to be sure that your notion of finite-digit arithmetic is correct.) Note that the three-digit chopping values simply retain the leading three digits, with no rounding involved, and differ from the three-digit rounding values. The exact result of the evaluation is

Exact: $f(4.71) = 104.487111 - 135.32301 + 15.072 + 1.5 = -14.263899,$

so the relative errors for the three-digit methods are

$$\text{Chopping:} \quad \left| \frac{-14.263899 + 13.5}{-14.263899} \right| \approx 0.05, \quad \text{and}$$

$$\text{Rounding:} \quad \left| \frac{-14.263899 + 13.4}{-14.263899} \right| \approx 0.06. \qquad \blacksquare$$

Illustration   As an alternative approach, the polynomial $f(x)$ in Example 1 can be written in a **nested** manner as

$$f(x) = x^3 - 6.1x^2 + 3.2x + 1.5 = ((x - 6.1)x + 3.2)x + 1.5.$$

Using three-digit chopping arithmetic now produces

$$f(4.71) = ((4.71 - 6.1)4.71 + 3.2)4.71 + 1.5 = ((-1.39)(4.71) + 3.2)4.71 + 1.5$$
$$= (-6.54 + 3.2)4.71 + 1.5 = (-3.34)4.71 + 1.5 = -15.7 + 1.5 = -14.2.$$

In a similar manner, we now obtain a three-digit rounding answer of -14.3. The new relative errors are

$$\text{Three-digit (chopping):} \quad \left| \frac{-14.263899 + 14.2}{-14.263899} \right| \approx 0.0045;$$

$$\text{Three-digit (rounding):} \quad \left| \frac{-14.263899 + 14.3}{-14.263899} \right| \approx 0.0025.$$

Nesting has reduced the relative error for the chopping approximation to less than 10% of that obtained initially. For the rounding approximation, the improvement has been even more dramatic; the error in this case has been reduced by more than 95%.   □

Characterizing Algorithms

We will be considering a variety of approximation problems throughout the text, and in each case we need to determine methods that produce dependably accurate results for a wide class of problems. Because of the differing ways in which the approximation methods are derived, we need a variety of conditions to categorize their accuracy. Not all of these conditions will be appropriate for any particular problem.

One criterion we will impose, whenever possible, is that of **stability**. A method is called **stable** if small changes in the initial data produce correspondingly small changes in the final results. When it is possible to have small changes in the initial data producing large changes in the final results, the method is **unstable**. Some methods are stable only for certain choices of initial data. These methods are called **conditionally stable**. We attempt to characterize stability properties whenever possible.

One of the most important topics affecting the stability of a method is the way in which the round-off error grows as the method is successively applied. Suppose an error with magnitude $E_0 > 0$ is introduced at some stage in the calculations and that the magnitude of the error after n subsequent operations is E_n. There are two distinct cases that often arise in practice.

- If a constant C exists independent of n, with $E_n \approx CnE_0$, the growth of error is **linear**.

- If a constant $C > 1$ exists independent of n, with $E_n \approx C^n E_0$, the growth of error is **exponential**.

It would be unlikely to have $E_n \approx C^n E_0$, with $C < 1$, because this implies that the error tends to zero.

Linear growth of error is usually unavoidable and, when C and E_0 are small, the results are generally acceptable. Methods having exponential growth of error should be avoided because the term C^n becomes large for even relatively small values of n and E_0. Consequently, a method that exhibits linear error growth is stable, while one exhibiting exponential error growth is unstable. (See Figure 1.10.)

Figure 1.10

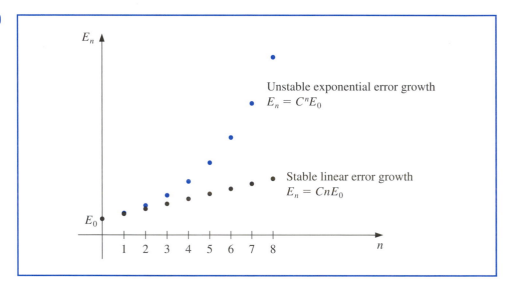

Rates of Convergence

Iterative techniques often involve sequences, and the section concludes with a brief discussion of some terminology used to describe the rate at which sequences converge when employing a numerical technique. In general, we would like to choose techniques that converge as rapidly as possible. The following definition is used to compare the convergence rates of various methods.

Suppose that $\{\alpha_n\}_{n=1}^{\infty}$ is a sequence that converges to a number α as n becomes large. If positive constants p and K exist with

$$|\alpha - \alpha_n| \leq \frac{K}{n^p}, \quad \text{for all large values of } n,$$

then we say that $\{\alpha_n\}$ **converges to α with rate, or order, of convergence $O(1/n^p)$** (read "big oh of $1/n^p$"). This is indicated by writing $\alpha_n = \alpha + O(1/n^p)$ and stated as "$\alpha_n \rightarrow \alpha$

There are numerous other ways of describing the growth of sequences and functions, some of which require bounds both above and below the sequence or function under consideration. Any good book that analyzes algorithms, for example [CLRS], will include this information.

with rate of convergence $1/n^p$." We are generally interested in the *largest* value of p for which $\alpha_n = \alpha + O(1/n^p)$.

We also use the "big oh" notation to describe how some divergent sequences grow as n becomes large. If positive constants p and K exist with

$$|\alpha_n| \leq Kn^p, \quad \text{for all large values of } n,$$

then we say that $\{\alpha_n\}$ **goes to ∞ with rate, or order, $O(n^p)$.** In the case of a sequence that becomes infinite, we are interested in the *smallest* value of p for which α_n is $O(n^p)$.

The "big oh" definition for sequences can be extended to incorporate more general sequences, but the definition as presented here is sufficient for our purposes.

Example 2 Suppose that, for $n \geq 1$,

$$\alpha_n = \frac{n+1}{n^2} \quad \text{and} \quad \hat{\alpha}_n = \frac{n+3}{n^3}.$$

Both of the sequences converge to 0, but the sequence $\{\hat{\alpha}_n\}$ converges much faster than the sequence $\{\alpha_n\}$. Using five-digit rounding arithmetic, we have the values shown in Table 1.2. Determine rates of convergence for these two sequences.

Table 1.2

n	1	2	3	4	5	6	7
α_n	2.00000	0.75000	0.44444	0.31250	0.24000	0.19444	0.16327
$\hat{\alpha}_n$	4.00000	0.62500	0.22222	0.10938	0.064000	0.041667	0.029155

Solution Define the sequences $\beta_n = 1/n$ and $\hat{\beta}_n = 1/n^2$. Then

$$|\alpha_n - 0| = \frac{n+1}{n^2} \leq \frac{n+n}{n^2} = 2 \cdot \frac{1}{n} = 2\beta_n$$

and

$$|\hat{\alpha}_n - 0| = \frac{n+3}{n^3} \leq \frac{n+3n}{n^3} = 4 \cdot \frac{1}{n^2} = 4\hat{\beta}_n.$$

Hence the rate of convergence of $\{\alpha_n\}$ to 0 is similar to the convergence of $\{1/n\}$ to 0, whereas $\{\hat{\alpha}_n\}$ converges to 0 at a rate similar to the more rapidly convergent sequence $\{1/n^2\}$. We express this by writing

$$\alpha_n = 0 + O\left(\frac{1}{n}\right) \quad \text{and} \quad \hat{\alpha}_n = 0 + O\left(\frac{1}{n^2}\right). \qquad \blacksquare$$

The "big oh" notation is also used to describe the rate of convergence of functions, particularly when the independent variable approaches 0.

Suppose that F is a function that converges to a number L as h goes to 0. If positive constants p and K exist with

$$|F(h) - L| \leq Kh^p, \quad \text{as } h \to 0,$$

then $F(h)$ **converges to L with rate, or order, of convergence $O(h^p)$.** This is written as $F(h) = L + O(h^p)$ and stated as "$F(h) \to L$ with rate of convergence h^p."

We are generally interested in the *largest* value of p for which $F(h) = L + O(h^p)$.

EXERCISE SET 1.4

1. (i) Use four-digit rounding arithmetic and Eqs. (1.2) and (1.3) to find the most accurate approximations to the roots of the following quadratic equations. (ii) Compute the absolute errors and relative errors for these approximations.

   a. $\dfrac{1}{3}x^2 - \dfrac{123}{4}x + \dfrac{1}{6} = 0$ 　　　　　　 b. $\dfrac{1}{3}x^2 + \dfrac{123}{4}x - \dfrac{1}{6} = 0$

   c. $1.002x^2 - 11.01x + 0.01265 = 0$ 　　 d. $1.002x^2 + 11.01x + 0.01265 = 0$

2. Repeat Exercise 1 using four-digit chopping arithmetic.

3. Let $f(x) = 1.013x^5 - 5.262x^3 - 0.01732x^2 + 0.8389x - 1.912$.

   a. Evaluate $f(2.279)$ by first calculating $(2.279)^2$, $(2.279)^3$, $(2.279)^4$, and $(2.279)^5$ using four-digit rounding arithmetic.

   b. Evaluate $f(2.279)$ using the formula

   $$f(x) = (((1.013x^2 - 5.262)x - 0.01732)x + 0.8389)x - 1.912$$

   and four-digit rounding arithmetic.

   c. Compute the absolute and relative errors in (a) and (b).

4. Repeat Exercise 3 using four-digit chopping arithmetic.

5. The fifth Maclaurin polynomials for e^{2x} and e^{-2x} are

   $$P_5(x) = \left(\left(\left(\left(\frac{4}{15}x + \frac{2}{3}\right)x + \frac{4}{3}\right)x + 2\right)x + 2\right)x + 1$$

   and

   $$\hat{P}_5(x) = \left(\left(\left(\left(-\frac{4}{15}x + \frac{2}{3}\right)x - \frac{4}{3}\right)x + 2\right)x - 2\right)x + 1$$

   a. Approximate $e^{-0.98}$ using $\hat{P}_5(0.49)$ and four-digit rounding arithmetic.

   b. Compute the absolute and relative error for the approximation in (a).

   c. Approximate $e^{-0.98}$ using $1/P_5(0.49)$ and four-digit rounding arithmetic.

   d. Compute the absolute and relative errors for the approximation in (c).

6. a. Show that the polynomial nesting technique described in the Illustration on page 25 can also be applied to the evaluation of

   $$f(x) = 1.01e^{4x} - 4.62e^{3x} - 3.11e^{2x} + 12.2e^x - 1.99.$$

   b. Use three-digit rounding arithmetic, the assumption that $e^{1.53} = 4.62$, and the fact that $e^{n(1.53)} = (e^{1.53})^n$ to evaluate $f(1.53)$ as given in (a).

   c. Redo the calculation in (b) by first nesting the calculations.

   d. Compare the approximations in (b) and (c) to the true three-digit result $f(1.53) = -7.61$.

7. Use three-digit chopping arithmetic to compute the sum $\sum_{i=1}^{10} 1/i^2$ first by $\frac{1}{1} + \frac{1}{4} + \cdots + \frac{1}{100}$ and then by $\frac{1}{100} + \frac{1}{81} + \cdots + \frac{1}{1}$. Which method is more accurate, and why?

8. The Maclaurin series for the arctangent function converges for $-1 < x \leq 1$ and is given by

   $$\arctan x = \lim_{n \to \infty} P_n(x) = \lim_{n \to \infty} \sum_{i=1}^{n} (-1)^{i+1} \frac{x^{2i-1}}{(2i-1)}.$$

   a. Use the fact that $\tan \pi/4 = 1$ to determine the number of terms of the series that need to be summed to ensure that $|4P_n(1) - \pi| < 10^{-3}$.

   b. The C programming language requires the value of π to be within 10^{-10}. How many terms of the series would we need to sum to obtain this degree of accuracy?

9. The number e is defined by $e = \sum_{n=0}^{\infty} 1/n!$, where $n! = n(n-1)\cdots 2 \cdot 1$, for $n \neq 0$ and $0! = 1$. (i) Use four-digit chopping arithmetic to compute the following approximations to e. (ii) Compute absolute and relative errors for these approximations.

   a. $\displaystyle\sum_{n=0}^{5} \frac{1}{n!}$ b. $\displaystyle\sum_{j=0}^{5} \frac{1}{(5-j)!}$

   c. $\displaystyle\sum_{n=0}^{10} \frac{1}{n!}$ d. $\displaystyle\sum_{j=0}^{10} \frac{1}{(10-j)!}$

10. Find the rates of convergence of the following sequences as $n \to \infty$.

   a. $\displaystyle\lim_{n\to\infty} \sin\left(\frac{1}{n}\right) = 0$ b. $\displaystyle\lim_{n\to\infty} \sin\left(\frac{1}{n^2}\right) = 0$

   c. $\displaystyle\lim_{n\to\infty} \left(\sin\left(\frac{1}{n}\right)\right)^2 = 0$ d. $\displaystyle\lim_{n\to\infty} [\ln(n+1) - \ln(n)] = 0$

11. Find the rates of convergence of the following functions as $h \to 0$.

   a. $\displaystyle\lim_{h\to 0} \frac{\sin h - h\cos h}{h} = 0$ b. $\displaystyle\lim_{h\to 0} \frac{1 - e^h}{h} = -1$

   c. $\displaystyle\lim_{h\to 0} \frac{\sin h}{h} = 1$ d. $\displaystyle\lim_{h\to 0} \frac{1 - \cos h}{h} = 0$

12. a. How many multiplications and additions are required to determine a sum of the form

$$\sum_{i=1}^{n} \sum_{j=1}^{i} a_i b_j ?$$

   b. Modify the sum in (a) to an equivalent form that reduces the number of computations.

13. The sequence $\{F_n\}$ described by $F_0 = 1$, $F_1 = 1$, and $F_{n+2} = F_n + F_{n+1}$, if $n \geq 0$, is called a *Fibonacci sequence*. Its terms occur naturally in many botanical species, particularly those with petals or scales arranged in the form of a logarithmic spiral. Consider the sequence $\{x_n\}$, where $x_n = F_{n+1}/F_n$. Assuming that $\lim_{n\to\infty} x_n = x$ exists, show that x is the *golden ratio* $(1 + \sqrt{5})/2$.

1.5   Computer Software

Computer software packages for approximating the numerical solutions to problems are available in many forms. On our website for the book

```
http://www.math.ysu.edu/~faires/Numerical-Methods/
```

we have provided programs written in C, FORTRAN, Maple, Mathematica, MATLAB, and Pascal, as well as JAVA applets. These can be used to solve the problems given in the examples and exercises, and will give satisfactory results for most problems that you may need to solve. However, they are what we call *special-purpose* programs. We use this term to distinguish these programs from those available in the standard mathematical subroutine libraries. The programs in these packages will be called *general purpose*.

General Purpose Algorithms

The programs in general-purpose software packages differ in their intent from the algo-rithms and programs provided with this book. General-purpose software packages consider ways to reduce errors due to machine rounding, underflow, and overflow. They also describe the range of input that will lead to results of a certain specified accuracy. These are machine-dependent characteristics, so general-purpose software packages use parameters that describe the floating-point characteristics of the machine being used for computations.

The system FORTRAN (FORmula TRANslator) was the original general-purpose scientific programming language. It is still in wide use in situations that require intensive scientific computations.

Many forms of general-purpose numerical software are available commercially and in the public domain. Most of the early software was written for mainframe computers, and a good reference for this is *Sources and Development of Mathematical Software*, edited by Wayne Cowell [Co].

Now that personal computers are sufficiently powerful, standard numerical software is available for them. Most of this numerical software is written in FORTRAN, although some packages are written in C, C++, and FORTRAN90.

ALGOL procedures were presented for matrix computations in 1971 in [WR]. A package of FORTRAN subroutines based mainly on the ALGOL procedures was then developed into the EISPACK routines. These routines are documented in the manuals published by Springer-Verlag as part of their Lecture Notes in Computer Science series [Sm,B] and [Gar]. The FORTRAN subroutines are used to compute eigenvalues and eigenvectors for a variety of different types of matrices.

The EISPACK project was the first large-scale numerical software package to be made available in the public domain and led the way for many packages to follow.

LINPACK is a package of FORTRAN subroutines for analyzing and solving systems of linear equations and solving linear least squares problems. The documentation for this package is contained in [DBMS]. A step-by-step introduction to LINPACK, EISPACK, and BLAS (Basic Linear Algebra Subprograms) is given in [CV].

The LAPACK package, first available in 1992, is a library of FORTRAN subroutines that supercedes LINPACK and EISPACK by integrating these two sets of algorithms into a unified and updated package. The software has been restructured to achieve greater efficiency on vector processors and other high-performance or shared-memory multiprocessors. LAPACK is expanded in depth and breadth in version 3.0, which is available in FORTRAN, FORTRAN90, C, C++, and JAVA. C and JAVA are only available as language interfaces or translations of the FORTRAN libraries of LAPACK. The package BLAS is not a part of LAPACK, but the code for BLAS is distributed with LAPACK.

Software engineering was established as a laboratory discipline during the 1970s and 1980s. EISPACK was developed at Argonne Labs and LINPACK shortly thereafter. By the early 1980s, Argonne was internationally recognized as a world leader in symbolic and numerical computation.

Other packages for solving specific types of problems are available in the public domain. As an alternative to netlib, you can use Xnetlib to search the database and retrieve software. More information can be found in the article *Software Distribution using Netlib* by Dongarra, Roman, and Wade [DRW].

These software packages are highly efficient, accurate, and reliable. They are thoroughly tested, and documentation is readily available. Although the packages are portable, it is a good idea to investigate the machine dependence and read the documentation thoroughly. The programs test for almost all special contingencies that might result in error and failures. At the end of each chapter, we will discuss some of the appropriate general-purpose packages.

Commercially available packages also represent the state of the art in numerical methods. Their contents are often based on the public-domain packages but include methods in libraries for almost every type of problem.

In 1970, IMSL became the first large-scale scientific library for mainframes. Since that time, the libraries have been made available for computer systems ranging from supercomputers to personal computers.

IMSL (International Mathematical and Statistical Libraries) consists of the libraries MATH, STAT, and SFUN for numerical mathematics, statistics, and special functions, respectively. These libraries contain more than 900 subroutines originally available in FORTRAN77 and now available in C, FORTRAN90, and JAVA. These subroutines solve the most common numerical analysis problems. The libraries are available commercially from Visual Numerics.

The packages are delivered in compiled form with extensive documentation. There is an example program for each routine as well as background reference information. IMSL contains methods for linear systems, eigensystem analysis, interpolation and approximation, integration and differentiation, differential equations, transforms, nonlinear equations, optimization, and basic matrix/vector operations. The library also contains extensive statistical routines.

The Numerical Algorithms Group (NAG) has been in existence in the United Kingdom since 1970. NAG offers more than 1000 subroutines in a FORTRAN77 library, about 400 subroutines in a C library, more than 200 subroutines in a FORTRAN90 library, and an MPI FORTRAN numerical library for parallel machines and clusters of workstations or personal computers. A useful introduction to the NAG routines is [Ph]. The NAG library contains routines to perform most standard numerical analysis tasks in a manner similar to those in the IMSL. It also includes some statistical routines and a set of graphic routines.

The IMSL and NAG packages are designed for the mathematician, scientist, or engineer who wishes to call high-quality C, Java, or FORTRAN subroutines from within a program. The documentation available with the commercial packages illustrates the typical driver program required to use the library routines. The next three software packages are stand-alone environments. When activated, the user enters commands to cause the package to solve a problem. However, each package allows programming within the command language.

MATLAB is a matrix laboratory that was originally a Fortran program published by Cleve Moler [Mo] in the 1980s. The laboratory is based mainly on the EISPACK and LINPACK subroutines, although functions such as nonlinear systems, numerical integration, cubic splines, curve fitting, optimization, ordinary differential equations, and graphical tools have been incorporated. The basic structure is to perform matrix operations, such as finding the eigenvalues of a matrix entered from the command line or from an external file via function calls. However, included with MATLAB is a Symbolic Toolbox. This is a computer algebra system based on the MuPAD system, which was developed in Germany in the 1990s. It has the ability to manipulate information in a symbolic manner, which permits the user to obtain exact answers in addition to numerical values. MATLAB is a powerful system that is especially useful and wide spread in engineering and science programs. This is why we have chosen it as the language of choice for this book.

Maple is a computer algebra system developed in 1980 by the Symbolic Computational Group at the University of Waterloo. The design for the original Maple system is presented in the paper by Char, Geddes, Gentlemen, and Gonnet [CGGG]. Maple, which is written in C, also has the ability to manipulate information in a symbolic manner which can give exact answers to mathematical problems such as integrals, differential equations, and linear systems. It contains a programming structure and permits text, as well as commands, to be saved in its worksheet files. These worksheets can then be loaded into Maple and the commands executed.

Numerous other systems are now quite widely available. Mathematica, from Wolfram Research, is a powerful computer algebra system that is used in many in colleges and universities. The open source system Sage is particularly useful for those who do not have access to a commercial product. Information about this system can be found at http://www.sagemath.org/ .

Additional information about software and software libraries can be found in the books by Cody and Waite [CW] and by Kockler [Ko], and in the 1995 article by Dongarra and Walker [DW]. More information about floating-point computation can be found in the book by Chaitini-Chatelin and Fraysse [CF] and the article by Goldberg [Go].

Books that address the application of numerical techniques on parallel computers include those by Schendell [Sche], Phillips and Freeman [PF], and Golub and Ortega [GO].

The Numerical Algorithms Group (NAG) developed its first mathematical software library in 1971. It now has over 10,000 users worldwide and contains over 1000 mathematical and statistical functions ranging from statistical, symbolic, visualization, and numerical simulation software, to compilers and application development tools.

MATLAB was originally written to provide easy access to matrix software developed in the LINPACK and EISPACK projects. The first version was written in the late 1970s for use in courses in matrix theory, linear algebra, and numerical analysis. There are currently more than 500,000 users of MATLAB in more than 100 countries.

Solutions of Equations of One Variable

2.1 Introduction

In this chapter we consider one of the most basic problems of numerical approximation, the root-finding problem. This process involves finding a **root**, or solution, of an equation of the form $f(x) = 0$. A root of this equation is also called a **zero** of the function f. This is one of the oldest known approximation problems, yet research continues in this area at the present time.

The problem of finding an approximation to the root of an equation can be traced at least as far back as 1700 B.C. A cuneiform table in the Yale Babylonian Collection dating from that period gives a sexagesimal (base-60) number equivalent to 1.414222 as an approximation to $\sqrt{2}$, a result that is accurate to within 10^{-5}. This approximation can be found by applying a technique given in Exercise 11 of Section 2.4.

2.2 The Bisection Method

The first and most elementary technique we consider is the **Bisection**, or *Binary-Search*, method. The Bisection method is used to determine, to any specified accuracy that your computer will permit, a solution to $f(x) = 0$ on an interval $[a, b]$, provided that f is continuous on the interval and that $f(a)$ and $f(b)$ are of opposite sign. Although the method will work for the case when more than one root is contained in the interval $[a, b]$, we assume for simplicity of our discussion that the root in this interval is unique.

Bisection Technique

To begin the Bisection method, set $a_1 = a$ and $b_1 = b$, as shown in Figure 2.1, and let p_1 be the midpoint of the interval $[a, b]$:

$$p_1 = a_1 + \frac{b_1 - a_1}{2} = \frac{a_1 + b_1}{2}.$$

If $f(p_1) = 0$, then the root p is given by $p = p_1$; if $f(p_1) \neq 0$, then $f(p_1)$ has the same sign as either $f(a_1)$ or $f(b_1)$.

Figure 2.1

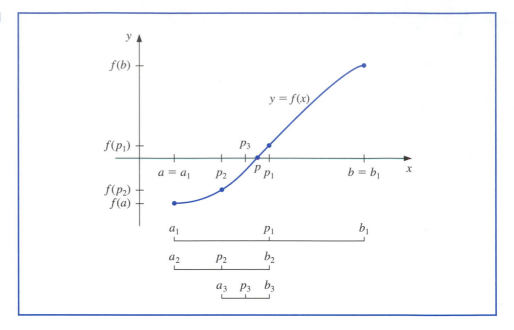

In computer science, the process of dividing a set continually in half to search for the solution to a problem, as the Bisection method does, is known as a *binary search* procedure.

- If $f(p_1)$ and $f(a_1)$ have opposite signs, then p is in the interval (a_1, p_1), and we set

$$a_2 = a_1 \quad \text{and} \quad b_2 = p_1.$$

- If $f(p_1)$ and $f(a_1)$ have the same sign, then p is in the interval (p_1, b_1), and we set

$$a_2 = p_1 \quad \text{and} \quad b_2 = b_1.$$

Reapply the process to the interval $[a_2, b_2]$, and continue forming $[a_3, b_3]$, $[a_4, b_4]$, ... Each new interval will contain p and have length one half of the length of the preceding interval.

Bisection Method

An interval $[a_{n+1}, b_{n+1}]$ containing an approximation to a root of $f(x) = 0$ is constructed from an interval $[a_n, b_n]$ containing the root by first letting

$$p_n = a_n + \frac{b_n - a_n}{2}.$$

Then set

$$a_{n+1} = a_n \quad \text{and} \quad b_{n+1} = p_n \quad \text{if} \quad f(a_n)f(p_n) < 0,$$

and

$$a_{n+1} = p_n \quad \text{and} \quad b_{n+1} = b_n \quad \text{otherwise.}$$

There are three stopping criteria commonly incorporated in the Bisection method, and incorporated within BISECT21.

- The method stops if one of the midpoints happens to coincide with the root.

- It also stops when the length of the search interval is less than some prescribed tolerance we call *TOL*.

- The procedure also stops if the number of iterations exceeds a preset bound N_0.

To start the Bisection method, an interval $[a, b]$ must be found with $f(a) \cdot f(b) < 0$; that is, $f(a)$ and $f(b)$ have opposite signs. At each step, the length of the interval known to contain a zero of f is reduced by a factor of 2. Since the midpoint p_1 must be within $(b-a)/2$ of the root p, and each succeeding iteration divides the interval under consideration by 2, we have

$$|p_n - p| \le \frac{b - a}{2^n}.$$

Consequently, it is easy to determine a bound for the number of iterations needed to ensure a given tolerance. If the root needs to be determined within the tolerance *TOL*, we need to determine the number of iterations, n, so that

$$\frac{b - a}{2^n} < TOL.$$

Using logarithms to solve for n in this inequality gives

$$\frac{b - a}{TOL} < 2^n, \quad \text{which implies that} \quad \log_2 \left(\frac{b - a}{TOL} \right) < n.$$

Since the number of required iterations to guarantee a given accuracy depends on the length of the initial interval $[a, b]$, we want to choose this interval as small as possible. For example, if $f(x) = 2x^3 - x^2 + x - 1$, we have both

$$f(-4) \cdot f(4) < 0 \quad \text{and} \quad f(0) \cdot f(1) < 0,$$

so the Bisection method could be used on either $[-4, 4]$ or $[0, 1]$. Starting the Bisection method on $[0, 1]$ instead of $[-4, 4]$ reduces by 3 the number of iterations required to achieve a specified accuracy.

Example 1 Show that $f(x) = x^3 + 4x^2 - 10 = 0$ has a root in $[1, 2]$ and use the Bisection method to determine an approximation to the root that is accurate to at least within 10^{-4}.

Solution Because $f(1) = -5$ and $f(2) = 14$, the Intermediate Value Theorem ensures that this continuous function has a root in $[1, 2]$. Since $f'(x) = 3x^2 + 8x$ is always positive on $[1, 2]$, the function f is increasing, and, as seen in Figure 2.2, the root is unique.

---

*These programs can be found at http://www.math.ysu.edu/~faires/Numerical-Methods/Programs/

Figure 2.2

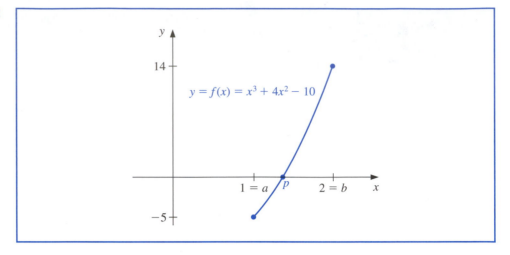

For the first iteration of the Bisection method we use the fact that at the midpoint of [1, 2] we have

$$f(1.5) = 2.375 > 0.$$

This indicates that we should select the interval [1, 1.5] for our second iteration. Then we find that

$$f(1.25) = -1.796875$$

so our new interval becomes [1.25, 1.5], whose midpoint is 1.375. Continuing in this manner gives the values in Table 2.1.

Table 2.1

n	a_n	b_n	p_n	$f(p_n)$
1	1.0	2.0	1.5	2.375
2	1.0	1.5	1.25	−1.79687
3	1.25	1.5	1.375	0.16211
4	1.25	1.375	1.3125	−0.84839
5	1.3125	1.375	1.34375	−0.35098
6	1.34375	1.375	1.359375	−0.09641
7	1.359375	1.375	1.3671875	0.03236
8	1.359375	1.3671875	1.36328125	−0.03215
9	1.36328125	1.3671875	1.365234375	0.000072
10	1.36328125	1.365234375	1.364257813	−0.01605
11	1.364257813	1.365234375	1.364746094	−0.00799
12	1.364746094	1.365234375	1.364990235	−0.00396
13	1.364990235	1.365234375	1.365112305	−0.00194

After 13 iterations, $p_{13} = 1.365112305$ approximates the root p with an error

$$|p - p_{13}| < |b_{14} - a_{14}| = |1.365234375 - 1.365112305| = 0.000122070.$$

Since $|a_{14}| < |p|$, we have

$$\frac{|p - p_{13}|}{|p|} < \frac{|b_{14} - a_{14}|}{|a_{14}|} \leq 9.0 \times 10^{-5},$$

so the approximation is correct to at least within 10^{-4}. The correct value of p to nine decimal places is $p = 1.365230013$. Note that p_9 is closer to p than is the final approximation p_{13}. You might suspect this is true because $|f(p_9)| < |f(p_{13})|$. ∎

The Bisection method, although conceptually clear, has serious drawbacks. It is slow to converge relative to the other techniques we will discuss, and a good intermediate approximation may be inadvertently discarded. This happened, for example, with p_9 in Example 1. However, the method has the important property that it always converges to a solution and it is easy to determine a bound for the number of iterations needed to ensure a given accuracy. For these reasons, the Bisection method is frequently used as a dependable starting procedure for the more efficient methods presented later in this chapter.

The bound for the number of iterations for the Bisection method assumes that the calculations are performed using infinite-digit arithmetic. When implementing the method on a computer, consideration must be given to the effects of round-off error. For example, the computation of the midpoint of the interval $[a_n, b_n]$ should be found from the equation

$$p_n = a_n + \frac{b_n - a_n}{2}$$

instead of from the algebraically equivalent equation

$$p_n = \frac{a_n + b_n}{2}.$$

The first equation adds a small correction, $(b_n - a_n)/2$, to the known value a_n. When $b_n - a_n$ is near the maximum precision of the machine, this correction might be in error, but the error would not significantly affect the computed value of p_n. However, in the second equation, if $b_n - a_n$ is near the maximum precision of the machine, it is possible for p_n to return a midpoint that is not even in the interval $[a_n, b_n]$.

A number of tests can be used to see if a root has been found. We would normally use a test of the form

$$|f(p_n)| < \epsilon,$$

where $\epsilon > 0$ would be a small number related in some way to the tolerance. However, it is also possible for the value $f(p_n)$ to be small when p_n is not near the root p.

The Latin word *signum* means "token" or "sign." So the signum function quite naturally returns the sign of a number (unless the number is 0).

As a final remark, to determine which subinterval of $[a_n, b_n]$ contains a root of f, it is better to make use of **signum** function, which is defined as

$$\text{sgn}(x) = \begin{cases} -1, & \text{if } x < 0, \\ 0, & \text{if } x = 0, \\ 1, & \text{if } x > 0. \end{cases}$$

The test

$$\text{sgn}(f(a_n))\,\text{sgn}(f(p_n)) < 0 \qquad \text{instead of} \qquad f(a_n)f(p_n) < 0$$

gives the same result but avoids the possibility of overflow or underflow in the multiplication of $f(a_n)$ and $f(p_n)$.

EXERCISE SET 2.2

1. Use the Bisection method to find p_3 for $f(x) = \sqrt{x} - \cos x$ on $[0, 1]$.

2. Let $f(x) = 3(x + 1)(x - \frac{1}{2})(x - 1)$. Use the Bisection method on the following intervals to find p_3.
    a. $[-2, 1.5]$
    b. $[-1.25, 2.5]$

3. Use the Bisection method to find solutions accurate to within 10^{-2} for $x^3 - 7x^2 + 14x - 6 = 0$ on each interval.
    a. $[0, 1]$
    b. $[1, 3.2]$
    c. $[3.2, 4]$

4. Use the Bisection method to find solutions accurate to within 10^{-2} for $x^4 - 2x^3 - 4x^2 + 4x + 4 = 0$ on each interval.
    a. $[-2, -1]$
    b. $[0, 2]$
    c. $[2, 3]$
    d. $[-1, 0]$

5. a. Sketch the graphs of $y = x$ and $y = 2 \sin x$.
    b. Use the Bisection method to find an approximation to within 10^{-2} to the first positive value of x with $x = 2 \sin x$.

6. a. Sketch the graphs of $y = x$ and $y = \tan x$.
    b. Use the Bisection method to find an approximation to within 10^{-2} to the first positive value of x with $x = \tan x$.

7. Let $f(x) = (x + 2)(x + 1)x(x - 1)^3(x - 2)$. To which zero of f does the Bisection method converge for the following intervals?
    a. $[-3, 2.5]$
    b. $[-2.5, 3]$
    c. $[-1.75, 1.5]$
    d. $[-1.5, 1.75]$

8. Let $f(x) = (x + 2)(x + 1)^2x(x - 1)^3(x - 2)$. To which zero of f does the Bisection method converge for the following intervals?
    a. $[-1.5, 2.5]$
    b. $[-0.5, 2.4]$
    c. $[-0.5, 3]$
    d. $[-3, -0.5]$

9. Use the Bisection method to find an approximation to $\sqrt{3}$ correct to within 10^{-4}. [*Hint:* Consider $f(x) = x^2 - 3$.]

10. Use the Bisection method to find an approximation to $\sqrt[3]{25}$ correct to within 10^{-4}.

11. Find a bound for the number of Bisection method iterations needed to achieve an approximation with accuracy 10^{-3} to the solution of $x^3 + x - 4 = 0$ lying in the interval $[1, 4]$. Find an approximation to the root with this degree of accuracy.

12. Find a bound for the number of Bisection method iterations needed to achieve an approximation with accuracy 10^{-4} to the solution of $x^3 - x - 1 = 0$ lying in the interval $[1, 2]$. Find an approximation to the root with this degree of accuracy.

13. The function defined by $f(x) = \sin \pi x$ has zeros at every integer. Show that when $-1 < a < 0$ and $2 < b < 3$, the Bisection method converges to
    a. 0,   if $a + b < 2$
    b. 2,   if $a + b > 2$
    c. 1,   if $a + b = 2$

2.3 The Secant Method

Although the Bisection method always converges, the speed of convergence is often too slow for general use. Figure 2.3 gives a graphical interpretation of the Bisection method that can be used to discover how improvements on this technique can be derived. It shows the graph of a continuous function that is negative at a_1 and positive at b_1. The first approximation p_1 to the root p is found by drawing the line joining the points $(a_1, \text{sgn}(f(a_1))) = (a_1, -1)$ and $(b_1, \text{sgn}(f(b_1))) = (b_1, 1)$ and letting p_1 be the point where this line intersects the x-axis. In essence, the line joining $(a_1, -1)$ and $(b_1, 1)$ has been used to approximate the graph of f on the interval $[a_1, b_1]$. Successive approximations apply this same process on subintervals of $[a_1, b_1]$, $[a_2, b_2]$, and so on. Notice that the Bisection method uses no information about the function f except the fact that $f(x)$ is positive and negative at certain values of x.

Figure 2.3

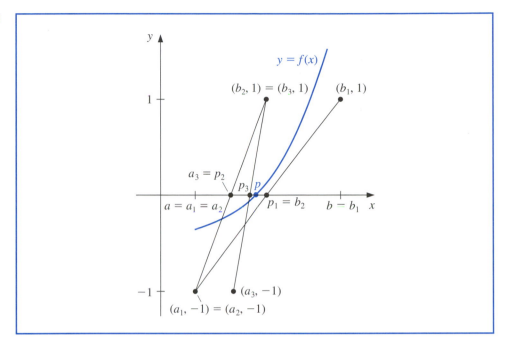

The Secant Method

Suppose that in the initial step we know that $|f(a_1)| < |f(b_1)|$. Then we would expect the root p to be closer to a_1 than to b_1. Alternatively, if $|f(b_1)| < |f(a_1)|$, p is likely to be closer to b_1 than to a_1. Instead of choosing the intersection of the line through $(a_1, \operatorname{sgn}(f(a_1))) = (a_1, -1)$ and $(b_1, \operatorname{sgn}(f(b_1))) = (b_1, 1)$ as the approximation to the root p, the *Secant method* chooses the x-intercept of the secant line to the curve, the line through $(a_1, f(a_1))$ and $(b_1, f(b_1))$. This places the approximation closer to the endpoint of the interval for which f has smaller absolute value, as shown in Figure 2.4.

Figure 2.4

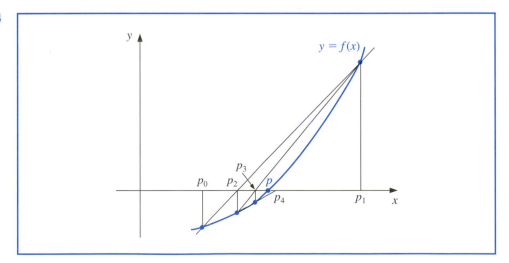

The sequence of approximations generated by the Secant method can be started by setting $p_0 = a$ and $p_1 = b$. The equation of the secant line through $(p_0, f(p_0))$ and $(p_1, f(p_1))$ is

$$y = f(p_1) + \frac{f(p_1) - f(p_0)}{p_1 - p_0}(x - p_1),$$

and the x-intercept $(p_2, 0)$ of this line satisfies

$$0 = f(p_1) + \frac{f(p_1) - f(p_0)}{p_1 - p_0}(p_2 - p_1).$$

Solving for p_2 gives

$$p_2 = p_1 - \frac{f(p_1)(p_1 - p_0)}{f(p_1) - f(p_0)}.$$

Secant Method

The approximation p_{n+1}, for $n > 1$, to a root of $f(x) = 0$ is computed from the approximations p_n and p_{n-1} using the equation

$$p_{n+1} = p_n - \frac{f(p_n)(p_n - p_{n-1})}{f(p_n) - f(p_{n-1})}.$$

The Secant method does not have the root-bracketing property of the Bisection method. That is, there is no assurance that the solution p is between p_n and p_{n+1}. As a consequence, the method does not always converge, but when it does converge, it generally does so much faster than the Bisection method.

We use two stopping conditions in the Secant method. First, we assume that p_n is sufficiently accurate when $|p_n - p_{n-1}|$ is within a given tolerance. Also, a safeguard exit based upon a maximum number of iterations is given in case the method fails to converge as expected.

The iteration equation should *not* be simplified algebraically to

$$p_{n+1} = p_n - \frac{f(p_n)(p_n - p_{n-1})}{f(p_n) - f(p_{n-1})} = \frac{f(p_{n-1})p_n - f(p_n)p_{n-1}}{f(p_{n-1}) - f(p_n)}.$$

Program SECANT22 implements the Secant method.

Although this is algebraically equivalent to the iteration equation, it could increase the significance of rounding error if the nearly equal numbers $f(p_{n-1})p_n$ and $f(p_n)p_{n-1}$ are subtracted.

Example 1   Use the Secant method to approximate a root of the equation $x^3 + 4x^2 - 10 = 0$.

Solution   We first need to choose two starting values p_0 and p_1. The Intermediate Value Theorem implies that a root lies in the interval $[1, 2]$ because $f(p_0) = -5$ and $f(p_1) = 14$, so a reasonable choice might be $p_0 = 1$ and $p_1 = 2$.

To employ MATLAB to find the approximations, we first define the function $f(x)$ and the numbers p_0 and p_1 with

```
f=inline('x^3+4*x^2-10','x')
p0=1, p1=2
```

The values of $f(p_0)$ and $f(p_1)$ are computed with

```
fp0=f(p0), fp1=f(p1)
```

The first Secant method approximation, denoted p2, is found by

```
p2=p1-fp1*(p1-p0)/(fp1-fp0)
```

which MATLAB gives as

$$p2 = 1.263157894736842$$

Continuing in the same manner, we then find that $fp2 \equiv f(p_2) = -1.602274384020994$ and that

$$p3 = p2 - \frac{fp2 * (p2 - p1)}{fp2 - fp1} = 1.338827838827839.$$

The program SECANT22 with inputs $p_0 = 1$, $p_1 = 2$, $TOL = 0.0005$, and $N_0 = 20$ produces the results in Table 2.2. About half the number of iterations are needed, compared to the Bisection method in Example 1 of Section 2.2. Further,

$$|p - p_6| = |1.3652300134 - 1.3652300011| < 1.3 \times 10^{-8}$$

is much smaller than the tolerance 0.0005. ■

Table 2.2

n	p_n	$f(p_n)$
2	1.2631578947	-1.6022743840
3	1.3388278388	-0.4303647480
4	1.3666163947	0.0229094308
5	1.3652119026	-0.0002990679
6	1.3652300011	-0.0000002032

The Method of False Position

There are other reasonable choices for generating a sequence of approximations based on the intersection of an approximating line and the x-axis. The **method of False Position** (or *Regula Falsi*) is a hybrid bisection-secant method that constructs approximating lines similar to those of the Secant method but always brackets the root in the manner of the Bisection method. As with the Bisection method, the method of False Position requires that an initial interval $[a, b]$ first be found, with $f(a)$ and $f(b)$ of opposite sign. With $a_1 = a$ and $b_1 = b$, the approximation, p_2, is given by

$$p_2 = a_1 - \frac{f(a_1)(b_1 - a_1)}{f(b_1) - f(a_1)}.$$

If $f(a_1) \cdot f(p_2) < 0$, then set $a_2 = a_1$ and $b_2 = p_2$. Alternatively, if $f(a_1) \cdot f(p_2) > 0$, then set $a_2 = p_2$ and $b_2 = b_1$. (See Figure 2.5.)

Figure 2.5

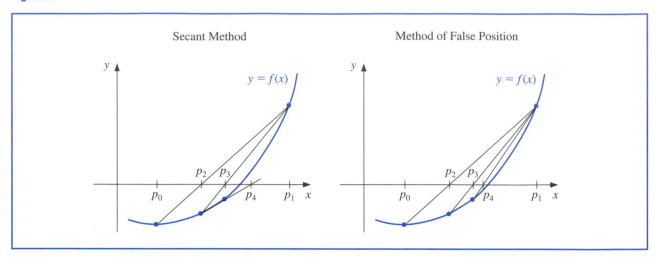

Method of False Position

An interval $[a_{n+1}, b_{n+1}]$, for $n > 1$, containing an approximation to a root of $f(x) = 0$ is found from an interval $[a_n, b_n]$ containing the root by first computing

$$p_{n+1} = a_n - \frac{f(a_n)(b_n - a_n)}{f(b_n) - f(a_n)}.$$

Then set

$$a_{n+1} = a_n \quad \text{and} \quad b_{n+1} = p_{n+1} \quad \text{if} \quad f(a_n)f(p_{n+1}) < 0,$$

and

$$a_{n+1} = p_{n+1} \quad \text{and} \quad b_{n+1} = b_n \quad \text{otherwise.}$$

Program FALPOS23 implements the method of False Position.

The False Position method might appear superior to the Secant method, but it generally converges more slowly. The results in Table 2.3 illustrate this for the problem we considered in Example 1. In fact, the method of False Position can even converge more slowly than the Bisection method (as the problem given in Exercise 14 shows), although this is not usually the case.

Table 2.3

n	a_n	b_n	p_{n+1}	$f(p_{n+1})$
1	1.00000000	2.00000000	1.26315789	−1.60227438
2	1.26315789	2.00000000	1.33882784	−0.43036475
3	1.33882784	2.00000000	1.35854634	−0.11000879
4	1.35854634	2.00000000	1.36354744	−0.02776209
5	1.36354744	2.00000000	1.36480703	−0.00698342
6	1.36480703	2.00000000	1.36512372	−0.00175521
7	1.36512372	2.00000000	1.36520330	−0.00044106

EXERCISE SET 2.3

1. Let $f(x) = x^2 - 6$, $p_0 = 3$, and $p_1 = 2$. Find p_3 using each method.

    a.   Secant method   **b.**   method of False Position

2. Let $f(x) = -x^3 - \cos x$, $p_0 = -1$, and $p_1 = 0$. Find p_3 using each method.

    a.   Secant method   **b.**   method of False Position

3. Use the Secant method to find solutions accurate to within 10^{-4} for the following problems.

    a. $x^3 - 2x^2 - 5 = 0$, on $[1, 4]$ **b.** $x^3 + 3x^2 - 1 = 0$, on $[-3, -2]$
    c. $x - \cos x = 0$, on $[0, \pi/2]$ **d.** $x - 0.8 - 0.2 \sin x = 0$, on $[0, \pi/2]$

4. Use the Secant method to find solutions accurate to within 10^{-5} for the following problems.

    a. $2x \cos 2x - (x - 2)^2 = 0$ on $[2, 3]$ and on $[3, 4]$

    b. $(x - 2)^2 - \ln x = 0$ on $[1, 2]$ and on $[e, 4]$

    c. $e^x - 3x^2 = 0$ on $[0, 1]$ and on $[3, 5]$

    d. $\sin x - e^{-x} = 0$ on $[0, 1]$, on $[3, 4]$ and on $[6, 7]$

5. Repeat Exercise 3 using the method of False Position.

6. Repeat Exercise 4 using the method of False Position.

7. Use the Secant method to find all four solutions of $4x \cos(2x) - (x - 2)^2 = 0$ in $[0, 8]$ accurate to within 10^{-5}.

8. Use the Secant method to find all solutions of $x^2 + 10 \cos x = 0$ accurate to within 10^{-5}.

9. Use the Secant method to find an approximation to $\sqrt{3}$ correct to within 10^{-4}, and compare the results to those obtained in Exercise 9 of Section 2.2.

10. Use the Secant method to find an approximation to $\sqrt[3]{25}$ correct to within 10^{-6}, and compare the results to those obtained in Exercise 10 of Section 2.2.

11. Approximate, to within 10^{-4}, the value of x that produces the point on the graph of $y = x^2$ that is closest to $(1, 0)$. [*Hint:* Minimize $[d(x)]^2$, where $d(x)$ represents the distance from (x, x^2) to $(1, 0)$.]

12. Approximate, to within 10^{-4}, the value of x that produces the point on the graph of $y = 1/x$ that is closest to $(2, 1)$.

13. The fourth-degree polynomial

$$f(x) = 230x^4 + 18x^3 + 9x^2 - 221x - 9$$

has two real zeros, one in $[-1, 0]$ and the other in $[0, 1]$. Attempt to approximate these zeros to within 10^{-6} using each method.

    a.   method of False Position   **b.**   Secant method

14. The function $f(x) = \tan \pi x - 6$ has a zero at $(1/\pi) \arctan 6 \approx 0.447431543$. Let $p_0 = 0$ and $p_1 = 0.48$ and use 10 iterations of each of the following methods to approximate this root. Which method is most successful and why?

    a.   Bisection method

    b.   method of False Position

    c.   Secant method

15. The sum of two numbers is 20. If each number is added to its square root, the product of the two sums is 155.55. Determine the two numbers to within 10^{-4}.

16. A trough of length L has a cross section in the shape of a semicircle with radius r. (See the accompanying figure.) When filled with water to within a distance h of the top, the volume, V, of water is

$$V = L \left[0.5\pi r^2 - r^2 \arcsin\left(\frac{h}{r}\right) - h(r^2 - h^2)^{1/2} \right]$$

Suppose $L = 10$ ft, $r = 1$ ft, and $V = 12.4$ ft^3. Find the depth of water in the trough to within 0.01 ft.

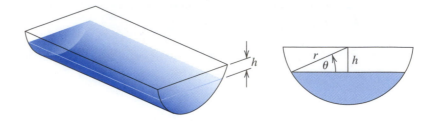

17.   The particle in the figure starts at rest on a smooth inclined plane whose angle θ is changing at a constant rate

$$\frac{d\theta}{dt} = \omega < 0.$$

At the end of t seconds, the position of the object is given by

$$x(t) = \frac{g}{2\omega^2}\left(\frac{e^{\omega t} - e^{-\omega t}}{2} - \sin\omega t\right).$$

Suppose the particle has moved 1.7 ft in 1 s. Find, to within 10^{-5}, the rate ω at which θ changes. Assume that $g = -32.17$ ft/s^2.

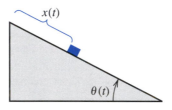

2.4 Newton's Method

Isaac Newton (1641–1727) was one of the most brilliant scientists of all time. The late 17th century was a vibrant period for science and mathematics and Newton's work touches nearly every aspect of mathematics. His method for solving was introduced to find a root of the equation $y^3 - 2y - 5 = 0$. Although he demonstrated the method only for polynomials, it is clear that he realized its broader applications.

The Bisection and Secant methods both have geometric representations that use the zero of an approximating line to the graph of a function f to approximate the solution to $f(x) = 0$. The increase in accuracy of the Secant method over the Bisection method is a consequence of the fact that the secant line to the curve better approximates the graph of f than does the line used to generate the approximations in the Bisection method.

The line that *best* approximates the graph of the function at a point on its graph is the tangent line to the graph at that point. Using this line instead of the secant line produces **Newton's method** (also called the *Newton–Raphson method*), the technique we consider in this section.

Newton's Method

Suppose that p_0 is an initial approximation to the root p of the equation $f(x) = 0$ and that f' exists in an interval containing all the approximations to p. The slope of the tangent line to the graph of f at the point $(p_0, f(p_0))$ is $f'(p_0)$, so the equation of this tangent line is

$$y - f(p_0) = f'(p_0)(x - p_0).$$

This tangent line crosses the x-axis when the y-coordinate of the point on the line is 0, so the next approximation, p_1, to p satisfies

$$0 - f(p_0) = f'(p_0)(p_1 - p_0),$$

which implies that

$$p_1 = p_0 - \frac{f(p_0)}{f'(p_0)},$$

provided that $f'(p_0) \neq 0$. Subsequent approximations are found for p in a similar manner, as shown in Figure 2.6.

Figure 2.6

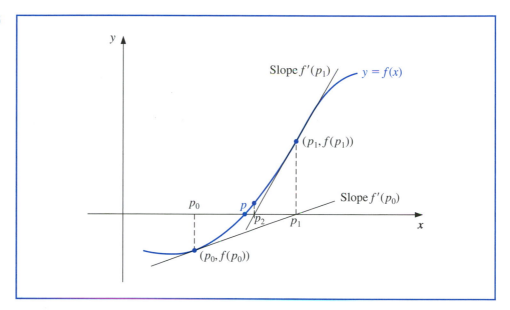

Newton's Method

The approximation p_{n+1} to a root of $f(x) = 0$ is computed from the approximation p_n using the equation

$$p_{n+1} = p_n - \frac{f(p_n)}{f'(p_n)},$$

provided that $f'(p_n) \neq 0$.

Example 1   Use Newton's method with $p_0 = 1$ to approximate the root of the equation $x^3 + 4x^2 - 10 = 0$.

Solution   We will use MATLAB to find the first two iterations of Newton's method with $p_0 = 1$. We first define $f(x)$, $f'(x)$, and $p0$ with

Program NEWTON24 implements Newton's method.

```
f=inline('x^3+4*x^2-10','x')
fp = inline('3*x^2+8*x','x')
p0=1
```

The first iteration of Newton's method gives $p1 = 1.454545454545455$ using the command

```
p1=p0-f(p0)/fp(p0)
```

The second iteration is $p2 = 1.368900401069519$ using

```
p2=p1-f(p1)/fp(p1)
```

and so on. The process can be continued to generate the entries in Table 2.4. This table was generated using $p_0 = 1$, $TOL = 0.0005$, and $N_0 = 20$ in the program NEWTON24. Note that we have

$$|p - p_4| \approx 10^{-10}.$$

If we compare the convergence of this method with those applied to this problem previously, we can see that the number of iterations needed to solve the problem by Newton's method is less than the number needed for the Secant method. Recall that the Secant method required less than half the iterations needed for the Bisection method. ▪

Table 2.4

n	p_n	$f(p_n)$
1	1.4545454545	1.5401953418
2	1.3689004011	0.0607196886
3	1.3652366002	0.0001087706
4	1.3652300134	0.0000000004

Convergence Using Newton's Method

Newton's method generally produces accurate results in just a few iterations. With the aid of Taylor polynomials we can see why this is true. Suppose p is the solution to $f(x) = 0$ and that f'' exists on an interval containing both p and the approximation p_n. Expanding f in the first Taylor polynomial at p_n gives

$$f(x) = f(p_n) + f'(p_n)(x - p_n) + \frac{f''(\xi)}{2}(x - p_n)^2,$$

and evaluating at $x = p$ gives

$$0 = f(p) = f(p_n) + f'(p_n)(p - p_n) + \frac{f''(\xi)}{2}(p - p_n)^2,$$

where ξ lies between p_n and p. Consequently, if $f'(p_n) \neq 0$, we have

$$p - p_n + \frac{f(p_n)}{f'(p_n)} = -\frac{f''(\xi)}{2f'(p_n)}(p - p_n)^2.$$

Since

$$p_{n+1} = p_n - \frac{f(p_n)}{f'(p_n)},$$

this implies that

$$p - p_{n+1} = p - p_n + \frac{f(p_n)}{f'(p_n)} = -\frac{f''(\xi)}{2f'(p_n)}(p - p_n)^2.$$

If a positive constant M exists with $|f''(x)| \le M$ on an interval about p, and if p_n is within this interval, then

$$|p - p_{n+1}| \le \frac{M}{2|f'(p_n)|}|p - p_n|^2.$$

The important feature of this inequality is that if f' is nonzero on the interval, then the error $|p - p_{n+1}|$ of the $(n + 1)$st approximation is bounded by approximately the square of the error of the nth approximation, $|p - p_n|$. This implies that Newton's method has the tendency to approximately double the number of digits of accuracy with each successive approximation. Newton's method is not, however, infallible, as we will see later in this section.

Example 2   Find an approximation to the solution of the equation $x = 3^{-x}$ that is accurate to within 10^{-8}.

Solution   A solution to $x = 3^{-x}$ corresponds to a solution of

$$0 = f(x) = x - 3^{-x}.$$

Since f is continuous with $f(0) = -1$ and $f(1) = \frac{2}{3}$, a solution of the equation lies in the interval $(0, 1)$. We have chosen the initial approximation to be the midpoint of this interval, $p_0 = 0.5$. Succeeding approximations are generated by applying the formula

$$p_{n+1} = p_n - \frac{f(p_n)}{f'(p_n)} = p_n - \frac{p_n - 3^{-p_n}}{1 + 3^{-p_n} \ln 3}.$$

These approximations are listed in Table 2.5, together with differences between successive approximations. Since Newton's method tends to double the number of decimal places of accuracy with each iteration, it is reasonable to suspect that p_3 is correct at least to the places listed.   ■

Table 2.5

n	p_n	$|p_n - p_{n-1}|$
0	0.500000000	
1	0.547329757	0.047329757
2	0.547808574	0.000478817
3	0.547808622	0.000000048

The success of Newton's method is predicated on the assumption that the derivative of f is nonzero at the approximations to the zero p. If f' is continuous, this means that the technique will be satisfactory provided that $f'(p) \ne 0$ and that a sufficiently accurate initial approximation is used. The condition $f'(p) \ne 0$ is not trivial; it is true precisely when p is a **simple zero**. A simple zero of a function f occurs at p if a function q exists with the property that, for $x \ne p$,

$$f(x) = (x - p)q(x), \qquad \text{where} \quad \lim_{x \to p} q(x) \ne 0.$$

In general, a **zero of multiplicity** m of a function f occurs at p if a function q exists with the property that, for $x \ne p$,

$$f(x) = (x - p)^m q(x), \qquad \text{where} \quad \lim_{x \to p} q(x) \ne 0.$$

So a simple zero is one that has multiplicity 1.

By taking consecutive derivatives and evaluating at p it can be shown that

- A function f with m derivatives at p has a zero of multiplicity m at p if and only if

$$0 = f(p) = f'(p) = \cdots = f^{(m-1)}(p), \text{ but } f^{(m)}(p) \neq 0.$$

When the zero is not simple, Newton's method might converge, but not with the speed we have seen in our previous examples.

Example 3   Let $f(x) = e^x - x - 1$. **(a)** Show that f has a zero of multiplicity 2 at $x = 0$. **(b)** Show that Newton's method with $p_0 = 1$ converges to this zero but not as rapidly as the convergence in Examples 1 and 2.

Table 2.6

n	p_n
0	1.0
1	0.58198
2	0.31906
3	0.16800
4	0.08635
5	0.04380
6	0.02206
7	0.01107
8	0.005545
9	2.7750×10^{-3}
10	1.3881×10^{-3}
11	6.9411×10^{-4}
12	3.4703×10^{-4}
13	1.7416×10^{-4}
14	8.8041×10^{-5}
15	4.2610×10^{-5}
16	1.9142×10^{-6}

Solution   **(a)** We have

$$f(x) = e^x - x - 1, \qquad f'(x) = e^x - 1 \qquad \text{and} \quad f''(x) = e^x,$$

so

$$f(0) = e^0 - 0 - 1 = 0, \qquad f'(0) = e^0 - 1 = 0 \qquad \text{and} \quad f''(0) = e^0 = 1.$$

This implies that f has a zero of multiplicity 2 at $x = 0$.

(b) The first two terms generated by Newton's method applied to f with $p_0 = 1$ are

$$p_1 = p_0 - \frac{f(p_0)}{f'(p_0)} = 1 - \frac{e-2}{e-1} \approx 0.58198,$$

and

$$p_2 = p_1 - \frac{f(p_1)}{f'(p_1)} \approx 0.58198 - \frac{0.20760}{0.78957} \approx 0.31906.$$

The first eight terms of the sequence generated by Newton's method are shown in Table 2.6. The sequence is clearly converging to 0, but not as rapidly as the convergence in Examples 1 and 2. The graph of f is shown in Figure 2.7.   ▪

Figure 2.7

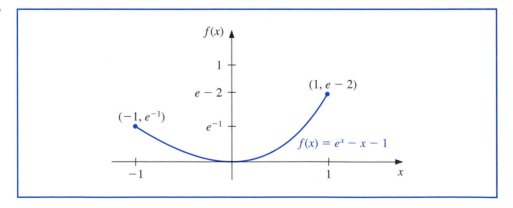

One method for improving the convergence to a multiple root is considered in Exercise 8.

EXERCISE SET 2.4

1. Let $f(x) = x^2 - 6$ and $p_0 = 1$. Use Newton's method to find p_2.

2. Let $f(x) = -x^3 - \cos x$ and $p_0 = -1$. Use Newton's method to find p_2. Could $p_0 = 0$ be used for this problem?

3. Use Newton's method to find solutions accurate to within 10^{-4} for the following problems.

    a. $x^3 - 2x^2 - 5 = 0$,   on $[1, 4]$

    b. $x^3 + 3x^2 - 1 = 0$,   on $[-3, -2]$

    c. $x - \cos x = 0$,   on $[0, \pi/2]$

    d. $x - 0.8 - 0.2 \sin x = 0$,   on $[0, \pi/2]$

4. Use Newton's method to find solutions accurate to within 10^{-5} for the following problems.

    a. $2x \cos 2x - (x - 2)^2 = 0$,   on $[2, 3]$ and $[3, 4]$

    b. $(x - 2)^2 - \ln x = 0$,   on $[1, 2]$ and $[e, 4]$

    c. $e^x - 3x^2 = 0$,   on $[0, 1]$ and $[3, 5]$

    d. $\sin x - e^{-x} = 0$,   on $[0, 1]$, $[3, 4]$, and $[6, 7]$

5. Use Newton's method to find all four solutions of $4x \cos(2x) - (x - 2)^2 = 0$ in $[0, 8]$ accurate to within 10^{-5}.

6. Use Newton's method to find all solutions of $x^2 + 10 \cos x = 0$ accurate to within 10^{-5}.

7. Use Newton's method to approximate the solutions of the following equations to within 10^{-5} in the given intervals. In these problems, the convergence will be slower than normal because the zeros are not simple.

    a. $x^2 - 2xe^{-x} + e^{-2x} = 0$,   on $[0, 1]$

    b. $\cos(x + \sqrt{2}) + x(x/2 + \sqrt{2}) = 0$,   on $[-2, -1]$

    c. $x^3 - 3x^2(2^{-x}) + 3x(4^{-x}) + 8^{-x} = 0$,   on $[0, 1]$

    d. $e^{6x} + 3(\ln 2)^2 e^{2x} - (\ln 8)e^{4x} - (\ln 2)^3$,   on $[-1, 0]$

8. The numerical method defined by

$$p_n = p_{n-1} - \frac{f(p_{n-1})f'(p_{n-1})}{[f'(p_{n-1})]^2 - f(p_{n-1})f''(p_{n-1})},$$

for $n = 1, 2, \ldots$, can be used instead of Newton's method for equations having multiple zeros. Repeat Exercise 7 using this method.

9. Use Newton's method to find an approximation to $\sqrt{3}$ correct to within 10^{-4}, and compare the results to those obtained in Exercise 9 of Sections 2.2 and 2.3.

10. Use Newton's method to find an approximation to $\sqrt[3]{25}$ correct to within 10^{-6}, and compare the results to those obtained in Exercise 10 of Sections 2.2 and 2.3.

11. Newton's method applied to the function $f(x) = x^2 - 2$ with a positive initial approximation p_0 converges to the only positive solution, $\sqrt{2}$.

    a. Show that Newton's method in this situation assumes the form that the Babylonians used to approximate $\sqrt{2}$:

$$p_{n+1} = \frac{1}{2} p_n + \frac{1}{p_n}.$$

    b. Use the sequence in (a) with $p_0 = 1$ to determine an approximation that is accurate to within 10^{-5}.

12. In Exercise 14 of Section 2.3, we found that for $f(x) = \tan \pi x - 6$, the Bisection method on $[0, 0.48]$ converges more quickly than the method of False Position with $p_0 = 0$ and $p_1 = 0.48$. Also, the Secant method with these values of p_0 and p_1 does not give convergence. Apply Newton's method to this problem with (a) $p_0 = 0$ and (b) $p_0 = 0.48$. (c) Explain the reason for any discrepancies.

13. Use Newton's method to determine the first positive solution to the equation $\tan x = x$, and explain why this problem can give difficulties.

14. Use Newton's method to solve the equation

$$0 = \frac{1}{2} + \frac{1}{4}x^2 - x \sin x - \frac{1}{2} \cos 2x, \quad \text{with } p_0 = \frac{\pi}{2}.$$

Iterate using Newton's method until an accuracy of 10^{-5} is obtained. Explain why the result seems unusual for Newton's method. Also, solve the equation with $p_0 = 5\pi$ and $p_0 = 10\pi$.

15. Player A will shut out (win by a score of 21–0) player B in a game of racquetball with probability

$$P = \frac{1+p}{2} \left(\frac{p}{1 - p + p^2} \right)^{21},$$

where p denotes the probability that A will win any specific rally (independent of the server). (See [K,J], p. 267.) Determine, to within 10^{-3}, the minimal value of p that will ensure that A will shut out B in at least half the matches they play.

16. The function described by $f(x) = \ln(x^2 + 1) - e^{0.4x} \cos \pi x$ has an infinite number of zeros.

    a. Determine, within 10^{-6}, the only negative zero.

    b. Determine, within 10^{-6}, the four smallest positive zeros.

    c. Determine a reasonable initial approximation to find the nth smallest positive zero of f. [*Hint:* Sketch an approximate graph of f.]

    d. Use (c) to determine, within 10^{-6}, the 25th smallest positive zero of f.

17. The accumulated value of a savings account based on regular periodic payments can be determined from the *annuity due equation,*

$$A = \frac{P}{i}[(1+i)^n - 1].$$

In this equation, A is the amount in the account, P is the amount regularly deposited, and i is the rate of interest per period for the n deposit periods. An engineer would like to have a savings account valued at \$750,000 upon retirement in 20 years and can afford to put \$1500 per month toward this goal. What is the minimum interest rate at which this amount can be invested, assuming that the interest is compounded monthly?

18. Problems involving the amount of money required to pay off a mortgage over a fixed period of time involve the formula

$$A = \frac{P}{i}[1 - (1+i)^{-n}],$$

known as an *ordinary annuity equation.* In this equation, A is the amount of the mortgage, P is the amount of each payment, and i is the interest rate per period for the n payment periods. Suppose that a 30-year home mortgage in the amount of \$135,000 is needed and that the borrower can afford house payments of at most \$1000 per month. What is the maximum interest rate the borrower can afford to pay?

19. A drug administered to a patient produces a concentration in the blood stream given by $c(t) = Ate^{-t/3}$ milligrams per milliliter t hours after A units have been injected. The maximum safe concentration is 1 mg/ml.

    a. What amount should be injected to reach this maximum safe concentration and when does this maximum occur?

    b. An additional amount of this drug is to be administered to the patient after the concentration falls to 0.25 mg/ml. Determine, to the nearest minute, when this second injection should be given.

    c. Assuming that the concentration from consecutive injections is additive and that 75% of the amount originally injected is administered in the second injection, when is it time for the third injection?

20. Let $f(x) = 3^{3x+1} - 7 \cdot 5^{2x}$.

    a. Use the MATLAB function `fzero` to try to find all zeros of f.

    b. Plot $f(x)$ to find initial approximations to roots of f.

    c. Use Newton's method to find roots of f to within 10^{-16}.

    d. Find the exact solutions of $f(x) = 0$ algebraically.

2.5 Error Analysis and Accelerating Convergence

In the previous section we found that Newton's method generally converges very rapidly if a sufficiently accurate initial approximation has been found. This rapid speed of convergence is due to the fact that Newton's method produces *quadratically* convergent approximations.

Order of Convergence

Suppose that a method produces a sequence $\{p_n\}$ of approximations that converge to a number p.

- The sequence converges **linearly** if, for large values of n, a constant $0 < M$ exists with

$$|p - p_{n+1}| \le M|p - p_n|.$$

- The sequence converges **quadratically** if, for large values of n, a constant $0 < M$ exists with

$$|p - p_{n+1}| \le M|p - p_n|^2.$$

The constant M is called an **asymptotic error constant**.

The following illustrates the advantage of quadratic over linear convergence.

Illustration Suppose that $\{p_n\}_{n=0}^{\infty}$ is linearly convergent to 0 with

$$\lim_{n \to \infty} \frac{|p_{n+1}|}{|p_n|} = 0.5$$

and that $\{\tilde{p}_n\}_{n=0}^{\infty}$ is quadratically convergent to 0 with the same asymptotic error constant, $M = 0.5$, so

$$\lim_{n \to \infty} \frac{|\tilde{p}_{n+1}|}{|\tilde{p}_n|^2} = 0.5.$$

For simplicity we assume that for each n we have

$$\frac{|p_{n+1}|}{|p_n|} \approx 0.5 \quad \text{and} \quad \frac{|\tilde{p}_{n+1}|}{|\tilde{p}_n|^2} \approx 0.5.$$

For the linearly convergent scheme, this means that

$$|p_n - 0| = |p_n| \approx 0.5|p_{n-1}| \approx (0.5)^2|p_{n-2}| \approx \cdots \approx (0.5)^n|p_0|,$$

whereas the quadratically convergent procedure has

$$|\tilde{p}_n - 0| = |\tilde{p}_n| \approx 0.5|\tilde{p}_{n-1}|^2 \approx (0.5)[0.5|\tilde{p}_{n-2}|^2]^2 = (0.5)^3|\tilde{p}_{n-2}|^4$$

$$\approx (0.5)^3[(0.5)|\tilde{p}_{n-3}|^2]^4 = (0.5)^7|\tilde{p}_{n-3}|^8$$

$$\approx \cdots \approx (0.5)^{2^n-1}|\tilde{p}_0|^{2^n}.$$

Table 2.7 illustrates the relative speed of convergence of the sequences to 0 if $|p_0| = |\tilde{p}_0| = 1$.

Table 2.7

n	Linear Convergence Sequence $\{p_n\}_{n=0}^{\infty}$ $(0.5)^n$	Quadratic Convergence Sequence $\{\tilde{p}_n\}_{n=0}^{\infty}$ $(0.5)^{2^n-1}$
1	5.0000×10^{-1}	5.0000×10^{-1}
2	2.5000×10^{-1}	1.2500×10^{-1}
3	1.2500×10^{-1}	7.8125×10^{-3}
4	6.2500×10^{-2}	3.0518×10^{-5}
5	3.1250×10^{-2}	4.6566×10^{-10}
6	1.5625×10^{-2}	1.0842×10^{-19}
7	7.8125×10^{-3}	5.8775×10^{-39}

The quadratically convergent sequence is within 10^{-38} of 0 by the seventh term. At least 126 terms are needed to ensure this accuracy for the linearly convergent sequence. □

Aitken's Δ^2 Method

As shown in the preceding Illustration, quadratically convergent sequences generally converge much more quickly than those that converge only linearly. However, linearly convergent methods are much more common than those that converge quadratically. **Aitken's Δ^2 method** is a technique that can be used to accelerate the convergence of a sequence that is linearly convergent, regardless of its origin or application.

Suppose $\{p_n\}_{n=0}^{\infty}$ is a linearly convergent sequence with limit p. To motivate the construction of a sequence $\{q_n\}$ that converges more rapidly to p than does $\{p_n\}$, let us first assume that the signs of $p_n - p$, $p_{n+1} - p$, and $p_{n+2} - p$ agree and that n is sufficiently large that

$$\frac{p_{n+1} - p}{p_n - p} \approx \frac{p_{n+2} - p}{p_{n+1} - p}.$$

Then

$$(p_{n+1} - p)^2 \approx (p_{n+2} - p)(p_n - p),$$

so

$$p_{n+1}^2 - 2p_{n+1}p + p^2 \approx p_{n+2}p_n - (p_n + p_{n+2})p + p^2$$

and

$$(p_{n+2} + p_n - 2p_{n+1})p \approx p_{n+2}p_n - p_{n+1}^2.$$

Solving for p gives

$$p \approx \frac{p_{n+2}p_n - p_{n+1}^2}{p_{n+2} - 2p_{n+1} + p_n}.$$

Adding and subtracting the terms p_n^2 and $2p_np_{n+1}$ in the numerator and grouping terms appropriately gives

$$p \approx \frac{p_np_{n+2} - 2p_np_{n+1} + p_n^2 - p_{n+1}^2 + 2p_np_{n+1} - p_n^2}{p_{n+2} - 2p_{n+1} + p_n}$$

$$= \frac{p_n(p_{n+2} - 2p_{n+1} + p_n) - (p_{n+1}^2 - 2p_np_{n+1} + p_n^2)}{p_{n+2} - 2p_{n+1} + p_n}$$

$$= p_n - \frac{(p_{n+1} - p_n)^2}{p_{n+2} - 2p_{n+1} + p_n}.$$

Alexander Aitken (1895–1967) used this technique in 1926 to accelerate the rate of convergence of a series in a paper on algebraic equations [Ai]. This process is similar to one used much earlier by Japanese mathematician Takakazu Seki Kowa (1642–1708).

Aitken's Δ^2 method uses the sequence $\{q_n\}_{n=0}^{\infty}$ defined by this approximation to p.

Aitken's Δ^2 Method

If $\{p_n\}_{n=0}^{\infty}$ is a sequence that converges linearly to p, and if

$$q_n = p_n - \frac{(p_{n+1} - p_n)^2}{p_{n+2} - 2p_{n+1} + p_n},$$

then $\{q_n\}_{n=0}^{\infty}$ also converges to p, and, in general, more rapidly.

Example 1   Apply Aitken's Δ^2 method to the linearly convergent sequence $\{\cos(1/n)\}_{n=1}^{\infty}$.

Table 2.8

n	p_n	$\hat{p}_n$
1	0.54030	0.96178
2	0.87758	0.98213
3	0.94496	0.98979
4	0.96891	0.99342
5	0.98007	0.99541
6	0.98614	
7	0.98981	

Solution   The first few terms of the sequences $\{p_n\}_{n=1}^{\infty}$ and $\{q_n\}_{n=1}^{\infty}$ are given in Table 2.8. It certainly appears that $\{q_n\}_{n=1}^{\infty}$ converges more rapidly to its limit $p = 1$ than does $\{p_n\}_{n=1}^{\infty}$. ∎

For a given sequence $\{p_n\}_{n=0}^{\infty}$, the **forward difference**, Δp_n (read "delta p_n"), is defined as

$$\Delta p_n = p_{n+1} - p_n, \quad \text{for } n \geq 0.$$

Higher powers of the operator Δ are defined recursively by

$$\Delta^k p_n = \Delta(\Delta^{k-1} p_n), \quad \text{for } k \geq 2.$$

The definition implies that

$$\Delta^2 p_n = \Delta(p_{n+1} - p_n) = \Delta p_{n+1} - \Delta p_n = (p_{n+2} - p_{n+1}) - (p_{n+1} - p_n),$$

so

$$\Delta^2 p_n = p_{n+2} - 2p_{n+1} + p_n.$$

Thus, the formula for q_n given in Aitken's Δ^2 method can be written as

$$q_n = p_n - \frac{(\Delta p_n)^2}{\Delta^2 p_n}, \quad \text{for all } n \geq 0.$$

The sequence $\{q_n\}_{n=1}^{\infty}$ converges to p more rapidly than does the original sequence $\{p_n\}_{n=0}^{\infty}$ in the following sense:

Aitken's Δ^2 Convergence

If $\{p_n\}$ is a sequence that converges linearly to the limit p and $(p_n - p)(p_{n+1} - p) > 0$ for large values of n, and

$$q_n = p_n - \frac{(\Delta p_n)^2}{\Delta^2 p_n}, \qquad \text{then} \qquad \lim_{n \to \infty} \frac{q_n - p}{p_n - p} = 0.$$

We will find occasion to apply this acceleration technique at various times in our study of approximation methods.

EXERCISE SET 2.5

1.   The following sequences are linearly convergent. Generate the first five terms of the sequence $\{q_n\}$ using Aitken's Δ^2 method.

   a. $p_0 = 0.5,$ $p_n = (2 - e^{p_{n-1}} + p_{n-1}^2)/3,$ for $n \geq 1$

   b. $p_0 = 0.75,$ $p_n = (e^{p_{n-1}}/3)^{1/2},$ for $n \geq 1$

   c. $p_0 = 0.5,$ $p_n = 3^{-p_{n-1}},$ for $n \geq 1$

   d. $p_0 = 0.5,$ $p_n = \cos p_{n-1},$ for $n \geq 1$

2.   Newton's method does not converge quadratically for the following problems. Accelerate the convergence using Aitken's Δ^2 method. Iterate until $|q_n - q_{n-1}| < 10^{-4}$.

   a. $x^2 - 2xe^{-x} + e^{-2x} = 0,$ $[0, 1]$

   b. $\cos(x + \sqrt{2}) + x(x/2 + \sqrt{2}) = 0,$ $[-2, -1]$

   c. $x^3 - 3x^2(2^{-x}) + 3x(4^{-x}) - 8^{-x} = 0,$ $[0, 1]$

   d. $e^{6x} - 27x^6 + 27x^4 e^x - 9x^2 e^{2x} = 0,$ $[4, 5]$

3.   Consider the function $f(x) = e^{6x} + 3(\ln 2)^2 e^{2x} - (\ln 8)e^{4x} - (\ln 2)^3$. Use Newton's method with $p_0 = 0$ to approximate a zero of f. Generate terms until $|p_{n+1} - p_n| < 0.0002$. Construct Aitken's Δ^2 sequence $\{q_n\}$. Is the convergence improved?

4.   Repeat Exercise 3 with the constants in $f(x)$ replaced by their four-digit approximations, that is, with $f(x) = e^{6x} + 1.441e^{2x} - 2.079e^{4x} - 0.3330$, and compare the solutions to the results in Exercise 3.

5.   (i) Show that the following sequences $\{p_n\}$ converge linearly to $p = 0$.   (ii) How large must n be before $|p_n - p| \leq 5 \times 10^{-2}$?   (iii) Use Aitken's Δ^2 method to generate a sequence $\{q_n\}$ until $|q_n - p| \leq 5 \times 10^{-2}$.

   a. $p_n = \dfrac{1}{n},$ for $n \geq 1$ **b.** $p_n = \dfrac{1}{n^2},$ for $n \geq 1$

6.   **a.**   Show that for any positive integer k, the sequence defined by $p_n = 1/n^k$ converges linearly to $p = 0$.

   b.   For each pair of integers k and m, determine a number N for which $1/N^k < 10^{-m}$.

7.   **a.**   Show that the sequence $p_n = 10^{-2^n}$ converges quadratically to zero.

   b.   Show that the sequence $p_n = 10^{-n^k}$ does not converge to zero quadratically, regardless of the size of the exponent $k > 1$.

8.   A sequence $\{p_n\}$ is said to be **superlinearly convergent** to p if a sequence $\{c_n\}$ converging to zero exists with

$$|p_{n+1} - p| \leq c_n |p_n - p|.$$

   a.   Show that if $\{p_n\}$ is superlinearly convergent to p, then $\{p_n\}$ is linearly convergent to p.

   b.   Show that $p_n = 1/n^n$ is superlinearly convergent to zero but is not quadratically convergent to zero.

2.6 Müller's Method

There are a number of root-finding problems for which the Secant, False Position, and Newton's methods will not give satisfactory results. They will not give rapid convergence, for example, when the function and its derivative are simultaneously close to zero. In addition, these methods cannot be used to approximate complex roots unless the initial approximation is a complex number whose imaginary part is nonzero. This often makes them a poor choice for use in approximating the roots of polynomials, which, even with real coefficients, commonly have complex roots occurring in conjugate pairs.

Müller's method is similar to the Secant method. The Secant method uses a line through two points on the curve to approximate the root. Müller's method uses a parabola through three points on the curve for the approximation.

Complex Zeros: Müller's Method

In this section we consider Müller's method, which is a generalization of the Secant method. The Secant method finds the zero of the line passing through points on the graph of the function that corresponds to the two immediately previous approximations, as shown in Figure 2.8(a). Müller's method uses the zero of the parabola through the three immediately previous points on the graph as the new approximation, as shown in Figure 2.8(b).

Figure 2.8

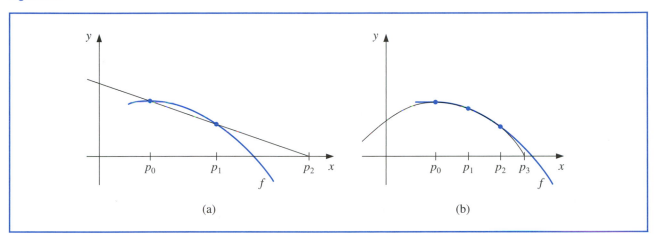

(a)   (b)

Suppose that three initial approximations, p_0, p_1, and p_2, are given for a solution of $f(x) = 0$. The derivation of Müller's method for determining the next approximation p_3 begins by considering the quadratic polynomial

$$P(x) = a(x - p_2)^2 + b(x - p_2) + c$$

that passes through $(p_0, f(p_0))$, $(p_1, f(p_1))$, and $(p_2, f(p_2))$. The constants $a, b,$ and c can be determined from the conditions

$$P(p_0) = a(p_0 - p_2)^2 + b(p_0 - p_2) + c,$$
$$P(p_1) = a(p_1 - p_2)^2 + b(p_1 - p_2) + c,$$

and

$$P(p_2) = a \cdot 0^2 + b \cdot 0 + c.$$

To determine p_3, the root of $P(x) = 0$, we apply the quadratic formula to $P(x)$. Because of round-off error problems caused by the subtraction of nearly equal numbers, however, we apply the formula in the manner prescribed in the Illustration on page 23:

$$p_3 - p_2 = \frac{-2c}{b \pm \sqrt{b^2 - 4ac}}.$$

This gives two possibilities for p_3, depending on the sign preceding the radical term. In Müller's method, the sign is chosen to agree with the sign of b. Chosen in this manner, the denominator will be the largest in magnitude, which avoids the possibility of subtracting nearly equal numbers, and results in p_3 being selected as the closest root of $P(x) = 0$ to p_2.

Müller's Method

Given initial approximations p_0, p_1, and p_2, generate

$$p_3 = p_2 - \frac{2c}{b + \text{sgn}(b)\sqrt{b^2 - 4ac}},$$

where

$$c = f(p_2),$$

$$b = \frac{(p_0 - p_2)^2[f(p_1) - f(p_2)] - (p_1 - p_2)^2[f(p_0) - f(p_2)]}{(p_0 - p_2)(p_1 - p_2)(p_0 - p_1)},$$

and

$$a = \frac{(p_1 - p_2)[f(p_0) - f(p_2)] - (p_0 - p_2)[f(p_1) - f(p_2)]}{(p_0 - p_2)(p_1 - p_2)(p_0 - p_1)}.$$

Then continue the iteration, with p_1, p_2, and p_3 replacing p_0, p_1, and p_2.

Program MULLER25 implements the Müller's method.

The method continues until a satisfactory approximation is obtained. The method involves the radical $\sqrt{b^2 - 4ac}$ at each step, so it approximates complex roots when $b^2 - 4ac < 0$, provided, of course, that complex arithmetic is used.

Illustration Consider the polynomial $f(x) = x^4 - 3x^3 + x^2 + x + 1$, part of whose graph is shown in Figure 2.9.

Figure 2.9

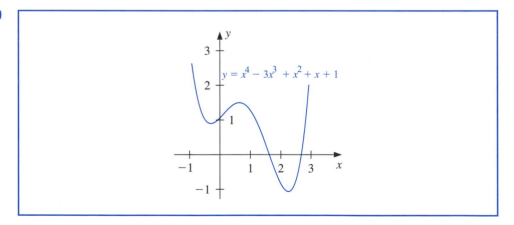

Three sets of three initial points will be used with program MULLER25 and $TOL = 10^{-5}$ to approximate the zeros of f. The first set will use $p_0 = 0.5$, $p_1 = -0.5$, and $p_2 = 0$. The parabola passing through these points has complex roots because it does not intersect the x-axis. Table 2.9 gives approximations to the corresponding complex zeros of f.

Table 2.10 gives the approximations to the two real zeros of f. The smallest of these uses $p_0 = 0.5$, $p_1 = 1.0$, and $p_2 = 1.5$, and the largest root is approximated when $p_0 = 1.5$, $p_1 = 2.0$, and $p_2 = 2.5$.

Table 2.9

	$p_0 = 0.5, \quad p_1 = -0.5, \quad p_2 = 0$	
i	p_i	$f(p_i)$
3	$-0.100000 + 0.888819i$	$-0.01120000 + 3.014875548i$
4	$-0.492146 + 0.447031i$	$-0.1691201 - 0.7367331502i$
5	$-0.352226 + 0.484132i$	$-0.1786004 + 0.0181872213i$
6	$-0.340229 + 0.443036i$	$0.01197670 - 0.0105562185i$
7	$-0.339095 + 0.446656i$	$-0.0010550 + 0.000387261i$
8	$-0.339093 + 0.446630i$	$0.000000 + 0.000000i$
9	$-0.339093 + 0.446630i$	$0.000000 + 0.000000i$

Table 2.10

$p_0 = 0.5, \quad p_1 = 1.0, \quad p_2 = 1.5$			$p_0 = 1.5, \quad p_1 = 2.0, \quad p_2 = 2.5$		
i	p_i	$f(p_i)$	i	p_i	$f(p_i)$
3	1.40637	-0.04851	3	2.24733	-0.24507
4	1.38878	0.00174	4	2.28652	-0.01446
5	1.38939	0.00000	5	2.28878	-0.00012
6	1.38939	0.00000	6	2.28880	0.00000
			7	2.28879	0.00000

The values in the tables are accurate approximations to the places listed.   □

We used MATLAB to generate the results in Tables 2.9 and 2.10. To find the first result in Table 2.9, define $f(x)$ and the initial approximations with

```
f=inline('x^4-3*x^3+ x^2+ x+1','x')
p0=0.5, p1=-0.5, p2=0.0
```

We evaluate the polynomial at the initial values

```
f0=f(p0), f1=f(p1), f2=f(p2)
```

and compute $c = 6$, $b = 10$, $a = 9$, and p_3 using the Müller's method formulas:

```
c=f2
b=((p0-p2)^2*(f1-f2)-(p1-p2)^2*(f0-f2))/((p0-p2)*(p1-p2)*(p0-p1))
a=((p1-p2)*(f0-f2)-(p0-p2)*(f1-f2))/((p0-p2)*(p1-p2)*(p0-p1))
p3=p2-(2*c)/(b+(b/abs(b))*sqrt(b^2-4*a*c))
```

The value p_3 was generated using complex arithmetic and found to be

$$p3 = -0.1000000 + 0.888819i$$

with $f(p_3) = -0.011200 + 3.01488i$. Reapplying the technique gives the entries in Table 2.9. These are accurate to the places listed.

The preceding Illustration shows that Müller's method can approximate the roots of polynomials with a variety of starting values. In fact, the technique generally converges to the root of a polynomial for any initial approximation choice. General-purpose software packages using Müller's method request only one initial approximation per root and, as an option, may even supply this approximation.

Although Müller's method is not quite as efficient as Newton's method, it is generally better than the Secant method. The relative efficiency, however, is not as important as the ease of implementation and the likelihood that a root will be found. Any of these methods will converge quite rapidly once a reasonable initial approximation is determined.

When a sufficiently accurate approximation p^* to a root has been found, $f(x)$ is divided by $x - p^*$ to produce what is called a *deflated* equation. If $f(x)$ is a polynomial of degree n, the deflated polynomial will be of degree $n - 1$, so the computations are simplified. After an approximation to the root of the deflated equation has been determined, either Müller's method or Newton's method can be used in the original function with this root as the initial approximation. This will ensure that the root being approximated is a solution to the true equation, not to the less accurate deflated equation.

EXERCISE SET 2.6

1. Find the approximations to within 10^{-4} to all the real zeros of the following polynomials using Newton's method.

   a. $P(x) = x^3 - 2x^2 - 5$

   b. $P(x) = x^3 + 3x^2 - 1$

   c. $P(x) = x^3 - x - 1$

   d. $P(x) = x^4 + 2x^2 - x - 3$

   e. $P(x) = x^3 + 4.001x^2 + 4.002x + 1.101$

   f. $P(x) = x^5 - x^4 + 2x^3 - 3x^2 + x - 4$

2. Find approximations to within 10^{-5} to all the zeros of each of the following polynomials by first finding the real zeros using Newton's method and then reducing to polynomials of lower degree to determine any complex zeros.

   a. $P(x) = x^4 + 5x^3 - 9x^2 - 85x - 136$

   b. $P(x) = x^4 - 2x^3 - 12x^2 + 16x - 40$

   c. $P(x) = x^4 + x^3 + 3x^2 + 2x + 2$

   d. $P(x) = x^5 + 11x^4 - 21x^3 - 10x^2 - 21x - 5$

   e. $P(x) = x^4 - 4x^2 - 3x + 5$

   f. $P(x) = x^3 - 7x^2 + 14x - 6$

3. Repeat Exercise 1 using Müller's method.

4. Repeat Exercise 2 using Müller's method.

5. Find, to within 10^{-3}, the zeros and critical points of the following functions. Use this information to sketch the graphs of P.

   a. $P(x) = x^3 - 9x^2 + 12$ b. $P(x) = x^4 - 2x^3 - 5x^2 + 12x - 5$

6. $P(x) = 10x^3 - 8.3x^2 + 2.295x - 0.21141 = 0$ has a root at $x = 0.29$.

   a. Use Newton's method with $p_0 = 0.28$ to attempt to find this root.

   b. Use Müller's method with $p_0 = 0.275$, $p_1 = 0.28$, and $p_2 = 0.285$ to attempt to find this root.

   c. Explain any discrepancies in (a) and (b).

7. Use the MATLAB function `roots` to find all zeros of the polynomial $P(x) = x^3 + 4x - 4$.

8. Use the MATLAB function `roots` to find all zeros of the polynomial $P(x) = x^3 - 2x - 5$.

9. Use each of the following methods to find a solution accurate to within 10^{-4} for the problem

   $$600x^4 - 550x^3 + 200x^2 - 20x - 1 = 0, \quad \text{for } 0.1 \le x \le 1.$$

   a. Bisection method

   b. Newton's method

   c.   Secant method

   d.   method of False Position

   e.   Müller's method

10.   Two ladders crisscross an alley of width W. Each ladder reaches from the base of one wall to some point on the opposite wall. The ladders cross at a height H above the pavement. Find W given that the lengths of the ladders are $x_1 = 20$ ft and $x_2 = 30$ ft and that $H = 8$ ft. (See the figure.)

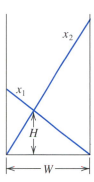

11.   A can in the shape of a right circular cylinder is to be constructed to contain 1000 cm^3. (See the figure.) The circular top and bottom of the can must have a radius of 0.25 cm more than the radius of the can so that the excess can be used to form a seal with the side. The sheet of material being formed into the side of the can must also be 0.25 cm longer than the circumference of the can so that a seal can be formed. Find, to within 10^{-4}, the minimal amount of material needed to construct the can.

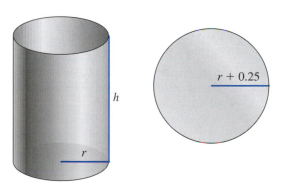

12.   In 1224, Leonardo of Pisa, better known as Fibonacci, answered a mathematical challenge of John of Palermo in the presence of Emperor Frederick II. His challenge was to find a root of the equation $x^3 + 2x^2 + 10x = 20$. He first showed that the equation had no rational roots and no Euclidean irrational root—that is, no root in one of the forms $a \pm \sqrt{b}$, $\sqrt{a} \pm \sqrt{b}$, $\sqrt{a \pm \sqrt{b}}$, or $\sqrt{\sqrt{a} \pm \sqrt{b}}$, where a and b are rational numbers. He then approximated the only real root, probably using an algebraic technique of Omar Khayyam involving the intersection of a circle and a parabola. His answer was given in the base-60 number system as

$$1 + 22\left(\frac{1}{60}\right) + 7\left(\frac{1}{60}\right)^2 + 42\left(\frac{1}{60}\right)^3 + 33\left(\frac{1}{60}\right)^4 + 4\left(\frac{1}{60}\right)^5 + 40\left(\frac{1}{60}\right)^6.$$

How accurate was his approximation?

2.7 Survey of Methods and Software

In this chapter we have considered the problem of solving the equation $f(x) = 0$, where f is a given continuous function. All the methods begin with an initial approximation and generate a sequence that converges to a root of the equation, if the method is successful. If $[a, b]$ is an interval on which $f(a)$ and $f(b)$ are of opposite sign, then the Bisection method and the method of False Position will converge. However, the convergence of these methods may be slow. Faster convergence is generally obtained using the Secant method or Newton's method. Good initial approximations are required for these methods, two for the Secant method and one for Newton's method, so the Bisection or the False Position method can be used as starter methods for the Secant or Newton's method.

Müller's method will give rapid convergence without a particularly good initial approximation. It is not quite as efficient as Newton's method, but it is better than the Secant method, and it has the added advantage of being able to approximate complex roots.

Deflation is generally used with Müller's method once an approximate root of a polynomial has been determined. After an approximation to the root of the deflated equation has been determined, use either Müller's method or Newton's method in the original polynomial with this root as the initial approximation. This procedure will ensure that the root being approximated is a solution to the true equation, not to the deflated equation. We recommended Müller's method for finding all the zeros of polynomials, real or complex. Müller's method can also be used for an arbitrary continuous function.

Other high-order methods are available for determining the roots of polynomials. If this topic is of particular interest, we recommend that consideration be given to Laguerre's method, which gives cubic convergence and also approximates complex roots (see [Ho, pp. 176–179] for a complete discussion), the Jenkins-Traub method (see [JT]), and Brent's method (see [Bre]). Both IMSL and NAG supply subroutines based on Brent's method. This technique uses a combination of linear interpolation, an inverse quadratic interpolation similar to Müller's method, and the bisection method.

The netlib FORTRAN subroutine fzero.f uses a combination of the Bisection and Secant method developed by Dekker to approximate a real zero of $f(x) = 0$ in the interval $[a, b]$. It requires specifying an interval $[a, b]$ that contains a root and returns an interval with a width that is within a specified tolerance. The FORTRAN subroutine sdzro.f uses a combination of the bisection method, interpolation, and extrapolation to find a real zero of $f(x) = 0$ in a given interval $[a, b]$. The routines rpzero and cpzero can be used to approximate all zeros of a real polynomial or complex polynomial, respectively. Both methods use Newton's method for systems, which will be considered in Chapter 10. All routines are given in single and double precision. These methods are available on the Internet from netlib at http://www.netlib.org/slatec/src.

Within MATLAB, the function `roots` is used to compute all the zeros, both real and complex, of a polynomial. For an arbitrary function, `fzero` computes a zero in a specified interval. For example, to find the zeros of

$$f(x) = x^4 - 3x^3 + x^2 + x + 1,$$

we define

```
f= inline ('x^4 - 3*x^3 + x^2 + x + 1','x')
```

Then the command

```
p = fzero(f,[1 2])
```

gives

$$p = 1.389390683334934$$

To find all the zeros of f we need a vector describing the coefficients of $f(x)$ in decreasing order:

```
ff = [1 -3 1 1 1]
```

Then the command

```
roots(ff)
```

gives

$$ans = 2.288794992188488$$
$$1.389390683334934$$
$$-0.339092837761710 + 0.446630099997518i$$
$$-0.339092837761710 - 0.446630099997518i$$

In spite of the diversity of methods, the professionally written packages are based primarily on the methods and principles discussed in this chapter. You should be able to use these packages by reading the manuals accompanying the packages to better understand the parameters and the specifications of the results that are obtained.

There are three books that we consider to be classics on the solution of nonlinear equations, those by Traub [Tr], by Ostrowski [Os], and by Householder [Ho]. In addition, the book by Brent [Bre] served as the basis for many of the currently used root-finding methods.

Interpolation and Polynomial Approximation

3.1 Introduction

Engineers and scientists commonly assume that relationships between variables in a physical problem can be approximately reproduced from data given by the problem. The ultimate goal might be to determine the values at intermediate points, to approximate the integral or derivative of the underlying function, or to simply give a smooth or continuous representation of the variables in the problem.

Interpolation refers to determining a function that exactly represents a collection of data. The most elementary type of interpolation consists of fitting a polynomial to a collection of data points. Polynomials have derivatives and integrals that are themselves polynomials, so they are a natural choice for approximating derivatives and integrals. In this chapter we will see that polynomials to approximate continuous functions are easily constructed. The following result implies that there are polynomials that are arbitrarily close to any continuous function.

Weierstrass Approximation Theorem

Suppose that f is defined and continuous on $[a, b]$. For each $\varepsilon > 0$, there exists a polynomial $P(x)$ defined on $[a, b]$, with the property that (see Figure 3.1)

$$|f(x) - P(x)| < \varepsilon, \qquad \text{for all } x \in [a, b].$$

Figure 3.1

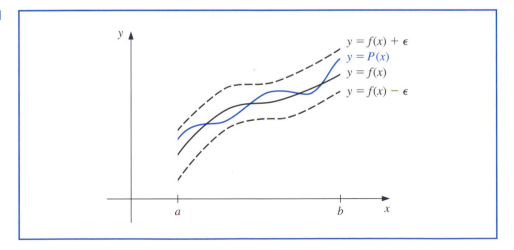

The Taylor polynomials were introduced in Chapter 1, where they were described as one of the fundamental building blocks of numerical analysis. Given this prominence, you might assume that polynomial interpolation makes heavy use of these functions. However, this is not the case. The Taylor polynomials agree as closely as possible with a given function at a specific point, but they concentrate their accuracy only near that point. A good interpolation polynomial needs to provide a relatively accurate approximation over an entire interval, and Taylor polynomials do not do that. For example, suppose we calculate the first six Taylor polynomials about $x_0 = 0$ for $f(x) = e^x$. Since the derivatives of $f(x)$ are all e^x, which evaluated at $x_0 = 0$ gives 1, the Taylor polynomials are

$$P_0(x) = 1, \quad P_1(x) = 1 + x, \quad P_2(x) = 1 + x + \frac{x^2}{2}, \quad P_3(x) = 1 + x + \frac{x^2}{2} + \frac{x^3}{6},$$

$$P_4(x) = 1 + x + \frac{x^2}{2} + \frac{x^3}{6} + \frac{x^4}{24}, \quad \text{and} \quad P_5(x) = 1 + x + \frac{x^2}{2} + \frac{x^3}{6} + \frac{x^4}{24} + \frac{x^5}{120}.$$

The graphs of these Taylor polynomials are shown in Figure 3.2. Notice that the error becomes progressively worse as we move away from zero.

Figure 3.2

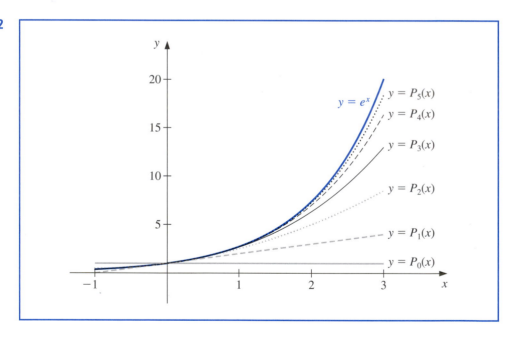

Although better approximations are obtained for this problem if higher-degree Taylor polynomials are used, this situation is not always true. Consider, as an extreme example, using Taylor polynomials of various degrees for $f(x) = 1/x$ expanded about $x_0 = 1$ to approximate $f(3) = \frac{1}{3}$. Since

$$f(x) = x^{-1}, f'(x) = -x^{-2}, f''(x) = (-1)^2 2 \cdot x^{-3},$$

and, in general,

$$f^{(n)}(x) = (-1)^n n! x^{-n-1},$$

the Taylor polynomials for $n \geq 0$ are

$$P_n(x) = \sum_{k=0}^{n} \frac{f^{(k)}(1)}{k!}(x-1)^k = \sum_{k=0}^{n} (-1)^k (x-1)^k.$$

When we approximate $f(3) = \frac{1}{3}$ by $P_n(3)$ for larger values of n, the approximations become increasingly inaccurate, as shown Table 3.1.

Table 3.1

n	0	1	2	3	4	5	6	7
$P_n(3)$	1	-1	3	-5	11	-21	43	-85

The Taylor polynomials have the property that all the information used in the approximation is concentrated at the single point x_0, so it is not uncommon for these polynomials to give inaccurate approximations as we move away from x_0. This limits Taylor polynomial approximation to the situation in which approximations are needed only at points close to x_0. For ordinary computational purposes, it is more efficient to use methods that include information at various points, which we will consider in the remainder of this chapter. The primary use of Taylor polynomials in numerical analysis is *not* for approximation purposes; instead it is for the derivation of numerical techniques.

3.2 Lagrange Polynomials

The interpolation formula named for Joseph Louis Lagrange (1736–1813) was likely known by Isaac Newton around 1675, but it appears to have been published first in 1779 by Edward Waring (1736–1798). Lagrange wrote extensively on the subject of interpolation and his work had significant influence on later mathematicians. He published this result in 1795.

In the previous section we discussed the general unsuitability of Taylor polynomials for approximation. These polynomials are useful only over small intervals for functions whose derivatives exist and are easily evaluated. In this section we find approximating polynomials that can be determined simply by specifying certain points on the plane through which they must pass.

Lagrange Interpolating Polynomials

Determining a polynomial of degree 1 that passes through the distinct points (x_0, y_0) and (x_1, y_1) is the same as approximating a function f for which $f(x_0) = y_0$ and $f(x_1) = y_1$ by means of a first-degree polynomial interpolating, or agreeing with, the values of f at the given points. We first define the functions

$$L_0(x) = \frac{x - x_1}{x_0 - x_1} \quad \text{and} \quad L_1(x) = \frac{x - x_0}{x_1 - x_0},$$

and note that these definitions imply that

$$L_0(x_0) = \frac{x_0 - x_1}{x_0 - x_1} = 1, \quad L_0(x_1) = \frac{x_1 - x_1}{x_0 - x_1} = 0, \quad L_1(x_0) = 0, \quad \text{and} \quad L_1(x_1) = 1.$$

We then define

$$P(x) = L_0(x)f(x_0) + L_1(x)f(x_1) = \frac{x - x_1}{x_0 - x_1}f(x_0) + \frac{x - x_0}{x_1 - x_0}f(x_1).$$

This gives

$$P(x_0) = 1 \cdot f(x_0) + 0 \cdot f(x_1) = f(x_0) = y_0$$

and

$$P(x_1) = 0 \cdot f(x_0) + 1 \cdot f(x_1) = f(x_1) = y_1.$$

So, P is the unique linear function passing through (x_0, y_0) and (x_1, y_1).

Example 1 Determine the linear Lagrange interpolating polynomial that passes through the points $(2, 4)$ and $(5, 1)$.

Solution In this case we have

$$L_0(x) = \frac{x - 5}{2 - 5} = -\frac{1}{3}(x - 5) \quad \text{and} \quad L_1(x) = \frac{x - 2}{5 - 2} = \frac{1}{3}(x - 2),$$

so

$$P(x) = -\frac{1}{3}(x - 5) \cdot 4 + \frac{1}{3}(x - 2) \cdot 1 = -\frac{4}{3}x + \frac{20}{3} + \frac{1}{3}x - \frac{2}{3} = -x + 6.$$

The graph of $y = P(x)$ is shown in Figure 3.3. ▪

Figure 3.3

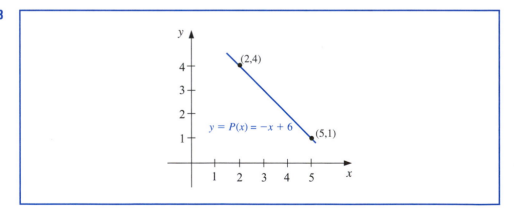

To generalize the concept of linear interpolation to higher-degree polynomials, consider the construction of a polynomial of degree at most n, shown in Figure 3.4, that passes through the $n + 1$ points

$$(x_0, f(x_0)), \ (x_1, f(x_1)), \ \ldots, \ (x_n, f(x_n)).$$

Figure 3.4

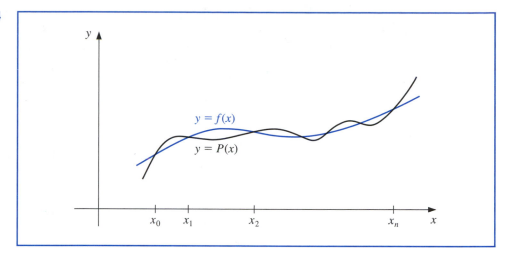

In this case, we construct, for each $k = 0, 1, \ldots, n$, a polynomial of degree n, which we will denote by $L_{n,k}(x)$, with the property that $L_{n,k}(x_i) = 0$ when $i \neq k$ and $L_{n,k}(x_k) = 1$.

To satisfy $L_{n,k}(x_i) = 0$ for each $i \neq k$, the numerator of $L_{n,k}(x)$ must contain the term

$$(x - x_0)(x - x_1) \cdots (x - x_{k-1})(x - x_{k+1}) \cdots (x - x_n).$$

To satisfy $L_{n,k}(x_k) = 1$, the denominator of $L_{n,k}(x)$ must be this term evaluated at $x = x_k$. Thus

$$L_{n,k}(x) = \frac{(x - x_0) \cdots (x - x_{k-1})(x - x_{k+1}) \cdots (x - x_n)}{(x_k - x_0) \cdots (x_k - x_{k-1})(x_k - x_{k+1}) \cdots (x_k - x_n)}.$$

A sketch of the graph of a typical $L_{n,k}$ when n is even is shown in Figure 3.5.

Figure 3.5

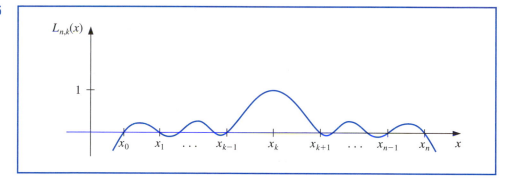

The interpolating polynomial is easily described now that the form of $L_{n,k}(x)$ is known. This polynomial is called the nth Lagrange interpolating polynomial.

_n_th Lagrange Interpolating Polynomial

$$P_n(x) = f(x_0)L_{n,0}(x) + \cdots + f(x_n)L_{n,n}(x) = \sum_{k=0}^{n} f(x_k)L_{n,k}(x),$$

where

$$L_{n,k}(x) = \frac{(x - x_0)(x - x_1) \cdots (x - x_{k-1})(x - x_{k+1}) \cdots (x - x_n)}{(x_k - x_0)(x_k - x_1) \cdots (x_k - x_{k-1})(x_k - x_{k+1}) \cdots (x_k - x_n)}$$

for each $k = 0, 1, \ldots, n$.

If $x_0, x_1, \ldots, x_n$ are $(n + 1)$ distinct numbers and f is a function whose values are given at these numbers, then $P_n(x)$ is the unique polynomial of degree at most n that agrees with $f(x)$ at $x_0, x_1, \ldots, x_n$. The notation for describing the Lagrange interpolating polynomial $P_n(x)$ is rather complicated because $P_n(x)$ is the sum of the $n + 1$ polynomials $f(x_k)L_{n,k}(x)$, for $k = 0, 1, \ldots, n$, each of which is of degree n, provided $f(x_k) \neq 0$. To reduce the notational complication, we will write $L_{n,k}(x)$ simply as $L_k(x)$ when there should be no confusion that its degree is n.

Example 2   (a) Use the numbers (called *nodes*) $x_0 = 2$, $x_1 = 2.75$, and $x_2 = 4$ to find the second Lagrange interpolating polynomial for $f(x) = 1/x$.

(b) Use this polynomial to approximate $f(3) = 1/3$.

Solution   (a) We first determine the coefficient polynomials $L_0(x)$, $L_1(x)$, and $L_2(x)$. They are

$$L_0(x) = \frac{(x - 2.75)(x - 4)}{(2 - 2.75)(2 - 4)} = \frac{2}{3}(x - 2.75)(x - 4),$$

$$L_1(x) = \frac{(x - 2)(x - 4)}{(2.75 - 2)(2.75 - 4)} = -\frac{16}{15}(x - 2)(x - 4),$$

and

$$L_2(x) = \frac{(x - 2)(x - 2.75)}{(4 - 2)(4 - 2.75)} = \frac{2}{5}(x - 2)(x - 2.75).$$

Also, $f(x_0) = f(2) = 1/2$, $f(x_1) = f(2.75) = 4/11$, and $f(x_2) = f(4) = 1/4$, so

$$P(x) = \sum_{k=0}^{2} f(x_k)L_k(x)$$

$$= \frac{1}{3}(x - 2.75)(x - 4) - \frac{64}{165}(x - 2)(x - 4) + \frac{1}{10}(x - 2)(x - 2.75)$$

$$= \frac{1}{22}x^2 - \frac{35}{88}x + \frac{49}{44}.$$

(b) An approximation to $f(3) = 1/3$ (see Figure 3.6) is

$$f(3) \approx P(3) = \frac{9}{22} - \frac{105}{88} + \frac{49}{44} = \frac{29}{88} \approx 0.32955.$$

Recall that in the Section 3.1 (see Table 3.1) we found that no Taylor polynomial expanded about $x_0 = 1$ could be used to reasonably approximate $f(x) = 1/x$ at $x = 3$.   ▪

Figure 3.6

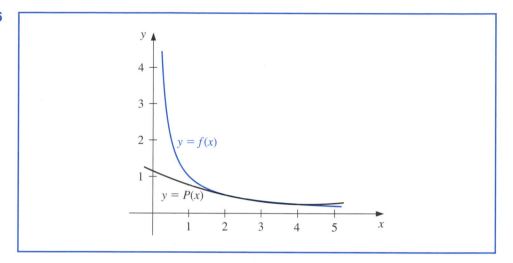

The Lagrange polynomials have remainder terms that are reminiscent of those for the Taylor polynomials. The nth Taylor polynomial about x_0 concentrates all the known information at x_0 and has an error term of the form

$$\frac{f^{(n+1)}(\xi(x))}{(n+1)!}(x-x_0)^{n+1},$$

where $\xi(x)$ is between x and x_0. The nth Lagrange polynomial uses information at the distinct numbers $x_0, x_1, \ldots, x_n$. In place of $(x-x_0)^{n+1}$, its error formula uses a product of the $n+1$ terms $(x-x_0), (x-x_1), \ldots, (x-x_n)$, and the number $\xi(x)$ can lie anywhere in the interval that contains the points $x_0, x_1, \ldots, x_n$, and x. Otherwise it has the same form as the error formula for the Taylor polynomials.

Lagrange Polynomial Error Formula

$$f(x) = P_n(x) + \frac{f^{(n+1)}(\xi(x))}{(n+1)!}(x-x_0)(x-x_1)\cdots(x-x_n),$$

for some (unknown) number $\xi(x)$ that lies in the smallest interval that contains $x_0, x_1, \ldots, x_n$ and x.

This error formula is an important theoretical result, because Lagrange polynomials are used extensively for deriving numerical differentiation and integration methods. Error bounds for these techniques are obtained from the Lagrange polynomial error formula. The specific use of this error formula, however, is restricted to those functions whose derivatives have known bounds. The next Illustration shows interpolation techniques for a situation in which the Lagrange error formula cannot be used. This shows that we should look for a more efficient way to obtain approximations via interpolation.

Illustration   Table 3.2 lists values of a function f at various points. The approximations to $f(1.5)$ obtained by various Lagrange polynomials that use this data will be compared to try to determine the accuracy of the approximation.

Table 3.2

x	$f(x)$
1.0	0.7651977
1.3	0.6200860
1.6	0.4554022
1.9	0.2818186
2.2	0.1103623

The most appropriate linear polynomial uses $x_0 = 1.3$ and $x_1 = 1.6$ because 1.5 is between 1.3 and 1.6. The value of the interpolating polynomial at 1.5 is

$$P_1(1.5) = \frac{(1.5 - 1.6)}{(1.3 - 1.6)} f(1.3) + \frac{(1.5 - 1.3)}{(1.6 - 1.3)} f(1.6)$$

$$= \frac{(1.5 - 1.6)}{(1.3 - 1.6)} (0.6200860) + \frac{(1.5 - 1.3)}{(1.6 - 1.3)} (0.4554022) = 0.5102968.$$

Two polynomials of degree 2 can reasonably be used, one with $x_0 = 1.3$, $x_1 = 1.6$, and $x_2 = 1.9$, which gives

$$P_2(1.5) = \frac{(1.5 - 1.6)(1.5 - 1.9)}{(1.3 - 1.6)(1.3 - 1.9)} (0.6200860) + \frac{(1.5 - 1.3)(1.5 - 1.9)}{(1.6 - 1.3)(1.6 - 1.9)} (0.4554022)$$

$$+ \frac{(1.5 - 1.3)(1.5 - 1.6)}{(1.9 - 1.3)(1.9 - 1.6)} (0.2818186) = 0.5112857,$$

and one with $x_0 = 1.0$, $x_1 = 1.3$, and $x_2 = 1.6$, which gives $\hat{P}_2(1.5) = 0.5124715$.

In the third-degree case, there are also two reasonable choices for the polynomial. One with $x_0 = 1.3$, $x_1 = 1.6$, $x_2 = 1.9$, and $x_3 = 2.2$, which gives $P_3(1.5) = 0.5118302$. The second third-degree approximation is obtained with $x_0 = 1.0$, $x_1 = 1.3$, $x_2 = 1.6$, and $x_3 = 1.9$, which gives $\hat{P}_3(1.5) = 0.5118127$.

The fourth-degree Lagrange polynomial uses all the entries in the table. With $x_0 = 1.0$, $x_1 = 1.3$, $x_2 = 1.6$, $x_3 = 1.9$, and $x_4 = 2.2$, the approximation is $P_4(1.5) = 0.5118200$.

Because $P_3(1.5)$, $\hat{P}_3(1.5)$, and $P_4(1.5)$ all agree to within 2×10^{-5} units, we expect this degree of accuracy for these approximations. We also expect $P_4(1.5)$ to be the most accurate approximation because it uses more of the given data.

The function we are approximating is actually the Bessel function of the first kind of order zero, whose value at 1.5 is known to be 0.5118277. Therefore, the true accuracies of the approximations are as follows:

$$|P_1(1.5) - f(1.5)| \approx 1.53 \times 10^{-3},$$

$$|P_2(1.5) - f(1.5)| \approx 5.42 \times 10^{-4},$$

$$|\hat{P}_2(1.5) - f(1.5)| \approx 6.44 \times 10^{-4},$$

$$|P_3(1.5) - f(1.5)| \approx 2.5 \times 10^{-6},$$

$$|\hat{P}_3(1.5) - f(1.5)| \approx 1.50 \times 10^{-5},$$

$$|P_4(1.5) - f(1.5)| \approx 7.7 \times 10^{-6}.$$

Although $P_3(1.5)$ is the most accurate approximation, if we had no knowledge of the actual value of $f(1.5)$, we would accept $P_4(1.5)$ as the best approximation because it includes the most data about the function. The Lagrange error term cannot be applied here because we have no knowledge of the fourth derivative of f. Unfortunately, this is generally the case.   ☐

Neville's Method

A practical difficulty with Lagrange interpolation is that because the error term is difficult to apply, the degree of the polynomial needed for the desired accuracy is generally not known until the computations are determined. The usual practice is to compute the results given from various polynomials until appropriate agreement is obtained, as was done in the previous example. However, the work done in calculating the approximation by the second polynomial does not lessen the work needed to calculate the third approximation; nor is the fourth approximation easier to obtain once the third approximation is known, and so on. To derive these approximating polynomials in a manner that uses the previous calculations to advantage, we need to introduce some new notation.

Let f be a function defined at $x_0, x_1, x_2, \ldots, x_n$ and suppose that $m_1, m_2, \ldots, m_k$ are k distinct integers with $0 \le m_i \le n$ for each i. The Lagrange polynomial that agrees with $f(x)$ at the k points $x_{m_1}, x_{m_2}, \ldots, x_{m_k}$ is denoted $P_{m_1,m_2,\ldots,m_k}(x)$.

Example 3 Suppose that $x_0 = 1$, $x_1 = 2$, $x_2 = 3$, $x_3 = 4$, $x_4 = 6$, and $f(x) = e^x$. Determine the interpolating polynomial denoted $P_{1,2,4}(x)$, and use this polynomial to approximate $f(5)$.

Solution This is the Lagrange polynomial that agrees with $f(x)$ at $x_1 = 2$, $x_2 = 3$, and $x_4 = 6$. Hence

$$P_{1,2,4}(x) = \frac{(x-3)(x-6)}{(2-3)(2-6)}e^2 + \frac{(x-2)(x-6)}{(3-2)(3-6)}e^3 + \frac{(x-2)(x-3)}{(6-2)(6-3)}e^6.$$

So

$$f(5) \approx P_{1,2,4}(5) = \frac{(5-3)(5-6)}{(2-3)(2-6)}e^2 + \frac{(5-2)(5-6)}{(3-2)(3-6)}e^3 + \frac{(5-2)(5-3)}{(6-2)(6-3)}e^6$$

$$= -\frac{1}{2}e^2 + e^3 + \frac{1}{2}e^6 \approx 218.105.$$ ∎

The next result describes a method for recursively generating Lagrange polynomial approximations.

Recursively Generated Lagrange Polynomials

Let f be defined at $x_0, x_1, \ldots, x_k$ and x_j, x_i be two numbers in this set. If

$$P(x) = \frac{(x-x_j)P_{0,1,\ldots,j-1,j+1,\ldots,k}(x) - (x-x_i)P_{0,1,\ldots,i-1,i+1,\ldots,k}(x)}{(x_i-x_j)},$$

then $P(x)$ is the kth Lagrange polynomial that interpolates, or agrees with, $f(x)$ at the $k+1$ points $x_0, x_1, \ldots, x_k$.

To see why this recursive formula is true, first let $Q \equiv P_{0,1,\ldots,i-1,i+1,\ldots,k}$ and $\hat{Q} \equiv P_{0,1,\ldots,j-1,j+1,\ldots,k}$. Since $Q(x)$ and $\hat{Q}(x)$ are polynomials of degree at most $k-1$,

$$P(x) = \frac{(x-x_j)\hat{Q}(x) - (x-x_i)Q(x)}{(x_i-x_j)}$$

must be of degree at most k. If $0 \le r \le k$ with $r \ne i$ and $r \ne j$, then $Q(x_r) = \hat{Q}(x_r) = f(x_r)$, so

$$P(x_r) = \frac{(x_r-x_j)\hat{Q}(x_r) - (x_r-x_i)Q(x_r)}{x_i-x_j} = \frac{(x_i-x_j)}{(x_i-x_j)}f(x_r) = f(x_r).$$

Moreover,

$$P(x_i) = \frac{(x_i - x_j)\hat{Q}(x_i) - (x_i - x_i)Q(x_i)}{x_i - x_j} = \frac{(x_i - x_j)}{(x_i - x_j)} f(x_i) = f(x_i),$$

and similarly, $P(x_j) = f(x_j)$. But there is only one polynomial of degree at most k that agrees with $f(x)$ at $x_0, x_1, \ldots, x_k$, and this polynomial by definition is $P_{0,1,\ldots,k}(x)$. Hence,

$$P_{0,1,\ldots,k}(x) = P(x) = \frac{(x - x_j)P_{0,1,\ldots,j-1,j+1,\ldots,k}(x) - (x - x_i)P_{0,1,\ldots,i-1,i+1,\ldots,k}(x)}{(x_i - x_j)}.$$

This result implies that the approximations from the interpolating polynomials can be generated recursively in the manner shown in Table 3.3. The row-by-row generation is performed to move across the rows as rapidly as possible, because these entries are given by successively higher-degree interpolating polynomials. This procedure is called **Neville's method**.

Table 3.3

x_0	P_0			
x_1	P_1	$P_{0,1}$		
x_2	P_2	$P_{1,2}$	$P_{0,1,2}$	
x_3	P_3	$P_{2,3}$	$P_{1,2,3}$	$P_{0,1,2,3}$
x_4	P_4	$P_{3,4}$	$P_{2,3,4}$	$P_{1,2,3,4}$	$P_{0,1,2,3,4}$

Program NEVLLE31 implements the Neville's method.

Eric Harold Neville (1889–1961) gave this modification of the Lagrange formula in a paper published in 1932 [N].

The P notation used in Table 3.3 is cumbersome because of the number of subscripts used to represent the entries. Note, however, that as an array is being constructed, only two subscripts are needed. Proceeding down the table corresponds to using consecutive points x_i with larger i, and proceeding to the right corresponds to increasing the degree of the interpolating polynomial. Since the points appear consecutively in each entry, we need to describe only a starting point and the number of additional points used in constructing the approximation. To avoid the cumbersome subscripts we let $Q_{i,j}(x)$, for $0 \leq j \leq i$, denote the jth interpolating polynomial on the $j + 1$ numbers $x_{i-j}, x_{i-j+1}, \ldots, x_{i-1}, x_i$; that is,

$$Q_{i,j} = P_{i-j,i-j+1,\ldots,i-1,i}.$$

Using this notation for Neville's method provides the Q notation in Table 3.4.

Table 3.4

x_0	$P_0 = Q_{0,0}$			
x_1	$P_1 = Q_{1,0}$	$P_{0,1} = Q_{1,1}$		
x_2	$P_2 = Q_{2,0}$	$P_{1,2} = Q_{2,1}$	$P_{0,1,2} = Q_{2,2}$	
x_3	$P_3 = Q_{3,0}$	$P_{2,3} = Q_{3,1}$	$P_{1,2,3} = Q_{3,2}$	$P_{0,1,2,3} = Q_{3,3}$
x_4	$P_4 = Q_{4,0}$	$P_{3,4} = Q_{4,1}$	$P_{2,3,4} = Q_{4,2}$	$P_{1,2,3,4} = Q_{4,3}$	$P_{0,1,2,3,4} = Q_{4,4}$

Example 4   Table 3.5 lists the values of $f(x) = \ln x$ accurate to the places given. Use Neville's method and four-digit rounding arithmetic to approximate $f(2.1) = \ln 2.1$ by completing the Neville table.

Table 3.5

i	x_i	$\ln x_i$
0	2.0	0.6931
1	2.2	0.7885
2	2.3	0.8329

Solution   Because $x - x_0 = 0.1$, $x - x_1 = -0.1$, $x - x_2 = -0.2$, and we are given $Q_{0,0} = 0.6931$, $Q_{1,0} = 0.7885$, and $Q_{2,0} = 0.8329$, we have

$$Q_{1,1} = \frac{1}{0.2}[(0.1)0.7885 - (-0.1)0.6931] = \frac{0.1482}{0.2} = 0.7410$$

and

$$Q_{2,1} = \frac{1}{0.1}[(-0.1)0.8329 - (-0.2)0.7885] = \frac{0.07441}{0.1} = 0.7441.$$

The final approximation we can obtain from this data is

$$Q_{2,2} = \frac{1}{0.3}[(0.1)0.7441 - (-0.2)0.7410] = \frac{0.2276}{0.3} = 0.7420.$$

These values are shown in Table 3.6.   ■

Table 3.6

i	x_i	$x - x_i$	Q_{i0}	Q_{i1}	Q_{i2}
0	2.0	0.1	0.6931		
1	2.2	−0.1	0.7885	0.7410	
2	2.3	−0.2	0.8329	0.7441	0.7420

If the latest approximation, $Q_{2,2}$, is not as accurate as desired, another node, x_3, can be selected and another row can be added to the table:

$$x_3 \qquad Q_{3,0} \qquad Q_{3,1} \qquad Q_{3,2} \qquad Q_{3,3}.$$

Then $Q_{2,2}$, $Q_{3,2}$, and $Q_{3,3}$ can be compared to determine further accuracy.

Using $x_3 = 2.4$ in Example 4 gives no improvement of accuracy because the additional row is

$$2.4 \quad 0.8755 \quad 0.7480 \quad 0.7420 \quad 0.7420.$$

Had the data been given with more digits of accuracy, there might have been an improvement.

EXERCISE SET 3.2

1. For the given functions $f(x)$, let $x_0 = 0$, $x_1 = 0.6$, and $x_2 = 0.9$. Construct the Lagrange interpolating polynomials of degree (i) at most 1 and (ii) at most 2 to approximate $f(0.45)$, and find the actual error.

   a. $f(x) = \cos x$
   b. $f(x) = \sqrt{1 + x}$
   c. $f(x) = \ln(x + 1)$
   d. $f(x) = \tan x$

2. Use the Lagrange polynomial error formula to find an error bound for the approximations in Exercise 1.

3. Use appropriate Lagrange interpolating polynomials of degrees 1, 2, and 3 to approximate each of the following:

   a. $f(8.4)$ if $f(8.1) = 16.94410$, $f(8.3) = 17.56492$, $f(8.6) = 18.50515$, $f(8.7) = 18.82091$

   b. $f\left(-\frac{1}{3}\right)$ if $f(-0.75) = -0.07181250$, $f(-0.5) = -0.02475000$, $f(-0.25) = 0.33493750$, $f(0) = 1.10100000$

   c. $f(0.25)$ if $f(0.1) = 0.62049958$, $f(0.2) = -0.28398668$, $f(0.3) = 0.00660095$, $f(0.4) = 0.24842440$

   d. $f(0.9)$ if $f(0.6) = -0.17694460$, $f(0.7) = 0.01375227$, $f(0.8) = 0.22363362$, $f(1.0) = 0.65809197$

4. Use Neville's method to obtain the approximations for Exercise 3.

5. Use Neville's method to approximate $\sqrt{3}$ with the function $f(x) = 3^x$ and the values $x_0 = -2$, $x_1 = -1$, $x_2 = 0$, $x_3 = 1$, and $x_4 = 2$.

6. Use Neville's method to approximate $\sqrt{3}$ with the function $f(x) = \sqrt{x}$ and the values $x_0 = 0, x_1 = 1,$ $x_2 = 2, x_3 = 4,$ and $x_4 = 5$. Compare the accuracy with that of Exercise 5.

7. The data for Exercise 3 were generated using the following functions. Use the error formula to find a bound for the error and compare the bound to the actual error for the cases $n = 1$ and $n = 2$.

    a. $f(x) = x \ln x$

    b. $f(x) = x^3 + 4.001x^2 + 4.002x + 1.101$

    c. $f(x) = x \cos x - 2x^2 + 3x - 1$

    d. $f(x) = \sin(e^x - 2)$

8. Use the Lagrange interpolating polynomial of degree 3 or less and four-digit chopping arithmetic to approximate $\cos 0.750$ using the following values. Find an error bound for the approximation.

$$\cos 0.698 = 0.7661 \quad \cos 0.733 = 0.7432$$
$$\cos 0.768 = 0.7193 \quad \cos 0.803 = 0.6946$$

The actual value of $\cos 0.750$ is 0.7317 (to four decimal places). Explain the discrepancy between the actual error and the error bound.

9. Use the following values and four-digit rounding arithmetic to construct a third Lagrange polynomial approximation to $f(1.09)$. The function being approximated is $f(x) = \log_{10}(\tan x)$. Use this knowledge to find a bound for the error in the approximation.

$$f(1.00) = 0.1924 \quad f(1.05) = 0.2414 \quad f(1.10) = 0.2933 \quad f(1.15) = 0.3492$$

10. Repeat Exercise 9 using MATLAB in `long format` mode.

11. Let $P_3(x)$ be the interpolating polynomial for the data $(0, 0), (0.5, y), (1, 3),$ and $(2, 2)$. Find y if the coefficient of x^3 in $P_3(x)$ is 6.

12. Neville's method is used to approximate $f(0.5)$, giving the following table.

$x_0 = 0$	$P_0 = 0$		
$x_1 = 0.4$	$P_1 = 2.8$	$P_{0,1} = 3.5$	
$x_2 = 0.7$	P_2	$P_{1,2}$	$P_{0,1,2} = \frac{27}{7}$

Determine $P_2 = f(0.7)$.

13. Suppose you need to construct eight-decimal-place tables for the common, or base-10, logarithm function from $x = 1$ to $x = 10$ in such a way that linear interpolation is accurate to within 10^{-6}. Determine a bound for the step size for this table. What choice of step size would you make to ensure that $x = 10$ is included in the table?

14. Suppose $x_j = j$ for $j = 0, 1, 2, 3$ and it is known that

$$P_{0,1}(x) = 2x + 1, \quad P_{0,2}(x) = x + 1, \quad \text{and} \quad P_{1,2,3}(2.5) = 3.$$

Find $P_{0,1,2,3}(2.5)$.

15. Neville's method is used to approximate $f(0)$ using $f(-2), f(-1), f(1),$ and $f(2)$. Suppose $f(-1)$ was overstated by 2 and $f(1)$ was understated by 3. Determine the error in the original calculation of the value of the interpolating polynomial to approximate $f(0)$.

16. The following table lists the population of the United States from 1960 to 2010.

Year	1960	1970	1980	1990	2000	2010
Population (thousands)	179,323	203,302	226,542	249,633	281,442	307,746

    a. Find the Lagrange polynomial of degree 5 fitting this data, and use this polynomial to estimate the population in the years 1950, 1975, and 2020.

    b. The population in 1950 was approximately 151,326,000. How accurate do you think your 1975 and 2020 figures are?

17.   In Exercise 15 of Section 1.2, a Maclaurin series was integrated to approximate erf(1), where erf(x) is the normal distribution error function defined by

$$\text{erf}(x) = \frac{2}{\sqrt{\pi}} \int_0^x e^{-t^2} dt.$$

a.   Use the Maclaurin series to construct a table for erf(x) that is accurate to within 10^{-4} for erf(x_i), where $x_i = 0.2i$, for $i = 0, 1, \ldots, 5$.

b.   Use both linear interpolation and quadratic interpolation to obtain an approximation to erf($\frac{1}{3}$). Which approach seems more feasible?

3.3 Divided Differences

Iterated interpolation was used in the previous section to generate successively higher degree polynomial approximations at a specific point. Divided-difference methods introduced in this section are used to successively generate the polynomials themselves.

Divided Differences

We first need to introduce the divided-difference notation, which should remind you of the Aitken's Δ^2 notation defined on page 53. Suppose we are given the $n + 1$ points $(x_0, f(x_0))$, $(x_1, f(x_1)), \ldots (x_n, f(x_n))$. There are $n + 1$ **zeroth divided differences** of the function f. For each $i = 0, 1, \ldots, n$ we define $f[x_i]$ simply as the value of f at x_i:

$$f[x_i] = f(x_i).$$

The remaining divided differences are defined inductively. There are n **first divided differences** of f, one for each $i = 0, 1, \ldots, n - 1$. The first divided difference relative to x_i and x_{i+1} is denoted $f[x_i, x_{i+1}]$ and is defined by

$$f[x_i, x_{i+1}] = \frac{f[x_{i+1}] - f[x_i]}{x_{i+1} - x_i}.$$

After the $(k - 1)$st divided differences,

$$f[x_i, x_{i+1}, x_{i+2}, \ldots, x_{i+k-1}] \quad \text{and} \quad f[x_{i+1}, x_{i+2}, \ldots, x_{i+k-1}, x_{i+k}],$$

have been determined, the **kth divided difference** relative to $x_i, x_{i+1}, x_{i+2}, \ldots, x_{i+k}$ is defined by

$$f[x_i, x_{i+1}, \ldots, x_{i+k-1}, x_{i+k}] = \frac{f[x_{i+1}, x_{i+2}, \ldots, x_{i+k}] - f[x_i, x_{i+1}, \ldots, x_{i+k-1}]}{x_{i+k} - x_i}.$$

The process ends with the single nth divided difference,

$$f[x_0, x_1, \ldots, x_n] = \frac{f[x_1, x_2, \ldots, x_n] - f[x_0, x_1, \ldots, x_{n-1}]}{x_n - x_0}.$$

With this notation, it can be shown that the nth Lagrange interpolation polynomial for f with respect to $x_0, x_1, \ldots, x_n$ can be expressed as

$$P_n(x) = f[x_0] + f[x_0, x_1](x - x_0) + f[x_0, x_1, x_2](x - x_0)(x - x_1) + \cdots$$
$$+ f[x_0, x_1, \ldots, x_n](x - x_0)(x - x_1) \cdots (x - x_{n-1})$$

which is called *Newton's divided-difference formula*. In compressed form we have the following.

As in so many areas, Isaac Newton is prominent in the study of difference equations. He developed interpolation formulas as early as 1675, using his Δ notation in tables of differences. He took a very general approach to the difference formulas, so explicit examples that he produced, including Lagrange's formulas, are often known by other names.

Newton's Interpolatory Divided-Difference Formula

$$P(x) = f[x_0] + \sum_{k=1}^{n} f[x_0, x_1, \ldots, x_k](x - x_0) \cdots (x - x_{k-1}).$$

The generation of the divided differences is outlined in Table 3.7. Two fourth differences and one fifth difference could also be determined from these data, but they have not been recorded in the table.

Table 3.7

x	$f(x)$	First divided differences	Second divided differences	Third divided differences
x_0	$f[x_0]$			
		$f[x_0, x_1] = \dfrac{f[x_1] - f[x_0]}{x_1 - x_0}$		
x_1	$f[x_1]$		$f[x_0, x_1, x_2] = \dfrac{f[x_1, x_2] - f[x_0, x_1]}{x_2 - x_0}$	
		$f[x_1, x_2] = \dfrac{f[x_2] - f[x_1]}{x_2 - x_1}$		$f[x_0, x_1, x_2, x_3] = \dfrac{f[x_1, x_2, x_3] - f[x_0, x_1, x_2]}{x_3 - x_0}$
x_2	$f[x_2]$		$f[x_1, x_2, x_3] = \dfrac{f[x_2, x_3] - f[x_1, x_2]}{x_3 - x_1}$	
		$f[x_2, x_3] = \dfrac{f[x_3] - f[x_2]}{x_3 - x_2}$		$f[x_1, x_2, x_3, x_4] = \dfrac{f[x_2, x_3, x_4] - f[x_1, x_2, x_3]}{x_4 - x_1}$
x_3	$f[x_3]$		$f[x_2, x_3, x_4] = \dfrac{f[x_3, x_4] - f[x_2, x_3]}{x_4 - x_2}$	
		$f[x_3, x_4] = \dfrac{f[x_4] - f[x_3]}{x_4 - x_3}$		$f[x_2, x_3, x_4, x_5] = \dfrac{f[x_3, x_4, x_5] - f[x_2, x_3, x_4]}{x_5 - x_2}$
x_4	$f[x_4]$		$f[x_3, x_4, x_5] = \dfrac{f[x_4, x_5] - f[x_3, x_4]}{x_5 - x_3}$	
		$f[x_4, x_5] = \dfrac{f[x_5] - f[x_4]}{x_5 - x_4}$		
x_5	$f[x_5]$			

> Program DIVDF32 implements the Divided Difference method.

Divided-difference tables are easily constructed by hand calculation. Alternatively, the program DIVDIF32 computes the interpolating polynomial for f at $x_0, x_1, \ldots, x_n$. The form of the output can be modified to produce all the divided differences, as is done in the following example.

Example 1 Complete the divided difference table for the data used in the Illustration on page 69 of Section 3.2, and reproduced in Table 3.8. Then construct the interpolating polynomial that uses all this data.

Table 3.8

x	$f(x)$
1.0	0.7651977
1.3	0.6200860
1.6	0.4554022
1.9	0.2818186
2.2	0.1103623

Solution The first divided difference involving x_0 and x_1 is

$$f[x_0, x_1] = \frac{f[x_1] - f[x_0]}{x_1 - x_0} = \frac{0.6200860 - 0.7651977}{1.3 - 1.0} = -0.4837057.$$

The remaining first divided differences are found in a similar manner and are shown in the fourth column in Table 3.9.

Table 3.9

i	x_i	$f[x_i]$	$f[x_{i-1}, x_i]$	$f[x_{i-2}, x_{i-1}, x_i]$	$f[x_{i-3}, \ldots, x_i]$	$f[x_{i-4}, \ldots, x_i]$
0	1.0	0.7651977				
			-0.4837057			
1	1.3	0.6200860		-0.1087339		
			-0.5489460		0.0658784	
2	1.6	0.4554022		-0.0494433		0.0018251
			-0.5786120		0.0680685	
3	1.9	0.2818186		0.0118183		
			-0.5715210			
4	2.2	0.1103623				

The second divided difference involving x_0, x_1, and x_2 is

$$f[x_0, x_1, x_2] = \frac{f[x_1, x_2] - f[x_0, x_1]}{x_2 - x_0} = \frac{-0.5489460 - (-0.4837057)}{1.6 - 1.0} = -0.1087339.$$

The remaining second divided differences are shown in the fifth column of Table 3.9. The third divided difference involving x_0, x_1, x_2, and x_3 and the fourth divided difference involving all the data points are, respectively,

$$f[x_0, x_1, x_2, x_3] = \frac{f[x_1, x_2, x_3] - f[x_0, x_1, x_2]}{x_3 - x_0} = \frac{-0.0494433 - (-0.1087339)}{1.9 - 1.0}$$
$$= 0.0658784,$$

and

$$f[x_0, x_1, x_2, x_3, x_4] = \frac{f[x_1, x_2, x_3, x_4] - f[x_0, x_1, x_2, x_3]}{x_4 - x_0} = \frac{0.0680685 - 0.0658784}{2.2 - 1.0}$$
$$= 0.0018251.$$

The coefficients of the Newton forward divided-difference form of the interpolating polynomial are along the diagonal in the table. This polynomial is

$$P_4(x) = 0.7651977 - 0.4837057(x - 1.0) - 0.1087339(x - 1.0)(x - 1.3)$$
$$+ 0.0658784(x - 1.0)(x - 1.3)(x - 1.6)$$
$$+ 0.0018251(x - 1.0)(x - 1.3)(x - 1.6)(x - 1.9).$$

Notice that the value $P_4(1.5) = 0.5118200$ agrees with the result in the Illustration on page 69 of Section 3.2, as it must because the polynomials are the same.   ■

Newton's interpolatory divided-difference formula has a simpler form when x_0, x_1, $\ldots$, x_n are arranged consecutively with equal spacing. In this case, we first introduce the notation $h = x_{i+1} - x_i$ for each $i = 0, 1, \ldots, n - 1$. Then we let a new variable s be defined by the equation $x = x_0 + sh$. The difference $x - x_i$ can then be written as $x - x_i = (s - i)h$, and the divided-difference formula becomes

$$P_n(x) = P_n(x_0 + sh) = f[x_0] + shf[x_0, x_1] + s(s - 1)h^2 f[x_0, x_1, x_2] + \cdots$$
$$+ s(s - 1) \cdots (s - n + 1)h^n f[x_0, x_1, \ldots, x_n]$$
$$= f[x_0] + \sum_{k=1}^{n} s(s - 1) \cdots (s - k + 1)h^k f[x_0, x_1, \ldots, x_k].$$

Using a generalization of the binomial-coefficient notation,

$$\binom{s}{k} = \frac{s(s-1)\cdots(s-k+1)}{k!},$$

where s need not be an integer, we can express $P_n(x)$ compactly as follows.

Newton Forward Divided-Difference Formula

$$P_n(x) = P_n(x_0 + sh) = f[x_0] + \sum_{k=1}^{n} \binom{s}{k} k!\, h^k f[x_0, x_1, \ldots, x_k].$$

Forward Differences

Another form is constructed by making use of the forward difference notation introduced in Section 2.5. With this notation

$$f[x_0, x_1] = \frac{f(x_1) - f(x_0)}{x_1 - x_0} = \frac{1}{h}\Delta f(x_0),$$

$$f[x_0, x_1, x_2] = \frac{1}{2h}\left[\frac{\Delta f(x_1) - \Delta f(x_0)}{h}\right] = \frac{1}{2h^2}\Delta^2 f(x_0),$$

and, in general,

$$f[x_0, x_1, \ldots, x_k] = \frac{1}{k!h^k}\Delta^k f(x_0).$$

This gives the following.

Newton Forward-Difference Formula

$$P_n(x) = f[x_0] + \sum_{k=1}^{n} \binom{s}{k} \Delta^k f(x_0).$$

Backward Differences

If the interpolating nodes are reordered from last to first as $x_n, x_{n-1}, \ldots, x_0$, we can write the interpolatory formula as

$$P_n(x) = f[x_n] + f[x_n, x_{n-1}](x - x_n) + f[x_n, x_{n-1}, x_{n-2}](x - x_n)(x - x_{n-1})$$

$$+ \cdots + f[x_n, \ldots, x_0](x - x_n)(x - x_{n-1})\cdots(x - x_1).$$

If the nodes are equally spaced with $x = x_n + sh$ and $x = x_i + (s + n - i)h$, then

$$P_n(x) = P_n(x_n + sh)$$

$$= f[x_n] + shf[x_n, x_{n-1}] + s(s + 1)h^2 f[x_n, x_{n-1}, x_{n-2}] + \cdots$$

$$+ s(s + 1) \cdots (s + n - 1)h^n f[x_n, \ldots, x_0].$$

This form is called the *Newton backward divided-difference formula*. It is used to derive a formula known as the *Newton backward-difference formula*. To discuss this formula, we first need to introduce some notation.

Given the sequence $\{p_n\}_{n=0}^{\infty}$, the **backward difference** ∇p_n (read "nabla p_n") is defined by

$$\nabla p_n \equiv p_n - p_{n-1}, \quad \text{for } n \geq 1$$

and higher powers are defined recursively by

$$\nabla^k p_n = \nabla \left(\nabla^{k-1} p_n \right), \quad \text{for } k \geq 2.$$

This implies that

$$f[x_n, x_{n-1}] = \frac{1}{h} \nabla f(x_n), \quad f[x_n, x_{n-1}, x_{n-2}] = \frac{1}{2h^2} \nabla^2 f(x_n),$$

and, in general,

$$f[x_n, x_{n-1}, \ldots, x_{n-k}] = \frac{1}{k! h^k} \nabla^k f(x_n).$$

Consequently,

$$P_n(x) = f[x_n] + s \nabla f(x_n) + \frac{s(s+1)}{2} \nabla^2 f(x_n) + \cdots$$
$$+ \frac{s(s+1) \cdots (s+n-1)}{n!} \nabla^n f(x_n).$$

Extending the binomial-coefficient notation to include all real values of s, we let

$$\binom{-s}{k} = \frac{-s(-s-1) \cdots (-s-k+1)}{k!} = (-1)^k \frac{s(s+1) \cdots (s+k-1)}{k!}$$

and

$$P_n(x) = f(x_n) + (-1)^1 \binom{-s}{1} \nabla f(x_n) + (-1)^2 \binom{-s}{2} \nabla^2 f(x_n) + \cdots$$
$$+ (-1)^n \binom{-s}{n} \nabla^n f(x_n),$$

which gives the following result.

Newton Backward-Difference Formula

$$P_n(x) = f[x_n] + \sum_{k=1}^{n} (-1)^k \binom{-s}{k} \nabla^k f(x_n).$$

Illustration   The divided-difference Table 3.10 corresponds to the data in Example 1.

Table 3.10

		First divided differences	Second divided differences	Third divided differences	Fourth divided differences
1.0	0.7651977				
		−0.4837057			
1.3	0.6200860		−0.1087339		
		−0.5489460		0.0658784	
1.6	0.4554022		−0.0494433		0.0018251
		−0.5786120		0.0680685	
1.9	0.2818186		0.0118183		
		−0.5715210			
2.2	0.1103623				

Only one interpolating polynomial of degree at most 4 uses these five data points, but we will organize the data points to obtain the best interpolation approximations of degrees 1, 2, and 3. This gives us a sense of the accuracy of the fourth-degree approximation for the given value of x.

If an approximation to $f(1.1)$ is required, the reasonable choice for the nodes would be $x_0 = 1.0, x_1 = 1.3, x_2 = 1.6, x_3 = 1.9$, and $x_4 = 2.2$ because this choice makes the earliest possible use of the data points closest to $x = 1.1$, and also makes use of the fourth divided difference. This implies that $h = 0.3$ and $s = \frac{1}{3}$, so the Newton forward divided-difference formula is used with the divided differences that have a *solid* underline (___) in Table 3.10:

$$P_4(1.1) = P_4\left(1.0 + \frac{1}{3}(0.3)\right)$$

$$= 0.7651977 + \frac{1}{3}(0.3)(-0.4837057) + \frac{1}{3}\left(-\frac{2}{3}\right)(0.3)^2(-0.1087339)$$

$$+ \frac{1}{3}\left(-\frac{2}{3}\right)\left(-\frac{5}{3}\right)(0.3)^3(0.0658784)$$

$$+ \frac{1}{3}\left(-\frac{2}{3}\right)\left(-\frac{5}{3}\right)\left(-\frac{8}{3}\right)(0.3)^4(0.0018251)$$

$$= 0.7196460.$$

To approximate a value when x is close to the end of the tabulated values, say, $x = 2.0$, we would again like to make the earliest use of the data points closest to x. This requires using the Newton backward divided-difference formula with $s = -\frac{2}{3}$ and the divided differences in Table 3.10 that have a *wavy* underline (︠︡). Notice that the fourth divided difference is used in both formulas.

$$P_4(2.0) = P_4\left(2.2 - \frac{2}{3}(0.3)\right)$$

$$= 0.1103623 - \frac{2}{3}(0.3)(-0.5715210) - \frac{2}{3}\left(\frac{1}{3}\right)(0.3)^2(0.0118183)$$

$$- \frac{2}{3}\left(\frac{1}{3}\right)\left(\frac{4}{3}\right)(0.3)^3(0.0680685) - \frac{2}{3}\left(\frac{1}{3}\right)\left(\frac{4}{3}\right)\left(\frac{7}{3}\right)(0.3)^4(0.0018251)$$

$$= 0.2238754. \qquad \square$$

Centered Differences

The Newton formulas are not appropriate for approximating $f(x)$ when x lies near the center of the table, since employing either the backward or forward method in such a way that the highest-order difference is involved will not allow x_0 to be close to x. A number of divided-difference formulas are available for this situation, each of which has situations when it can be used to maximum advantage. These methods are known as *centered-difference formulas*. There are a number of such methods, but we do not discuss any of these techniques. They can be found in many classical numerical analysis books, including the book by Hildebrand [Hi] that is listed in the bibliography.

EXERCISE SET 3.3

1. Use Newton's interpolatory divided-difference formula to construct interpolating polynomials of degrees 1, 2, and 3 for the following data. Approximate the specified value using each of the polynomials.

   a. $f(8.4)$ if $f(8.1) = 16.94410$, $f(8.3) = 17.56492$, $f(8.6) = 18.50515$, $f(8.7) = 18.82091$

   b. $f(0.9)$ if $f(0.6) = -0.17694460$, $f(0.7) = 0.01375227$, $f(0.8) = 0.22363362$, $f(1.0) = 0.65809197$

2. Use Newton's forward-difference formula to construct interpolating polynomials of degrees 1, 2, and 3 for the following data. Approximate the specified value using each of the polynomials.

   a. $f\left(-\frac{1}{3}\right)$ if $f(-0.75) = -0.07181250$, $f(-0.5) = -0.02475000$, $f(-0.25) = 0.33493750$, $f(0) = 1.10100000$

   b. $f(0.25)$ if $f(0.1) = -0.62049958$, $f(0.2) = -0.28398668$, $f(0.3) = 0.00660095$, $f(0.4) = 0.24842440$

3. Use Newton's backward-difference formula to construct interpolating polynomials of degrees 1, 2, and 3 for the following data. Approximate the specified value using each of the polynomials.

   a. $f\left(-\frac{1}{3}\right)$ if $f(-0.75) = -0.07181250$, $f(-0.5) = -0.02475000$, $f(-0.25) = 0.33493750$, $f(0) = 1.10100000$

   b. $f(0.25)$ if $f(0.1) = -0.62049958$, $f(0.2) = -0.28398668$, $f(0.3) = 0.00660095$, $f(0.4) = 0.24842440$

4. a. Construct the fourth-degree interpolating polynomial for the unequally spaced points given in the following table:

x	0.0	0.1	0.3	0.6	1.0
$f(x)$	-6.00000	-5.89483	-5.65014	-5.17788	-4.28172

   b. Suppose $f(1.1) = -3.99583$ is added to the table. Construct the fifth interpolating polynomial.

5. a. Use the following data and the Newton forward divided-difference formula to approximate $f(0.05)$.

x	0.0	0.2	0.4	0.6	0.8
$f(x)$	1.00000	1.22140	1.49182	1.82212	2.22554

   b. Use the Newton backward divided-difference formula to approximate $f(0.65)$.

6. The following population table was given in Exercise 16 of Section 3.2.

Year	1960	1970	1980	1990	2000	2010
Population (thousands)	179,323	203,302	226,542	249,633	281,442	307,746

a. Use an appropriate divided difference method to estimate the population in the years 1950, 1975, and 2020.

b. The population in 1950 was approximately 151,326,000. How accurate do you think your 1975 and 2020 figures are?

7. Show that the polynomial interpolating the following data has degree 3.

x	-2	-1	0	1	2	3
$f(x)$	1	4	11	16	13	-4

8. a. Show that the Newton forward divided-difference polynomials

$$P(x) = 3 - 2(x + 1) + 0(x + 1)(x) + (x + 1)(x)(x - 1)$$

and

$$Q(x) = -1 + 4(x + 2) - 3(x + 2)(x + 1) + (x + 2)(x + 1)(x)$$

both interpolate the data

x	-2	-1	0	1	2
$f(x)$	-1	3	1	-1	3

b. Why does (a) not violate the uniqueness property of interpolating polynomials?

9. A fourth-degree polynomial $P(x)$ satisfies $\Delta^4 P(0) = 24$, $\Delta^3 P(0) = 6$, and $\Delta^2 P(0) = 0$, where $\Delta P(x) = P(x + 1) - P(x)$. Compute $\Delta^2 P(10)$.

10. The following data are given for a polynomial $P(x)$ of unknown degree.

x	0	1	2	3
$P(x)$	4	9	15	18

Determine the coefficient of x^2 in $P(x)$ if all third-order forward differences are 1.

11. The Newton forward divided-difference formula is used to approximate $f(0.3)$ given the following data.

x	0.0	0.2	0.4	0.6
$f(x)$	15.0	21.0	30.0	51.0

Suppose it is discovered that $f(0.4)$ was understated by 10 and $f(0.6)$ was overstated by 5. By what amount should the approximation to $f(0.3)$ be changed?

12. For a function f, the Newton's interpolatory divided-difference formula gives the interpolating polynomial

$$P_3(x) = 1 + 4x + 4x(x - 0.25) + \frac{16}{3}x(x - 0.25)(x - 0.5)$$

on the nodes $x_0 = 0$, $x_1 = 0.25$, $x_2 = 0.5$ and $x_3 = 0.75$. Find $f(0.75)$.

13. For a function f, the forward divided differences are given by

$x_0 = 0.0$	$f[x_0]$		
		$f[x_0, x_1]$	
$x_1 = 0.4$	$f[x_1]$		$f[x_0, x_1, x_2] = \frac{50}{7}$
		$f[x_1, x_2] = 10$	
$x_2 = 0.7$	$f[x_2] = 6$		

Determine the missing entries in the table.

3.4   Hermite Interpolation

Charles Hermite (1822–1901) made significant mathematical discoveries throughout his life in areas such as complex analysis and number theory, particularly involving the theory of equations. He is perhaps best known for proving in 1873 that e is transcendental, that is, it is not the solution to any algebraic equation having integer coefficients. This led in 1882 to Lindemann's proof that π is also transcendental, which demonstrated that it is impossible to use the standard geometry tools of Euclid to construct a square that has the same area as a unit circle.

The Lagrange polynomials agree with a function f at specified points. The values of f are often determined from observation, and in some situations it is possible to determine the derivative of f as well. This is likely to be the case, for example, if the independent variable is time and the function describes the position of an object. The derivative of the function in this case is the velocity, which might be available.

Hermite Polynomials

In this section we will consider Hermite interpolation, which determines a polynomial that agrees with the function and its first derivative at specified points. Suppose that $n + 1$ data points and values of the function, $(x_0, f(x_0)), (x_1, f(x_1)), \ldots, (x_n, f(x_n))$, are given, and that f has a continuous first derivative on an interval $[a, b]$ that contains $x_0, x_1, \ldots, x_n$. (Recall from Section 1.2 that this is denoted $f \in C^1[a, b]$.) The Lagrange polynomial agreeing with f at these points will generally have degree n. If we require, in addition, that the derivative of the Hermite polynomial agrees with the derivative of f at $x_0, x_1, \ldots, x_n$, then the additional $n + 1$ conditions raise the expected degree of the Hermite polynomial to $2n + 1$.

Hermite Polynomial

Suppose that $f \in C^1[a, b]$ and that $x_0, \ldots, x_n$ in $[a, b]$ are distinct. The unique polynomial of least degree agreeing with f and f' at $x_0, \ldots, x_n$ is the polynomial of degree at most $2n + 1$ given by

$$H_{2n+1}(x) = \sum_{j=0}^{n} f(x_j) H_{n,j}(x) + \sum_{j=0}^{n} f'(x_j) \hat{H}_{n,j}(x),$$

where

$$H_{n,j}(x) = [1 - 2(x - x_j) L'_{n,j}(x_j)] L^2_{n,j}(x)$$

and

$$\hat{H}_{n,j}(x) = (x - x_j) L^2_{n,j}(x).$$

Here, $L_{n,j}(x)$ denotes the jth Lagrange coefficient polynomial of degree n.

The error term for the Hermite polynomial is similar to that of the Lagrange polynomial, with the only modifications being those needed to accommodate the increased amount of data used in the Hermite polynomial.

Hermite Polynomial Error Formula

If $f \in C^{2n+2}[a, b]$, then

$$f(x) = H_{2n+1}(x) + \frac{f^{(2n+2)}(\xi(x))}{(2n+2)!}(x - x_0)^2 \cdots (x - x_n)^2$$

for some $\xi(x)$ (unknown) in the interval (a, b).

Hermite Polynomials Using Divided Differences

Although the Hermite formula provides a complete description of the Hermite polynomials, the need to determine and evaluate the Lagrange polynomials and their derivatives makes the procedure tedious even for small values of n. An alternative method for generating Hermite approximations is based on the connection between the nth divided difference and the nth derivative of f.

Divided-Difference Relationship to the Derivative

If $f \in C^n[a, b]$ and $x_0, x_1, \ldots, x_n$ are distinct in $[a, b]$, then some number ξ in (a, b) exists with

$$f[x_0, x_1, \ldots, x_n] = \frac{f^{(n)}(\xi)}{n!}.$$

To use this result to generate the Hermite polynomial, first suppose that the distinct numbers $x_0, x_1, \ldots, x_n$ are given together with the values of f and f' at these numbers. Define a new sequence $z_0, z_1, \ldots, z_{2n+1}$ by

$$z_{2k} = z_{2k+1} = x_k, \quad \text{for each } k = 0, 1, \ldots, n.$$

Now construct the divided-difference table using the variables $z_0, z_1, \ldots, z_{2n+1}$.

Since $z_{2k} = z_{2k+1} = x_k$ for each k, we cannot define $f[z_{2k}, z_{2k+1}]$ by the basic divided-difference relation:

$$f[z_{2k}, z_{2k+1}] = \frac{f[z_{2k+1}] - f[z_{2k}]}{z_{2k+1} - z_{2k}}.$$

But, for each k we have $f[x_k, x_{k+1}] = f'(\xi_k)$ for some number ξ_k in (x_k, x_{k+1}), and $\lim_{x_{k+1} \to x_k} f[x_k, x_{k+1}] = f'(x_k)$. So a reasonable substitution in this situation is $f[z_{2k}, z_{2k+1}] = f'(x_k)$, and we use the entries

$$f'(x_0), f'(x_1), \ldots, f'(x_n)$$

in place of the first divided differences

$$f[z_0, z_1], f[z_2, z_3], \ldots, f[z_{2n}, z_{2n+1}].$$

The remaining divided differences are produced as usual, and the appropriate divided differences are employed in Newton's interpolatory divided-difference formula. This provides us with an alternative, and more easily evaluated, method for determining coefficients of the Hermite polynomial.

Divided-Difference Form of the Hermite Polynomial

If $f \in C^1[a, b]$ and $x_0, x_1, \ldots, x_n$ are distinct in $[a, b]$, then

$$H_{2n+1}(x) = f[z_0] + \sum_{k=1}^{2n+1} f[z_0, z_1, \ldots, z_k](x - z_0), \ldots, (x - z_{k-1}),$$

where $z_{2k} = z_{2k+1} = x_k$ and $f[z_{2k}, z_{2k+1}] = f'(x_k)$ for each $k = 0, 1, \ldots, n$.

Table 3.11 shows the entries that are used for the first three divided-difference columns when determining the Hermite polynomial $H_5(x)$ for x_0, x_1, and x_2. The remaining entries are generated in the usual divided-difference manner.

Table 3.11

z	$f(z)$	First divided differences	Second divided differences
$z_0 = x_0$	$f[z_0] = f(x_0)$		
		$f[z_0, z_1] = f'(x_0)$	
$z_1 = x_0$	$f[z_1] = f(x_0)$		$f[z_0, z_1, z_2] = \dfrac{f[z_1, z_2] - f[z_0, z_1]}{z_2 - z_0}$
		$f[z_1, z_2] = \dfrac{f[z_2] - f[z_1]}{z_2 - z_1}$	
$z_2 = x_1$	$f[z_2] = f(x_1)$		$f[z_1, z_2, z_3] = \dfrac{f[z_2, z_3] - f[z_1, z_2]}{z_3 - z_1}$
		$f[z_2, z_3] = f'(x_1)$	
$z_3 = x_1$	$f[z_3] = f(x_1)$		$f[z_2, z_3, z_4] = \dfrac{f[z_3, z_4] - f[z_2, z_3]}{z_4 - z_2}$
		$f[z_3, z_4] = \dfrac{f[z_4] - f[z_3]}{z_4 - z_3}$	
$z_4 = x_2$	$f[z_4] = f(x_2)$		$f[z_3, z_4, z_5] = \dfrac{f[z_4, z_5] - f[z_3, z_4]}{z_5 - z_3}$
		$f[z_4, z_5] = f'(x_2)$	
$z_5 = x_2$	$f[z_5] = f(x_2)$		

Program HERMIT33 implements the Hermite's method.

Example 1   Use the data given in the Illustration on page 80 and the divided difference method to determine the Hermite polynomial approximation at $x = 1.5$.

Solution   The underlined entries in the first three columns of Table 3.12 are the data given in the Illustration and the appropriate derivatives. The remaining entries in this table are generated by the standard divided-difference formula, as shown in Table 3.11.

Table 3.12

1.3	0.6200860				
		−0.5220232			
1.3	0.6200860		−0.0897427		
		−0.5489460		0.0663657	
1.6	0.4554022		−0.0698330		0.0026663
		−0.5698959		0.0679655	−0.0027738
1.6	0.4554022		−0.0290537		0.0010020
		−0.5786120		0.0685667	
1.9	0.2818186		−0.0084837		
		−0.5811571			
1.9	0.2818186				

For example, for the second entry in the third column we use the second 1.3 entry in the second column and the first 1.6 entry in that column to obtain

$$\frac{0.4554022 - 0.6200860}{1.6 - 1.3} = -0.5489460.$$

For the first entry in the fourth column we use the first 1.3 entry in the third column and the first 1.6 entry in that column to obtain

$$\frac{-0.5489460 - (-0.5220232)}{1.6 - 1.3} = -0.0897427.$$

The value of the Hermite polynomial at 1.5 is

$$
\begin{aligned}
H_5(1.5) = {} & f[1.3] + f'(1.3)(1.5 - 1.3) + f[1.3, 1.3, 1.6](1.5 - 1.3)^2 \\
& + f[1.3, 1.3, 1.6, 1.6](1.5 - 1.3)^2(1.5 - 1.6) \\
& + f[1.3, 1.3, 1.6, 1.6, 1.9](1.5 - 1.3)^2(1.5 - 1.6)^2 \\
& + f[1.3, 1.3, 1.6, 1.6, 1.9, 1.9](1.5 - 1.3)^2(1.5 - 1.6)^2(1.5 - 1.9) \\
= {} & 0.6200860 + (-0.5220232)(0.2) + (-0.0897427)(0.2)^2 \\
& + 0.0663657(0.2)^2(-0.1) + 0.0026663(0.2)^2(-0.1)^2 \\
& + (-0.0027738)(0.2)^2(-0.1)^2(-0.4) \\
= {} & 0.5118277. \qquad \blacksquare
\end{aligned}
$$

The program HERMIT33 generates the coefficients for the Hermite polynomials using this modified Newton interpolatory divided-difference formula. The program is structured slightly differently from the discussion to take advantage of efficiency of computation.

EXERCISE SET 3.4

1. Use Hermite interpolation to construct an approximating polynomial for the following data.

a.

x	$f(x)$	$f'(x)$
8.3	17.56492	3.116256
8.6	18.50515	3.151762

b.

x	$f(x)$	$f'(x)$
0.8	0.22363362	2.1691753
1.0	0.65809197	2.0466965

c.

x	$f(x)$	$f'(x)$
−0.5	−0.0247500	0.7510000
−0.25	0.3349375	2.1890000
0	1.1010000	4.0020000

d.

x	$f(x)$	$f'(x)$
0.1	−0.62049958	3.58502082
0.2	−0.28398668	3.14033271
0.3	0.00660095	2.66668043
0.4	0.24842440	2.16529366

2. The data in Exercise 1 were generated using the following functions. For the given value of x, use the polynomials constructed in Exercise 1 to approximate $f(x)$, and calculate the actual error.

a. $f(x) = x \ln x$;   approximate $f(8.4)$.

b. $f(x) = \sin(e^x - 2)$;   approximate $f(0.9)$.

c. $f(x) = x^3 + 4.001x^2 + 4.002x + 1.101$;   approximate $f(-\frac{1}{3})$.

d. $f(x) = x \cos x - 2x^2 + 3x - 1$;   approximate $f(0.25)$.

3. **a.** Use the following values and five-digit rounding arithmetic to construct the Hermite interpolating polynomial to approximate $\sin 0.34$.

x	$f(x) = \sin x$	$f'(x) = \cos x$
0.30	0.29552	0.95534
0.32	0.31457	0.94924
0.35	0.34290	0.93937

b. Determine an error bound for the approximation in part (a) and compare to the actual error.

c. Add $\sin 0.33 = 0.32404$ and $\cos 0.33 = 0.94604$ to the data and redo the calculations.

4. Let $f(x) = 3xe^x - e^{2x}$.

a. Approximate $f(1.03)$ by the Hermite interpolating polynomial of degree at most 3 using $x_0 = 1$ and $x_1 = 1.05$. Compare the actual error to the error bound.

b. Repeat (a) with the Hermite interpolating polynomial of degree at most 5, using $x_0 = 1$, $x_1 = 1.05$, and $x_2 = 1.07$.

5. Use the error formula and MATLAB to find a bound for the errors in the approximations of $f(x)$ in (a) and (c) of Exercise 2.

6. The following table lists data for the function described by $f(x) = e^{0.1x^2}$. Approximate $f(1.25)$ by using $H_5(1.25)$ and $H_3(1.25)$, where H_5 uses the nodes $x_0 = 1$, $x_1 = 2$, and $x_2 = 3$ and H_3 uses the nodes $\bar{x}_0 = 1$ and $\bar{x}_1 = 1.5$. Find error bounds for these approximations.

x	$f(x) = e^{0.1x^2}$	$f'(x) = 0.2xe^{0.1x^2}$
$x_0 = \bar{x}_0 = 1$	1.105170918	0.2210341836
$\bar{x}_1 = 1.5$	1.252322716	0.3756968148
$x_1 = 2$	1.491824698	0.5967298792
$x_2 = 3$	2.459603111	1.475761867

7. A car traveling along a straight road is clocked at a number of points. The data from the observations arc given in the following table, where the time is in seconds, the distance is in feet, and the speed is in feet per second.

Time	0	3	5	8	13
Distance	0	225	383	623	993
Speed	75	77	80	74	72

a. Use a Hermite polynomial to predict the position of the car and its speed when $t = 10$ s.

b. Use the derivative of the Hermite polynomial to determine whether the car ever exceeds a 55-mi/h speed limit on the road. If so, what is the first time the car exceeds this speed?

c. What is the predicted maximum speed for the car?

8. Let $z_0 = x_0$, $z_1 = x_0$, $z_2 = x_1$, and $z_3 = x_1$. Form the following divided-difference table.

$z_0 = x_0$	$f[z_0] = f(x_0)$			
		$f[z_0, z_1] = f'(x_0)$		
$z_1 = x_0$	$f[z_1] = f(x_0)$		$f[z_0, z_1, z_2]$	
		$f[z_1, z_2]$		$f[z_0, z_1, z_2, z_3]$
$z_2 = x_1$	$f[z_2] = f(x_1)$		$f[z_1, z_2, z_3]$	
		$f[z_2, z_3] = f'(x_1)$		
$z_3 = x_1$	$f[z_3] = f(x_1)$			

Show if

$$P(x) = f[z_0] + f[z_0, z_1](x - x_0) + f[z_0, z_1, z_2](x - x_0)^2 + f[z_0, z_1, z_2, z_3](x - x_0)^2(x - x_1),$$

then

$$P(x_0) = f(x_0), \quad P(x_1) = f(x_1), \quad P'(x_0) = f'(x_0), \quad \text{and} \quad P'(x_1) = f'(x_1),$$

which implies that $P(x) \equiv H_3(x)$.

3.5 Spline Interpolation

The previous sections use polynomials to approximate arbitrary functions. However, relatively high-degree polynomials are needed for accurate approximation and these have some serious disadvantages. They can have an oscillatory nature, and a fluctuation over a small portion of the interval can induce large fluctuations over the entire range. We will see an example of this later in this section.

An alternative approach is to divide the interval into a collection of subintervals and construct a different approximating polynomial on each subinterval. This is called **piecewise polynomial approximation**.

Piecewise-Polynomial Approximation

The simplest piecewise polynomial approximation joins the data points $(x_0, f(x_0))$, $(x_1, f(x_1)), \ldots, (x_n, f(x_n))$ by a series of straight lines, such as those shown in Figure 3.7.

A disadvantage of linear approximation is that the approximation is generally not differentiable at the endpoints of the subintervals, so the interpolating function is not "smooth" at these points. It is often clear from physical conditions that smoothness is required, and the approximating function must be continuously differentiable.

Figure 3.7

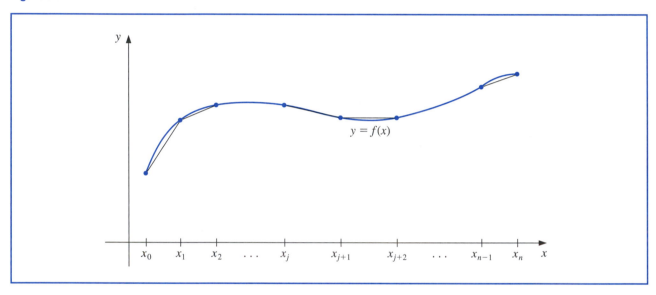

One remedy for this problem is to use a piecewise polynomial of Hermite type. For example, if the values of f and f' are known at each of the points $x_0 < x_1 < \cdots < x_n$, a cubic Hermite polynomial can be used on each of the subintervals $[x_0, x_1], [x_1, x_2], \ldots, [x_{n-1}, x_n]$ to obtain an approximating function that has a continuous derivative on the interval $[x_0, x_n]$. To determine the appropriate cubic Hermite polynomial on a given interval, we simply compute the function $H_3(x)$ for that interval.

The Hermite polynomials are commonly used in application problems to study the motion of particles in space. The difficulty with using Hermite piecewise polynomials for general interpolation problems concerns the need to know the derivative of the function being approximated. The remainder of this section considers approximations using piecewise polynomials that require no derivative information, except perhaps at the endpoints of the interval on which the function is being approximated.

Isaac Jacob Schoenberg (1903–1990) developed his work on splines during World War II at the Army's Ballistic Research Laboratory in Aberdeen, Maryland, while on leave from the University of Pennsylvania. His original work involved numerical procedures for solving differential equations. The much broader application of splines to the areas of data fitting and computer-aided geometric design became evident with the widespread availability of computers in the 1960s.

Cubic Splines

The most common piecewise-polynomial approximation uses cubic polynomials between pairs of nodes and is called **cubic spline** interpolation. A general cubic polynomial involves four constants, so there is sufficient flexibility in the cubic spline procedure to ensure that the interpolant has two continuous derivatives on the interval. The derivatives of the cubic spline do not, in general, however, agree with the derivatives of the function, even at the nodes. (See Figure 3.8.)

Figure 3.8

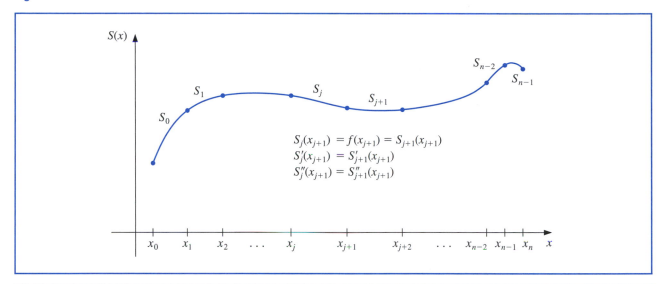

Cubic Spline Interpolation

Given a function f defined on $[a, b]$ and a set of nodes, $a = x_0 < x_1 < \cdots < x_n = b$, a cubic spline interpolant, S, for f is a function that satisfies the following conditions:

(a) For each $j = 0, 1, \ldots, n - 1$, $S(x)$ is a cubic polynomial, denoted by $S_j(x)$, on the subinterval $[x_j, x_{j+1}]$.

(b) $S_j(x_j) = f(x_j)$ and $S_j(x_{j+1}) = f(x_{j+1})$ for each $j = 0, 1, \ldots, n - 1$.

(c) $S_{j+1}(x_{j+1}) = S_j(x_{j+1})$ for each $j = 0, 1, \ldots, n - 2$ (Implied by (b).)

(d) $S'_{j+1}(x_{j+1}) = S'_j(x_{j+1})$ for each $j = 0, 1, \ldots, n - 2$.

(e) $S''_{j+1}(x_{j+1}) = S''_j(x_{j+1})$ for each $j = 0, 1, \ldots, n - 2$.

(f) One of the following sets of boundary conditions is satisfied:

   (i) $S''(x_0) = S''(x_n) = 0$ (natural or free boundary);

   (ii) $S'(x_0) = f'(x_0)$ and $S'(x_n) = f'(x_n)$ (clamped boundary).

The root of the word "spline" is the same as that of splint. It was originally a small strip of wood that could be used to join two boards. Later the word was used to refer to a long flexible strip, generally of metal, that could be used to draw continuous smooth curves by forcing the strip to pass through specified points and tracing along the curve.

Although cubic splines are defined with other boundary conditions, the conditions given in (f) are sufficient for our purposes. When the natural boundary conditions are used, the spline assumes the shape that a long flexible rod would take if forced to go through the points $\{(x_0, f(x_0)), (x_1, f(x_1)), \ldots, (x_n, f(x_n))\}$. This spline continues linearly when $x \le x_0$ and when $x \ge x_n$.

Example 1   Construct a natural cubic spline that passes through the points $(1, 2)$, $(2, 3)$, and $(3, 5)$.

Solution   This spline consists of two cubics. The first for the interval $[1, 2]$, denoted

$$S_0(x) = a_0 + b_0(x - 1) + c_0(x - 1)^2 + d_0(x - 1)^3,$$

A natural spline has no conditions imposed for the direction at its endpoints, so the curve takes the shape of a straight line after it passes through the interpolation points nearest its endpoints. The name derives from the fact that this is the natural shape a flexible strip assumes if forced to pass through specified interpolation points with no additional constraints. (See Figure 3.9.)

and the other for [2, 3], denoted

$$S_1(x) = a_1 + b_1(x - 2) + c_1(x - 2)^2 + d_1(x - 2)^3.$$

There are 8 constants to be determined, which requires 8 conditions. Four conditions come from the fact that the splines must agree with the data at the nodes. Hence

$$2 = f(1) = a_0, \quad 3 = f(2) = a_0 + b_0 + c_0 + d_0, \quad 3 = f(2) = a_1, \quad \text{and}$$

$$5 = f(3) = a_1 + b_1 + c_1 + d_1.$$

Two more come from the fact that $S_0'(2) = S_1'(2)$ and $S_0''(2) = S_1''(2)$. These are

$$S_0'(2) = S_1'(2): \quad b_0 + 2c_0 + 3d_0 = b_1 \quad \text{and} \quad S_0''(2) = S_1''(2): \quad 2c_0 + 6d_0 = 2c_1.$$

The final two come from the natural boundary conditions:

$$S_0''(1) = 0: \quad 2c_0 = 0 \quad \text{and} \quad S_1''(3) = 0: \quad 2c_1 + 6d_1 = 0.$$

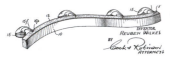

Figure 3.9

Solving this system of equations gives the spline

$$S(x) = \begin{cases} 2 + \dfrac{3}{4}(x - 1) + \dfrac{1}{4}(x - 1)^3, & \text{for } x \in [1, 2] \\[2ex] 3 + \dfrac{3}{2}(x - 2) + \dfrac{3}{4}(x - 2)^2 - \dfrac{1}{4}(x - 2)^3, & \text{for } x \in [2, 3]. \end{cases} \qquad \blacksquare$$

Construction of a Cubic Spline

Clamping a spline indicates that the ends of the flexible strip are fixed so that it is forced to take a specific direction at each of its endpoints. This is important, for example, when two spline functions should match at their endpoints. This is done mathematically by specifying the values of the derivative of the curve at the endpoints of the spline.

In general, clamped boundary conditions lead to more accurate approximations because they include more information about the function. However, for this type of boundary condition, we need values of the derivative at the endpoints or an accurate approximation to those values.

To construct the cubic spline interpolant for a given function f, the conditions in the definition are applied to the cubic polynomials

$$S_j(x) = a_j + b_j(x - x_j) + c_j(x - x_j)^2 + d_j(x - x_j)^3$$

for each $j = 0, 1, \ldots, n - 1$.

Since

$$S_j(x_j) = a_j = f(x_j),$$

condition (c) can be applied to obtain

$$a_{j+1} = S_{j+1}(x_{j+1}) = S_j(x_{j+1}) = a_j + b_j(x_{j+1} - x_j) + c_j(x_{j+1} - x_j)^2 + d_j(x_{j+1} - x_j)^3$$

for each $j = 0, 1, \ldots, n - 2$.

Since the term $x_{j+1} - x_j$ is used repeatedly in this development, it is convenient to introduce the simpler notation

$$h_j = x_{j+1} - x_j,$$

for each $j = 0, 1, \ldots, n - 1$. If we also define $a_n = f(x_n)$, then the equation

$$a_{j+1} = a_j + b_j h_j + c_j h_j^2 + d_j h_j^3 \qquad (3.1)$$

holds for each $j = 0, 1, \ldots, n - 1$.

In a similar manner, define $b_n = S'(x_n)$ and observe that

$$S'_j(x) = b_j + 2c_j(x - x_j) + 3d_j(x - x_j)^2 \tag{3.2}$$

implies that $S'_j(x_j) = b_j$ for each $j = 0, 1, \ldots, n - 1$. Applying condition (d) gives

$$b_{j+1} = b_j + 2c_j h_j + 3d_j h_j^2, \tag{3.3}$$

for each $j = 0, 1, \ldots, n - 1$.

Another relation between the coefficients of S_j is obtained by defining $c_n = S''(x_n)/2$ and applying condition (e). In this case,

$$c_{j+1} = c_j + 3d_j h_j, \tag{3.4}$$

for each $j = 0, 1, \ldots, n - 1$.

Solving for d_j in Eq. (3.4) and substituting this value into Eqs. (3.1) and (3.3) gives the new equations

$$a_{j+1} = a_j + b_j h_j + \frac{h_j^2}{3}(2c_j + c_{j+1}) \tag{3.5}$$

and

$$b_{j+1} = b_j + h_j(c_j + c_{j+1}) \tag{3.6}$$

for each $j = 0, 1, \ldots, n - 1$.

The final relationship involving the coefficients is obtained by solving the appropriate equation in the form of Eq. (3.5) for b_j,

$$b_j = \frac{1}{h_j}(a_{j+1} - a_j) - \frac{h_j}{3}(2c_j + c_{j+1}), \tag{3.7}$$

and then, with a reduction of the index, for b_{j-1}, which gives

$$b_{j-1} = \frac{1}{h_{j-1}}(a_j - a_{j-1}) - \frac{h_{j-1}}{3}(2c_{j-1} + c_j).$$

Substituting these values into the equation derived from Eq. (3.6), when the index is reduced by 1, gives the linear system of equations

$$h_{j-1}c_{j-1} + 2(h_{j-1} + h_j)c_j + h_j c_{j+1} = \frac{3}{h_j}(a_{j+1} - a_j) - \frac{3}{h_{j-1}}(a_j - a_{j-1}) \tag{3.8}$$

for each $j = 1, 2, \ldots, n - 1$. This system involves only $\{c_j\}_{j=0}^n$ as unknowns since the values of $\{h_j\}_{j=0}^{n-1}$ and $\{a_j\}_{j=0}^n$ are given by the spacing of the nodes $\{x_j\}_{j=0}^n$ and the values $\{f(x_j)\}_{j=0}^n$.

Once the values of $\{c_j\}_{j=0}^n$ are determined, it is a simple matter to find the remainder of the constants $\{b_j\}_{j=0}^{n-1}$ from Eq. (3.7) and $\{d_j\}_{j=0}^{n-1}$ from Eq. (3.4) and to construct the cubic polynomials $\{S_j(x)\}_{j=0}^{n-1}$. In the case of the clamped spline, we also need equations involving the $\{c_j\}$ that ensure that $S'(x_0) = f'(x_0)$ and $S'(x_n) = f'(x_n)$. In Eq. (3.2) we have $S'_j(x)$ in terms of b_j, c_j, and d_j. Since we now know b_j and d_j in terms of c_j, we can use this equation to show that the appropriate equations are

$$2h_0 c_0 + h_0 c_1 = \frac{3}{h_0}(a_1 - a_0) - 3f'(x_0) \tag{3.9}$$

Program NCUBSP34
creates a Natural Cubic
Spline

and

$$h_{n-1}c_{n-1} + 2h_{n-1}c_n = 3f'(x_n) - \frac{3}{h_{n-1}}(a_n - a_{n-1}). \tag{3.10}$$

The solution to the cubic spline problem with the natural boundary conditions $S''(x_0) = S''(x_n) = 0$ can be obtained by applying the program NCUBSP34. The program CCUBSP35 determines the cubic spline with the clamped boundary conditions $S'(x_0) = f'(x_0)$ and $S'(x_n) = f'(x_n)$.

Example 2   Determine the clamped cubic spline for $f(x) = x \sin 4x$ using the nodes $x_0 = 0, x_1 = 0.25$, $x_2 = 0.4$, and $x_3 = 0.6$.

Program CCUBSP35
creates a Clamped Cubic
Spline

Solution   We first define the function $f(x)$ and its derivative $fp(x) \equiv f'(x)$ in MATLAB with

```
f = inline('x*sin(4*x)','x')
fp = inline('sin(4*x)+4*x*cos(4*x)','x')
```

We define the nodes using the MATLAB capability of defining subscripted variables within square brackets, where a blank is used to separate entries. The subscripts in MATLAB begin with 1 so $x(1) = 0, x(2) = 0.25, x(3) = 0.4$, and $x(4) = 0.6$. This is entered in MATLAB as

```
x = [0 0.25 0.4 0.6]
```

The step sizes are defined by

```
h = [x(2)-x(1)     x(3)-x(2)     x(4)-x(3)]
```

and the values of the function at the nodes by

```
a = [f(x(1))     f(x(2))     f(x(3))     f(x(4))]
```

MATLAB responds to this last command with

$$a = 0 \quad 0.210367746201974 \quad 0.399829441216602 \quad 0.405277908330691$$

A 4×4 array A is defined whose rows are defined by the system of equations used to determine the quadratic coefficients, that is, the c's. These are given in Eqs. (3.8), (3.9), and (3.10), but Eq. (3.9) involves an index of 0, which MATLAB does not permit. So the indices must all be increased by 1 to compensate. So A is defined as follows:

Row 1: the left-hand side of Eq. (3.9), with all the indices increased by 1; that is:

$$2h_1c_1 + h_1c_2 + 0 \cdot c_3 + 0 \cdot c_4.$$

Row 2: the left-hand side of Eq. (3.8) when $j = 2$:

$$h_1c_1 + 2(h_1 + h_2)c_2 + h_2 \cdot c_3 + 0 \cdot c_4.$$

Row 3: the left-hand side of Eq. (3.8) when $j = 3$:

$$0 \cdot c_1 + h_2c_2 + 2(h_2 + h_3)c_3 + h_3c_4.$$

Row 4: the left-hand side of Eq. (3.10) when $n = 4$:

$$0 \cdot c_1 + 0 \cdot c_2 + h_3 c_3 + 2 h_3 c_4.$$

This gives

```
A = [ 2*h(1)  h(1)  0  0; h(1)  2*(h(1)+h(2))  h(2) 0;  0  h(2)
      2*(h(2)+h(3))  h(3);  0  0  h(3)  2*h(3)]
```

The right-hand sides of the same equations are stored in the 4×1 array B.

Row 1: the right-hand side of Eq. (3.9), with all the indices increased by 1; that is:

$$\frac{3}{h_1}(a_2 - a_1) - 3 f'(x_1).$$

Row 2: the right-hand side of Eq. (3.8) when $j = 2$:

$$\frac{3}{h_2}(a_3 - a_2) - \frac{3}{h_1}(a_2 - a_1).$$

Row 3: the right-hand side of Eq. (3.8) when $j = 3$:

$$\frac{3}{h_3}(a_4 - a_3) - \frac{3}{h_2}(a_3 - a_2).$$

Row 4: the right-hand side of Eq. (3.10) when $n = 4$:

$$3 f'(x_4) - \frac{3}{h_3}(a_4 - a_3).$$

So B is defined by

```
B = [3*(a(2)-a(1))/h(1) - 3*fp(x(1));
     3*(a(3)-a(2))/h(2)-3*(a(2)-a(1))/h(1);
     3*(a(4)-a(3))/h(3)-3*(a(3)-a(2))/h(2);
     3*fp(x(4))-3*(a(4)-a(3))/h(3)]
```

MATLAB gives these as

$$A = \begin{bmatrix} 0.500000000000000 & 0.250000000000000 & 0 & 0 \\ 0.250000000000000 & 0.800000000000000 & 0.150000000000000 & 0 \\ 0 & 0.150000000000000 & 0.700000000000000 & 0.200000000000000 \\ 0 & 0 & 0.200000000000000 & 0.400000000000000 \end{bmatrix}$$

and

$$B = \begin{bmatrix} 2.524412954423689 \\ 1.264820945868869 \\ -3.707506893581232 \\ -3.364572216954841 \end{bmatrix}$$

We now can have MATLAB solve the system using the `linsolve` command.

```
c = linsolve(A,B)
```

MATLAB responds with solution consisting of $c(1)$, $c(2)$, $c(3)$, and $c(4)$ as

$$c = \begin{bmatrix} 4.649673230468573 \\ 0.798305356757612 \\ -3.574944314362422 \\ -6.623958385205892 \end{bmatrix}$$

Now use Eq. (3.7) to obtain the values of $b(1)$, $b(2)$, and $b(3)$ with the command

```
b = [(a(2)-a(1))/h(1) - h(1)*(2*c(1)+c(2))/3;
     (a(3)-a(2))/h(2) - h(2)*(2*c(2)+c(3))/3;
     (a(4)-a(3))/h(3) - h(3)*(2*c(3)+c(4))/3]
```

which MATLAB gives as

$$b = \begin{bmatrix} 0.000000000000000 \\ 1.361994646806546 \\ 0.945498803165825 \end{bmatrix}$$

Finally, the values of $d(1)$, $d(2)$, and $d(3)$ are obtained using Eq. (3.4) and the command

```
d = [(c(2)-c(1))/(3*h(1));  (c(3)-c(2))/(3*h(2));
     (c(4)-c(3))/(3*h(3))]
```

producing

$$d = \begin{bmatrix} -5.135157164947948 \\ -9.718332602488962 \\ -5.081690118072451 \end{bmatrix}$$

This implies that the cubic spline, to three decimal places, is as shown in Table 3.13.

Table 3.13

j	x_j	a_j	b_j	c_j	d_j
0	0.000	0.000	0.000	4.650	−5.135
1	0.250	0.210	1.362	0.798	−9.718
2	0.400	0.400	0.945	−3.575	−5.082
3	0.600	0.405		−6.624	

■

In the following Illustration the shape of the curve is much more complex. Placing a minimal number of data points along the curve to get a good representation would require some experimentation.

Illustration   Figure 3.10 shows a ruddy duck in flight. To approximate the top profile of the duck, we have chosen points along the curve through which we want the approximating curve to pass. Table 3.14 lists the coordinates of 21 data points relative to the superimposed coordinate system shown in Figure 3.11. Notice that more points are used when the curve is changing rapidly than when it is changing more slowly.

Figure 3.10

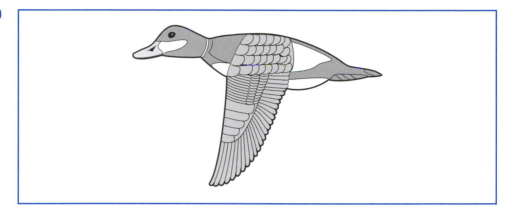

Figure 3.11

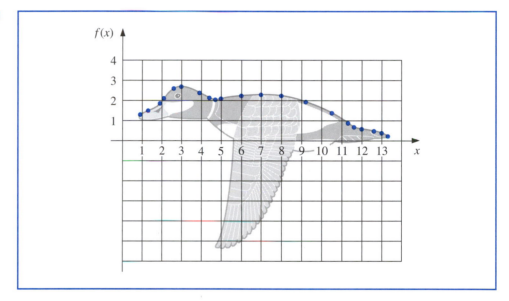

Table 3.14

x	0.9	1.3	1.9	2.1	2.6	3.0	3.9	4.4	4.7	5.0	6.0	7.0	8.0	9.2	10.5	11.3	11.6	12.0	12.6	13.0	13.3
$f(x)$	1.3	1.5	1.85	2.1	2.6	2.7	2.4	2.15	2.05	2.1	2.25	2.3	2.25	1.95	1.4	0.9	0.7	0.6	0.5	0.4	0.25

Using the program NCUBSP34 to generate the natural cubic spline for this data produces the coefficients shown in Table 3.15. This spline curve is nearly identical to the profile, as shown in Figure 3.12.

For comparison purposes, Figure 3.13 gives an illustration of the curve that is generated using a Lagrange interpolating polynomial to fit the data given in Table 3.14. The interpolating polynomial in this case is of degree 20 and oscillates wildly. It produces a very strange illustration of the back of a duck, in flight or otherwise.

Table 3.15

j	x_j	a_j	b_j	c_j	d_j
0	0.9	1.3	5.40	0.00	−0.25
1	1.3	1.5	0.42	−0.30	0.95
2	1.9	1.85	1.09	1.41	−2.96
3	2.1	2.1	1.29	−0.37	−0.45
4	2.6	2.6	0.59	−1.04	0.45
5	3.0	2.7	−0.02	−0.50	0.17
6	3.9	2.4	−0.50	−0.03	0.08
7	4.4	2.15	−0.48	0.08	1.31
8	4.7	2.05	−0.07	1.27	−1.58
9	5.0	2.1	0.26	−0.16	0.04
10	6.0	2.25	0.08	−0.03	0.00
11	7.0	2.3	0.01	−0.04	−0.02
12	8.0	2.25	−0.14	−0.11	0.02
13	9.2	1.95	−0.34	−0.05	−0.01
14	10.5	1.4	−0.53	−0.10	−0.02
15	11.3	0.9	−0.73	−0.15	1.21
16	11.6	0.7	−0.49	0.94	−0.84
17	12.0	0.6	−0.14	−0.06	0.04
18	12.6	0.5	−0.18	0.00	−0.45
19	13.0	0.4	−0.39	−0.54	0.60
20	13.3	0.25			

Figure 3.12

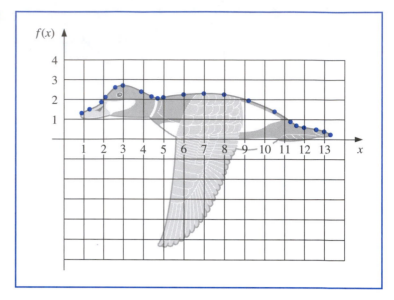

Figure 3.13

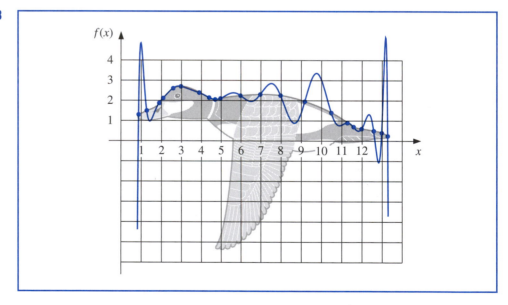

To use a clamped spline to approximate this curve we would need derivative approximations for the endpoints. Even if these approximations were available, we could expect little improvement because of the close agreement of the natural cubic spline to the curve of the top profile.   ☐

Cubic splines generally agree quite well with the function being approximated, provided that the points are not too far apart and the fourth derivative of the function is well behaved. For example, suppose that f has four continuous derivatives on $[a, b]$ and that the fourth derivative on this interval has a magnitude bounded by M. Then the clamped cubic spline

$S(x)$ agreeing with $f(x)$ at the points $a = x_0 < x_1 < \cdots < x_n = b$ has the property that for all x in $[a, b]$,

$$|f(x) - S(x)| \leq \frac{5M}{384} \max_{0 \leq j \leq n-1} (x_{j+1} - x_j)^4.$$

A similar—but more complicated—result holds for the natural cubic splines.

EXERCISE SET 3.5

1. Determine the natural cubic spline S that interpolates the data $f(0) = 0$, $f(1) = 1$, and $f(2) = 2$.

2. Determine the clamped cubic spline s that interpolates the data $f(0) = 0$, $f(1) = 1$, $f(2) = 2$ and satisfies $s'(0) = s'(2) = 1$.

3. Construct the natural cubic spline for the following data.

   a.

x	$f(x)$
8.3	17.56492
8.6	18.50515

   b.

x	$f(x)$
0.8	0.22363362
1.0	0.65809197

   c.

x	$f(x)$
-0.5	-0.0247500
-0.25	0.3349375
0	1.1010000

   d.

x	$f(x)$
0.1	-0.62049958
0.2	-0.28398668
0.3	0.00660095
0.4	0.24842440

4. The data in Exercise 3 were generated using the following functions. Use the cubic splines constructed in Exercise 3 for the given value of x to approximate $f(x)$ and $f'(x)$, and calculate the actual error.

   a. $f(x) = x \ln x$; approximate $f(8.4)$ and $f'(8.4)$.

   b. $f(x) = \sin(e^x - 2)$; approximate $f(0.9)$ and $f'(0.9)$.

   c. $f(x) = x^3 + 4.001x^2 + 4.002x + 1.101$; approximate $f(-\frac{1}{3})$ and $f'(-\frac{1}{3})$.

   d. $f(x) = x \cos x - 2x^2 + 3x - 1$; approximate $f(0.25)$ and $f'(0.25)$.

5. Construct the clamped cubic spline using the data of Exercise 3 and the fact that

   a. $f'(8.3) = 3.116256$ and $f'(8.6) = 3.151762$

   b. $f'(0.8) = 2.1691753$ and $f'(1.0) = 2.0466965$

   c. $f'(-0.5) = 0.7510000$ and $f'(0) = 4.0020000$

   d. $f'(0.1) = 3.58502082$ and $f'(0.4) = 2.16529366$

6. Repeat Exercise 4 using the clamped cubic splines constructed in Exercise 5.

7. a. Construct a natural cubic spline to approximate $f(x) = \cos \pi x$ by using the values given by $f(x)$ at $x = 0, 0.25, 0.5, 0.75$, and 1.0.

   b. Integrate the spline over $[0, 1]$, and compare the result to $\int_0^1 \cos \pi x \, dx = 0$.

   c. Use the derivatives of the spline to approximate $f'(0.5)$ and $f''(0.5)$, and compare these approximations to the actual values.

8. a. Construct a natural cubic spline to approximate $f(x) = e^{-x}$ by using the values given by $f(x)$ at $x = 0, 0.25, 0.75$, and 1.0.

   b. Integrate the spline over $[0, 1]$, and compare the result to $\int_0^1 e^{-x} \, dx = 1 - 1/e$.

   c. Use the derivatives of the spline to approximate $f'(0.5)$ and $f''(0.5)$, and compare the approximations to the actual values.

9. Repeat Exercise 7, constructing instead the clamped cubic spline with $f'(0) = f'(1) = 0$.

10. Repeat Exercise 8, constructing instead the clamped cubic spline with $f'(0) = -1$, $f'(1) = -e^{-1}$.

11.   A natural cubic spline S on $[0, 2]$ is defined by

$$S(x) = \begin{cases} S_0(x) = 1 + 2x - x^3, & \text{if } 0 \le x < 1, \\ S_1(x) = a + b(x-1) + c(x-1)^2 + d(x-1)^3, & \text{if } 1 \le x \le 2. \end{cases}$$

Find a, b, c, and d.

12.   A clamped cubic spline s for a function f is defined on $[1, 3]$ by

$$s(x) = \begin{cases} s_0(x) = 3(x-1) + 2(x-1)^2 - (x-1)^3, & \text{if } 1 \le x < 2, \\ s_1(x) = a + b(x-2) + c(x-2)^2 + d(x-2)^3, & \text{if } 2 \le x \le 3. \end{cases}$$

Given $f'(1) = f'(3)$, find a, b, c, and d.

13.   A natural cubic spline S is defined by

$$S(x) = \begin{cases} S_0(x) = 1 + B(x-1) - D(x-1)^3, & \text{if } 1 \le x < 2, \\ S_1(x) = 1 + b(x-2) - \frac{3}{4}(x-2)^2 + d(x-2)^3, & \text{if } 2 \le x \le 3. \end{cases}$$

If S interpolates the data $(1, 1)$, $(2, 1)$, and $(3, 0)$, find B, D, b, and d.

14.   A clamped cubic spline s for a function f is defined by

$$s(x) = \begin{cases} s_0(x) = 1 + Bx + 2x^2 - 2x^3, & \text{if } 0 \le x < 1, \\ s_1(x) = 1 + b(x-1) - 4(x-1)^2 + 7(x-1)^3, & \text{if } 1 \le x \le 2. \end{cases}$$

Find $f'(0)$ and $f'(2)$.

15.   Suppose that $f(x)$ is a polynomial of degree 3. Show that $f(x)$ is its own clamped cubic spline but that it cannot be its own natural cubic spline.

16.   Suppose the data $\{x_i, f(x_i)\}_{i=1}^n$ lie on a straight line. What can be said about the natural and clamped cubic splines for the function f? [*Hint:* Take a cue from the results of Exercises 1 and 2.]

17.   The data in the following table give the population of the United States for the years 1960 to 2010 and were considered in Exercise 16 of Section 3.2 and Exercise 6 of Section 3.3.

Year	1960	1970	1980	1990	2000	2010
Population (thousands)	179,323	203,302	226,542	249,633	281,442	307,746

   a.   Find a natural cubic spline agreeing with these data, and use the spline to predict the population in the years 1950, 1975, and 2020.

   b.   Compare your approximations with those previously obtained. If you had to make a choice, which interpolation procedure would you choose?

18.   A car traveling along a straight road is clocked at a number of points. The data from the observations are given in the following table, where the time is in seconds, the distance is in feet, and the speed is in feet per second.

Time	0	3	5	8	13
Distance	0	225	383	623	993
Speed	75	77	80	74	72

   a.   Use a clamped cubic spline to predict the position of the car and its speed when $t = 10$ s.

   b.   Use the derivative of the spline to determine whether the car ever exceeds a 55-mi/h speed limit on the road; if so, what is the first time the car exceeds this speed?

   c.   What is the predicted maximum speed for the car?

19.   The 2011 Kentucky Derby was won by a horse named Animal Kingdom (at 20:1 odds) in a time of 2:02.04 (2 minutes and 2.04 seconds) for the $1\frac{1}{4}$-mile race. Times at the quarter-mile, half-mile, and mile poles were 0:24.26, 0:59.68, and 1:47.95.

a. Use these values together with the starting time to construct a natural cubic spline for Animal Kingdom's race.

b. Use the spline to predict the time at the three-quarter-mile pole, and compare this to the actual time of 1:24.40.

c. Use the spline to approximate Animal Kingdom's speed at the finish line.

20. It is suspected that the high amounts of tannin in mature oak leaves inhibit the growth of the winter moth (*Operophtera bromata L., Geometridae*) larvae that extensively damage these trees in certain years. The following table lists the average weight of two samples of larvae at times in the first 28 days after birth. The first sample was reared on young oak leaves, whereas the second sample was reared on mature leaves from the same tree.

a. Use a natural cubic spline to approximate the average weight curve for each sample.

b. Find an approximate maximum average weight for each sample by determining the maximum of the spline.

Day	0	6	10	13	17	20	28
Sample 1 average weight (mg)	6.67	17.33	42.67	37.33	30.10	29.31	28.74
Sample 2 average weight (mg)	6.67	16.11	18.89	15.00	10.56	9.44	8.89

3.6 Parametric Curves

None of the techniques we have developed can be used to generate curves of the form shown in Figure 3.14, because this curve cannot be expressed as a function of one coordinate variable in terms of the other. In this section we will see how to represent general curves by using a parameter to express both the x- and y-coordinate variables. This technique can be extended to represent general curves and surfaces in space.

Figure 3.14

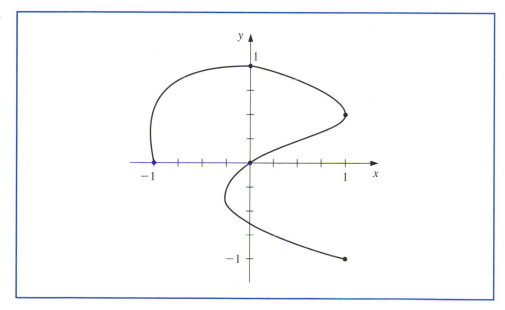

A straightforward parametric technique for determining a polynomial or piecewise polynomial to connect the points $(x_0, y_0), (x_1, y_1), \ldots, (x_n, y_n)$ is to use a parameter t on an interval $[t_0, t_n]$, with $t_0 < t_1 < \cdots < t_n$, and construct approximation functions with

$$x_i = x(t_i) \quad \text{and} \quad y_i = y(t_i) \quad \text{for each } i = 0, 1, \ldots, n.$$

The following example demonstrates the technique when both approximating functions are Lagrange interpolating polynomials.

Example 1 Construct a pair of Lagrange polynomials to approximate the curve shown in Figure 3.15, using the data points shown on the curve.

Table 3.16

i	0	1	2	3	4
t_i	0	0.25	0.5	0.75	1
x_i	−1	0	1	0	1
y_i	0	1	0.5	0	−1

Solution There is flexibility in choosing the parameter, and we will choose the points $\{t_i\}_{i=0}^{\infty}$ equally spaced in [0,1], which gives the data in Table 3.16.

This produces the interpolating polynomials

$$x(t) = \left(\left(\left(64t - \tfrac{352}{3}\right)t + 60\right)t - \tfrac{14}{3}\right)t - 1 \quad \text{and} \quad y(t) = \left(\left(\left(-\tfrac{64}{3}t + 48\right)t - \tfrac{116}{3}\right)t + 11\right)t.$$

Plotting this parametric system produces the graph shown in blue in Figure 3.15. Although it passes through the required points and has the same basic shape, it is quite a crude approximation to the original curve. A more accurate approximation would require additional nodes, with the accompanying increase in computation. ▪

Figure 3.15

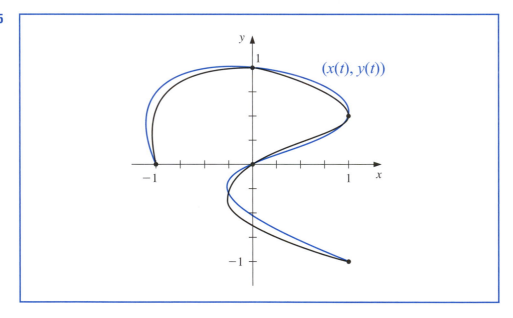

A successful computer design system needs to be based on a formal mathematical theory so that the results are predictable, but this theory should be performed in the background so that the artist can base the design on aesthetics.

Hermite and spline curves can be generated in a similar manner, but these also require extensive computation.

Applications in computer graphics require the rapid generation of smooth curves that can be easily and quickly modified. Also, for both aesthetic and computational reasons, changing one portion of the curves should have little or no effect on other portions. This eliminates the use of interpolating polynomials and splines because, for these, changing a single data point affects the whole curve.

The choice of curve for use in computer graphics is generally a form of the piecewise cubic Hermite polynomial. Each portion of a cubic Hermite polynomial is completely determined by specifying its endpoints and the derivatives at these endpoints. As a consequence, one portion of the curve can be changed while leaving most of the curve the same. Only the adjacent portions need to be modified if we want to ensure smoothness at the endpoints. The computations can be performed quickly, and the curve can be modified a section at a time.

The problem with Hermite interpolation is the need to specify the derivatives at the endpoints of each section of the curve. Suppose the curve has $n + 1$ data points $(x_0, y_0), \ldots, (x_n, y_n)$, and we wish to parameterize the cubic to allow complex features. Then if $(x_i, y_i) = (x(t_i), y(t_i))$ for each $i = 0, 1, \ldots, n$, we must specify $x'(t_i)$ and $y'(t_i)$. This is not as difficult as it would first appear, however, since each portion of the curve is generated independently. Essentially, then, we can simplify the process to one of determining a pair of cubic Hermite polynomials in the parameter t, where $t_0 = 0$, $t_1 = 1$, given the endpoint data $(x(0), y(0))$, and $(x(1), y(1))$ and the derivatives dy/dx (at $t = 0$) and dy/dx (at $t = 1$).

Notice that we are specifying only six conditions, and each cubic polynomial has four parameters, for a total of eight. This provides considerable flexibility in choosing the pair of cubic Hermite polynomials to satisfy these conditions, because the natural form for determining $x(t)$ and $y(t)$ requires that we specify $x'(0)$, $x'(1)$, $y'(0)$, and $y'(1)$. The explicit Hermite curve in x and y requires specifying only the quotients

$$\frac{dy}{dx}(\text{at } t = 0) = \frac{y'(0)}{x'(0)} \quad \text{and} \quad \frac{dy}{dx}(\text{at } t = 1) = \frac{y'(1)}{x'(1)}.$$

By multiplying $x'(0)$ and $y'(0)$ by a common scaling factor, the tangent line to the curve at $(x(0), y(0))$ remains the same, but the shape of the curve varies. The larger the scaling factor, the closer the curve comes to approximating the tangent line near $(x(0), y(0))$. A similar situation exists at the other endpoint $(x(1), y(1))$.

To further simplify the process, the derivative at an endpoint is specified graphically by describing a second point, called a *guide point*, on the desired tangent line, as shown in Figure 3.16. The farther the guide point is from the node, the larger the scaling factor and the more closely the curve approximates the tangent line near the node.

Figure 3.16

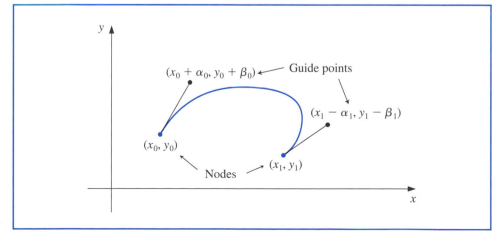

In Figure 3.16, the nodes occur at (x_0, y_0) and (x_1, y_1), the guide point for (x_0, y_0) is $(x_0 + \alpha_0, y_0 + \beta_0)$, and the guide point for (x_1, y_1) is $(x_1 - \alpha_1, y_1 - \beta_1)$. The cubic Hermite polynomial $x(t)$ on $[0, 1]$ must satisfy

$$x(0) = x_0, \qquad x(1) = x_1, \qquad x'(0) = \alpha_0, \quad \text{and} \quad x'(1) = \alpha_1.$$

The unique cubic polynomial satisfying these conditions is

$$x(t) = [2(x_0 - x_1) + (\alpha_0 + \alpha_1)]t^3 + [3(x_1 - x_0) - (\alpha_1 + 2\alpha_0)]t^2 + \alpha_0 t + x_0. \quad (3.11)$$

In a similar manner, the unique cubic polynomial for y satisfying $y(0) = y_0$, $y(1) = y_1$, $y'(0) = \beta_0$, and $y'(1) = \beta_1$ is

$$y(t) = [2(y_0 - y_1) + (\beta_0 + \beta_1)]t^3 + [3(y_1 - y_0) - (\beta_1 + 2\beta_0)]t^2 + \beta_0 t + y_0. \quad (3.12)$$

Example 2 Determine the graph of the parametric curve generated Eq. (3.11) and (3.12) when the end points are $(x_0, y_0) = (0, 0)$ and $(x_1, y_1) = (1, 0)$, and respective guide points, as shown in Figure 3.17 are $(1, 1)$ and $(0, 1)$.

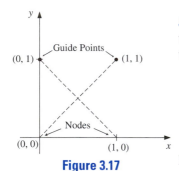

Figure 3.17

Solution The endpoint information implies that $x_0 = 0$, $x_1 = 1$, $y_0 = 0$, and $y_1 = 0$, and the guide points at $(1, 1)$ and $(0, 1)$ imply that $\alpha_0 = 1$, $\alpha_1 = 1$, $\beta_0 = 1$, and $\beta_1 = -1$. Note that the slopes of the guide lines at $(0, 0)$ and $(1, 0)$ are, respectively

$$\frac{\beta_0}{\alpha_0} = \frac{1}{1} = 1 \quad \text{and} \quad \frac{\beta_1}{\alpha_1} = \frac{-1}{1} = -1.$$

Equations (3.11) and (3.12) imply that for $t \in [0, 1]$ we have

$$x(t) = [2(0 - 1) + (1 + 1)]t^3 + [3(0 - 0) - (1 + 2 \cdot 1)]t^2 + 1 \cdot t + 0 = t$$

and

$$y(t) = [2(0 - 0) + (1 + (-1))]t^3 + [3(0 - 0) - (-1 + 2 \cdot 1)]t^2 + 1 \cdot t + 0 = -t^2 + t.$$

This graph is shown as (a) in Figure 3.18, together with some other possibilities of curves produced by Eqs. (3.11) and (3.12) when the nodes are $(0, 0)$ and $(1, 0)$ and the slopes at these nodes are 1 and -1, respectively. ▪

Figure 3.18

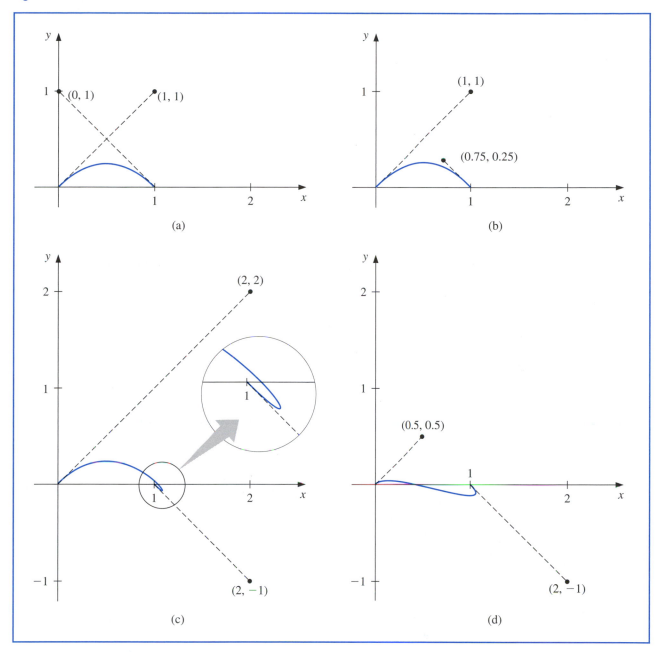

The standard procedure for determining curves in an interactive graphics mode is to first use a mouse to set the nodes and guide points to generate a first approximation to the curve. These can be set manually, but most graphics systems permit you to use your input device to draw the curve on the screen freehand and will select appropriate nodes and guide points for your freehand curve. The nodes and guide points can then be manipulated into a position that produces an aesthetically satisfying curve. Since the computation is minimal, the curve is determined so quickly that the resulting change can be seen almost immediately.

Pierre Etienne Bézier (1910–1999) was head of design and production for Renault motorcars for most of his professional life. He began his research into computer-aided design and manufacturing in 1960, developing interactive tools for curve and surface design, and initiated computer-generated milling for automobile modeling.

The Bézier curves that bear his name have the advantage of being based on a rigorous mathematical theory that does not need to be explicitly recognized by the practitioner who simply wants to make an aesthetically pleasing curve or surface. These are the curves that are the basis of the powerful Adobe Postscript system, and produce the freehand curves that are generated in most sufficiently powerful computer graphics packages.

Moreover, all the data needed to compute the curves are imbedded in coordinates of the nodes and guide points, so no analytical knowledge is required of the user of the system.

Popular graphics programs use this type of system for their freehand graphic representations in a slightly modified form. The Hermite cubics are described as **Bézier polynomials**, which incorporate a scaling factor of 3 when computing the derivatives at the endpoints. This modifies the parametric equations to

$$x(t) = [2(x_0 - x_1) + 3(\alpha_0 + \alpha_1)]t^3 + [3(x_1 - x_0) - 3(\alpha_1 + 2\alpha_0)]t^2 + 3\alpha_0 t + x_0,$$

and

$$y(t) = [2(y_0 - y_1) + 3(\beta_0 + \beta_1)]t^3 + [3(y_1 - y_0) - 3(\beta_1 + 2\beta_0)]t^2 + 3\beta_0 t + y_0,$$

for $0 \le t \le 1$, but this change is transparent to the user of the system.

Three-dimensional curves can be generated in a similar manner by additionally specifying third components z_0 and z_1 for the nodes and $z_0 + \gamma_0$ and $z_1 - \gamma_1$ for the guide points. The more difficult problem involving the representation of three-dimensional curves concerns the loss of the third dimension when the curve is projected onto a two-dimensional medium such as a computer screen or printer paper. Various projection techniques are used, but this topic lies within the realm of computer graphics. For an introduction to this topic and ways that the technique can be modified for surface representations, see one of the many books on computer graphics methods.

Program BEZIER36
generates Bézier curves

EXERCISE SET 3.6

1. Let $(x_0, y_0) = (0, 0)$ and $(x_1, y_1) = (5, 2)$ be the endpoints of a curve. Use the given guide points to construct parametric cubic Hermite approximations $(x(t), y(t))$ to the curve and graph the approximations.

   a. (1, 1) and (6, 1) **b.** (0.5, 0.5) and (5.5, 1.5)

   c. (1, 1) and (6, 3) **d.** (2, 2) and (7, 0)

2. Repeat Exercise 1 using cubic Bézier polynomials.

3. Construct and graph the cubic Bézier polynomials given the following points and guide points.

   a. Point (1, 1) with guide point (1.5, 1.25) to point (6, 2) with guide point (7, 3)

   b. Point (1, 1) with guide point (1.25, 1.5) to point (6, 2) with guide point (5, 3)

   c. Point (0, 0) with guide point (0.5, 0.5) to point (4, 6) with entering guide point (3.5, 7) and exiting guide point (4.5, 5) to point (6, 1) with guide point (7, 2)

   d. Point (0, 0) with guide point (0.5, 0.25) to point (2, 1) with entering guide point (3, 1) and exiting guide point (3, 1) to point (4, 0) with entering guide point (5, 1) and exiting guide point (3, −1) to point (6, −1) with guide point (6.5, −0.25)

4. Use the data in the following table to approximate the letter 𝑛.

i	x_i	y_i	α_i	β_i	α_i'	β_i'
0	3	6	3.3	6.5		
1	2	2	2.8	3.0	2.5	2.5
2	6	6	5.8	5.0	5.0	5.8
3	5	2	5.5	2.2	4.5	2.5
4	6.5	3			6.4	2.8

3.7 Survey of Methods and Software

In this chapter we have considered approximating a function using polynomials and piecewise polynomials. The function can be specified by a given defining equation or by providing points in the plane through which the graph of the function passes. A set of nodes $x_0, x_1, \ldots, x_n$ is given in each case, and more information, such as the value of various derivatives, may also be required. We need to find an approximating function that satisfies the conditions specified by these data.

The interpolating polynomial $P(x)$ is the polynomial of least degree that satisfies, for a function f,

$$P(x_i) = f(x_i) \quad \text{for each } i = 0, 1, \ldots, n.$$

Although there is a unique interpolating polynomial, it can take many different forms. The Lagrange form is most often used for interpolating tables when n is small and for deriving formulas for approximating derivatives and integrals. Neville's method is used for evaluating several interpolating polynomials at the same value of x. Newton's forms of the polynomial are more appropriate for computation and are also used extensively for deriving formulas for solving differential equations. However, polynomial interpolation has the inherent weaknesses of oscillation, particularly if the number of nodes is large. In this case there are other methods that can be better applied.

The Hermite polynomials interpolate a function and its derivative at the nodes. They can be very accurate but require more information about the function being approximated. When you have a large number of nodes, the Hermite polynomials also exhibit oscillation weaknesses.

The most commonly used form of interpolation is piecewise polynomial interpolation. If function and derivative values are available, piecewise cubic Hermite interpolation is recommended. This is the preferred method for interpolating values of a function that is the solution to a differential equation. When only the function values are available, natural cubic spline interpolation could be used. This spline forces the second derivative of the spline to be zero at the endpoints. Some of the other cubic splines require additional data. For example, the clamped cubic spline needs values of the derivative of the function at the endpoints of the interval.

Other methods of interpolation are commonly used. Trigonometric interpolation, in particular, the Fast Fourier Transform discussed in Chapter 8, is used with large amounts of data when the function has a periodic nature. Interpolation by rational functions is also used. If the data are suspected to be inaccurate, smoothing techniques can be applied, and some form of least squares fit of data is recommended. Polynomials, trigonometric functions, rational functions, and splines can be used in least squares fitting of data. We consider these topics in Chapter 8.

Interpolation routines included in the IMSL and the NAG Library are based on the book *A Practical Guide to Splines* by de Boor [Deb2] and use interpolation by cubic splines. The libraries contain subroutines for spline interpolation with user supplied end conditions, periodic end conditions, and the not-a-knot condition. There are also cubic splines to minimize oscillations or to preserve concavity. Methods for two-dimensional interpolation by bicubic splines are also included.

The netlib Library contains the subroutines to compute the cubic spline with various endpoint conditions. One package produces Newton's divided-difference coefficients for a discrete set of data points, and there are various routines for evaluating Hermite piecewise polynomials.

The MATLAB function `interp1` can be used to interpolate a discrete set of data points using either the nearest neighbor interpolation, linear interpolation, cubic spline interpolation, or piecewise cubic Hermite interpolation. The function `interp1` outputs the polynomial evaluated at a discrete set of points. The function `polyfit`, based on a least squares approximation (see Section 8.2), can be used to find an interpolating function of degree at most n that passes through $n + 1$ specified points. Cubic splines can be produced with the function `spline`.

General references to the methods in this chapter are the books by Powell [Po] and by Davis [Da]. The seminal paper on splines is due to Schoenberg [Scho]. Important books on splines are by Schultz [Schul], De Boor [Deb2], and Schumaker [Schum]. The book by Diercx [Di] is also recommended for those needing more information about splines.

Numerical Integration and Differentiation

4.1 Introduction

Many techniques are described in calculus courses for the exact evaluation of integrals, but exact techniques fail to solve many problems that arise in the physical world. For these we need approximation methods of the type we consider in this chapter. The basic techniques are discussed in Section 4.2, and refinements and special applications of these procedures are given in the next six sections.

Section 4.9 considers approximating the derivatives of functions. Methods of this type will be needed in Chapters 11 and 12 for approximating the solutions to ordinary and partial differential equations. You might wonder why there is so much more emphasis on approximating integrals than on approximating derivatives. Determining the actual derivative of a function is a constructive process that leads to straightforward rules for evaluation. Although the definition of the integral is also constructive, the principal tool for evaluating a definite integral is the Fundamental Theorem of Calculus. To apply this theorem, we must determine the antiderivative of the function we wish to evaluate. This is not generally a constructive process, and it leads to the need for accurate approximation procedures.

In this chapter we will also discover one of the more interesting facts in the study of numerical methods. The approximation of integrals—a task that is frequently needed—can usually be accomplished very accurately and often with little effort. The accurate approximation of derivatives—which is needed far less frequently—is a more difficult problem. We think that there is something satisfying about a subject that provides good approximation methods for problems that need them, but is less successful for problems that don't.

4.2 Basic Quadrature Rules

The basic procedure for approximating the definite integral of a function f on the interval $[a, b]$ is to determine an interpolating polynomial that approximates f and then integrate this polynomial. In this section we determine approximations that arise when some basic polynomials are used for the approximations and determine error bounds for these approximations.

The approximations we consider use interpolating polynomials at equally spaced points in the interval $[a, b]$. The first of these is the *Midpoint rule*, which uses the midpoint of $[a, b]$, $\frac{1}{2}(a + b)$, as its only interpolation point. The Midpoint rule approximation is easy to generate geometrically, as shown in Figure 4.1, but to establish the pattern for the higher-order methods and to determine an error formula for the technique, we will use a basic tool for these derivations, the Newton interpolatory divided-difference formula which we discussed on page 76.

Figure 4.1

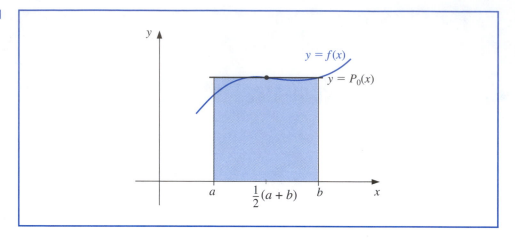

Suppose that $f \in C^{n+1}[a, b]$, where $[a, b]$ is an interval that contains all the nodes $x_0, x_1, \ldots, x_n$. The Newton interpolatory divided-difference formula states that the interpolating polynomial for the function f using the nodes $x_0, x_1, \ldots, x_n$ can be expressed in the form

$$P_{0,1,\ldots,n}(x) = f[x_0] + f[x_0, x_1](x - x_0) + f[x_0, x_1, x_2](x - x_0)(x - x_1) + \cdots$$

$$+ f[x_0, x_1, \ldots, x_n](x - x_0)(x - x_1) \cdots (x - x_{n-1}).$$

Since this is equivalent to the nth Lagrange polynomial, the error formula has the form

$$f(x) - P_{0,1,\ldots,n}(x) = \frac{f^{(n+1)}(\xi(x))}{(n + 1)!}(x - x_0)(x - x_1) \cdots (x - x_n),$$

where $\xi(x)$ is a number, depending on x, that lies in the smallest interval that contains all of $x, x_0, x_1, \ldots, x_n$.

To derive the Midpoint rule we could use the constant interpolating polynomial with $x_0 = \frac{1}{2}(a + b)$ to produce

$$\int_a^b f(x)dx \approx \int_a^b f[x_0]dx = f[x_0](b - a) = f\left(\frac{a + b}{2}\right)(b - a).$$

But we could also use a linear interpolating polynomial with this value of x_0 and an arbitrary value of x_1. This is because the integral of the second term in the Newton interpolatory divided-difference formula is zero for our choice of x_0, independent of the value of x_1, and as such does not contribute to the approximation:

$$\int_a^b f[x_0, x_1](x - x_0)dx = \frac{f[x_0, x_1]}{2}(x - x_0)^2 \Big|_a^b$$

$$= \frac{f[x_0, x_1]}{2}\left(x - \frac{a + b}{2}\right)^2 \Big|_a^b$$

$$= \frac{f[x_0, x_1]}{2}\left[\left(b - \frac{a + b}{2}\right)^2 - \left(a - \frac{a + b}{2}\right)^2\right]$$

$$= \frac{f[x_0, x_1]}{2}\left[\left(\frac{b - a}{2}\right)^2 - \left(\frac{a - b}{2}\right)^2\right] = 0.$$

We would like to derive approximation methods that have high powers of $b - a$ in the error term. In general, the higher the degree of the approximation, the higher the power of $b - a$ in the error term, so we will integrate the error for the linear interpolation polynomial instead of the constant polynomial to determine an error formula for the Midpoint rule.

Suppose that the arbitrary x_1 was chosen to be the same value as x_0. (In fact, this is the only value that we *cannot* have for x_1, but we will ignore this problem for the moment.) Then the integral of the error formula for the interpolating polynomial $P_{0,1}(x)$ has the form

$$\int_a^b \frac{(x - x_0)(x - x_1)}{2} f''(\xi(x))dx = \int_a^b \frac{(x - x_0)^2}{2} f''(\xi(x))dx,$$

where, for each x, the number $\xi(x)$ lies in the interval (a, b).

The term $(x - x_0)^2$ does not change sign on the interval (a, b), so the Mean Value Theorem for Integrals (see page 8) implies that a number ξ, independent of x, exists in (a, b) with

$$\int_a^b \frac{(x - x_0)^2}{2} f''(\xi(x))dx = f''(\xi) \int_a^b \frac{(x - x_0)^2}{2} dx = \frac{f''(\xi)}{6}(x - x_0)^3 \Big]_a^b$$

$$= \frac{f''(\xi)}{6} \left[\left(b - \frac{b + a}{2} \right)^3 - \left(a - \frac{b + a}{2} \right)^3 \right]$$

$$= \frac{f''(\xi)}{6} \frac{(b - a)^3}{4} = \frac{f''(\xi)}{24}(b - a)^3.$$

As a consequence, the Midpoint rule with its error formula has the following form:

Midpoint Rule

If $f \in C^2[a, b]$, then a number ξ in (a, b) exists with

$$\int_a^b f(x)dx = (b - a)f\left(\frac{a + b}{2} \right) + \frac{f''(\xi)}{24}(b - a)^3.$$

The invalid assumption, $x_1 = x_0$, that leads to this result can be avoided by taking x_1 close, but not equal, to x_0 and using limits to show that the error formula is still valid.

The Trapezoidal Rule

The Midpoint rule uses a constant interpolating polynomial disguised as a linear interpolating polynomial. The next method we consider uses a true linear interpolating polynomial, one with the distinct nodes $x_0 = a$ and $x_1 = b$. This approximation is also easy to generate geometrically, as shown in Figure 4.2 on the following page, and is aptly called the *Trapezoidal*, or *Trapezium*, rule. If we integrate the linear interpolating polynomial with $x_0 = a$ and $x_1 = b$, we also produce this formula:

$$\int_a^b f[x_0] + f[x_0, x_1](x - x_0)dx = \left[f[a]x + f[a, b]\frac{(x - a)^2}{2} \right]_a^b$$

$$= f(a)(b - a) + \frac{f(b) - f(a)}{b - a} \left[\frac{(b - a)^2}{2} - \frac{(a - a)^2}{2} \right]$$

$$= \frac{f(a) + f(b)}{2}(b - a).$$

Figure 4.2

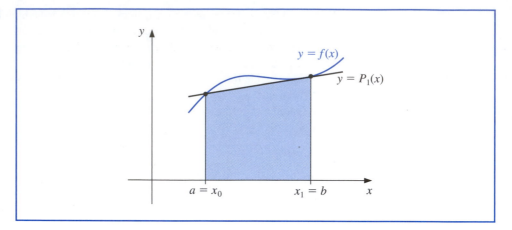

The error for the Trapezoidal rule follows from integrating the error term for $P_{0,1}(x)$ when $x_0 = a$ and $x_1 = b$. Since $(x - x_0)(x - x_1) = (x - a)(x - b)$ is always negative in the interval (a, b), we can again apply the Mean Value Theorem for Integrals. In this case it implies that a number ξ in (a, b) exists with

$$\int_a^b \frac{(x - a)(x - b)}{2} f''(\xi(x)) dx = \frac{f''(\xi)}{2} \int_a^b (x - a)[(x - a) - (b - a)] dx$$

$$= \frac{f''(\xi)}{2} \left[\frac{(x - a)^3}{3} - \frac{(x - a)^2}{2}(b - a) \right]_a^b$$

$$= \frac{f''(\xi)}{2} \left[\frac{(b - a)^3}{3} - \frac{(b - a)^2}{2}(b - a) \right]$$

$$= -\frac{f''(\xi)}{12}(b - a)^3.$$

This gives the Trapezoidal rule with its error formula.

Trapezoidal Rule

If $f \in C^2[a, b]$, then a number ξ in (a, b) exists with

$$\int_a^b f(x) dx = \frac{f(a) + f(b)}{2}(b - a) - \frac{f''(\xi)}{12}(b - a)^3.$$

When we use the term *trapezoid* we mean a four-sided figure that has at least two of its sides parallel. The European term for this figure is *trapezium*. To further confuse the issue, the European word trapezoidal refers to a four-sided figure with no sides equal, and the American word for this type of figure is trapezium.

We cannot improve on the power of $b - a$ in the error formula for the Trapezoidal rule, as we did in the case of the Midpoint rule, because the integral of the next higher term in the Newton interpolatory divided-difference formula is

$$\int_a^b f[x_0, x_1, x_2](x - x_0)(x - x_1) dx = f[x_0, x_1, x_2] \int_a^b (x - a)(x - b) dx.$$

Since $(x - a)(x - b) < 0$ for all x in (a, b), this term will not be zero unless $f[x_0, x_1, x_2] = 0$. As a consequence, the error formulas for the Midpoint and the Trapezoidal rules both involve $(b - a)^3$, even though they are derived from interpolation formulas with error formulas that involve $b - a$ and $(b - a)^2$, respectively.

Simpson's Rule

Next we consider an integration formula based on approximating the function f by a quadratic polynomial that agrees with f at the equally spaced points $x_0 = a, x_1 = (a+b)/2$, and $x_2 = b$. This formula is not easy to generate geometrically, although the approximation is illustrated in Figure 4.3.

Figure 4.3

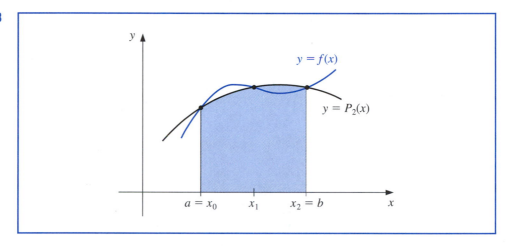

To derive the formula, we integrate $P_{0,1,2}(x)$.

$$\int_a^b P_{0,1,2}(x)\,dx$$

$$= \int_a^b \left\{ f(a) + f\left[a, \frac{a+b}{2}\right](x-a) + f\left[a, \frac{a+b}{2}, b\right](x-a)\left(x - \frac{a+b}{2}\right)\right\} dx$$

$$= \left[f(a)x + f\left[a, \frac{a+b}{2}\right]\frac{(x-a)^2}{2}\right]_a^b$$

$$\quad + f\left[a, \frac{a+b}{2}, b\right]\int_a^b (x-a)\left[(x-a) + \left(a - \frac{a+b}{2}\right)\right] dx$$

$$= f(a)(b-a) + \frac{f(\frac{a+b}{2}) - f(a)}{\frac{a+b}{2} - a}\frac{(b-a)^2}{2}$$

$$\quad + \frac{f[\frac{a+b}{2}, b] - f[a, \frac{a+b}{2}]}{b-a}\left[\frac{(x-a)^3}{3} + \frac{(x-a)^2}{2}\left(\frac{a-b}{2}\right)\right]_a^b$$

$$= (b-a)\left[f(a) + f\left(\frac{a+b}{2}\right) - f(a)\right]$$

$$\quad + \left(\frac{1}{b-a}\right)\left[\frac{f(b) - f(\frac{a+b}{2})}{\frac{b-a}{2}} - \frac{f(\frac{a+b}{2}) - f(a)}{\frac{b-a}{2}}\right]\left[\frac{(b-a)^3}{3} - \frac{(b-a)^3}{4}\right]$$

$$= (b-a)f\left(\frac{a+b}{2}\right) + \frac{2}{(b-a)^2}\left[f(b) - 2f\left(\frac{a+b}{2}\right) + f(a)\right]\frac{(b-a)^3}{12}.$$

Thomas Simpson (1710–1761) was a self-taught mathematician who supported himself as a weaver during his early years. His primary interest was probability theory, although in 1750 he published a two-volume calculus book entitled *The Doctrine and Application of Fluxions.*

Simplifying this equation gives the approximation method known as Simpson's rule:

$$\int_a^b f(x)dx \approx \frac{(b-a)}{6}\left[f(a) + 4f\left(\frac{a+b}{2}\right) + f(b)\right].$$

An error formula for Simpson's rule involving $(b-a)^4$ can be derived by using the error formula for the quadratic interpolating polynomial $P_{0,1,2}(x)$. However, similar to the case of the Midpoint rule, the integral of the next term in the Newton interpolatory divided-difference formula is zero. This implies that the error formula for the cubic interpolating polynomial $P_{0,1,2,3}(x)$ can be used to produce an error formula that involves $(b-a)^5$. When simplified, Simpson's rule with this error formula is as follows:

Simpson's Rule

If $f \in C^4[a, b]$, then a number ξ in (a, b) exists with

$$\int_a^b f(x)dx = \frac{(b-a)}{6}\left[f(a) + 4f\left(\frac{a+b}{2}\right) + f(b)\right] - \frac{f^{(4)}(\xi)}{2880}(b-a)^5.$$

This higher power of $b-a$ in the error term makes Simpson's rule significantly superior to the Midpoint and Trapezoidal rules in almost all situations, provided that $b - a$ is small. This is illustrated in the following example.

Example 1 Compare the Midpoint, Trapezoidal, and Simpson's rules approximations to $\int_0^2 f(x)dx$ when $f(x)$ is

    (a) x^2 **(b)** x^4 **(c)** $(x + 1)^{-1}$

    (d) $\sqrt{1 + x^2}$ **(e)** $\sin x$ **(f)** e^x

Solution On [0, 2] the Midpoint, Trapezoidal, and Simpson's rules have the forms

Midpoint: $\int_0^2 f(x)dx \approx 2f(1),$ Trapezoidal: $\int_0^2 f(x)dx \approx f(0) + f(2),$

and

Simpson's: $\int_0^2 f(x)dx \approx \frac{1}{3}[f(0) + 4f(1) + f(2)].$

When $f(x) = x^2$, they give

Midpoint: $\int_0^2 f(x)dx \approx 2 \cdot 1 = 2,$ Trapezoidal: $\int_0^2 f(x)dx \approx 0^2 + 2^2 = 4,$

and

Simpson's: $\int_0^2 f(x)dx \approx \frac{1}{3}[f(0) + 4f(1) + f(2)] = \frac{1}{3}(0^2 + 4 \cdot 1^2 + 2^2) = \frac{8}{3}.$

The approximation from Simpson's rule is exact because its truncation error involves $f^{(4)}$, which is identically 0 when $f(x) = x^2$.

    The results to three places for the functions are summarized in Table 4.1. Notice that, in each instance, Simpson's rule is significantly superior.

Table 4.1

	(a)	**(b)**	**(c)**	**(d)**	**(e)**	**(f)**
$f(x)$	x^2	x^4	$(x+1)^{-1}$	$\sqrt{1+x^2}$	$\sin x$	e^x
Exact value	2.667	6.400	1.099	2.958	1.416	6.389
Midpoint	2.000	2.000	1.000	2.818	1.682	5.436
Trapezoidal	4.000	16.000	1.333	3.326	0.909	8.389
Simpson's	2.667	6.667	1.111	2.964	1.425	6.421

To demonstrate the error terms for the Midpoint, Trapezoidal, and Simpson's methods, we will find bounds for the errors in approximating $\int_0^2 \sqrt{1+x^2}\,dx$. With $f(x) = (1+x^2)^{1/2}$, we have

$$f'(x) = \frac{x}{(1+x^2)^{1/2}}, \quad f''(x) = \frac{1}{(1+x^2)^{3/2}}, \quad \text{and} \quad f'''(x) = \frac{-3x}{(1+x^2)^{5/2}}.$$

To bound the error of the Midpoint method, we need to determine $\max_{0 \le x \le 2} |f''(x)|$. This maximum will occur at either the maximum or the minimum value of f'' on $[0, 2]$. Maximum and minimum values for f'' on $[0, 2]$ can occur only when $x = 0$, $x = 2$, or when $f'''(x) = 0$. Since $f'''(x) = 0$ only when $x = 0$, we have

$$\max_{0 \le x \le 2} |f''(x)| = \max\{|f''(0)|, |f''(2)|\} = \max\left\{1, 5^{-3/2}\right\} = 1.$$

So a bound for the error in the Midpoint method is

$$\left| \frac{f''(\xi)}{24}(b-a)^3 \right| \le \frac{1}{24}(2-0)^3 = \frac{1}{3} = 0.\overline{3}.$$

The actual error is within this bound, since $|2.958 - 2.818| = 0.14$. For the Trapezoidal method, we have the error bound

$$\left| -\frac{f''(\xi)}{12}(b-a)^3 \right| \le \frac{1}{12}(2-0)^3 = \frac{2}{3} = 0.\overline{6},$$

and the actual error is $|2.958 - 3.326| = 0.368$. We need more derivatives for Simpson's rule:

$$f^{(4)}(x) = \frac{12x^2 - 3}{(1+x^2)^{7/2}} \quad \text{and} \quad f^{(5)}(x) = \frac{45x - 60x^3}{(1+x^2)^{9/2}}.$$

Since $f^{(5)}(x) = 0$ implies

$$0 = 45x - 60x^3 = 15x(3 - 4x^2),$$

$f^{(4)}(x)$ has critical points $0, \pm\sqrt{3}/2$. Evaluating the fourth derivative at the critical points and endpoints we have

$$|f^{(4)}(\xi)| \le \max_{0 \le x \le 2} |f^{(4)}(x)| = \max\{|f^{(4)}(0)|, |f^{(4)}(\sqrt{3}/2)|, |f^{(4)}(2)|\}$$

$$= \max\left\{|-3|, \frac{768\sqrt{7}}{2401}, \frac{9\sqrt{5}}{125}\right\} = 3.$$

The error for Simpson's rule is consequently bounded by

$$\left| -\frac{f^{(4)}(\xi)}{2880}(b-a)^5 \right| \le \frac{3}{2880}(2-0)^5 = \frac{96}{2880} = 0.0\overline{3},$$

and the actual error is $|2.958 - 2.964| = 0.006$.

The error formulas all contain $b - a$ to a power, so they are most effective when the interval $[a, b]$ is small, so that $b - a$ is much smaller than one. There are formulas that can be used to improve the accuracy when integrating over large intervals, some of which are considered in the exercises. However, a better solution to the problem is considered in the next section.

EXERCISE SET 4.2

1. Use the Midpoint rule to approximate the following integrals.

   a. $\displaystyle\int_{0.5}^{1} x^4 dx$

   b. $\displaystyle\int_{0}^{0.5} \frac{2}{x-4} dx$

   c. $\displaystyle\int_{1}^{1.5} x^2 \ln x \, dx$

   d. $\displaystyle\int_{0}^{1} x^2 e^{-x} dx$

   e. $\displaystyle\int_{1}^{1.6} \frac{2x}{x^2-4} dx$

   f. $\displaystyle\int_{0}^{0.35} \frac{2}{x^2-4} dx$

   g. $\displaystyle\int_{0}^{\pi/4} x \sin x \, dx$

   h. $\displaystyle\int_{0}^{\pi/4} e^{3x} \sin 2x \, dx$

2. Use the error formula to find a bound for the error in Exercise 1, and compare the bound to the actual error.

3. Repeat Exercise 1 using the Trapezoidal rule.

4. Repeat Exercise 2 using the Trapezoidal rule and the results of Exercise 3.

5. Repeat Exercise 1 using Simpson's rule.

6. Repeat Exercise 2 using Simpson's rule and the results of Exercise 5.

   Other quadrature formulas with error terms are given by

   (i) $\displaystyle\int_{a}^{b} f(x)dx = \frac{3h}{8}[f(a) + 3f(a+h) + 3f(a+2h) + f(b)] - \frac{3h^5}{80} f^{(4)}(\xi)$, where $h = \frac{b-a}{3}$;

   (ii) $\displaystyle\int_{a}^{b} f(x)dx = \frac{3h}{2}[f(a+h) + f(a+2h)] + \frac{3h^3}{4} f''(\xi)$, where $h = \frac{b-a}{3}$.

7. Repeat Exercises 1 using (a) Formula (i) and (b) Formula (ii).

8. Repeat Exercises 2 using (a) Formula (i) and (b) Formula (ii).

9. The Trapezoidal rule applied to $\int_{0}^{2} f(x)dx$ gives the value 4, and Simpson's rule gives the value 2. What is $f(1)$?

10. The Trapezoidal rule applied to $\int_{0}^{2} f(x)dx$ gives the value 5, and the Midpoint rule gives the value 4. What value does Simpson's rule give?

11. Find the constants c_0, c_1, and x_1 so that the quadrature formula

    $$\int_{0}^{1} f(x)dx = c_0 f(0) + c_1 f(x_1)$$

    gives exact results for all polynomials of degree at most 2.

12. Find the constants x_0, x_1, and c_1 so that the quadrature formula

    $$\int_{0}^{1} f(x)dx = \frac{1}{2} f(x_0) + c_1 f(x_1)$$

    gives exact results for all polynomials of degree at most 3.

13. Given the function f at the following values:

x	1.8	2.0	2.2	2.4	2.6
$f(x)$	3.12014	4.42569	6.04241	8.03014	10.46675

a. Approximate $\int_{1.8}^{2.6} f(x)dx$ using each of the following.

(i) the Midpoint rule **(ii)** the Trapezoidal rule **(iii)** Simpson's rule

b. Suppose the data have round-off errors given by the following table:

x	1.8	2.0	2.2	2.4	2.6
Error in $f(x)$	2×10^{-6}	-2×10^{-6}	-0.9×10^{-6}	-0.9×10^{-6}	2×10^{-6}

Calculate the errors due to round-off in each of the approximation methods.

4.3 Composite Quadrature Rules

The basic notions underlying numerical integration were derived in the previous section, but the techniques given there are not satisfactory for most problems. We saw an example of this at the end of that section, where the approximations were poor for integrals of functions on the interval $[0, 2]$. To see why this occurs, let us consider Simpson's method, generally the most accurate of these techniques. Assuming that $f \in C^4[a, b]$, Simpson's method with its error formula is given by

Piecewise approximation is often effective. Recall that this was used for spline interpolation.

$$\int_a^b f(x)dx = \frac{b-a}{6}\left[f(a) + 4f\left(\frac{a+b}{2}\right) + f(b)\right] - \frac{(b-a)^5}{2880}f^{(4)}(\xi)$$

$$= \frac{h}{3}[f(a) + 4f(a+h) + f(b)] - \frac{h^5}{90}f^{(4)}(\xi)$$

where $h = (b-a)/2$ and ξ lies somewhere in the interval (a, b). Since $f \in C^4[a, b]$ implies that $f^{(4)}$ is bounded on $[a, b]$, there exists a constant M such that $|f^{(4)}(x)| \le M$ for all x in $[a, b]$. As a consequence,

$$\left|\frac{h}{3}[f(a) + 4f(a+h) + f(b)] - \int_a^b f(x)dx\right| = \left|\frac{h^5}{90}f^{(4)}(\xi)\right| \le \frac{M}{90}h^5.$$

The error term in this formula involves M, a bound for the fourth derivative of f, and h^5, so we can expect the error to be small provided that

- the fourth derivative of f is not erratic, and

- the value of $h = b - a$ is small.

The first assumption we will need to live with, but the second might be quite unreasonable. There is no reason, in general, to expect that the interval $[a, b]$ over which the integration is performed is small, and if it is not, the h^5 portion in the error term will likely dominate the calculations.

We circumvent the problem involving a large interval of integration by subdividing the interval $[a, b]$ into a collection of intervals that are sufficiently small so that the error over each is kept under control.

Example 1   Use Simpson's rule to approximate $\int_0^4 e^x\,dx$ and compare this to the results obtained by adding the Simpson's rule approximations for **(a)** $\int_0^2 e^x\,dx$ and $\int_2^4 e^x\,dx$, and **(b)** $\int_0^1 e^x\,dx$, $\int_1^2 e^x\,dx$, $\int_2^3 e^x\,dx$, and $\int_3^4 e^x\,dx$.

Solution   **(a)** Simpson's rule on $[0, 4]$ uses $h = 2$ and gives

$$\int_0^4 e^x\,dx \approx \frac{2}{3}(e^0 + 4e^2 + e^4) = 56.76958.$$

The exact answer in this case is $e^4 - e^0 = 53.59815$, and the error -3.17143 is far larger than we would normally accept.

Applying Simpson's rule on each of the intervals $[0, 2]$ and $[2, 4]$ uses $h = 1$ and gives

$$\int_0^4 e^x\,dx = \int_0^2 e^x\,dx + \int_2^4 e^x\,dx \approx \frac{1}{3}(e^0 + 4e + e^2) + \frac{1}{3}(e^2 + 4e^3 + e^4)$$

$$= \frac{1}{3}(e^0 + 4e + 2e^2 + 4e^3 + e^4) = 53.86385.$$

The error has been reduced to -0.26570.

(b) For the integrals on $[0, 1], [1, 2], [3, 4]$, and $[3, 4]$ we use Simpson's rule four times with $h = \frac{1}{2}$ giving

$$\int_0^4 e^x\,dx = \int_0^1 e^x\,dx + \int_1^2 e^x\,dx + \int_2^3 e^x\,dx + \int_3^4 e^x\,dx$$

$$\approx \frac{1}{6}(e_0 + 4e^{1/2} + e) + \frac{1}{6}(e + 4e^{3/2} + e^2)$$

$$+ \frac{1}{6}(e^2 + 4e^{5/2} + e^3) + \frac{1}{6}(e^3 + 4e^{7/2} + e^4)$$

$$= \frac{1}{6}(e^0 + 4e^{1/2} + 2e + 4e^{3/2} + 2e^2 + 4e^{5/2} + 2e^3 + 4e^{7/2} + e^4)$$

$$= 53.61622.$$

The error for this approximation has been reduced to -0.01807. If we had used eight subintervals, the error would have been reduced to 1.15×10^{-3}.   ▪

The generalization of the procedure considered in Example 1 is called the *Composite Simpson's rule* and is described as follows.

Choose an even integer n, subdivide the interval $[a, b]$ into n subintervals, and use Simpson's rule on each consecutive pair of subintervals. Let $h = (b - a)/n$ and $a = x_0 < x_1 < \cdots < x_n = b$, where $x_j = x_0 + jh$ for each $j = 0, 1, \ldots, n$. Then

$$\int_a^b f(x)\,dx = \sum_{j=1}^{n/2} \int_{x_{2j-2}}^{x_{2j}} f(x)\,dx$$

$$= \sum_{j=1}^{n/2} \left\{ \frac{h}{3}[f(x_{2j-2}) + 4f(x_{2j-1}) + f(x_{2j})] - \frac{h^5}{90} f^{(4)}(\xi_j) \right\}$$

for some ξ_j with $x_{2j-2} < \xi_j < x_{2j}$, provided that $f \in C^4[a, b]$. To simplify this formula, first note that for each $j = 1, 2, \ldots, \frac{n}{2} - 1$, we have $f(x_{2j})$ appearing in the term corresponding to the interval $[x_{2j-2}, x_{2j}]$ and also in the term corresponding to the interval

$[x_{2j}, x_{2j+2}]$. Combining these terms gives a simplified form for the approximation (see Figure 4.4):

$$\int_a^b f(x)dx = \frac{h}{3}\left[f(x_0) + 2\sum_{j=1}^{(n/2)-1} f(x_{2j}) + 4\sum_{j=1}^{n/2} f(x_{2j-1}) + f(x_n)\right] - \frac{h^5}{90}\sum_{j=1}^{n/2} f^{(4)}(\xi_j).$$

Figure 4.4

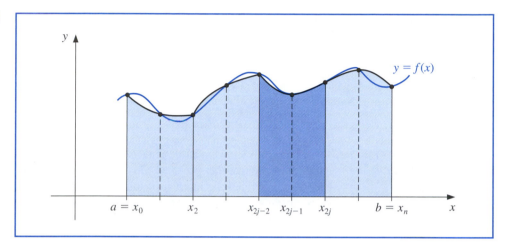

We can also simplify the error term

$$E(f) = -\frac{h^5}{90}\sum_{j=1}^{n/2} f^{(4)}(\xi_j),$$

where $x_{2j-2} < \xi_j < x_{2j}$ for each $j = 1, 2, \ldots, \frac{n}{2}$. If $f \in C^4[a,b]$, the Extreme Value Theorem implies that $f^{(4)}$ assumes its maximum and minimum in $[a,b]$. Since

$$\min_{x\in[a,b]} f^{(4)}(x) \le f^{(4)}(\xi_j) \le \max_{x\in[a,b]} f^{(4)}(x),$$

we have

$$\frac{n}{2}\min_{x\in[a,b]} f^{(4)}(x) \le \sum_{j=1}^{n/2} f^{(4)}(\xi_j) \le \frac{n}{2}\max_{x\in[a,b]} f^{(4)}(x),$$

and

$$\min_{x\in[a,b]} f^{(4)}(x) \le \frac{2}{n}\sum_{j=1}^{n/2} f^{(4)}(\xi_j) \le \max_{x\in[a,b]} f^{(4)}(x).$$

The term in the middle lies between values of the continuous function $f^{(4)}$, so the Intermediate Value Theorem implies that a number μ in (a,b) exists with

$$f^{(4)}(\mu) = \frac{2}{n}\sum_{j=1}^{n/2} f^{(4)}(\xi_j).$$

But $n = (b - a)/h$, so the error formula simplifies to

$$E(f) = -\frac{h^5}{90} \sum_{j=1}^{n/2} f^{(4)}(\xi_j) = -\frac{h^5}{90} \left(\frac{(b-a)}{2h} f^{(4)}(\mu) \right) = -\frac{h^4(b-a)}{180} f^{(4)}(\mu).$$

Summarizing these results, we have the following: If $f \in C^4[a, b]$ and n is even, the Composite Simpson's rule for n subintervals of $[a, b]$ can be expressed with error term as follows:

Composite Simpson's Rule

Suppose that $f \in C^4[a, b]$. Then for some μ in (a, b) we have

$$\int_a^b f(x)dx = \frac{h}{3} \left[f(a) + 2 \sum_{j=1}^{(n/2)-1} f(x_{2j}) + 4 \sum_{j=1}^{n/2} f(x_{2j-1}) + f(b) \right] - \frac{(b-a)h^4}{180} f^{(4)}(\mu).$$

The program CSIMPR41 implements Composite Simpson's rule.

Composite Simpson's rule is $O(h^4)$, so the approximations converge to $\int_a^b f(x)dx$ at about the same rate that $h^4 \to 0$. This rate of convergence is sufficient for most common problems, provided the interval of integration is subdivided so that h is small. Composite Simpson's rule on n subintervals is the most frequently used general-purpose integral approximation technique.

The subdivision approach can be applied to any of the formulas we saw in the preceding section. Since the Trapezoidal rule (see Figure 4.5) uses only interval endpoint values to determine its approximation, the integer n can be odd or even.

Figure 4.5

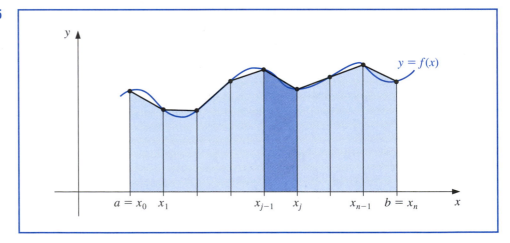

The Composite Trapezoidal rule for n subintervals is $O(h^2)$ if $f \in C^2[a, b]$. In this case $h = (b - a)/n$, and that $x_j = a + jh$ for each $j = 0, 1, \ldots, n$.

Composite Trapezoidal Rule

Suppose that $f \in C^2[a, b]$. Then for some μ in (a, b) we have

$$\int_a^b f(x)dx = \frac{h}{2} \left[f(a) + f(b) + 2 \sum_{j=1}^{n-1} f(x_j) \right] - \frac{(b-a)h^2}{12} f''(\mu).$$

The Composite Midpoint rule for $n + 2$ subintervals is also $O(h^2)$ if $f \in C^2[a, b]$ when n is an even integer. Because we do not use the endpoints in this case, we define $h = (b - a)/(n + 2)$ and $x_j = a + (j + 1)h$ for each $j = -1, 0, \ldots, n + 1$ (see Figure 4.6).

Composite Midpoint Rule

Suppose that $f \in C^2[a, b]$. Then for some μ in (a, b) we have

$$\int_a^b f(x)\,dx = 2h \sum_{j=0}^{n/2} f(x_{2j}) + \frac{(b - a)h^2}{6} f''(\mu).$$

Figure 4.6

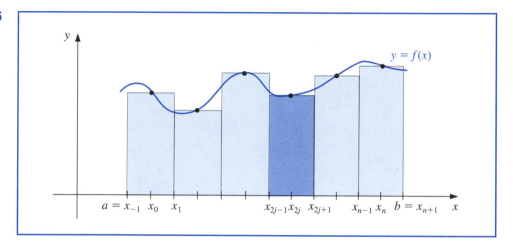

Example 2   Determine values of h that will ensure an approximation error of less than 0.00002 when approximating $\int_0^\pi \sin x \, dx$ and employing **(a)** Composite Trapezoidal rule and **(b)** Composite Simpson's rule.

Solution   **(a)** The error form for the Composite Trapezoidal rule for $f(x) = \sin x$ on $[0, \pi]$ is

$$\left| \frac{\pi h^2}{12} f''(\mu) \right| = \left| \frac{\pi h^2}{12} (-\sin \mu) \right| = \frac{\pi h^2}{12} |\sin \mu|.$$

To ensure sufficient accuracy with this technique we need to have

$$\frac{\pi h^2}{12} |\sin \mu| \leq \frac{\pi h^2}{12} < 0.00002.$$

Since $h = \pi/n$ implies that $n = \pi/h$, we need

$$\frac{\pi^3}{12n^2} < 0.00002, \quad \text{which implies that} \quad n > \left(\frac{\pi^3}{12(0.00002)} \right)^{1/2} \approx 359.44.$$

So the Composite Trapezoidal rule requires $n \geq 360$.

(b) The error form for the Composite Simpson's rule for $f(x) = \sin x$ on $[0, \pi]$ is

$$\left| \frac{\pi h^4}{180} f^{(4)}(\mu) \right| = \left| \frac{\pi h^4}{180} \sin \mu \right| = \frac{\pi h^4}{180} |\sin \mu|.$$

To ensure sufficient accuracy with this technique we need to have

$$\frac{\pi h^4}{180}|\sin\mu| \le \frac{\pi h^4}{180} < 0.00002.$$

Using again the fact that $n = \pi/h$ gives

$$\frac{\pi^5}{180n^4} < 0.00002, \quad \text{which implies that} \quad n > \left(\frac{\pi^5}{180(0.00002)}\right)^{1/4} \approx 17.07.$$

So Composite Simpson's rule requires only $n \ge 18$. In fact, Composite Simpson's rule with $n = 18$ gives

$$\int_0^\pi \sin x\, dx \approx \frac{\pi}{54}\left[2\sum_{j=1}^{8}\sin\left(\frac{j\pi}{9}\right) + 4\sum_{j=1}^{9}\sin\left(\frac{(2j-1)\pi}{18}\right)\right] = 2.0000104.$$

This is accurate to within about 10^{-5} because the true value is $-\cos(\pi) - (-\cos(0)) = 2.$ ∎

MATLAB uses a function quad to approximate integrals using a modification of Simpson's method. For example, to approximate

$$\int_0^\pi \sin x\, dx = 2$$

we first define the function being integrated with

f = inline('sin(x)','x')

and give the interval of integration with

a=0,b=pi

To perform the integration we use

q=quad(f,a,b)

and MATLAB responds with

$$q = 1.999999996398431$$

MATLAB can also be used to code integration methods directly. To give a brief description of this we will illustrate the Composite Midpoint rule using loops. This is the easiest of the common composite methods to code because the coefficients in the formula are all the same. However the looping procedure is common to all the composite rules.

First we define $f(x)$ and the endpoints a and b, and the values of n and h.

```
f = inline('sin(x)','x')
a = 0, b = pi, n = 18, h = (b-a)/(n+2)
```

We need a variable, which we will call Tot, to calculate the running sum, and we initialize it to 0.

```
Tot = 0
```

In MATLAB the counter-controlled loop is defined by

```
for loop control variable = initial-value : terminal value
    statement;
    statement;
    .
    .
    .
    statement;
end
```

In our example we have called the loop control variable j, which begins at 0 and goes to $n/2 = 9$ in steps of 1. For each value of j= 0, 1, ... , 9 the loop is traversed, and each calculation inside the loop is performed until the word end is encountered. The reserved words involved are for and end. Note that no semicolon follows the for statement.

```
for j = 0 : n/2
    xj = a + (2*j + 1)*h;
    Tot = Tot + f(xj);
end;
```

This produces a series of results culminating in the final summation, which is found with the command

```
Tot
```

to be

$$Tot = 6.392453221499663$$

We then multiply by $2h$ to finish the Composite Midpoint method and obtain the final result, which we call integral

```
integral = 2*h*Tot
```

MATLAB calculates this as

$$integral = 2.008248407907975$$

Round-Off Error Stability

An important property shared by all the composite rules is stability with respect to round-off error. To demonstrate this, suppose we apply the Composite Simpson's rule with n subintervals to a function on $[a, b]$ and determine the maximum bound for the round-off error. Assume that the true values $f(x_i)$ are approximated by $\tilde{f}(x_i)$ and that

$$f(x_i) = \tilde{f}(x_i) + e_i, \qquad \text{for each } i = 0, 1, \ldots, n,$$

where e_i denotes the round-off error associated with using $\tilde{f}(x_i)$ to approximate $f(x_i)$. Then the accumulated round-off error, $e(h)$, in the Composite Simpson's rule is

$$e(h) = \left| \frac{h}{3} \left[e_0 + 2 \sum_{j=1}^{(n/2)-1} e_{2j} + 4 \sum_{j=1}^{n/2} e_{2j-1} + e_n \right] \right|$$

$$\leq \frac{h}{3} \left[|e_0| + 2 \sum_{j=1}^{(n/2)-1} |e_{2j}| + 4 \sum_{j=1}^{n/2} |e_{2j-1}| + |e_n| \right].$$

If the round-off errors are uniformly bounded by some known tolerance ε, then

$$e(h) \leq \frac{h}{3}\left[\varepsilon + 2\left(\frac{n}{2} - 1\right)\varepsilon + 4\left(\frac{n}{2}\right)\varepsilon + \varepsilon\right] = \frac{h}{3}3n\varepsilon = nh\varepsilon.$$

But $nh = b - a$, so $e(h) \leq (b-a)\varepsilon$, a bound independent of h and n. This means that even though we may need to divide an interval into more parts to ensure accuracy, the increased computation that this requires does not increase the round-off error bound.

EXERCISE SET 4.3

1.   Use the Composite Trapezoidal rule with the indicated values of n to approximate the following integrals.

   a. $\displaystyle\int_1^2 x \ln x \, dx, \quad n = 4$

   b. $\displaystyle\int_{-2}^2 x^3 e^x \, dx, \quad n = 4$

   c. $\displaystyle\int_0^2 \frac{2}{x^2 + 4} dx, \quad n = 6$

   d. $\displaystyle\int_0^{\pi} x^2 \cos x \, dx, \quad n = 6$

   e. $\displaystyle\int_0^2 e^{2x} \sin 3x \, dx, \quad n = 8$

   f. $\displaystyle\int_1^3 \frac{x}{x^2 + 4} dx, \quad n = 8$

   g. $\displaystyle\int_3^5 \frac{1}{\sqrt{x^2 - 4}} dx, \quad n = 8$

   h. $\displaystyle\int_0^{3\pi/8} \tan x \, dx, \quad n = 8$

2.   Use the Composite Simpson's rule to approximate the integrals in Exercise 1.

3.   Use the Composite Midpoint rule with $n + 2$ subintervals to approximate the integrals in Exercise 1.

4.   Approximate $\displaystyle\int_0^2 x^2 e^{-x^2} \, dx$ using $h = 0.25$.

   a. Use the Composite Trapezoidal rule.

   b. Use the Composite Simpson's rule.

   c. Use the Composite Midpoint rule.

5.   Determine the values of n and h required to approximate $\displaystyle\int_0^2 e^{2x} \sin 3x \, dx$ to within 10^{-4}.

   a. Use the Composite Trapezoidal rule.

   b. Use the Composite Simpson's rule.

   c. Use the Composite Midpoint rule.

6.   Repeat Exercise 5 for the integral $\displaystyle\int_0^{\pi} x^2 \cos x \, dx$.

7.   Determine the values of n and h required to approximate $\displaystyle\int_0^2 \frac{1}{x + 4} dx$ to within 10^{-5} and compute the approximation.

   a. Use the Composite Trapezoidal rule.

   b. Use the Composite Simpson's rule.

   c. Use the Composite Midpoint rule.

8.   Repeat Exercise 7 for the integral $\int_1^2 x \ln x \, dx$.

9.   Suppose that $f(0) = 1$, $f(0.5) = 2.5$, $f(1) = 2$, and $f(0.25) = f(0.75) = \alpha$. Find α if the Composite Trapezoidal rule with $n = 4$ gives the value 1.75 for $\int_0^1 f(x)dx$.

10.   The Midpoint rule for approximating $\int_{-1}^1 f(x)dx$ gives the value 12, the Composite Midpoint rule with $n = 2$ gives 5, and Composite Simpson's rule gives 6. Use the fact that $f(-1) = f(1)$ and $f(-0.5) = f(0.5) - 1$ to determine $f(-1)$, $f(-0.5)$, $f(0)$, $f(0.5)$, and $f(1)$.

11. In multivariable calculus and in statistics courses, it is shown that

$$\int_{-\infty}^{\infty} \frac{1}{\sigma\sqrt{2\pi}} e^{-(1/2)(x/\sigma)^2} dx = 1$$

for any positive σ. The function

$$f(x) = \frac{1}{\sigma\sqrt{2\pi}} e^{-(1/2)(x/\sigma)^2}$$

is the *normal density function* with *mean* $\mu = 0$ and *standard deviation* σ. The probability that a randomly chosen value described by this distribution lies in $[a, b]$ is given by $\int_a^b f(x)dx$. Approximate to within 10^{-5} the probability that a randomly chosen value described by this distribution will lie in

    a. $[-2\sigma, 2\sigma]$ **b.** $[-3\sigma, 3\sigma]$

12. Determine to within 10^{-6} the length of the graph of the ellipse with equation $4x^2 + 9y^2 = 36$.

13. A car laps a race track in 84 seconds. The speed of the car at each 6-second interval is determined using a radar gun and is given from the beginning of the lap, in feet per second, by the entries in the following table.

Time	0	6	12	18	24	30	36	42	48	54	60	66	72	78	84
Speed	124	134	148	156	147	133	121	109	99	85	78	89	104	116	123

How long is the track?

14. A particle of mass m moving through a fluid is subjected to a viscous resistance R, which is a function of the velocity v. The relationship between the resistance R, velocity v, and time t is given by the equation

$$t = \int_{v(t_0)}^{v(t)} \frac{m}{R(u)} du.$$

Suppose that $R(v) = -v\sqrt{v}$ for a particular fluid, where R is in newtons and v is in meters per second. If $m = 10$ kg and $v(0) = 10$ m/s, approximate the time required for the particle to slow to $v = 5$ m/s.

15. The equation

$$\int_0^x \frac{1}{\sqrt{2\pi}} e^{-t^2/2} dt = 0.45$$

can be solved for x by using Newton's method with

$$f(x) = \int_0^x \frac{1}{\sqrt{2\pi}} e^{-t^2/2} dt - 0.45$$

and

$$f'(x) = \frac{1}{\sqrt{2\pi}} e^{-x^2/2}.$$

To evaluate f at the approximation p_k, we need a quadrature formula to approximate

$$\int_0^{p_k} \frac{1}{\sqrt{2\pi}} e^{-t^2/2} dt.$$

    a. Find a solution to $f(x) = 0$ accurate to within 10^{-5} using Newton's method with $p_0 = 0.5$ and the Composite Simpson's rule.

    b. Repeat (a) using the Composite Trapezoidal rule in place of the Composite Simpson's rule.

16. To determine the thermal characteristics of disk brakes, you need the "area averaged lining temperature," T, of the brake pad from the equation

$$T = \frac{\displaystyle\int_{r_e}^{r_0} T(r) r \theta_p \, dr}{\displaystyle\int_{r_e}^{r_0} r \theta_p \, dr},$$

where r_e represents the radius at which the pad-disk contact begins, r_0 represents the outside radius of the pad-disk contact, and θ_p represents the angle subtended by the sector brake pads. The temperature $T(r)$ at each point of the pad is obtained numerically from analyzing the heat equation (see Section 12.2). Suppose $r_e = 0.308$ ft, $r_0 = 0.478$ ft, $\theta_p = 0.7051$ radians, and the temperatures given in the following table have been calculated at the various points on the disk. Approximate T.

r (ft)	$T(r)$ (°F)	r (ft)	$T(r)$ (°F)	r (ft)	$T(r)$ (°F)
0.308	640	0.376	1034	0.444	1204
0.325	794	0.393	1064	0.461	1222
0.342	885	0.410	1114	0.478	1239
0.359	943	0.427	1152		

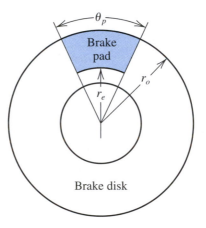

Brake disk

4.4 Romberg Integration

Extrapolation is used to accelerate the convergence of many approximation techniques. It can be applied whenever it is known that the approximation technique has an error term with a particular form, which depends on a parameter, usually the step size h. In this section we will illustrate how extrapolation can be applied to results from the Composite Trapezoidal rule to obtain high accuracy approximations with little computational cost.

Extrapolation

Suppose that $N(h)$ is a formula involving a step size h that approximates an unknown value M, and that it is known that the error for $N(h)$ has the form

$$M = N(h) + K_1 h + K_2 h^2 + K_3 h^3 + \cdots. \tag{4.1}$$

for some unspecified, and often unknown, collection of constants, $K_1, K_2, K_3 \ldots..$

We assume here that $h > 0$ can be arbitrarily chosen and that improved approximations occur as h becomes small. The objective of extrapolation is to improve the formula from one of order $O(h)$ to one of higher order. Do not be misled by the relative simplicity of Eq. (4.1). It might be quite difficult to obtain the approximation $N(h)$, particularly for small values of h.

Since Eq. (4.1) is assumed to hold for all positive h, consider the result when we replace the parameter h by half its value. This gives the formula

$$M = N\left(\frac{h}{2}\right) + K_1\frac{h}{2} + K_2\frac{h^2}{4} + K_3\frac{h^3}{8} + \cdots. \tag{4.2}$$

Subtracting Eq. (4.1) from twice Eq. (4.2) eliminates the term involving K_1 and gives

$$M = \left[N\left(\frac{h}{2}\right) + \left(N\left(\frac{h}{2}\right) - N(h)\right)\right] + K_2\left(\frac{h^2}{2} - h^2\right) + K_3\left(\frac{h^3}{4} - h^3\right) + \cdots.$$

To simplify the notation, define $N_1(h) \equiv N(h)$ and

$$N_2(h) = N_1\left(\frac{h}{2}\right) + \left[N_1\left(\frac{h}{2}\right) - N_1(h)\right].$$

Then the two original $O(h)$ approximations combine to produce an $O(h^2)$ approximation formula for M:

$$M = N_2(h) - \frac{K_2}{2}h^2 - \frac{3K_3}{4}h^3 - \cdots. \tag{4.3}$$

If we now replace h by $h/2$ in Eq. (4.3), we have

$$M = N_2\left(\frac{h}{2}\right) - \frac{K_2}{8}h^2 - \frac{3K_3}{32}h^3 - \cdots. \tag{4.4}$$

This can be combined with Eq. (4.3) to eliminate the h^2 term. Specifically, subtracting Eq. (4.3) from 4 times Eq. (4.4) gives

$$3M = 4N_2\left(\frac{h}{2}\right) - N_2(h) + \frac{3K_3}{4}\left(-\frac{h^3}{2} + h^3\right) + \cdots,$$

and

$$M = \left[N_2\left(\frac{h}{2}\right) + \frac{N_2(h/2) - N_2(h)}{3}\right] + \frac{K_3}{8}h^3 + \cdots.$$

Defining

$$N_3(h) \equiv N_2\left(\frac{h}{2}\right) + \frac{N_2(h/2) - N_2(h)}{3}$$

simplifies this to the $O(h^3)$ approximation formula for M:

$$M = N_3(h) + \frac{K_3}{8}h^3 + \cdots.$$

The process is continued by constructing the $O(h^4)$ approximation

$$N_4(h) = N_3\left(\frac{h}{2}\right) + \frac{N_3(h/2) - N_3(h)}{7},$$

the $O(h^5)$ approximation

$$N_5(h) = N_4\left(\frac{h}{2}\right) + \frac{N_4(h/2) - N_4(h)}{15},$$

and so on. In general, if M can be written in the form

$$M = N(h) + \sum_{j=1}^{m-1} K_j h^j + O(h^m),$$

then for each $j = 2, 3, \ldots, m$, we have an $O(h^j)$ approximation of the form

$$N_j(h) = N_{j-1}\left(\frac{h}{2}\right) + \frac{N_{j-1}(h/2) - N_{j-1}(h)}{2^{j-1} - 1}.$$

These approximations are generated by rows to take advantage of the highest order formulas. The first four rows are shown in Table 4.2. The entries are numbered by the manner in which they are most efficiently generated.

Table 4.2

$O(h)$	$O(h^2)$	$O(h^3)$	$O(h^4)$
1: $N_1(h)$			
2: $N_1(\frac{h}{2})$	**3:** $N_2(h)$		
4: $N_1(\frac{h}{4})$	**5:** $N_2(\frac{h}{2})$	**6:** $N_3(h)$	
7: $N_1(\frac{h}{8})$	**8:** $N_2(\frac{h}{4})$	**9:** $N_3(\frac{h}{2})$	**10:** $N_4(h)$

Extrapolation with the Composite Trapezoidal Rule

In Section 4.3 we found that the Composite Trapezoidal rule has a truncation error of order $O(h^2)$. Specifically, we showed that for $h = (b - a)/n$ and $x_j = a + jh$ we have

$$\int_a^b f(x)dx = \frac{h}{2}\left[f(a) + 2\sum_{j=1}^{n-1} f(x_j) + f(b)\right] - \frac{(b-a)f''(\mu)}{12}h^2.$$

for some number μ in (a, b). By an alternative method it can be shown that if $f \in C^\infty[a, b]$, the Composite Trapezoidal rule can also be written with an error term in the form

$$\int_a^b f(x)dx = \frac{h}{2}\left[f(a) + 2\sum_{j=1}^{n-1} f(x_j) + f(b)\right] + K_1 h^2 + K_2 h^4 + K_3 h^6 + \cdots, \quad (4.5)$$

where each K_i is a constant that depends only on $f^{(2i-1)}(a)$ and $f^{(2i-1)}(b)$. Because the Composite Trapezoidal rule has this form, it is an obvious candidate for extrapolation. This results in a technique known as **Romberg integration**.

Romberg integration is even more effective than the extrapolation method we described earlier because there are no odd powers in the error term. In this case we have

$$N_2(h) = N_1\left(\frac{h}{2}\right) + \frac{1}{3}[N_1(h/2) - N_1(h)], \quad N_3(h) = N_2\left(\frac{h}{2}\right) + \frac{1}{15}[N_2(h/2) - N_2(h)],$$

and, in general, for each $j = 2, 3, \ldots$, we have the $O(h^{2j})$ approximation

$$N_j(h) = N_{j-1}\left(\frac{h}{2}\right) + \frac{N_{j-1}(h/2) - N_{j-1}(h)}{4^{j-1} - 1}$$

and the approximations have the improved error terms given in Table 4.3.

Table 4.3

$O(h^2)$	$O(h^4)$	$O(h^6)$	$O(h^8)$
$N_1(h)$			
$N_1(\frac{h}{2})$	$N_2(h)$		
$N_1(\frac{h}{4})$	$N_2(\frac{h}{2})$	$N_3(h)$	
$N_1(\frac{h}{8})$	$N_2(\frac{h}{4})$	$N_3(\frac{h}{2})$	$N_4(h)$

Werner Romberg (1909–2003) devised this procedure for improving the accuracy of the trapezoidal rule by eliminating the successive terms in the asymptotic expansion in 1955.

To approximate the integral $\int_a^b f(x)\,dx$ we use the results of the Composite Trapezoidal rule with $n = 1, 2, 4, 8, 16, \ldots$, and denote the resulting approximations, respectively, by $R_{1,1}, R_{2,1}, R_{3,1}$, etc. We then apply extrapolation to obtain $O(h^4)$ approximations $R_{2,2}, R_{3,2}, R_{4,2}$, etc. by

$$R_{k,2} = R_{k,1} + \frac{1}{3}(R_{k,1} - R_{k-1,1}), \quad \text{for } k = 2, 3, \ldots$$

Then we obtain $O(h^6)$ approximations $R_{3,3}, R_{4,3}, R_{5,3}$, etc. by

$$R_{k,3} = R_{k,2} + \frac{1}{15}(R_{k,2} - R_{k-1,2}), \quad \text{for } k = 3, 4, \ldots.$$

In general, after the appropriate $R_{k,j-1}$ approximations have been obtained, we determine the $O(h^{2j})$ approximations from

$$R_{k,j} = R_{k,j-1} + \frac{1}{4^{j-1} - 1}(R_{k,j-1} - R_{k-1,j-1}), \quad \text{for } k = j, j+1, \ldots$$

This gives approximation of the form shown in Table 4.4.

Table 4.4

k	$O(h^2)$	$O(h^4)$	$O(h^6)$	$O(h^8)$		$O(h^{2n})$
1	$R_{1,1}$					
2	$R_{2,1}$	$R_{2,2}$				
3	$R_{3,1}$	$R_{3,2}$	$R_{3,3}$			
4	$R_{4,1}$	$R_{4,2}$	$R_{4,3}$	$R_{4,4}$		
$\vdots$	$\vdots$	$\vdots$	$\vdots$	$\vdots$	$\ddots$	
n	$R_{n,1}$	$R_{n,2}$	$R_{n,3}$	$R_{n,4}$	$\cdots$	$R_{n,n}$

Example 1   Use the Composite Trapezoidal rule to find approximations to $\displaystyle\int_0^\pi \sin x\,dx$ with $n = 1, 2, 4, 8,$ and 16.

Solution The Composite Trapezoidal rule for the various values of n gives the following approximations to the true value 2.

$$R_{1,1} = \frac{\pi}{2}[\sin 0 + \sin \pi] = 0;$$

$$R_{2,1} = \frac{\pi}{4}\left[\sin 0 + 2\sin \frac{\pi}{2} + \sin \pi\right] = 1.57079633;$$

$$R_{3,1} = \frac{\pi}{8}\left[\sin 0 + 2\left(\sin \frac{\pi}{4} + \sin \frac{\pi}{2} + \sin \frac{3\pi}{4}\right) + \sin \pi\right] = 1.89611890;$$

$$R_{4,1} = \frac{\pi}{16}\left[\sin 0 + 2\left(\sin \frac{\pi}{8} + \sin \frac{\pi}{4} + \cdots + \sin \frac{3\pi}{4} + \sin \frac{7\pi}{8}\right) + \sin \pi\right] = 1.97423160;$$

$$R_{5,1} = \frac{\pi}{32}\left[\sin 0 + 2\left(\sin \frac{\pi}{16} + \sin \frac{\pi}{8} + \cdots + \sin \frac{7\pi}{8} + \sin \frac{15\pi}{16}\right) + \sin \pi\right] = 1.99357034.$$

■

Consecutive applications of the Composite Trapezoidal rule include the evaluations of the function at the previous step, but use only half their values. To efficiently use the Composite Trapezoidal rule to approximate $\int_a^b f(x)dx$, we incorporate this observation and express

$$R_{k,1} = \frac{1}{2}\left[R_{k-1,1} + h_{k-1}\sum_{i=1}^{2^{k-2}} f(a + (2i-1)h_k)\right], \quad \text{where} \quad h_k = \frac{b-a}{2^{k-1}}, \qquad (4.6)$$

for each $k = 2, 3, \ldots, n$.

Extrapolation is then used to produce $O(h_k^{2j})$ approximations by

$$R_{k,j} = R_{k,j-1} + \frac{1}{4^{j-1}-1}(R_{k,j-1} - R_{k-1,j-1}), \quad \text{for } k = j, j+1, \ldots.$$

Example 2 In Example 1 we found Composite Trapezoidal approximations to $\int_0^\pi \sin x \, dx$ with $n = 1$, 2, 4, 8, and 16:

$$R_{1,1} = 0, \quad R_{2,1} = 1.57079633, \quad R_{3,1} = 1.89611890, \quad R_{4,1} = 1.97423160, \quad \text{and}$$

$$R_{5,1} = 1.99357034.$$

Perform Romberg extrapolation on these results.

Solution The four $O(h^4)$ approximations are

$$R_{2,2} = R_{2,1} + \frac{1}{3}(R_{2,1} - R_{1,1}) = 2.09439511;$$

$$R_{3,2} = R_{3,1} + \frac{1}{3}(R_{3,1} - R_{2,1}) = 2.00455976;$$

$$R_{4,2} = R_{4,1} + \frac{1}{3}(R_{4,1} - R_{3,1}) = 2.00026917;$$

$$R_{5,2} = R_{5,1} + \frac{1}{3}(R_{5,1} - R_{4,1}) = 2.00001659;$$

The three $O(h^6)$ approximations are

$$R_{3,3} = R_{3,2} + \frac{1}{15}(R_{3,2} - R_{2,2}) = 1.99857073;$$

$$R_{4,3} = R_{4,2} + \frac{1}{15}(R_{4,2} - R_{3,2}) = 1.99998313;$$

$$R_{5,3} = R_{5,2} + \frac{1}{15}(R_{5,2} - R_{4,2}) = 1.99999975.$$

The two $O(h^8)$ approximations are

$$R_{4,4} = R_{4,3} + \frac{1}{63}(R_{4,3} - R_{3,3}) = 2.00000555;$$

$$R_{5,4} = R_{5,3} + \frac{1}{63}(R_{5,3} - R_{4,3}) = 2.00000001,$$

and the final $O(h^{10})$ approximation is

$$R_{5,5} = R_{5,4} + \frac{1}{255}(R_{5,4} - R_{4,4}) = 1.99999999.$$

These results are shown in Table 4.5. ■

Table 4.5

0				
1.57079633	2.09439511			
1.89611890	2.00455976	1.99857073		
1.97423160	2.00026917	1.99998313	2.00000555	
1.99357034	2.00001659	1.99999975	2.00000001	1.99999999

The effective method to construct the Romberg table makes use of the highest order of approximation at each step. That is, it calculates the entries row by row, in the order $R_{1,1}, R_{2,1}, R_{2,2}, R_{3,1}, R_{3,2}, R_{3,3}$, etc. This also permits an entire new row in the table to be calculated by doing only one additional application of the Composite Trapezoidal rule. It then uses a simple averaging on the previously calculated values to obtain the remaining entries in the row.

Example 3 If the Trapezoidal rule with $n = 32$ is applied to $\int_0^\pi \sin x \, dx$, we obtain

$$R_{6,1} = \frac{1}{2}\left[R_{5,1} + \frac{\pi}{16}\sum_{k=1}^{2^4}\sin\frac{(2k-1)\pi}{32}\right] = 1.99839336.$$

Use this approximation and the results in Example 2 to add an additional extrapolation row to Table 4.5.

Solution   Using the values in Table 4.5 we have

$$R_{6,2} = R_{6,1} + \frac{1}{3}(R_{6,1} - R_{5,1}) = 1.99839336 + \frac{1}{3}(1.99839336 - 1.99357035)$$

$$= 2.00000103;$$

$$R_{6,3} = R_{6,2} + \frac{1}{15}(R_{6,2} - R_{5,2}) = 2.00000103 + \frac{1}{15}(2.00000103 - 2.00001659)$$

$$= 2.00000000;$$

$$R_{6,4} = R_{6,3} + \frac{1}{63}(R_{6,3} - R_{5,3}) = 2.00000000; \quad R_{6,5} = R_{6,4} + \frac{1}{255}(R_{6,4} - R_{5,4})$$

$$= 2.00000000;$$

and $R_{6,6} = R_{6,5} + \frac{1}{1023}(R_{6,5} - R_{5,5}) = 2.00000000$. The new extrapolation table is shown in Table 4.6.   ▪

Table 4.6

0					
1.57079633	2.09439511				
1.89611890	2.00455976	1.99857073			
1.97423160	2.00026917	1.99998313	2.00000555		
1.99357034	2.00001659	1.99999975	2.00000001	1.99999999	
1.99839336	2.00000103	2.00000000	2.00000000	2.00000000	2.00000000

Notice that all the extrapolated values except for the first (in the first row of the second column) are more accurate than the best Composite Trapezoidal approximation (in the last row of the first column). Although there are 21 entries in Table 4.6, only the six in the left column require function evaluations, those generated by the Composite Trapezoidal rule. The other entries are obtained by an averaging process. In fact, because of the recurrence relationship of the terms in the left column, the only function evaluations needed are those to compute the final Composite Trapezoidal rule approximation. In general, $R_{k,1}$ requires $1 + 2^{k-1}$ function evaluations, so in this case $1 + 2^5 = 33$ are needed.

The program ROMBRG32 requires a preset integer n to determine the number of rows to be generated. It is often also useful to prescribe an error tolerance for the approximation and generate R_{nn}, for n within some upper bound, until consecutive diagonal entries $R_{n,n}$ agree to within the tolerance.

Romberg integration applied to a function f on the interval $[a, b]$ relies on the assumption that the Composite Trapezoidal rule has an error term that can be expressed in the form of Eq. (4.5); that is, we must have $f \in C^{2k+2}[a, b]$ for the kth row to be generated. General-purpose algorithms using Romberg integration include a check at each stage to ensure that this assumption is fulfilled. These methods are known as *cautious Romberg algorithms* and are described in [Joh]. This reference also describes methods for using the Romberg technique as an adaptive procedure, similar to the adaptive Simpson's rule that will be discussed in Section 4.6.

> The program ROMBRG42 implements Romberg integration.

EXERCISE SET 4.4

1.   Use Romberg integration to compute $R_{3,3}$ for the following integrals.

   a. $\displaystyle\int_{1}^{1.5} x^2 \ln x \, dx$

   b. $\displaystyle\int_{0}^{1} x^2 e^{-x} dx$

   c. $\displaystyle\int_0^{0.35} \frac{2}{x^2-4}\,dx$ **d.** $\displaystyle\int_0^{\pi/4} x^2 \sin x\,dx$

   e. $\displaystyle\int_0^{\pi/4} e^{3x} \sin 2x\,dx$ **f.** $\displaystyle\int_1^{1.6} \frac{2x}{x^2-4}\,dx$

   g. $\displaystyle\int_3^{3.5} \frac{x}{\sqrt{x^2-4}}\,dx$ **h.** $\displaystyle\int_0^{\pi/4} (\cos x)^2\,dx$

2.   Calculate $R_{4,4}$ for the integrals in Exercise 1.

3.   Use Romberg integration to approximate the integrals in Exercise 1 to within 10^{-6}. Compute the Romberg table until $|R_{n-1,n-1} - R_{n,n}| < 10^{-6}$, or until $n = 10$. Compare your results to the exact values of the integrals.

4.   Use Romberg integration to approximate the following integrals to within 10^{-6}. Compute the Romberg table until either $|R_{n-1,n-1} - R_{n,n}| < 10^{-6}$, or $n = 10$.

   a. $\displaystyle\int_1^1 (\cos x)^2\,dx$ **b.** $\displaystyle\int_{-0.75}^{0.75} x \ln(x+1)\,dx$

   c. $\displaystyle\int_1^4 \left((\sin x)^2 - 2x \sin x + 1\right)dx$ **d.** $\displaystyle\int_e^{2e} \frac{1}{x \ln x}\,dx$

5.   Use Romberg integration and the following data to approximate $\int_1^5 f(x)dx$ as accurately as possible.

x	1	2	3	4	5
$f(x)$	2.4142	2.6734	2.8974	3.0976	3.2804

6.   Romberg integration is used to approximate $\displaystyle\int_0^1 \frac{x^2}{1+x^3}\,dx.$

   If $R_{11} = 0.250$ and $R_{22} = 0.2315$, what is R_{21}?

7.   Romberg integration is used to approximate $\displaystyle\int_2^3 f(x)dx.$

   If $f(2) = 0.51342$, $f(3) = 0.36788$, $R_{31} = 0.43687$, and $R_{33} = 0.43662$, find $f(2.5)$.

8.   Romberg integration for approximating $\int_0^1 f(x)dx$ gives $R_{11} = 4$ and $R_{22} = 5$. Find $f(\frac{1}{2})$.

9.   Romberg integration for approximating $\int_a^b f(x)dx$ gives $R_{11} = 8$, $R_{22} = \frac{16}{3}$, and $R_{33} = \frac{208}{45}$. Find R_{31}.

10.   Suppose that $N(h)$ is an approximation to M for every $h > 0$ and that

$$M = N(h) + K_1 h + K_2 h^2 + K_3 h^3 + \cdots$$

   for some constants $K_1, K_2, K_3, \ldots$. Use the values $N(h)$, $N\left(\frac{h}{3}\right)$, and $N\left(\frac{h}{9}\right)$ to produce an $O(h^3)$ approximation to M.

11.   Suppose that $N(h)$ is an approximation to M for every $h > 0$ and that

$$M = N(h) + K_1 h^2 + K_2 h^4 + K_3 h^6 + \cdots$$

   for some constants $K_1, K_2, K_3, \ldots$. Use the values $N(h)$, $N\left(\frac{h}{3}\right)$, and $N\left(\frac{h}{9}\right)$ to produce an $O(h^6)$ approximation to M.

12.   We learn in calculus that $e = \lim_{h \to 0}(1+h)^{1/h}$.

   a.   Determine approximations to e corresponding to $h = 0.04, 0.02$, and 0.01.

   b.   Use extrapolation on the approximations, assuming that constants $K_1, K_2, \ldots$, exist with

$$e = (1+h)^{1/h} + K_1 h + K_2 h^2 + K_3 h^3 + \cdots$$

   to produce an $O(h^3)$ approximation to e, where $h = 0.04$.

   c.   Do you think that the assumption in (b) is correct?

13. **a.** Show that

$$\lim_{h \to 0} \left(\frac{2+h}{2-h} \right)^{1/h} = e.$$

b. Compute approximations to e using the formula

$$N(h) = \left(\frac{2+h}{2-h} \right)^{1/h}$$

for $h = 0.04, 0.02$, and 0.01.

c. Assume that $e = N(h) + K_1 h + K_2 h^2 + K_3 h^3 + \cdots$. Use extrapolation, with at least 16 digits of precision, to compute an $O(h^3)$ approximation to e with $h = 0.04$. Do you think the assumption is correct?

d. Show that $N(-h) = N(h)$.

e. Use (d) to show that $K_1 = K_3 = K_5 = \cdots = 0$ in the formula

$$e = N(h) + K_1 h + K_2 h^2 + K_3 h^3 + K_4 h^4 + K_5^5 + \cdots,$$

so that the formula reduces to

$$e = N(h) + K_2 h^2 + K_4 h^4 + K_6 h^6 + \cdots.$$

f. Use the results of (e) and extrapolation to compute an $O(h^6)$ approximation to e with $h = 0.04$.

14. Suppose the following extrapolation table has been constructed to approximate the number M with $M = N_1(h) + K_1 h^2 + K_2 h^4 + K_3 h^6$:

$N_1(h)$		
$N_1\left(\dfrac{h}{2}\right)$	$N_2(h)$	
$N_1\left(\dfrac{h}{4}\right)$	$N_2\left(\dfrac{h}{2}\right)$	$N_3(h)$

a. Show that the linear interpolating polynomial $P_{0,1}(h)$ through $(h^2, N_1(h))$ and $(h^2/4, N_1(h/2))$ satisfies $P_{0,1}(0) = N_2(h)$. Similarly, show that $P_{1,2}(0) = N_2(h/2)$.

b. Show that the linear interpolating polynomial $P_{0,2}(h)$ through $(h^4, N_2(h))$ and $(h^4/16, N_2(h/2))$ satisfies $P_{0,2}(0) = N_3(h)$.

4.5 Gaussian Quadrature

The formulas in Section 4.2 for approximating the integral of a function were derived by integrating interpolating polynomials, and the error term in the interpolating polynomial of degree n involves the $(n + 1)$st derivative of the function being approximated. Every polynomial of degree less than or equal to n has zero for its $(n+1)$st derivative, so applying a formula of this type to such polynomials gives an exact result.

All the formulas in Section 4.2 use functional values at equally spaced points. This is convenient when the formulas are combined to form the composite rules we considered in Section 4.3, but this restriction can significantly decrease the accuracy of the approximation. Consider, for example, the Trapezoidal rule applied to determine the integrals of the functions shown in Figure 4.7. The Trapezoidal rule approximates the integral of the function by integrating the linear function that joins the endpoints of the graph of the function.

Figure 4.7

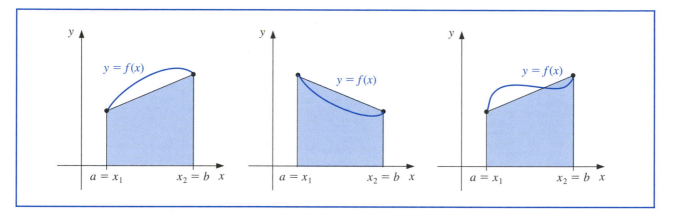

But this is not likely the best line to use for approximating the integral. Lines such as those shown in Figure 4.8 would give better approximations in most cases.

Figure 4.8

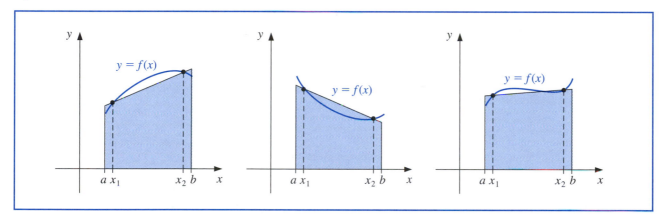

Gaussian quadrature uses evaluation points, or nodes, that are not equally spaced in the interval. The nodes $x_1, x_2, \ldots, x_n$ in the interval $[a, b]$ and coefficients $c_1, c_2, \ldots, c_n$ are chosen to minimize the expected error obtained in the approximation

$$\int_a^b f(x)dx \approx \sum_{i=1}^n c_i f(x_i).$$

To minimize the expected error, we assume that the best choice of these values is that which produces the exact result for the largest class of polynomials.

The coefficients $c_1, c_2, \ldots, c_n$ in the approximation formula are arbitrary, and the nodes $x_1, x_2, \ldots, x_n$ are restricted only by the fact that they must lie in $[a, b]$, the interval of integration. This gives $2n$ parameters to choose. If the coefficients of a polynomial are considered parameters, the class of polynomials of degree at most $2n - 1$ also contains $2n$ parameters. This, then, is the largest class of polynomials for which it is reasonable to expect the formula to be exact. For the proper choice of the nodes and coefficients, exactness on this set can be obtained.

To illustrate the procedure for choosing the appropriate constants, we will show how to select the coefficients and nodes when $n = 2$ and the interval of integration is $[-1, 1]$.

Illustration   Suppose we want to determine c_1, c_2, x_1, and x_2 so that the integration formula

$$\int_{-1}^{1} f(x)dx \approx c_1 f(x_1) + c_2 f(x_2)$$

gives the exact result whenever $f(x)$ is a polynomial of degree $2(2) - 1 = 3$ or less, that is, when

$$f(x) = a_0 + a_1 x + a_2 x^2 + a_3 x^3,$$

for any collection of constants, a_0, a_1, a_2, and a_3. Because

$$\int_{-1}^{1} (a_0 + a_1 x + a_2 x^2 + a_3 x^3)dx = a_0 \int_{-1}^{1} 1dx + a_1 \int_{-1}^{1} xdx + a_2 \int_{-1}^{1} x^2 dx + a_3 \int_{-1}^{1} x^3 dx,$$

this is equivalent to showing that the formula gives exact results for all the integrals on the right side of this equation. Hence, we need c_1, c_2, x_1, and x_2, so that

$$c_1 \cdot 1 + c_2 \cdot 1 = \int_{-1}^{1} 1dx = 2, \qquad c_1 \cdot x_1 + c_2 \cdot x_2 = \int_{-1}^{1} x\, dx = 0,$$

$$c_1 \cdot x_1^2 + c_2 \cdot x_2^2 = \int_{-1}^{1} x^2 dx = \frac{2}{3}, \quad \text{and} \quad c_1 \cdot x_1^3 + c_2 \cdot x_2^3 = \int_{-1}^{1} x^3 dx = 0.$$

A little algebra shows that this system of equations has the unique solution

$$c_1 = 1, \quad c_2 = 1, \quad x_1 = -\frac{\sqrt{3}}{3}, \quad \text{and} \quad x_2 = \frac{\sqrt{3}}{3},$$

which gives the approximation formula

$$\int_{-1}^{1} f(x)dx \approx f\left(\frac{-\sqrt{3}}{3}\right) + f\left(\frac{\sqrt{3}}{3}\right). \tag{4.7}$$

This formula has degree of precision 3; that is, it produces the exact result for every polynomial of degree 3 or less. ▪

Legendre Polynomials

The technique in the Illustration can be used to determine the nodes and coefficients for formulas that give exact results for higher-degree polynomials, but an alternative method obtains them more easily. In Section 8.3 we will consider various collections of orthogonal polynomials, sets of functions that have the property that a particular definite integral of the product of any two of them is zero. The set that is relevant to our problem is the set of *Legendre polynomials*, a collection $\{P_0(x), P_1(x), \ldots, P_n(x), \ldots\}$ that has the following properties:

- For each n, $P_n(x)$ is a monic polynomial of degree n.

Monic polynomials have leading coefficient 1.

- $\int_{-1}^{1} P_i(x)P_j(x)dx = 0$ whenever $i \neq j$.

The first few Legendre polynomials are

$$P_0(x) = 1, \qquad P_1(x) = x, \qquad P_2(x) = x^2 - \frac{1}{3},$$

$$P_3(x) = x^3 - \frac{3}{5}x, \qquad \text{and} \qquad P_4(x) = x^4 - \frac{6}{7}x^2 + \frac{3}{35}.$$

Adrien-Marie Legendre (1752–1833) introduced this set of polynomials in 1785. He had numerous priority disputes with Gauss, generally due to Gauss's failure to publish many of his original results until long after he had discovered them.

The roots of these polynomials are distinct, lie in the interval $(-1, 1)$, have a symmetry with respect to the origin, and, most importantly, are the nodes to use to solve our problem.

The nodes $x_1, x_2, \ldots, x_n$ needed to produce an integral approximation formula to give exact results for any polynomial of degree $2n - 1$ or less are the roots of the nth-degree Legendre polynomial. In addition, once the roots are known, the appropriate coefficients for the function evaluations at these nodes can be found from the fact that for each $i = 1, 2, \ldots, n$, we have

$$c_i = \int_{-1}^{1} \frac{(x - x_1)(x - x_2) \cdots (x - x_{i-1})(x - x_{i+1}) \cdots (x - x_n)}{(x_i - x_1)(x_i - x_2) \cdots (x_i - x_{i-1})(x_i - x_{i+1}) \cdots (x_i - x_n)} dx.$$

However, both the roots of the Legendre polynomials and the coefficients are extensively tabulated, so it is not necessary to perform these evaluations. A small sample is given in Table 4.7, and listings for higher-degree polynomials are given on the Internet or, for example, in the book by Stroud and Secrest [StS].

Table 4.7

n	Roots $r_{n,i}$	Coefficients $c_{n,i}$
2	0.5773502692	1.0000000000
	−0.5773502692	1.0000000000
3	0.7745966692	0.5555555556
	0.0000000000	0.8888888889
	−0.7745966692	0.5555555556
4	0.8611363116	0.3478548451
	0.3399810436	0.6521451549
	−0.3399810436	0.6521451549
	−0.8611363116	0.3478548451
5	0.9061798459	0.2369268850
	0.5384693101	0.4786286705
	0.0000000000	0.5688888889
	−0.5384693101	0.4786286705
	−0.9061798459	0.2369268850

Example 1 Approximate $\int_{-1}^{1} e^x \cos x \, dx$ using Gaussian quadrature with $n = 3$.

Solution The entries in Table 4.7 give

$$\int_{-1}^{1} e^x \cos x \, dx \approx 0.\overline{5} e^{0.774596692} \cos 0.774596692$$

$$+ 0.\overline{8} \cos 0 + 0.\overline{5} e^{-0.774596692} \cos(-0.774596692)$$

$$= 1.9333904.$$

Integration by parts can be used to show that the true value of the integral is 1.9334214, so the absolute error is less than 3.2×10^{-5}.

We now have a solution to the approximation problem, but only for definite integrals of functions on the interval $[-1, 1]$. However, this solution is sufficient for any closed interval because the simple linear relation

$$t = \frac{2x - a - b}{b - a}$$

transforms the variable x in the interval $[a, b]$ into the variable t in the interval $[-1, 1]$. Then the Legendre polynomials can be used to approximate

$$\int_a^b f(x)dx = \int_{-1}^1 f\left(\frac{(b - a)t + b + a}{2}\right)\frac{(b - a)}{2}dt.$$

Using the roots $r_{n,1}, r_{n,2}, \ldots, r_{n,n}$ and the coefficients $c_{n,1}, c_{n,2}, \ldots, c_{n,n}$ given in Table 4.7 produces the following approximation formula, which gives the exact result for a polynomial of degree $2n - 1$ or less.

Gaussian Quadrature

The approximation

$$\int_a^b f(x)dx \approx \frac{b - a}{2} \sum_{j=1}^n c_{n,j} f\left(\frac{(b - a)r_{n,j} + b + a}{2}\right)$$

is exact whenever $f(x)$ is a polynomial of degree $2n - 1$ or less.

Example 2   Consider the integral $\int_1^3 x^6 - x^2 \sin(2x)dx \approx 317.3442466$.

(a) Compare the results for the Trapezoidal rule, the Composite Midpoint rule with $n = 2$, and Gaussian quadrature when $n = 2$.

(b) Compare the results for the Composite Trapezoidal rule with $n = 2$, Simpson's rule, and Gaussian quadrature when $n = 3$.

Solution   (a) Each of the formulas in this part requires two evaluations of the function $f(x) = x^6 - x^2 \sin(2x)$. The Newton-Cotes approximations are

Trapezoidal rule: $[f(1) + f(3)] = 731.6054420;$

Composite Midpoint rule: $[f(3/2) + f(5/2)] = 261.2070067.$

The wide variation in these approximations indicates that at least one of them is very inaccurate.

Gaussian quadrature applied to this problem requires that the integral first be transformed into a problem whose interval of integration is $[-1, 1]$. Using the Gaussian quadrature formula gives

$$\int_1^3 x^6 - x^2 \sin(2x)dx = \int_{-1}^1 (t + 2)^6 - (t + 2)^2 \sin(2(t + 2))dt.$$

Because $c_{2,1} = c_{2,2} = 1$, $r_{2,1} = -0.5773502692$, and $r_{2,2} = 0.5773502692$ we have

$$\int_1^3 x^6 - x^2 \sin(2x)dx \approx f(-0.5773502692 + 2) + f(0.5773502692 + 2) = 306.8199344.$$

(b) Each of these formulas requires three function evaluations. The Newton-Cotes approximations are

$$\textbf{Composite Trapezoidal rule:} \quad \frac{1}{2}[f(1) + 2f(2) + f(3)] = 432.8299311.$$

$$\textbf{Simpson's rule:} \quad \frac{1}{3}[f(1) + 4f(2) + f(3)] = 333.2380940.$$

Gaussian quadrature, once the transformation has been done, gives

$$\int_1^3 x^6 - x^2 \sin(2x)\,dx \approx 0.\overline{5}f(-0.7745966692 + 2)$$

$$+ 0.\overline{8}f(2) + 0.\overline{5}f(0.7745966692 + 2) = 317.2641516.$$

The Gaussian quadrature results are clearly superior in each instance. ∎

For small problems, Composite Simpson's rule might be acceptable to avoid the computational complexity of Gaussian quadrature, but for problems requiring expensive function evaluations, the Gaussian procedure should certainly be considered. Gaussian quadrature is particularly important for approximating multiple integrals because the number of function evaluations increases as a power of the number of integrals being evaluated. This topic is considered in Section 4.7.

MATLAB has a command quadgk that uses Gaussian quadrature by way of the Gauss-Kronrod method, which can be used to approximate a wide variety of integrals. This procedure requires the integrand to be defined as an M-file. For the function in the example we have

```
function Y = ourfunction2(x)
Y = x^6-x^2*sin(2*x);
```

which is saved with the filename ourfunction2.m. This script file is made known to MATLAB using the command

```
F2 = @ourfunction2
```

The limits of integration a and b are defined, as usual, with

```
a=1, b=3
```

The routine quadgk is called using the following command, which returns an integral approximation that is accurate to the places listed, as well as an error bound.

```
[q,errbnd]=quadgk(F2,a,b)
```

$$q = 3.173442466738263e + 002 \quad \text{and} \quad errbnd = 2.164518564384821e - 014$$

EXERCISE SET 4.5

1. Approximate the following integrals using Gaussian quadrature with $n = 2$ and compare your results to the exact values of the integrals.

a. $\displaystyle\int_1^{1.5} x^2 \ln x\, dx$
  b. $\displaystyle\int_0^1 x^2 e^{-x}\, dx$

c. $\displaystyle\int_0^{0.35} \frac{2}{x^2 - 4}\,dx$

d. $\displaystyle\int_0^{\pi/4} x^2 \sin x\,dx$

e. $\displaystyle\int_0^{\pi/4} e^{3x} \sin 2x\,dx$

f. $\displaystyle\int_1^{1.6} \frac{2x}{x^2 - 4}\,dx$

g. $\displaystyle\int_3^{3.5} \frac{x}{\sqrt{x^2 - 4}}\,dx$

h. $\displaystyle\int_0^{\pi/4} (\cos x)^2\,dx$

2. Repeat Exercise 1 with $n = 3$.

3. Repeat Exercise 1 with $n = 4$.

4. Repeat Exercise 1 with $n = 5$.

5. Determine constants a, b, c, and d that will produce a quadrature formula

$$\int_{-1}^1 f(x)\,dx = af(-1) + bf(1) + cf'(-1) + df'(1)$$

that gives exact results for polynomials of degree 3 or less.

6. Determine constants a, b, c, and d that will produce a quadrature formula

$$\int_{-1}^1 f(x)\,dx = af(-1) + bf(0) + cf(1) + df'(-1) + ef'(1)$$

that gives exact results for polynomials of degree 4 or less.

7. Verify the entries for the values of $n = 2$ and 3 in Table 4.7 by finding the roots of the respective Legendre polynomials and use the equations preceding this table to find the coefficients associated with the values.

8. Use the recurrence relation

$$P_{n+1}(x) = x P_n(x) - \frac{n^2}{4n^2 - 1} P_{n-1}(x), \quad \text{for } n \geq 1$$

to generate the Legendre polynomials $P_2(x)$, $P_3(x)$, and $P_4(x)$, given that $P_0(x) = 1$ and $P_1(x) = x$.

4.6 Adaptive Quadrature

In Section 4.3 we used composite methods of approximation to break an integral over a large interval into integrals over smaller subintervals. The approach used in that section involved equally-sized subintervals, which permitted us to combine the individual approximations into a convenient form. Although this is satisfactory for most problems, it leads to increased computation when the function being integrated varies widely on some, but not all, parts of the interval of integration. In this case, techniques that adapt to the differing accuracy needs of the interval are superior. Adaptive numerical methods are particularly popular for inclusion in professional software packages because, in addition to being efficient, they generally provide approximations that are within a given specified tolerance.

In this section we consider an Adaptive quadrature method and see how it can be used not only to reduce approximation error, but also to predict an error estimate that does not rely on knowledge of higher derivatives of the function.

Adaptive Simpson's Rule

Suppose that we want to approximate $\int_a^b f(x)dx$ to within a specified tolerance $\varepsilon > 0$. We first apply Simpson's rule with step size $h = \frac{1}{2}(b-a)$, which gives (see Figure 4.9)

$$\int_a^b f(x)dx = S(a,b) - \frac{(b-a)^5}{2880} f^{(4)}(\xi) = S(a,b) - \frac{h^5}{90} f^{(4)}(\xi), \qquad (4.8)$$

for some number ξ in (a,b), where we denote the Simpson's rule approximation on $[a,b]$ by

$$S(a,b) = \frac{h}{3}[f(a) + 4f(a+h) + f(b)].$$

Figure 4.9

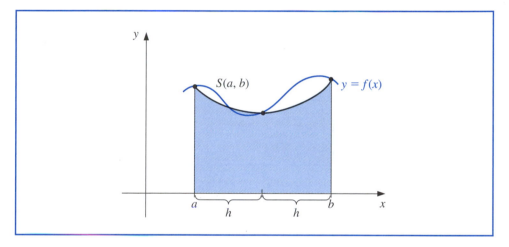

Next we determine an estimate for the accuracy of our approximation, in particular, one that does not require determining $f^{(4)}(\xi)$. To do this, we apply the Composite Simpson's rule to the problem with $n = 4$ and step size $(b-a)/4 = h/2$. Thus

$$\int_a^b f(x)dx = \frac{h}{6}\left[f(a) + 4f\left(a + \frac{h}{2}\right) + 2f(a+h) + 4f\left(a + \frac{3h}{2}\right) + f(b)\right]$$

$$- \left(\frac{h}{2}\right)^4 \frac{(b-a)}{180} f^{(4)}(\tilde{\xi}), \qquad (4.9)$$

for some number $\tilde{\xi}$ in (a,b). To simplify notation, let the Simpson's rule approximation on $[a, (a+b)/2]$ be denoted

$$S\left(a, \frac{a+b}{2}\right) = \frac{h}{6}\left[f(a) + 4f\left(a + \frac{h}{2}\right) + f(a+h)\right]$$

and the Simpson's rule approximation on $[(a+b)/2, b]$ be denoted

$$S\left(\frac{a+b}{2}, b\right) = \frac{h}{6}\left[f(a+h) + 4f\left(a + \frac{3h}{2}\right) + f(b)\right].$$

Then Eq. (4.9) can be rewritten (see Figure 4.10) as

$$\int_a^b f(x)dx = S\left(a, \frac{a+b}{2}\right) + S\left(\frac{a+b}{2}, b\right) - \frac{1}{16}\left(\frac{h^5}{90}\right)f^{(4)}(\tilde{\xi}). \tag{4.10}$$

Figure 4.10

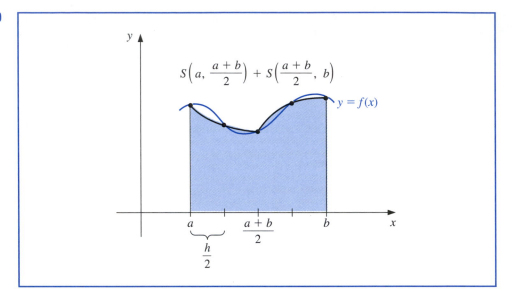

The error estimate is derived by assuming that $f^{(4)}(\xi) \approx f^{(4)}(\tilde{\xi})$, and the success of the technique depends on the accuracy of this assumption. If it is accurate, then equating the integrals in Eqs. (4.8) and (4.10) implies that

$$S\left(a, \frac{a+b}{2}\right) + S\left(\frac{a+b}{2}, b\right) - \frac{1}{16}\left(\frac{h^5}{90}\right)f^{(4)}(\xi) \approx S(a, b) - \frac{h^5}{90}f^{(4)}(\xi),$$

so

$$\frac{h^5}{90}f^{(4)}(\xi) \approx \frac{16}{15}\left[S(a, b) - S\left(a, \frac{a+b}{2}\right) - S\left(\frac{a+b}{2}, b\right)\right].$$

Using this estimate in Eq. (4.10) produces the error estimation

$$\left|\int_a^b f(x)dx - S\left(a, \frac{a+b}{2}\right) - S\left(\frac{a+b}{2}, b\right)\right|$$

$$\approx \frac{1}{15}\left|S(a, b) - S\left(a, \frac{a+b}{2}\right) - S\left(\frac{a+b}{2}, b\right)\right|.$$

Thus, $S(a, (a+b)/2) + S((a+b)/2, b)$ approximates $\int_a^b f(x)dx$ about 15 times better than it agrees with the known value $S(a, b)$. As a consequence, $S(a, (a+b)/2) + S((a+b)/2, b)$ approximates $\int_a^b f(x)dx$ to within ε provided that the two approximations $S(a, (a+b)/2) + S((a+b)/2, b)$ and $S(a, b)$ differ by less than 15ε.

Adaptive Quadrature Error Estimate

If

$$\left| S(a,b) - S\left(a, \frac{a+b}{2}\right) - S\left(\frac{a+b}{2}, b\right) \right| \leq 15\varepsilon,$$

then

$$\left| \int_a^b f(x)\,dx - S\left(a, \frac{a+b}{2}\right) - S\left(\frac{a+b}{2}, b\right) \right| \lessapprox \varepsilon.$$

Example 1   Check the accuracy of the error estimate given in the result when applied to the integral $\int_0^{\pi/2} \sin x \, dx = 1$ by comparing

$$\frac{1}{15}\left| S\left(0, \frac{\pi}{2}\right) - S\left(0, \frac{\pi}{4}\right) - S\left(\frac{\pi}{4}, \frac{\pi}{2}\right) \right| \quad \text{to} \quad \left| \int_0^{\pi/2} \sin x \, dx - S\left(0, \frac{\pi}{4}\right) - S\left(\frac{\pi}{4}, \frac{\pi}{2}\right) \right|.$$

Solution   We have

$$S\left(0, \frac{\pi}{2}\right) = \frac{\pi/4}{3}\left[\sin 0 + 4\sin\frac{\pi}{4} + \sin\frac{\pi}{2}\right] = \frac{\pi}{12}(2\sqrt{2}+1) = 1.002279878$$

and

$$S\left(0, \frac{\pi}{4}\right) + S\left(\frac{\pi}{4}, \frac{\pi}{2}\right) = \frac{\pi/8}{3}\left[\sin 0 + 4\sin\frac{\pi}{8} + 2\sin\frac{\pi}{4} + 4\sin\frac{3\pi}{8} + \sin\frac{\pi}{2}\right]$$

$$= 1.000134585.$$

So

$$\left| S\left(0, \frac{\pi}{2}\right) - S\left(0, \frac{\pi}{4}\right) - S\left(\frac{\pi}{4}, \frac{\pi}{2}\right) \right| = |1.002279878 - 1.000134585| = 0.002145293.$$

The estimate for the error obtained when using $S(a, (a+b)) + S((a+b), b)$ to approximate $\int_a^b f(x)\,dx$ is consequently

$$\frac{1}{15}\left| S\left(0, \frac{\pi}{2}\right) - S\left(0, \frac{\pi}{4}\right) - S\left(\frac{\pi}{4}, \frac{\pi}{2}\right) \right| = 0.000143020,$$

which closely approximates the actual error

$$\left| \int_0^{\pi/2} \sin x \, dx - 1.000134585 \right| = 0.000134585,$$

even though $D_x^4 \sin x = \sin x$ varies significantly in the interval $(0, \pi/2)$.   ∎

   Example 1 gives a demonstration of the error estimate for the Adaptive quadrature method, but the real value of the method comes from being able to reasonably predict the correct steps to ensure that a given approximation accuracy is obtained. Suppose that an error tolerance $\varepsilon > 0$ is specified for the integral of f on $[a, b]$, and

$$\left| S(a,b) - S\left(a, \frac{a+b}{2}\right) - S\left(\frac{a+b}{2}, b\right) \right| < 15\varepsilon.$$

Then we assume that $S\left(a, \frac{a+b}{2}\right) + S\left(\frac{a+b}{2}, b\right)$ is within ε of the value of $\int_a^b f(x)\,dx$.

If the approximation is not sufficiently accurate, the error-estimation procedure can be applied individually to the subintervals $[a, (a + b)/2]$ and $[(a + b)/2, b]$ to determine if the approximation to the integral on each subinterval is within a tolerance of $\varepsilon/2$. If so, the sum of the approximations agrees with $\int_a^b f(x)dx$ to within the tolerance ε.

If the approximation on one of the subintervals fails to be within the tolerance $\varepsilon/2$, then that subinterval is itself subdivided, and each of its subintervals is analyzed to determine if the integral approximation on each subinterval is accurate to within $\varepsilon/4$.

The halving procedure is continued until each portion is within the required tolerance. Although problems can be constructed for which this tolerance will never be met, the technique is generally successful because each subdivision increases the accuracy of the approximation by a factor of approximately 15 while requiring an increased accuracy factor of only 2.

Some technical difficulties require an implementation of the method in program ADAPQR43 that differs from the preceding discussion. For example, the tolerance between successive approximations has been conservatively set at 10ε rather than the derived value of 15ε to compensate for possible error in the assumption $f^{(4)}(\xi) \approx f^{(4)}(\tilde{\xi})$. In problems when $f^{(4)}$ is known to be widely varying, it would be reasonable to lower this bound further.

It is always a good idea to include a margin of safety when it is impossible to verify accuracy assumptions.

The program ADAPQR43 implements Simpson's Adaptive quadrature method.

Illustration The graph of the function $f(x) = (100/x^2)\sin(10/x)$ for x in $[1, 3]$ is shown in Figure 4.11. Using the program ADAPQR43 with tolerance 10^{-4} to approximate $\int_1^3 f(x)dx$ produces -1.426014, a result that is accurate to within 1.1×10^{-5}. The approximation required that Simpson's rule with $n = 4$ be performed on the 23 subintervals whose endpoints are shown on the horizontal axis in Figure 4.11. The total number of functional evaluations required for this approximation using ADAPQR43 is 93.

Figure 4.11

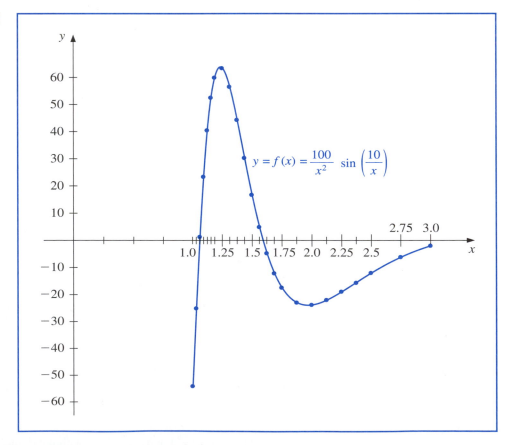

The largest value of h for which the standard Composite Simpson's rule gives 10^{-4} accuracy is $h = 1/88$. This application requires 177 function evaluations, nearly twice as many as Adaptive quadrature. ☐

The MATLAB function quad is based on an adaptive Simpson's method that is even more efficient than the method we have described. The call

```
[q,fcnt] = quad(f,a,b,tol)
```

gives the user the ability to select an error tolerance and obtain both the approximation and the number of function evaluations the method required to obtain that accuracy.

For our example we define the function, limits of integration, and the accuracy of 0.00001 with the following code.

```
f = inline('(100./x.^2).*sin(10./x)','x')
a=1, b=3
[q,fcnt]=quad(f,a,b,0.00001)
```

MATLAB responds with the approximation, q, and the number of function evaluations fcnt:

$$q = -1.426024570836794 \quad \text{and} \quad fcnt = 85$$

EXERCISE SET 4.6

1. For the following integrals, compute the Simpson's rule approximations $S(a, b)$, $S(a, (a+b)/2)$, and $S((a + b)/2, b)$, and verify the estimate given in the Adaptive Quadrature Error Estimate result on page 141.

   a. $\displaystyle\int_1^{1.5} x^2 \ln x \, dx$

   b. $\displaystyle\int_0^1 x^2 e^{-x} dx$

   c. $\displaystyle\int_0^{0.35} \frac{2}{x^2 - 4} dx$

   d. $\displaystyle\int_0^{\pi/4} x^2 \sin x \, dx$

   e. $\displaystyle\int_0^{\pi/4} e^{3x} \sin 2x \, dx$

   f. $\displaystyle\int_1^{1.6} \frac{2x}{x^2 - 4} dx$

   g. $\displaystyle\int_3^{3.5} \frac{x}{\sqrt{x^2 - 4}} dx$

   h. $\displaystyle\int_0^{\pi/4} (\cos x)^2 dx$

2. Use Adaptive quadrature to find approximations to within 10^{-3} for the integrals in Exercise 1. Do not use a computer program to generate these results.

3. Use Adaptive quadrature to approximate the following integrals to within 10^{-5}.

   a. $\displaystyle\int_1^3 e^{2x} \sin 3x \, dx$

   b. $\displaystyle\int_1^3 e^{3x} \sin 2x \, dx$

   c. $\displaystyle\int_0^5 [2x \cos(2x) - (x - 2)^2] dx$

   d. $\displaystyle\int_0^5 [4x \cos(2x) - (x - 2)^2] dx$

4. Use Composite Simpson's rule with $n = 4, 6, 8, \ldots$, until successive approximations to the following integrals agree to within 10^{-6}. Determine the number of nodes required. Use Adaptive quadrature

to approximate the integral to within 10^{-6} and count the number of nodes. Did Adaptive quadrature produce any improvement?

a. $\displaystyle\int_0^\pi x \cos x^2 dx$ **b.** $\displaystyle\int_0^\pi x \sin x^2 dx$

c. $\displaystyle\int_0^\pi x^2 \cos x \, dx$ **d.** $\displaystyle\int_0^\pi x^2 \sin x \, dx$

5.   Sketch the graphs of $\sin(1/x)$ and $\cos(1/x)$ on $[0.1, 2]$. Use Adaptive quadrature to approximate the integrals

$$\int_{0.1}^2 \sin\frac{1}{x}dx \quad \text{and} \quad \int_{0.1}^2 \cos\frac{1}{x}dx$$

to within 10^{-3}.

6.   Let $T(a, b)$ and $T(a, \frac{a+b}{2}) + T(\frac{a+b}{2}, b)$ be the single and double applications of the Trapezoidal rule to $\int_a^b f(x)dx$. Determine a relationship between

$$\left| T(a, b) - \left(T\left(a, \frac{a+b}{2}\right) + T\left(\frac{a+b}{2}, b\right) \right) \right|$$

and

$$\left| \int_a^b f(x)dx - \left(T\left(a, \frac{a+b}{2}\right) + T\left(\frac{a+b}{2}, b\right) \right) \right|.$$

7.   The differential equation

$$mu''(t) + ku(t) = F_0 \cos \omega t$$

describes a spring-mass system with mass m, spring constant k, and no applied damping. The term $F_0 \cos \omega t$ describes a periodic external force applied to the system. The solution to the equation when the system is initially at rest $(u'(0) = u(0) = 0)$ is

$$u(t) = \frac{2F_0}{m(\omega_0^2 - \omega^2)} \sin\frac{(\omega_0 - \omega)}{2} t \sin\frac{(\omega_0 + \omega)}{2} t, \quad \text{where} \quad \omega_0 = \sqrt{\frac{k}{m}} \neq \omega.$$

Sketch the graph of u when $m = 1$, $k = 9$, $F_0 = 1$, $\omega = 2$, and $t \in [0, 2\pi]$. Approximate $\int_0^{2\pi} u(t)dt$ to within 10^{-4}.

8.   If the term $cu'(t)$ is added to the left side of the motion equation in Exercise 7, the resulting differential equation describes a spring-mass system that is damped with damping constant c. The solution to this equation when the solution is initially at rest is

$$u(t) = c_1 e^{r_1 t} + c_2 e^{r_2 t} + \frac{F_0}{\sqrt{m^2(\omega_0^2 - \omega^2)^2 + c^2\omega^2}} \cos(\omega t - \delta),$$

where

$$\delta = \arctan\left(\frac{c\omega}{m(\omega_0^2 - \omega^2)}\right), \qquad r_1 = \frac{-c + \sqrt{c^2 - 4\omega_0^2 m^2}}{2m},$$

and

$$r_2 = \frac{-c - \sqrt{c^2 - 4\omega_0^2 m^2}}{2m}.$$

Sketch the graph of u when $m = 1$, $k = 9$, $F_0 = 1$, $c = 10$, $\omega = 2$, and $t \in [0, 2\pi]$. Approximate $\int_0^{2\pi} u(t)dt$ to within 10^{-4}.

9. The study of light diffraction at a rectangular aperture involves the Fresnel integrals

$$c(t) = \int_0^t \cos\frac{\pi}{2}w^2\,dw \quad \text{and} \quad s(t) = \int_0^t \sin\frac{\pi}{2}w^2\,dw.$$

Construct a table of values for $c(t)$ and $s(t)$ that is accurate to within 10^{-4} for values of $t = 0.1, 0.2, \ldots, 1.0$.

4.7 Multiple Integrals

The techniques discussed in the previous sections can be modified in a straightforward manner for use in the approximation of multiple integrals. Let us first consider the double integral

$$\iint_R f(x, y)\,dA,$$

where R is a rectangular region in the plane:

$$R = \{(x, y) \mid a \le x \le b, \quad c \le y \le d\}$$

for some constants a, b, c, and d (see Figure 4.12).

Figure 4.12

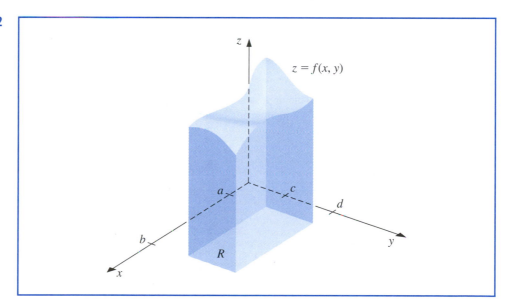

Illustration Writing the double integral as an iterated integral gives

$$\iint_R f(x, y)\,dA = \int_a^b \left(\int_c^d f(x, y)\,dy \right) dx.$$

To simplify notation, let $k = (d-c)/2$ and $h = (b-a)/2$. Apply the Composite Trapezoidal rule to the interior integral to obtain

$$\int_c^d f(x, y)\,dy \approx \frac{k}{2}\left[f(x, c) + f(x, d) + 2f\left(x, \frac{c+d}{2}\right) \right].$$

This approximation is of order $O((d-c)^3)$. Then apply the Composite Trapezoidal rule again to approximate the integral of this function of x:

$$\int_a^b \left(\int_c^d f(x,y)dy \right) dx \approx \int_a^b \left(\frac{d-c}{4} \right) \left[f(x,c) + 2f\left(x, \frac{c+d}{2} \right) + f(d) \right] dx$$

$$= \frac{b-a}{4} \left(\frac{d-c}{4} \right) \left[f(a,c) + 2f\left(a, \frac{c+d}{2} \right) + f(a,d) \right]$$

$$+ \frac{b-a}{4} \left(2\left(\frac{d-c}{4} \right) \left[f\left(\frac{a+b}{2}, c \right) \right. \right.$$

$$\left. + 2f\left(\frac{a+b}{2}, \frac{c+d}{2} \right) + f\left(\frac{a+b}{2}, d \right) \right] \right)$$

$$+ \frac{b-a}{4} \left(\frac{d-c}{4} \right) \left[f(b,c) + 2f\left(b, \frac{c+d}{2} \right) + f(b,d) \right]$$

$$= \frac{(b-a)(d-c)}{16} \left[f(a,c) + f(a,d) + f(b,c) + f(b,d) \right.$$

$$+ 2\left(f\left(\frac{a+b}{2}, c \right) + f\left(\frac{a+b}{2}, d \right) + f\left(a, \frac{c+d}{2} \right) \right.$$

$$\left. + f\left(b, \frac{c+d}{2} \right) \right) + 4f\left(\frac{a+b}{2}, \frac{c+d}{2} \right) \right]$$

This approximation is of order $O((b-a)(d-c)[(b-a)^2 + (d-c)^2])$. Figure 4.13 shows a grid with the number of functional evaluations at each of the nodes used in the approximation. ☐

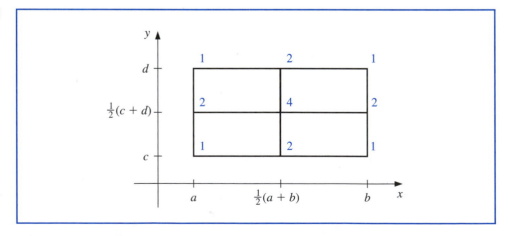

Figure 4.13

We will now employ the Composite Simpson's rule to illustrate the approximation technique, although any other approximation formula could be used without major modifications.

Suppose that even integers n and m are chosen to determine the step sizes $h = (b-a)/n$ and $k = (d-c)/m$. We first write the double integral as an iterated integral,

$$\iint_R f(x, y)\, dA = \int_a^b \left(\int_c^d f(x, y)dy \right) dx,$$

and use the Composite Simpson's rule to approximate

$$\int_c^d f(x, y)dy,$$

treating x as a constant. Let $y_j = c + jk$ for each $j = 0, 1, \ldots, m$. Then

$$\int_c^d f(x, y)dy = \frac{k}{3}\left[f(x, y_0) + 2\sum_{j=1}^{(m/2)-1} f(x, y_{2j}) + 4\sum_{j=1}^{m/2} f(x, y_{2j-1}) + f(x, y_m) \right]$$
$$- \frac{(d-c)k^4}{180}\frac{\partial^4 f}{\partial y^4}(x, \mu)$$

for some μ in (c, d). Thus

$$\int_a^b \int_c^d f(x, y)dy\, dx = \frac{k}{3}\int_a^b f(x, y_0)dx + \frac{2k}{3}\sum_{j=1}^{(m/2)-1}\int_a^b f(x, y_{2j})dx$$
$$+ \frac{4k}{3}\sum_{j=1}^{m/2}\int_a^b f(x, y_{2j-1})dx + \frac{k}{3}\int_a^b f(x, y_m)dx$$
$$- \frac{(d-c)k^4}{180}\int_a^b \frac{\partial^4 f}{\partial y^4}(x, \mu)dx.$$

Composite Simpson's rule is now employed on each of the first four integrals in this equation. Let $x_i = a + ih$ for each $i = 0, 1, \ldots, n$. Then for each $j = 0, 1, \ldots, m$, we have

$$\int_a^b f(x, y_j)dx = \frac{h}{3}\left[f(x_0, y_j) + 2\sum_{i=1}^{(n/2)-1} f(x_{2i}, y_j) + 4\sum_{i=1}^{n/2} f(x_{2i-1}, y_j) + f(x_n, y_j) \right]$$
$$- \frac{(b-a)h^4}{180}\frac{\partial^4 f}{\partial x^4}(\xi_j, y_j)$$

for some ξ_j in (a, b).

Substituting the Composite Simpson's rule approximations into this equation gives the approximation as

$$
\int_a^b \int_c^d f(x, y)dy\, dx \approx \frac{hk}{9} \Bigg\{ \Bigg[f(x_0, y_0) + 2 \sum_{i=1}^{(n/2)-1} f(x_{2i}, y_0)
$$

$$
+ 4 \sum_{i=1}^{n/2} f(x_{2i-1}, y_0) + f(x_n, y_0) \Bigg]
$$

$$
+ 2 \Bigg[\sum_{j=1}^{(m/2)-1} f(x_0, y_{2j}) + 2 \sum_{j=1}^{(m/2)-1} \sum_{i=1}^{(n/2)-1} f(x_{2i}, y_{2j})
$$

$$
+ 4 \sum_{j=1}^{(m/2)-1} \sum_{i=1}^{n/2} f(x_{2i-1}, y_{2j}) + \sum_{j=1}^{(m/2)-1} f(x_n, y_{2j}) \Bigg]
$$

$$
+ 4 \Bigg[\sum_{j=1}^{m/2} f(x_0, y_{2j-1}) + 2 \sum_{j=1}^{m/2} \sum_{i=1}^{(n/2)-1} f(x_{2i}, y_{2j-1})
$$

$$
+ 4 \sum_{j=1}^{m/2} \sum_{i=1}^{n/2} f(x_{2i-1}, y_{2j-1}) + \sum_{j=1}^{m/2} f(x_n, y_{2j-1}) \Bigg]
$$

$$
+ \Bigg[f(x_0, y_m) + 2 \sum_{i=1}^{(n/2)-1} f(x_{2i}, y_m) + 4 \sum_{i=1}^{n/2} f(x_{2i-1}, y_m)
$$

$$
+ f(x_n, y_m) \Bigg] \Bigg\}.
$$

The error term, E, is given by

$$
E = \frac{-k(b-a)h^4}{540} \Bigg[\frac{\partial^4 f}{\partial x^4}(\xi_0, y_0) + 2 \sum_{j=1}^{(m/2)-1} \frac{\partial^4 f}{\partial x^4}(\xi_{2j}, y_{2j}) + 4 \sum_{j=1}^{m/2} \frac{\partial^4 f}{\partial x^4}(\xi_{2j-1}, y_{2j-1})
$$

$$
+ \frac{\partial^4 f}{\partial x^4}(\xi_m, y_m) \Bigg] - \frac{(d-c)k^4}{180} \int_a^b \frac{\partial^4 f}{\partial y^4}(x, \mu)dx.
$$

If the fourth partial derivatives $\partial^4 f/\partial x^4$ and $\partial^4 f/\partial y^4$ are continuous, the Intermediate Value Theorem and Mean Value Theorem for Integrals can be used to show that the error formula can be simplified to

$$
E = \frac{-(d-c)(b-a)}{180} \Bigg[h^4 \frac{\partial^4 f}{\partial x^4}(\bar{\eta}, \bar{\mu}) + k^4 \frac{\partial^4 f}{\partial y^4}(\hat{\eta}, \hat{\mu}) \Bigg]
$$

for some $(\bar{\eta}, \bar{\mu})$ and $(\hat{\eta}, \hat{\mu})$ in R.

Example 1 Use Composite Simpson's rule with $n = 4$ and $m = 2$ to approximate

$$
\int_{1.4}^{2.0} \int_{1.0}^{1.5} \ln(x + 2y)dy\, dx,
$$

Solution   The step sizes for this application are $h = (2.0 - 1.4)/4 = 0.15$ and $k = (1.5 - 1.0)/2 = 0.25$. The region of integration R is shown in Figure 4.14, together with the nodes (x_i, y_j), where $i = 0, 1, 2, 3, 4$ and $j = 0, 1, 2$. It also shows the coefficients $w_{i,j}$ of $f(x_i, y_i) = \ln(x_i + 2y_i)$ in the sum that gives the Composite Simpson's rule approximation to the integral.

Figure 4.14

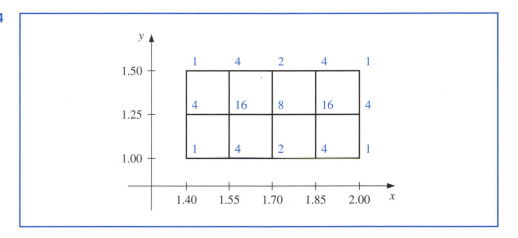

The approximation is

$$\int_{1.4}^{2.0} \int_{1.0}^{1.5} \ln(x + 2y) \, dy \, dx \approx \frac{(0.15)(0.25)}{9} \sum_{i=0}^{4} \sum_{j=0}^{2} w_{i,j} \ln(x_i + 2y_j) = 0.4295524387.$$

We have

$$\frac{\partial^4 f}{\partial x^4}(x, y) = \frac{-6}{(x + 2y)^4} \quad \text{and} \quad \frac{\partial^4 f}{\partial y^4}(x, y) = \frac{-96}{(x + 2y)^4},$$

and the maximum values of the absolute values of these partial derivatives occur on R when $x = 1.4$ and $y = 1.0$. So the error is bounded by

$$|E| \leq \frac{(0.5)(0.6)}{180} \left[(0.15)^4 \max_{(x,y)\text{in}R} \frac{6}{(x + 2y)^4} + (0.25)^4 \max_{(x,y)\text{in}R} \frac{96}{(x + 2y)^4} \right] \leq 4.72 \times 10^{-6}.$$

The actual value of the integral to ten decimal places is

$$\int_{1.4}^{2.0} \int_{1.0}^{1.5} \ln(x + 2y) \, dy \, dx = 0.4295545265,$$

so the approximation is accurate to within 2.1×10^{-6}.   ∎

The same techniques can be applied for the approximation of triple integrals, as well as higher integrals for functions of more than three variables. The number of functional evaluations required for the approximation is the product of the number required when the method is applied to each variable.

Gaussian Quadrature for Double Integral Approximation

The reduced calculation makes it generally worthwhile to apply Gaussian quadrature rather than a Simpson's technique when approximating double integrals.

To reduce the number of functional evaluations, more efficient methods such as Gaussian quadrature, Romberg integration, or Adaptive quadrature can be incorporated in place of Simpson's formula. The following example illustrates the use of Gaussian quadrature for the integral considered in Example 1. In this example, the Gaussian nodes $r_{3,j}$ and coefficients $c_{3,j}$, for $j = 1, 2, 3$ are given in Table 4.7 on page 135.

Example 2   Use Gaussian quadrature with $n = 3$ in both dimensions to approximate the integral

$$\int_{1.4}^{2.0} \int_{1.0}^{1.5} \ln(x + 2y)dy\,dx.$$

Solution   Before employing Gaussian quadrature to approximate this integral, we need to transform the region of integration

$$R = \{(x, y) \mid 1.4 \le x \le 2.0, 1.0 \le y \le 1.5\} \quad \text{into}$$

$$\hat{R} = \{(u, v) \mid -1 \le u \le 1, -1 \le v \le 1\}.$$

The linear transformations that accomplish this are

$$u = \frac{1}{2.0 - 1.4}(2x - 1.4 - 2.0) \quad \text{and} \quad v = \frac{1}{1.5 - 1.0}(2y - 1.0 - 1.5),$$

or, equivalently, $x = 0.3u + 1.7$ and $y = 0.25v + 1.25$. Employing this change of variables gives an integral on which Gaussian quadrature can be applied:

$$\int_{1.4}^{2.0} \int_{1.0}^{1.5} \ln(x + 2y)dy\,dx = 0.075 \int_{-1}^{1} \int_{-1}^{1} \ln(0.3u + 0.5v + 4.2)\,dv\,du.$$

The Gaussian quadrature formula for $n = 3$ in both u and v requires that we use the nodes

$$u_1 = v_1 = r_{3,2} = 0, \quad u_0 = v_0 = r_{3,1} = -0.7745966692, \quad \text{and}$$

$$u_2 = v_2 = r_{3,3} = 0.7745966692.$$

The associated weights are $c_{3,2} = 0.\overline{8}$ and $c_{3,1} = c_{3,3} = 0.\overline{5}$. (These are also given in Table 4.7 on page 135.) The resulting approximation is

$$\int_{1.4}^{2.0} \int_{1.0}^{1.5} \ln(x + 2y)dy\,dx \approx 0.075 \sum_{i=1}^{3} \sum_{j=1}^{3} c_{3,i}c_{3,j} \ln(0.3r_{3,i} + 0.5r_{3,j} + 4.2)$$

$$= 0.4295545313.$$

Although this result requires only 9 functional evaluations compared to 15 for the Composite Simpson's rule considered in Example 1, it is accurate to within 4.8×10^{-9}, compared to 2.1×10^{-6} accuracy in Example 1.   ▪

Non-Rectangular Regions

The use of approximation methods for double integrals is not limited to integrals with rectangular regions of integration. The techniques previously discussed can be modified to approximate double integrals with variable inner limits—that is, integrals of the form

$$\int_{a}^{b} \int_{c(x)}^{d(x)} f(x, y)dy\,dx.$$

For this type of integral, we begin as before by applying Simpson's rule to integrate with respect to both variables. The step size for the variable x is $h = (b-a)/2$, but the step size $k(x)$ for y varies with x (see Figure 4.15):

$$k(x) = \frac{d(x) - c(x)}{2}.$$

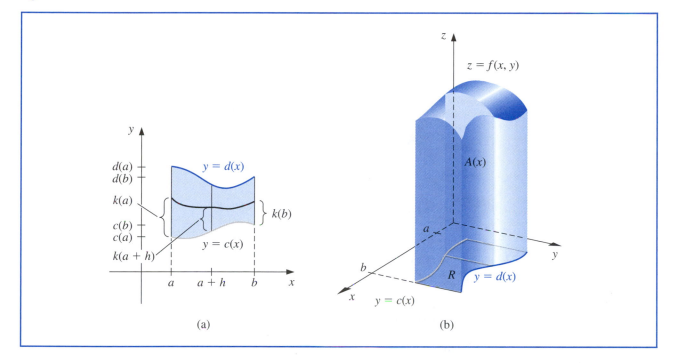

(a)   (b)

Consequently

$$\int_a^b \int_{c(x)}^{d(x)} f(x,y)\,dy\,dx \approx \int_a^b \frac{k(x)}{3}[f(x,c(x)) + 4f(x,c(x)+k(x)) + f(x,d(x))]dx$$

$$\approx \frac{h}{3}\left\{ \frac{k(a)}{3}[f(a,c(a)) + 4f(a,c(a)+k(a)) + f(a,d(a))] \right.$$

$$+ \frac{4k(a+h)}{3}[f(a+h,c(a+h)) + 4f(a+h,c(a+h)$$

$$+ k(a+h)) + f(a+h,d(a+h))]$$

$$+ \left. \frac{k(b)}{3}[f(b,c(b)) + 4f(b,c(b)+k(b)) + f(b,d(b))] \right\}.$$

The program DINTGL44 implements Composite Simpson's Rule for Double Integrals.

The program DINTGL44 applies the Composite Simpson's rule to a double integral in this form on a general region, and is also appropriate, of course, when $c(x) \equiv c$ and $d(x) \equiv d$ are constants.

Double Integrals using Gaussian Quadrature

To apply Gaussian quadrature to the double integral

$$\int_a^b \int_{c(x)}^{d(x)} f(x, y)dy\, dx$$

first requires transforming, for each x in $[a, b]$, the variable y in the interval $[c(x), d(x)]$ into the variable t in the interval $[-1, 1]$. This linear transformation gives

$$f(x, y) = f\left(x, \frac{(d(x) - c(x))t + d(x) + c(x)}{2}\right) \quad \text{and} \quad dy = \frac{d(x) - c(x)}{2}dt.$$

Then, for each x in $[a, b]$, we apply Gaussian quadrature to the resulting integral

$$\int_{c(x)}^{d(x)} f(x, y)dy = \int_{-1}^1 f\left(x, \frac{(d(x) - c(x))t + d(x) + c(x)}{2}\right)dt$$

to produce

$$\int_a^b \int_{c(x)}^{d(x)} f(x, y)dy\, dx \approx \int_a^b \frac{d(x) - c(x)}{2} \sum_{j=1}^n c_{n,j} f\left(x, \frac{(d(x) - c(x))r_{n,j} + d(x) + c(x)}{2}\right)dx.$$

The constants $r_{n,j}$ in this formula are the roots of the nth Legendre polynomial. These are tabulated, for $n = 2, 3, 4,$ and 5, in Table 4.7 on page 135. The values of $c_{n,j}$ are also given in that table.

After this, the interval $[a, b]$ is transformed to $[-1, 1]$, and Gaussian quadrature is applied to approximate the integral on the right side of this equation. The program DGQINT45 uses this technique.

> The program DGQINT45 implements Gaussian Quadrature for Double Integrals.

Illustration The volume of the solid in Figure 4.16 is approximated by applying Composite Simpson's rule using program DINTGL44 with $n = m = 10$ to

$$\int_{0.1}^{0.5} \int_{x^3}^{x^2} e^{y/x}dy\, dx.$$

This requires 121 evaluations of the function $f(x, y) = e^{y/x}$ and produces the value 0.0333054, which approximates the volume of the solid shown in Figure 4.16 to nearly seven decimal places. Applying the Gaussian quadrature program DGQINT45 with $n = m = 5$ requires only 25 function evaluations and gives the approximation 0.03330556611, which is accurate to 11 decimal places. ☐

Figure 4.16

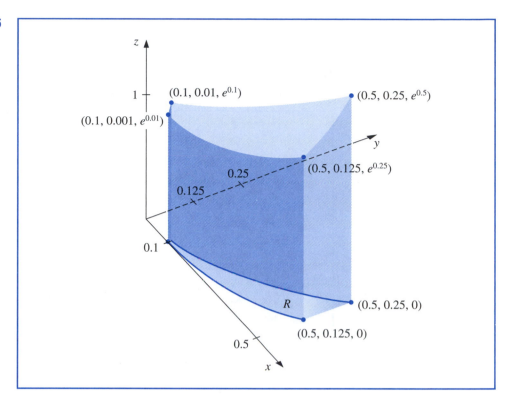

Triple Integral Approximation

The program TINTGL46 implements Gaussian Quadrature for Triple Integrals.

Triple integrals of the form

$$\int_a^b \int_{c(x)}^{d(x)} \int_{\alpha(x,y)}^{\beta(x,y)} f(x, y, z) dz \, dy \, dx$$

are approximated in a manner similar to double integration. Because of the number of calculations involved, Gaussian quadrature is the method of choice.

The following Illustration requires the evaluation of four triple integrals.

Illustration   The center of a mass of a solid region D with density function σ occurs at

The reduced calculation makes it almost always worthwhile to apply Gaussian quadrature rather than a Simpson's technique when approximating triple or higher integrals.

$$(\bar{x}, \bar{y}, \bar{z}) = \left(\frac{M_{yz}}{M}, \frac{M_{xz}}{M}, \frac{M_{xy}}{M} \right),$$

where

$$M_{yz} = \iiint_D x\sigma(x, y, z) dV, \quad M_{xz} = \iiint_D y\sigma(x, y, z) dV, \quad \text{and}$$

$$M_{xy} = \iiint_D z\sigma(x, y, z) dV$$

are the moments about the coordinate planes and the mass of D is

$$M = \iiint_D \sigma(x, y, z) dV.$$

The solid shown in Figure 4.17 is bounded by the upper nappe of the cone $z^2 = x^2 + y^2$ and the plane $z = 2$. Suppose that this solid has the density function

$$\sigma(x, y, z) = \sqrt{x^2 + y^2}.$$

Figure 4.17

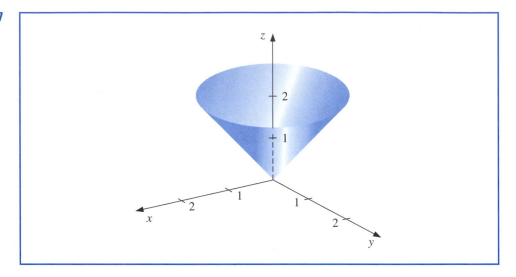

Applying the program TINTGL46 with $n = m = p = 5$ requires 125 function evaluations per integral and gives the following approximations:

$$M = \int_{-2}^{2} \int_{-\sqrt{4-x^2}}^{\sqrt{4-x^2}} \int_{\sqrt{x^2+y^2}}^{2} \sqrt{x^2 + y^2} \, dz \, dy \, dx$$

$$= 4 \int_{0}^{2} \int_{0}^{\sqrt{4-x^2}} \int_{\sqrt{x^2+y^2}}^{2} \sqrt{x^2 + y^2} \, dz \, dy \, dx \approx 8.37504476,$$

$$M_{yz} = \int_{-2}^{2} \int_{-\sqrt{4-x^2}}^{\sqrt{4-x^2}} \int_{\sqrt{x^2+y^2}}^{2} x \sqrt{x^2 + y^2} \, dz \, dy \, dx \approx -5.55111512 \times 10^{-17},$$

$$M_{xz} = \int_{-2}^{2} \int_{-\sqrt{4-x^2}}^{\sqrt{4-x^2}} \int_{\sqrt{x^2+y^2}}^{2} y \sqrt{x^2 + y^2} \, dz \, dy \, dx \approx -8.01513675 \times 10^{-17},$$

$$M_{xy} = \int_{-2}^{2} \int_{-\sqrt{4-x^2}}^{\sqrt{4-x^2}} \int_{\sqrt{x^2+y^2}}^{2} z \sqrt{x^2 + y^2} \, dz \, dy \, dx \approx 13.40038156.$$

This implies that the approximate location of the center of mass is

$$(\overline{x}, \overline{y}, \overline{z}) = (0, 0, 1.60003701).$$

These integrals are quite easy to evaluate directly. If you do this, you will find that the exact center of mass occurs at $(0, 0, 1.6)$. ☐

MATLAB has commands

- `dblquad` for computing a double integral over a rectangle,

- `quad2d` for computing a double integral over a more general region, and

- `triplequad` for computing a triple integral over a rectangular parallelepiped.

All these commands can be used to compute the integral to within a specified tolerance. The input and output of these commands is similar to that of `quad`, but you should consult the help facility of MATLAB before using them.

EXERCISE SET 4.7

1. Use Composite Simpson's rule for double integrals with $n = m = 4$ to approximate the following double integrals. Compare the results to the exact answer.

   a. $\displaystyle\int_{2.1}^{2.5}\int_{1.2}^{1.4} xy^2 dy\, dx$

   b. $\displaystyle\int_{0}^{0.5}\int_{0}^{0.5} e^{y-x} dy\, dx$

   c. $\displaystyle\int_{2}^{2.2}\int_{x}^{2x} (x^2 + y^3) dy\, dx$

   d. $\displaystyle\int_{1}^{1.5}\int_{0}^{x} (x^2 + \sqrt{y}) dy\, dx$

2. Find the smallest values for $n = m$ so that Composite Simpson's rule for double integrals can be used to approximate the integrals in Exercise 1 to within 10^{-6} of the actual value.

3. Use Composite Simpson's rule for double integrals with $n = 4, m = 8$ and $n = 8, m = 4$ and $n = m = 6$ to approximate the following double integrals. Compare the results to the exact answer.

   a. $\displaystyle\int_{0}^{\pi/4}\int_{\sin x}^{\cos x} (2y \sin x + \cos^2 x) dy\, dx$

   b. $\displaystyle\int_{1}^{e}\int_{1}^{x} \ln xy\, dy\, dx$

   c. $\displaystyle\int_{0}^{1}\int_{x}^{2x} (x^2 + y^3) dy\, dx$

   d. $\displaystyle\int_{0}^{1}\int_{x}^{2x} (y^2 + x^3) dy\, dx$

   e. $\displaystyle\int_{0}^{\pi}\int_{0}^{x} \cos x\, dy\, dx$

   f. $\displaystyle\int_{0}^{\pi}\int_{0}^{x} \cos y\, dy\, dx$

   g. $\displaystyle\int_{0}^{\pi/4}\int_{0}^{\sin x} \frac{1}{\sqrt{1 - y^2}} dy\, dx$

   h. $\displaystyle\int_{-\pi}^{3\pi/2}\int_{0}^{2\pi} (y \sin x + x \cos y) dy\, dx$

4. Find the smallest values for $n = m$ so that Composite Simpson's rule for double integrals can be used to approximate the integrals in Exercise 3 to within 10^{-6} of the actual value.

5. Use Gaussian quadrature for double integrals with $n = m = 2$ to approximate the integrals in Exercise 1 and compare the results to those obtained in Exercise 1.

6. Find the smallest values of $n = m$ so that Gaussian quadrature for double integrals may be used to approximate the integrals in Exercise 1 to within 10^{-6}. Do not continue beyond $n = m = 5$. Compare the number of functional evaluations required to the number required in Exercise 2.

7. Use Gaussian quadrature for double integrals with $n = m = 3$; $n = 3, m = 4$; $n = 4, m = 3$ and $n = m = 4$ to approximate the integrals in Exercise 3.

8. Use Gaussian quadrature for double integrals with $n = m = 5$ to approximate the integrals in Exercise 3. Compare the number of functional evaluations required to the number required in Exercise 4.

9. Use Gaussian quadrature for triple integrals with $n = m = p = 2$ to approximate the following triple integrals, and compare the results to the exact answer.

   a. $\displaystyle\int_{0}^{1}\int_{1}^{2}\int_{0}^{0.5} e^{x+y+z} dz\, dy\, dx$

   b. $\displaystyle\int_{0}^{1}\int_{x}^{1}\int_{0}^{y} y^2 z\, dz\, dy\, dx$

c. $\displaystyle\int_0^1 \int_{x^2}^x \int_{x-y}^{x+y} y\,dz\,dy\,dx$

d. $\displaystyle\int_0^1 \int_{x^2}^x \int_{x-y}^{x+y} z\,dz\,dy\,dx$

e. $\displaystyle\int_0^\pi \int_0^x \int_0^{xy} \frac{1}{y}\sin\frac{z}{y}\,dz\,dy\,dx$

f. $\displaystyle\int_0^1 \int_0^1 \int_{-xy}^{xy} e^{x^2+y^2}\,dz\,dy\,dx$

10. Repeat Exercise 9 using $n = m = p = 3$.

11. Use Composite Simpson's rule for double integrals with $n = m = 14$ and Gaussian quadrature for double integrals with $n = m = 4$ to approximate

$$\iint_R e^{-(x+y)}\,dA$$

for the region R in the plane bounded by the curves $y = x^2$ and $y = \sqrt{x}$.

12. Use Composite Simpson's rule for double integrals to approximate

$$\iint_R \sqrt{xy + y^2}\,dA,$$

where R is the region in the plane bounded by the lines $x + y = 6$, $3y - x = 2$, and $3x - y = 2$. First partition R into two regions, R_1 and R_2, on which Composite Simpson's rule for double integrals can be applied. Use $n = m = 6$ on both R_1 and R_2.

13. The area of the surface described by $z = f(x, y)$ for (x, y) in R is given by

$$\iint_R \sqrt{[f_x(x, y)]^2 + [f_y(x, y)]^2 + 1}\,dA.$$

Find an approximation to the area of the surface on the hemisphere $x^2 + y^2 + z^2 = 9$, $z \geq 0$ that lies above the region $R = \{(x, y) \mid 0 \leq x \leq 1,\ 0 \leq y \leq 1\}$, using each program.

a. DINTGL44 with $n = m = 8$ **b.** DGQINT45 with $n = m = 4$

14. A plane lamina is defined to be a thin sheet of continuously distributed mass. If σ is a function describing the density of a lamina having the shape of a region R in the xy-plane, then the center of the mass of the lamina $(\bar{x}, \bar{y})$ is defined by

$$\bar{x} = \frac{\iint_R x\sigma(x, y)\,dA}{\iint_R \sigma(x, y)\,dA}, \qquad \bar{y} = \frac{\iint_R y\sigma(x, y)\,dA}{\iint_R \sigma(x, y)\,dA}.$$

Find the center of mass of the lamina described by $R = \{(x, y) \mid 0 \leq x \leq 1,\ 0 \leq y \leq \sqrt{1 - x^2}\}$ with the density function $\sigma(x, y) = e^{-(x^2+y^2)}$ using each program.

a. DINTGL44 with $n = m = 14$ **b.** DGQINT45 with $n = m = 5$

15. Use Gaussian quadrature for triple integrals with $n = m = p = 4$ to approximate

$$\iiint_S xy\sin(yz)\,dV,$$

where S is the solid bounded by the coordinate planes and the planes $x = \pi$, $y = \pi/2$, $z = \pi/3$. Compare this approximation to the exact result.

16. Use Gaussian quadrature for triple integrals with $n = m = p = 5$ to approximate

$$\iiint_S \sqrt{xyz}\,dV,$$

when S is the region in the first octant bounded by the cylinder $x^2 + y^2 = 4$, the sphere $x^2 + y^2 + z^2 = 4$, and the plane $x + y + z = 8$. How many functional evaluations are required for the approximation?

4.8   Improper Integrals

Improper integrals result when the notion of integration is extended either to an interval of integration on which the function is unbounded or to an interval with one or both infinite endpoints. In either circumstance, the normal rules of integral approximation must be modified.

Left Endpoint Singularity

We will first handle the situation when the integrand is unbounded at the left endpoint of the interval of integration, as shown in Figure 4.18. In this case, we say that f has a **singularity** at the endpoint a. We will then show that, by a suitable manipulation, the other improper integrals can be reduced to problems of this form.

Figure 4.18

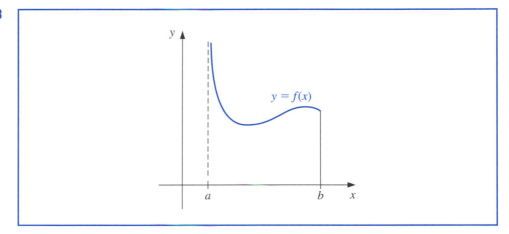

The improper integral with a singularity at the left endpoint,

$$\int_a^b \frac{1}{(x-a)^p}dx,$$

converges if and only if $0 < p < 1$, and in this case,

$$\int_a^b \frac{1}{(x-a)^p}dx = \frac{(x-a)^{1-p}}{1-p}\Big|_a^b = \frac{(b-a)^{1-p}}{1-p}.$$

Example 1   Show that the improper integral $\int_0^1 \frac{1}{\sqrt{x}}dx$ converges but $\int_0^1 \frac{1}{x^2}dx$ diverges.

Solution   For the first integral we have

$$\int_0^1 \frac{1}{\sqrt{x}}dx = \lim_{M\to 0^+}\int_M^1 x^{-1/2}dx = \lim_{M\to 0^+} 2x^{1/2}\Big|_{x=M}^{x=1} = 2-0 = 2,$$

but the second integral

$$\int_0^1 \frac{1}{x^2}dx = \lim_{M\to 0^+}\int_M^1 x^{-2}dx = \lim_{M\to 0^+} -x^{-1}\Big|_{x=M}^{x=1}$$

is unbounded.   ■

If f is a function that can be written in the form

$$f(x) = \frac{g(x)}{(x-a)^p},$$

where $0 < p < 1$ and g is continuous on $[a, b]$, then the improper integral

$$\int_a^b f(x)dx$$

also exists. We will approximate this integral using Composite Simpson's rule. If $g \in C^5[a, b]$, we can construct the fourth Taylor polynomial, $P_4(x)$, for g about a:

$$P_4(x) = g(a) + g'(a)(x-a) + \frac{g''(a)}{2!}(x-a)^2 + \frac{g'''(a)}{3!}(x-a)^3 + \frac{g^{(4)}(a)}{4!}(x-a)^4,$$

and write

$$\int_a^b f(x)dx = \int_a^b \frac{g(x) - P_4(x)}{(x-a)^p}dx + \int_a^b \frac{P_4(x)}{(x-a)^p}dx.$$

We can exactly determine the value of

$$\int_a^b \frac{P_4(x)}{(x-a)^p}dx = \sum_{k=0}^4 \int_a^b \frac{g^{(k)}(a)}{k!}(x-a)^{k-p}dx = \sum_{k=0}^4 \frac{g^{(k)}(a)}{k!(k+1-p)}(b-a)^{k+1-p}.$$

$$(4.11)$$

This is generally the dominant portion of the approximation, especially when the Taylor polynomial $P_4(x)$ agrees closely with the function g throughout the interval $[a, b]$.

To approximate the integral of f, we need to add this value to the approximation of

$$\int_a^b \frac{g(x) - P_4(x)}{(x-a)^p}dx.$$

To determine this, we first define

$$G(x) = \begin{cases} \frac{g(x) - P_4(x)}{(x-a)^p}, & \text{if } a < x \le b, \\ 0, & \text{if } x = a. \end{cases}$$

Since $0 < p < 1$ and $P_4^{(k)}(a)$ agrees with $g^{(k)}(a)$ for each $k = 0, 1, 2, 3, 4$, we have $G \in C^4[a, b]$. This implies that Composite Simpson's rule can be applied to approximate the integral of G on $[a, b]$. Adding this approximation to the value from Eq. (4.11) gives an approximation to the improper integral of f on $[a, b]$, within the accuracy of the Composite Simpson's rule approximation.

Example 2   Use Composite Simpson's rule with $h = 0.25$ to approximate the value of the improper integral

$$\int_0^1 \frac{e^x}{\sqrt{x}}dx.$$

Solution   In Example 1 we showed that $\int_0^1 \frac{1}{\sqrt{x}}dx$ is convergent, so the integral in this example is also convergent.

The fourth Taylor polynomial for e^x about $x = 0$ is

$$P_4(x) = 1 + x + \frac{x^2}{2} + \frac{x^3}{6} + \frac{x^4}{24},$$

so the dominant portion of the approximation to $\int_0^1 \frac{e^x}{\sqrt{x}} dx$ is

$$\int_0^1 \frac{P_4(x)}{\sqrt{x}} dx = \int_0^1 \left(x^{-1/2} + x^{1/2} + \frac{1}{2}x^{3/2} + \frac{1}{6}x^{5/2} + \frac{1}{24}x^{7/2} \right) dx$$

$$= \lim_{M \to 0^+} \left[2x^{1/2} + \frac{2}{3}x^{3/2} + \frac{1}{5}x^{5/2} + \frac{1}{21}x^{7/2} + \frac{1}{108}x^{9/2} \right]_M^1$$

$$= 2 + \frac{2}{3} + \frac{1}{5} + \frac{1}{21} + \frac{1}{108} \approx 2.9235450.$$

For the second portion of the approximation to $\int_0^1 \frac{e^x}{\sqrt{x}} dx$ we need to approximate $\int_0^1 G(x) dx$, where

$$G(x) = \begin{cases} \dfrac{1}{\sqrt{x}}(e^x - P_4(x)), & \text{if } 0 < x \leq 1, \\ 0, & \text{if } x = 0. \end{cases}$$

Table 4.8

x	$G(x)$
0.00	0
0.25	0.0000170
0.50	0.0004013
0.75	0.0026026
1.00	0.0099485

Table 4.8 lists the values needed for the Composite Simpson's rule for this approximation. Using these data and the Composite Simpson's rule gives

$$\int_0^1 G(x) dx \approx \frac{0.25}{3}[0 + 4(0.0000170) + 2(0.0004013) + 4(0.0026026) + 0.0099485]$$

$$= 0.0017691.$$

Hence

$$\int_0^1 \frac{e^x}{\sqrt{x}} dx \approx 2.9235450 + 0.0017691 = 2.9253141.$$

This result is accurate to within the accuracy of the Composite Simpson's rule approximation for the function G. Because $|G^{(4)}(x)| < 1$ on $[0, 1]$, the error is bounded by

$$\frac{1 - 0}{180}(0.25)^4 = 0.0000217. \qquad \blacksquare$$

Right Endpoint Singularity

To approximate the improper integral with a singularity at the right endpoint, we could apply the technique we used previously but expand in terms of the right endpoint b instead of the left endpoint a. Alternatively, we can make the substitution $z = -x$, $dz = -dx$ to change the improper integral into one of the form

$$\int_a^b f(x) dx = \int_{-b}^{-a} f(-z) dz,$$

which has its singularity at the left endpoint. (See Figure 4.19.) We can now approximate $\int_{-b}^{-a} f(-z)\,dz$ as we did earlier, and this gives us our approximation for $\int_a^b f(x)dx$.

Figure 4.19

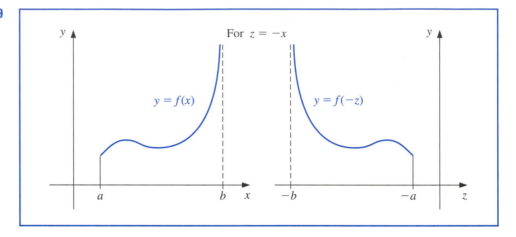

An improper integral with a singularity at c, where $a < c < b$, is treated as the sum of improper integrals with endpoint singularities since

$$\int_a^b f(x)dx = \int_a^c f(x)dx + \int_c^b f(x)dx.$$

Infinite Singularity

The other type of improper integral involves infinite limits of integration. A basic convergent integral of this type has the form

$$\int_a^\infty \frac{1}{x^p}dx,$$

for $p > 1$. This integral is converted to an integral with left-endpoint singularity by making the integration substitution

$$t = x^{-1}, \quad dt = -x^{-2}dx, \quad \text{so} \quad dx = -x^2 dt = -t^{-2}dt.$$

Then

$$\int_a^\infty \frac{1}{x^p}dx = \int_{1/a}^0 -\frac{t^p}{t^2}dt = \int_0^{1/a} \frac{1}{t^{2-p}}dt.$$

In a similar manner, the variable change $t = x^{-1}$ converts the improper integral $\int_a^\infty f(x)dx$ into one that has a left-endpoint singularity at zero:

$$\int_a^\infty f(x)dx = \int_0^{1/a} t^{-2} f\left(\frac{1}{t}\right) dt.$$

It can now be approximated using a quadrature formula of the type described earlier.

Example 3 Approximate the value of the improper integral

$$I = \int_1^\infty x^{-3/2} \sin \frac{1}{x} dx.$$

Solution We first make the variable change $t = x^{-1}$, which converts the infinite singularity into one with a left endpoint singularity. Then

$$dt = -x^{-2} dx, \quad \text{so} \quad dx = -x^2 dt = -\frac{1}{t^2} dt,$$

and

$$I = \int_{x=1}^{x=\infty} x^{-3/2} \sin \frac{1}{x} dx = \int_{t=1}^{t=0} \left(\frac{1}{t}\right)^{-3/2} \sin t \left(-\frac{1}{t^2} dt\right) = \int_0^1 \frac{\sin t}{t^{1/2}} dt.$$

The fourth Taylor polynomial, $P_4(t)$, for $\sin t$ about 0 is

$$P_4(t) = t - \frac{1}{6} t^3, \quad \text{so} \quad G(t) = \begin{cases} \dfrac{\sin t - t + \frac{1}{6} t^3}{t^{1/2}}, & \text{if } 0 < t \le 1 \\ 0, & \text{if } t = 0 \end{cases}$$

is in $C^4[0, 1]$, and we have

$$I = \int_0^1 t^{-1/2} \left(t - \frac{1}{6} t^3\right) dt + \int_0^1 \frac{\sin t - t + \frac{1}{6} t^3}{t^{1/2}} dt$$

$$= \left[\frac{2}{3} t^{3/2} - \frac{1}{21} t^{7/2}\right]_0^1 + \int_0^1 \frac{\sin t - t + \frac{1}{6} t^3}{t^{1/2}} dt$$

$$= 0.61904761 + \int_0^1 \frac{\sin t - t + \frac{1}{6} t^3}{t^{1/2}} dt.$$

The result from the Composite Simpson's rule with $n = 16$ for the remaining integral is 0.0014890097. This gives a final approximation of

$$I = 0.0014890097 + 0.61904761 = 0.62053661,$$

which is accurate to within 4.0×10^{-8}. ∎

The function `quadgk` can be used when the integrand has a singularity or when the limits of integration are infinite. Consider first the problem in Example 2. We define the integrand using the `M-file` called `ourfunction3.m` as

```
function Y = ourfunction3(x)
Y=exp(x)./sqrt(x)
```

which is presented to MATLAB using

```
F3=@ourfunction3
```

To obtain the MATLAB approximation we use

```
[q,errbnd]=quadgk(F3,0,1)
```

which gives the MATLAB response

$$q = 2.925303491814127 \quad \text{and} \quad errbnd = 2.561659218081046e - 013$$

For the integral in Example 3 we use an `M-file` called `ourfunction4` to define the integrand:

```
function Y = ourfunction4(x)
Y=1./(x.^1.5).*sin(1./x)
```

In the call to quadgk we use Inf to represent infinity. (If we wanted minus infinity we would use −Inf.)

```
[q,errbnd]=quadgk(F4,1,Inf)
```

The results are

$$q = 0.620536603446762 \quad \text{and} \quad errbnd = 5.247538514829841e - 017$$

EXERCISE SET 4.8

1. Use Composite Simpson's rule and the given values of n to approximate the following improper integrals.

    a. $\displaystyle\int_0^1 x^{-1/4} \sin x \, dx$ with $n = 4$ **b.** $\displaystyle\int_0^1 \frac{e^{2x}}{\sqrt[5]{x^2}} dx$ with $n = 6$

    c. $\displaystyle\int_1^2 \frac{\ln x}{(x-1)^{1/5}} dx$ with $n = 8$ **d.** $\displaystyle\int_0^1 \frac{\cos 2x}{x^{1/3}} dx$ with $n = 6$

2. Use the Composite Simpson's rule and the given values of n to approximate the following improper integrals.

    a. $\displaystyle\int_0^1 \frac{e^{-x}}{\sqrt{1-x}} dx$ with $n = 6$ **b.** $\displaystyle\int_0^2 \frac{xe^x}{\sqrt[3]{(x-1)^2}} dx$ with $n = 8$

3. Use the transformation $t = x^{-1}$ and then the Composite Simpson's rule and the given values of n to approximate the following improper integrals.

    a. $\displaystyle\int_1^\infty \frac{1}{x^2+9} dx$ with $n = 4$ **b.** $\displaystyle\int_1^\infty \frac{1}{1+x^4} dx$ with $n = 4$

    c. $\displaystyle\int_1^\infty \frac{\cos x}{x^3} dx$ with $n = 6$ **d.** $\displaystyle\int_1^\infty x^{-4} \sin x \, dx$ with $n = 6$

4. The improper integral $\int_0^\infty f(x)dx$ cannot be converted into an integral with finite limits using the substitution $t = 1/x$ because the limit at zero becomes infinite. The problem is resolved by first writing $\int_0^\infty f(x)dx = \int_0^1 f(x)dx + \int_1^\infty f(x)dx$. Apply this technique to approximate the following improper integrals to within 10^{-6}.

    a. $\displaystyle\int_0^\infty \frac{1}{1+x^4} dx$ **b.** $\displaystyle\int_0^\infty \frac{1}{(1+x^2)^3} dx$

5. Suppose a body of mass m is travelling vertically upward starting at the surface of the earth. If all resistance except gravity is neglected, the escape velocity v is given by

$$v^2 = 2gR \int_1^\infty z^{-2} \, dz, \quad \text{where } z = \frac{x}{R},$$

$R = 3960$ mi is the radius of the earth, and $g = 0.00609$ mi/s^2 is the force of gravity at the earth's surface. Approximate the escape velocity v.

4.9 Numerical Differentiation

At the beginning of this chapter we stated that derivative approximations are not as frequently needed as integral approximations. This is true for the approximation of single derivatives, but derivative approximation formulas are used extensively for approximating the solutions to ordinary and partial differential equations, a subject we consider in Chapters 11 and 12.

The derivative of the function f at x_0 is defined as

$$f'(x_0) = \lim_{h \to 0} \frac{f(x_0 + h) - f(x_0)}{h}.$$

This formula gives an obvious way to generate an approximation to $f'(x_0)$; simply compute

$$\frac{f(x_0 + h) - f(x_0)}{h}$$

for small values of h. Although this may be obvious, it is not very successful, due to our old nemesis, round-off error. But it is certainly the place to start.

To approximate $f'(x_0)$, suppose first that $x_0 \in (a, b)$, where $f \in C^2[a, b]$, and that $x_1 = x_0 + h$ for some $h \neq 0$ that is sufficiently small to ensure that $x_1 \in [a, b]$. We construct the first Lagrange polynomial, $P_{0,1}$, for f determined by x_0 and x_1 with its error term

$$f(x) = P_{0,1}(x) + \frac{(x - x_0)(x - x_1)}{2!} f''(\xi(x))$$

$$= \frac{f(x_0)(x - x_0 - h)}{-h} + \frac{f(x_0 + h)(x - x_0)}{h} + \frac{(x - x_0)(x - x_0 - h)}{2} f''(\xi(x))$$

for some number $\xi(x)$ in $[a, b]$. Differentiating this equation gives

$$f'(x) = \frac{f(x_0 + h) - f(x_0)}{h} + D_x \left[\frac{(x - x_0)(x - x_0 - h)}{2} f''(\xi(x)) \right]$$

$$= \frac{f(x_0 + h) - f(x_0)}{h} + \frac{2(x - x_0) - h}{2} f''(\xi(x))$$

$$+ \frac{(x - x_0)(x - x_0 - h)}{2} D_x(f''(\xi(x))),$$

so

$$f'(x) \approx \frac{f(x_0 + h) - f(x_0)}{h},$$

Difference equations were used and popularized by Isaac Newton in the last quarter of the 17th century, but many of these techniques had previously been developed by Thomas Harriot (1561–1621) and Henry Briggs (1561–1630). Harriot made significant advances in navigation techniques, and Briggs was the person most responsible for the acceptance of logarithms as an aid to computation.

with error

$$\frac{2(x - x_0) - h}{2} f''(\xi(x)) + \frac{(x - x_0)(x - x_0 - h)}{2} D_x(f''(\xi(x))).$$

There are two terms for the error in this approximation. The first term involves $f''(\xi(x))$, which can be bounded if we have a bound for the second derivative of f. The second part of the truncation error involves $D_x f''(\xi(x)) = f'''(\xi(x)) \cdot \xi'(x)$, which generally cannot be estimated because it contains the unknown term $\xi'(x)$. However, when x is x_0, the coefficient of $D_x f''(\xi(x))$ is zero. In this case, the formula simplifies to the following:

Two-Point Formula

If f'' exists on the interval containing x_0 and $x_0 + h$, then

$$f'(x_0) = \frac{f(x_0 + h) - f(x_0)}{h} - \frac{h}{2} f''(\xi),$$

for some number ξ between x_0 and $x_0 + h$.

Suppose that M is a bound on $|f''(x)|$ for $x \in [a, b]$. Then for small values of h, the difference quotient $[f(x_0 + h) - f(x_0)]/h$ can be used to approximate $f'(x_0)$ with an error bounded by $M|h|/2$. This is a two-point formula known as the **forward-difference formula** if $h > 0$ (see Figure 4.20) and the **backward-difference formula** if $h < 0$.

Figure 4.20

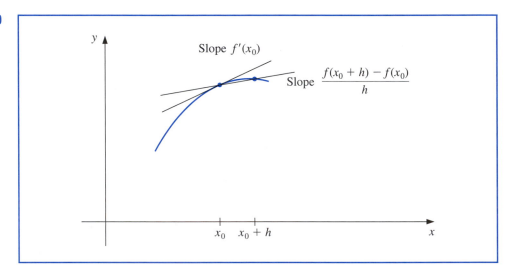

Example 1   Use the forward-difference formula to approximate the derivative of $f(x) = \ln x$ at $x_0 = 1.8$ using $h = 0.1, h = 0.05$, and $h = 0.01$, and determine bounds for the approximation errors.

Solution   The forward-difference formula

$$\frac{f(1.8 + h) - f(1.8)}{h}$$

with $h = 0.1$ gives

$$\frac{\ln 1.9 - \ln 1.8}{0.1} = \frac{0.64185389 - 0.58778667}{0.1} = 0.5406722.$$

Because $f''(x) = -1/x^2$ and $1.8 < \xi < 1.9$, a bound for this approximation error is

$$\frac{|hf''(\xi)|}{2} = \frac{|h|}{2\xi^2} < \frac{0.1}{2(1.8)^2} = 0.0154321.$$

The approximation and error bounds when $h = 0.05$ and $h = 0.01$ are found in a similar manner and the results are shown in Table 4.9.

Table 4.9

h	$f(1.8+h)$	$\dfrac{f(1.8+h)-f(1.8)}{h}$	$\dfrac{\lvert h\rvert}{2(1.8)^2}$
0.1	0.64185389	0.5406722	0.0154321
0.05	0.61518564	0.5479795	0.0077160
0.01	0.59332685	0.5540180	0.0015432

Since $f'(x) = 1/x$, the exact value of $f'(1.8)$ is $0.55\overline{5}$, and in this case the error bounds are quite close to the true approximation error.   ∎

To obtain general derivative approximation formulas, suppose that $x_0, x_1, \ldots, x_n$ are $(n+1)$ distinct numbers in some interval I and that $f \in C^{n+1}(I)$. Then

$$f(x) = \sum_{j=0}^{n} f(x_j)L_j(x) + \frac{(x-x_0)\cdots(x-x_n)}{(n+1)!}f^{(n+1)}(\xi(x))$$

for some $\xi(x)$ in I, where $L_j(x)$ denotes the jth Lagrange coefficient polynomial for f at $x_0, x_1, \ldots, x_n$. Differentiating this expression gives

$$f'(x) = \sum_{j=0}^{n} f(x_j)L_j'(x) + D_x\left[\frac{(x-x_0)\cdots(x-x_n)}{(n+1)!}\right]f^{(n+1)}(\xi(x))$$

$$+ \frac{(x-x_0)\cdots(x-x_n)}{(n+1)!}D_x[f^{(n+1)}(\xi(x))].$$

Again we have a problem with the second part of the truncation error unless x is one of the numbers x_k. In this case, the multiplier of $D_x[f^{(n+1)}(\xi(x))]$ is zero, and the formula becomes

$$f'(x_k) = \sum_{j=0}^{n} f(x_j)L_j'(x_k) + \frac{f^{(n+1)}(\xi(x_k))}{(n+1)!}\prod_{\substack{j=0\\j\neq k}}^{n}(x_k - x_j).$$

Three-Point Formulas

Applying this technique using the second Lagrange polynomial at x_0, $x_1 = x_0 + h$, and $x_2 = x_0 + 2h$ produces the following formula.

Three-Point Endpoint Formula

If f''' exists on the interval containing x_0 and $x_0 + 2h$, then

$$f'(x_0) = \frac{1}{2h}[-3f(x_0) + 4f(x_0 + h) - f(x_0 + 2h)] + \frac{h^2}{3}f'''(\xi),$$

for some number ξ between x_0 and $x_0 + 2h$.

This formula is useful when approximating the derivative at the endpoint of an interval. This situation occurs, for example, when approximations are needed for the derivatives used for the clamped cubic splines. Left endpoint approximations are found using $h > 0$, and right endpoint approximations using $h < 0$.

When approximating the derivative of a function at an interior point of an interval, it is better to use the formula that is produced from the second Lagrange polynomial at $x_0 - h$, x_0, and $x_0 + h$.

Three-Point Midpoint Formula

If f''' exists on the interval containing $x_0 - h$ and $x_0 + h$, then

$$f'(x_0) = \frac{1}{2h}[f(x_0 + h) - f(x_0 - h)] - \frac{h^2}{6}f'''(\xi),$$

for some number ξ between $x_0 - h$ and $x_0 + h$.

The error in the Midpoint formula is approximately half the error in the Endpoint formula and f needs to be evaluated at only two points whereas in the Endpoint formula three evaluations are required. Figure 4.21 gives an illustration of the approximation produced from the Midpoint formula.

Figure 4.21

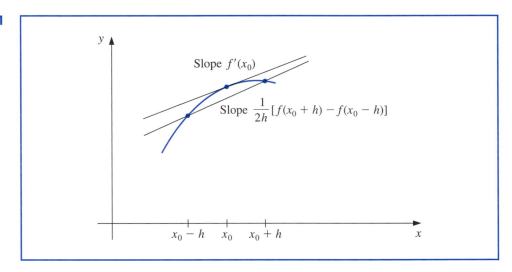

These methods are called *three-point formulas* (even though the third point, $f(x_0)$, does not appear in the Midpoint formula). Similarly, there are *five-point formulas* that involve evaluating the function at two additional points, whose error term is $O(h^4)$. These formulas are generated by differentiating fourth Lagrange polynomials that pass through the evaluation points. The most useful is the Midpoint formula.

Five-Point Formulas

Five-Point Midpoint Formula

If $f^{(5)}$ exists on the interval containing $x_0 - 2h$ and $x_0 + 2h$, then

$$f'(x_0) = \frac{1}{12h}[f(x_0 - 2h) - 8f(x_0 - h) + 8f(x_0 + h) - f(x_0 + 2h)] + \frac{h^4}{30}f^{(5)}(\xi),$$

for some number ξ between $x_0 - 2h$ and $x_0 + 2h$.

There is another five-point formula that is useful, particularly with regard to the clamped cubic spline interpolation.

Five-Point Endpoint Formula

If $f^{(5)}$ exists on the interval containing x_0 and $x_0 + 4h$, then

$$f'(x_0) = \frac{1}{12h}[-25f(x_0) + 48f(x_0 + h) - 36f(x_0 + 2h)$$

$$+ 16f(x_0 + 3h) - 3f(x_0 + 4h)] + \frac{h^4}{5}f^{(5)}(\xi),$$

for some number ξ between x_0 and $x_0 + 4h$.

Left-endpoint approximations are found using $h > 0$, and right-endpoint approximations are found using $h < 0$.

Example 2   Values for $f(x) = xe^x$ are given in Table 4.10. Use all the applicable three-point and five-point formulas to approximate $f'(2.0)$.

Solution   The data in the table permit us to find four different three-point approximations:

Table 4.10

x	$f(x)$
1.8	10.889365
1.9	12.703199
2.0	14.778112
2.1	17.148957
2.2	19.855030

Three-Point Endpoint Formula with $h = 0.1$:

$$\frac{1}{0.2}[-3f(2.0) + 4f(2.1) - f(2.2)] = 5[-3(14.778112) + 4(17.148957) - 19.855030)]$$

$$= 22.032310,$$

Three-Point Endpoint Formula with $h = -0.1$:

$$\frac{1}{-0.2}[-3f(2.0) + 4f(1.9) - f(1.8)] = -5[-3(14.778112) + 4(12.703199)$$

$$-10.889365)] = 22.054525,$$

Three-Point Midpoint Formula with $h = 0.1$:

$$\frac{1}{0.2}[f(2.1) - f(1.9)] = 5(17.148957 - 12.703199) = 22.228790,$$

Three-Point Midpoint Formula with $h = 0.2$:

$$\frac{1}{0.4}[f(2.2) - f(1.8)] = \frac{5}{2}(19.855030 - 10.889365) = 22.414163.$$

The only five-point formula for which the table gives sufficient data is the midpoint formula with $h = 0.1$.

Five-Point Midpoint Formula with $h = 0.1$:

$$\frac{1}{1.2}[f(1.8) - 8f(1.9) + 8f(2.1) - f(2.2)] = \frac{1}{1.2}[10.889365 - 8(12.703199)$$

$$+ 8(17.148957) - 19.855030]$$

$$= 22.166999.$$

If we had no other information, we would accept the five-point midpoint approximation using $h = 0.1$ as the most accurate. The true value for this problem is $f'(2.0) = (2+1)e^2 = 22.167168$. ∎

Round-Off Error Instability

It is particularly important to pay attention to round-off error when approximating derivatives. When approximating integrals in Section 4.3, we found that reducing the step size in the Composite Simpson's rule reduced the truncation error, and, even though the amount of calculation increased, the total round-off error remained bounded. This is not the case when approximating derivatives.

When applying a numerical differentiation technique, the truncation error will also decrease if the step size is reduced, but only at the expense of increased round-off error. To see why this occurs, let us examine more closely the Three-Point Midpoint formula:

$$f'(x_0) = \frac{1}{2h}[f(x_0 + h) - f(x_0 - h)] - \frac{h^2}{6}f'''(\xi).$$

Suppose that, in evaluating $f(x_0 + h)$ and $f(x_0 - h)$, we encounter round-off errors $e(x_0 + h)$ and $e(x_0 - h)$. Then our computed values $\tilde{f}(x_0 + h)$ and $\tilde{f}(x_0 - h)$ are related to the true values $f(x_0 + h)$ and $f(x_0 - h)$ by

$$f(x_0 + h) = \tilde{f}(x_0 + h) + e(x_0 + h) \quad \text{and} \quad f(x_0 - h) = \tilde{f}(x_0 - h) + e(x_0 - h).$$

In this case, the total error in the approximation,

$$f'(x_0) - \frac{\tilde{f}(x_0 + h) - \tilde{f}(x_0 - h)}{2h} = \frac{e(x_0 + h) - e(x_0 - h)}{2h} - \frac{h^2}{6}f'''(\xi),$$

is due in part to round-off and in part to truncating. If we assume that the round-off errors, $e(x_0 \pm h)$, for the function evaluations are bounded by some number $\varepsilon > 0$ and that the third derivative of f is bounded by a number $M > 0$, then

$$\left| f'(x_0) - \frac{\tilde{f}(x_0 + h) - \tilde{f}(x_0 + h)}{2h} \right| \leq \frac{\varepsilon}{h} + \frac{h^2}{6}M.$$

To reduce the truncation portion of the error, $h^2 M/6$, we must reduce h. But as h is reduced, the round-off portion of the error, ε/h, grows. In practice, then, it is seldom advantageous to let h be too small, since the round-off error will dominate the calculations.

Illustration   Consider using the values in Table 4.11 to approximate $f'(0.900)$, where $f(x) = \sin x$. The true value is $\cos 0.900 = 0.62161$. The formula

$$f'(0.900) \approx \frac{f(0.900 + h) - f(0.900 - h)}{2h},$$

with different values of h, gives the approximations in Table 4.12.

Table 4.11

x	$\sin x$	x	$\sin x$
0.800	0.71736	0.901	0.78395
0.850	0.75128	0.902	0.78457
0.880	0.77074	0.905	0.78643
0.890	0.77707	0.910	0.78950
0.895	0.78021	0.920	0.79560
0.898	0.78208	0.950	0.81342
0.899	0.78270	1.000	0.84147

Table 4.12

h	Approximation to $f'(0.900)$	Error
0.001	0.62500	0.00339
0.002	0.62250	0.00089
0.005	0.62200	0.00039
0.010	0.62150	-0.00011
0.020	0.62150	-0.00011
0.050	0.62140	-0.00021
0.100	0.62055	-0.00106

The optimal choice for h appears to lie between 0.005 and 0.05. We can use calculus to verify (see Exercise 13) that a minimum for

$$e(h) = \frac{\varepsilon}{h} + \frac{h^2}{6}M,$$

occurs at $h = \sqrt[3]{3\varepsilon/M}$, where

$$M = \max_{x \in [0.800, 1.00]} |f'''(x)| = \max_{x \in [0.800, 1.00]} |\cos x| = \cos 0.8 \approx 0.69671.$$

Because values of f are given to five decimal places, we will assume that the round-off error is bounded by $\varepsilon = 5 \times 10^{-6}$. Therefore, the optimal choice of h is approximately

$$h = \sqrt[3]{\frac{3(0.000005)}{0.69671}} \approx 0.028,$$

which is consistent with the results in Table 4.12.

In practice, we cannot compute an optimal h to use in approximating the derivative, because we have no knowledge of the third derivative of the function. But we must remain aware that reducing the step size will not always improve the approximation. $\square$

We have considered only the round-off error problems that are presented by the Three-Point Midpoint formula, but similar difficulties occur with all the differentiation formulas. The reason for the problems can be traced to the need to divide by a power of h. As we found in Section 1.4 (see, in particular, Example 1), division by small numbers tends to exaggerate round-off error, and this operation should be avoided if possible. In the case of numerical differentiation, it is impossible to avoid the problem entirely, although the higher-order methods reduce the difficulty.

Keep in mind that, as an approximation method, numerical differentiation is unstable, because the small values of h needed to reduce truncation error cause the round-off error to grow. This is the first class of unstable methods that we have encountered, and these techniques would be avoided if it were possible. However it is not, because these formulas are needed in Chapters 11 and 12 for approximating the solutions of ordinary and partial-differential equations.

Keep in mind that difference method approximations can be unstable.

Methods for approximating higher derivatives of functions using Taylor polynomials can be derived as was done when approximating the first derivative or by using an averaging technique that is similar to that used for extrapolation. These techniques, of course, suffer from the same stability weaknesses as the approximation methods for first derivatives, but they are needed for approximating the solution to boundary value problems in differential equations. The only one we will need is a Three-Point Midpoint formula, which has the following form.

Three-Point Midpoint Formula for Approximating f''

If $f^{(4)}$ exists on the interval containing $x_0 - h$ and $x_0 + h$, then

$$f''(x_0) = \frac{1}{h^2}[f(x_0 - h) - 2f(x_0) + f(x_0 + h)] - \frac{h^2}{12}f^{(4)}(\xi),$$

for some number ξ between $x_0 - h$ and $x_0 + h$.

EXERCISE SET 4.9

1. Use the forward-difference formulas and backward-difference formulas to determine each missing entry in the following tables.

a.

x	$f(x)$	$f'(x)$
0.5	0.4794	
0.6	0.5646	
0.7	0.6442	

b.

x	$f(x)$	$f'(x)$
0.0	0.00000	
0.2	0.74140	
0.4	1.3718	

2. The data in Exercise 1 were taken from the following functions. Compute the actual errors in Exercise 1, and find error bounds using the error formulas.

a. $f(x) = \sin x$ 　　　　　　　　**b.** $f(x) = e^x - 2x^2 + 3x - 1$

3. Use the most accurate three-point formula to determine each missing entry in the following tables.

a.

x	$f(x)$	$f'(x)$
1.1	9.025013	
1.2	11.02318	
1.3	13.46374	
1.4	16.44465	

b.

x	$f(x)$	$f'(x)$
8.1	16.94410	
8.3	17.56492	
8.5	18.19056	
8.7	18.82091	

c.

x	$f(x)$	$f'(x)$
2.9	−4.827866	
3.0	−4.240058	
3.1	−3.496909	
3.2	−2.596792	

d.

x	$f(x)$	$f'(x)$
2.0	3.6887983	
2.1	3.6905701	
2.2	3.6688192	
2.3	3.6245909	

4. The data in Exercise 3 were taken from the following functions. Compute the actual errors in Exercise 3, and find error bounds using the error formulas.

a. $f(x) = e^{2x}$ 　　　　　　　　　**b.** $f(x) = x \ln x$
c. $f(x) = x \cos x - x^2 \sin x$ 　　　**d.** $f(x) = 2(\ln x)^2 + 3 \sin x$

5. Use the formulas given in this section to determine, as accurately as possible, approximations for each missing entry in the following tables.

a.

x	$f(x)$	$f'(x)$
2.1	−1.709847	
2.2	−1.373823	
2.3	−1.119214	
2.4	−0.9160143	
2.5	−0.7470223	
2.6	−0.6015966	

b.

x	$f(x)$	$f'(x)$
−3.0	9.367879	
−2.8	8.233241	
−2.6	7.180350	
−2.4	6.209329	
−2.2	5.320305	
−2.0	4.513417	

6. The data in Exercise 5 were taken from the following functions. Compute the actual errors in Exercise 5, and find error bounds using the error formulas.

a. $f(x) = \tan x$ 　　　　　　　　**b.** $f(x) = e^{x/3} + x^2$

7. Let $f(x) = \cos \pi x$. Use the Three-Point Midpoint formula for f'' and the values of $f(x)$ at $x = 0.25$, 0.5, and 0.75 to approximate $f''(0.5)$. Compare this result to the exact value and to the approximation found in Exercise 7 of Section 3.5. Explain why this method is particularly accurate for this problem.

8. Let $f(x) = 3xe^x - \cos x$. Use the following data and the Three-Point Midpoint formula for f'' to approximate $f''(1.3)$ with $h = 0.1$ and with $h = 0.01$.

x	1.20	1.29	1.30	1.31	1.40
$f(x)$	11.59006	13.78176	14.04276	14.30741	16.86187

Compare your results to $f''(1.3)$.

9. Use the following data and the knowledge that the first five derivatives of f were bounded on $[1, 5]$ by 2, 3, 6, 12, and 23, respectively, to approximate $f'(3)$ as accurately as possible. Find a bound for the error.

x	1	2	3	4	5
$f(x)$	2.4142	2.6734	2.8974	3.0976	3.2804

10. Repeat Exercise 9, assuming instead that the third derivative of f is bounded on $[1, 5]$ by 4.

11. Analyze the round-off errors for the formula

$$f'(x_0) = \frac{f(x_0 + h) - f(x_0)}{h} - \frac{h}{2} f''(\xi_0).$$

Find an optimal $h > 0$ in terms of a bound M for f'' on $(x_0, x_0 + h)$.

12. All calculus students know that the derivative of a function f at x can be defined as

$$f'(x) = \lim_{h \to 0} \frac{f(x + h) - f(x)}{h}.$$

Choose your favorite function f, nonzero number x, and computer or calculator. Generate approximations $f'_n(x)$ to $f'(x)$ by

$$f'_n(x) = \frac{f(x + 10^{-n}) - f(x)}{10^{-n}}$$

for $n = 1, 2, \ldots, 20$ and describe what happens.

13. Consider the function

$$e(h) = \frac{\varepsilon}{h} + \frac{h^2}{6} M,$$

where M is a bound for the third derivative of a function. Show that $e(h)$ has a minimum at $\sqrt[3]{3\varepsilon/M}$.

14. The forward-difference formula can be expressed as

$$f'(x_0) = \frac{1}{h}[f(x_0 + h) - f(x_0)] - \frac{h}{2} f''(x_0) - \frac{h^2}{6} f'''(x_0) + O(h^3).$$

Use extrapolation on this formula to derive an $O(h^3)$ formula for $f'(x_0)$.

15. In Exercise 7 of Section 3.4, data were given describing a car traveling on a straight road. That problem asked to predict the position and speed of the car when $t = 10$ s. Use the following times and positions to predict the speed at each time listed.

Time	0	3	5	8	10	13
Distance	0	225	383	623	742	993

16. In a circuit with impressed voltage $\mathcal{E}(t)$ and inductance L, Kirchhoff's first law gives the relationship

$$\mathcal{E}(t) = L\frac{di}{dt} + Ri,$$

where R is the resistance in the circuit and i is the current. Suppose we measure the current for several values of t and obtain:

t	1.00	1.01	1.02	1.03	1.0
i	3.10	3.12	3.14	3.18	3.24

where t is measured in seconds, i is in amperes, the inductance L is a constant 0.98 henries, and the resistance is 0.142 ohms. Approximate the voltage $\mathcal{E}(t)$ when $t = 1.00, 1.01, 1.02, 1.03$, and 1.04.

17. Derive a method for approximating $f''(x_0)$ whose error term is of order h^2 by expanding the function f in a third Taylor polynomial about x_0 and evaluating at $x_0 + h$ and $x_0 - h$.

4.10 Survey of Methods and Software

In this chapter we considered approximating integrals of functions of one, two, or three variables and approximating the derivatives of a function of a single real variable.

The Midpoint rule, Trapezoidal rule, and Simpson's rule were studied to introduce the techniques and error analysis of quadrature methods. Composite Simpson's rule is easy to use and produces accurate approximations unless the function oscillates in a subinterval of the interval of integration. Adaptive quadrature can be used if the function is suspected of oscillatory behavior. To minimize the number of nodes and also increase the accuracy, we studied Gaussian quadrature. Romberg integration was introduced to take advantage of the easily-applied Composite Trapezoidal rule and extrapolation.

Most software for integrating a function of a single real variable is based on the adaptive approach or extremely accurate Gaussian formulas. Cautious Romberg integration is an adaptive technique that includes a check to make sure that the integrand is smoothly behaved over subintervals of the integral of integration. This method has been successfully used in software libraries. Multiple integrals are generally approximated by extending good adaptive methods to higher dimensions. Gaussian-type quadrature is also recommended to decrease the number of function evaluations.

The main routines in both the IMSL and NAG Libraries are based on QUADPACK: *A Subroutine Package for Automatic Integration* by R. Piessens, E. de Doncker-Kapenga, C. W. Uberhuber, and D. K. Kahaner published by Springer-Verlag in 1983 [PDUK]. The routines are also available as public domain software, at http://www.netlib.org/quadpack. The main technique is an adaptive integration scheme based on the 21-point Gaussian-Kronrod rule using the 10-point Gaussian rule for error estimation. The Gaussian rule uses the 10 points $x_1, \ldots, x_{10}$ and weights $w_1, \ldots, w_{10}$ to give the quadrature formula $\sum_{i=1}^{10} w_i f(x_i)$ to approximate $\int_a^b f(x)dx$. The additional points $x_{11}, \ldots, x_{21}$ and the new weights $v_1, \ldots, v_{21}$ are then used in the Kronrod formula, $\sum_{i=1}^{21} v_i f(x_i)$. The results of the two formulas are compared to eliminate error. The advantage in using $x_1, \ldots, x_{10}$ in each formula is that f needs to be evaluated at only 21 points. If independent 10- and 21-point Gaussian rules were used, 31 function evaluations would be needed. This procedure also permits endpoint singularities in the integrand. Other subroutines allow user specified singularities and infinite intervals of integration. Methods are also available for multiple integrals.

Although numerical differentiation is unstable, derivative approximation formulas are needed for solving differential equations. The NAG Library includes a subroutine for the numerical differentiation of a function of one real variable, with differentiation to the fourteenth derivative being possible. An IMSL function uses an adaptive change in step size for finite differences to approximate a derivative of f at x to within a given tolerance. Both packages allow the differentiation and integration of interpolatory cubic splines.

For further reading on numerical integration, we recommend the books by Engels [E] and by Davis and Rabinowitz [DR]. For more information on Gaussian quadrature, see Stroud and Secrest [StS]. Books on multiple integrals include those by Stroud [Stro] and by Sloan and Joe [SJ].

Numerical Solution of Initial-Value Problems

5.1 Introduction

Differential equations are used to model problems that involve the change of some variable with respect to another. These problems require the solution to an initial-value problem— that is, the solution to a differential equation that satisfies a given initial condition.

In many real-life situations, the differential equation that models the problem is too complicated to solve exactly, and one of two approaches is taken to approximate the solution. The first approach is to simplify the differential equation to one that can be solved exactly, and then use the solution of the simplified equation to approximate the solution to the original equation. The other approach, the one we examine in this chapter, involves finding methods for directly approximating the solution of the original problem. This is the approach commonly taken because more accurate results and realistic error information can be obtained.

The methods we consider in this chapter do not produce a continuous approximation to the solution of the initial-value problem. Rather, approximations are found at certain specified, and often equally-spaced, points. Some method of interpolation, commonly cubic Hermite, is used if intermediate values are needed.

The first part of the chapter concerns approximating the solution $y(t)$ to a problem of the form

$$\frac{dy}{dt} = f(t, y), \quad \text{for } a \leq t \leq b,$$

subject to an initial condition

$$y(a) = \alpha.$$

These techniques form the core of the study because more general procedures use these as a base. Later in the chapter we deal with the extension of these methods to a system of first-order differential equations in the form

$$\frac{dy_1}{dt} = f_1(t, y_1, y_2, \ldots, y_n),$$

$$\frac{dy_2}{dt} = f_2(t, y_1, y_2, \ldots, y_n),$$

$$\vdots$$

$$\frac{dy_n}{dt} = f_n(t, y_1, y_2, \ldots, y_n),$$

for $a \leq t \leq b$, subject to the initial conditions

$$y_1(a) = \alpha_1, \quad y_2(a) = \alpha_2, \quad \ldots, \quad y_n(a) = \alpha_n.$$

We also examine the relationship of a system of this type to the general nth-order initial-value problem of the form

$$y^{(n)} = f\left(t, y, y', y'', \dots, y^{(n-1)}\right)$$

for $a \le t \le b$, subject to the multiple initial conditions

$$y(a) = \alpha_0, \quad y'(a) = \alpha_1, \dots, y^{(n-1)}(a) = \alpha_{n-1}.$$

Well-Posed Problems

Before describing the methods for approximating the solution to our basic problem, we consider some situations that ensure the solution will exist. In fact, because we will not be solving the given problem, only an approximation to the problem, we need to know when problems that are close to the given problem have solutions that accurately approximate the solution to the given problem. This property of an initial-value problem is called **well-posed**, and these are the problems for which numerical methods are appropriate. The following result shows that the class of well-posed problems is quite broad.

Well-Posed Condition

Suppose that f and f_y, its first partial derivative with respect to y, are continuous for t in $[a, b]$ and for all y. Then the initial-value problem

$$y' = f(t, y), \quad \text{for } a \le t \le b, \quad \text{with } y(a) = \alpha,$$

has a unique solution $y(t)$ for $a \le t \le b$, and the problem is well-posed.

Example 1   Consider the initial-value problem

$$y' = 1 + t\sin(ty), \quad \text{for } 0 \le t \le 2, \quad \text{with } y(0) = 0.$$

Since the functions

$$f(t, y) = 1 + t\sin(ty) \qquad \text{and} \qquad f_y(t, y) = t^2\cos(ty)$$

are both continuous for $0 \le t \le 2$ and for all y, a unique solution exists to this well-posed initial-value problem.

If you have taken a course in differential equations, you might attempt to determine the solution to this problem by using one of the techniques you learned in that course. ■

5.2 Taylor Methods

The methods in this section use Taylor polynomials and the knowledge of the derivative at a node to approximate the value of the function at a new node.

Many of the numerical methods we saw in the first four chapters have an underlying derivation from Taylor's Theorem. The approximation of the solution to initial-value problems is no exception. In this case, the function we need to expand in a Taylor polynomial is the (unknown) solution to the problem, $y(t)$. In its most elementary form, this leads to **Euler's Method**. Although Euler's method is seldom used in practice, the simplicity of its derivation illustrates the technique used for more advanced procedures, without the cumbersome algebra that accompanies these constructions.

The use of elementary difference methods to approximate the solution to differential equations was one of the numerous mathematical topics that was first presented to the mathematical public by the most prolific of mathematicians, Leonhard Euler (1707–1783).

The objective of Euler's method is to find, for a given positive integer N, an approximation to the solution of a problem of the form

$$\frac{dy}{dt} = f(t, y), \quad \text{for } a \leq t \leq b, \quad \text{with } y(a) = \alpha$$

at the $N + 1$ equally-spaced **mesh points** $\{t_0, t_1, t_2, \ldots, t_N\}$ (see Figure 5.1). The common distance between the points, $h = (b - a)/N$, is called the **step size**, and

$$t_i = a + ih, \quad \text{for each } i = 0, 1, \ldots N.$$

Approximations at other values of t in $[a, b]$ can then be found using interpolation.

Figure 5.1

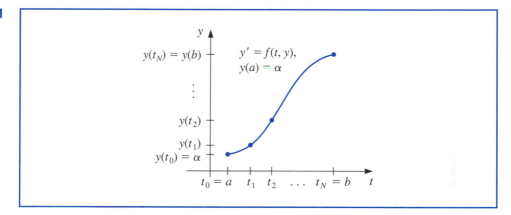

Suppose that $y(t)$, the solution to the problem, has two continuous derivatives on $[a, b]$, so that for each $i = 0, 1, 2, \ldots, N - 1$, Taylor's Theorem implies that

$$y(t_{i+1}) = y(t_i) + (t_{i+1} - t_i)y'(t_i) + \frac{(t_{i+1} - t_i)^2}{2}y''(\xi_i),$$

for some number ξ_i in (t_i, t_{i+1}). Letting $h = (b - a)/N = t_{i+1} - t_i$, we have

$$y(t_{i+1}) = y(t_i) + hy'(t_i) + \frac{h^2}{2}y''(\xi_i),$$

and, since $y(t)$ satisfies the differential equation $y'(t) = f(t, y(t))$,

$$y(t_{i+1}) = y(t_i) + hf(t_i, y(t_i)) + \frac{h^2}{2}y''(\xi_i).$$

Euler's method constructs the approximation w_i to $y(t_i)$ for each $i = 1, 2, \ldots, N$ by deleting the error term in this equation. This produces a *difference equation* that approximates the differential equation. The term **local error** refers to the error at the given step if it is assumed that all the previous results are exact. The true, or accumulated, error of the method is called **global error**.

Euler's Method

$$w_0 = \alpha,$$

$$w_{i+1} = w_i + hf(t_i, w_i),$$

where $i = 0, 1, \ldots, N - 1$, with local error $\frac{1}{2}y''(\xi_i)h^2$ for some ξ_i in (t_i, t_{i+1}).

Illustration   In Example 1 we will use Euler's method to approximate the solution to

$$y' = y - t^2 + 1, \quad 0 \le t \le 2, \quad y(0) = 0.5,$$

at $t = 2$. Here we will simply illustrate the steps in the technique when we have $h = 0.5$. For this problem $f(t, y) = y - t^2 + 1$, so

$$w_0 = y(0) = 0.5;$$

$$w_1 = w_0 + 0.5(w_0 - (0.0)^2 + 1) = 0.5 + 0.5(1.5) = 1.25;$$

$$w_2 = w_1 + 0.5(w_1 - (0.5)^2 + 1) = 1.25 + 0.5(2.0) = 2.25;$$

$$w_3 = w_2 + 0.5(w_2 - (1.0)^2 + 1) = 2.25 + 0.5(2.25) = 3.375;$$

and

$$y(2) \approx w_4 = w_3 + 0.5(w_3 - (1.5)^2 + 1) = 3.375 + 0.5(2.125) = 4.4375. \qquad \square$$

To interpret Euler's method geometrically, note that when w_i is a close approximation to $y(t_i)$, the assumption that the problem is well-posed implies that

$$f(t_i, w_i) \approx y'(t_i) = f(t_i, y(t_i)).$$

> The program EULERM51 implements Euler's method.

The first step of Euler's method appears in Figure 5.2(a), and a series of steps appears in Figure 5.2(b).

Figure 5.2

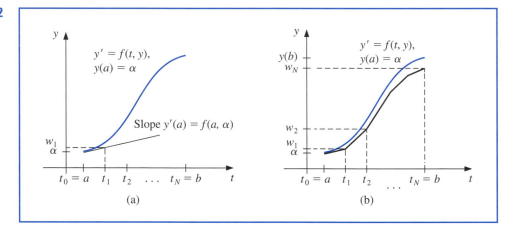

(a)   (b)

Example 1   Euler's method was used in the Illustration with $h = 0.5$ to approximate the solution to the initial-value problem

$$y' = y - t^2 + 1, \quad 0 \le t \le 2, \quad y(0) = 0.5.$$

Use the program EULERM51 with $N = 10$ to determine approximations, and compare these with the exact values given by $y(t) = (t + 1)^2 - 0.5e^t$.

Solution   With $N = 10$ we have $h = 0.2$, $t_i = 0.2i$, $w_0 = 0.5$, and

$$w_{i+1} = w_i + h\left(w_i - t_i^2 + 1\right) = w_i + 0.2\left[w_i - 0.04i^2 + 1\right] = 1.2w_i - 0.008i^2 + 0.2,$$

for $i = 0, 1, \ldots, 9$. So

$$w_1 = 1.2(0.5) - 0.008(0)^2 + 0.2 = 0.8; \quad w_2 = 1.2(0.8) - 0.008(1)^2 + 0.2 = 1.152;$$

and so on. Table 5.1 shows the comparison between the approximate values at t_i and the actual values. ∎

Table 5.1

t_i	w_i	$y_i = y(t_i)$	$\lvert y_i - w_i \rvert$
0.0	0.5000000	0.5000000	0.0000000
0.2	0.8000000	0.8292986	0.0292986
0.4	1.1520000	1.2140877	0.0620877
0.6	1.5504000	1.6489406	0.0985406
0.8	1.9884800	2.1272295	0.1387495
1.0	2.4581760	2.6408591	0.1826831
1.2	2.9498112	3.1799415	0.2301303
1.4	3.4517734	3.7324000	0.2806266
1.6	3.9501281	4.2834838	0.3333557
1.8	4.4281538	4.8151763	0.3870225
2.0	4.8657845	5.3054720	0.4396874

Error Bounds for Euler's Method

Euler's method is derived from a Taylor polynomial whose error term involves the square of the step size h, so the local error at each step is proportional to h^2, so it is $O(h^2)$. However, the total error, or global error, accumulates these local errors, so it generally grows at a much faster rate.

Euler's Method Error Bound

Let $y(t)$ denote the unique solution to the initial-value problem

$$y' = f(t, y), \quad \text{for } a \le t \le b, \quad \text{with } y(a) = \alpha,$$

and $w_0, w_1, \ldots, w_N$ be the approximations generated by Euler's method for some positive integer N. Suppose that f is continuous for all t in $[a, b]$ and all y in $(-\infty, \infty)$, and constants L and M exist with

$$\left\lvert \frac{\partial f}{\partial y}(t, y(t)) \right\rvert \le L \quad \text{and} \quad \lvert y''(t) \rvert \le M.$$

Then, for each $i = 0, 1, 2, \ldots, N$,

$$\lvert y(t_i) - w_i \rvert \le \frac{hM}{2L}[e^{L(t_i - a)} - 1].$$

An important point to notice is that, although the local error of Euler's method, that is, the error at an individual step, is $O(h^2)$, the global error, which is the error over the entire interval, is only $O(h)$. The reduction of one power of h from local to global error is typical of initial-value techniques. Even though we have a reduction in order from local to global errors, the formula shows that the error tends to zero with h.

Example 2   The solution to the initial-value problem

$$y' = y - t^2 + 1, \quad 0 \le t \le 2, \quad y(0) = 0.5,$$

was approximated in Example 1 using Euler's method with $h = 0.2$. Find bounds for the approximation errors and compare these to the actual errors.

Solution   Because $f(t, y) = y - t^2 + 1$, we have $\partial f(t, y)/\partial y = 1$ for all y, so $L = 1$. For this problem, the exact solution is $y(t) = (t + 1)^2 - 0.5e^t$, so $y''(t) = 2 - 0.5e^t$ and

$$|y''(t)| \le 0.5e^2 - 2, \quad \text{for all } t \in [0, 2].$$

Using the inequality in the error bound for Euler's method with $h = 0.2$, $L = 1$, and $M = 0.5e^2 - 2$ gives

$$|y_i - w_i| \le 0.1(0.5e^2 - 2)(e^{t_i} - 1).$$

Hence

$$|y(0.2) - w_1| \le 0.1(0.5e^2 - 2)(e^{0.2} - 1) = 0.03752;$$

$$|y(0.4) - w_2| \le 0.1(0.5e^2 - 2)(e^{0.4} - 1) = 0.08334;$$

and so on. Table 5.2 lists the actual error found in Example 1, together with this error bound. Note that even though the true bound for the second derivative of the solution was used, the error bound is considerably larger than the actual error, especially for increasing values of t.   ▪

Table 5.2

t_i	0.2	0.4	0.6	0.8	1.0	1.2	1.4	1.6	1.8	2.0
Actual Error	0.02930	0.06209	0.09854	0.13875	0.18268	0.23013	0.28063	0.33336	0.38702	0.43969
Error Bound	0.03752	0.08334	0.13931	0.20767	0.29117	0.39315	0.51771	0.66985	0.85568	1.08264

Higher Order Taylor Methods

Euler's method was derived using Taylor's Theorem with $n = 1$, so the first attempt to find methods for improving the accuracy of difference methods is to extend this technique of derivation to larger values of n. Suppose the solution $y(t)$ to the initial-value problem

$$y' = f(t, y), \quad \text{for } a \le t \le b, \quad \text{with } y(a) = \alpha,$$

has $n + 1$ continuous derivatives. If we expand the solution $y(t)$ in terms of its nth Taylor polynomial about t_i, we obtain

$$y(t_{i+1}) = y(t_i) + hy'(t_i) + \frac{h^2}{2}y''(t_i) + \cdots + \frac{h^n}{n!}y^{(n)}(t_i) + \frac{h^{n+1}}{(n+1)!}y^{(n+1)}(\xi_i)$$

for some number ξ_i in (t_i, t_{i+1}). Successive differentiation of the solution $y(t)$ gives

$$y'(t) = f(t, y(t)), \quad y''(t) = f'(t, y(t)), \quad \text{and, generally,} \quad y^{(k)}(t) = f^{(k-1)}(t, y(t)).$$

Substituting these results into the Taylor expansion gives

$$y(t_{i+1}) = y(t_i) + hf(t_i, y(t_i)) + \frac{h^2}{2}f'(t_i, y(t_i)) + \cdots$$

$$+ \frac{h^n}{n!}f^{(n-1)}(t_i, y(t_i)) + \frac{h^{n+1}}{(n+1)!}f^{(n)}(\xi_i, y(\xi_i)).$$

The difference-equation method corresponding to this equation is obtained by deleting the remainder term involving ξ_i.

Taylor Method of Order *n*

$$w_0 = \alpha,$$

$$w_{i+1} = w_i + hT^{(n)}(t_i, w_i)$$

for each $i = 0, 1, \ldots, N - 1$, where

$$T^{(n)}(t_i, w_i) = f(t_i, w_i) + \frac{h}{2} f'(t_i, w_i) + \cdots + \frac{h^{n-1}}{n!} f^{(n-1)}(t_i, w_i).$$

The local error is $\frac{1}{(n+1)!} y^{(n+1)}(\xi_i) h^{n+1}$ for some ξ_i in (t_i, t_{i+1}).

The formula for $T^{(n)}$ is easily expressed but difficult to use because it requires the derivatives of f with respect to t. Since f is described as a multivariable function of both t and y, the chain rule implies that the total derivative of f with respect to t, which we denoted $f'(t, y(t))$, is obtained by

$$f'(t, y(t)) = \frac{\partial f}{\partial t}(t, y(t)) \cdot \frac{dt}{dt} + \frac{\partial f}{\partial y}(t, y(t)) \frac{dy(t)}{dt} = \frac{\partial f}{\partial t}(t, y(t)) + \frac{\partial f}{\partial y}(t, y(t)) y'(t)$$

or, since $y'(t) = f(t, y(t))$, by

$$f'(t, y(t)) = \frac{\partial f}{\partial t}(t, y(t)) + f(t, y(t)) \frac{\partial f}{\partial y}(t, y(t)).$$

Higher derivatives can be obtained in a similar manner, but they might become increasingly complicated. For example, $f''(t, y(t))$ involves the partial derivatives of all the terms on the right side of this equation with respect to both t and y.

Example 3 Apply Taylor's method of orders **(a)** 2 and **(b)** 4 with $N = 10$ to the initial-value problem

$$y' = y - t^2 + 1, \quad 0 \le t \le 2, \quad y(0) = 0.5.$$

Solution **(a)** For the method of order 2 we need the first derivative of $f(t, y(t)) = y(t) - t^2 + 1$ with respect to the variable t. Because $y' = y - t^2 + 1$ we have

$$f'(t, y(t)) = \frac{d}{dt}(y - t^2 + 1) = y' - 2t = y - t^2 + 1 - 2t,$$

so

$$T^{(2)}(t_i, w_i) = f(t_i, w_i) + \frac{h}{2} f'(t_i, w_i) = w_i - t_i^2 + 1 + \frac{h}{2}(w_i - t_i^2 + 1 - 2t_i)$$

$$= \left(1 + \frac{h}{2}\right)(w_i - t_i^2 + 1) - ht_i.$$

Because $N = 10$ we have $h = 0.2$, and $t_i = 0.2i$ for each $i = 1, 2, \ldots, 10$. Thus the second-order method becomes

$$w_0 = 0.5,$$

$$w_{i+1} = w_i + h\left[\left(1 + \frac{h}{2}\right)(w_i - t_i^2 + 1) - ht_i\right]$$

$$= w_i + 0.2\left[\left(1 + \frac{0.2}{2}\right)(w_i - 0.04i^2 + 1) - 0.04i\right]$$

$$= 1.22w_i - 0.0088i^2 - 0.008i + 0.22.$$

Table 5.3

t_i	Taylor Order 2 w_i	Error $|y(t_i) - w_i|$
0.0	0.500000	0
0.2	0.830000	0.000701
0.4	1.215800	0.001712
0.6	1.652076	0.003135
0.8	2.132333	0.005103
1.0	2.648646	0.007787
1.2	3.191348	0.011407
1.4	3.748645	0.016245
1.6	4.306146	0.022663
1.8	4.846299	0.031122
2.0	5.347684	0.042212

The first two steps give the approximations

$$y(0.2) \approx w_1 = 1.22(0.5) - 0.0088(0)^2 - 0.008(0) + 0.22 = 0.83;$$

$$y(0.4) \approx w_2 = 1.22(0.83) - 0.0088(0.2)^2 - 0.008(0.2) + 0.22 = 1.2158.$$

All the approximations and their errors are shown in Table 5.3.

(b) For Taylor's method of order 4 we need the first three derivatives of $f(t, y(t))$ with respect to t. Again using $y' = y - t^2 + 1$ we have

$$f'(t, y(t)) = y - t^2 + 1 - 2t,$$

$$f''(t, y(t)) = \frac{d}{dt}(y - t^2 + 1 - 2t) = y' - 2t - 2 = y - t^2 + 1 - 2t - 2$$

$$= y - t^2 - 2t - 1,$$

and

$$f'''(t, y(t)) = \frac{d}{dt}(y - t^2 - 2t - 1) = y' - 2t - 2 = y - t^2 - 2t - 1,$$

so

$$T^{(4)}(t_i, w_i) = f(t_i, w_i) + \frac{h}{2}f'(t_i, w_i) + \frac{h^2}{6}f''(t_i, w_i) + \frac{h^3}{24}f'''(t_i, w_i)$$

$$= w_i - t_i^2 + 1 + \frac{h}{2}\left(w_i - t_i^2 + 1 - 2t_i\right) + \frac{h^2}{6}\left(w_i - t_i^2 - 2t_i - 1\right)$$

$$+ \frac{h^3}{24}\left(w_i - t_i^2 - 2t_i - 1\right)$$

$$= \left(1 + \frac{h}{2} + \frac{h^2}{6} + \frac{h^3}{24}\right)\left(w_i - t_i^2\right) - \left(1 + \frac{h}{3} + \frac{h^2}{12}\right)(ht_i) + 1 + \frac{h}{2} - \frac{h^2}{6} - \frac{h^3}{24}.$$

Hence Taylor's method of order 4 is

$$w_0 = 0.5,$$

$$w_{i+1} = w_i + h\left[\left(1 + \frac{h}{2} + \frac{h^2}{6} + \frac{h^3}{24}\right)\left(w_i - t_i^2\right) - \left(1 + \frac{h}{3} + \frac{h^2}{12}\right)ht_i + 1 + \frac{h}{2} - \frac{h^2}{6} - \frac{h^3}{24}\right],$$

for $i = 0, 1, \ldots, N - 1$.

Because $N = 10$ and $h = 0.2$ the method becomes

$$w_{i+1} = w_i + 0.2\left[\left(1 + \frac{0.2}{2} + \frac{0.04}{6} + \frac{0.008}{24}\right)\left(w_i - 0.04i^2\right)\right.$$

$$- \left(1 + \frac{0.2}{3} + \frac{0.04}{12}\right)(0.04i) + 1 + \frac{0.2}{2} - \frac{0.04}{6} - \frac{0.008}{24}\right]$$

$$= 1.2214w_i - 0.008856i^2 - 0.00856i + 0.2186,$$

for each $i = 0, 1, \ldots, 9$. The first two steps give the approximations

$$y(0.2) \approx w_1 = 1.2214(0.5) - 0.008856(0)^2 - 0.00856(0) + 0.2186 = 0.8293;$$

$$y(0.4) \approx w_2 = 1.2214(0.8293) - 0.008856(0.2)^2 - 0.00856(0.2) + 0.2186 = 1.214091.$$

All the approximations and their errors are shown in Table 5.4. ▪

Table 5.4

t_i	Taylor Order 4 w_i	Error $|y(t_i) - w_i|$
0.0	0.500000	0
0.2	0.829300	0.000001
0.4	1.214091	0.000003
0.6	1.648947	0.000006
0.8	2.127240	0.000010
1.0	2.640874	0.000015
1.2	3.179964	0.000023
1.4	3.732432	0.000032
1.6	4.283529	0.000045
1.8	4.815238	0.000062
2.0	5.305555	0.000083

Compare these results with those of Taylor's method of order 2 in Table 5.3 and you will see that the order 4 results are vastly superior.

Approximating Intermediate Results

Hermite interpolation requires both the value of the function and its derivative at each node. This makes it a natural interpolation method for approximating differential equations because these data are all available.

The results from Table 5.4 indicate the Taylor's method of order 4 results are quite accurate at the nodes 0.2, 0.4, etc. But suppose we need to determine an approximation to an intermediate point in the table, for example, at $t = 1.25$. If we use linear interpolation on the Taylor method of order four approximations at $t = 1.2$ and $t = 1.4$, we have

$$y(1.25) \approx \left(\frac{1.25 - 1.4}{1.2 - 1.4}\right)3.179964 + \left(\frac{1.25 - 1.2}{1.4 - 1.2}\right)3.732432 = 3.318081.$$

The true value is $y(1.25) = 3.317329$, so this approximation has an error of 0.000752, which is nearly 30 times the average of the approximation errors at 1.2 and 1.4.

We can significantly improve the approximation by using cubic Hermite interpolation. To determine this approximation for $y(1.25)$ requires approximations to $y'(1.2)$ and $y'(1.4)$ as well as approximations to $y(1.2)$ and $y(1.4)$. However, the approximations for $y(1.2)$ and $y(1.4)$ are in the table, and the derivative approximations are available from the differential equation because $y'(t) = f(t, y(t))$. In our example $y'(t) = y(t) - t^2 + 1$, so

$$y'(1.2) = y(1.2) - (1.2)^2 + 1 \approx 3.179964 - 1.44 + 1 = 2.739964$$

and

$$y'(1.4) = y(1.4) - (1.4)^2 + 1 \approx 3.732433 - 1.96 + 1 = 2.772432.$$

The divided-difference procedure in Section 3.4 gives the information in Table 5.5. The underlined entries come from the data, and the other entries use the divided-difference formulas.

Table 5.5

1.2	<u>3.179964</u>			
		2.739964		
1.2	<u>3.179964</u>		0.111880	
		2.762340		−0.307100
1.4	<u>3.732432</u>		0.050460	
		2.772432		
1.4	<u>3.732432</u>			

The cubic Hermite polynomial is

$$y(t) \approx 3.179964 + (t - 1.2)2.739964 + (t - 1.2)^2 0.111880$$
$$+ (t - 1.2)^2(t - 1.4)(-0.307100),$$

so

$$y(1.25) \approx 3.179964 + 0.136998 + 0.000280 + 0.000115 = 3.317357,$$

a result that is accurate to within 0.000028. This is about the average of the errors at 1.2 and at 1.4, and only 4% of the error obtained using linear interpolation. This improvement in accuracy certainly justifies the added computation required for the Hermite method.

Error estimates for the Taylor methods are similar to those for Euler's method. If sufficient differentiability conditions are met, an nth-order Taylor method will have local error $O(h^{n+1})$ and global error $O(h^n)$.

EXERCISE SET 5.2

1. Use Euler's method to approximate the solutions for each of the following initial-value problems.

    a. $y' = te^{3t} - 2y$,   for $0 \le t \le 1$,   with $y(0) = 0$ and $h = 0.5$

    b. $y' = 1 + (t - y)^2$,   for $2 \le t \le 3$,   with $y(2) = 1$ and $h = 0.5$

    c. $y' = 1 + \dfrac{y}{t}$,   for $1 \le t \le 2$,   with $y(1) = 2$ and $h = 0.25$

    d. $y' = \cos 2t + \sin 3t$,   for $0 \le t \le 1$,   with $y(0) = 1$ and $h = 0.25$

2. The actual solutions to the initial-value problems in Exercise 1 are given here. Compare the actual error at each step to the error bound.

    a. $y(t) = \dfrac{1}{5}te^{3t} - \dfrac{1}{25}e^{3t} + \dfrac{1}{25}e^{-2t}$ b. $y(t) = t + (1 - t)^{-1}$

    c. $y(t) = t \ln t + 2t$ d. $y(t) = \dfrac{1}{2}\sin 2t - \dfrac{1}{3}\cos 3t + \dfrac{4}{3}$

3. Use Euler's method to approximate the solutions for each of the following initial-value problems.

    a. $y' = \dfrac{y}{t} - \left(\dfrac{y}{t}\right)^2$,   for $1 \le t \le 2$,   with $y(1) = 1$ and $h = 0.1$

    b. $y' = 1 + \dfrac{y}{t} + \left(\dfrac{y}{t}\right)^2$,   for $1 \le t \le 3$,   with $y(1) = 0$ and $h = 0.2$

    c. $y' = -(y + 1)(y + 3)$,   for $0 \le t \le 2$,   with $y(0) = -2$ and $h = 0.2$

    d. $y' = -5y + 5t^2 + 2t$,   for $0 \le t \le 1$,   with $y(0) = 1/3$ and $h = 0.1$

4. The actual solutions to the initial-value problems in Exercise 3 are given here. Compute the actual error in the approximations of Exercise 3.

    a. $y(t) = t(1 + \ln t)^{-1}$

    b. $y(t) = t \tan(\ln t)$

    c. $y(t) = -3 + 2(1 + e^{-2t})^{-1}$

    d. $y(t) = t^2 + \dfrac{1}{3}e^{-5t}$

5. Repeat Exercise 1 using Taylor's method of order 2.

6. Repeat Exercise 3 using Taylor's method of order 2.

7. Repeat Exercise 3 using Taylor's method of order 4.

8. Given the initial-value problem

    $$y' = \frac{2}{t}y + t^2e^t, \quad 1 \le t \le 2, \quad y(1) = 0$$

    with the exact solution $y(t) = t^2(e^t - e)$:

    a. Use Euler's method with $h = 0.1$ to approximate the solution and compare it with the actual values of y.

    b. Use the answers generated in (a) and linear interpolation to approximate the following values of y and compare them to the actual values.

    i. $y(1.04)$ ii. $y(1.55)$ iii. $y(1.97)$

    c. Use Taylor's method of order 2 with $h = 0.1$ to approximate the solution and compare it with the actual values of y.

    d. Use the answers generated in (c) and linear interpolation to approximate y at the following values and compare them to the actual values of y.

    i. $y(1.04)$ ii. $y(1.55)$ iii. $y(1.97)$

    e. Use Taylor's method of order 4 with $h = 0.1$ to approximate the solution and compare it with the actual values of y.

f. Use the answers generated in (e) and piecewise cubic Hermite interpolation to approximate y at the following values and compare them to the actual values of y.

 i. $y(1.04)$ **ii.** $y(1.55)$ **iii.** $y(1.97)$

9. Given the initial-value problem

$$y' = \frac{1}{t^2} - \frac{y}{t} - y^2, \quad 1 \le t \le 2, \quad y(1) = -1$$

with the exact solution $y(t) = -1/t$.

a. Use Euler's method with $h = 0.05$ to approximate the solution and compare it with the actual values of y.

b. Use the answers generated in (a) and linear interpolation to approximate the following values of y and compare them to the actual values.

 i. $y(1.052)$ **ii.** $y(1.555)$ **iii.** $y(1.978)$

c. Use Taylor's method of order 2 with $h = 0.05$ to approximate the solution and compare it with the actual values of y.

d. Use the answers generated in (c) and linear interpolation to approximate the following values of y and compare them to the actual values.

 i. $y(1.052)$ **ii.** $y(1.555)$ **iii.** $y(1.978)$

e. Use Taylor's method of order 4 with $h = 0.05$ to approximate the solution and compare it with the actual values of y.

f. Use the answers generated in (e) and piecewise cubic Hermite interpolation to approximate the following values of y and compare them to the actual values.

 i. $y(1.052)$ **ii.** $y(1.555)$ **iii.** $y(1.978)$

10. In an electrical circuit with impressed voltage $\mathcal{E}$, having resistance R, inductance L, and capacitance C in parallel, the current i satisfies the differential equation

$$\frac{di}{dt} = C\frac{d^2\mathcal{E}}{dt^2} + \frac{1}{R}\frac{d\mathcal{E}}{dt} + \frac{1}{L}\mathcal{E}.$$

Suppose $i(0) = 0$, $C = 0.3$ farads, $R = 1.4$ ohms, $L = 1.7$ henries, and the voltage is given by

$$\mathcal{E}(t) = e^{-0.06\pi t}\sin(2t - \pi).$$

Use Euler's method to find the current i for the values $t = 0.1j$, $j = 0, 1, \ldots, 100$.

11. A projectile of mass $m = 0.11$ kg shot vertically upward with initial velocity $v(0) = 8$ m/s is slowed due to the force of gravity $F_g = mg$ and due to air resistance $F_r = -kv|v|$, where $g = -9.8$ m/s^2 and $k = 0.002$ kg/m. The differential equation for the velocity v is given by

$$mv' = mg - kv|v|.$$

a. Find the velocity after $0.1, 0.2, \ldots, 1.0$ s.

b. To the nearest tenth of a second, determine when the projectile reaches its maximum height and begins falling.

5.3 Runge-Kutta Methods

In the last section we saw how Taylor methods of arbitrary high order can be generated. However, the application of these high-order methods to a specific problem is complicated by the need to determine and evaluate high-order derivatives with respect to t on the right side of the differential equation. The widespread use of computer algebra systems has simplified this process, but it still remains cumbersome.

In the later 1800s, Carl Runge (1856–1927) used methods similar to those in this section to derive numerous formulas for approximating the solution to initial-value problems.

In this section we consider Runge-Kutta methods, which modify the Taylor methods so that the high-order error bounds are preserved, but the need to determine and evaluate the high-order partial derivatives is eliminated. The strategy behind these techniques involves approximating a Taylor method with a method that is easier to evaluate. This approximation might increase the error, but the increase does not exceed the order of the truncation error that is already present in the Taylor method. As a consequence, the new error does not significantly influence the calculations.

Runge-Kutta Methods of Order Two

In 1901, Martin Wilhelm Kutta (1867–1944) generalized the methods that Runge developed in 1895 to incorporate systems of first-order differential equations. These techniques differ slightly from those we currently call Runge-Kutta methods.

The Runge-Kutta techniques make use of the Taylor expansion of f, the function on the right side of the differential equation. Since f is a function of two variables, t and y, we must first consider the generalization of Taylor's Theorem to functions of this type. This generalization appears more complicated than the single-variable form, but this is only because of all the partial derivatives of the function f.

Taylor's Theorem for Two Variables

If f and all its partial derivatives of order less than or equal to $n + 1$ are continuous on $D = \{(t, y) | a \leq t \leq b, \ c \leq y \leq d\}$ and (t, y) and $(t + \alpha, y + \beta)$ both belong to D, then

$$f(t + \alpha, y + \beta) \approx f(t, y) + \left[\alpha \frac{\partial f}{\partial t}(t, y) + \beta \frac{\partial f}{\partial y}(t, y) \right]$$

$$+ \left[\frac{\alpha^2}{2} \frac{\partial^2 f}{\partial t^2}(t, y) + \alpha\beta \frac{\partial^2 f}{\partial t \, \partial y}(t, y) + \frac{\beta^2}{2} \frac{\partial^2 f}{\partial y^2}(t, y) \right] + \cdots$$

$$+ \frac{1}{n!} \sum_{j=0}^{n} \binom{n}{j} \alpha^{n-j} \beta^j \frac{\partial^n f}{\partial t^{n-j} \partial y^j}(t, y).$$

The error term in this approximation is similar to that given in Taylor's Theorem, with the added complications that arise because of the incorporation of all the partial derivatives of order $n + 1$.

To illustrate the use of this formula in developing the Runge-Kutta methods, let us consider a Runge-Kutta method of order 2. We saw in the previous section that the Taylor method of order 2 comes from

$$y(t_{i+1}) = y(t_i) + hy'(t_i) + \frac{h^2}{2} y''(t_i) + \frac{h^3}{3!} y'''(\xi)$$

$$= y(t_i) + hf(t_i, y(t_i)) + \frac{h^2}{2} f'(t_i, y(t_i)) + \frac{h^3}{3!} y'''(\xi),$$

or, since

$$f'(t_i, y(t_i)) = \frac{\partial f}{\partial t}(t_i, y(t_i)) + \frac{\partial f}{\partial y}(t_i, y(t_i)) \cdot y'(t_i)$$

and $y'(t_i) = f(t_i, y(t_i))$, we have

$$y(t_{i+1}) = y(t_i) + h \left[f(t_i, y(t_i)) + \frac{h}{2} \frac{\partial f}{\partial t}(t_i, y(t_i)) + \frac{h}{2} \frac{\partial f}{\partial y}(t_i, y(t_i)) \cdot f(t_i, y(t_i)) \right]$$

$$+ \frac{h^3}{3!} y'''(\xi).$$

Taylor's Theorem of two variables permits us to replace the term in the braces with a multiple of a function evaluation of f of the form $a_1 f(t_i + \alpha, y(t_i) + \beta)$. If we expand this term using Taylor's Theorem with $n = 1$, we have

$$a_1 f(t_i + \alpha, y(t_i) + \beta) \approx a_1 \left[f(t_i, y(t_i)) + \alpha \frac{\partial f}{\partial t}(t_i, y(t_i)) + \beta \frac{\partial f}{\partial y}(t_i, y(t_i)) \right]$$

$$= a_1 f(t_i, y(t_i)) + a_1 \alpha \frac{\partial f}{\partial t}(t_i, y(t_i)) + a_1 \beta \frac{\partial f}{\partial y}(t_i, y(t_i)).$$

Equating this expression with the terms enclosed in the braces in the preceding equation implies that $a_1, \alpha,$ and β should be chosen so that

$$1 = a_1, \quad \frac{h}{2} = a_1 \alpha, \quad \text{and} \quad \frac{h}{2} f(t_i, y(t_i)) = a_1 \beta;$$

that is,

$$a_1 = 1, \quad \alpha = \frac{h}{2}, \quad \text{and} \quad \beta = \frac{h}{2} f(t_i, y(t_i)).$$

The error introduced by replacing the term in the Taylor method with its approximation has the same order as the error term for the method, so the Runge-Kutta method produced in this way, called the **Midpoint method**, is also a second-order method. As a consequence, the local error of the method is proportional to h^3, and the global error is proportional to h^2.

Midpoint Method

$$w_0 = \alpha$$

$$w_{i+1} = w_i + h \left[f \left(t_i + \frac{h}{2}, w_i + \frac{h}{2} f(t_i, w_i) \right) \right],$$

where $i = 0, 1, \ldots, N - 1$, with local error $O(h^3)$ and global error $O(h^2)$.

Using $a_1 f(t + \alpha, y + \beta)$ to replace the term in the Taylor method is the easiest choice, but it is not the only one. If we instead use a term of the form

$$a_1 f(t, y) + a_2 f(t + \alpha, y + \beta f(t, y)),$$

the extra parameter in this formula provides an infinite number of second-order Runge-Kutta formulas. When $a_1 = a_2 = \frac{1}{2}$ and $\alpha = \beta = h$, we have the **Modified Euler method**.

Modified Euler Method

$$w_0 = \alpha$$

$$w_{i+1} = w_i + \frac{h}{2} [f(t_i, w_i) + f(t_{i+1}, w_i + h f(t_i, w_i))]$$

where $i = 0, 1, \ldots, N - 1$, with local error $O(h^3)$ and global error $O(h^2)$.

Example 1   Use the Midpoint method and the Modified Euler method with $N = 10, h = 0.2, t_i = 0.2i$, and $w_0 = 0.5$ to approximate the solution to our usual example,

$$y' = y - t^2 + 1, \quad 0 \leq t \leq 2, \quad y(0) = 0.5.$$

Solution   The difference equations produced from the various formulas are

$$\text{Midpoint method:}\quad w_{i+1} = 1.22w_i - 0.0088i^2 - 0.008i + 0.218;$$

$$\text{Modified Euler method:}\quad w_{i+1} = 1.22w_i - 0.0088i^2 - 0.008i + 0.216,$$

for each $i = 0, 1, \ldots, 9$. The first two steps of these methods give

$$\text{Midpoint method:}\quad w_1 = 1.22(0.5) - 0.0088(0)^2 - 0.008(0) + 0.218 = 0.828;$$

$$\text{Modified Euler method:}\quad w_1 = 1.22(0.5) - 0.0088(0)^2 - 0.008(0) + 0.216 = 0.826,$$

and

$$\text{Midpoint method:}\quad w_2 = 1.22(0.828) - 0.0088(0.2)^2 - 0.008(0.2) + 0.218$$
$$= 1.21136;$$

$$\text{Modified Euler method:}\quad w_2 = 1.22(0.826) - 0.0088(0.2)^2 - 0.008(0.2) + 0.216$$
$$= 1.20692,$$

Table 5.6 lists all the results of the calculations. For this problem, the Midpoint method is superior to the Modified Euler method.

Table 5.6

		Midpoint		Modified Euler	
t_i	$y(t_i)$	Method	Error	Method	Error
0.0	0.5000000	0.5000000	0	0.5000000	0
0.2	0.8292986	0.8280000	0.0012986	0.8260000	0.0032986
0.4	1.2140877	1.2113600	0.0027277	1.2069200	0.0071677
0.6	1.6489406	1.6446592	0.0042814	1.6372424	0.0116982
0.8	2.1272295	2.1212842	0.0059453	2.1102357	0.0169938
1.0	2.6408591	2.6331668	0.0076923	2.6176876	0.0231715
1.2	3.1799415	3.1704634	0.0094781	3.1495789	0.0303627
1.4	3.7324000	3.7211654	0.0112346	3.6936862	0.0387138
1.6	4.2834838	4.2706218	0.0128620	4.2350972	0.0483866
1.8	4.8151763	4.8009586	0.0142177	4.7556185	0.0595577
2.0	5.3054720	5.2903695	0.0151025	5.2330546	0.0724173

Higher-Order Runge-Kutta Methods

Karl Heun (1859–1929) was a professor at the Technical University of Karlsruhe. He introduced this technique in a paper published in 1900 [Heu].

The term $T^{(3)}(t, y)$ can be approximated with global error $O(h^3)$ by an expression of the form

$$f(t + \alpha_1, y + \delta_1 f(t + \alpha_2, y + \delta_2 f(t, y))),$$

involving four parameters, but the algebra involved in the determination of $\alpha_1, \delta_1, \alpha_2,$ and δ_2 is quite involved. The most common $O(h^3)$ is Heun's method.

Heun's Method

$$w_0 = \alpha$$

$$w_{i+1} = w_i + \frac{h}{4}\left(f(t_i, w_i) + 3\left(f\left(t_i + \frac{2h}{3}, w_i + \frac{2h}{3}f\left(t_i + \frac{h}{3}, w_i + \frac{h}{3}f(t_i, w_i)\right)\right)\right)\right),$$

for $i = 0, 1, \ldots, N - 1$, with local error $O(h^4)$ and global error $O(h^3)$.

Illustration Applying Heun's method with $N = 10$, $h = 0.2$, $t_i = 0.2i$, and $w_0 = 0.5$ to approximate the solution to our usual example,

$$y' = y - t^2 + 1, \quad 0 \le t \le 2, \quad y(0) = 0.5$$

gives the values in Table 5.7. Note the decreased error throughout the range over the Midpoint and Modified Euler approximations. □

Table 5.7

t_i	$y(t_i)$	Heun's Method	Error
0.0	0.5000000	0.5000000	0
0.2	0.8292986	0.8292444	0.0000542
0.4	1.2140877	1.2139750	0.0001127
0.6	1.6489406	1.6487659	0.0001747
0.8	2.1272295	2.1269905	0.0002390
1.0	2.6408591	2.6405555	0.0003035
1.2	3.1799415	3.1795763	0.0003653
1.4	3.7324000	3.7319803	0.0004197
1.6	4.2834838	4.2830230	0.0004608
1.8	4.8151763	4.8146966	0.0004797
2.0	5.3054720	5.3050072	0.0004648

The program RKO4M52 implements the Runge-Kutta Order 4 method.

Runge-Kutta Order Four

Runge-Kutta methods of order 3 are not generally used. The most common Runge-Kutta method in use is of order 4. It is given by the following.

Runge-Kutta Method of Order 4

$$w_0 = \alpha,$$

$$k_1 = hf(t_i, w_i),$$

$$k_2 = hf\left(t_i + \frac{h}{2}, w_i + \frac{1}{2}k_1\right),$$

$$k_3 = hf\left(t_i + \frac{h}{2}, w_i + \frac{1}{2}k_2\right),$$

$$k_4 = hf(t_{i+1}, w_i + k_3),$$

$$w_{i+1} = w_i + \frac{1}{6}(k_1 + 2k_2 + 2k_3 + k_4),$$

where $i = 0, 1, \ldots, N - 1$, with local error $O(h^5)$ and global error $O(h^4)$.

Example 2 Use the Runge-Kutta method of order 4 with $h = 0.2$, $N = 10$, and $t_i = 0.2i$ to obtain approximations to the solution of the initial-value problem

$$y' = y - t^2 + 1, \quad 0 \le t \le 2, \quad y(0) = 0.5.$$

Solution   The approximation to $y(0.2)$ is obtained by

$$w_0 = 0.5$$

$$k_1 = 0.2 f(0, 0.5) = 0.2(1.5) = 0.3$$

$$k_2 = 0.2 f(0.1, 0.65) = 0.328$$

$$k_3 = 0.2 f(0.1, 0.664) = 0.3308$$

$$k_4 = 0.2 f(0.2, 0.8308) = 0.35816$$

$$w_1 = 0.5 + \frac{1}{6}(0.3 + 2(0.328) + 2(0.3308) + 0.35816) = 0.8292933.$$

The remaining results and their errors are listed in Table 5.8.   ▪

Table 5.8

t_i	Exact $y_i = y(t_i)$	Runge-Kutta Order 4 w_i	Error $|y_i - w_i|$
0.0	0.5000000	0.5000000	0
0.2	0.8292986	0.8292933	0.0000053
0.4	1.2140877	1.2140762	0.0000114
0.6	1.6489406	1.6489220	0.0000186
0.8	2.1272295	2.1272027	0.0000269
1.0	2.6408591	2.6408227	0.0000364
1.2	3.1799415	3.1798942	0.0000474
1.4	3.7324000	3.7323401	0.0000599
1.6	4.2834838	4.2834095	0.0000743
1.8	4.8151763	4.8150857	0.0000906
2.0	5.3054720	5.3053630	0.0001089

Computational Comparisons

The main computational effort in applying the Runge-Kutta methods involves the function evaluations of f. In the second-order methods, the local error is $O(h^3)$ and the cost is two functional evaluations per step. The Runge-Kutta method of order 4 requires four evaluations per step and the local error is $O(h^5)$. The relationship between the number of evaluations per step and the order of the global error is shown in Table 5.9. Because of the relative decrease in the order for n greater than 4, the methods of order less than 5 with smaller step size are used in preference to the higher-order methods using a larger step size.

Table 5.9

Evaluations per step:	2	3	4	$5 \leq n \leq 7$	$8 \leq n \leq 9$	$10 \leq n$
Best possible global error:	$O(h^2)$	$O(h^3)$	$O(h^4)$	$O(h^{n-1})$	$O(h^{n-2})$	$O(h^{n-3})$

One way to compare the lower-order Runge-Kutta methods is described as follows: The Runge-Kutta method of order 4 requires four evaluations per step, so to be superior to Euler's method, which requires only one evaluation per step, it should give more accurate answers than when Euler's method uses one-fourth the Runge-Kutta step size. Similarly, if the Runge-Kutta method of order 4 is to be superior to the second-order Runge-Kutta methods, which require two evaluations per step, it should give more accuracy with step size h than a second-order method with step size $\frac{1}{2}h$. The following Illustration indicates the superiority of the Runge-Kutta method of order 4 by this measure.

Illustration   For the problem

$$y' = y - t^2 + 1, \quad 0 \le t \le 2, \quad y(0) = 0.5,$$

Euler's method with $h = 0.025$, the Midpoint method with $h = 0.05$, and the Runge-Kutta fourth-order method with $h = 0.1$ are compared at the common mesh points of these methods 0.1, 0.2, 0.3, 0.4, and 0.5. Each of these techniques requires 20 function evaluations to determine the values listed in Table 5.10 to approximate $y(0.5)$. In this example, the fourth-order method is clearly superior.   □

Table 5.10

t_i	Exact	Euler $h = 0.025$	Modified Euler $h = 0.05$	Runge-Kutta Order 4 $h = 0.1$
0.0	0.5000000	0.5000000	0.5000000	0.5000000
0.1	0.6574145	0.6554982	0.6573085	0.6574144
0.2	0.8292986	0.8253385	0.8290778	0.8292983
0.3	1.0150706	1.0089334	1.0147254	1.0150701
0.4	1.2140877	1.2056345	1.2136079	1.2140869
0.5	1.4256394	1.4147264	1.4250141	1.4256384

MATLAB uses methods to approximate the solutions to ordinary differential equations that are more sophisticated than the standard Runge-Kutta techniques. An introduction to methods of this type is considered in Section 5.6.

EXERCISE SET 5.3

1. Use the Modified Euler method to approximate the solutions to each of the following initial-value problems, and compare the results to the actual values.

    a. $y' = te^{3t} - 2y, \quad 0 \le t \le 1, \quad y(0) = 0$, with $h = 0.5$; actual solution $y(t) = \frac{1}{5}te^{3t} - \frac{1}{25}e^{3t} + \frac{1}{25}e^{-2t}$.

    b. $y' = 1 + (t - y)^2, \quad 2 \le t \le 3, \quad y(2) = 1$, with $h = 0.5$; actual solution $y(t) = t + \frac{1}{1-t}$.

    c. $y' = 1 + y/t, \quad 1 \le t \le 2, \quad y(1) = 2$, with $h = 0.25$; actual solution $y(t) = t \ln t + 2t$.

    d. $y' = \cos 2t + \sin 3t, \quad 0 \le t \le 1, \quad y(0) = 1$, with $h = 0.25$; actual solution $y(t) = \frac{1}{2}\sin 2t - \frac{1}{3}\cos 3t + \frac{4}{3}$.

2. Use the Modified Euler method to approximate the solutions to each of the following initial-value problems, and compare the results to the actual values.

    a. $y' = y/t - (y/t)^2, \quad 1 \le t \le 2, \quad y(1) = 1$, with $h = 0.1$; actual solution $y(t) = t/(1 + \ln t)$.

    b. $y' = 1 + y/t + (y/t)^2, \quad 1 \le t \le 3, \quad y(1) = 0$, with $h = 0.2$; actual solution $y(t) = t \tan(\ln t)$.

    c. $y' = -(y + 1)(y + 3), \quad 0 \le t \le 2, \quad y(0) = -2$, with $h = 0.2$; actual solution $y(t) = -3 + 2(1 + e^{-2t})^{-1}$.

    d. $y' = -5y + 5t^2 + 2t, \quad 0 \le t \le 1, \quad y(0) = \frac{1}{3}$, with $h = 0.1$; actual solution $y(t) = t^2 + \frac{1}{3}e^{-5t}$.

3. Use the Modified Euler method to approximate the solutions to each of the following initial-value problems, and compare the results to the actual values.

    a. $y' = \dfrac{2 - 2ty}{t^2 + 1}, \quad 0 \le t \le 1, \quad y(0) = 1$, with $h = 0.1$; actual solution $y(t) = \dfrac{2t + 1}{t^2 + 1}$.

    b. $y' = \dfrac{y^2}{1 + t}, \quad 1 \le t \le 2, \quad y(1) = \frac{-1}{\ln 2}$, with $h = 0.1$; actual solution $y(t) = \dfrac{-1}{\ln(t + 1)}$.

    c. $y' = (y^2 + y)/t$, $1 \le t \le 3$, $y(1) = -2$, with $h = 0.2$; actual solution $y(t) = \dfrac{2t}{1-t}$.

    d. $y' = -ty + 4t/y$, $0 \le t \le 1$, $y(0) = 1$, with $h = 0.1$; actual solution $y(t) = \sqrt{4 - 3e^{-t^2}}$.

4.   Repeat Exercise 1 using the Midpoint method.

5.   Repeat Exercise 2 using the Midpoint method.

6.   Repeat Exercise 3 using the Midpoint method.

7.   Repeat Exercise 1 using Heun's method.

8.   Repeat Exercise 2 using Heun's method.

9.   Repeat Exercise 3 using Heun's method.

10.   Repeat Exercise 1 using the Runge-Kutta method of order 4.

11.   Repeat Exercise 2 using the Runge-Kutta method of order 4.

12.   Repeat Exercise 3 using the Runge-Kutta method of order 4.

13.   Use the results of Exercise 2 and linear interpolation to approximate values of $y(t)$, and compare the results to the actual values.

    a. $y(1.25)$ and $y(1.93)$ **b.** $y(2.1)$ and $y(2.75)$

    c. $y(1.3)$ and $y(1.93)$ **d.** $y(0.54)$ and $y(0.94)$

14.   Use the results of Exercise 3 and linear interpolation to approximate values of $y(t)$, and compare the results to the actual values.

    a. $y(0.54)$ and $y(0.94)$ **b.** $y(1.25)$ and $y(1.93)$

    c. $y(1.3)$ and $y(2.93)$ **d.** $y(0.54)$ and $y(0.94)$

15.   Use the results of Exercise 11 and Cubic Hermite interpolation to approximate values of $y(t)$, and compare the approximations to the actual values.

    a. $y(1.25)$ and $y(1.93)$ **b.** $y(2.1)$ and $y(2.75)$

    c. $y(1.3)$ and $y(1.93)$ **d.** $y(0.54)$ and $y(0.94)$

16.   Use the results of Exercise 12 and Cubic Hermite interpolation to approximate values of $y(t)$, and compare the approximations to the actual values.

    a. $y(0.54)$ and $y(0.94)$ **b.** $y(1.25)$ and $y(1.93)$

    c. $y(1.3)$ and $y(2.93)$ **d.** $y(0.54)$ and $y(0.94)$

17.   Show that the Midpoint method and the Modified Euler method give the same approximations to the initial-value problem

$$y' = -y + t + 1, \quad 0 \le t \le 1, \quad y(0) = 1,$$

    for any choice of h. Why is this true?

18.   Water flows from an inverted conical tank with a circular orifice at the rate

$$\frac{dx}{dt} = -0.6\pi r^2 \sqrt{2g}\, \frac{\sqrt{x}}{A(x)},$$

    where r is the radius of the orifice, x is the height of the liquid level from the vertex of the cone, and $A(x)$ is the area of the cross-section of the tank x units above the orifice. Suppose $r = 0.1$ ft, $g = 32.1$ ft/s^2, and the tank has an initial water level of 8 ft and initial volume of $512(\pi/3)$ ft^3. Use the Runge-Kutta method of order 4 to find the following.

    a.   The water level after 10 min with $h = 20$ s

    b.   When the tank will be empty, to within 1 min.

19.   The irreversible chemical reaction in which two molecules of solid potassium dichromate ($K_2Cr_2O_7$), two molecules of water (H_2O), and three atoms of solid sulfur (S) combine to yield three molecules of the gas sulfur dioxide (SO_2), four molecules of solid potassium hydroxide (KOH), and two molecules of solid chromic oxide (Cr_2O_3) can be represented symbolically by the *stoichiometric equation*:

$$2K_2Cr_2O_7 + 2H_2O + 3S \longrightarrow 4KOH + 2Cr_2O_3 + 3SO_2.$$

If n_1 molecules of $K_2Cr_2O_7$, n_2 molecules of H_2O, and n_3 molecules of S are originally available, the following differential equation describes the amount $x(t)$ of KOH after time t:

$$\frac{dx}{dt} = k \left(n_1 - \frac{x}{2} \right)^2 \left(n_2 - \frac{x}{2} \right)^2 \left(n_3 - \frac{3x}{4} \right)^3,$$

where k is the velocity constant of the reaction. If $k = 6.22 \times 10^{-19}$, $n_1 = n_2 = 2 \times 10^3$, and $n_3 = 3 \times 10^3$, use the Runge-Kutta method of order 4 to determine how many units of potassium hydroxide will have been formed after 0.2 s.

20. Show that Heun's Method can be expressed in difference form, similar to that of the Runge-Kutta method of order 4, as

$$w_0 = \alpha,$$

$$k_1 = hf(t_i, w_i),$$

$$k_2 = hf \left(t_i + \frac{h}{3}, w_i + \frac{1}{3}k_1 \right),$$

$$k_3 = hf \left(t_i + \frac{2h}{3}, w_i + \frac{2}{3}k_2 \right),$$

$$w_{i+1} = w_i + \frac{1}{4}(k_1 + 3k_3),$$

for each $i = 0, 1, \ldots, N - 1$.

5.4 Predictor-Corrector Methods

The Taylor and Runge-Kutta methods are examples of **one-step methods** for approximating the solution to initial-value problems. These methods use w_i in the approximation w_{i+1} to $y(t_{i+1})$ but do not involve any of the prior approximations $w_0, w_1, \ldots, w_{i-1}$. Generally some functional evaluations of f are required at intermediate points, but these are discarded as soon as w_{i+1} is obtained.

Since $|y(t_j) - w_j|$ decreases in accuracy as j increases, better approximation methods can be derived if, when approximating $y(t_{i+1})$, we include in the method some of the approximations prior to w_i. Methods developed using this philosophy are called **multistep methods**. In brief, one-step methods consider what occurred at only one previous step; multistep methods consider what happened at more than one previous step.

To derive a multistep method, suppose that the solution to the initial-value problem

$$\frac{dy}{dt} = f(t, y), \quad \text{for } a \le t \le b, \quad \text{with } y(a) = \alpha,$$

is integrated over the interval $[t_i, t_{i+1}]$. Then

$$y(t_{i+1}) - y(t_i) = \int_{t_i}^{t_{i+1}} y'(t)dt = \int_{t_i}^{t_{i+1}} f(t, y(t))dt,$$

and

$$y(t_{i+1}) = y(t_i) + \int_{t_i}^{t_{i+1}} f(t, y(t))dt.$$

Since we cannot integrate $f(t, y(t))$ without knowing $y(t)$, which is the solution to the problem, we instead integrate an interpolating polynomial, $P(t)$, for $f(t, y(t))$ determined by some of the previously obtained data points $(t_0, w_0), (t_1, w_1), \ldots, (t_i, w_i)$. When we assume, in addition, that $y(t_i) \approx w_i$, we have

$$y(t_{i+1}) \approx w_i + \int_{t_i}^{t_{i+1}} P(t)dt.$$

If w_{m+1} is the first approximation generated by the multistep method, then we need to supply starting values $w_0, w_1, \ldots, w_m$ for the method. These starting values are generated using a one-step Runge-Kutta method with the same error characteristics as the multistep method.

There are two distinct classes of multistep methods. In an **explicit method**, w_{i+1} does not involve the function evaluation $f(t_{i+1}, w_{i+1})$. A method that does depend in part on $f(t_{i+1}, w_{i+1})$ is an **implicit** method.

Adams-Bashforth Explicit Methods

Some of the explicit multistep methods, together with their required starting values and local error terms, are given next.

Adams-Bashforth Two-Step Explicit Method

$$w_0 = \alpha, \ w_1 = \alpha_1,$$

$$w_{i+1} = w_i + \frac{h}{2}[3f(t_i, w_i) - f(t_{i-1}, w_{i-1})],$$

where $i = 1, 2, \ldots, N - 1$, with local error $\frac{5}{12}y'''(\mu_i)h^3$ for some μ_i in (t_{i-1}, t_{i+1}).

Adams-Bashforth Three-Step Explicit Method

$$w_0 = \alpha, \ w_1 = \alpha_1, \ w_2 = \alpha_2,$$

$$w_{i+1} = w_i + \frac{h}{12}[23f(t_i, w_i) - 16f(t_{i-1}, w_{i-1}) + 5f(t_{i-2}, w_{i-2})]$$

where $i = 2, 3, \ldots, N - 1$, with local error $\frac{3}{8}y^{(4)}(\mu_i)h^4$ for some μ_i in (t_{i-2}, t_{i+1}).

Adams-Bashforth Four-Step Explicit Method

$$w_0 = \alpha, \ w_1 = \alpha_1, \ w_2 = \alpha_2, \ w_3 = \alpha_3,$$

$$w_{i+1} = w_i + \frac{h}{24}[55f(t_i, w_i) - 59f(t_{i-1}, w_{i-1}) + 37f(t_{i-2}, w_{i-2}) - 9f(t_{i-3}, w_{i-3})]$$

where $i = 3, 4, \ldots, N - 1$, with local error $\frac{251}{720}y^{(5)}(\mu_i)h^5$ for some μ_i in (t_{i-3}, t_{i+1}).

Adams-Bashforth Five-Step Explicit Method

$$w_0 = \alpha, \ w_1 = \alpha_1, \ w_2 = \alpha_2, \ w_3 = \alpha_3, \ w_4 = \alpha_4$$

$$w_{i+1} = w_i + \frac{h}{720}[1901 f(t_i, w_i) - 2774 f(t_{i-1}, w_{i-1})$$

$$+ 2616 f(t_{i-2}, w_{i-2}) - 1274 f(t_{i-3}, w_{i-3}) + 251 f(t_{i-4}, w_{i-4})]$$

where $i = 4, 5, \ldots, N - 1$, with local error $\frac{95}{288} y^{(6)}(\mu_i) h^6$ for some μ_i in (t_{i-4}, t_{i+1}).

Adams-Moulton Implicit Methods

Implicit methods use $(t_{i+1}, f(t_{i+1}, w_{i+1}))$ as an additional interpolation node in the approximation of the integral

$$\int_{t_i}^{t_{i+1}} f(t, y(t)) dt.$$

Some of the more common implicit methods are listed next. Notice that the local error of an $(m - 1)$-step implicit method is $O(h^{m+1})$, the same as that of an m-step explicit method. They both use m function evaluations, however, because the implicit methods use $f(t_{i+1}, w_{i+1})$, but the explicit methods do not.

Adams-Moulton Two-Step Implicit Method

$$w_0 = \alpha, \ w_1 = \alpha_1$$

$$w_{i+1} = w_i + \frac{h}{12}[5 f(t_{i+1}, w_{i+1}) + 8 f(t_i, w_i) - f(t_{i-1}, w_{i-1})]$$

where $i = 1, 2, \ldots, N - 1$, with local error $-\frac{1}{24} y^{(4)}(\mu_i) h^4$ for some μ_i in (t_{i-1}, t_{i+1}).

Adams-Moulton Three-Step Implicit Method

$$w_0 = \alpha, \ w_1 = \alpha_1, \ w_2 = \alpha_2,$$

$$w_{i+1} = w_i + \frac{h}{24}[9 f(t_{i+1}, w_{i+1}) + 19 f(t_i, w_i) - 5 f(t_{i-1}, w_{i-1}) + f(t_{i-2}, w_{i-2})],$$

where $i = 2, 3, \ldots, N - 1$, with local error $-\frac{19}{720} y^{(5)}(\mu_i) h^5$ for some μ_i in (t_{i-2}, t_{i+1}).

Adams-Moulton Four-Step Implicit Method

$$w_0 = \alpha, \ w_1 = \alpha_1, \ w_2 = \alpha_2, \ w_3 = \alpha_3,$$

$$w_{i+1} = w_i + \frac{h}{720}[251 f(t_{i+1}, w_{i+1}) + 646 f(t_i, w_i) - 264 f(t_{i-1}, w_{i-1})$$

$$+ 106 f(t_{i-2}, w_{i-2}) - 19 f(t_{i-3}, w_{i-3})]$$

where $i = 3, 4, \ldots, N - 1$, with local error $-\frac{3}{160} y^{(6)}(\mu_i) h^6$ for some μ_i in (t_{i-3}, t_{i+1}).

It is interesting to compare an m-step Adams-Bashforth explicit method to an $(m-1)$-step Adams-Moulton implicit method. Both require m evaluations of f per step, and both have the terms $y^{(m+1)}(\mu_i)h^{m+1}$ in their local errors. In general, the coefficients of the terms involving f in the approximation and those in the local error are smaller for the implicit methods than for the explicit methods. This leads to smaller truncation and round-off errors for the implicit methods.

Example 1   In Example 2 of Section 5.3 (see Table 5.8 on page 188) we used the Runge-Kutta method of order 4 with $h = 0.2$ to approximate the solutions to the initial value problem

$$y' = y - t^2 + 1, \quad 0 \le t \le 2, \quad y(0) = 0.5.$$

The first approximations were found to be $y(0) = w_0 = 0.5$, $y(0.2) \approx w_1 = 0.8292933$, $y(0.4) \approx w_2 = 1.2140762$, and $y(0.6) \approx w_3 = 1.6489220$. Use these as starting values for the fourth-order Adams-Bashforth method to compute new approximations for $y(0.8)$ and $y(1.0)$, and compare these new approximations to those produced by the Runge-Kutta method of order 4.

Solution   For the fourth-order Adams-Bashforth we have

$$
\begin{aligned}
y(0.8) \approx w_4 &= w_3 + \frac{0.2}{24}(55f(0.6, w_3) - 59f(0.4, w_2) + 37f(0.2, w_1) - 9f(0, w_0)) \\
&= 1.6489220 + \frac{0.2}{24}(55f(0.6, 1.6489220) - 59f(0.4, 1.2140762) \\
&\quad + 37f(0.2, 0.8292933) - 9f(0, 0.5)) \\
&= 1.6489220 + 0.0083333(55(2.2889220) - 59(2.0540762) \\
&\quad + 37(1.7892933) - 9(1.5)) \\
&= 2.1272892,
\end{aligned}
$$

and

$$
\begin{aligned}
y(1.0) \approx w_5 &= w_4 + \frac{0.2}{24}(55f(0.8, w_4) - 59f(0.6, w_3) + 37f(0.4, w_2) - 9f(0.2, w_1)) \\
&= 2.1272892 + \frac{0.2}{24}(55f(0.8, 2.1272892) - 59f(0.6, 1.6489220) \\
&\quad + 37f(0.4, 1.2140762) - 9f(0.2, 0.8292933)) \\
&= 2.1272892 + 0.0083333(55(2.4872892) - 59(2.2889220) \\
&\quad + 37(2.0540762) - 9(1.7892933)) \\
&= 2.6410533,
\end{aligned}
$$

The errors for these approximations at $t = 0.8$ and $t = 1.0$ are, respectively,

$$|2.1272295 - 2.1272892| = 5.97 \times 10^{-5} \quad \text{and} \quad |2.6410533 - 2.6408591| = 1.94 \times 10^{-4}.$$

The corresponding Runge-Kutta approximations had errors

$$|2.1272027 - 2.1272892| = 2.69 \times 10^{-5} \quad \text{and} \quad |2.6408227 - 2.6408591| = 3.64 \times 10^{-5}.$$

■

The implicit Adams-Moulton methods generally give considerably better results than the explicit Adams-Bashforth method of the same order. However, the implicit methods have the inherent weakness of first having to convert the method algebraically to an explicit representation for w_{i+1}. That this procedure can become difficult, if not impossible, can be seen by considering the elementary initial-value problem

$$y' = e^y, \quad \text{for } 0 \leq t \leq 0.25, \quad \text{with } y(0) = 1.$$

Since $f(t, y) = e^y$, the Adams-Moulton Three-Step method has

$$w_{i+1} = w_i + \frac{h}{24}[9e^{w_{i+1}} + 19e^{w_i} - 5e^{w_{i-1}} + e^{w_{i-2}}]$$

as its difference equation, and this equation cannot be solved explicitly for w_{i+1}. We could use Newton's method or the Secant method to approximate w_{i+1}, but this complicates the procedure considerably.

Predictor-Corrector Methods

In practice, implicit multistep methods are not used alone. Rather, they are used to improve approximations obtained by explicit methods. The combination of an explicit and implicit technique is called a **predictor-corrector method**. The explicit method predicts an approximation, and the implicit method corrects this prediction.

Consider the following fourth-order method for solving an initial-value problem. The first step is to calculate the starting values $w_0, w_1, w_2,$ and w_3 for the explicit Adams-Bashforth Four-Step method. To do this, we use a fourth-order one-step method, specifically, the Runge-Kutta method of order 4. The next step is to calculate an approximation, w_{4p}, to $y(t_4)$ using the explicit Adams-Bashforth Four-Step method as predictor:

$$w_{4p} = w_3 + \frac{h}{24}[55f(t_3, w_3) - 59f(t_2, w_2) + 37f(t_1, w_1) - 9f(t_0, w_0)].$$

This approximation is improved by use of the implicit Adams-Moulton Three-Step method as corrector:

$$w_4 = w_3 + \frac{h}{24}[9f(t_4, w_{4p}) + 19f(t_3, w_3) - 5f(t_2, w_2) + f(t_1, w_1)].$$

The value w_4 is now used as the approximation to $y(t_4)$. Then the technique of using the Adams-Bashforth method as a predictor and the Adams-Moulton method as a corrector is repeated to find w_{5p} and w_5, the initial and final approximations to $y(t_5)$. This process is continued until we obtain an approximation to $y(t_N) = y(b)$.

Program PRCORM53 is based on the Adams-Bashforth Four-Step method as predictor and one iteration of the Adams-Moulton Three-Step method as corrector, with the starting values obtained from the Runge-Kutta method of order 4.

The program PRCORM53 implements the Adams Predictor-Corrector method.

Example 2   Apply the Adams fourth-order predictor-corrector method with $h = 0.2$ and starting values from the Runge-Kutta fourth-order method to the initial-value problem

$$y' = y - t^2 + 1, \quad 0 \leq t \leq 2, \quad y(0) = 0.5.$$

Solution   This is a continuation and modification of the problem considered in Example 1 at the beginning of the section. In that example, we found that the starting approximations from Runge-Kutta are

$$y(0) = w_0 = 0.5, \ y(0.2) \approx w_1 = 0.8292933, \ y(0.4) \approx w_2 = 1.2140762, \quad \text{and}$$

$$y(0.6) \approx w_3 = 1.6489220.$$

and the fourth-order Adams-Bashforth method gave

$$y(0.8) \approx w_{4p} = w_3 + \frac{0.2}{24}[55f(0.6, w_3) - 59f(0.4, w_2) + 37f(0.2, w_1) - 9f(0, w_0)]$$

$$= 1.6489220 + \frac{0.2}{24}[55f(0.6, 1.6489220) - 59f(0.4, 1.2140762)$$

$$+ 37f(0.2, 0.8292933) - 9f(0, 0.5)]$$

$$= 1.6489220 + 0.0083333[55(2.2889220) - 59(2.0540762)$$

$$+ 37(1.7892933) - 9(1.5)]$$

$$= 2.1272892.$$

We will now use w_{4p} as the predictor of the approximation to $y(0.8)$ and determine the corrected value w_4, from the implicit Adams-Moulton method. This gives

$$y(0.8) \approx w_4 = w_3 + \frac{0.2}{24}[9f(0.8, w_{4p}) + 19f(0.6, w_3) - 5f(0.4, w_2) + f(0.2, w_1)]$$

$$= 1.6489220 + \frac{0.2}{24}[9f(0.8, 2.1272892) + 19f(0.6, 1.6489220)$$

$$- 5f(0.4, 1.2140762) + f(0.2, 0.8292933)]$$

$$= 1.6489220 + 0.0083333[9(2.4872892) + 19(2.2889220)$$

$$- 5(2.0540762) + (1.7892933)]$$

$$= 2.1272056.$$

Now we use this approximation to determine the predictor, w_{5p}, for $y(1.0)$ as

$$y(1.0) \approx w_{5p} = w_4 + \frac{0.2}{24}[55f(0.8, w_4) - 59f(0.6, w_3) + 37f(0.4, w_2) - 9f(0.2, w_1)]$$

$$= 2.1272056 + \frac{0.2}{24}[55f(0.8, 2.1272056) - 59f(0.6, 1.6489220)$$

$$+ 37f(0.4, 1.2140762) - 9f(0.2, 0.8292933)]$$

$$= 2.1272056 + 0.0083333[55(2.4872056) - 59(2.2889220)$$

$$+ 37(2.0540762) - 9(1.7892933)]$$

$$= 2.6409314,$$

and correct this with

$$y(1.0) \approx w_5 = w_4 + \frac{0.2}{24}[9f(1.0, w_{5p}) + 19f(0.8, w_4) - 5f(0.6, w_3) + f(0.4, w_2)]$$

$$= 2.1272056 + \frac{0.2}{24}[9f(1.0, 2.6409314) + 19f(0.8, 2.1272892)$$

$$- 5f(0.6, 1.6489220) + f(0.4, 1.2140762)]$$

$$= 2.1272056 + 0.0083333[9(2.6409314) + 19(2.4872056)$$

$$- 5(2.2889220) + (2.0540762)]$$

$$= 2.6408286.$$

In Example 1 we found that using the explicit Adams-Bashforth method alone produced results that were inferior to those of Runge-Kutta. However, these approximations to $y(0.8)$ and $y(1.0)$ are accurate to within

$$|2.1272295 - 2.1272056| = 2.39 \times 10^{-5} \quad \text{and}$$

$$|2.6408286 - 2.6408591| = 3.05 \times 10^{-5},$$

respectively, compared to those of Runge-Kutta, which were accurate, respectively, to within

$$|2.1272027 - 2.1272892| = 2.69 \times 10^{-5} \quad \text{and}$$

$$|2.6408227 - 2.6408591| = 3.64 \times 10^{-5}. \qquad \blacksquare$$

Other multistep methods can be derived using integration of interpolating polynomials over intervals of the form $[t_j, t_{i+1}]$ for $j \le i - 1$, where some of the data points are omitted. Milne's method is an explicit technique that results when a Newton Backward-Difference interpolating polynomial is integrated over $[t_{i-3}, t_{i+1}]$.

Milne's Method

$$w_{i+1} = w_{i-3} + \frac{4h}{3}[2f(t_i, w_i) - f(t_{i-1}, w_{i-1}) + 2f(t_{i-2}, w_{i-2})],$$

where $i = 3, 4, \ldots, N - 1$, with local error $\frac{14}{45}h^5 y^{(5)}(\mu_i)$ for some μ_i in (t_{i-3}, t_{i+1}).

This method is used as a predictor for an implicit method called Simpson's method. Its name comes from the fact that it can be derived using Simpson's rule for approximating integrals.

Simpson's Method

$$w_{i+1} = w_{i-1} + \frac{h}{3}[f(t_{i+1}, w_{i+1}) + 4f(t_i, w_i) + f(t_{i-1}, w_{i-1})],$$

where $i = 1, 2, \ldots, N - 1$, with local error $-\frac{1}{90}h^5 y^{(5)}(\mu_i)$ for some μ_i in (t_{i-1}, t_{i+1}).

Although the local error involved with a predictor-corrector method of the Milne-Simpson type is generally smaller than that of the Adams-Bashforth-Moulton method, the technique has limited use because of round-off error problems, which do not occur with the Adams procedure.

MATLAB uses methods that are more sophisticated than the standard Adams-Bashforth-Moulton techniques to approximate the solutions to ordinary differential equations. An introduction to methods of this type is considered in Section 5.6.

EXERCISE SET 5.4

1. Use all the Adams-Bashforth methods to approximate the solutions to the following initial-value problems. In each case, use exact starting values and compare the results to the actual values.

   a. $y' = te^{3t} - 2y$, for $0 \le t \le 1$, with $y(0) = 0$ and $h = 0.2$; actual solution $y(t) = \frac{1}{5}te^{3t} - \frac{1}{25}e^{3t} + \frac{1}{25}e^{-2t}$.

   b. $y' = 1 + (t - y)^2$, for $2 \le t \le 3$, with $y(2) = 1$ and $h = 0.2$; actual solution $y(t) = t + 1/(1 - t)$.

   c. $y' = 1 + \frac{y}{t}$, for $1 \le t \le 2$, with $y(1) = 2$ and $h = 0.2$; actual solution $y(t) = t \ln t + 2t$.

   d. $y' = \cos 2t + \sin 3t$, for $0 \le t \le 1$ with $y(0) = 1$ and $h = 0.2$; actual solution $y(t) = \frac{1}{2}\sin 2t - \frac{1}{3}\cos 3t + \frac{4}{3}$.

2. Use all the Adams-Moulton methods to approximate the solutions to Exercises 1(a), 1(c), and 1(d). In each case, use exact starting values and explicitly solve for w_{i+1}. Compare the results to the actual values.

3. Use each of the Adams-Bashforth methods to approximate the solutions to the following initial-value problems. In each case, use starting values obtained from the Runge-Kutta method of order 4. Compare the results to the actual values.

   a. $y' = \frac{y}{t} - \left(\frac{y}{t}\right)^2$, for $1 \le t \le 2$, with $y(1) = 1$ and $h = 0.1$; actual solution $y(t) = t/(1 + \ln t)$.

   b. $y' = 1 + \frac{y}{t} + \left(\frac{y}{t}\right)^2$, for $1 \le t \le 3$, with $y(1) = 0$ and $h = 0.2$; actual solution $y(t) = t \tan(\ln t)$.

   c. $y' = -(y + 1)(y + 3)$, for $0 \le t \le 2$, with $y(0) = -2$ and $h = 0.1$; actual solution $y(t) = -3 + 2/(1 + e^{-2t})$.

   d. $y' = -5y + 5t^2 + 2t$, for $0 \le t \le 1$, with $y(0) = 1/3$ and $h = 0.1$; actual solution $y(t) = t^2 + \frac{1}{3}e^{-5t}$.

4. Use the predictor-corrector method based on the Adams-Bashforth Four-Step method and the Adams-Moulton Three-Step method to approximate the solutions to the initial-value problems in Exercise 1.

5. Use the predictor-corrector method based on the Adams-Bashforth Four-Step method and the Adams-Moulton Three-Step method to approximate the solutions to the initial-value problem in Exercise 3.

6. The initial-value problem

$$y' = e^y, \quad \text{for } 0 \le t \le 0.20, \quad \text{with } y(0) = 1$$

   has the exact solution

$$y(t) = 1 - \ln(1 - et).$$

   Applying the Adams-Moulton Three-Step method to the problem requires solving for w_{i+1} in the equation

$$g(w_{i+1}) = w_{i+1} - \left(w_i + \frac{h}{24}\left(9e^{w_{i+1}} + 19e^{w_i} - 5e^{w_{i-1}} + e^{w_{i-2}}\right)\right) = 0.$$

   Suppose that $h = 0.01$ and we use the exact starting values for w_0, w_1, and w_2.

   a. Apply Newton's method to this equation with the starting value w_2 to approximate w_3 to within 10^{-6}.

   b. Repeat the calculations in (a) using the starting value w_i to determine approximations accurate to within 10^{-6} for each w_{i+1} for $i = 2, \ldots, 19$.

7. Use the Milne-Simpson Predictor-Corrector method to approximate the solutions to the initial-value problems in Exercise 3.

5.5 Extrapolation Methods

Extrapolation was used in Romberg integration for the approximation of definite integrals, where we found that, by correctly averaging relatively inaccurate trapezoidal approximations, we could produce new approximations that are exceedingly accurate. In this section we will apply extrapolation to increase the accuracy of approximations to the solution of initial-value problems. As we have previously seen, the original approximations must have an error expansion of a specific form for the procedure to be successful.

To apply extrapolation to solve initial-value problems, we use a technique based on the Midpoint method:

$$w_{i+1} = w_{i-1} + 2hf(t_i, w_i), \quad \text{for } i \geq 1. \tag{5.1}$$

This technique requires two starting values because both w_0 and w_1 are needed before the first midpoint approximation, w_2, can be determined. As usual, we use the initial condition for $w_0 = y(a) = \alpha$. To determine the second starting value, w_1, we apply Euler's method. Subsequent approximations are obtained from Eq. (5.1). After a series of approximations of this type are generated ending at a value t, an endpoint correction is performed that involves the final two midpoint approximations. This produces an approximation $w(t, h)$ to $y(t)$ that has the form

$$y(t) = w(t, h) + \sum_{k=1}^{\infty} \delta_k h^{2k}, \tag{5.2}$$

where the δ_k are constants related to the derivatives of the solution $y(t)$. The important point is that the δ_k do not depend on the step size h.

To illustrate the extrapolation technique for solving

$$y'(t) = f(t, y), \quad \text{for } a \leq t \leq b, \quad \text{with } y(a) = \alpha,$$

let us assume that we have a fixed step size h and that we wish to approximate $y(a + h)$.

As the first step let $h_0 = h/2$ and use Euler's method with $w_0 = \alpha$ to approximate $y(a + h_0) = y(a + h/2)$ as

$$w_1 = w_0 + h_0 f(a, w_0).$$

Then apply the Midpoint method with $t_{i-1} = a$ and $t_i = a + h_0 = a + h/2$ to produce a first approximation to $y(a + h) = y(a + 2h_0)$,

$$w_2 = w_0 + 2h_0 f(a + h_0, w_1).$$

The endpoint correction is applied to obtain the final approximation to $y(a + h)$ for the step size h_0. This results in an $O(h_0^2)$ approximation to $y(t_1)$ given by

$$y_{1,1} = \frac{1}{2}[w_2 + w_1 + h_0 f(a + 2h_0, w_2)].$$

We save the first approximation $y_{1,1}$ to $y(t_1)$ and discard the intermediate results, w_1 and w_2.

To obtain a second approximation, $y_{2,1}$, to $y(t_1)$, let $h_1 = h/4$ and use Euler's method with $w_0 = \alpha$ to obtain an approximation w_1 to $y(a + h_1) = y(a + h/4)$; that is,

$$w_1 = w_0 + h_1 f(a, w_0).$$

Then use the Midpoint method to produce approximations w_2 to $y(a + 2h_1) = y(a + h/2)$, w_3 to $y(a + 3h_1) = y(a + 3h/4)$, and w_4 to $y(a + 4h_1) = y(t_1)$. They are

$$w_2 = w_0 + 2h_1 f(a + h_1, w_1)$$

$$w_3 = w_1 + 2h_1 f(a + 2h_1, w_2)$$

and

$$w_4 = w_2 + 2h_1 f(a + 3h_1, w_3).$$

The endpoint correction is now applied to w_3 and w_4 to produce the improved $O(h_1^2)$ approximation to $y(t_1)$,

$$y_{2,1} = \frac{1}{2}[w_4 + w_3 + h_1 f(a + 4h_1, w_4)].$$

Because of the form of the error given in Eq. (5.2), the two approximations to $y(a + h)$ have the property that

$$y(a + h) = y_{1,1} + \delta_1 \left(\frac{h}{2}\right)^2 + \delta_2 \left(\frac{h}{2}\right)^4 + \cdots = y_{1,1} + \delta_1 \frac{h^2}{4} + \delta_2 \frac{h^4}{16} + \cdots$$

and

$$y(a + h) = y_{2,1} + \delta_1 \left(\frac{h}{4}\right)^2 + \delta_2 \left(\frac{h}{4}\right)^4 + \cdots = y_{2,1} + \delta_1 \frac{h^2}{16} + \delta_2 \frac{h^4}{256} + \cdots .$$

We can eliminate the $O(h^2)$ portion of this truncation error by averaging these two formulas appropriately. Specifically, if we subtract the first equation from 4 times the second equation and divide the result by 3, we have

$$y(a + h) = y_{2,1} + \frac{1}{3}(y_{2,1} - y_{1,1}) - \delta_2 \frac{h^4}{64} + \cdots .$$

So the approximation to $y(t_1)$ given by

$$y_{2,2} = y_{2,1} + \frac{1}{3}(y_{2,1} - y_{1,1})$$

has error of order $O(h^4)$.

We next let $h_2 = h/6$ and apply Euler's method once, followed by the Midpoint method five times. Then we use the endpoint correction to determine the h^2 approximation, $y_{3,1}$, to $y(a + h) = y(t_1)$. This approximation can be averaged with $y_{2,1}$ to produce a second $O(h^4)$ approximation that we denote $y_{3,2}$. Then $y_{3,2}$ and $y_{2,2}$ are averaged to eliminate the $O(h^4)$ error terms and produce an approximation with error of order $O(h^6)$. Higher-order formulas are generated by continuing the process.

The only difference between the extrapolation performed here and that used for Romberg integration in Section 4.4 results from the way the subdivisions are chosen. In Romberg integration there is a convenient formula for representing the Composite Trapezoidal rule approximations that uses consecutive divisions of the step size by the integers 1, 2, 4, 8, 16, 32, 64, ... This permits the averaging process to proceed in a straightforward manner. We do not have a means for easily producing consecutive approximations for initial-value problems, so the divisions for the extrapolation technique are chosen to minimize the number of required function evaluations. The averaging procedure arising from this choice of subdivision is not as elementary as that used for Romberg integration, but other than that, the process is the same.

Program EXTRAP54 uses the extrapolation technique with the sequence of integers

> The program EXTRAP54 implements the Extrapolation method.

$$q_0 = 2, \, q_1 = 4, \, q_2 = 6, \, q_3 = 8, \, q_4 = 12, \, q_5 = 16, \, q_6 = 24, \, q_7 = 32.$$

A basic step size, h, is selected, and the method progresses by using $h_j = h/q_j$, for each $j = 0, \ldots, 7$, to approximate $y(t + h)$. The error is controlled by requiring that the approximations $y_{1,1}$, $y_{2,2}$, ... be computed until $|y_{i,i} - y_{i-1,i-1}|$ is less than a given tolerance. If the tolerance is not achieved by $i = 8$, then h is reduced, and the process is reapplied.

Minimum and maximum values of h, $hmin$, and $hmax$, respectively, are specified in the program to ensure control over the method. If $y_{i,i}$ is found to be acceptable, then w_1 is set to $y_{i,i}$, and computations begin again to determine w_2, which will approximate $y(t_2) = y(a+2h)$. The process is repeated until the approximation w_N to $y(b)$ is determined.

Example 1   Use the Extrapolation method with maximum step size $hmax = 0.2$, minimum step size $hmin = 0.01$, and tolerance $TOL = 10^{-9}$ to approximate the solution of the initial-value problem

$$y' = y - t^2 + 1, \quad 0 \le t \le 2, \quad y(0) = 0.5.$$

Solution   For the first step of the extrapolation method we let $w_0 = 0.5$, $t_0 = 0$, and $h = 0.2$. Then we compute

$$h_0 = h/2 = 0.1;$$
$$w_1 = w_0 + h_0 f(t_0, w_0) = 0.5 + 0.1(1.5) = 0.65;$$
$$w_2 = w_0 + 2h_0 f(t_0 + h_0, w_1) = 0.5 + 0.2(1.64) = 0.828;$$

and the first approximation to $y(0.2)$ is

$$y_{1,1} = \frac{1}{2}(w_2 + w_1 + h_0 f(t_0 + 2h_0, w_2)) = \frac{1}{2}(0.828 + 0.65 + 0.1 f(0.2, 0.828))$$
$$= 0.8284.$$

For the second approximation to $y(0.2)$ we compute

$$h_1 = h/4 = 0.05;$$
$$w_1 = w_0 + h_1 f(t_0, w_0) = 0.5 + 0.05(1.5) = 0.575;$$
$$w_2 = w_0 + 2h_1 f(t_0 + h_1, w_1) = 0.5 + 0.1(1.5725) = 0.65725;$$
$$w_3 = w_1 + 2h_1 f(t_0 + 2h_1, w_2) = 0.575 + 0.1(1.64725) = 0.739725;$$
$$w_4 = w_2 + 2h_1 f(t_0 + 3h_1, w_3) = 0.65725 + 0.1(1.717225) = 0.8289725.$$

Then the endpoint correction approximation is

$$y_{2,1} = \frac{1}{2}(w_4 + w_3 + h_1 f(t_0 + 4h_1, w_4))$$
$$= \frac{1}{2}(0.8289725 + 0.739725 + 0.05 f(0.2, 0.8289725)) = 0.8290730625.$$

This gives the first extrapolation approximation

$$y_{2,2} = y_{2,1} + \left(\frac{(1/4)^2}{(1/2)^2 - (1/4)^2} \right) (y_{21} - y_{11}) = 0.8292974167.$$

The third approximation is found by computing

$$h_2 = h/6 = 0.0\overline{3};$$

$$w_1 = w_0 + h_2 f(t_0, w_0) = 0.55;$$

$$w_2 = w_0 + 2h_2 f(t_0 + h_2, w_1) = 0.6032592593;$$

$$w_3 = w_1 + 2h_2 f(t_0 + 2h_2, w_2) = 0.6565876543;$$

$$w_4 = w_2 + 2h_2 f(t_0 + 3h_2, w_3) = 0.7130317696;$$

$$w_5 = w_3 + 2h_2 f(t_0 + 4h_2, w_4) = 0.7696045871;$$

$$w_6 = w_4 + 2h_2 f(t_0 + 5h_2, w_5) = 0.8291535569;$$

and the end-point correction approximation

$$y_{3,1} = \frac{1}{2}(w_6 + w_5 + h_2 f(t_0 + 6h_2, w_6)) = 0.8291982979.$$

We can now find two extrapolated approximations,

$$y_{3,2} = y_{3,1} + \left(\frac{(1/6)^2}{(1/4)^2 - (1/6)^2} \right)(y_{3,1} - y_{2,1}) = 0.8292984862,$$

and

$$y_{3,3} = y_{3,2} + \left(\frac{(1/6)^2}{(1/2)^2 - (1/6)^2} \right)(y_{3,2} - y_{2,2}) = 0.8292986199.$$

Because

$$|y_{3,3} - y_{2,2}| = 1.2 \times 10^{-6}$$

does not satisfy the tolerance, we need to compute at least one more row of the extrapolation table. We use $h_3 = h/8 = 0.025$ and calculate w_1 by Euler's method, $w_2, \cdots, w_8$ by the midpoint method and apply the endpoint correction. This will give us the new approximation $y_{4,1}$, which permits us to compute the new extrapolation row

$$y_{4,1} = 0.8292421745 \quad y_{4,2} = 0.8292985873 \quad y_{4,3} = 0.8292986210 \quad y_{4,4} = 0.8292986211.$$

Comparing $|y_{4,4} - y_{3,3}| = 1.2 \times 10^{-9}$ we find that the accuracy tolerance still has not been reached.

To obtain the entries in the next row, we use $h_4 = h/12 = 0.0\overline{6}$. First calculate w_1 by Euler's method, then w_2 through w_{12} by the Midpoint method. Finally use the endpoint correction to obtain $y_{5,1}$. The remaining entries in the fifth row are obtained using extrapolation, and are shown in Table 5.11.

Table 5.11

$y_{1,1} = 0.8284000000$				
$y_{2,1} = 0.8290730625$	$y_{2,2} = 0.8292974167$			
$y_{3,1} = 0.8291982979$	$y_{3,2} = 0.8292984862$	$y_{3,3} = 0.8292986199$		
$y_{4,1} = 0.8292421745$	$y_{4,2} = 0.8292985873$	$y_{4,3} = 0.8292986210$	$y_{4,4} = 0.8292986211$	
$y_{5,1} = 0.8292735291$	$y_{5,2} = 0.8292986128$	$y_{5,3} = 0.8292986213$	$y_{5,4} = 0.8292986213$	$y_{5,5} = 0.8292986213$

Because $y_{5,5} = 0.8292986213$ is within 10^{-9} of $y_{4,4}$ it is accepted as the approximation to $y(0.2)$. The procedure begins anew to approximate $y(0.4)$. The complete set of approximations accurate to the places listed is given in Table 5.12. The numbers in the final column indicate the level of extrapolation needed to obtain the required accuracy. Note that we chose $h = 0.05$ in the last line of Table 5.12 so that the final approximation is to $y(2.0)$. ■

Table 5.12

t_i	$y_i = y(t_i)$	w_i	h_i	k
0.200	0.8292986210	0.8292986213	0.200	5
0.400	1.2140876512	1.2140876510	0.200	4
0.600	1.6489405998	1.6489406000	0.200	4
0.700	1.8831236462	1.8831236460	0.100	5
0.800	2.1272295358	2.1272295360	0.100	4
0.900	2.3801984444	2.3801984450	0.100	7
0.925	2.4446908698	2.4446908710	0.025	8
0.950	2.5096451704	2.5096451700	0.025	3
1.000	2.6408590858	2.6408590860	0.050	3
1.100	2.9079169880	2.9079169880	0.100	7
1.200	3.1799415386	3.1799415380	0.100	6
1.300	3.4553516662	3.4553516610	0.100	8
1.400	3.7324000166	3.7324000100	0.100	5
1.450	3.8709427424	3.8709427340	0.050	7
1.475	3.9401071136	3.9401071050	0.025	3
1.525	4.0780532154	4.0780532060	0.050	4
1.575	4.2152541820	4.2152541820	0.050	3
1.675	4.4862274254	4.4862274160	0.100	4
1.775	4.7504844318	4.7504844210	0.100	4
1.825	4.8792274904	4.8792274790	0.050	3
1.875	5.0052154398	5.0052154290	0.050	3
1.925	5.1280506670	5.1280506570	0.050	4
1.975	5.2473151731	5.2473151660	0.050	8
2.000	5.3054719506	5.3054719440	0.025	3

EXERCISE SET 5.5

1. The initial-value problem

$$y' = \sqrt{2 - y^2}\, e^t, \quad \text{for } 0 \leq t \leq 0.8, \quad \text{with } y(0) = 0$$

has the exact solution $y(t) = \sqrt{2}\sin(e^t - 1)$. Use extrapolation with $h = 0.1$ to find an approximation for $y(0.1)$ to within a tolerance of $TOL = 10^{-5}$. Compare the approximation to the actual value.

2. The initial-value problem

$$y' = -y + 1 - \frac{y}{t}, \quad \text{for } 1 \leq t \leq 2, \quad \text{with } y(1) = 1$$

has the exact solution $y(t) = 1 + (e^{1-t} - 1)/t$. Use extrapolation with $h = 0.2$ to find an approximation for $y(1.2)$ to within a tolerance of $TOL = 0.00005$. Compare the approximation to the actual value.

3. Use the extrapolation program EXTRAP54 with $TOL = 10^{-4}$ to approximate the solutions to the following initial-value problems:

   a. $y' = \left(\frac{y}{t}\right)^2 + \left(\frac{y}{t}\right)$, for $1 \leq t \leq 1.2$, with $y(1) = 1$, $hmax = 0.05$, and $hmin = 0.02$.

   b. $y' = \sin t + e^{-t}$, for $0 \leq t \leq 1$, with $y(0) = 0$, $hmax = 0.25$, and $hmin = 0.02$.

   c. $y' = (y^2 + y)t^{-1}$, for $1 \leq t \leq 3$, with $y(1) = -2$, $hmax = 0.5$, and $hmin = 0.02$.

   d. $y' = -ty + 4ty^{-1}$, for $0 \leq t \leq 1$, with $y(0) = 1$, $hmax = 0.25$, and $hmin = 0.02$.

4. Use the extrapolation program EXTRAP54 with tolerance $TOL = 10^{-6}$, $hmax = 0.5$, and $hmin = 0.05$ to approximate the solutions to the following initial-value problems. Compare the results to the actual values.

   a. $y' = \frac{y}{t} - \frac{y^2}{t^2}$, for $1 \leq t \leq 4$, with $y(1) = 1$; actual solution $y(t) = t/(1 + \ln t)$.

   b. $y' = 1 + \frac{y}{t} + \left(\frac{y}{t}\right)^2$, for $1 \leq t \leq 3$, with $y(1) = 0$; actual solution $y(t) = t\tan(\ln t)$.

    c. $y' = -(y+1)(y+3)$, for $0 \le t \le 3$, with $y(0) = -2$; actual solution $y(t) = -3 + 2(1 + e^{-2t})^{-1}$.

    d. $y' = (t + 2t^3)y^3 - ty$, for $0 \le t \le 2$, with $y(0) = \frac{1}{3}$; actual solution $y(t) = (3 + 2t^2 + 6e^{t^2})^{-1/2}$.

5. Let $P(t)$ be the number of individuals in a population at time t, measured in years. If the average birth rate b is constant and the average death rate d is proportional to the size of the population (due to overcrowding), then the growth rate of the population is given by the *logistic equation*

$$\frac{dP(t)}{dt} = bP(t) - k[P(t)]^2$$

where $d = kP(t)$. Suppose $P(0) = 50,976$, $b = 2.9 \times 10^{-2}$, and $k = 1.4 \times 10^{-7}$. Find the population after 5 years.

5.6 Adaptive Techniques

The appropriate use of varying step size was seen in Section 4.6 to produce integral approximating methods that are efficient in the amount of computation required. This might not be sufficient to favor these methods due to the increased complication of applying them, but they have another important feature. The step-size selection procedure produces an estimate of the local error that does not require the approximation of the higher derivatives of the function. These methods are called *adaptive* because they adapt the number and position of the nodes used in the approximation to keep the local error within a specified bound.

You might like to review the Adaptive Quadrature material beginning on page 138 before considering this material.

There is a close connection between the problem of approximating the value of a definite integral and that of approximating the solution to an initial-value problem. It is not surprising, then, that there are adaptive methods for approximating the solutions to initial-value problems, and that these methods are not only efficient but incorporate the control of error.

Any one-step method for approximating the solution, $y(t)$, of the initial-value problem

$$y' = f(t, y), \quad \text{for } a \le t \le b, \quad \text{with } y(a) = \alpha$$

can be expressed in the form

$$w_{i+1} = w_i + h_i\phi(t_i, w_i, h_i), \quad \text{for } i = 0, 1, \ldots, N-1,$$

for some function ϕ. An ideal difference-equation method would have the property that given a tolerance $\varepsilon > 0$, the minimal number of mesh points would be used to ensure that the global error, $|y(t_i) - w_i|$, would not exceed ε for any $i = 0, 1, \ldots, N$. Having a minimal number of mesh points and also controlling the global error of a difference method is, not surprisingly, inconsistent with the points being equally spaced in the interval. In this section we examine techniques used to control the error of a difference-equation method in an efficient manner by the appropriate choice of mesh points.

Although we cannot generally determine the global error of a method, there is often a close connection between the local error and the global error. If a method has local error $O(h^{n+1})$, then the global error of the method is $O(h^n)$. By using methods of differing order we can predict the local error and, using this prediction, choose a step size that will keep the global error in check.

To illustrate the technique, suppose that we have two approximation techniques. The first is an nth-order method obtained from an nth-order Taylor method of the form

$$y(t_{i+1}) = y(t_i) + h\phi(t_i, y(t_i), h) + O(h^{n+1}).$$

This method produces approximations

$$w_0 = \alpha$$

$$w_{i+1} = w_i + h\phi(t_i, w_i, h) \quad \text{for} \quad i > 0,$$

which satisfy, for some $\hat{K}$ and all relevant h and i,

$$|y(t_i) - w_i| < \hat{K}h^n.$$

In general, the method is generated by applying a Runge-Kutta modification to the Taylor method, but the specific derivation is unimportant.

The second method is similar but of higher order. For example, let us suppose it comes from an $(n + 1)$st-order Taylor method of the form

$$y(t_{i+1}) = y(t_i) + h\tilde{\phi}(t_i, y(t_i), h) + O(h^{n+2}),$$

producing approximations

$$\tilde{w}_0 = \alpha$$

$$\tilde{w}_{i+1} = \tilde{w}_i + h\tilde{\phi}(t_i, \tilde{w}_i, h) \quad \text{for} \quad i > 0,$$

which satisfy, for some $\tilde{K}$ and all relevant h and i,

$$|y(t_i) - \tilde{w}_i| < \tilde{K}h^{n+1}.$$

We assume now that at the point t_i we have

$$w_i = \tilde{w}_i = z(t_i),$$

where $z(t)$ is the solution to the differential equation that does not satisfy the original initial condition but instead satisfies the condition $z(t_i) = w_i$. The typical difference between $y(t)$ and $z(t)$ is shown in Figure 5.3.

Figure 5.3

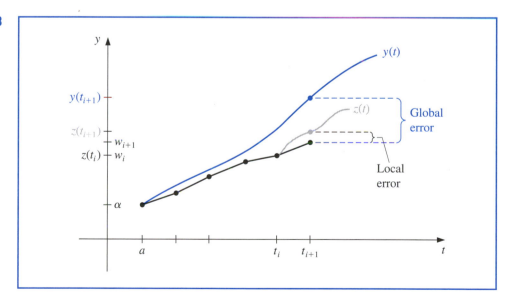

Applying the two methods to the differential equation with the fixed step size h produces two approximations, w_{i+1} and $\tilde{w}_{i+1}$, whose differences from $y(t_i + h)$ represent global errors but whose differences from $z(t_i + h)$ represent local errors.

Now consider

$$z(t_i + h) - w_{i+1} = (\tilde{w}_{i+1} - w_{i+1}) + (z(t_i + h) - \tilde{w}_{i+1}).$$

The term on the left side of this equation is $O(h^{n+1})$, the local error of the nth order method, but the second term on the right side is $O(h^{n+2})$, the local error of the $(n+1)$st order method. This implies that the dominant portion on the right side comes from the first term; that is,

$$z(t_i + h) - w_{i+1} \approx \tilde{w}_{i+1} - w_{i+1} = O(h^{n+1}).$$

So, a constant K exists with

$$Kh^{n+1} = |z(t_i + h) - w_{i+1}| \approx |\tilde{w}_{i+1} - w_{i+1}|,$$

and K can be approximated by

$$K \approx \frac{|\tilde{w}_{i+1} - w_{i+1}|}{h^{n+1}}. \tag{5.3}$$

Let us now return to the global error associated with the problem we really want to solve,

$$y' = f(t, y), \quad \text{for } a \leq t \leq b, \quad \text{with } y(a) = \alpha,$$

and consider the adjustment to the step size needed if this global error is expected to be bounded by the tolerance ε. Using a multiple q of the original step size implies that we need to ensure that

$$|y(t_i + qh) - w_{i+1}(\text{using the new step size } qh)| < K(qh)^n < \varepsilon.$$

This implies, from Eq. (5.3), that for the approximations w_{i+1} and $\tilde{w}_{i+1}$ using the original step size h we have

$$Kq^n h^n \approx \frac{|\tilde{w}_{i+1} - w_{i+1}|}{h^{n+1}} q^n h^n = \frac{q^n |\tilde{w}_{i+1} - w_{i+1}|}{h} < \varepsilon.$$

Solving this inequality for q tells us how we should choose the new step size, qh, to ensure that the global error is within the bound ε:

$$q < \left[\frac{\varepsilon h}{|\tilde{w}_{i+1} - w_{i+1}|} \right]^{1/n}.$$

Erwin Fehlberg developed this and other error control techniques while working for the NASA facility in Huntsville, Alabama during the 1960s. In 1969 he received NASA's Exceptional Scientific Achievement Medal for his work.

The Runge-Kutta-Fehlberg Method

One popular technique that uses this inequality for error control is the **Runge-Kutta-Fehlberg method** (see [Fe]). It uses a Runge-Kutta method of order 5,

$$\tilde{w}_{i+1} = w_i + \frac{16}{135}k_1 + \frac{6656}{12825}k_3 + \frac{28561}{56430}k_4 - \frac{9}{50}k_5 + \frac{2}{55}k_6,$$

to estimate the local error in a Runge-Kutta method of order 4,

$$w_{i+1} = w_i + \frac{25}{216}k_1 + \frac{1408}{2565}k_3 + \frac{2197}{4104}k_4 - \frac{1}{5}k_5,$$

where the coefficient equations are

$$k_1 = hf(t_i, w_i),$$

$$k_2 = hf\left(t_i + \frac{h}{4}, w_i + \frac{1}{4}k_1\right),$$

$$k_3 = hf\left(t_i + \frac{3h}{8}, w_i + \frac{3}{32}k_1 + \frac{9}{32}k_2\right),$$

$$k_4 = hf\left(t_i + \frac{12h}{13}, w_i + \frac{1932}{2197}k_1 - \frac{7200}{2197}k_2 + \frac{7296}{2197}k_3\right),$$

$$k_5 = hf\left(t_i + h, w_i + \frac{439}{216}k_1 - 8k_2 + \frac{3680}{513}k_3 - \frac{845}{4104}k_4\right),$$

$$k_6 = hf\left(t_i + \frac{h}{2}, w_i - \frac{8}{27}k_1 + 2k_2 - \frac{3544}{2565}k_3 + \frac{1859}{4104}k_4 - \frac{11}{40}k_5\right).$$

An advantage of this method is that only six evaluations of f are required per step, whereas arbitrary Runge-Kutta methods of order 4 and 5 used together would require (see Table 5.9 on page 188) at least four evaluations of f for the fourth-order method and an additional six for the fifth-order method.

In the theory of error control, an initial value of h at the ith step was used to find the first values of w_{i+1} and $\tilde{w}_{i+1}$, which led to the determination of q for that step. Then the calculations were repeated with the step size h replaced by qh. This procedure requires twice the number of functional evaluations per step as without error control. In practice, the value of q to be used is chosen somewhat differently in order to make the increased functional-evaluation cost worthwhile. The value of q determined at the ith step is used for two purposes:

- When $q < 1$, to reject the initial choice of h at the ith step and repeat the calculations using qh, and

- When $q \geq 1$, to accept the computed value at the ith step using the step size h and to increase the step size to qh for the $(i+1)$st step.

Because of the penalty in terms of functional evaluations that must be paid if many steps are repeated, q tends to be chosen conservatively. In fact, for the Runge-Kutta-Fehlberg method with $n = 4$, the usual choice is

$$q = \left(\frac{\varepsilon h}{2|\tilde{w}_{i+1} - w_{i+1}|}\right)^{1/4} \approx 0.84\left(\frac{\varepsilon h}{|\tilde{w}_{i+1} - w_{i+1}|}\right)^{1/4}.$$

The program RKFVSM55 implements the Runge-Kutta-Fehlberg method.

Program RKFVSM55 incorporates a technique to eliminate large modifications in step size. This is done to avoid spending too much time with very small step sizes in regions with irregularities in the derivatives of y, and to avoid large step sizes, which might result in skipping sensitive regions nearby. In some instances the step-size-increase procedure is omitted completely and the step-size-decrease procedure is modified to be incorporated only when needed to bring the error under control.

Example 1   Use the Runge-Kutta-Fehlberg method with a tolerance $TOL = 10^{-5}$, a maximum step size $hmax = 0.25$, and a minimum step size $hmin = 0.01$ to approximate the solution to the initial-value problem

$$y' = y - t^2 + 1, \quad 0 \le t \le 2, \quad y(0) = 0.5,$$

and compare the results with the exact solution $y(t) = (t+1)^2 - 0.5e^t$.

Solution   We will work through the first step of the calculations and then apply the program RKFVSM55 to determine the remaining results. The initial condition gives $t_0 = 0$ and $w_0 = 0.5$. To determine w_1 using $h = 0.25$, the maximum allowable stepsize, we compute

$$k_1 = hf(t_0, w_0) = 0.25(0.5 - 0^2 + 1) = 0.375;$$

$$k_2 = hf\left(t_0 + \frac{1}{4}h, w_0 + \frac{1}{4}k_1\right) = 0.25f\left(\frac{1}{4}0.25, 0.5 + \frac{1}{4}0.375\right) = 0.3974609;$$

$$k_3 = hf\left(t_0 + \frac{3}{8}h, w_0 + \frac{3}{32}k_1 + \frac{9}{32}k_2\right)$$

$$= 0.25f\left(0.09375, 0.5 + \frac{3}{32}0.375 + \frac{9}{32}0.3974609\right) = 0.4095383;$$

$$k_4 = hf\left(t_0 + \frac{12}{13}h, w_0 + \frac{1932}{2197}k_1 - \frac{7200}{2197}k_2 + \frac{7296}{2197}k_3\right)$$

$$= 0.25f\left(0.2307692, 0.5 + \frac{1932}{2197}0.375 - \frac{7200}{2197}0.3974609 + \frac{7296}{2197}0.4095383\right)$$

$$= 0.4584971;$$

$$k_5 = hf\left(t_0 + h, w_0 + \frac{439}{216}k_1 - 8k_2 + \frac{3680}{513}k_3 - \frac{845}{4104}k_4\right)$$

$$= 0.25f\left(0.25, 0.5 + \frac{439}{216}0.375 - 8(0.3974609) + \frac{3680}{513}0.4095383 - \frac{845}{4104}0.4584971\right)$$

$$= 0.4658452;$$

$$k_6 = hf\left(t_0 + \frac{1}{2}h, w_0 - \frac{8}{27}k_1 + 2k_2 - \frac{3544}{2565}k_3 + \frac{1859}{4104}k_4 - \frac{11}{40}k_5\right)$$

$$= 0.25f\left(0.125, 0.5 - \frac{8}{27}0.375 + 2(0.3974609) - \frac{3544}{2565}0.4095383\right.$$

$$\left. + \frac{1859}{4104}0.4584971 - \frac{11}{40}0.4658452\right)$$

$$= 0.4204789.$$

The two approximations to $y(0.25)$ are then found to be

$$\tilde{w}_1 = w_0 + \frac{16}{135}k_1 + \frac{6656}{12825}k_3 + \frac{28561}{56430}k_4 - \frac{9}{50}k_5 + \frac{2}{55}k_6$$

$$= 0.5 + \frac{16}{135}0.375 + \frac{6656}{12825}0.4095383 + \frac{28561}{56430}0.4584971 - \frac{9}{50}0.4658452$$

$$+ \frac{2}{55}0.4204789$$

$$= 0.9204870,$$

and

$$w_1 = w_0 + \frac{25}{216}k_1 + \frac{1408}{2565}k_3 + \frac{2197}{4104}k_4 - \frac{1}{5}k_5$$

$$= 0.5 + \frac{25}{216}0.375 + \frac{1408}{2565}0.4095383 + \frac{2197}{4104}0.4584971 - \frac{1}{5}0.4658452$$

$$= 0.9204886.$$

The approximations with $h = 0.25$ and $\varepsilon = 10^{-5}$ give

$$q = \left(\frac{h\varepsilon}{2|\tilde{w}_1 - w_1|}\right)^{1/4} = 0.9462100.$$

Since $q < 1$, we need to recompute the values of $\tilde{w}_1$ and w_1 with the new step size

$$h = 0.9462100(0.05) = 0.2365525.$$

Recalculating with this step size gives us $q = 0.9986023$ which is again less than 1. Redoing the calculations with the step size

$$h = (0.9986023)(0.2365525) = 0.2362221.$$

gives a value of $q = 0.9999643$ which is still less than one. This requires a change in step size to

$$h = (0.9999643)(0.236222) = 0.2362137.$$

For this step size we get $q = 1.000002 > 1$. So, we accept that, with the step size 0.2362137, the approximation w_1 is sufficiently accurate approximation to $y(0.2362137)$.

The results from program RKFVSM55 are shown in Table 5.13. Notice that the step size on the last line has been adjusted so that an approximation is given for $y(2)$. Keep in mind that the errors listed are global errors, and the Runge-Kutta-Fehlberg method controls local errors.   ∎

Table 5.13

			RKF-4		RKF-5	
t_i	$y(t_i)$	h_i	w_i	$|y(t_i) - w_i|$	$\hat{w}_i$	$|y(t_i) - \hat{w}_i|$
0.2362137	0.8950018	0.2362137	0.8950028	1.0×10^{-6}	0.8950016	2.0×10^{-7}
0.4724278	1.3661020	0.2362142	1.3661042	2.2×10^{-6}	1.3661031	1.1×10^{-6}
0.7147675	1.9185718	0.2423397	1.9185755	3.7×10^{-6}	1.9185745	2.7×10^{-6}
0.9647675	2.5482226	0.2500000	2.5482282	5.6×10^{-6}	2.5482272	4.6×10^{-6}
1.2147675	3.2204398	0.2500000	3.2204475	7.7×10^{-6}	3.2204469	7.1×10^{-6}
1.4647675	3.9118102	0.2500000	3.9118204	1.02×10^{-5}	3.9118202	1.00×10^{-5}
1.7147675	4.5922707	0.2500000	4.5922836	1.29×10^{-5}	4.5922839	1.32×10^{-5}
1.9647675	5.2232194	0.2500000	5.2232351	1.57×10^{-5}	5.2232362	1.68×10^{-5}
2.0000000	5.3054720	0.0352325	5.3054883	1.63×10^{-5}	5.3054883	1.63×10^{-5}

An adaptive Runge-Kutta method is available in MATLAB using the command ode45. To use this command for our sample initial value problem, we first need to define the function $f(t, y)$ using an M-file which we name odefunction2.m.

```
function YY = odefunction2(t,y)
YY=y-t^2+1
```

We then need to make the function known to MATLAB using

```
f = @odefunction2
```

We define a range of t values for which we want to approximate the $y(t)$ values beginning at the initial value $a = 0$ and ending at the value $b = 2$. The range of t values are placed in the structure called `tspan`

```
tspan = [0.0 0.2 0.4 0.6 0.8 1.0 1.2 1.4 1.6 1.8 2.0]
```

and we define the initial value $y(0) = 0.5$ with

```
y0=0.5
```

We invoke the function `ode45` using the following command

```
[T,YY] = ode45(f,tspan,y0)
```

which produces the values of t and corresponding approximations to $y(t)$ in the two 11×1 arrays

$$
T = \begin{bmatrix} 0 \\ 0.200000000000000 \\ 0.400000000000000 \\ 0.600000000000000 \\ 0.800000000000000 \\ 1.000000000000000 \\ 1.200000000000000 \\ 1.400000000000000 \\ 1.600000000000000 \\ 1.800000000000000 \\ 2.000000000000000 \end{bmatrix} \quad \text{and} \quad YY = \begin{bmatrix} 0.500000000000000 \\ 0.829298806102213 \\ 1.214087852350423 \\ 1.648940818452941 \\ 2.127229773242783 \\ 2.640859343211129 \\ 3.179941816703143 \\ 3.732400315281681 \\ 4.283484106130412 \\ 4.815176603278762 \\ 5.305472351203669 \end{bmatrix}
$$

Variable Step-Size Predictor-Corrector Methods

The Runge-Kutta-Fehlberg method is popular for error control because at each step it provides, at little additional cost, *two* approximations that can be compared and related to the local error. Predictor-corrector techniques always generate two approximations at each step, so they are natural candidates for error-control adaptation.

To demonstrate the procedure, we construct a *variable-step-size predictor-corrector* method using the explicit Adams-Bashforth Four-Step method as predictor and the implicit Adams-Moulton Three-Step method as corrector.

The Adams-Bashforth Four-Step method comes from the equation

$$
y(t_{i+1}) = y(t_i) + \frac{h}{24}[55f(t_i, y(t_i)) - 59f(t_{i-1}, y(t_{i-1}))
$$

$$
+ 37f(t_{i-2}, y(t_{i-2})) - 9f(t_{i-3}, y(t_{i-3}))] + \frac{251}{720}y^{(5)}(\hat{\mu}_i)h^5
$$

for some $\hat{\mu}_i$ in (t_{i-3}, t_{i+1}). Suppose we assume that the approximations $w_0, w_1, \ldots, w_i$ are all exact and, as in the case of the one-step methods, we let z represent the solution to the differential equation satisfying the initial condition $z(t_i) = w_i$. Then the difference between the actual solution and the predicted solution $w_{i+1,p}$ is

$$
z(t_{i+1}) - w_{i+1,p} = \frac{251}{720}z^{(5)}(\hat{\mu}_i)h^5 \quad \text{for some } \hat{\mu}_i \text{ in } (t_{i-3}, t_i). \tag{5.4}
$$

A similar analysis of the Adams-Moulton Three-Step corrector method leads to the local error

$$z(t_{i+1}) - w_{i+1} = -\frac{19}{720} z^{(5)}(\tilde{\mu}_i) h^5 \quad \text{for some } \tilde{\mu}_i \text{ in } (t_{i-2}, t_{i+1}). \tag{5.5}$$

To proceed further, we must make the assumption that for small values of h,

$$z^{(5)}(\hat{\mu}_i) \approx z^{(5)}(\tilde{\mu}_i),$$

and the effectiveness of the error-control technique depends directly on this assumption.

If we subtract Eq. (5.5) from Eq. (5.4), and assume that $z^{(5)}(\hat{\mu}_i) \approx z^{(5)}(\tilde{\mu}_i)$, we have

$$w_{i+1} - w_{i+1,p} = \frac{h^5}{720} \left[251 z^{(5)}(\hat{\mu}_i) + 19 z^{(5)}(\tilde{\mu}_i) \right] \approx \frac{3}{8} h^5 z^{(5)}(\tilde{\mu}_i),$$

so

$$z^{(5)}(\tilde{\mu}_i) \approx \frac{8}{3h^5} (w_{i+1} - w_{i+1,p}).$$

Using this result to eliminate the term involving $h^5 z^{(5)}(\tilde{\mu}_i)$ from Eq. (5.5) gives the approximation to the error

$$|z(t_{i+1}) - w_{i+1}| \approx \frac{19h^5}{720} \cdot \frac{8}{3h^5} |w_{i+1} - w_{i+1,p}| = \frac{19|w_{i+1} - w_{i+1,p}|}{270}.$$

This expression was derived assuming that $w_0, w_1, \dots, w_i$ are the exact values of $y(t_0), y(t_1), \dots, y(t_i)$, respectively, which means that this is an approximation to the local error. As in the case of the one-step methods, the global error is of order one degree less, so for the function y that is the solution to the original initial-value problem,

$$y' = f(t, y), \quad \text{for } a \le t \le b, \quad \text{with } y(a) = \alpha,$$

the global error can be estimated by

$$|y(t_{i+1}) - w_{i+1}| \approx \frac{|z(t_{i+1}) - w_{i+1}|}{h} \approx \frac{19|w_{i+1} - w_{i+1,p}|}{270h}.$$

Suppose that we now reconsider the situation with a new step size qh generating new approximations $\hat{w}_{i+1,p}$ and $\hat{w}_{i+1}$. To control the global error to within ε, we want to choose q so that

$$\frac{|z(t_i + qh) - \hat{w}_{i+1} \text{ (using the step size } qh)|}{qh} < \varepsilon.$$

But from Eq. (5.5),

$$\frac{|z(t_i + qh) - \hat{w}_{i+1}(\text{using } qh)|}{qh} = \frac{19}{720} |z^{(5)}(\tilde{\mu}_i)| q^4 h^4 \approx \frac{19}{720} \left[\frac{8}{3h^5} |w_{i+1} - w_{i+1,p}| \right] q^4 h^4,$$

so we need to choose q with

$$\frac{19}{720} \left[\frac{8}{3h^5} |w_{i+1} - w_{i+1,p}| \right] q^4 h^4 = \frac{19}{270} \frac{|w_{i+1} - w_{i+1,p}|}{h} q^4 < \varepsilon.$$

Consequently, we need the change in step size from h to qh, where q satisfies

$$q < \left(\frac{270}{19} \frac{h\varepsilon}{|w_{i+1} - w_{i+1,p}|} \right)^{1/4} \approx 2 \left(\frac{h\varepsilon}{|w_{i+1} - w_{i+1,p}|} \right)^{1/4}.$$

A number of approximation assumptions have been made in this development, so in practice q is chosen conservatively, usually as

$$q = 1.5 \left(\frac{h\varepsilon}{|w_{i+1} - w_{i+1,p}|} \right)^{1/4}.$$

A change in step size for a multistep method is more costly in terms of functional evaluations than for a one-step method because new equally-spaced starting values must be computed. As a consequence, it is also common practice to ignore the step-size change whenever the global error is between $\varepsilon/10$ and ε; that is, when

$$\frac{\varepsilon}{10} < \frac{|y(t_{i+1}) - w_{i+1}|}{h} \approx \frac{19}{270h} |w_{i+1} - w_{i+1,p}| < \varepsilon.$$

The program VPRCOR56 implements the Adams Variable Predictor-Corrector method.

In addition, q is generally given an upper bound to ensure that a single unusually accurate approximation does not result in too large a step size. Program VPRCOR56 incorporates this safeguard with an upper bound of four times the previous step size.

It should be emphasized that because the multistep methods require equal step sizes for the starting values, any change in step size necessitates recalculating new starting values at that point. In VPRCOR56, this is done by incorporating as a subroutine the program RKO4M53, which is the Runge-Kutta method of order 4.

Example 2   Use the Adams Variable Step-Size Predictor-Corrector method with maximum step size $hmax = 0.2$, minimum step size $hmin = 0.01$, and tolerance $TOL = 10^{-5}$ to approximate the solution of the initial-value problem

$$y' = y - t^2 + 1, \quad 0 \le t \le 2, \quad y(0) = 0.5.$$

Solution   We begin with $h = hmax = 0.2$, and obtain w_0, w_1, w_2, and w_3 using Runge-Kutta, then find w_{4p} and w_4 by applying the predictor-corrector method. These calculations were done in Example 2 of Section 5.3 where it was determined that the Runge-Kutta approximations are

$$y(0) = w_0 = 0.5, \ y(0.2) \approx w_1 = 0.8292933, \ y(0.4) \approx w_2 = 1.2140762, \ \text{and}$$

$$y(0.6) \approx w_3 = 1.6489220.$$

The predictor and corrector method in Example 2 of Section 5.4 gave

$$y(0.8) \approx w_{4p} = w_3 + \frac{0.2}{24}(55f(0.6, w_3) - 59f(0.4, w_2) + 37f(0.2, w_1) - 9f(0, w_0))$$

$$= 2.1272892,$$

and

$$y(0.8) \approx w_4 = w_3 + \frac{0.2}{24}(9f(0.8, w_{4p}) + 19f(0.6, w_3) - 5f(0.42, w_2) + f(0.2, w_1))$$

$$= 2.1272056.$$

We now need to determine if these approximations are sufficiently accurate or if there needs to be a change in the step size. First we find

$$\sigma = \frac{19}{270h}|w_4 - w_{4p}| = \frac{19}{270(0.2)}|2.1272056 - 2.1272892| = 2.941 \times 10^{-5}.$$

Because this exceeds the tolerance of 10^{-5} a new step size is needed and the new step size is

$$qh = \left(\frac{10^{-5}}{2\sigma}\right)^{1/4} = \left(\frac{10^{-5}}{2(2.941 \times 10^{-5})}\right)^{1/4}(0.2) = 0.642(0.2) \approx 0.128.$$

As a consequence, we need to begin the procedure again computing the Runge-Kutta values with this step size, and then use the predictor-corrector method with this same step size to compute the new values of w_{4p} and w_4. We then need to run the accuracy check on these approximations to see that we have been successful. Table 5.14 shows that this second run is successful and lists all the results obtained using the Adams Variable Predictor-Corrector method. ∎

Table 5.14

t_i	$y(t_i)$	w_i	h_i	σ_i	$|y(t_i) - w_i|$
0	0.5	0.5			
0.1257017	0.7002323	0.7002318	0.1257017	4.051×10^{-6}	0.0000005
0.2514033	0.9230960	0.9230949	0.1257017	4.051×10^{-6}	0.0000011
0.3771050	1.1673894	1.1673877	0.1257017	4.051×10^{-6}	0.0000017
0.5028066	1.4317502	1.4317480	0.1257017	4.051×10^{-6}	0.0000022
0.6285083	1.7146334	1.7146306	0.1257017	4.610×10^{-6}	0.0000028
0.7542100	2.0142869	2.0142834	0.1257017	5.210×10^{-6}	0.0000035
0.8799116	2.3287244	2.3287200	0.1257017	5.913×10^{-6}	0.0000043
1.0056133	2.6556930	2.6556877	0.1257017	6.706×10^{-6}	0.0000054
1.1313149	2.9926385	2.9926319	0.1257017	7.604×10^{-6}	0.0000066
1.2570166	3.3366642	3.3366562	0.1257017	8.622×10^{-6}	0.0000080
1.3827183	3.6844857	3.6844761	0.1257017	9.777×10^{-6}	0.0000097
1.4857283	3.9697541	3.9697433	0.1030100	7.029×10^{-6}	0.0000108
1.5887383	4.2527830	4.2527711	0.1030100	7.029×10^{-6}	0.0000120
1.6917483	4.5310269	4.5310137	0.1030100	7.029×10^{-6}	0.0000133
1.7947583	4.8016639	4.8016488	0.1030100	7.029×10^{-6}	0.0000151
1.8977683	5.0615660	5.0615488	0.1030100	7.760×10^{-6}	0.0000172
1.9233262	5.1239941	5.1239764	0.0255579	3.918×10^{-8}	0.0000177
1.9488841	5.1854932	5.1854751	0.0255579	3.918×10^{-8}	0.0000181
1.9744421	5.2460056	5.2459870	0.0255579	3.918×10^{-8}	0.0000186
2.0000000	5.3054720	5.3054529	0.0255579	3.918×10^{-8}	0.0000191

MATLAB uses the command `ode113` to implement a variable-step size and variable-order Adams-Bashforth-Moulton method. It can be more efficient than `ode45` for smaller error tolerances because it can use higher-order methods if needed.

EXERCISE SET 5.6

1. The initial-value problem

$$y' = \sqrt{2 - y^2}e^t, \quad \text{for } 0 \le t \le 0.8, \quad \text{with } y(0) = 0$$

has the exact solution $y(t) = \sqrt{2}\sin(e^t - 1)$.

a. Use the Runge-Kutta-Fehlberg method with tolerance $TOL = 10^{-4}$ to find w_1. Compare the approximate solution to the actual solution.

b. Use the Adams Variable-Step-Size Predictor-Corrector method with tolerance $TOL = 10^{-4}$ and starting values from the Runge-Kutta method of order 4 to find w_4. Compare the approximate solution to the actual solution.

2.   The initial-value problem

$$y' = -y + 1 - \frac{y}{t}, \quad \text{for } 1 \le t \le 2, \quad \text{with } y(1) = 1$$

has the exact solution $y(t) = 1 + (e^{1-t} - 1)t^{-1}$.

    a.   Use the Runge-Kutta-Fehlberg method with tolerance $TOL = 10^{-3}$ to find w_1 and w_2. Compare the approximate solutions to the actual values.

    b.   Use the Adams Variable-Step-Size Predictor-Corrector method with tolerance $TOL = 0.002$ and starting values from the Runge-Kutta method of order 4 to find w_4 and w_5. Compare the approximate solutions to the actual values.

3.   Use the Runge-Kutta-Fehlberg method with tolerance $TOL = 10^{-4}$ to approximate the solution to the following initial-value problems.

    a. $y' = \left(\frac{y}{t}\right)^2 + \frac{y}{t}$, for $1 \le t \le 1.2$, with $y(1) = 1$, $hmax = 0.05$, and $hmin = 0.02$.

    b. $y' = \sin t + e^{-t}$, for $0 \le t \le 1$, with $y(0) = 0$, $hmax = 0.25$, and $hmin = 0.02$.

    c. $y' = (y^2 + y)t^{-1}$, for $1 \le t \le 3$, with $y(1) = -2$, $hmax = 0.5$, and $hmin = 0.02$.

    d. $y' = -ty + 4ty^{-1}$, for $0 \le t \le 1$, with $y(0) = 1$, $hmax = 0.2$, and $hmin = 0.01$.

4.   Use the Runge-Kutta-Fehlberg method with tolerance $TOL = 10^{-6}$, $hmax = 0.5$, and $hmin = 0.05$ to approximate the solutions to the following initial-value problems. Compare the results to the actual values.

    a. $y' = \frac{y}{t} - \frac{y^2}{t^2}$, for $1 \le t \le 4$, with $y(1) = 1$; actual solution $y(t) = t/(1 + \ln t)$.

    b. $y' = 1 + \frac{y}{t} + \left(\frac{y}{t}\right)^2$, for $1 \le t \le 3$, with $y(1) = 0$; actual solution $y(t) = t \tan(\ln t)$.

    c. $y' = -(y+1)(y+3)$, for $0 \le t \le 3$, with $y(0) = -2$; actual solution $y(t) = -3 + 2(1 + e^{-2t})^{-1}$.

    d. $y' = (t + 2t^3)y^3 - ty$, for $0 \le t \le 2$, with $y(0) = \frac{1}{3}$; actual solution $y(t) = (3 + 2t^2 + 6e^{t^2})^{-1/2}$.

5.   Use the Adams Variable-Step-Size Predictor-Corrector method with $TOL = 10^{-4}$ to approximate the solutions to the initial-value problems in Exercise 3.

6.   Use the Adams Variable-Step-Size Predictor-Corrector method with tolerance $TOL = 10^{-6}$, $hmax = 0.5$, and $hmin = 0.02$ to approximate the solutions to the initial-value problems in Exercise 4.

7.   An electrical circuit consists of a capacitor of constant capacitance $C = 1.1$ farads in series with a resistor of constant resistance $R_0 = 2.1$ ohms. A voltage $\mathcal{E}(t) = 110 \sin t$ is applied at time $t = 0$. When the resistor heats up, the resistance becomes a function of the current i,

$$R(t) = R_0 + ki, \quad \text{where } k = 0.9,$$

and the differential equation for i becomes

$$\left(1 + \frac{2k}{R_0}i\right)\frac{di}{dt} + \frac{1}{R_0 C}i = \frac{1}{R_0 C}\frac{d\mathcal{E}}{dt}.$$

Find the current i after 2 s, assuming $i(0) = 0$.

5.7 Methods for Systems of Equations

The most common application of numerical methods for approximating the solution of initial-value problems concerns not a single problem, but a linked system of differential equations. Why, then, have we spent the majority of this chapter considering the solution of a single equation? The answer is simple: to approximate the solution of a system of initial-value problems, we successively apply the techniques that we used to solve problems involving a single equation. As is so often the case in mathematics, the key to the methods for systems can be found by examining the easier problem and then logically modifying it to treat the more complicated situation.

An *m***th-order system** of first-order initial-value problems has the form

$$\frac{du_1}{dt} = f_1(t, u_1, u_2, \ldots, u_m),$$

$$\frac{du_2}{dt} = f_2(t, u_1, u_2, \ldots, u_m),$$

$$\vdots$$

$$\frac{du_m}{dt} = f_m(t, u_1, u_2, \ldots, u_m),$$

for $a \leq t \leq b$, with the initial conditions

$$u_1(a) = \alpha_1, \quad u_2(a) = \alpha_2, \quad \ldots, \quad u_m(a) = \alpha_m.$$

The object is to find m functions $u_1, u_2, \ldots, u_m$ that satisfy the system of differential equations together with all the initial conditions.

Methods to solve systems of first-order differential equations are generalizations of the methods for a single first-order equation presented earlier in this chapter. For example, the classical Runge-Kutta method of order 4 given by

$$w_0 = \alpha,$$

$$k_1 = hf(t_i, w_i),$$

$$k_2 = hf\left(t_i + \frac{h}{2}, w_i + \frac{1}{2}k_1\right),$$

$$k_3 = hf\left(t_i + \frac{h}{2}, w_i + \frac{1}{2}k_2\right),$$

$$k_4 = hf(t_{i+1}, w_i + k_3),$$

and

$$w_{i+1} = w_i + \frac{1}{6}[k_1 + 2k_2 + 2k_3 + k_4],$$

for each $i = 0, 1, \ldots, N - 1$, is used to solve the first-order initial-value problem

$$y' = f(t, y), \quad \text{for } a \leq t \leq b, \quad \text{with } y(a) = \alpha.$$

It is generalized for systems as follows.

Let an integer $N > 0$ be chosen and set $h = (b - a)/N$. Partition the interval $[a, b]$ into N subintervals with the mesh points

$$t_j = a + jh \quad \text{for each } j = 0, 1, \ldots, N.$$

Use the notation w_{ij} for each $j = 0, 1, \ldots, N$ and $i = 1, 2, \ldots, m$ to denote an approximation to $u_i(t_j)$; that is, w_{ij} approximates the ith solution $u_i(t)$ of the system at the jth mesh point t_j. For the initial conditions, set

$$w_{1,0} = \alpha_1, \quad w_{2,0} = \alpha_2, \quad \ldots, \quad w_{m,0} = \alpha_m.$$

Figure 5.4 gives an illustration of this notation.

Figure 5.4

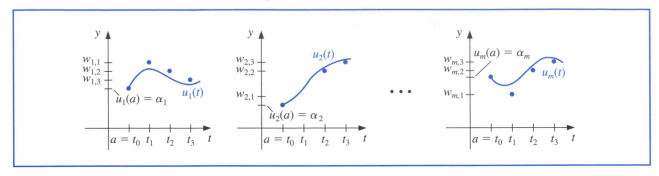

Suppose that the values $w_{1,j}, w_{2,j}, \ldots, w_{m,j}$ have been computed. We obtain $w_{1,j+1}$, $w_{2,j+1}, \ldots, w_{m,j+1}$ by first calculating, for each $i = 1, 2, \ldots, m$,

$$k_{1,i} = hf_i(t_j, w_{1,j}, w_{2,j}, \ldots, w_{m,j}),$$

and then finding, for each i,

$$k_{2,i} = hf_i\left(t_j + \frac{h}{2}, w_{1,j} + \frac{1}{2}k_{1,1}, w_{2,j} + \frac{1}{2}k_{1,2}, \ldots, w_{m,j} + \frac{1}{2}k_{1,m}\right).$$

We next determine all the terms

$$k_{3,i} = hf_i\left(t_j + \frac{h}{2}, w_{1,j} + \frac{1}{2}k_{2,1}, w_{2,j} + \frac{1}{2}k_{2,2}, \ldots, w_{m,j} + \frac{1}{2}k_{2,m}\right)$$

and, finally, calculate all the terms

$$k_{4,i} = hf_i(t_j + h, w_{1,j} + k_{3,1}, w_{2,j} + k_{3,2}, \ldots, w_{m,j} + k_{3,m}).$$

Combining these values gives

$$w_{i,j+1} = w_{i,j} + \frac{1}{6}[k_{1,i} + 2k_{2,i} + 2k_{3,i} + k_{4,i}]$$

The program RKO4SY57 implements the Runge-Kutta method of order 4 for systems.

for each $i = 1, 2, \ldots m$.

Note that all the values $k_{1,1}, k_{1,2}, \ldots, k_{1,m}$ must be computed before any of the terms of the form $k_{2,i}$ can be determined. In general, each $k_{l,1}, k_{l,2}, \ldots, k_{l,m}$ must be computed before any of the expressions $k_{l+1,i}$.

Example 1 Kirchhoff's Law states that the sum of all instantaneous voltage changes around a closed electrical circuit is zero. This implies that the current, $I(t)$, in a closed circuit containing a resistance of R ohms, a capacitance of C farads, an inductance of L henrys, and a voltage source of $E(t)$ volts must satisfy the equation

$$LI'(t) + RI(t) + \frac{1}{C}\int I(t)dt = E(t).$$

The currents $I_1(t)$ and $I_2(t)$ in the left and right loops, respectively, of the circuit shown in Figure 5.5 are the solutions to the system of equations

$$2I_1(t) + 6[I_1(t) - I_2(t)] + 2I_1'(t) = 12,$$

$$\frac{1}{0.5}\int I_2(t)dt + 4I_2(t) + 6[I_2(t) - I_1(t)] = 0.$$

Figure 5.5

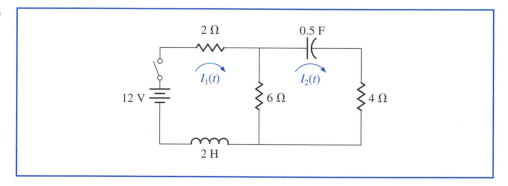

Suppose that the switch in the circuit is closed at time $t = 0$. This implies that $I_1(0)$ and $I_2(0) = 0$. Solve for $I_1'(t)$ in the first equation, differentiate the second equation, and substitute for $I_1'(t)$ to get

$$I_1' = f_1(t, I_1, I_2) = -4I_1 + 3I_2 + 6, \quad \text{with } I_1(0) = 0,$$

$$I_2' = f_2(t, I_1, I_2) = 0.6I_1' - 0.2I_2 = -2.4I_1 + 1.6I_2 + 3.6, \quad \text{with } I_2(0) = 0.$$

The exact solution to this system is

$$I_1(t) = -3.375e^{-2t} + 1.875e^{-0.4t} + 1.5 \quad \text{and} \quad I_2(t) = -2.25e^{-2t} + 2.25e^{-0.4t}.$$

We will apply the Runge-Kutta method of order 4 to this system with $h = 0.1$. Since $w_{1,0} = I_1(0) = 0$ and $w_{2,0} = I_2(0) = 0$,

$$k_{1,1} = hf_1(t_0, w_{1,0}, w_{2,0}) = 0.1 \ f_1(0, 0, 0) = 0.1[-4(0) + 3(0) + 6] = 0.6,$$

$$k_{1,2} = hf_2(t_0, w_{1,0}, w_{2,0}) = 0.1 \ f_2(0, 0, 0) = 0.1[-2.4(0) + 1.6(0) + 3.6] = 0.36,$$

$$k_{2,1} = hf_1\left(t_0 + \frac{1}{2}h, w_{1,0} + \frac{1}{2}k_{1,1}, w_{2,0} + \frac{1}{2}k_{1,2}\right) = 0.1 \ f_1(0.05, 0.3, 0.18)$$

$$= 0.1[-4(0.3) + 3(0.18) + 6] = 0.534,$$

$$k_{2,2} = hf_2\left(t_0 + \frac{1}{2}h, w_{1,0} + \frac{1}{2}k_{1,1}, w_{2,0} + \frac{1}{2}k_{1,2}\right) = 0.1 \ f_2(0.05, 0.3, 0.18)$$

$$= 0.1[-2.4(0.3) + 1.6(0.18) + 3.6] = 0.3168.$$

Generating the remaining entries in a similar manner produces

$$k_{3,1} = (0.1) f_1(0.05, 0.267, 0.1584) = 0.54072,$$

$$k_{3,2} = (0.1) f_2(0.05, 0.267, 0.1584) = 0.321264,$$

$$k_{4,1} = (0.1) f_1(0.1, 0.54072, 0.321264) = 0.4800912,$$

and

$$k_{4,2} = (0.1) f_2(0.1, 0.54072, 0.321264) = 0.28162944.$$

As a consequence,

$$I_1(0.1) \approx w_{1,1} = w_{1,0} + \frac{1}{6}[k_{1,1} + 2k_{2,1} + 2k_{3,1} + k_{4,1}]$$

$$= 0 + \frac{1}{6}[0.6 + 2(0.534) + 2(0.54072) + 0.4800912] = 0.5382552$$

and

$$I_2(0.1) \approx w_{2,1} = w_{2,0} + \frac{1}{6}[k_{1,2} + 2k_{2,2} + 2k_{3,2} + k_{4,2}] = 0.3196263.$$

The remaining entries in Table 5.15 are generated in a similar manner.   ▪

Table 5.15

t_j	$w_{1,j}$	$w_{2,j}$	$|I_1(t_j) - w_{1,j}|$	$|I_2(t_j) - w_{2,j}|$
0.0	0	0	0	0
0.1	0.5382550	0.3196263	0.8285×10^{-5}	0.5803×10^{-5}
0.2	0.9684983	0.5687817	0.1514×10^{-4}	0.9596×10^{-5}
0.3	1.310717	0.7607328	0.1907×10^{-4}	0.1216×10^{-4}
0.4	1.581263	0.9063208	0.2098×10^{-4}	0.1311×10^{-4}
0.5	1.793505	1.014402	0.2193×10^{-4}	0.1240×10^{-4}

Any of the methods implemented in MATLAB can be used for systems of differential equations. For example, to use ode45 to solve our system given in Example 1 we first define the right-hand sides using an M-file called F.m that contains the statements

```
function dy = F(t,y)
dy = zeros(2,1);
dy(1) = -4*y(1)+3*y(2)+6;
dy(2) = -2.4*y(1)+1.6*y(2)+3.6;
```

Then make the right-hand side of the system of differential equations known to MATLAB with

```
FF = @F
```

We now define the t values at which we want to approximate the solutions

```
tspan = [0 0.1 0.2 0.3 0.4 0.5]
```

The following command computes the solution to the system at the given values of t. The initial conditions $I_1(0) = 0$ and $I_2(0) = 0$ are given as [0 0].

```
[T,YY]=ode45(FF,tspan,[0 0])
```

The MATLAB response places the t values in the array T and the approximate solution values in YY, with the approximations for $I_1(t)$ in the first column and $I_2(t)$ in the second.

$$T = \begin{bmatrix} 0 \\ 0.100000000000000 \\ 0.200000000000000 \\ 0.300000000000000 \\ 0.400000000000000 \\ 0.500000000000000 \end{bmatrix} \text{ and } YY = \begin{bmatrix} 0 & 0 \\ 0.538263922676270 & 0.319632054268176 \\ 0.968513005638230 & 0.568791683477228 \\ 1.310736555252393 & 0.760744806883952 \\ 1.581284356153020 & 0.906333359733513 \\ 1.793527044389029 & 1.014415449337470 \end{bmatrix}$$

Higher-Order Differential Equations

Many important physical problems—for example, electrical circuits and vibrating systems—involve initial-value problems whose equations have order higher than 1. New techniques are not required for solving these problems. By relabeling the variables we can reduce a higher-order differential equation into a system of first-order differential equations and then apply one of the methods we have already discussed.

A general ***m*th-order initial-value problem** has the form

$$y^{(m)}(t) = f\left(t, y, y', \ldots, y^{(m-1)}\right),$$

for $a \leq t \leq b$, with initial conditions

$$y(a) = \alpha_1, \, y'(a) = \alpha_2, \ldots, \, y^{(m-1)}(a) = \alpha_m.$$

To convert this into a system of first-order differential equations, define

$$u_1(t) = y(t), \, u_2(t) = y'(t), \ldots, \, u_m(t) = y^{(m-1)}(t).$$

Using this notation, we obtain the first-order system

$$\frac{du_1}{dt} = \frac{dy}{dt} = u_2,$$

$$\frac{du_2}{dt} = \frac{dy'}{dt} = u_3,$$

$$\vdots$$

$$\frac{du_{m-1}}{dt} = \frac{dy^{(m-2)}}{dt} = u_m,$$

and

$$\frac{du_m}{dt} = \frac{dy^{(m-1)}}{dt} = y^{(m)} = f\left(t, y, y', \ldots, y^{(m-1)}\right) = f(t, u_1, u_2, \ldots, u_m),$$

with initial conditions

$$u_1(a) = y(a) = \alpha_1, \, u_2(a) = y'(a) = \alpha_2, \ldots, \, u_m(a) = y^{(m-1)}(a) = \alpha_m.$$

Example 2   Transform the second-order initial-value problem

$$y'' - 2y' + 2y = e^{2t} \sin t, \qquad \text{for } 0 \leq t \leq 1, \quad \text{with } y(0) = -0.4, \; y'(0) = -0.6.$$

into a system of first order initial-value problems, and use the Runge-Kutta method of order 4 with $h = 0.1$ to approximate the solution.

Solution   Let $u_1(t) = y(t)$ and $u_2(t) = y'(t)$. This transforms the second-order equation into the system

$$u'_1(t) = u_2(t),$$

$$u'_2(t) = e^{2t} \sin t - 2u_1(t) + 2u_2(t),$$

with initial conditions $u_1(0) = -0.4$, $u_2(0) = -0.6$.

The initial conditions give $w_{1,0} = -0.4$ and $w_{2,0} = -0.6$. The Runge-Kutta method of order 4 for systems described on page 216 with $j = 0$ give

$$k_{1,1} = hf_1(t_0, w_{1,0}, w_{2,0}) = hw_{2,0} = -0.06,$$

$$k_{1,2} = hf_2(t_0, w_{1,0}, w_{2,0}) = h[e^{2t_0} \sin t_0 - 2w_{1,0} + 2w_{2,0}] = -0.04,$$

$$k_{2,1} = hf_1\left(t_0 + \frac{h}{2}, w_{1,0} + \frac{1}{2}k_{1,1}, w_{2,0} + \frac{1}{2}k_{1,2}\right) = h\left[w_{2,0} + \frac{1}{2}k_{1,2}\right] = -0.062,$$

$$k_{2,2} = hf_2\left(t_0 + \frac{h}{2}, w_{1,0} + \frac{1}{2}k_{1,1}, w_{2,0} + \frac{1}{2}k_{1,2}\right)$$

$$= h\left[e^{2(t_0+0.05)} \sin(t_0 + 0.05) - 2\left(w_{1,0} + \frac{1}{2}k_{1,1}\right) + 2\left(w_{2,0} + \frac{1}{2}k_{1,2}\right)\right]$$

$$= -0.03247644757,$$

$$k_{3,1} = h\left[w_{2,0} + \frac{1}{2}k_{2,2}\right] = -0.06162832238,$$

$$k_{3,2} = h\left[e^{2(t_0+0.05)} \sin(t_0 + 0.05) - 2\left(w_{1,0} + \frac{1}{2}k_{2,1}\right) + 2\left(w_{2,0} + \frac{1}{2}k_{2,2}\right)\right]$$

$$= -0.03152409237,$$

$$k_{4,1} = h\left[w_{2,0} + k_{3,2}\right] = -0.06315240924,$$

and

$$k_{4,2} = h\left[e^{2(t_0+0.1)} \sin(t_0 + 0.1) - 2(w_{1,0} + k_{3,1}) + 2(w_{2,0} + k_{3,2})\right] = -0.02178637298.$$

So

$$w_{1,1} = w_{1,0} + \frac{1}{6}(k_{1,1} + 2k_{2,1} + 2k_{3,1} + k_{4,1}) = -0.4617333423 \quad \text{and}$$

$$w_{2,1} = w_{2,0} + \frac{1}{6}(k_{1,2} + 2k_{2,2} + 2k_{3,2} + k_{4,2}) = -0.6316312421.$$

The value $w_{1,1}$ approximates $u_1(0.1) = y(0.1) = 0.2e^{2(0.1)}(\sin 0.1 - 2\cos 0.1)$, and $w_{2,1}$ approximates $u_2(0.1) = y'(0.1) = 0.2e^{2(0.1)}(4\sin 0.1 - 3\cos 0.1)$.

The set of values $w_{1,j}$ and $w_{2,j}$, for $j = 0, 1, \ldots, 10$, are presented in Table 5.16 and are compared to the actual values of $u_1(t) = 0.2e^{2t}(\sin t - 2\cos t)$ and $u_2(t) = u_1'(t) = 0.2e^{2t}(4\sin t - 3\cos t)$.   ■

Table 5.16

t_j	$y(t_j) = u_1(t_j)$	$w_{1,j}$	$y'(t_j) = u_2(t_j)$	$w_{2,j}$	$\lvert y(t_j) - w_{1,j}\rvert$	$\lvert y'(t_j) - w_{2,j}\rvert$
0.0	−0.40000000	−0.40000000	−0.60000000	−0.60000000	0	0
0.1	−0.46173297	−0.46173334	−0.63163105	−0.63163124	3.72×10^{-7}	1.92×10^{-7}
0.2	−0.52555905	−0.52555988	−0.64014866	−0.64014895	8.36×10^{-7}	2.84×10^{-7}
0.3	−0.58860005	−0.58860144	−0.61366361	−0.61366381	1.39×10^{-6}	1.99×10^{-7}
0.4	−0.64661028	−0.64661231	−0.53658220	−0.53658203	2.02×10^{-6}	1.68×10^{-7}
0.5	−0.69356395	−0.69356666	−0.38873906	−0.38873810	2.71×10^{-6}	9.58×10^{-7}
0.6	−0.72114849	−0.72115190	−0.14438322	−0.14438087	3.41×10^{-6}	2.35×10^{-6}
0.7	−0.71814890	−0.71815295	0.22899243	0.22899702	4.06×10^{-6}	4.59×10^{-6}
0.8	−0.66970677	−0.66971133	0.77198383	0.77199180	4.55×10^{-6}	7.97×10^{-6}
0.9	−0.55643814	−0.55644290	1.5347686	1.5347815	4.77×10^{-6}	1.29×10^{-5}
1.0	−0.35339436	−0.35339886	2.5787466	2.5787663	4.50×10^{-6}	1.97×10^{-5}

Other one-step approximation methods can be extended to systems. If the Runge-Kutta-Fehlberg method is extended, then each component of the numerical solution $w_{1j}, w_{2j}, \ldots,$ w_{mj} must be examined for accuracy. If any of the components fail to be sufficiently accurate, the entire numerical solution must be recomputed.

The multistep methods and predictor-corrector techniques can also be extended easily to systems. Again, if error control is used, each component must be accurate. The extension of the extrapolation technique to systems can also be done, but the notation becomes quite involved.

EXERCISE SET 5.7

1. Use the Runge-Kutta method of order 4 for systems to approximate the solutions of the following systems of first-order differential equations and compare the results to the actual solutions.

a. $u_1' = 3u_1 + 2u_2 - (2t^2 + 1)e^{2t},$ for $0 \le t \le 1$ with $u_1(0) = 1;$
  $u_2' = 4u_1 + u_2 + (t^2 + 2t - 4)e^{2t},$ for $0 \le t \le 1$ with $u_2(0) = 1;$
  $h = 0.2;$ actual solutions $u_1(t) = \frac{1}{3}e^{5t} - \frac{1}{3}e^{-t} + e^{2t}$ and $u_2(t) = \frac{1}{3}e^{5t} + \frac{2}{3}e^{-t} + t^2e^{2t}.$

b. $u_1' = -4u_1 - 2u_2 + \cos t + 4\sin t,$ for $0 \le t \le 2$ with $u_1(0) = 0;$
  $u_2' = 3u_1 + u_2 - 3\sin t,$ for $0 \le t \le 2$ with $u_2(0) = -1;$
  $h = 0.1;$ actual solutions $u_1(t) = 2e^{-t} - 2e^{-2t} + \sin t$ and $u_2(t) = -3e^{-t} + 2e^{-2t}.$

c. $u_1' = u_2,$ for $0 \le t \le 2$ with $u_1(0) = 1;$
  $u_2' = -u_1 - 2e^t + 1,$ for $0 \le t \le 2$ with $u_2(0) = 0;$
  $u_3' = -u_1 - e^t + 1,$ for $0 \le t \le 2$ with $u_3(0) = 1;$
  $h = 0.5;$ actual solutions $u_1(t) = \cos t + \sin t - e^t + 1,$ $u_2(t) = -\sin t + \cos t - e^t,$ and
  $u_3(t) = -\sin t + \cos t.$

d. $u_1' = u_2 - u_3 + t,$ for $0 \le t \le 1$ with $u_1(0) = 1;$
  $u_2' = 3t^2,$ for $0 \le t \le 1$ with $u_2(0) = 1;$
  $u_3' = u_2 + e^{-t},$ for $0 \le t \le 1$ with $u_3(0) = -1;$
  $h = 0.1;$ actual solutions $u_1(t) = -0.05t^5 + 0.25t^4 + t + 2 - e^{-t},$ $u_2(t) = t^3 + 1,$ and
  $u_3(t) = 0.25t^4 + t - e^{-t}.$

2. Use the Runge-Kutta method for systems to approximate the solutions of the following higher-order differential equations and compare the results to the actual solutions.

a. $y'' - 2y' + y = te^t - t,$ for $0 \le t \le 1$ with $y(0) = y'(0) = 0$ and $h = 0.1;$ actual solution $y(t) = \frac{1}{6}t^3e^t - te^t + 2e^t - t - 2.$

b. $t^2y'' - 2ty' + 2y = t^3 \ln t,$ for $1 \le t \le 2$ with $y(1) = 1,$ $y'(1) = 0,$ and $h = 0.1;$ actual solution $y(t) = \frac{7}{4}t + \frac{1}{2}t^3 \ln t - \frac{3}{4}t^3.$

c. $y''' + 2y'' - y' - 2y = e^t,$ for $0 \le t \le 3$ with $y(0) = 1,$ $y'(0) = 2,$ $y''(0) = 0,$ and $h = 0.2;$ actual solution $y(t) = \frac{43}{36}e^t + \frac{1}{4}e^{-t} - \frac{4}{9}e^{-2t} + \frac{1}{6}te^t.$

d. $t^3y''' - t^2y'' + 3ty' - 4y = 5t^3 \ln t + 9t^3,$ for $1 \le t \le 2$ with $y(1) = 0,$ $y'(1) = 1,$ $y''(1) = 3,$ and $h = 0.1;$ actual solution $y(t) = -t^2 + t\cos(\ln t) + t\sin(\ln t) + t^3 \ln t.$

3. Change the Adams Fourth-Order Predictor-Corrector method to obtain approximate solutions to systems of first-order equations.

4. Repeat Exercise 1 using the method developed in Exercise 3.

5. The study of mathematical models for predicting the population dynamics of competing species has its origin in independent works published in the early part of this century by A. J. Lotka and V. Volterra. Consider the problem of predicting the population of two species, one of which is a predator, whose population at time t is $x_2(t),$ feeding on the other, which is the prey, whose population is $x_1(t).$ We will assume that the prey always has an adequate food supply and that its birth rate at any time is proportional to the number of prey alive at that time; that is, birth rate (prey) is $k_1x_1(t).$ The death rate of the prey depends on both the number of prey and predators alive at that time. For simplicity, we assume death rate (prey) $= k_2x_1(t)x_2(t).$ The birth rate of the predator, on the other hand, depends on its food supply, $x_1(t),$ as well as on the number of predators available for reproduction purposes.

For this reason, we assume that the birth rate (predator) is $k_3x_1(t)x_2(t)$. The death rate of the predator will be taken as simply proportional to the number of predators alive at the time; that is, death rate (predator) $= k_4x_2(t)$.

Since $x_1'(t)$ and $x_2'(t)$ represent the change in the prey and predator populations, respectively, with respect to time, the problem is expressed by the system of nonlinear differential equations

$$x_1'(t) = k_1x_1(t) - k_2x_1(t)x_2(t) \quad \text{and} \quad x_2'(t) = k_3x_1(t)x_2(t) - k_4x_2(t).$$

Use Runge-Kutta of order 4 for systems to solve this system for $0 \le t \le 4$, assuming that the initial population of the prey is 1000 and of the predators is 500 and that the constants are $k_1 = 3, k_2 = 0.002, k_3 = 0.0006$, and $k_4 = 0.5$. Is there a stable solution to this population model? If so, for what values x_1 and x_2 is the solution stable?

6. In Exercise 5 we considered the problem of predicting the population in a predator-prey model. Another problem of this type is concerned with two species competing for the same food supply. If the numbers of species alive at time t are denoted by $x_1(t)$ and $x_2(t)$, it is often assumed that, although the birth rate of each of the species is simply proportional to the number of species alive at that time, the death rate of each species depends on the population of both species. We will assume that the population of a particular pair of species is described by the equations

$$\frac{dx_1(t)}{dt} = x_1(t)[4 - 0.0003x_1(t) - 0.0004x_2(t)]$$

and

$$\frac{dx_2(t)}{dt} = x_2(t)[2 - 0.0002x_1(t) - 0.0001x_2(t)].$$

If it is known that the initial population of each species is 10,000, find the solution to this system for $0 \le t \le 4$. Is there a stable solution to this population model? If so, for what values of x_1 and x_2 is the solution stable?

5.8 Stiff Differential Equations

All the methods for approximating the solution to initial-value problems have error terms that involve a higher derivative of the solution of the equation. If the derivative can be reasonably bounded, then the method will have a predictable error bound that can be used to estimate the accuracy of the approximation. Even if the derivative grows as the step sizes increase, the error can be kept in relative control, provided that the solution also grows in magnitude. Problems frequently arise, however, where the magnitude of the derivative increases, but the solution does not. In this situation, the error can grow so large that it dominates the calculations. Initial-value problems for which this is likely to occur are called **stiff equations** and are quite common, particularly in the study of vibrations, chemical reactions, and electrical circuits.

Stiff differential equations have an exact solution with a term of the form e^{-ct}, where c is a large positive constant. This is usually only a part of the solution, called the *transient* solution; the more important portion of the solution is called the *steady-state* solution. A transient portion of a stiff equation will rapidly decay to zero as t increases, but since the nth derivative of this term has magnitude c^ne^{-ct}, the derivative does not decay as quickly, and for large values of c it can grow very large. In addition, the derivative in the error term is evaluated not at t, but at a number between zero and t, so the derivative terms may increase as t increases—and very rapidly indeed. Fortunately, stiff equations can generally be predicted from the physical problem from which the equation is derived, and with care the error can be kept under control. The manner in which this is done is considered in this section.

Stiff systems derive their name from the motion of spring and mass systems that have large spring constants.

Illustration   The system of initial-value problems

$$u_1' = 9u_1 + 24u_2 + 5\cos t - \frac{1}{3}\sin t, \quad u_1(0) = \frac{4}{3}$$

$$u_2' = -24u_1 - 51u_2 - 9\cos t + \frac{1}{3}\sin t, \quad u_2(0) = \frac{2}{3}$$

has the unique solution

$$u_1(t) = 2e^{-3t} - e^{-39t} + \frac{1}{3}\cos t, \quad u_2(t) = -e^{-3t} + 2e^{-39t} - \frac{1}{3}\cos t.$$

The transient term e^{-39t} in the solution causes this system to be stiff. Applying the Runge-Kutta Fourth-Order Method for Systems program RKO4SY57 gives results listed in Table 5.17. When $h = 0.05$, stability results and the approximations are accurate. Increasing the step size to $h = 0.1$, however, leads to the disastrous results shown in the table. ☐

Table 5.17

t	Exact $u_1(t)$	$w_1(t)$ $h = 0.05$	$w_1(t)$ $h = 0.1$	Exact $u_2(t)$	$w_2(t)$ $h = 0.05$	$w_2(t)$ $h = 0.1$
0.1	1.793061	1.712219	−2.645169	−1.032001	−0.8703152	7.844527
0.2	1.423901	1.414070	−18.45158	−0.8746809	−0.8550148	38.87631
0.3	1.131575	1.130523	−87.47221	−0.7249984	−0.7228910	176.4828
0.4	0.9094086	0.9092763	−934.0722	−0.6082141	−0.6079475	789.3540
0.5	0.7387877	9.7387506	−1760.016	−0.5156575	−0.5155810	3520.00
0.6	0.6057094	0.6056833	−7848.550	−0.4404108	−0.4403558	15697.84
0.7	0.4998603	0.4998361	−34989.63	−0.3774038	−0.3773540	69979.87
0.8	0.4136714	0.4136490	−155979.4	−0.3229535	−0.3229078	311959.5
0.9	0.3416143	0.3415939	−695332.0	−0.2744088	−0.2743673	1390664.
1.0	0.2796748	0.2796568	−3099671.	−0.2298877	−0.2298511	6199352.

Although stiffness is usually associated with systems of differential equations, the approximation characteristics of a particular numerical method applied to a stiff system can be predicted by examining the error produced when the method is applied to a simple *test equation*,

$$y' = \lambda y, \quad \text{with } y(0) = \alpha,$$

where λ is a negative real number. The solution to this equation contains the transient solution $e^{\lambda t}$ and the steady-state solution is zero, so the approximation characteristics of a method are easy to determine. (A more complete discussion of the round-off error associated with stiff systems requires an examination of the test equation when λ is a complex number with negative real part.)

One-Step Methods

Suppose that we apply Euler's method to the test equation. Letting $h = (b-a)/N$ and $t_j = jh$, for $j = 0, 1, 2, \ldots, N$, implies that

$$w_0 = \alpha$$

and, for $j = 0, 1, \ldots, N-1,$

$$w_{j+1} = w_j + h(\lambda w_j) = (1 + h\lambda)w_j,$$

so

$$w_{j+1} = (1 + h\lambda)^{j+1} w_0 = (1 + h\lambda)^{j+1} \alpha, \quad \text{for } j = 0, 1, \ldots, N - 1. \qquad (5.6)$$

Since the exact solution is $y(t) = \alpha e^{\lambda t}$, the absolute error is

$$|y(t_j) - w_j| = |e^{jh\lambda} - (1 + h\lambda)^j| \, |\alpha| = |(e^{h\lambda})^j - (1 + h\lambda)^j| \, |\alpha|,$$

and the accuracy depends on how well the term $1 + h\lambda$ approximates $e^{h\lambda}$. When $\lambda < 0$, the exact solution, $(e^{h\lambda})^j$, decays to zero as j increases, but, by Eq. (5.6), the approximation will have this property only if $|1 + h\lambda| < 1$. This effectively restricts the step size h for Euler's method to satisfy $|1 + h\lambda| < 1$, which in turn implies that $h < 2/|\lambda|$.

Suppose now that a round-off error δ_0 is introduced in the initial condition for Euler's method,

$$w_0 = \alpha + \delta_0.$$

At the jth step the round-off error is

$$\delta_j = (1 + h\lambda)^j \delta_0.$$

If $\lambda < 0$, the condition for the control of the growth of round-off error, $|1 + h\lambda| < 1$, is the same as the condition for controlling the absolute error, so we need to have $h < 2/|\lambda|$.

Illustration   The test differential equation

$$y' = -30y, \quad 0 \le t \le 1.5, \quad y(0) = \frac{1}{3}$$

has exact solution $y = \frac{1}{3} e^{-30t}$. Using $h = 0.1$ for Euler's method, the Runge-Kutta method of order 4, and the Adams Predictor-Corrector method, gives the results at $t = 1.5$ in Table 5.18.   □

Table 5.18

Exact solution	9.54173×10^{-21}
Euler's method	-1.09225×10^4
Runge-Kutta method	3.95730×10^1
Predictor-Corrector method	8.03840×10^5

The situation is similar for other one-step methods. In general, a function Q exists with the property that the difference method, when applied to the test equation, gives

$$w_{j+1} = Q(h\lambda) w_j.$$

The accuracy of the method depends upon how well $Q(h\lambda)$ approximates $e^{h\lambda}$, and the error will grow without bound if $|Q(h\lambda)| > 1$.

Multistep Methods

The problem of determining when a method is stable is more complicated in the case of multistep methods, due to the interplay of previous approximations at each step. Explicit multistep methods tend to have stability problems, as do predictor-corrector methods, because they involve explicit techniques. In practice, the techniques used for stiff systems are implicit multistep methods. Generally, w_{i+1} is obtained by iteratively solving a nonlinear equation or nonlinear system, often by Newton's method. To illustrate the procedure, consider the following implicit technique.

Implicit Trapezoidal Method

$$w_0 = \alpha$$

$$w_{j+1} = w_j + \frac{h}{2}[f(t_j, w_j) + f(t_{j+1}, w_{j+1})]$$

where $j = 0, 1, \ldots, N - 1$.

To determine w_1 using this technique, we apply Newton's method to find the root of the equation

$$0 = F(w) = w - w_0 - \frac{h}{2}[f(t_0, w_0) + f(t_1, w)] = w - \alpha - \frac{h}{2}[f(a, \alpha) + f(t_1, w)].$$

To approximate this solution, select $w_1^{(0)}$ (usually as w_0) and generate $w_1^{(k)}$ by applying Newton's method to obtain

$$w_1^{(k)} = w_1^{(k-1)} - \frac{F\left(w_1^{(k-1)}\right)}{F'\left(w_1^{(k-1)}\right)} = w_1^{(k-1)} - \frac{w_1^{(k-1)} - \alpha - \frac{h}{2}\left[f(a, \alpha) + f\left(t_1, w_1^{(k-1)}\right)\right]}{1 - \frac{h}{2}f_y\left(t_1, w_1^{(k-1)}\right)}$$

The program TRAPNT58 implements the Implicit Trapezoidal method.

until $|w_1^{(k)} - w_1^{(k-1)}|$ is sufficiently small. Normally only three or four iterations are required. Once a satisfactory approximation for w_1 has been determined, the method is repeated to find w_2 and so on.

Alternatively, the Secant method can be used in the Implicit Trapezoidal method in place of Newton's method, but then two distinct initial approximations to w_{j+1} are required. To determine these, the usual practice is to let $w_{j+1}^{(0)} = w_j$ and obtain $w_{j+1}^{(1)}$ from some explicit multistep method. When a system of stiff equations is involved, a generalization is required for either Newton's or the Secant method. These topics are considered in Chapter 10.

Illustration   The stiff initial-value problem

$$y' = 5e^{5t}(y - t)^2 + 1, \quad 0 \le t \le 1, \quad y(0) = -1$$

has solution $y(t) = t - e^{-5t}$. To show the effects of stiffness, the Implicit Trapezoidal method and the Runge-Kutta method of order 4 are applied both with $N = 4$, giving $h = 0.25$, and with $N = 5$, giving $h = 0.20$.

The Trapezoidal method performs well in both cases using a maximum of 10 iterations per step and $TOL = 10^{-6}$, as does Runge-Kutta with $h = 0.2$. However, $h = 0.25$ is outside the region of absolute stability of the Runge-Kutta method, which is evident from the results in Table 5.19. ☐

Table 5.19

	Runge–Kutta Method		Trapezoidal Method	
	$h = 0.2$		$h = 0.2$	
t_i	w_i	$\|y(t_i) - w_i\|$	w_i	$\|y(t_i) - w_i\|$
0.0	-1.0000000	0	-1.0000000	0
0.2	-0.1488521	1.9027×10^{-2}	-0.1414969	2.6383×10^{-2}
0.4	0.2684884	3.8237×10^{-3}	0.2748614	1.0197×10^{-2}
0.6	0.5519927	1.7798×10^{-3}	0.5539828	3.7700×10^{-3}
0.8	0.7822857	6.0131×10^{-4}	0.7830720	1.3876×10^{-3}
1.0	0.9934905	2.2845×10^{-4}	0.9937726	5.1050×10^{-4}
	$h = 0.25$		$h = 0.25$	
t_i	w_i	$\|y(t_i) - w_i\|$	w_i	$\|y(t_i) - w_i\|$
0.0	-1.0000000	0	-1.0000000	0
0.25	0.4014315	4.37936×10^{-1}	0.0054557	4.1961×10^{-2}
0.5	3.4374753	3.01956×10^{0}	0.4267572	8.8422×10^{-3}
0.75	1.44639×10^{23}	1.44639×10^{23}	0.7291528	2.6706×10^{-3}
1.0	Overflow		0.9940199	7.5790×10^{-4}

MATLAB has specific routines to solve stiff systems. To use the variable order multistep method ode15s to solve the stiff system in the Illustration at the beginning of the section:

$$u_1' = 9u_1 + 24u_2 + 5\cos t - \frac{1}{3}\sin t, \quad u_1(0) = \frac{4}{3},$$

$$u_2' = -24u_1 - 51u_2 - 9\cos t + \frac{1}{3}\sin t, \quad u_2(0) = \frac{2}{3},$$

we first define the right-hand side of the system by creating an M-file called F1.m.

```
function dy = F1(t,y)
dy = zeros(2,1);
dy(1) = 9*y(1)+24*y(2)+5*cos(t)-sin(t)/3;
dy(2) = -24*y(1)-51*y(2)-9*cos(t)+sin(t)/3;
```

and make F1.m known to MATLAB with

```
F3 = @F1
```

The initial conditions $u_1(0) = 4/3$ and $u_2(0) = 2/3$ are defined by

```
y10 = 4/3, y20 = 2/3
```

and the t values where we want the approximations by

```
tspan = [0 0.1 0.2 0.3 0.4 0.5 0.6 0.7 0.8 0.9 1.0]
```

The MATLAB response to

```
[T,YY]=ode15s(F3,tspan,[y10 y20])
```

gives the t values in the array T and in the array YY are the approximations to $u_1(t)$ in the first column and to $u_2(t)$ in the second.

$$T = \begin{bmatrix} 0 \\ 0.100000000000000 \\ 0.200000000000000 \\ 0.300000000000000 \\ 0.400000000000000 \\ 0.500000000000000 \\ 0.600000000000000 \\ 0.700000000000000 \\ 0.800000000000000 \\ 0.900000000000000 \\ 1.000000000000000 \end{bmatrix} \text{ and } YY = \begin{bmatrix} 1.333333333333333 & 0.666666666666667 \\ 1.792822173096780 & -1.031510588260698 \\ 1.423841705229490 & -0.874543162822479 \\ 1.131606544984823 & -0.725026524390037 \\ 0.909737405881405 & -0.608356645400698 \\ 0.739488715596804 & -0.515984454184929 \\ 0.606600245598731 & -0.440862764215503 \\ 0.500693300085545 & -0.377829713594847 \\ 0.414374560936528 & -0.323304488389232 \\ 0.342214086384561 & -0.274706988032098 \\ 0.280123638387467 & -0.230112384428887 \end{bmatrix}$$

EXERCISE SET 5.8

1. Solve the following stiff initial-value problems using Euler's method and compare the results with the actual solution.

   a. $y' = -9y$, for $0 \le t \le 1$, with $y(0) = e$ and $h = 0.1$; actual solution $y(t) = e^{1-9t}$.

   b. $y' = -20(y - t^2) + 2t$, for $0 \le t \le 1$, with $y(0) = \frac{1}{3}$ and $h = 0.1$; actual solution $y(t) = t^2 + \frac{1}{3}e^{-20t}$.

   c. $y' = -20y + 20\sin t + \cos t$, for $0 \le t \le 2$, with $y(0) = 1$ and $h = 0.25$; actual solution $y(t) = \sin t + e^{-20t}$.

   d. $y' = \dfrac{50}{y} - 50y$, for $0 \le t \le 1$, with $y(0) = \sqrt{2}$ and $h = 0.1$; actual solution $y(t) = (1 + e^{-100t})^{1/2}$.

2. Repeat Exercise 1 using the Runge-Kutta Fourth-Order method.

3. Repeat Exercise 1 using the Adams Fourth-Order Predictor-Corrector method.

4. Repeat Exercise 1 using the Trapezoidal method with a tolerance of 10^{-5}.

5. The Backward Euler One-Step method is defined by

   $$w_{i+1} = w_i + hf(t_{i+1}, w_{i+1}) \quad \text{for } i = 0, 1, \ldots, N - 1.$$

   Repeat Exercise 1 using the Backward Euler method incorporating Newton's method to solve for w_{i+1}.

6. The differential equation

   $$\frac{dp(t)}{dt} = rb(1 - p(t))$$

   can be used as a model for studying the proportion $p(t)$ of nonconformists in a society whose birth rate is b, and where r represents the rate at which an offspring becomes nonconformist when at least one of the parents is a conformist. Use the Trapezoidal method to find an approximation for $p(50)$ when $p(0) = 0.01$, $b = 0.02$, $r = 0.1$, and t takes on the integral values from 1 to 50.

5.9 Survey of Methods and Software

In this chapter we have considered methods to approximate the solutions to initial-value problems for ordinary differential equations. We began with a discussion of the most elementary numerical technique, Euler's method. This procedure is not sufficiently accurate to be of use in applications, but it illustrates the general behavior of the more powerful techniques, without the accompanying algebraic difficulties. The Taylor methods were then considered as generalizations of Euler's method. They were found to be accurate but cumbersome because of the need to determine extensive partial derivatives of the defining function of the differential equation. The Runge-Kutta formulas simplified the Taylor methods, while not significantly increasing the error. To this point we had considered only one-step methods, techniques that use only data at the most recently computed point.

Multistep methods were discussed in Section 5.4, where explicit methods of Adams-Bashforth type and implicit methods of Adams-Moulton type were considered. These culminate in predictor-corrector methods, which use an explicit method, such as an Adams-Bashforth, to predict the solution and then apply a corresponding implicit method, like an Adams-Moulton, to correct the approximation.

The more accurate adaptive methods are based on the relatively uncomplicated one-step and multistep techniques. In particular, we saw in Section 5.6 that the Runge-Kutta-Fehlberg method is a one-step procedure that seeks to select mesh spacing to keep the local error of the approximation under control. The Variable Step-Size Predictor-Corrector method also presented in Section 5.6 is based on the four-step Adams-Bashforth method and three-step Adams-Moulton method. It also changes the step size to keep the local error within a given tolerance. The Extrapolation method discussed in Section 5.5 is based on a modification of the Midpoint method and incorporates extrapolation to maintain a desired accuracy of approximation.

Section 5.7 illustrated how the techniques for a single first-order initial-value problem can be adapted to solve systems of initial-value problems and higher-order initial-value problems.

The final topic in the chapter concerned the difficulty that is inherent in the approximation of the solution to a stiff equation, a differential equation whose exact solution contains a portion of the form $e^{\lambda t}$, where λ is a complex number with a negative real part. Special caution must be taken with problems of this type, or the results can be overwhelmed by round-off error.

Methods of the Runge-Kutta-Fehlberg type are generally sufficient for non-stiff problems when moderate accuracy is required. The extrapolation procedures are recommended for non-stiff problems where high accuracy is required. Extensions of the Implicit Trapezoidal method to variable-order and variable step-size implicit Adams-type methods are used for stiff initial-value problems.

The IMSL Library includes two subroutines for approximating the solutions of initial-value problems. One is a variable step-size subroutine similar to the Runge-Kutta-Fehlberg method but based on fifth- and sixth-order formulas. The other subroutine is designed for stiff systems and uses implicit multistep methods of order up to 12. The NAG Library contains a Runge-Kutta type formula with a variable step size. A variable order, variable step-size backward-difference method for stiff systems is also available.

The netlib Library includes several subroutines for approximating the solutions of initial-value problems in the package ODE, located on the Internet at http://www.netlib.org/ode. The subroutine dverk.f is based on the Runge-Kutta-Verner fifth- and sixth-order methods. The subroutine rkf45.f is based on the Runge-Kutta-Fehlberg fourth- and fifth-order methods as described in Section 5.6. For stiff ordinary differential equation initial-value problems, the subroutine epsode.f based on a variable coefficient backward difference formula can be used.

Many books specialize in the numerical solution of initial-value problems. Two classics are by Henrici [He1] and Gear [Gea1]. Other books that survey the field are by Botha and Pinder [BP], Ortega and Poole [OP], Golub and Ortega [GO], Shampine [Sh], and Dormand [Do].

Two books by Hairer, Nörsett, and Warner provide comprehensive discussions on nonstiff [HNW1] and stiff [HNW2] problems. The book by Burrage [Bur] describes parallel and sequential methods.

CHAPTER

6

Direct Methods for Solving Linear Systems

6.1 Introduction

Systems of equations are used to represent physical problems that involve the interaction of various properties. The variables in the system represent the properties being studied, and the equations describe the interaction between the variables. The system is easiest to study when the equations are all linear. Often the number of equations is the same as the number of variables, for only in this case is it likely that a unique solution will exist.

Not all physical problems can be reasonably represented using a linear system with the same number of equations as unknowns, but the solutions to many problems either have this form or can be approximated by such a system. In fact, this is quite often the only approach that can give quantitative information about a physical problem.

In this chapter we consider *direct methods* for approximating the solution of a system of n linear equations in n unknowns. A direct method is one that gives the exact solution to the system, if it is assumed that all calculations can be performed without round-off error effects. However, we cannot generally avoid round-off error and we need to consider quite carefully the role of finite-digit arithmetic error in the approximation to the solution to the system, and how to arrange the calculations to minimize its effect.

6.2 Gaussian Elimination

If you have studied linear algebra or matrix theory, you probably have been introduced to Gaussian elimination, the most elementary method for systematically determining the solution of a system of linear equations. Variables are eliminated from the equations until one equation involves only one variable, a second equation involves only that variable and one other, a third has only these two and one additional, and so on. The solution is found by solving for the variable in the single equation, using this to reduce the second equation to one that now contains a single variable, and so on, until values for all the variables are found.

Three operations are permitted on a system of equations $E_1, E_2, \ldots, E_n$.

Operations on Systems of Equations

- Equation E_i can be multiplied by any nonzero constant λ, with the resulting equation used in place of E_i. This operation is denoted $(\lambda E_i) \rightarrow (E_i)$.

- Equation E_j can be multiplied by any constant λ, and added to equation E_i, with the resulting equation used in place of E_i. This operation is denoted $(E_i + \lambda E_j) \rightarrow (E_i)$.

- Equations E_i and E_j can be transposed in order. This operation is denoted $(E_i) \leftrightarrow (E_j)$.

By a sequence of the operations just given, a linear system can be transformed to a more easily-solved linear system with the same solutions. The sequence of operations is shown in the next Illustration.

Illustration   The four equations

$$
\begin{aligned}
E_1: & \quad x_1 + x_2 \qquad\;\; + 3x_4 = \quad 4, \\
E_2: & \quad 2x_1 + x_2 - x_3 + x_4 = \quad 1, \\
E_3: & \quad 3x_1 - x_2 - x_3 + 2x_4 = -3, \\
E_4: & \; -x_1 + 2x_2 + 3x_3 - x_4 = \quad 4,
\end{aligned}
\tag{6.1}
$$

will be solved for x_1, x_2, x_3, and x_4. We first use equation E_1 to eliminate the unknown x_1 from equations E_2, E_3, and E_4 by performing $(E_2 - 2E_1) \to (E_2)$, $(E_3 - 3E_1) \to (E_3)$, and $(E_4 + E_1) \to (E_4)$. For example, in the second equation

$$
(E_2 - 2E_1) \to (E_2)
$$

produces

$$
(2x_1 + x_2 - x_3 + x_4) - 2(x_1 + x_2 + 3x_4) = 1 - 2(4),
$$

which simplifies to the result shown as E_2 in

$$
\begin{aligned}
E_1: & \quad x_1 + x_2 \qquad\;\; + 3x_4 = \quad 4, \\
E_2: & \qquad\;\; - x_2 - x_3 - 5x_4 = \quad -7, \\
E_3: & \qquad\;\; - 4x_2 - x_3 - 7x_4 = -15, \\
E_4: & \qquad\qquad\;\; 3x_2 + 3x_3 + 2x_4 = \quad 8.
\end{aligned}
$$

For simplicity, the new equations are again labeled E_1, E_2, E_3, and E_4.

In the new system, E_2 is used to eliminate the unknown x_2 from E_3 and E_4 by performing $(E_3 - 4E_2) \to (E_3)$ and $(E_4 + 3E_2) \to (E_4)$. This results in

$$
\begin{aligned}
E_1: & \quad x_1 + x_2 \qquad\; + 3x_4 = \quad 4, \\
E_2: & \qquad\; - x_2 - x_3 - 5x_4 = \quad -7, \\
E_3: & \qquad\qquad\quad 3x_3 + 13x_4 = \quad 13, \\
E_4: & \qquad\qquad\qquad\; - 13x_4 = -13.
\end{aligned}
\tag{6.2}
$$

The system of equations (6.2) is now in **triangular** (or **reduced**) **form** and can be solved for the unknowns by a **backward-substitution process**. Since E_4 implies $x_4 = 1$, we can solve E_3 for x_3 to give

$$
x_3 = \frac{1}{3}(13 - 13x_4) = \frac{1}{3}(13 - 13) = 0.
$$

Continuing, E_2 gives

$$
x_2 = -(-7 + 5x_4 + x_3) = -(-7 + 5 + 0) = 2,
$$

and E_1 gives

$$
x_1 = 4 - 3x_4 - x_2 = 4 - 3 - 2 = -1.
$$

The solution to system (6.2), and consequently to system (6.1), is therefore $x_1 = -1$, $x_2 = 2$, $x_3 = 0$, and $x_4 = 1$.   □

Matrices and Vectors

When performing the calculations of the Illustration, we did not need to write out the full equations at each step or to carry the variables x_1, x_2, x_3, and x_4 through the calculations because they always remained in the same column. The only variation from system to system occurred in the coefficients of the unknowns and in the values on the right side of the equations. For this reason, a linear system is often replaced by a **matrix**, a rectangular array of elements in which not only is the value of an element important, but also its position in the array. The matrix contains all the information about the system that is necessary to determine its solution in a compact form.

The notation for an $n \times m$ (n by m) matrix will be a capital letter, such as A, for the matrix, and lowercase letters with double subscripts, such as a_{ij}, to refer to the entry at the intersection of the ith row and jth column; that is,

$$A = [a_{ij}] = \begin{bmatrix} a_{11} & a_{12} & \cdots & a_{1m} \\ a_{21} & a_{22} & \cdots & a_{2m} \\ \vdots & \vdots & & \vdots \\ a_{n1} & a_{n2} & \cdots & a_{nm} \end{bmatrix}.$$

Example 1   Determine the size and respective entries of the matrix

$$A = \begin{bmatrix} 2 & -1 & 7 \\ 3 & 1 & 0 \end{bmatrix}.$$

Solution   The matrix has two rows and three columns, so it is of size 2×3. Its entries are described by $a_{11} = 2$, $a_{12} = -1$, $a_{13} = 7$, $a_{21} = 3$, $a_{22} = 1$, and $a_{23} = 0$.   ∎

The $1 \times n$ matrix $A = [a_{11} \ a_{12} \ \cdots a_{1n}]$ is called an ***n*-dimensional row vector**, and an $n \times 1$ matrix

$$A = \begin{bmatrix} a_{11} \\ a_{21} \\ \vdots \\ a_{n1} \end{bmatrix}$$

is called an ***n*-dimensional column vector**. Usually the unnecessary subscript is omitted for vectors and a boldface lowercase letter is used for notation. So,

$$\mathbf{x} = \begin{bmatrix} x_1 \\ x_2 \\ \vdots \\ x_n \end{bmatrix}$$

denotes a column vector, and $\mathbf{y} = [y_1 \ y_2 \ \cdots y_n]$ denotes a row vector.

A system of n linear equations in the n unknowns $x_1, x_2, \ldots, x_n$ has the form

$$a_{11}x_1 + a_{12}x_2 + \cdots + a_{1n}x_n = b_1,$$

$$a_{21}x_1 + a_{22}x_2 + \cdots + a_{2n}x_n = b_2,$$

$$\vdots$$

$$a_{n1}x_1 + a_{n2}x_2 + \cdots + a_{nn}x_n = b_n.$$

An $n \times (n+1)$ matrix can be used to represent this linear system by first constructing

$$A = [a_{ij}] = \begin{bmatrix} a_{11} & a_{12} & \cdots & a_{1n} \\ a_{21} & a_{22} & \cdots & a_{2n} \\ \vdots & \vdots & & \vdots \\ a_{n1} & a_{n2} & \cdots & a_{nn} \end{bmatrix} \quad \text{and} \quad \mathbf{b} = \begin{bmatrix} b_1 \\ b_2 \\ \vdots \\ b_n \end{bmatrix}$$

Augmented refers to the fact that the right-hand side of the system has been included in the matrix.

and then combining these matrices to form the *augmented matrix*:

$$[A, \mathbf{b}] = \begin{bmatrix} a_{11} & a_{12} & \cdots & a_{1n} & \vdots & b_1 \\ a_{21} & a_{22} & \cdots & a_{2n} & \vdots & b_2 \\ \vdots & \vdots & & \vdots & \vdots & \vdots \\ a_{n1} & a_{n2} & \cdots & a_{nn} & \vdots & b_n \end{bmatrix},$$

where the vertical dotted line before the last column is used to separate the coefficients of the unknowns from the values on the right-hand side of the equations.

Repeating the operations involved in the Illustration on page 230 with the matrix notation results in first considering the augmented matrix:

$$\begin{bmatrix} 1 & 1 & 0 & 3 & \vdots & 4 \\ 2 & 1 & -1 & 1 & \vdots & 1 \\ 3 & -1 & -1 & 2 & \vdots & -3 \\ -1 & 2 & 3 & -1 & \vdots & 4 \end{bmatrix}.$$

Performing the operations

$$(E_2 - 2E_1) \rightarrow (E_2), \quad (E_3 - 3E_1) \rightarrow (E_3), \quad \text{and} \quad (E_4 + E_1) \rightarrow (E_4)$$

produces

$$\begin{bmatrix} 1 & 1 & 0 & 3 & \vdots & 4 \\ 0 & -1 & -1 & -5 & \vdots & -7 \\ 0 & -4 & -1 & -7 & \vdots & -15 \\ 0 & 3 & 3 & 2 & \vdots & 8 \end{bmatrix}$$

Then

$$(E_3 - 4E_2) \rightarrow (E_3) \quad \text{and} \quad (E_4 + 3E_2) \rightarrow (E_4),$$

A technique similar to Gaussian elimination first appeared during the Han dynasty in China in the text Nine Chapters on the Mathematical Art, *which was written in approximately 200* BCE. *Joseph Louis Lagrange (1736–1813) described a technique similar to this procedure in 1778 for the case when the value of each equation is 0. Gauss gave a more general description in* Theoria Motus corporum coelestium sectionibus solem ambientium, *which described the least squares technique he used in 1801 to determine the orbit of the dwarf planet Ceres.*

produces the final matrix

$$\begin{bmatrix} 1 & 1 & 0 & 3 & \vdots & 4 \\ 0 & -1 & -1 & -5 & \vdots & -7 \\ 0 & 0 & 3 & 13 & \vdots & 13 \\ 0 & 0 & 0 & -13 & \vdots & -13 \end{bmatrix}.$$

This final matrix can be transformed into its corresponding linear system and solutions for $x_1, x_2, x_3,$ and x_4 obtained. The procedure involved in this process is called **Gaussian Elimination with Backward Substitution**.

The general Gaussian elimination procedure applied to the linear system

$$E_1: \quad a_{11}x_1 + a_{12}x_2 + \cdots + a_{1n}x_n = b_1,$$

$$E_2: \quad a_{21}x_1 + a_{22}x_2 + \cdots + a_{2n}x_n = b_2,$$

$$\vdots$$

$$E_n: \quad a_{n1}x_1 + a_{n2}x_2 + \cdots + a_{nn}x_n = b_n,$$

is handled in a similar manner. First form the augmented matrix $\tilde{A}$:

$$\tilde{A} = [A, \mathbf{b}] = \begin{bmatrix} a_{11} & a_{12} & \cdots & a_{1n} & \vdots & a_{1,n+1} \\ a_{21} & a_{22} & \cdots & a_{2n} & \vdots & a_{2,n+1} \\ \vdots & \vdots & & \vdots & \vdots & \vdots \\ a_{n1} & a_{n2} & \cdots & a_{nn} & \vdots & a_{n,n+1} \end{bmatrix},$$

where A denotes the matrix formed by the coefficients and the entries in the $(n + 1)$st column are the values of $\mathbf{b}$; that is, $a_{i,n+1} = b_i$ for each $i = 1, 2, \ldots, n$.

Suppose that $a_{11} \neq 0$. To convert the entries in the first column, below a_{11}, to zero, we perform the operations $(E_k - m_{k1}E_1) \rightarrow (E_k)$ for each $k = 2, 3 \ldots, n$ for an appropriate multiplier m_{k1}. We first designate the diagonal element in the column, a_{11} as the **pivot element**. The *multiplier* for the kth row is defined by $m_{k1} = a_{k1}/a_{11}$. Performing the operations $(E_k - m_{k1}E_1) \rightarrow (E_k)$ for each $k = 2, 3, \ldots, n$ eliminates (that is, changes to zero) the coefficient of x_1 in each of these rows:

$$\begin{bmatrix} a_{11} & a_{12} & \cdots & a_{1n} & \vdots & b_1 \\ a_{21} & a_{22} & \cdots & a_{2n} & \vdots & b_2 \\ \vdots & \vdots & & \vdots & \vdots & \vdots \\ a_{n1} & a_{n2} & \cdots & a_{nn} & \vdots & b_n \end{bmatrix} \begin{matrix} E_2 - m_{21}E_1 \rightarrow E_2 \\ E_3 - m_{31}E_1 \rightarrow E_3 \\ \vdots \\ E_n - m_{n1}E_1 \rightarrow E_n \end{matrix} \begin{bmatrix} a_{11} & a_{12} & \cdots & a_{1n} & \vdots & b_1 \\ 0 & a_{22} & \cdots & a_{2n} & \vdots & b_2 \\ \vdots & \vdots & & \vdots & \vdots & \vdots \\ 0 & a_{n2} & \cdots & a_{nn} & \vdots & b_n \end{bmatrix}.$$

Although the entries in rows $2, 3, \ldots, n$ are expected to change, for ease of notation, we again denote the entry in the ith row and the jth column by a_{ij}.

If the pivot element $a_{22} \neq 0$, we form the multipliers $m_{k2} = a_{k2}/a_{22}$ and perform the operations $(E_k - m_{k2}E_2) \rightarrow E_k$ for each $k = 3, \ldots, n$ obtaining

$$\begin{bmatrix} a_{11} & a_{12} & \cdots & a_{1n} & \vdots & b_1 \\ 0 & a_{22} & \cdots & a_{2n} & \vdots & b_2 \\ \vdots & \vdots & & \vdots & \vdots & \vdots \\ 0 & a_{n2} & \cdots & a_{nn} & \vdots & b_n \end{bmatrix} \begin{matrix} E_3 - m_{32}E_2 \rightarrow E_3 \\ \vdots \\ E_n - m_{n2}E_2 \rightarrow E_n \end{matrix} \begin{bmatrix} a_{11} & a_{12} & \cdots & a_{1n} & \vdots & b_1 \\ 0 & a_{22} & \cdots & a_{2n} & \vdots & b_2 \\ \vdots & \vdots & & \vdots & \vdots & \vdots \\ 0 & 0 & \cdots & a_{nn} & \vdots & b_n \end{bmatrix}$$

We then follow this sequential procedure for the rows $i = 3 \ldots, n - 1$. Define the multiplier $m_{ki} = a_{ki}/a_{ii}$ and perform the operation

$$(E_k - m_{ki}E_i) \rightarrow (E_k)$$

for each $k = i + 1, i + 2, \ldots, n$, provided the pivot element a_{ii} is nonzero. This eliminates x_i in each row below the ith for all values of $i = 1, 2, \ldots, n - 1$. The resulting matrix has the form

$$\tilde{\tilde{A}} = \begin{bmatrix} a_{11} & a_{12} & \cdots & a_{1n} & \vdots & a_{1,n+1} \\ 0 & a_{22} & \cdots & a_{2n} & \vdots & a_{2,n+1} \\ \vdots & & \ddots & \vdots & \vdots & \vdots \\ 0 & \cdots & 0 & a_{nn} & \vdots & a_{n,n+1} \end{bmatrix},$$

where, except in the first row, the values of a_{ij} are not expected to agree with those in the original matrix $\tilde{A}$. The matrix $\tilde{\tilde{A}}$ represents a linear system with the same solution set as the original system, that is,

$$a_{11}x_1 + a_{12}x_2 + \cdots + a_{1n}x_n = a_{1,n+1},$$
$$a_{22}x_2 + \cdots + a_{2n}x_n = a_{2,n+1},$$
$$\vdots \qquad \vdots$$
$$a_{nn}x_n = a_{n,n+1},$$

Backward substitution can be performed on this system. Solving the nth equation for x_n gives

$$x_n = \frac{a_{n,n+1}}{a_{nn}}.$$

Then solving the $(n-1)$st equation for x_{n-1} and using the known value for x_n yields

$$x_{n-1} = \frac{a_{n-1,n+1} - a_{n-1,n}x_n}{a_{n-1,n-1}}.$$

Continuing this process, we obtain

$$x_i = \frac{a_{i,n+1} - (a_{i,i+1}x_{i+1} + \cdots + a_{i,n}x_n)}{a_{ii}} = \frac{a_{i,n+1} - \sum_{j=i+1}^{n} a_{ij}x_j}{a_{ii}}$$

for each $i = n-1, n-2, \ldots, 2, 1$.

The procedure will fail if at the ith step the pivot element a_{ii} is zero, for then either the multipliers $m_{ki} = a_{ki}/a_{ii}$ are not defined (this occurs if $a_{ii} = 0$ for some $i < n$) or the backward substitution cannot be performed (if $a_{nn} = 0$). This does not necessarily mean that the system has no solution, but rather that the technique for finding the solution must be altered by interchanging rows when a pivot is 0.

> The program GAUSEL61 implements Gaussian Elimination with Backward Substitution.

Program GAUSEL61 incorporates row interchanges when required. An illustration is given in the following example.

Example 2 Represent the linear system

$$
\begin{aligned}
E_1: & \quad x_1 - x_2 + 2x_3 - x_4 = -8, \\
E_2: & \quad 2x_1 - 2x_2 + 3x_3 - 3x_4 = -20, \\
E_3: & \quad x_1 + x_2 + x_3 \qquad\quad = -2, \\
E_4: & \quad x_1 - x_2 + 4x_3 + 3x_4 = \quad 4,
\end{aligned}
$$

as an augmented matrix and use Gaussian elimination to find its solution.

Solution The augmented matrix is

$$
\tilde{A} = \left[
\begin{array}{cccc:c}
1 & -1 & 2 & -1 & -8 \\
2 & -2 & 3 & -3 & -20 \\
1 & 1 & 1 & 0 & -2 \\
1 & -1 & 4 & 3 & 4
\end{array}
\right].
$$

Performing the operations

$$(E_2 - 2E_1) \rightarrow (E_2), \quad (E_3 - E_1) \rightarrow (E_3), \quad \text{and} \quad (E_4 - E_1) \rightarrow (E_4),$$

gives the matrix

$$
\left[
\begin{array}{cccc:c}
1 & -1 & 2 & -1 & -8 \\
0 & 0 & -1 & -1 & -4 \\
0 & 2 & -1 & 1 & 6 \\
0 & 0 & 2 & 4 & 12
\end{array}
\right].
$$

> The pivot element for a specific column is the entry that is used to place zeros in the other entries in that column.

The new diagonal entry a_{22}, called the **pivot element**, is 0, so the procedure cannot continue in its present form. But operations $(E_i) \leftrightarrow (E_j)$ are permitted, so a search is made of the elements a_{32} and a_{42} for the first nonzero element. Since $a_{32} \neq 0$, the operation $(E_2) \leftrightarrow (E_3)$ is performed to obtain a new matrix,

$$
\begin{bmatrix}
1 & -1 & 2 & -1 & : & -8 \\
0 & 2 & -1 & 1 & : & 6 \\
0 & 0 & -1 & -1 & : & -4 \\
0 & 0 & 2 & 4 & : & 12
\end{bmatrix}.
$$

Since x_2 is already eliminated from E_3 and E_4, the computations continue with the operation $(E_4 + 2E_3) \rightarrow (E_4)$, giving

$$
\tilde{A}^{(4)} =
\begin{bmatrix}
1 & -1 & 2 & -1 & : & -8 \\
0 & 2 & -1 & 1 & : & 6 \\
0 & 0 & -1 & -1 & : & -4 \\
0 & 0 & 0 & 2 & : & 4
\end{bmatrix}.
$$

The matrix is now converted back into a linear system that has a solution equivalent to the solution of the original system and the backward substitution is applied:

$$
x_4 = \frac{4}{2} = 2,
$$

$$
x_3 = \frac{[-4 - (-1)x_4]}{-1} = 2,
$$

$$
x_2 = \frac{[6 - x_4 - (-1)x_3]}{2} = 3,
$$

$$
x_1 = \frac{[-8 - (-1)x_4 - 2x_3 - (-1)x_2]}{1} = -7.
$$

To define the initial augmented matrix in MATLAB, which we will call `AA`, we enter the matrix row by row. A space is placed between each entry in a row, and the rows in `AA` are separated by a colon. So, for the matrix in Example 2 we have

`AA = [1 -1 2 -1 -8; 2 -2 3 -3 -20; 1 1 1 0 -2; 1 -1 4 3 4]`

MATLAB responds with

$$
AA =
\begin{bmatrix}
1 & -1 & 2 & -1 & -8 \\
2 & -2 & 3 & -3 & -20 \\
1 & 1 & 1 & 0 & -2 \\
1 & -1 & 4 & 3 & 4
\end{bmatrix}
$$

To perform the operation $(E_j + mE_i) \rightarrow (E_j)$ in MATLAB we use the command

`AA(j,:) = AA(j,:) + m * AA(i,:)`

The notation `AA(k,l)` refers to entry in the kth row and lth column. The use of `:` in MATLAB refers to multiplying an entire row or column. For example, multiplying the kth row by m is done with `m *AA(k,:)`. Similarly, multiplying the lth column by m would be done with `m*AA(:,l)`. So the next command subtracts twice the first row of `AA` from the second row.

`AA(2,:) = AA(2,:)-2*AA(1,:)`

which gives

$$AA = \begin{bmatrix} 1 & -1 & 2 & -1 & -8 \\ 0 & 0 & -1 & -1 & -4 \\ 1 & 1 & 1 & 0 & -2 \\ 1 & -1 & 4 & 3 & 4 \end{bmatrix}$$

We then subtract the first row of AA from the third row, followed by the subtraction of the first row from the fourth row with

```
AA(3,:) = AA(3,:)-AA(1,:)
```

and

```
AA(4,:) = AA(4,:)-AA(1,:)
```

This gives

$$AA = \begin{bmatrix} 1 & -1 & 2 & -1 & -8 \\ 0 & 0 & -1 & -1 & -4 \\ 0 & 2 & -1 & 1 & 6 \\ 0 & 0 & 2 & 4 & 12 \end{bmatrix}$$

The variable x_1 has now been eliminated from the rows corresponding to the second, third, and fourth equations. Since a_{22} is zero, we need to interchange rows to move a nonzero entry to a_{22}. To interchange rows 2 and 3, we store row 2 in a temporary row vector B, move row 3 to row 2, and then more the temporary row vector B to row 3. This is done with

```
B = AA(2,:)
AA(2,:) = AA(3,:)
AA(3,:) = B
```

The result is

$$AA = \begin{bmatrix} 1 & -1 & 2 & -1 & -8 \\ 0 & 2 & -1 & 1 & 6 \\ 0 & 0 & -1 & -1 & -4 \\ 0 & 0 & 2 & 4 & 12 \end{bmatrix}$$

The final operation in Gaussian elimination for this matrix is to add 2 times the third row to the fourth row with

```
AA(4,:) = AA(4,:)+2*AA(3,:)
```

This produces

$$AA = \begin{bmatrix} 1 & -1 & 2 & -1 & -8 \\ 0 & 2 & -1 & 1 & 6 \\ 0 & 0 & -1 & -1 & -4 \\ 0 & 0 & 0 & 2 & 4 \end{bmatrix}$$

To perform the backward substitution we need to define the vector **x** that will contain the solution. We initialize a vector **x** as the **0** vector and will replace these entries as we progress through the backward substitution.

```
x = [0 0 0 0]
```

Now we replace the 0 in the fourth column of **x** with

```
x(4) = AA(4,5)/AA(4,4)
```

which gives

$$x = 0\ 0\ 0\ 2$$

Then

```
x(3) = (AA(3,5) - AA(3,4)*x(4))/AA(3,3)
```

gives

$$x = 0\ 0\ 2\ 2$$

```
x(2) = (AA(2,5) - (AA(2,3)*x(3)+AA(2,4)*x(4)))/AA(2,2)
```

gives

$$x = 0\ 3\ 2\ 2$$

and

```
x(1) = (AA(1,5) - (AA(1,2)*x(2)+AA(1,3)*x(3)+AA(1,4)*x(4)))/AA(1,1)
```

gives the final solution

$$x = -7\ 3\ 2\ 2$$

which corresponds to $x_1 = -7$, $x_2 = 3$, $x_3 = 2$, and $x_4 = 2$.

Example 2 illustrates what is done if one of the pivot elements is zero. If the ith pivot element is zero, the ith column of the matrix is searched from the ith row downward for the first nonzero entry, and a row interchange is performed to obtain the new matrix. Then the procedure continues as before. If no nonzero entry is found the procedure stops, and the linear system does not have a unique solution. It might have no solution or an infinite number of solutions.

Operation Counts

The computations in the program are performed using only one $n \times (n + 1)$ array for storage. This is done by replacing, at each step, the previous value of a_{ij} by the new one. In addition, the multipliers are stored in the locations of a_{ki} known to have zero values—that is, when $i < n$ and $k = i + 1, i + 2, \ldots, n$. Thus, the original matrix A is overwritten by the multipliers below the main diagonal and by the nonzero entries of the final reduced matrix on and above the main diagonal. We will see in Section 6.5 that these values can be used to solve other linear systems involving the original matrix A.

Both the amount of time required to complete the calculations and the subsequent round-off error depend on the number of floating-point arithmetic operations needed to solve a routine problem. In general, the amount of time required to perform a multiplication or division on a computer is approximately the same and is considerably greater than that required to perform an addition or subtraction. Even though the actual differences in execution time depend on the particular computing system being used, the count of the additions/subtractions are kept separate from the count of the multiplications/divisions because of the time differential. The total number of arithmetic operations depends on the size n, as follows:

$$\text{Multiplications/divisions:} \quad \frac{n^3}{3} + n^2 - \frac{n}{3}.$$

$$\text{Additions/subtractions:} \quad \frac{n^3}{3} + \frac{n^2}{2} - \frac{5n}{6}.$$

For large n, the total number of multiplications and divisions is approximately $n^3/3$, that is, $O(n^3)$, as is the total number of additions and subtractions. The amount of computation, and the time required to perform it, increases with n in approximate proportion to $n^3/3$, as shown in Table 6.1.

Table 6.1

n	Multiplications/Divisions	Additions/Subtractions
3	17	11
10	430	375
50	44,150	42,875
100	343,300	338,250

EXERCISE SET 6.2

1. Obtain a solution by graphical methods of the following linear systems, if possible.

   a. $x_1 + 2x_2 = 3,$
   $x_1 - x_2 = 0.$

   b. $x_1 + 2x_2 = 0,$
   $x_1 - x_2 = 0.$

   c. $x_1 + 2x_2 = 3,$
   $2x_1 + 4x_2 = 6.$

   d. $x_1 + 2x_2 = 3,$
   $-2x_1 - 4x_2 = 6.$

   e. $x_1 + 2x_2 = 0,$
   $2x_1 + 4x_2 = 0.$

   f. $2x_1 + x_2 = -1,$
   $x_1 + x_2 = 2,$
   $x_1 - 3x_2 = 5.$

   g. $2x_1 + x_2 = -1,$
   $4x_1 + 2x_2 = -2,$
   $x_1 - 3x_2 = 5.$

   h. $2x_1 + x_2 + x_3 = 1,$
   $2x_1 + 4x_2 - x_3 = -1.$

2. Use Gaussian elimination and two-digit rounding arithmetic to solve the following linear systems. Do not reorder the equations. (The exact solution to each system is $x_1 = 1, x_2 = -1, x_3 = 3$.)

   a. $4x_1 - x_2 + x_3 = 8,$
   $2x_1 + 5x_2 + 2x_3 = 3,$
   $x_1 + 2x_2 + 4x_3 = 11.$

   b. $4x_1 + x_2 + 2x_3 = 9,$
   $2x_1 + 4x_2 - x_3 = -5,$
   $x_1 + x_2 - 3x_3 = -9.$

3. Use Gaussian elimination to solve the following linear systems, if possible, and determine whether row interchanges are necessary:

   a. $x_1 - x_2 + 3x_3 = 2,$
   $3x_1 - 3x_2 + x_3 = -1,$
   $x_1 + x_2 = 3.$

   b. $2x_1 - 1.5x_2 + 3x_3 = 1,$
   $-x_1 + 2x_3 = 3,$
   $4x_1 - 4.5x_2 + 5x_3 = 1.$

c.
$$2x_1 \qquad\qquad\qquad = 3,$$
$$x_1 + 1.5x_2 \qquad\qquad = 4.5,$$
$$-3x_2 + 0.5x_3 \qquad = -6.6,$$
$$2x_1 - \quad 2x_2 + \quad x_3 + x_4 = 0.8.$$

d.
$$x_1 - \tfrac{1}{2}x_2 + x_3 \qquad = 4,$$
$$2x_1 - \quad x_2 - x_3 + x_4 = 5,$$
$$x_1 + \quad x_2 \qquad\qquad = 2,$$
$$x_1 - \tfrac{1}{2}x_2 + x_3 + x_4 = 5.$$

e.
$$x_1 + x_2 \qquad + \quad x_4 = 2,$$
$$2x_1 + x_2 - \quad x_3 + \quad x_4 = 1,$$
$$4x_1 - x_2 - 2x_3 + 2x_4 = 0,$$
$$3x_1 - x_2 - \quad x_3 + 2x_4 = -3.$$

f.
$$x_1 + \quad x_2 \qquad + \quad x_4 = 2,$$
$$2x_1 + \quad x_2 - \quad x_3 + \quad x_4 = 1,$$
$$-x_1 + 2x_2 + 3x_3 - \quad x_4 = 4,$$
$$3x_1 - \quad x_2 - \quad x_3 + 2x_4 = -3.$$

4. Use MATLAB with `long format` and Gaussian elimination to solve the following linear systems.

a.
$$\tfrac{1}{4}x_1 + \tfrac{1}{5}x_2 + \tfrac{1}{6}x_3 = 9,$$
$$\tfrac{1}{3}x_1 + \tfrac{1}{4}x_2 + \tfrac{1}{5}x_3 = 8,$$
$$\tfrac{1}{2}x_1 + \quad x_2 + 2x_3 = 8.$$

b.
$$3.333x_1 + \quad 15920x_2 - 10.333x_3 = 15913,$$
$$2.222x_1 + \quad 16.71x_2 + \quad 9.612x_3 = 28.544,$$
$$1.5611x_1 + 5.1791x_2 + 1.6852x_3 = 8.4254.$$

c.
$$x_1 + \tfrac{1}{2}x_2 + \tfrac{1}{3}x_3 + \tfrac{1}{4}x_4 = \tfrac{1}{6},$$
$$\tfrac{1}{2}x_1 + \tfrac{1}{3}x_2 + \tfrac{1}{4}x_3 + \tfrac{1}{5}x_4 = \tfrac{1}{7},$$
$$\tfrac{1}{3}x_1 + \tfrac{1}{4}x_2 + \tfrac{1}{5}x_3 + \tfrac{1}{6}x_4 = \tfrac{1}{8},$$
$$\tfrac{1}{4}x_1 + \tfrac{1}{5}x_2 + \tfrac{1}{6}x_3 + \tfrac{1}{7}x_4 = \tfrac{1}{9}.$$

d.
$$2x_1 + \quad x_2 - \quad x_3 + x_4 - 3x_5 = 7,$$
$$x_1 \qquad\qquad + 2x_3 - x_4 + \quad x_5 = 2,$$
$$- 2x_2 - \quad x_3 + x_4 - \quad x_5 = -5,$$
$$3x_1 + \quad x_2 - 4x_3 \qquad + 5x_5 = 6,$$
$$x_1 - \quad x_2 - \quad x_3 - x_4 + \quad x_5 = 3.$$

5. Given the linear system
$$2x_1 - 6\alpha x_2 = 3,$$
$$3\alpha x_1 - \quad x_2 = \tfrac{3}{2}.$$

a. Find value(s) of α for which the system has no solutions.

b. Find value(s) of α for which the system has an infinite number of solutions.

c. Assuming a unique solution exists for a given α, find the solution.

6. Given the linear system
$$x_1 - \quad x_2 + \alpha x_3 = -2,$$
$$-x_1 + 2x_2 - \alpha x_3 = 3,$$
$$\alpha x_1 + \quad x_2 + \quad x_3 = 2.$$

a. Find value(s) of α for which the system has no solutions.

b. Find value(s) of α for which the system has an infinite number of solutions.

c. Assuming a unique solution exists for a given α, find the solution.

7. Suppose that in a biological system there are n species of animals and m sources of food. Let x_j represent the population of the jth species for each $j = 1, \ldots, n$; b_i represent the available daily supply of the ith food; and a_{ij} represent the amount of the ith food consumed on average by a member of the jth species. The linear system

$$
\begin{aligned}
a_{11}x_1 + a_{12}x_2 + \cdots + a_{1n}x_n &= b_1, \\
a_{21}x_1 + a_{22}x_2 + \cdots + a_{2n}x_n &= b_2, \\
&\vdots \\
a_{m1}x_1 + a_{m2}x_2 + \cdots + a_{mn}x_n &= b_m
\end{aligned}
$$

represents an equilibrium where there is a daily supply of food to precisely meet the average daily consumption of each species.

a.   Let

$$
A = [a_{ij}] = \begin{bmatrix} 1 & 2 & 0 & 3 \\ 1 & 0 & 2 & 2 \\ 0 & 0 & 1 & 1 \end{bmatrix},
$$

$\mathbf{x} = (x_j) = [1000, 500, 350, 400]$, and $\mathbf{b} = (b_i) = [3500, 2700, 900]$. Is there sufficient food to satisfy the average daily consumption?

b.   What is the maximum number of animals of each species that could be individually added to the system with the supply of food still meeting the consumption?

c.   If species 1 became extinct, how much of an individual increase of each of the remaining species could be supported?

d.   If species 2 became extinct, how much of an individual increase of each of the remaining species could be supported?

8.   A Fredholm integral equation of the second kind is an equation of the form

$$
u(x) = f(x) + \int_a^b K(x, t)u(t)\,dt,
$$

where a and b and the functions f and K are given. To approximate the function u on the interval $[a, b]$, a partition $x_0 = a < x_1 < \cdots < x_{m-1} < x_m = b$ is selected and the equations

$$
u(x_i) = f(x_i) + \int_a^b K(x_i, t)u(t)\,dt, \quad \text{for each } i = 0, \ldots, m,
$$

are solved for $u(x_0), u(x_1), \ldots, u(x_m)$. The integrals are approximated using quadrature formulas based on the nodes $x_0, \ldots, x_m$. In our problem, $a = 0$, $b = 1$, $f(x) = x^2$, and $K(x, t) = e^{|x-t|}$.

a.   Show that the linear system

$$
u(0) = f(0) + \frac{1}{2}[K(0, 0)u(0) + K(0, 1)u(1)],
$$

$$
u(1) = f(1) + \frac{1}{2}[K(1, 0)u(0) + K(1, 1)u(1)]
$$

must be solved when the Trapezoidal rule is used.

b.   Set up and solve the linear system that results when the Composite Trapezoidal rule is used with $n = 4$.

c.   Repeat (b) using the Composite Simpson's rule.

6.3 Pivoting Strategies

If all the calculations could be done using exact arithmetic, we could almost end the chapter with the previous section. We now know how many calculations are needed to perform Gaussian elimination on a system, and from this we should be able to determine whether our computational device can solve our problem in reasonable time. In a practical situation, however, we do not have exact arithmetic, and the large number of arithmetic computations, on the order of $O(n^3)$, makes the consideration of computational round-off error necessary. In fact, as we will see in our first example, even for certain very small systems, round-off error can dominate the calculations. In this section we will see how the calculations in Gaussian elimination can be arranged to reduce the effect of this error.

In deriving the Gaussian elimination method, we found that a row interchange is needed when one of the pivot elements, a_{ii}, is zero. This row interchange has the form $(E_i) \leftrightarrow (E_p)$, where p is the smallest integer greater than i with $a_{pi} \neq 0$. To reduce the round-off error associated with finite-digit arithmetic, it is often necessary to perform row interchanges even when the pivot elements are not zero.

Suppose that a_{ii} is small in magnitude compared to a_{ki}. Then the magnitude of the multiplier

with a small value of a_{ii}, any round-off error in the numerator is dramatically when dividing by a_{ii}. An illustration of this difficulty is given in the following example.

Example 1 Apply Gaussian elimination to the system

$$E_1: \quad 0.003000x_1 + 59.14x_2 = 59.17$$
$$E_2: \quad 5.291x_1 - 6.130x_2 = 46.78,$$

using four-digit arithmetic with rounding, and compare the results to the exact solution $x_1 = 10.00$ and $x_2 = 1.000$.

Solution The first pivot element, $a_{11} = 0.003000$, is small, and its associated multiplier,

$$m_{21} = \frac{5.291}{0.003000} = 1763.6\overline{6},$$

rounds to the large number 1764. Performing $(E_2 - m_{21}E_1) \rightarrow (E_2)$ and the appropriate rounding gives the system

$$0.003000x_1 + 59.14x_2 \approx 59.17$$
$$-104300x_2 \approx -104400,$$

instead of the exact system, which is

$$0.003000x_1 + 59.14x_2 = 59.17$$
$$-104309.37\overline{6}x_2 = -104309.37\overline{6}.$$

The disparity in the magnitudes of $m_{21}a_{13}$ and a_{23} has introduced round-off error, but the round-off error has not yet been propagated. Backward substitution yields

$$x_2 \approx 1.001,$$

which is a close approximation to the actual value, $x_2 = 1.000$. However, because of the small pivot $a_{11} = 0.003000$,

$$x_1 \approx \frac{59.17 - (59.14)(1.001)}{0.003000} = -10.00$$

contains the small error of 0.001 multiplied by

$$\frac{59.14}{0.003000} \approx 20000.$$

This ruins the approximation to the actual value $x_1 = 10.00$.

This is clearly a contrived example and the graph in Figure 6.1 shows why the error can so easily occur. For larger systems it is much more difficult to predict in advance when devastating round-off error might occur. ▪

Figure 6.1

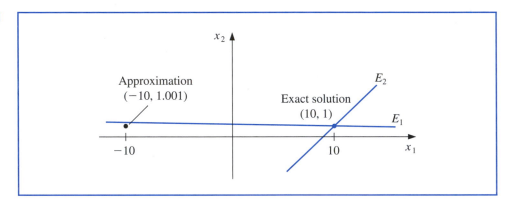

Partial Pivoting

Example 1 shows how difficulties can arise when the pivot element a_{ii} is small relative to the entries a_{kj} for $i \le k \le n$ and $i \le j \le n$. To avoid this problem, we first select a more appropriate pivot element a_{pi} and interchange the ith and pth rows.

The simplest strategy is to select, at the ith step, the element in the same column that is below the diagonal and has the largest absolute value; that is, to determine the smallest $p \ge i$ such that

$$|a_{pi}| = \max_{i \le k \le n} |a_{ki}|$$

and perform $(E_i) \leftrightarrow (E_p)$. In this case no interchange of columns is used. This technique is called **partial pivoting**, or *maximal column pivoting*.

Example 2 Apply Gaussian elimination to the system

$$\begin{aligned} E_1: \quad & 0.003000 x_1 + 59.14 x_2 = 59.17 \\ E_2: \quad & 5.291 x_1 - 6.130 x_2 = 46.78, \end{aligned}$$

using partial pivoting and four-digit arithmetic with rounding, and compare the results to the exact solution $x_1 = 10.00$ and $x_2 = 1.000$.

Solution The partial-pivoting procedure first requires finding

$$\max \{ |a_{11}|, |a_{21}| \} = \max \{ |0.003000|, |5.291| \} = |5.291| = |a_{21}|.$$

This requires that the operation $(E_2) \leftrightarrow (E_1)$ be performed to produce the equivalent system

$$E_1: \qquad 5.291x_1 - 6.130x_2 = 46.78,$$
$$E_2: \quad 0.003000x_1 + 59.14x_2 = 59.17.$$

The multiplier for this system is

$$m_{21} = \frac{a_{21}}{a_{11}} = 0.0005670,$$

and the operation $(E_2 - m_{21}E_1) \rightarrow (E_2)$ reduces the system to

$$5.291x_1 - 6.130x_2 \approx 46.78,$$
$$59.14x_2 \approx 59.14.$$

The four-digit answers resulting from the backward subst
$x_1 = 10.00$ and $x_2 = 1.000$.

> The program GAUPP62 implements Gaussian Elimination with Partial Pivoting.

Scaled Partial Pivoting

Although partial pivoting is sufficient for many linear syste
is inadequate. For example, the linear system

$$E_1: \quad 30.00x_1 + 591400x_2 = 5$$
$$E_2: \quad 5.291x_1 - \quad 6.130x_2 = 46.78$$

is the same as that in Examples 1 and 2 except that all entries in the first equation have been multiplied by 10^4. Partial pivoting with four-digit arithmetic leads to the same results as obtained in Example 1 because no row interchange would be performed. We need a more sophisticated method before we can accurately approximate solutions of systems like this; one that takes into account the relative sizes of the coefficients of the variables in the equations.

A technique known as **scaled partial pivoting** is needed for this system. The first step in this procedure is to define a scale factor s_k for each row:

$$s_k = \max_{1 \le j \le n} |a_{kj}|.$$

The appropriate row interchange to place zeros in the first column is determined by choosing the first integer p with

$$\frac{|a_{p1}|}{s_p} = \max_{1 \le k \le n} \frac{|a_{k1}|}{s_k}$$

and performing $(E_1) \leftrightarrow (E_p)$. The effect of scaling is to ensure that the largest element in each row has a relative magnitude of 1 before the comparison for row interchange is performed.

In a similar manner, before eliminating variable x_i using the operations

$$E_k - m_{ki}E_i \rightarrow E_k \quad \text{for } k = i + 1, \dots, n$$

we select the smallest integer $p \ge i$ with

$$\frac{|a_{pi}|}{s_p} = \max_{i \le k \le n} \frac{|a_{ki}|}{s_k}$$

The program GAUSPP62 implements Gaussian Elimination with Scaled Partial Pivoting.

and perform the row interchange $E_i \leftrightarrow E_p$ if $i \neq p$. The scale factors $s_1, \ldots, s_n$ are computed only once at the start of the procedure and must also be interchanged when row interchanges are performed.

In GAUSPP63 the scaling is done only for comparison purposes, so the division by scaling factors produces no round-off error in the system.

Illustration   The linear system

$$E_1: \quad 30.00x_1 + 591400x_2 = 591700,$$

$$E_2: \quad 5.291x_1 - \quad 6.130x_2 = 46.78,$$

is the same as that in Examples 1 and 2 except that all the entries in the first equation have been multiplied by 10^4. The partial pivoting procedure with four-digit arithmetic leads to the same results as obtained in Example 1. The maximal value in the first column is 30.00, and the multiplier

$$m_{21} = \frac{5.291}{30.00} = 0.1764$$

leads to the system

$$30.00x_1 + 591400x_2 \approx 591700,$$

$$-104300x_2 \approx -104400,$$

which has the same inaccurate solutions as in Example 1: $x_2 \approx 1.001$ and $x_1 \approx -10.00$.

Applying scaled partial pivoting to the system gives

$$s_1 = \max\{|30.00|, |591400|\} = 591400 \quad \text{and} \quad s_2 = \max\{|5.291|, |-6.130|\} = 6.130.$$

Consequently

$$\frac{|a_{11}|}{s_1} = \frac{30.00}{591400} = 0.5073 \times 10^{-4} \quad \text{and} \quad \frac{|a_{21}|}{s_2} = \frac{5.291}{6.130} = 0.8631$$

and the interchange $(E_1) \leftrightarrow (E_2)$ is made. Applying Gaussian elimination to the new system produces the correct results: $x_1 = 10.00$ and $x_2 = 1.000$. $\quad\square$

Example 3   Use scaled partial pivoting to solve the linear system using three-digit rounding arithmetic.

$$2.11x_1 - \quad 4.21x_2 + 0.921x_3 = 2.01,$$
$$4.01x_1 + \quad 10.2x_2 - \quad 1.12x_3 = -3.09,$$
$$1.09x_1 + 0.987x_2 + 0.832x_3 = 4.21.$$

Solution   The coefficients that are largest in magnitude give the scaling factors, so $s_1 = 4.21$, $s_2 = 10.2$, and $s_3 = 1.09$. Using three-digit rounding arithmetic gives

$$\frac{|a_{11}|}{s_1} = \frac{2.11}{4.21} = 0.501, \quad \frac{|a_{21}|}{s_2} = \frac{4.01}{10.2} = 0.393, \quad \text{and} \quad \frac{|a_{31}|}{s_3} = \frac{1.09}{1.09} = 1.$$

and the augmented matrix $\hat{A}$ is defined by

$$\hat{A} = \begin{bmatrix} 2.11 & -4.21 & 0.921 & : & 2.01 \\ 4.01 & 10.2 & -1.12 & : & -3.09 \\ 1.09 & 0.987 & 0.832 & : & 4.21 \end{bmatrix}.$$

Since $|a_{31}|/s_3$ is largest, we perform $(E_1) \leftrightarrow (E_3)$ to change the augmented matrix to

$$\hat{A} = \begin{bmatrix} 1.09 & 0.987 & 0.832 & \vdots & 4.21 \\ 4.01 & 10.2 & -1.12 & \vdots & -3.09 \\ 2.11 & -4.21 & 0.921 & \vdots & 2.01 \end{bmatrix}.$$

The multipliers for the first column of this matrix are $m_{21} = 3.68$ and $m_{31} = 1.94$, and performing $(E_2) - m_{21}(E_1)$ and $(E_3) - m_{31}(E_1)$ using three-digit arithmetic gives the new augmented matrix

$$\hat{A} = \begin{bmatrix} 1.09 & 0.987 & 0.832 & \vdots & 4.21 \\ 0 & 6.57 & -4.18 & \vdots & -18.6 \\ 0 & -6.12 & -0.689 & \vdots & -6.16 \end{bmatrix}.$$

Since

$$\frac{|a_{22}|}{s_2} = \frac{6.57}{10.2} = 0.644 < \frac{|a_{32}|}{s_3} = \frac{6.12}{4.21} = 1.45,$$

we perform $(E_2) \leftrightarrow (E_3)$ giving

$$\hat{A} = \begin{bmatrix} 1.09 & 0.987 & 0.832 & \vdots & 4.21 \\ 0 & -6.12 & -0.689 & \vdots & -6.16 \\ 0 & 6.57 & -4.18 & \vdots & -18.6 \end{bmatrix}.$$

Now we compute the multiplier $m_{32} = 6.57/(-6.12) = -1.07$ and perform $(E_3) - m_{32}(E_2)$ using three-digit arithmetic to obtain

$$\hat{A} := \begin{bmatrix} 1.09 & 0.987 & 0.832 & \vdots & 4.21 \\ 0 & -6.12 & -0.689 & \vdots & -6.16 \\ 0 & 0 & -4.92 & \vdots & -25.2 \end{bmatrix}.$$

Performing backward substitution using three-digit rounding arithmetic on this final matrix gives

$$x_1 = -0.431, \quad x_2 = 0.430, \quad \text{and} \quad x_3 = 5.12. \qquad \blacksquare$$

The scaled partial pivoting procedure adds a total of

$$\frac{3}{2}n(n-1) \quad \text{comparisons} \quad \text{and} \quad \frac{n(n+1)}{2} - 1 \quad \text{divisions}$$

to the Gaussian elimination procedure. The time required to perform a comparison is slightly more than that of an addition/subtraction. The total time to perform the basic Gaussian elimination procedure is the time required for approximately $n^3/3$ multiplications/divisions and $n^3/3$ additions/subtractions. So scaled partial pivoting does not add significantly to the computational time required to solve a system for large values of n.

Complete Pivoting

If a system has computation difficulties that scaled partial pivoting cannot resolve, *complete* (also called *total* or *maximal*) pivoting can be used. Complete pivoting at the ith step searches all the entries a_{kj}, for $k = i, i+1, \ldots, n$ and $j = i, i+1, \ldots, n$, to find the entry with the largest magnitude. Both row and column interchanges are performed to bring this entry to the pivot position.

The additional time required to incorporate complete pivoting into Gaussian elimination is

$$\frac{n(n-1)(2n+5)}{6} \quad \text{comparisons.}$$

This approximately doubles the amount of addition/subtraction time over ordinary Gaussian elimination.

EXERCISE SET 6.3

1. Use standard Gaussian elimination to find the row interchanges that are required to solve the following linear systems.

   a.
   $$\begin{aligned} x_1 - 5x_2 + x_3 &= 7 \\ 10x_1 + 20x_3 &= 6 \\ 5x_1 - x_3 &= 4 \end{aligned}$$

   b.
   $$\begin{aligned} x_1 + x_2 - x_3 &= 1 \\ x_1 + x_2 + 4x_3 &= 2 \\ 2x_1 - x_2 + 2x_3 &= 3 \end{aligned}$$

   c.
   $$\begin{aligned} 2x_1 - 3x_2 + 2x_3 &= 5 \\ -4x_1 + 2x_2 - 6x_3 &= 14 \\ 2x_1 + 2x_2 + 4x_3 &= 8 \end{aligned}$$

   d.
   $$\begin{aligned} x_2 + x_3 &= 6 \\ x_1 - 2x_2 - x_3 &= 4 \\ x_1 - x_2 + x_3 &= 5 \end{aligned}$$

2. Repeat Exercise 1 using Gaussian elimination with partial pivoting.

3. Repeat Exercise 1 using Gaussian elimination with scaled partial pivoting.

4. Repeat Exercise 1 using Gaussian elimination with complete pivoting.

5. Use Gaussian elimination and three-digit chopping arithmetic to solve the following linear systems, and compare the approximations to the actual solution.

   a.
   $$\begin{aligned} 0.03x_1 + 58.9x_2 &= 59.2 \\ 5.31x_1 - 6.10x_2 &= 47.0 \end{aligned}$$
   Actual solution $x_1 = 10, x_2 = 1$.

   b.
   $$\begin{aligned} 58.9x_1 + 0.03x_2 &= 59.2 \\ -6.10x_1 + 5.31x_2 &= 47.0 \end{aligned}$$
   Actual solution $x_1 = 1, x_2 = 10$.

   c.
   $$\begin{aligned} 3.03x_1 - 12.1x_2 + 14x_3 &= -119 \\ -3.03x_1 + 12.1x_2 - 7x_3 &= 120 \\ 6.11x_1 - 14.2x_2 + 21x_3 &= -139 \end{aligned}$$
   Actual solution $x_1 = 0, x_2 = 10, x_3 = \frac{1}{7}$.

   d.
   $$\begin{aligned} 3.3330x_1 + 15920x_2 + 10.333x_3 &= 7953 \\ 2.2220x_1 + 16.710x_2 + 9.6120x_3 &= 0.965 \\ -1.5611x_1 + 5.1792x_2 - 1.6855x_3 &= 2.714 \end{aligned}$$
   Actual solution $x_1 = 1, x_2 = 0.5, x_3 = -1$.

   e.
   $$\begin{aligned} 1.19x_1 + 2.11x_2 - 100x_3 + x_4 &= 1.12 \\ 14.2x_1 - 0.122x_2 + 12.2x_3 - x_4 &= 3.44 \\ 100x_2 - 99.9x_3 + x_4 &= 2.15 \\ 15.3x_1 + 0.110x_2 - 13.1x_3 - x_4 &= 4.16 \end{aligned}$$
   Actual solution $x_1 = 0.17682530, x_2 = 0.01269269, x_3 = -0.02065405, x_4 = -1.18260870$.

   f.
   $$\begin{aligned} \pi x_1 - ex_2 + \sqrt{2}x_3 - \sqrt{3}x_4 &= \sqrt{11} \\ \pi^2 x_1 + ex_2 - e^2 x_3 + \tfrac{3}{7}x_4 &= 0 \\ \sqrt{5}x_1 - \sqrt{6}x_2 + x_3 - \sqrt{2}x_4 &= \pi \\ \pi^3 x_1 + e^2 x_2 - \sqrt{7}x_3 + \tfrac{1}{9}x_4 &= \sqrt{2} \end{aligned}$$
   Actual solution $x_1 = 0.78839378, x_2 = -3.12541367, x_3 = 0.16759660, x_4 = 4.55700252$.

6. Repeat Exercise 5 using three-digit rounding arithmetic.

7. Repeat Exercise 5 using Gaussian elimination with partial pivoting.

8. Repeat Exercise 5 using Gaussian elimination with scaled partial pivoting.

9. Suppose that

$$
\begin{aligned}
2x_1 + x_2 + 3x_3 &= 1 \\
4x_1 + 6x_2 + 8x_3 &= 5 \\
6x_1 + \alpha x_2 + 10x_3 &= 5
\end{aligned}
$$

with $|\alpha| < 10$. For which of the following values of α will there be no row interchange required when solving this system using scaled partial pivoting?

a. $\alpha = 6$ **b.** $\alpha = 9$ **c.** $\alpha = -3$

6.4 Linear Algebra and Matrix Inversion

Early in this chapter we illustrated the convenience of matrix notation for the study of linear systems of equations, but there is a wealth of additional material in linear algebra that finds application in the study of approximation techniques. In this section we introduce some basic notation and results that are needed for both theory and application. All the topics discussed here should be familiar to anyone who has studied matrix theory at the undergraduate level. This section could then be omitted, but it is advisable to read the section to see the results from linear algebra that will be frequently called upon for service, and to be aware of the notation being used.

Two matrices A and B are **equal** if both are of the same size, say, $n \times m$, and if $a_{ij} = b_{ij}$ for each $i = 1, 2, \ldots, n$ and $j = 1, 2, \ldots, m$.

This definition means, for example, that

$$
\begin{bmatrix} 2 & -1 & 7 \\ 3 & 1 & 0 \end{bmatrix} \neq \begin{bmatrix} 2 & 3 \\ -1 & 1 \\ 7 & 0 \end{bmatrix}
$$

because they differ in dimension.

Matrix Arithmetic

If A and B are $n \times m$ matrices and λ is a real number, then

- the **sum** of A and B, denoted $A + B$, is the $n \times m$ matrix whose entries are $a_{ij} + b_{ij}$,

- the **scalar product** of λ and A, denoted λA, is the $n \times m$ matrix whose entries are λa_{ij}.

Example 1 Determine $A + B$ and λA when

$$
A = \begin{bmatrix} 2 & -1 & 7 \\ 3 & 1 & 0 \end{bmatrix}, \quad B = \begin{bmatrix} 4 & 2 & -8 \\ 0 & 1 & 6 \end{bmatrix}, \quad \text{and } \lambda = -2.
$$

Solution We have

$$
A + B = \begin{bmatrix} 2+4 & -1+2 & 7-8 \\ 3+0 & 1+1 & 0+6 \end{bmatrix} = \begin{bmatrix} 6 & 1 & -1 \\ 3 & 2 & 6 \end{bmatrix},
$$

and

$$
\lambda A = \begin{bmatrix} -2(2) & -2(-1) & -2(7) \\ -2(3) & -2(1) & -2(0) \end{bmatrix} = \begin{bmatrix} -4 & 2 & -14 \\ -6 & -2 & 0 \end{bmatrix}. \quad \blacksquare
$$

Matrix-Matrix Products

If A is an $n \times m$ matrix and B is an $m \times p$ matrix,

- the **matrix product** of A and B, denoted AB, is an $n \times p$ matrix C whose entries c_{ij} are given by

$$c_{ij} = \sum_{k=1}^{m} a_{ik}b_{kj} = a_{i1}b_{1j} + a_{i2}b_{2j} + \cdots + a_{im}b_{mj},$$

for each $i = 1, 2, \ldots n$ and $j = 1, 2, \ldots, p$.

The computation of c_{ij} can be viewed as the multiplication of the entries of the ith row of A with corresponding entries in the jth column of B, followed by a summation; that is,

$$[a_{i1}, a_{i2}, \ldots, a_{im}] \begin{bmatrix} b_{1j} \\ b_{2j} \\ \vdots \\ b_{mj} \end{bmatrix} = [c_{ij}],$$

where

$$c_{ij} = a_{i1}b_{1j} + a_{i2}b_{2j} + \cdots + a_{im}b_{mj} = \sum_{k=1}^{m} a_{ik}b_{kj}.$$

This explains why the number of columns of A must equal the number of rows of B for the product AB to be defined.

The following example illustrates a common matrix multiplying operation in the case when the matrix on the right of the multiplication has only one column, that is, it is a column vector.

Example 2 Determine the product $A\mathbf{b}$ if $A = \begin{bmatrix} 3 & 2 \\ -1 & 1 \\ 6 & 4 \end{bmatrix}$ and $\mathbf{b} = \begin{bmatrix} 3 \\ -1 \end{bmatrix}$.

Solution Because A has dimension 3×2 and $\mathbf{b}$ has dimension 2×1, the product is defined and is 3×1, that is, a vector with three rows. These are

$$3(3) + 2(-1) = 7, \quad (-1)(3) + 1(-1) = -4, \quad \text{and} \quad 6(3) + 4(-1) = 14.$$

So

$$A\mathbf{b} = \begin{bmatrix} 3 & 2 \\ -1 & 1 \\ 6 & 4 \end{bmatrix} \begin{bmatrix} 3 \\ -1 \end{bmatrix} = \begin{bmatrix} 7 \\ -4 \\ 14 \end{bmatrix} \qquad ■$$

In the next example we consider product operations in various situations.

Example 3 Determine all possible products of the matrices

$$A = \begin{bmatrix} 3 & 2 \\ -1 & 1 \\ 1 & 4 \end{bmatrix}, \quad B = \begin{bmatrix} 2 & 1 & -1 \\ 3 & 1 & 2 \end{bmatrix}, \quad C = \begin{bmatrix} 2 & 1 & 0 & 1 \\ -1 & 3 & 2 & 1 \\ 1 & 1 & 2 & 0 \end{bmatrix}, \quad \text{and}$$

$$D = \begin{bmatrix} 1 & -1 \\ 2 & -1 \end{bmatrix}.$$

Solution The sizes of the matrices are

$$A: 3 \times 2, \quad B: 2 \times 3, \quad C: 3 \times 4, \quad \text{and} \quad D: 2 \times 2.$$

The products that can be defined, and their dimensions, are:

$$AB: 3 \times 3, \quad BA: 2 \times 2, \quad AD: 3 \times 2, \quad BC: 2 \times 4, \quad DB: 2 \times 3, \quad \text{and} \quad DD: 2 \times 2.$$

These products are

$$AB = \begin{bmatrix} 12 & 5 & 1 \\ 1 & 0 & 3 \\ 14 & 5 & 7 \end{bmatrix}, \qquad BA = \begin{bmatrix} 4 & 1 \\ 10 & 15 \end{bmatrix}, \qquad AD = \begin{bmatrix} 7 & -5 \\ 1 & 0 \\ 9 & -5 \end{bmatrix},$$

$$BC = \begin{bmatrix} 2 & 4 & 0 & 3 \\ 7 & 8 & 6 & 4 \end{bmatrix}, \qquad DB = \begin{bmatrix} -1 & 0 & -3 \\ 1 & 1 & -4 \end{bmatrix}, \quad \text{and} \quad DD = \begin{bmatrix} -1 & 0 \\ 0 & -1 \end{bmatrix}.$$

■

Square Matrices

We have some special names and notation for matrices with the same number of rows and columns.

> The term *diagonal* applied to a matrix refers to the entries in the diagonal that run from the top left entry to the bottom right entry.

- A **square** matrix has the same number of rows as columns.

- A **diagonal** matrix is a square matrix whose only nonzero elements are along the main diagonal. So if $D = [d_{ij}]$ is a diagonal matrix, then $d_{ij} = 0$ whenever $i \neq j$.

- The **identity** matrix of order n, $I_n = [\delta_{ij}]$, is a diagonal matrix with 1s along the diagonal. That is,

$$I_n = \begin{bmatrix} 1 & 0 & \cdots\cdots & 0 \\ 0 & 1 & & \vdots \\ \vdots & & \ddots & \\ \vdots & & & 0 \\ 0 & \cdots\cdots & 0 & 1 \end{bmatrix}.$$

When the size of I_n is clear, this matrix is generally written simply as I. For example, the identity matrix of order three is

$$\begin{bmatrix} 1 & 0 & 0 \\ 0 & 1 & 0 \\ 0 & 0 & 1 \end{bmatrix}.$$

If A is any $n \times n$ matrix and $I = I_n$, then $AI = IA = A$.

Illustration Consider the identity matrix of order three,

$$I_3 = \begin{bmatrix} 1 & 0 & 0 \\ 0 & 1 & 0 \\ 0 & 0 & 1 \end{bmatrix}.$$

If A is any 3×3 matrix, then

$$AI_3 = \begin{bmatrix} a_{11} & a_{12} & a_{13} \\ a_{21} & a_{22} & a_{23} \\ a_{31} & a_{32} & a_{33} \end{bmatrix} \begin{bmatrix} 1 & 0 & 0 \\ 0 & 1 & 0 \\ 0 & 0 & 1 \end{bmatrix} = \begin{bmatrix} a_{11} & a_{12} & a_{13} \\ a_{21} & a_{22} & a_{23} \\ a_{31} & a_{32} & a_{33} \end{bmatrix} = A. \qquad \square$$

A triangular matrix is one that has all its nonzero entries either on and above (upper) or on and below (lower) the main diagonal. A matrix is diagonal precisely when it is both upper- and lower-triangular.

An $n \times n$ **upper-triangular** matrix $U = [u_{ij}]$ has all its nonzero entries on or above the main diagonal, that is, for each $j = 1, 2, \ldots, n$, the entries

$$u_{ij} = 0, \quad \text{for each } i = j + 1, j + 2, \ldots, n.$$

In a similar manner, a **lower-triangular** matrix $L = [l_{ij}]$ has all its nonzero entries on or below the main diagonal, that is, for each $j = 1, 2, \ldots, n$, the entries

$$l_{ij} = 0, \quad \text{for each } i = 1, 2, \ldots, j - 1.$$

(A diagonal matrix is both upper and lower triangular.)

In Example 3 we found that, in general, $AB \neq BA$, even when both products are defined. However, the other arithmetic properties associated with multiplication do hold. For example, when A, B, and C are matrices of the appropriate size and λ is a scalar, we have

- $A(BC) = (AB)C$,

- $A(B + C) = AB + AC$, and

- $\lambda(AB) = (\lambda A)B = A(\lambda B)$.

Inverse Matrices

The word *singular* means something that deviates from the ordinary. Hence a singular matrix does *not* have an inverse.

Certain $n \times n$ matrices have the property that another $n \times n$ matrix, which we will denote A^{-1}, exists with $AA^{-1} = A^{-1}A = I$. In this case A is said to be **nonsingular**, or *invertible*, and the matrix A^{-1} is called the **inverse** of A. A matrix without an inverse is called **singular**, or *noninvertible*.

Example 4 Let

$$A = \begin{bmatrix} 1 & 2 & -1 \\ 2 & 1 & 0 \\ -1 & 1 & 2 \end{bmatrix} \quad \text{and} \quad B = \begin{bmatrix} -\frac{2}{9} & \frac{5}{9} & -\frac{1}{9} \\ \frac{4}{9} & -\frac{1}{9} & \frac{2}{9} \\ -\frac{1}{3} & \frac{1}{3} & \frac{1}{3} \end{bmatrix}.$$

Show that $B = A^{-1}$, and that the solution to the linear system described by

$$x_1 + 2x_2 - x_3 = 2,$$
$$2x_1 + x_2 = 3,$$
$$-x_1 + x_2 + 2x_3 = 4.$$

is given by the entries in $B\mathbf{b}$, where $\mathbf{b}$ is the column vector with entries 2, 3, and 4.

Solution First note that

$$AB = \begin{bmatrix} 1 & 2 & -1 \\ 2 & 1 & 0 \\ -1 & 1 & 2 \end{bmatrix} \cdot \begin{bmatrix} -\frac{2}{9} & \frac{5}{9} & -\frac{1}{9} \\ \frac{4}{9} & -\frac{1}{9} & \frac{2}{9} \\ -\frac{1}{3} & \frac{1}{3} & \frac{1}{3} \end{bmatrix} = \begin{bmatrix} 1 & 0 & 0 \\ 0 & 1 & 0 \\ 0 & 0 & 1 \end{bmatrix} = I_3.$$

In a similar manner, $BA = I_3$, so A and B are both nonsingular with $B = A^{-1}$ and $A = B^{-1}$.

Now convert the given linear system to the matrix equation

$$\begin{bmatrix} 1 & 2 & -1 \\ 2 & 1 & 0 \\ -1 & 1 & 2 \end{bmatrix} \begin{bmatrix} x_1 \\ x_2 \\ x_3 \end{bmatrix} = \begin{bmatrix} 2 \\ 3 \\ 4 \end{bmatrix},$$

and multiply both sides by B, the inverse of A. Then we have

$$B(A\mathbf{x}) = B\mathbf{b},$$

with

$$B(A\mathbf{x}) = (BA)\mathbf{x} = \left(\begin{bmatrix} -\frac{2}{9} & \frac{5}{9} & -\frac{1}{9} \\ \frac{4}{9} & -\frac{1}{9} & \frac{2}{9} \\ -\frac{3}{9} & \frac{3}{9} & \frac{3}{9} \end{bmatrix} \begin{bmatrix} 1 & 2 & -1 \\ 2 & 1 & 0 \\ -1 & 1 & 2 \end{bmatrix} \right) \mathbf{x} = \mathbf{x}$$

and

$$B\mathbf{b} = \begin{bmatrix} -\frac{2}{9} & \frac{5}{9} & -\frac{1}{9} \\ \frac{4}{9} & -\frac{1}{9} & \frac{2}{9} \\ -\frac{1}{3} & \frac{1}{3} & \frac{1}{3} \end{bmatrix} \begin{bmatrix} 2 \\ 3 \\ 4 \end{bmatrix} = \begin{bmatrix} \frac{7}{9} \\ \frac{13}{9} \\ \frac{5}{3} \end{bmatrix}.$$

This implies that $\mathbf{x} = B\mathbf{b}$ and gives the solution $x_1 = 7/9$, $x_2 = 13/9$, and $x_3 = 5/3$.   ∎

The reason for introducing this matrix operation at this time is that the linear system

$$a_{11}x_1 + a_{12}x_2 + \cdots + a_{1n}x_n = b_1,$$

$$a_{21}x_1 + a_{22}x_2 + \cdots + a_{2n}x_n = b_2,$$

$$\vdots \qquad\qquad \vdots$$

$$a_{n1}x_1 + a_{n2}x_2 + \cdots + a_{nn}x_n = b_n,$$

can be viewed as the matrix equation $A\mathbf{x} = \mathbf{b}$, where

$$A = \begin{bmatrix} a_{11} & a_{12} & \cdots & a_{1n} \\ a_{21} & a_{22} & \cdots & a_{2n} \\ \vdots & \vdots & & \vdots \\ a_{n1} & a_{n2} & \cdots & a_{nn} \end{bmatrix}, \quad \mathbf{x} = \begin{bmatrix} x_1 \\ x_2 \\ \vdots \\ x_n \end{bmatrix}, \quad \text{and} \quad \mathbf{b} = \begin{bmatrix} b_1 \\ b_2 \\ \vdots \\ b_n \end{bmatrix}.$$

If A is a nonsingular matrix, then the solution $\mathbf{x}$ to the linear system $A\mathbf{x} = \mathbf{b}$ is given by

$$\mathbf{x} = A^{-1}(A\mathbf{x}) = A^{-1}\mathbf{b}.$$

In general, however, it is more difficult to determine A^{-1} than it is to solve the system $A\mathbf{x} = \mathbf{b}$ because the number of operations involved in determining A^{-1} is larger. Even so, it is useful from a conceptual standpoint to describe a method for determining the inverse of a matrix.

Illustration   To determine the inverse of the matrix

$$A = \begin{bmatrix} 1 & 2 & -1 \\ 2 & 1 & 0 \\ -1 & 1 & 2 \end{bmatrix},$$

let us first consider the product AB, where B is an arbitrary 3×3 matrix.

$$AB = \begin{bmatrix} 1 & 2 & -1 \\ 2 & 1 & 0 \\ -1 & 1 & 2 \end{bmatrix} \begin{bmatrix} b_{11} & b_{12} & b_{13} \\ b_{21} & b_{22} & b_{23} \\ b_{31} & b_{32} & b_{33} \end{bmatrix}$$

$$= \begin{bmatrix} b_{11} + 2b_{21} - b_{31} & b_{12} + 2b_{22} - b_{32} & b_{13} + 2b_{23} - b_{33} \\ 2b_{11} + b_{21} & 2b_{12} + b_{22} & 2b_{13} + b_{23} \\ -b_{11} + b_{21} + 2b_{31} & -b_{12} + b_{22} + 2b_{32} & -b_{13} + b_{23} + 2b_{33} \end{bmatrix}.$$

If $B = A^{-1}$, then $AB = I$, so we must have

$$\begin{aligned} b_{11} + 2b_{21} - b_{31} &= 1, & b_{12} + 2b_{22} - b_{32} &= 0, \\ 2b_{11} + b_{21} &= 0, & 2b_{12} + b_{22} &= 1, \\ -b_{11} + b_{21} + 2b_{31} &= 0, & -b_{12} + b_{22} + 2b_{32} &= 0, \end{aligned}$$

$$\begin{aligned} b_{13} + 2b_{23} - b_{33} &= 0 \\ 2b_{13} + b_{23} &= 0 \\ -b_{13} + b_{23} + 2b_{33} &= 1 \end{aligned}$$

Notice that the coefficients in each of the systems of equations are the same; the only change in the systems occurs on the right side of the equations. As a consequence, the computations can be performed on the larger augmented matrix, which is formed by combining the matrices for each of the systems

$$\begin{bmatrix} 1 & 2 & -1 & \vdots & 1 & 0 & 0 \\ 2 & 1 & 0 & \vdots & 0 & 1 & 0 \\ -1 & 1 & 2 & \vdots & 0 & 0 & 1 \end{bmatrix}.$$

First, performing $(E_2 - 2E_1) \rightarrow (E_1)$ and $(E_3 + E_1) \rightarrow (E_3)$ gives

$$\begin{bmatrix} 1 & 2 & -1 & \vdots & 1 & 0 & 0 \\ 0 & -3 & 2 & \vdots & -2 & 1 & 0 \\ 0 & 3 & 1 & \vdots & 1 & 0 & 1 \end{bmatrix}.$$

Next, performing $(E_3 + E_2) \rightarrow (E_3)$ produces

$$\begin{bmatrix} 1 & 2 & -1 & \vdots & 1 & 0 & 0 \\ 0 & -3 & 2 & \vdots & -2 & 1 & 0 \\ 0 & 0 & 3 & \vdots & -1 & 1 & 1 \end{bmatrix}.$$

Backward substitution is performed on each of the three augmented matrices,

$$\begin{bmatrix} 1 & 2 & -1 & \vdots & 1 \\ 0 & -3 & 2 & \vdots & -2 \\ 0 & 0 & 3 & \vdots & -1 \end{bmatrix}, \begin{bmatrix} 1 & 2 & -1 & \vdots & 0 \\ 0 & -3 & 2 & \vdots & 1 \\ 0 & 0 & 3 & \vdots & 1 \end{bmatrix}, \begin{bmatrix} 1 & 2 & -1 & \vdots & 0 \\ 0 & -3 & 2 & \vdots & 0 \\ 0 & 0 & 3 & \vdots & 1 \end{bmatrix},$$

to eventually give

$$\begin{aligned} b_{11} &= -\tfrac{2}{9}, & b_{12} &= \tfrac{5}{9}, & b_{13} &= -\tfrac{1}{9}, \\ b_{21} &= \tfrac{4}{9}, & b_{22} &= -\tfrac{1}{9}, & \text{and} \quad b_{23} &= \tfrac{2}{9}, \\ b_{31} &= -\tfrac{1}{3}, & b_{32} &= \tfrac{1}{3}, & b_{33} &= \tfrac{1}{3}. \end{aligned}$$

These are the entries of A^{-1}:

$$A^{-1} = \begin{bmatrix} -\tfrac{2}{9} & \tfrac{5}{9} & -\tfrac{1}{9} \\ \tfrac{4}{9} & -\tfrac{1}{9} & \tfrac{2}{9} \\ -\tfrac{1}{3} & \tfrac{1}{3} & \tfrac{1}{3} \end{bmatrix} = \frac{1}{9} \begin{bmatrix} -2 & 5 & -1 \\ 4 & -1 & 2 \\ -3 & 3 & 3 \end{bmatrix}. \tag{6.3}$$

□

Transpose of a Matrix

- The **transpose** of an $n \times m$ matrix $A = [a_{ij}]$ is the $m \times n$ matrix $A^t = [a_{ji}]$.

- A square matrix A is **symmetric** if $A = A^t$.

Illustration The matrices

$$
A = \begin{bmatrix} 7 & 2 & 0 \\ 3 & 5 & -1 \\ 0 & 5 & -6 \end{bmatrix}, \quad
B = \begin{bmatrix} 2 & 4 & 7 \\ 3 & -5 & -1 \end{bmatrix}, \quad
C = \begin{bmatrix} 6 & 4 & -3 \\ 4 & -2 & 0 \\ -3 & 0 & 1 \end{bmatrix}
$$

have transposes

$$
A^t = \begin{bmatrix} 7 & 3 & 0 \\ 2 & 5 & 5 \\ 0 & -1 & -6 \end{bmatrix}, \quad
B^t = \begin{bmatrix} 2 & 3 \\ 4 & -5 \\ 7 & -1 \end{bmatrix}, \quad
C^t = \begin{bmatrix} 6 & 4 & -3 \\ 4 & -2 & 0 \\ -3 & 0 & 1 \end{bmatrix}.
$$

The matrix C is symmetric because $C^t = C$. The matrices A and B are not symmetric. $\square$

The transpose notation is convenient for expressing column vectors in a more compact manner. Because the transpose of a column vector is a row vector,

the column vector $\mathbf{x} = \begin{bmatrix} x_1 \\ x_2 \\ \vdots \\ x_n \end{bmatrix}$ is generally written in text as $\mathbf{x} = (x_1, x_2, \ldots, x_n)^t$.

The following operations involving the transpose of a matrix hold whenever the operation is possible.

Transpose Facts

(i) $(A^t)^t = A$.

(ii) $(A + B)^t = A^t + B^t$.

(iii) $(AB)^t = B^t A^t$.

(iv) If A^{-1} exists, $(A^{-1})^t = (A^t)^{-1}$.

Matrix Determinants

The determinant of a square matrix is a number that can be useful in determining the existence and uniqueness of solutions to linear systems. We will denote the determinant of a matrix A by $\det A$, but it is also common to use the notation $|A|$.

Determinant of a Matrix

(i) If $A = [a]$ is a 1×1 matrix, then $\det A = a$.

(ii) If A is an $n \times n$ matrix, the **minor** M_{ij} is the determinant of the $(n - 1) \times (n - 1)$ submatrix of A obtained by deleting the ith row and jth column of the matrix A.

Then the determinant of A is given either by

$$\det A = \sum_{j=1}^{n} (-1)^{i+j} a_{ij} M_{ij} \quad \text{for any } i = 1, 2, \ldots, n,$$

or by

$$\det A = \sum_{i=1}^{n} (-1)^{i+j} a_{ij} M_{ij} \quad \text{for any } j = 1, 2, \ldots, n.$$

The notion of a determinant appeared independently in 1683 in Japan and Europe, although neither Takakazu Seki Kowa (1642–1708) nor Gottfried Leibniz (1646–1716) appear to have used the term determinant.

To calculate the determinant of a general $n \times n$ matrix by expanding by minors requires $O(n!)$ multiplications/divisions and additions/subtractions. Even for relatively small values of n, the number of calculations becomes unwieldy. Fortunately, the precise value of the determinant is seldom needed, and there are efficient ways to approximate its value.

Although it appears that there are $2n$ different definitions of $\det A$, depending on which row or column is chosen, all definitions give the same numerical result. The flexibility in the definition is used in the following example. It is most convenient to compute $\det A$ across the row or down the column with the most zeros.

Example 5   Find the determinant of the matrix

$$A = \begin{bmatrix} 2 & -1 & 3 & 0 \\ 4 & -2 & 7 & 0 \\ -3 & -4 & 1 & 5 \\ 6 & -6 & 8 & 0 \end{bmatrix}$$

using the row or column with the most zero entries.

Solution   To compute $\det A$, it is easiest to use the fourth column because three of its entries are 0.

$$\det A = a_{14}(-1)^5 M_{14} + a_{24}(-1)^6 M_{24} + a_{34}(-1)^7 M_{34} + a_{44}(-1)^8 M_{44} = -5M_{34}.$$

Eliminating the third row and the fourth column of A and expanding the resulting 3×3 matrix by its first row gives

$$\det A = -5 \det \begin{bmatrix} 2 & -1 & 3 \\ 4 & -2 & 7 \\ 6 & -6 & 8 \end{bmatrix}$$

$$= -5 \left\{ 2 \det \begin{bmatrix} -2 & 7 \\ -6 & 8 \end{bmatrix} - (-1) \det \begin{bmatrix} 4 & 7 \\ 6 & 8 \end{bmatrix} + 3 \det \begin{bmatrix} 4 & -2 \\ 6 & -6 \end{bmatrix} \right\}$$

$$= -5 [2(-16 + 42) + (32 - 42) + 3(-24 + 12)] = -30. \quad ▪$$

The following properties of determinants are useful in relating linear systems and Gaussian elimination to determinants.

Determinant Facts

Suppose A is an $n \times n$ matrix:

(i) If any row or column of A has only zero entries, then $\det A = 0$.

(ii) If $\tilde{A}$ is obtained from A by the operation $(E_i) \leftrightarrow (E_k)$, with $i \neq k$, then $\det \tilde{A} = -\det A$.

(iii) If A has two rows or two columns the same, then $\det A = 0$.

(iv) If $\tilde{A}$ is obtained from A by the operation $(\lambda E_i) \rightarrow (E_i)$, then $\det \tilde{A} = \lambda \det A$.

(v) If $\tilde{A}$ is obtained from A by the operation $(E_i + \lambda E_k) \rightarrow (E_i)$ with $i \neq k$, then $\det \tilde{A} = \det A$.

(vi) If B is also an $n \times n$ matrix, then $\det AB = \det A \cdot \det B$.

(vii) $\det A^t = \det A$.

(viii) If A^{-1} exists, then $\det A^{-1} = \dfrac{1}{\det A}$.

(ix) If A is an upper triangular, lower triangular, or diagonal matrix, then

$$\det A = a_{11} \cdot a_{22} \cdots a_{nn}.$$

Example 6 Compute the determinant of the matrix

$$A = \begin{bmatrix} 2 & 1 & -1 & 1 \\ 1 & 1 & 0 & 3 \\ -1 & 2 & 3 & -1 \\ 3 & -1 & -1 & 2 \end{bmatrix}$$

using Determinant Facts (ii), (iv), (v), and (ix), and doing the computations in MATLAB.

Solution Matrix A is defined in MATLAB by

```
A=[2 1 -1 1; 1 1 0 3; -1 2 3 -1; 3 -1 -1 2]
```

We will use the operations in Table 6.2 to first place the matrix in upper-triangular form. Then we can use the final determinant fact to contain the result from the entries on the diagonal. We have used some of these steps to illustrate the commands in MATLAB. For example, the first operation would not normally be performed when placing the matrix in upper-triangular form.

Table 6.2

Operation	MATLAB Command	Effect
$\frac{1}{2}E_1 \rightarrow E_1$	A1(1,:) = A(1,:)/2	$\det A1 = \frac{1}{2}\det A$
$E_2 - E_1 \rightarrow E_2$	A2(2,:)=A1(2,:)-A1(1,:)	$\det A2 = \det A1 = \frac{1}{2}\det A$
$E_3 + E_1 \rightarrow E_3$	A3(3,:)=A2(3,:)+A2(1,:)	$\det A3 = \det A2 = \frac{1}{2}\det A$
$E_4 - 3E_1 \rightarrow E_4$	A4(4,:)=A3(4,:)-3*A3(1,:)	$\det A4 = \det A3 = \frac{1}{2}\det A$
$2E_2 \rightarrow E_2$	A5(2,:)=2*A4(2,:)	$\det A5 = 2\det A4 = \det A$
$E_3 - \frac{5}{2}E_2 \rightarrow E_3$	A6(3,:)=A5(3,:)-2.5*A5(2,:)	$\det A6 = \det A5 = \det A$
$E_4 + \frac{5}{2}E_2 \rightarrow E_4$	A7(4,:)=A6(4,:)+2.5*A6(2,:)	$\det A7 = \det A6 = \det A$
$E_3 \leftrightarrow E_4$	B=A7(3,:),A8(3,:)=A7(4,:),A8(4,:)=B	$\det A8 = -\det A7 = -\det A$

After these operations are performed, the matrix will have the form

$$
A8 = \begin{bmatrix} 1 & \frac{1}{2} & -\frac{1}{2} & \frac{1}{2} \\ 0 & 1 & 1 & 5 \\ 0 & 0 & 3 & 13 \\ 0 & 0 & 0 & -13 \end{bmatrix}
$$

By (ix), det $A8 = 1 \cdot 1 \cdot 3(-13) = -39$, so det $A = -$ det $A8 = 39$. ▪

The key result relating nonsingularity, Gaussian elimination, linear systems, and determinants is that the following statements are equivalent.

Equivalent Statements about an $n \times n$ Matrix A

(i) The equation $A\mathbf{x} = \mathbf{0}$ has the unique solution $\mathbf{x} = \mathbf{0}$.

(ii) The system $A\mathbf{x} = \mathbf{b}$ has a unique solution for any n-dimensional column vector $\mathbf{b}$.

(iii) The matrix A is nonsingular; that is, A^{-1} exists.

(iv) det $A \neq 0$.

(v) Gaussian elimination with row interchanges can be performed on the system $A\mathbf{x} = \mathbf{b}$ for any n-dimensional column vector $\mathbf{b}$ to find the unique solution $\mathbf{x}$.

MATLAB has numerous commands that can directly perform operations on matrices. For example, for matrices A and B and scalar a, when the operations are possible, the following MATLAB commands can be used.

- To add A and B: A+B
- To multiply A and B: A*B
- To multiply A by a: a*A
- To obtain the transpose of A: A'
- To obtain the inverse of A: inv(A)
- To find the determinant of A: det(A)

EXERCISE SET 6.4

1. Compute the following matrix products.

a.
$$
\begin{bmatrix} 1 & 0 & 0 \\ -1 & 1 & 0 \\ 2 & 3 & 1 \end{bmatrix} \cdot \begin{bmatrix} 1 & 0 & 0 \\ 2 & 2 & 0 \\ 1 & -1 & 1 \end{bmatrix}
$$

b.
$$
\begin{bmatrix} 1 & 0 & 0 \\ 2 & 1 & 0 \\ -2 & -1 & 1 \end{bmatrix} \cdot \begin{bmatrix} 1 & -1 & 2 \\ 0 & 1 & 3 \\ 0 & 0 & 2 \end{bmatrix}
$$

c.
$$
\begin{bmatrix} 1 & 0 & 0 \\ 0 & 1 & 0 \\ 0 & -2 & 1 \end{bmatrix} \cdot \begin{bmatrix} 1 & 0 & 0 \\ 2 & 1 & 0 \\ -3 & 0 & 1 \end{bmatrix}
$$

d.
$$
\begin{bmatrix} 2 & -1 & 4 \\ 0 & -1 & 2 \\ 0 & 0 & 3 \end{bmatrix} \cdot \begin{bmatrix} 3 & -3 & 4 \\ 0 & 1 & 1 \\ 0 & 0 & 2 \end{bmatrix}
$$

2. For the following matrices:

i. Find the transpose of the matrix.

ii. Determine which matrices are nonsingular and compute their inverses.

a.
$$
\begin{bmatrix} 4 & 2 & 6 \\ 3 & 0 & 7 \\ -2 & -1 & -3 \end{bmatrix}
$$

b.
$$
\begin{bmatrix} 1 & 2 & 0 \\ 2 & 1 & -1 \\ 3 & 1 & 1 \end{bmatrix}
$$

c. $\begin{bmatrix} 4 & 0 & 0 \\ 0 & 0 & 0 \\ 0 & 0 & 3 \end{bmatrix}$

d. $\begin{bmatrix} 1 & 1 & -1 & 1 \\ 1 & 2 & -4 & -2 \\ 2 & 1 & 1 & 5 \\ -1 & 0 & -2 & -4 \end{bmatrix}$

e. $\begin{bmatrix} 4 & 0 & 0 & 0 \\ 6 & 7 & 0 & 0 \\ 9 & 11 & 1 & 0 \\ 5 & 4 & 1 & 1 \end{bmatrix}$

f. $\begin{bmatrix} 2 & 0 & 1 & 2 \\ 1 & 1 & 0 & 2 \\ 2 & -1 & 3 & 1 \\ 3 & -1 & 4 & 3 \end{bmatrix}$

3. Compute the determinants of the matrices in Exercise 2 and the determinants of the inverse matrices of those that are nonsingular.

4. Consider the four 3×3 linear systems having the same coefficient matrix:

$$\begin{aligned} 2x_1 - 3x_2 + x_3 &= 2, & 2x_1 - 3x_2 + x_3 &= 6, \\ x_1 + x_2 - x_3 &= -1, & x_1 + x_2 - x_3 &= 4, \\ -x_1 + x_2 - 3x_3 &= 0, & -x_1 + x_2 - 3x_3 &= 5, \end{aligned}$$

$$\begin{aligned} 2x_1 - 3x_2 + x_3 &= 0, & 2x_1 - 3x_2 + x_3 &= -1, \\ x_1 + x_2 - x_3 &= 1, & x_1 + x_2 - x_3 &= 0, \\ -x_1 + x_2 - 3x_3 &= -3, & -x_1 + x_2 - 3x_3 &= 0. \end{aligned}$$

a. Solve the linear systems by applying Gaussian elimination to the augmented matrix

$$\begin{bmatrix} 2 & -3 & 1 & : & 2 & 6 & 0 & -1 \\ 1 & 1 & -1 & : & -1 & 4 & 1 & 0 \\ -1 & 1 & -3 & : & 0 & 5 & -3 & 0 \end{bmatrix}.$$

b. Solve the linear systems by finding and multiplying by the inverse of

$$A = \begin{bmatrix} 2 & -3 & 1 \\ 1 & 1 & -1 \\ -1 & 1 & -3 \end{bmatrix}.$$

c. Which method requires more operations?

5. Show that the following statements are true or provide counterexamples to show they are not.

a. The product of two symmetric matrices is symmetric.

b. The inverse of a nonsingular symmetric matrix is a nonsingular symmetric matrix.

c. If A and B are $n \times n$ matrices, then $(AB)^t = A^t B^t$.

6. **a.** Show that the product of two $n \times n$ lower triangular matrices is lower triangular.

b. Show that the product of two $n \times n$ upper triangular matrices is upper triangular.

c. Show that the inverse of a nonsingular $n \times n$ lower triangular matrix is lower triangular.

7. The solution by **Cramer's rule** to the linear system

$$\begin{aligned} a_{11}x_1 + a_{12}x_2 + a_{13}x_3 &= b_1, \\ a_{21}x_1 + a_{22}x_2 + a_{23}x_3 &= b_2, \\ a_{31}x_1 + a_{32}x_2 + a_{33}x_3 &= b_3 \end{aligned}$$

has

$$x_1 = \frac{1}{D} \det \begin{bmatrix} b_1 & a_{12} & a_{13} \\ b_2 & a_{22} & a_{23} \\ b_3 & a_{32} & a_{33} \end{bmatrix} \equiv \frac{D_1}{D},$$

$$x_2 = \frac{1}{D} \det \begin{bmatrix} a_{11} & b_1 & a_{13} \\ a_{21} & b_2 & a_{23} \\ a_{31} & b_3 & a_{33} \end{bmatrix} \equiv \frac{D_2}{D},$$

and

$$x_3 = \frac{1}{D} \det \begin{bmatrix} a_{11} & a_{12} & b_1 \\ a_{21} & a_{22} & b_2 \\ a_{31} & a_{32} & b_3 \end{bmatrix} \equiv \frac{D_3}{D},$$

where

$$D = \det \begin{bmatrix} a_{11} & a_{12} & a_{13} \\ a_{21} & a_{22} & a_{23} \\ a_{31} & a_{32} & a_{33} \end{bmatrix}.$$

a.   Use Cramer's rule to find the solution to the linear system

$$\begin{aligned} 2x_1 + 3x_2 - x_3 &= 4, \\ x_1 - 2x_2 + x_3 &= 6, \\ x_1 - 12x_2 + 5x_3 &= 10. \end{aligned}$$

b.   Show that the linear system

$$\begin{aligned} 2x_1 + 3x_2 - x_3 &= 4, \\ x_1 - 2x_2 + x_3 &= 6, \\ -x_1 - 12x_2 + 5x_3 &= 9 \end{aligned}$$

does not have a solution. Compute D_1, D_2, and D_3.

c.   Show that the linear system

$$\begin{aligned} 2x_1 + 3x_2 - x_3 &= 4, \\ x_1 - 2x_2 + x_3 &= 6, \\ -x_1 - 12x_2 + 5x_3 &= 10 \end{aligned}$$

has an infinite number of solutions. Compute D_1, D_2, and D_3.

d.   Suppose that a 3×3 linear system with $D = 0$ has solutions. Explain why we must also have $D_1 = D_2 = D_3 = 0$.

8.   In a paper entitled "Population Waves," Bernadelli [Ber] hypothesizes a type of simplified beetle, which has a natural life span of 3 years. The female of this species has a survival rate of $\frac{1}{2}$ in the first year of life, has a survival rate of $\frac{1}{3}$ from the second to third years, and gives birth to an average of six new females before expiring at the end of the third year. A matrix can be used to show the contribution an individual female beetle makes, in a probabilistic sense, to the female population of the species by letting a_{ij} in the matrix $A = [a_{ij}]$ denote the contribution that a single female beetle of age j will make to the next year's female population of age i; that is,

$$A = \begin{bmatrix} 0 & 0 & 6 \\ \frac{1}{2} & 0 & 0 \\ 0 & \frac{1}{3} & 0 \end{bmatrix}.$$

a.   The contribution that a female beetle makes to the population 2 years hence is determined from the entries of A^2, of 3 years hence from A^3, and so on. Construct A^2 and A^3, and try to make a general statement about the contribution of a female beetle to the population in n years' time for any positive integral value of n.

b.   Use your conclusions from part (a) to describe what will occur in future years to a population of these beetles that initially consists of 6000 female beetles in each of the three age groups.

c.   Construct A^{-1} and describe its significance regarding the population of this species.

9. The study of food chains is an important topic in the determination of the spread and accumulation of environmental pollutants in living matter. Suppose that a food chain has three links. The first link consists of vegetation of types $v_1, v_2, \ldots, v_n$, which provide all the food requirements for herbivores of species $h_1, h_2, \ldots, h_m$ in the second link. The third link consists of carnivorous animals $c_1, c_2, \ldots, c_k$, which depend entirely on the herbivores in the second link for their food supply. The coordinate a_{ij} of the matrix

$$A = \begin{bmatrix} a_{11} & a_{12} & \cdots & a_{1m} \\ a_{21} & a_{22} & \cdots & a_{2m} \\ \vdots & \vdots & & \vdots \\ a_{n1} & a_{n2} & \cdots & a_{nm} \end{bmatrix}$$

represents the total number of plants of type v_i eaten by the herbivores in the species h_j, whereas b_{ij} in

$$B = \begin{bmatrix} b_{11} & b_{12} & \cdots & b_{1k} \\ b_{21} & b_{22} & \cdots & b_{2k} \\ \vdots & \vdots & & \vdots \\ b_{m1} & b_{m2} & \cdots & b_{mk} \end{bmatrix}$$

describes the number of herbivores in species h_i that are devoured by the animals of type c_j.

  a. Show that the number of plants of type v_i that eventually end up in the animals of species c_j is given by the entry in the ith row and jth column of the matrix AB.

  b. What physical significance is associated with the matrices A^{-1}, B^{-1}, and $(AB)^{-1} = B^{-1}A^{-1}$?

10. In Section 3.6 we found that the parametric form $(x(t), y(t))$ of the cubic Hermite polynomials through $(x(0), y(0)) = (x_0, y_0)$ and $(x(1), y(1)) = (x_1, y_1)$ with guidepoints $(x_0 + \alpha_0, y_0 + \beta_0)$ and $(x_1 - \alpha_1, y_1 - \beta_1)$, respectively, is given by

$$x(t) = [2(x_0 - x_1) + (\alpha_0 + \alpha_1)]t^3 + [3(x_1 - x_0) - \alpha_1 - 2\alpha_0]t^2 + \alpha_0 t + x_0$$

and

$$y(t) = [2(y_0 - y_1) + (\beta_0 + \beta_1)]t^3 + [3(y_1 - y_0) - \beta_1 - 2\beta_0]t^2 + \beta_0 t + y_0.$$

The Bézier cubic polynomials have the form

$$\hat{x}(t) = [2(x_0 - x_1) + 3(\alpha_0 + \alpha_1)]t^3 + [3(x_1 - x_0) - 3(\alpha_1 + 2\alpha_0)]t^2 + 3\alpha_0 t + x_0$$

and

$$\hat{y}(t) = [2(y_0 - y_1) + 3(\beta_0 + \beta_1)]t^3 + [3(y_1 - y_0) - 3(\beta_1 + 2\beta_0)]t^2 + 3\beta_0 t + y_0.$$

  a. Show that the matrix

$$A = \begin{bmatrix} 7 & 4 & 4 & 0 \\ -6 & -3 & -6 & 0 \\ 0 & 0 & 3 & 0 \\ 0 & 0 & 0 & 1 \end{bmatrix}$$

  maps the Hermite polynomial coefficients onto the Bézier polynomial coefficients.

  b. Determine a matrix B that maps the Bézier polynomial coefficients onto the Hermite polynomial coefficients.

11. Consider the 2×2 linear system $(A + iB)(\mathbf{x} + i\mathbf{y}) = \mathbf{c} + i\mathbf{d}$ with complex entries in component form:

$$(a_{11} + ib_{11})(x_1 + iy_1) + (a_{12} + ib_{12})(x_2 + iy_2) = c_1 + id_1,$$
$$(a_{21} + ib_{21})(x_1 + iy_1) + (a_{22} + ib_{22})(x_2 + iy_2) = c_2 + id_2.$$

a. Use the properties of complex numbers to convert this system to the equivalent 4×4 real linear system

$$\begin{array}{lc} \text{Real part:} & A\mathbf{x} - B\mathbf{y} = \mathbf{c}, \\ \text{Imaginary part:} & B\mathbf{x} + A\mathbf{y} = \mathbf{d}. \end{array}$$

b. Solve the linear system

$$(1 - 2i)(x_1 + iy_1) + (3 + 2i)(x_2 + iy_2) = 5 + 2i,$$
$$(2 + i)(x_1 + iy_1) + (4 + 3i)(x_2 + iy_2) = 4 - i.$$

6.5 Matrix Factorization

Matrix factorization is another of the important techniques that Gauss seems to be the first to have discovered. It is included in his two-volume treatise on celestial mechanics *Theoria motus corporum coelestium in sectionibus conicis Solem ambientium*, which was published in 1809.

Gaussian elimination is the principal tool in the direct solution of linear systems of equations, so it should be no surprise that it appears in other guises. In this section we will see that the steps used to solve a system of the form $A\mathbf{x} = \mathbf{b}$ can be used to factor a matrix. The factorization is particularly useful when it has the form $A = LU$, where L is lower triangular and U is upper triangular. Although not all matrices have this type of representation, many do that occur frequently in the application of numerical techniques.

In Section 6.2 we found that Gaussian elimination applied to an arbitrary linear system $A\mathbf{x} = \mathbf{b}$ requires $O(n^3/3)$ arithmetic operations to determine $\mathbf{x}$. However, to solve a linear system that involves an upper-triangular system requires only backward substitution, which takes $O(n^2)$ operations. The number of operations required to solve a lower-triangular system is similar.

Suppose that A has been factored into the triangular form $A = LU$, where L is lower triangular and U is upper triangular. Then we can easily solve for $\mathbf{x}$ using a two-step process.

- First define the temporary vector $\mathbf{y} = U\mathbf{x}$ and solve the lower triangular system $L\mathbf{y} = \mathbf{b}$ for $\mathbf{y}$. Since L is triangular, determining $\mathbf{y}$ from this equation requires only $O(n^2)$ operations.

- Once $\mathbf{y}$ is known, the upper triangular system $U\mathbf{x} = \mathbf{y}$ requires only an additional $O(n^2)$ operations to determine the solution $\mathbf{x}$.

Solving a linear system $A\mathbf{x} = \mathbf{b}$ in factored form means that the number of operations needed to solve the system $A\mathbf{x} = \mathbf{b}$ is reduced from $O(n^3/3)$ to $O(2n^2)$.

Example 1 Compare the approximate number of operations required to determine the solution to a linear system using a technique requiring $O(n^3/3)$ operations and one requiring $O(2n^2)$ when $n = 20$, $n = 100$, and $n = 1000$.

Solution Table 6.3 gives the results of these calculations. ▪

Table 6.3

n	$n^3/3$	$2n^2$	Reduction
10	$3.\overline{3} \times 10^2$	2×10^2	40%
100	$3.\overline{3} \times 10^5$	2×10^4	94%
1000	$3.\overline{3} \times 10^8$	2×10^6	99.4%

As the example illustrates, the reduction factor increases dramatically with the size of the matrix. Not surprisingly, the reductions from the factorization come at a cost; determining the specific matrices L and U requires $O(n^3/3)$ operations. But once the factorization is determined, systems involving the matrix A can be solved in this simplified manner for any number of vectors **b**.

To obtain the LU factorization of an $n \times n$ matrix A:

- use Gaussian elimination to solve a linear system of the form $A\mathbf{x} = \mathbf{b}$.

- if Gaussian elimination can be performed without row interchanges, then

  - the upper triangular matrix U is the matrix that results when Gaussian elimination is complete,

  - the lower triangular matrix L has 1s on its main diagonal, and each entry below the main diagonal is the multiplier that was needed to place a zero in that entry when Gaussian elimination was performed.

The process is outlined in the following example.

Example 2 Determine the LU factorization for matrix

$$A = \begin{bmatrix} 1 & 1 & 0 & 3 \\ 2 & 1 & -1 & 1 \\ 3 & -1 & -1 & 2 \\ -1 & 2 & 3 & -1 \end{bmatrix}$$

Solution A system involving the matrix A system was considered in the Illustration of Section 6.2 (see page 230), where we saw that the sequence of operations

$$(E_2 - 2E_1) \to (E_2), \quad (E_3 - 3E_1) \to (E_3), \quad (E_4 - (-1)E_1) \to (E_4,)$$

followed by

$$(E_3 - 4E_2) \to (E_3) \quad \text{and} \quad (E_4 - (-3)E_2) \to (E_4)$$

converts the system to the triangular system

$$\begin{aligned} x_1 + x_2 \quad\quad + 3x_4 &= 4, \\ -x_2 - x_3 - 5x_4 &= -7, \\ 3x_3 + 13x_4 &= 13, \\ -13x_4 &= -13. \end{aligned}$$

As a consequence, the upper triangular matrix in the factorization is

$$U = \begin{bmatrix} 1 & 1 & 0 & 3 \\ 0 & -1 & -1 & -5 \\ 0 & 0 & 3 & 13 \\ 0 & 0 & 0 & -13 \end{bmatrix}.$$

The multipliers used in Gaussian elimination were

$$m_{21} = 2, \quad m_{31} = 3, \quad m_{41} = -1, \quad m_{32} = 4, \quad m_{42} = -3, \quad \text{and} \quad m_{43} = 0,$$

so the lower triangular matrix is

$$L = \begin{bmatrix} 1 & 0 & 0 & 0 \\ 2 & 1 & 0 & 0 \\ 3 & 4 & 1 & 0 \\ -1 & -3 & 0 & 1 \end{bmatrix}$$

Hence the LU factorization of A is

$$A = \begin{bmatrix} 1 & 1 & 0 & 3 \\ 2 & 1 & -1 & 1 \\ 3 & -1 & -1 & 2 \\ -1 & 2 & 3 & -1 \end{bmatrix} = \begin{bmatrix} 1 & 0 & 0 & 0 \\ 2 & 1 & 0 & 0 \\ 3 & 4 & 1 & 0 \\ -1 & -3 & 0 & 1 \end{bmatrix} \begin{bmatrix} 1 & 1 & 0 & 3 \\ 0 & -1 & -1 & -5 \\ 0 & 0 & 3 & 13 \\ 0 & 0 & 0 & -13 \end{bmatrix} = LU.$$

■

The next example uses the factorization in the previous to solve a linear system.

Example 3   Use the factorization found in Example 2 to solve the system

$$\begin{aligned} x_1 + x_2 \quad\quad + 3x_4 &= 8, \\ 2x_1 + x_2 - x_3 + x_4 &= 7, \\ 3x_1 - x_2 - x_3 + 2x_4 &= 14, \\ -x_1 + 2x_2 + 3x_3 - x_4 &= -7. \end{aligned}$$

Solution   To solve

$$A\mathbf{x} = LU\mathbf{x} = \begin{bmatrix} 1 & 0 & 0 & 0 \\ 2 & 1 & 0 & 0 \\ 3 & 4 & 1 & 0 \\ -1 & -3 & 0 & 1 \end{bmatrix} \begin{bmatrix} 1 & 1 & 0 & 3 \\ 0 & -1 & -1 & -5 \\ 0 & 0 & 3 & 13 \\ 0 & 0 & 0 & -13 \end{bmatrix} \begin{bmatrix} x_1 \\ x_2 \\ x_3 \\ x_4 \end{bmatrix} = \begin{bmatrix} 8 \\ 7 \\ 14 \\ -7 \end{bmatrix},$$

we first introduce the temporary vector $\mathbf{y} = U\mathbf{x}$. Then $\mathbf{b} = L(U\mathbf{x}) = L\mathbf{y}$. That is,

$$L\mathbf{y} = \begin{bmatrix} 1 & 0 & 0 & 0 \\ 2 & 1 & 0 & 0 \\ 3 & 4 & 1 & 0 \\ -1 & -3 & 0 & 1 \end{bmatrix} \begin{bmatrix} y_1 \\ y_2 \\ y_3 \\ y_4 \end{bmatrix} = \begin{bmatrix} 8 \\ 7 \\ 14 \\ -7 \end{bmatrix}.$$

This system is solved for $\mathbf{y}$ by a simple forward-substitution process:

$$\begin{aligned} y_1 &= 8; \\ 2y_1 + y_2 &= 7, \quad\text{so } y_2 = 7 - 2y_1 = -9; \\ 3y_1 + 4y_2 + y_3 &= 14, \quad\text{so } y_3 = 14 - 3y_1 - 4y_2 = 26; \\ -y_1 - 3y_2 + y_4 &= -7, \quad\text{so } y_4 = -7 + y_1 + 3y_2 = -26. \end{aligned}$$

We then solve $U\mathbf{x} = \mathbf{y}$ for $\mathbf{x}$, the solution of the original system; that is,

$$\begin{bmatrix} 1 & 1 & 0 & 3 \\ 0 & -1 & -1 & -5 \\ 0 & 0 & 3 & 13 \\ 0 & 0 & 0 & -13 \end{bmatrix} \begin{bmatrix} x_1 \\ x_2 \\ x_3 \\ x_4 \end{bmatrix} = \begin{bmatrix} 8 \\ -9 \\ 26 \\ -26 \end{bmatrix}.$$

Using backward substitution we obtain $x_4 = 2$, $x_3 = 0$, $x_2 = -1$, $x_1 = 3$.   ■

The program LUFACT64 performs the LU Factorization.

Although new matrices $L = [l_{ij}]$ and $U = [u_{ij}]$ are constructed by the program LUFACT64, the values generated replace the corresponding entries of A that are no longer needed. Thus, the new matrix has entries $a_{ij} = l_{ij}$ for each $i = 2, 3, \ldots, n$ and $j = 1, 2, \ldots, i - 1$ and $a_{ij} = u_{ij}$ for each $i = 1, 2, \ldots, n$ and $j = i + 1, i + 2, \ldots, n$.

The factorization is particularly useful when a number of linear systems involving A must be solved because most of the operations, those involving the Gaussian Elimination, need to be performed only once.

Permutation Matrices

In the previous discussion we assumed that A is such that a linear system of the form $A\mathbf{x} = \mathbf{b}$ can be solved using Gaussian elimination that does not require row interchanges. From a practical standpoint, this factorization is useful only when row interchanges are not required to control the round-off error resulting from the use of finite-digit arithmetic. Although many systems we encounter when using approximation methods are of this type, factorization modifications must be made when row interchanges are required. We begin the discussion with the introduction of a class of matrices that are used to rearrange, or permute, rows of a given matrix.

An $n \times n$ **permutation matrix** P has precisely one entry in each column and each row whose value is 1, and all of whose other entries are 0.

Illustration   The matrix

$$P = \begin{bmatrix} 1 & 0 & 0 \\ 0 & 0 & 1 \\ 0 & 1 & 0 \end{bmatrix}$$

is a 3×3 permutation matrix. For any 3×3 matrix A, multiplying on the left by P has the effect of interchanging the second and third rows of A:

$$PA = \begin{bmatrix} 1 & 0 & 0 \\ 0 & 0 & 1 \\ 0 & 1 & 0 \end{bmatrix} \begin{bmatrix} a_{11} & a_{12} & a_{13} \\ a_{21} & a_{22} & a_{23} \\ a_{31} & a_{32} & a_{33} \end{bmatrix} = \begin{bmatrix} a_{11} & a_{12} & a_{13} \\ a_{31} & a_{32} & a_{33} \\ a_{21} & a_{22} & a_{23} \end{bmatrix}.$$

Similarly, multiplying A on the right by P interchanges the second and third columns of A.   ☐

The matrix multiplication PA permutes rows of A.

There are two useful properties of permutation matrices that relate to Gaussian elimination. The first of these was shown in the Illustration.

The matrix multiplication AP permutes columns of A.

- If $k_1, \ldots, k_n$ is a permutation of the integers $1, \ldots, n$ and the permutation matrix $P = [p_{ij}]$ is defined by

$$p_{ij} = \begin{cases} 1, & \text{if } j = k_i, \\ 0, & \text{otherwise,} \end{cases} \quad \text{then} \quad PA = \begin{bmatrix} a_{k_1,1} & a_{k_1,2} & \cdots & a_{k_1,n} \\ a_{k_2,1} & a_{k_2,2} & \cdots & a_{k_2,n} \\ \vdots & \vdots & & \vdots \\ a_{k_n,1} & a_{k_n,2} & \cdots & a_{k_n,n} \end{bmatrix}.$$

The second is

- If P is a permutation matrix, then P^{-1} exists and $P^{-1} = P^t$.

Example 4   Determine a factorization in the form $A = (P^t L)U$ for the matrix

$$A = \begin{bmatrix} 0 & 0 & -1 & 1 \\ 1 & 1 & -1 & 2 \\ -1 & -1 & 2 & 0 \\ 1 & 2 & 0 & 2 \end{bmatrix}.$$

Solution The matrix A does not have an LU factorization because $a_{11} = 0$. However, using the row interchange $(E_1) \leftrightarrow (E_2)$, followed by $(E_3 + E_1) \rightarrow (E_3)$ and $(E_4 - E_1) \rightarrow (E_4)$, produces

$$
\begin{bmatrix}
1 & 1 & -1 & 2 \\
0 & 0 & -1 & 1 \\
0 & 0 & 1 & 2 \\
0 & 1 & 1 & 0
\end{bmatrix}.
$$

Then the row interchange $(E_2) \leftrightarrow (E_4)$, followed by $(E_4 + E_3) \rightarrow (E_4)$, gives the matrix

$$
U = \begin{bmatrix}
1 & 1 & -1 & 2 \\
0 & 1 & 1 & 0 \\
0 & 0 & 1 & 2 \\
0 & 0 & 0 & 3
\end{bmatrix}.
$$

The permutation matrix associated with the row interchanges $(E_1) \leftrightarrow (E_2)$ and $(E_2) \leftrightarrow (E_4)$ is

$$
P = \begin{bmatrix}
1 & 0 & 0 & 0 \\
0 & 0 & 0 & 1 \\
0 & 0 & 1 & 0 \\
0 & 1 & 0 & 0
\end{bmatrix} \cdot \begin{bmatrix}
0 & 1 & 0 & 0 \\
1 & 0 & 0 & 0 \\
0 & 0 & 1 & 0 \\
0 & 0 & 0 & 1
\end{bmatrix} = \begin{bmatrix}
0 & 1 & 0 & 0 \\
0 & 0 & 0 & 1 \\
0 & 0 & 1 & 0 \\
1 & 0 & 0 & 0
\end{bmatrix},
$$

and

$$
PA = \begin{bmatrix}
1 & 1 & -1 & 2 \\
1 & 2 & 0 & 2 \\
-1 & -1 & 2 & 0 \\
0 & 0 & -1 & 1
\end{bmatrix}.
$$

Gaussian elimination is performed on PA using the same operations as on A, except without the row interchanges. That is, $(E_2 - E_1) \rightarrow (E_2)$, $(E_3 + E_1) \rightarrow (E_3)$, followed by $(E_4 + E_3) \rightarrow (E_4)$. The nonzero multipliers for PA are, consequently,

$$ m_{21} = 1, \quad m_{31} = -1, \quad \text{and} \quad m_{43} = -1, $$

and the LU factorization of PA is

$$
PA = \begin{bmatrix}
1 & 0 & 0 & 0 \\
1 & 1 & 0 & 0 \\
-1 & 0 & 1 & 0 \\
0 & 0 & -1 & 1
\end{bmatrix} \begin{bmatrix}
1 & 1 & -1 & 2 \\
0 & 1 & 1 & 0 \\
0 & 0 & 1 & 2 \\
0 & 0 & 0 & 3
\end{bmatrix} = LU.
$$

Multiplying by $P^{-1} = P^t$ produces the factorization

$$
A = P^{-1}(LU) = P^t(LU) = (P^t L)U = \begin{bmatrix}
0 & 0 & -1 & 1 \\
1 & 0 & 0 & 0 \\
-1 & 0 & 1 & 0 \\
1 & 1 & 0 & 0
\end{bmatrix} \begin{bmatrix}
1 & 1 & -1 & 2 \\
0 & 1 & 1 & 0 \\
0 & 0 & 1 & 2 \\
0 & 0 & 0 & 3
\end{bmatrix}.
$$

■

MATLAB has the command `lu(A)` to obtain an LU factorization of a matrix A in the form $A = PLU$, where U is upper triangular, L is lower triangular, and P is a permutation matrix. Notice that the permutation matrix P that MATLAB constructs is the matrix that we would call P^t. We apply this command to the matrix in Example 4 by first defining

```
A = [0 0 -1 1; 1 1 -1 2; -1 -1 1 0; 1 2 0 2]
```

and then calling

```
[L, U, P] = lu(A)
```

MATLAB responds with

$$L = \begin{bmatrix} 1 & 0 & 0 & 0 \\ 1 & 1 & 0 & 0 \\ -1 & 0 & 1 & 0 \\ 0 & 0 & -1 & 1 \end{bmatrix}, \quad U = \begin{bmatrix} 1 & 1 & -1 & 2 \\ 0 & 1 & 1 & 0 \\ 0 & 0 & 1 & 2 \\ 0 & 0 & 0 & 3 \end{bmatrix}, \quad \text{and } P = \begin{bmatrix} 0 & 0 & 0 & 1 \\ 1 & 0 & 0 & 0 \\ 0 & 0 & 1 & 0 \\ 0 & 1 & 0 & 0 \end{bmatrix}.$$

For these matrices L, U, and P, we have $A = PLU$.

EXERCISE SET 6.5

1. Solve the following linear systems.

    a. $\begin{bmatrix} 1 & 0 & 0 \\ 2 & 1 & 0 \\ -1 & 0 & 1 \end{bmatrix} \begin{bmatrix} 2 & 3 & -1 \\ 0 & -2 & 1 \\ 0 & 0 & 3 \end{bmatrix} \begin{bmatrix} x_1 \\ x_2 \\ x_3 \end{bmatrix} = \begin{bmatrix} 2 \\ -1 \\ 1 \end{bmatrix}$

    b. $\begin{bmatrix} 2 & 0 & 0 \\ -1 & 1 & 0 \\ 3 & 2 & -1 \end{bmatrix} \begin{bmatrix} 1 & 1 & 1 \\ 0 & 1 & 2 \\ 0 & 0 & 1 \end{bmatrix} \begin{bmatrix} x_1 \\ x_2 \\ x_3 \end{bmatrix} = \begin{bmatrix} -1 \\ 3 \\ 0 \end{bmatrix}$

2. Factor the following matrices into the LU decomposition with $l_{ii} = 1$ for all i.

    a. $\begin{bmatrix} 2 & -1 & 1 \\ 3 & 3 & 9 \\ 3 & 3 & 5 \end{bmatrix}$
        b. $\begin{bmatrix} 1.012 & -2.132 & 3.104 \\ -2.132 & 4.096 & -7.013 \\ 3.104 & -7.013 & 0.014 \end{bmatrix}$

    c. $\begin{bmatrix} 2 & 0 & 0 & 0 \\ 1 & 1.5 & 0 & 0 \\ 0 & -3 & 0.5 & 0 \\ 2 & -2 & 1 & 1 \end{bmatrix}$

    d. $\begin{bmatrix} 2.1756 & 4.0231 & -2.1732 & 5.1967 \\ -4.0231 & 6.0000 & 0 & 1.1973 \\ -1.0000 & -5.2107 & 1.1111 & 0 \\ 6.0235 & 7.0000 & 0 & -4.1561 \end{bmatrix}$

3. Obtain factorizations of the form $A = P^t LU$ for the following matrices.

    a. $A = \begin{bmatrix} 0 & 2 & 3 \\ 1 & 1 & -1 \\ 0 & -1 & 1 \end{bmatrix}$
        b. $A = \begin{bmatrix} 1 & 2 & -1 \\ 1 & 2 & 3 \\ 2 & -1 & 4 \end{bmatrix}$

    c. $A = \begin{bmatrix} 1 & -2 & 3 & 0 \\ 3 & -6 & 9 & 3 \\ 2 & 1 & 4 & 1 \\ 1 & -2 & 2 & -2 \end{bmatrix}$
    d. $A = \begin{bmatrix} 1 & -2 & 3 & 0 \\ 1 & -2 & 3 & 1 \\ 1 & -2 & 2 & -2 \\ 2 & 1 & 3 & -1 \end{bmatrix}$

4. Suppose $A = P^t LU$, where P is a permutation matrix, L is a lower-triangular matrix with 1s on the diagonal, and U is an upper-triangular matrix.

   a. Count the number of operations needed to compute $P^t LU$ for a given matrix A.

   b. Show that if P contains k row interchanges, then

   $$\det P = \det P^t = (-1)^k.$$

   c. Use $\det A = \det P^t \cdot \det L \cdot \det U = (-1)^k \det U$ to count the number of operations for determining $\det A$ by factoring.

   d. Factor A as $P^t LU$ and use this factorization to compute $\det A$ and to count the number of operations when

   $$A = \begin{bmatrix} 0 & 2 & 1 & 4 & -1 & 3 \\ 1 & 2 & -1 & 3 & 4 & 0 \\ 0 & 1 & 1 & -1 & 2 & -1 \\ 2 & 3 & -4 & 2 & 0 & 5 \\ 1 & 1 & 1 & 3 & 0 & 2 \\ -1 & -1 & 2 & -1 & 2 & 0 \end{bmatrix}.$$

5. Use the LU factorization obtained in Exercise 2 to solve the following linear systems.

   a. $2x_1 - x_2 + x_3 = -1,$
      $3x_1 + 3x_2 + 9x_3 = 0,$
      $3x_1 + 3x_2 + 5x_3 = 4.$

   b. $1.012x_1 - 2.132x_2 + 3.104x_3 = 1.984,$
      $-2.132x_1 + 4.096x_2 - 7.013x_3 = -5.049,$
      $3.104x_1 - 7.013x_2 + 0.014x_3 = -3.895.$

   c. $2x_1 = 3,$
      $x_1 + 1.5x_2 = 4.5,$
      $- 3x_2 + 0.5x_3 = -6.6,$
      $2x_1 - 2x_2 + x_3 + x_4 = 0.8.$

   d. $2.1756x_1 + 4.0231x_2 - 2.1732x_3 + 5.1967x_4 = 17.102,$
      $-4.0231x_1 + 6.0000x_2 + 1.1973x_4 = -6.1593,$
      $-1.0000x_1 - 5.2107x_2 + 1.1111x_3 = 3.0004,$
      $6.0235x_1 + 7.0000x_2 - 4.1561x_4 = 0.0000.$

6.6 Techniques for Special Matrices

Although this chapter has been concerned primarily with the effective application of Gaussian elimination for finding the solution to a linear system of equations, many of the results have wider application. It might be said that Gaussian elimination is the hub about which the chapter revolves, but the wheel itself is of equal interest and has application in many forms in the study of numerical methods. In this section we consider some matrices that are of special types, forms that will be used in other chapters of the book.

Strict Diagonal Dominance

Each main diagonal entry in a strictly diagonally dominant matrix has a magnitude that is strictly greater than the sum of the magnitudes of all the other entries in that row.

The $n \times n$ matrix A is said to be **strictly diagonally dominant** when

$$|a_{ii}| > \sum_{\substack{j=1, \\ j \neq i}}^{n} |a_{ij}| \quad \text{holds for each } i = 1, 2, \ldots, n.$$

Illustration   Consider the matrices

$$A = \begin{bmatrix} 7 & 2 & 0 \\ 3 & 5 & -1 \\ 0 & 5 & -6 \end{bmatrix} \quad \text{and} \quad B = \begin{bmatrix} 6 & 4 & -3 \\ 4 & -2 & 0 \\ -3 & 0 & 1 \end{bmatrix}.$$

The nonsymmetric matrix A is strictly diagonally dominant because

$$|7| > |2| + |0|, \quad |5| > |3| + |-1|, \quad \text{and} \quad |-6| > |0| + |5|.$$

The symmetric matrix B is not strictly diagonally dominant because, for example, in the first row the absolute value of the diagonal element is $|6| < |4| + |-3| = 7$. It is interesting to note that A^t is not strictly diagonally dominant because the middle row of A^t is [2 5 5], nor, of course, is B^t because $B^t = B$.   □

Strictly Diagonally Dominant Matrices

A strictly diagonally dominant matrix A has an inverse. Moreover, Gaussian elimination can be performed on any linear system of the form $A\mathbf{x} = \mathbf{b}$ to obtain its unique solution without row or column interchanges, and the computations are stable with respect to the growth of round-off error.

Positive Definite Matrices

The name positive definite refers to the fact that the number $\mathbf{x}^t A\mathbf{x}$ must be positive whenever $\mathbf{x} \neq \mathbf{0}$.

A matrix A is **positive definite** if it is symmetric and if $\mathbf{x}^t A\mathbf{x} > 0$ for every n-dimensional column vector $\mathbf{x} \neq \mathbf{0}$.

Using the definition to determine whether a matrix is positive definite can be difficult. Fortunately, there are more easily verified criteria for identifying members that are and are not of this important class.

Positive Definite Matrix Properties

If A is an $n \times n$ positive definite matrix, then

   (i) A has an inverse;

   (ii) $a_{ii} > 0$ for each $i = 1, 2, \ldots, n$;

   (iii) $\max_{1 \leq k, j \leq n} |a_{kj}| \leq \max_{1 \leq i \leq n} |a_{ii}|$;

   (iv) $(a_{ij})^2 < a_{ii} a_{jj}$ for each $i \neq j$.

Our definition of positive definite requires the matrix to be symmetric, but not all authors make this requirement. For example, Golub and Van Loan [GV], a standard reference in matrix methods, requires only that $\mathbf{x}^t A\mathbf{x} > 0$ for each nonzero vector $\mathbf{x}$. Matrices that we call positive definite are called symmetric positive definite in [GV]. Keep this discrepancy in mind if you are using material from other sources.

The next result parallels the strictly diagonally dominant result presented previously.

Positive Definite Matrix Equivalences

The following are equivalent for any $n \times n$ symmetric matrix A:

(i) A is positive definite.

(ii) Gaussian elimination without row interchanges can be performed on the linear system $A\mathbf{x} = \mathbf{b}$ with all pivot elements positive. (This ensures that the computations are stable with respect to the growth of round-off error.)

(iii) A can be factored in the form LL^t, where L is lower triangular with positive diagonal entries.

(iv) A can be factored in the form LDL^t, where L is lower triangular with 1s on its diagonal and D is a diagonal matrix with positive diagonal entries.

(v) For each $i = 1, 2, \ldots, n$, we have

$$\det \begin{bmatrix} a_{11} & a_{12} & \cdots & a_{1i} \\ a_{21} & a_{22} & \cdots & a_{2i} \\ \vdots & \vdots & & \vdots \\ a_{i1} & a_{i2} & \cdots & a_{ii} \end{bmatrix} > 0.$$

The next examples illustrate portions of this result. First we will consider (v).

Example 1   Show that the symmetric matrix

$$A = \begin{bmatrix} 2 & -1 & 0 \\ -1 & 2 & -1 \\ 0 & -1 & 2 \end{bmatrix}$$

is positive definite.

Solution   We have

$$\det[2] = 2 > 0, \quad \det \begin{bmatrix} 2 & -1 \\ -1 & 2 \end{bmatrix} = 4 - 1 = 3 > 0,$$

and

$$\det \begin{bmatrix} 2 & -1 & 0 \\ -1 & 2 & -1 \\ 0 & -1 & 2 \end{bmatrix} = 2 \det \begin{bmatrix} 2 & -1 \\ -1 & 2 \end{bmatrix} - (-1) \det \begin{bmatrix} -1 & -1 \\ 0 & 2 \end{bmatrix}$$

$$= 2(4 - 1) + (-2 + 0) = 4 > 0,$$

so, by (v), A is positive definite.   ■

> The program LDLFCT65 performs the LDL^t Factorization.

The next example illustrates how the LDL^t factorization of a positive definite matrix described in (iv) of the result is formed.

Example 2   Determine the LDL^t factorization of the positive definite matrix

$$A = \begin{bmatrix} 4 & -1 & 1 \\ -1 & 4.25 & 2.75 \\ 1 & 2.75 & 3.5 \end{bmatrix}.$$

Solution The LDL^t factorization has 1s on the diagonal of the lower triangular matrix L so we need to have

$$A = \begin{bmatrix} a_{11} & a_{21} & a_{31} \\ a_{21} & a_{22} & a_{32} \\ a_{31} & a_{32} & a_{33} \end{bmatrix} = \begin{bmatrix} 1 & 0 & 0 \\ l_{21} & 1 & 0 \\ l_{31} & l_{32} & 1 \end{bmatrix} \begin{bmatrix} d_1 & 0 & 0 \\ 0 & d_2 & 0 \\ 0 & 0 & d_3 \end{bmatrix} \begin{bmatrix} 1 & l_{21} & l_{31} \\ 0 & 1 & l_{32} \\ 0 & 0 & 1 \end{bmatrix}$$

$$= \begin{bmatrix} d_1 & d_1 l_{21} & d_1 l_{31} \\ d_1 l_{21} & d_2 + d_1 l_{21}^2 & d_2 l_{32} + d_1 l_{21} l_{31} \\ d_1 l_{31} & d_1 l_{21} l_{31} + d_2 l_{32} & d_1 l_{31}^2 + d_2 l_{32}^2 + d_3 \end{bmatrix}$$

Thus

a_{11}: $4 = d_1 \implies d_1 = 4,$ a_{21}: $-1 = d_1 l_{21} \implies l_{21} = -0.25$

a_{31}: $1 = d_1 l_{31} \implies l_{31} = 0.25,$ a_{22}: $4.25 = d_2 + d_1 l_{21}^2 \implies d_2 = 4$

a_{32}: $2.75 = d_1 l_{21} l_{31} + d_2 l_{32} \implies l_{32} = 0.75,$ a_{33}: $3.5 = d_1 l_{31}^2 + d_2 l_{32}^2 + d_3 \implies d_3 = 1,$

and we have

$$A = LDL^t = \begin{bmatrix} 1 & 0 & 0 \\ -0.25 & 1 & 0 \\ 0.25 & 0.75 & 1 \end{bmatrix} \begin{bmatrix} 4 & 0 & 0 \\ 0 & 4 & 0 \\ 0 & 0 & 1 \end{bmatrix} \begin{bmatrix} 1 & -0.25 & 0.25 \\ 0 & 1 & 0.75 \\ 0 & 0 & 1 \end{bmatrix}.$$ ∎

Andre-Louis Cholesky (1875–1918) was a French military officer involved in geodesy and surveying in the early 1900s. He developed this factorization method to compute solutions to least squares problems.

Any symmetric matrix A for which Gaussian elimination can be applied without row interchanges can be factored into the form LDL^t. In this general case, L is lower triangular with 1s on its diagonal, and D is the diagonal matrix with the Gaussian elimination pivots on its diagonal. This result is widely applied because symmetric matrices are common and easily recognized.

The factorization in part (iii) of the positive definite matrix equivalences, that is, $A = LL^t$, is known as Cholesky's factorization. The next example shows how this is done.

Example 3 Determine the Cholesky LL^t factorization of the positive definite matrix

$$A = \begin{bmatrix} 4 & -1 & 1 \\ -1 & 4.25 & 2.75 \\ 1 & 2.75 & 3.5 \end{bmatrix}.$$

The program CHOLFC66 performs the LL^t Factorization.

Solution The LL^t factorization does not necessarily have 1s on the diagonal of the lower triangular matrix L so we need to have

$$A = \begin{bmatrix} a_{11} & a_{21} & a_{31} \\ a_{21} & a_{22} & a_{32} \\ a_{31} & a_{32} & a_{33} \end{bmatrix} = \begin{bmatrix} l_{11} & 0 & 0 \\ l_{21} & l_{22} & 0 \\ l_{31} & l_{32} & l_{33} \end{bmatrix} \begin{bmatrix} l_{11} & l_{21} & l_{31} \\ 0 & l_{22} & l_{32} \\ 0 & 0 & l_{33} \end{bmatrix}$$

$$= \begin{bmatrix} l_{11}^2 & l_{11} l_{21} & l_{11} l_{31} \\ l_{11} l_{21} & l_{21}^2 + l_{22}^2 & l_{21} l_{31} + l_{22} l_{32} \\ l_{11} l_{31} & l_{21} l_{31} + l_{22} l_{32} & l_{31}^2 + l_{32}^2 + l_{33}^2 \end{bmatrix}$$

Thus

a_{11}: $4 = l_{11}^2 \implies l_{11} = 2,$ a_{21}: $-1 = l_{11} l_{21} \implies l_{21} = -0.5$

a_{31}: $1 = l_{11} l_{31} \implies l_{31} = 0.5,$ a_{22}: $4.25 = l_{21}^2 + l_{22}^2 \implies l_{22} = 2$

a_{32}: $2.75 = l_{21} l_{31} + l_{22} l_{32} \implies l_{32} = 1.5,$ a_{33}: $3.5 = l_{31}^2 + l_{32}^2 + l_{33}^2 \implies l_{33} = 1,$

and we have

$$A = LL^t = \begin{bmatrix} 2 & 0 & 0 \\ -0.5 & 2 & 0 \\ 0.5 & 1.5 & 1 \end{bmatrix} \begin{bmatrix} 2 & -0.5 & 0.5 \\ 0 & 2 & 1.5 \\ 0 & 0 & 1 \end{bmatrix}.$$ ▪

MATLAB commands are available for computing both the LDL^t (ldl) and Cholesky (LL^t) (chol) factorizations. For example, for the matrix A defined by

```
A = [4 -1 1; -1 4.25 2.75; 1 2.75 3.5]
```

we have

```
[L, D] = ldl(A)
```

giving

$$L = \begin{bmatrix} 1.000000000000000 & 0 & 0 \\ -0.250000000000000 & 1.000000000000000 & 0 \\ 0.250000000000000 & 0.750000000000000 & 1.000000000000000 \end{bmatrix} \quad \text{and}$$

$$D = \begin{bmatrix} 4 & 0 & 0 \\ 0 & 4 & 0 \\ 0 & 0 & 1 \end{bmatrix}$$

and

```
L=chol(A)
```

giving

$$L = \begin{bmatrix} 2.000000000000000 & -0.500000000000000 & 0.500000000000000 \\ 0 & 2.000000000000000 & 1.500000000000000 \\ 0 & 0 & 1.000000000000000 \end{bmatrix}$$

Band Matrices

The last special matrices considered are *band matrices*. In many applications, band matrices are also strictly diagonally dominant or positive definite. This combination of properties is very useful.

The name for a band matrix comes from the fact that all the nonzero entries lie in a band that is centered on the main diagonal.

An $n \times n$ matrix is called a **band matrix** if integers p and q, with $1 < p, q < n$, exist with the property that $a_{ij} = 0$ whenever $p \le j - i$ or $q \le i - j$. The number p describes the number of diagonals above, and including, the main diagonal on which nonzero entries may lie. The number q describes the number of diagonals below, and including, the main diagonal on which nonzero entries may lie. In most applications, $p = q$ and the nonzero entries are evenly banded about the main diagonal.

The **bandwidth** of the band matrix is $w = p + q - 1$, which tells us how many of the diagonals can contain nonzero entries. The 1 is subtracted from the sum of p and q because both of these numbers count the main diagonal.

For example, the matrix

$$A = \begin{bmatrix} 7 & 2 & 1 & 0 \\ 3 & 5 & -3 & -2 \\ 0 & 4 & 6 & -1 \\ 0 & 0 & 5 & 8 \end{bmatrix}$$

is a band matrix with $p = 3$ and $q = 2$, so it has bandwidth $3 + 2 - 1 = 4$.

Tridiagonal Matrices

Band matrices concentrate all their nonzero entries about the diagonal. Two special cases of band matrices that occur often have $p = q = 2$ and $p = q = 4$. Matrices of bandwidth 3 that occur when $p = q = 2$ are called **tridiagonal** because they have the form

$$A = \begin{bmatrix} a_{11} & a_{12} & 0 & \cdots & \cdots & 0 \\ a_{21} & a_{22} & a_{23} & & & \vdots \\ 0 & a_{32} & a_{33} & a_{34} & & \vdots \\ \vdots & & & & & 0 \\ \vdots & & & & & a_{n-1,n} \\ 0 & \cdots & \cdots & 0 & a_{n-1,n} & a_{nn} \end{bmatrix}.$$

Tridiagonal matrices will appear in Chapter 11 in connection with the study of piecewise linear approximations to boundary-value problems. The case of $p = q = 4$ will also be used in that chapter for the solution of boundary-value problems, when the approximating functions assume the form of cubic splines.

The factorization methods can be simplified considerably in the case of band matrices because a large number of zeros appear in regular patterns. Of particular interest is the Crout factorization, where $A = LU$ with U having all 1s on its diagonal.

Crout factorization is illustrated in the following example.

> The program CRTRLS67 performs Crout Factorization.

Example 4 Determine the Crout factorization of the symmetric tridiagonal matrix

$$\begin{bmatrix} 2 & -1 & 0 & 0 \\ -1 & 2 & -1 & 0 \\ 0 & -1 & 2 & -1 \\ 0 & 0 & -1 & 2 \end{bmatrix}.$$

Solution The LU factorization of A has the form

$$A = \begin{bmatrix} a_{11} & 0 & 0 & 0 \\ a_{21} & a_{22} & a_{23} & 0 \\ 0 & a_{32} & a_{33} & a_{34} \\ 0 & 0 & a_{43} & a_{44} \end{bmatrix} = \begin{bmatrix} l_{11} & 0 & 0 & 0 \\ l_{21} & l_{22} & 0 & 0 \\ 0 & l_{32} & l_{33} & 0 \\ 0 & 0 & l_{43} & l_{44} \end{bmatrix} \begin{bmatrix} 1 & u_{12} & 0 & 0 \\ 0 & 1 & u_{23} & 0 \\ 0 & 0 & 1 & u_{34} \\ 0 & 0 & 0 & 1 \end{bmatrix}$$

$$= \begin{bmatrix} l_{11} & l_{11}u_{12} & 0 & 0 \\ l_{21} & l_{22} + l_{21}u_{12} & l_{22}u_{23} & 0 \\ 0 & l_{32} & l_{33} + l_{32}u_{23} & l_{33}u_{34} \\ 0 & 0 & l_{43} & l_{44} + l_{43}u_{34} \end{bmatrix}.$$

Thus

$$a_{11}: \quad 2 = l_{11} \implies l_{11} = 2, \qquad\qquad a_{12}: \quad -1 = l_{11}u_{12} \implies u_{12} = -\tfrac{1}{2},$$

$$a_{21}: \quad -1 = l_{21} \implies l_{21} = -1, \qquad a_{22}: \quad 2 = l_{22} + l_{21}u_{12} \implies l_{22} = \tfrac{3}{2},$$

$$a_{23}: \quad -1 = l_{22}u_{23} \implies u_{23} = -\tfrac{2}{3}, \qquad a_{32}: \quad -1 = l_{32} \implies l_{32} = -1,$$

$$a_{33}: \quad 2 = l_{33} + l_{32}u_{23} \implies l_{33} = \tfrac{4}{3}, \qquad a_{34}: \quad -1 = l_{33}u_{34} \implies u_{34} = -\tfrac{3}{4},$$

$$a_{43}: \quad -1 = l_{43} \implies l_{43} = -1, \qquad a_{44}: \quad 2 = l_{44} + l_{43}u_{34} \implies l_{44} = \tfrac{5}{4}.$$

This gives the Crout factorization

$$A = \begin{bmatrix} 2 & -1 & 0 & 0 \\ -1 & 2 & -1 & 0 \\ 0 & -1 & 2 & -1 \\ 0 & 0 & -1 & 2 \end{bmatrix} = \begin{bmatrix} 2 & 0 & 0 & 0 \\ -1 & \frac{3}{2} & 0 & 0 \\ 0 & -1 & \frac{4}{3} & 0 \\ 0 & 0 & -1 & \frac{5}{4} \end{bmatrix} \begin{bmatrix} 1 & -\frac{1}{2} & 0 & 0 \\ 0 & 1 & -\frac{2}{3} & 0 \\ 0 & 0 & 1 & -\frac{3}{4} \\ 0 & 0 & 0 & 1 \end{bmatrix} = LU.$$

■

The next example shows how a linear system is solved once the Crout factorization is known.

Example 5   Use this Crout factorization found in Example 4 to solve the linear system

$$\begin{aligned} 2x_1 - x_2 &= 1, \\ -x_1 + 2x_2 - x_3 &= 0, \\ -x_2 + 2x_3 - x_4 &= 0, \\ -x_3 + 2x_4 &= 1. \end{aligned}$$

Solution   First we introduce a temporary vector $\mathbf{y} = U\mathbf{x}$ and use forward substitution to solve the system

$$L\mathbf{y} = \begin{bmatrix} 2 & 0 & 0 & 0 \\ -1 & \frac{3}{2} & 0 & 0 \\ 0 & -1 & \frac{4}{3} & 0 \\ 0 & 0 & -1 & \frac{5}{4} \end{bmatrix} \begin{bmatrix} y_1 \\ y_2 \\ y_3 \\ y_4 \end{bmatrix} = \begin{bmatrix} 1 \\ 0 \\ 0 \\ 1 \end{bmatrix}.$$

This gives

$$2y_1 = 1 \implies y_1 = \frac{1}{2},$$

$$-1y_1 + \frac{3}{2}y_2 = 0 \implies y_2 = \frac{1}{3}$$

$$-y_2 + \frac{4}{3}y_3 = 0 \implies y_3 = \frac{1}{4},$$

$$-y_3 + \frac{5}{4}y_4 = 1 \implies y_4 = 1,$$

so $\mathbf{y} = \left(\frac{1}{2}, \frac{1}{3}, \frac{1}{4}, 1 \right)^t.$

Then using backward substitution to solve $U\mathbf{x} = \mathbf{y}$,

$$U\mathbf{x} = \begin{bmatrix} 1 & -\frac{1}{2} & 0 & 0 \\ 0 & 1 & -\frac{2}{3} & 0 \\ 0 & 0 & 1 & -\frac{3}{4} \\ 0 & 0 & 0 & 1 \end{bmatrix} \begin{bmatrix} x_1 \\ x_2 \\ x_3 \\ x_4 \end{bmatrix} = \begin{bmatrix} \frac{1}{2} \\ \frac{1}{3} \\ \frac{1}{4} \\ 1 \end{bmatrix}$$

gives

$$x_4 = 1,$$

$$x_3 - \frac{3}{4}x_4 = \frac{1}{4} \implies x_3 = 1,$$

$$x_2 - \frac{2}{3}x_3 = \frac{1}{3} \implies x_2 = 1,$$

$$x_1 - \frac{1}{2}x_2 = \frac{1}{2} \implies x_1 = 1$$

and $\mathbf{x} = (1, 1, 1, 1)^t$.   ∎

The tridiagonal factorization can be applied whenever $l_i \neq 0$ for each $i = 1, 2, \ldots, n$. Two conditions, either of which ensure that this is true, are that the coefficient matrix of the system is positive definite or that it is strictly diagonally dominant. An additional condition that ensures this method can be applied is as follows.

Nonsingular Tridiagonal Matrices

Suppose that A is tridiagonal with $a_{i,i-1} \neq 0$ and $a_{i,i+1} \neq 0$ for each $i = 2, 3, \ldots, n-1$. If $|a_{11}| > |a_{12}|$, $|a_{nn}| > |a_{n,n-1}|$, and $|a_{ii}| \geq |a_{i,i-1}| + |a_{i,i+1}|$ for each $i = 2, 3, \ldots, n-1$, then A is nonsingular, and the values of l_i are nonzero for each $i = 1, 2, \ldots, n$.

EXERCISE SET 6.6

1. Determine which of the following matrices are (i) symmetric, (ii) singular, (iii) strictly diagonally dominant, and (iv) positive definite.

   a. $\begin{bmatrix} 2 & 1 \\ 1 & 3 \end{bmatrix}$

   b. $\begin{bmatrix} -2 & 1 \\ 1 & -3 \end{bmatrix}$

   c. $\begin{bmatrix} 2 & 1 & 0 \\ 0 & 3 & 0 \\ 1 & 0 & 4 \end{bmatrix}$

   d. $\begin{bmatrix} 2 & 1 & 0 \\ 0 & 3 & 2 \\ 1 & 2 & 4 \end{bmatrix}$

   e. $\begin{bmatrix} 4 & 2 & 6 \\ 3 & 0 & 7 \\ -2 & -1 & -3 \end{bmatrix}$

   f. $\begin{bmatrix} 2 & -1 & 0 \\ -1 & 4 & 2 \\ 0 & 2 & 2 \end{bmatrix}$

   g. $\begin{bmatrix} 4 & 0 & 0 & 0 \\ 6 & 7 & 0 & 0 \\ 9 & 11 & 1 & 0 \\ 5 & 4 & 1 & 1 \end{bmatrix}$

   h. $\begin{bmatrix} 2 & 3 & 1 & 2 \\ -2 & 4 & -1 & 5 \\ 3 & 7 & 1.5 & 1 \\ 6 & -9 & 3 & 7 \end{bmatrix}$

2. Find a factorizaton of the form $A = LDL^t$ for the following symmetric matrices:

   a. $A = \begin{bmatrix} 2 & -1 & 0 \\ -1 & 2 & -1 \\ 0 & -1 & 2 \end{bmatrix}$

   b. $A = \begin{bmatrix} 4 & 1 & 1 & 1 \\ 1 & 3 & -1 & 1 \\ 1 & -1 & 2 & 0 \\ 1 & 1 & 0 & 2 \end{bmatrix}$

   c. $A = \begin{bmatrix} 4 & 1 & -1 & 0 \\ 1 & 3 & -1 & 0 \\ -1 & -1 & 5 & 2 \\ 0 & 0 & 2 & 4 \end{bmatrix}$

   d. $A = \begin{bmatrix} 6 & 2 & 1 & -1 \\ 2 & 4 & 1 & 0 \\ 1 & 1 & 4 & -1 \\ -1 & 0 & -1 & 3 \end{bmatrix}$

3. Find a factorization of the form $A = LL^t$ for the matrices in Exercise 2.

4. Use the factorization in Exercise 2 to solve the following linear systems.

a.
$$\begin{aligned}
2x_1 - x_2 &= 3, \\
-x_1 + 2x_2 - x_3 &= -3, \\
-x_2 + 2x_3 &= 1.
\end{aligned}$$

b.
$$\begin{aligned}
4x_1 + x_2 + x_3 + x_4 &= 0.65, \\
x_1 + 3x_2 - x_3 + x_4 &= 0.05, \\
x_1 - x_2 + 2x_3 &= 0, \\
x_1 + x_2 + 2x_4 &= 0.5.
\end{aligned}$$

c.
$$\begin{aligned}
4x_1 + x_2 - x_3 &= 7, \\
x_1 + 3x_2 - x_3 &= 8, \\
-x_1 - x_2 + 5x_3 + 2x_4 &= -4, \\
2x_3 + 4x_4 &= 6.
\end{aligned}$$

d.
$$\begin{aligned}
6x_1 + 2x_2 + x_3 - x_4 &= 0, \\
2x_1 + 4x_2 + x_3 &= 7, \\
x_1 + x_2 + 4x_3 - x_4 &= -1, \\
-x_1 - x_3 + 3x_4 &= -2.
\end{aligned}$$

5. Use Crout factorization for tridiagonal systems to solve the following linear systems.

a.
$$\begin{aligned}
x_1 - x_2 &= 0, \\
-2x_1 + 4x_2 - 2x_3 &= -1, \\
-x_2 + 2x_3 &= 1.5.
\end{aligned}$$

b.
$$\begin{aligned}
3x_1 + x_2 &= -1, \\
2x_1 + 4x_2 + x_3 &= 7, \\
2x_2 + 5x_3 &= 9.
\end{aligned}$$

c.
$$\begin{aligned}
2x_1 - x_2 &= 3, \\
-x_1 + 2x_2 - x_3 &= -3, \\
-x_2 + 2x_3 &= 1.
\end{aligned}$$

d.
$$\begin{aligned}
0.5x_1 + 0.25x_2 &= 0.35, \\
0.35x_1 + 0.8x_2 + 0.4x_3 &= 0.77, \\
0.25x_2 + x_3 + 0.5x_4 &= -0.5, \\
x_3 - 2x_4 &= -2.25.
\end{aligned}$$

6. Let A be the 10×10 tridiagonal matrix given by $a_{ii} = 2, a_{i,i+1} = a_{i,i-1} = -1$, for each $i = 2, \ldots, 9$, and $a_{11} = a_{10,10} = 2, a_{12} = a_{10,9} = -1$. Let **b** be the 10-dimensional column vector given by $b_1 = b_{10} = 1$ and $b_i = 0$ for each $i = 2, 3, \ldots, 9$. Solve $A\mathbf{x} = \mathbf{b}$ using the Crout factorization for tridiagonal systems.

7. Suppose that A and B are positive definite $n \times n$ matrices.

a. Must $-A$ also be positive definite?

b. Must A^t also be positive definite?

c. Must $A + B$ also be positive definite?

d. Must A^2 also be positive definite?

e. Must $A - B$ also be positive definite?

8. Let

$$A = \begin{bmatrix} 1 & 0 & -1 \\ 0 & 1 & 1 \\ -1 & 1 & \alpha \end{bmatrix}.$$

Find all values of α for which

a. A is singular.

b. A is strictly diagonally dominant.

c. A is symmetric.

d. A is positive definite.

9. Let

$$A = \begin{bmatrix} \alpha & 1 & 0 \\ \beta & 2 & 1 \\ 0 & 1 & 2 \end{bmatrix}.$$

Find all values of α and β for which

a. A is singular.

b. A is strictly diagonally dominant.

    c. A is symmetric.

    d. A is positive definite.

10. Suppose A and B commute; that is, $AB = BA$. Must A^t and B^t also commute?

11. In a paper by Dorn and Burdick [DoB], it is reported that the average wing length that resulted from mating three mutant varieties of fruit flies (*Drosophila melanogaster*) can be expressed in the symmetric matrix form

$$A = \begin{bmatrix} 1.59 & 1.69 & 2.13 \\ 1.69 & 1.31 & 1.72 \\ 2.13 & 1.72 & 1.85 \end{bmatrix},$$

where a_{ij} denotes the average wing length of an offspring resulting from the mating of a male of type i with a female of type j.

    a. What physical significance is associated with the symmetry of this matrix?

    b. Is this matrix positive definite? If so, prove it; if not, find a nonzero vector $\mathbf{x}$ for which $\mathbf{x}^t A \mathbf{x} \leq 0$.

6.7 Survey of Methods and Software

In this chapter we have looked at direct methods for solving linear systems. A linear system consists of n equations in n unknowns expressed in matrix notation as $A\mathbf{x} = \mathbf{b}$. These techniques use a finite sequence of arithmetic operations to determine the exact solution of the system subject only to round-off error. We found that the linear system $A\mathbf{x} = \mathbf{b}$ has a unique solution if and only if A^{-1} exists, which is equivalent to $\det A \neq 0$. The solution of the linear system is the vector $\mathbf{x} = A^{-1}\mathbf{b}$.

Pivoting techniques were introduced to minimize the effects of round-off error, which can dominate the solution when using direct methods. We studied partial pivoting, scaled partial pivoting, and total pivoting. We recommend the partial or scaled partial pivoting methods for most problems because these decrease the effects of round-off error without adding much extra computation. Total pivoting should be used if round-off error is suspected to be large. In Section 7.6 we will see some procedures for estimating this round-off error.

Gaussian elimination was shown to yield a factorization of the matrix A into LU, where L is lower triangular with 1s on the diagonal and U is upper triangular. (This process is sometimes called Doolittle's factorization.) Not all nonsingular matrices can be factored this way, but a permutation of the rows will always give a factorization of the form $PA = LU$, where P is the permutation matrix used to rearrange the rows of A. The advantage of the factorization is that the work is reduced when solving linear systems $A\mathbf{x} = \mathbf{b}$ with the same coefficient matrix A and different vectors $\mathbf{b}$.

Factorizations take a simpler form when the matrix A is positive definite. For example, the Cholesky factorization has the form $A = LL^t$, where L is lower triangular. A symmetric matrix that has an LU factorization can also be factored in the form $A = LDL^t$, where D is diagonal and L is lower triangular with 1s on the diagonal. With these factorizations, manipulations involving A can be simplified. If A is tridiagonal, the LU factorization takes a particularly simple form, with U having 1s on the main diagonal and its only other nonzero entries on the diagonal immediately above. In addition, L has its only nonzero entries on the main diagonal and on the diagonal immediately below. Another important matrix factorization technique is the singular value decomposition considered in Section 9.6.

The direct methods are the methods of choice for most linear systems. For tridiagonal, banded, and positive definite matrices, the special methods are recommended. For the general case, Gaussian elimination or LU factorization methods, which allow pivoting, are recommended. In these cases, the effects of round-off error should be monitored. In Section 7.6 we discuss estimating errors in direct methods.

Large linear systems with primarily 0 entries occurring in regular patterns can be solved efficiently using an iterative procedure such as those discussed in Chapter 7. Systems of this type arise naturally, for example, when finite-difference techniques are used to solve boundary-value problems, a common application in the numerical solution of partial-differential equations.

It can be difficult to solve a large linear system that has primarily nonzero entries or one where the 0 entries are not in a predictable pattern. The matrix associated with the system can be placed in secondary storage in partitioned form and portions read into main memory only as needed for calculation. Methods that require secondary storage can be either iterative or direct, but they generally require techniques from the fields of data structures and graph theory. The reader is referred to [BuR] and [RW] for a discussion of the current techniques.

The software for matrix operations and the direct solution of linear systems implemented in IMSL and NAG is based on LAPACK, a subroutine package in the public domain. There is excellent documentation available with it and from the books written about it.

Accompanying LAPACK is a set of lower-level operations called Basic Linear Algebra Subprograms (BLAS). Level 1 of BLAS generally consists of vector-vector operations with input data and operation counts of $O(n)$. Level 2 consists of the matrix-vector operations with input data and operation counts of $O(n^2)$. Level 3 consists of the matrix-matrix operations with input data and operation counts of $O(n^3)$.

The subroutines in LAPACK for solving linear systems first factor the matrix A. The factorization depends on the type of matrix in the following way:

- General matrix $PA = LU$;

- Positive definite matrix $A = LL^t$;

- Symmetric matrix $A = LDL^t$;

- Tridiagonal matrix $A = LU$ (in banded form).

Linear systems are then solved based on the factorization. It is also possible to compute determinants and inverses and to estimate the round-off error involved.

The IMSL Library includes counterparts to almost all the LAPACK subroutines and some extensions as well. The NAG Library has numerous subroutines for direct methods of solving linear systems similar to those in LAPACK and IMSL.

Further information on the numerical solution of linear systems and matrices can be found in Golub and Van Loan [GV], Forsythe and Moler [FM], and Stewart [Stew]. The use of direct techniques for solving large sparse systems is discussed in detail in George and Liu [GL] and in Pissanetzky [Pi]. Coleman and Van Loan [CV] consider the use of BLAS, LINPACK, and MATLAB.

Iterative Methods for Solving Linear Systems

7.1 Introduction

The previous chapter considered the approximation of the solution of a linear system using *direct* methods, techniques that would produce the exact solution if all the calculations were performed using exact arithmetic. In this chapter we describe some popular *iterative* techniques, which require an initial approximation to the solution. These methods are not expected to return the exact solution even if all the calculations could be performed using exact arithmetic. In many instances, however, they are more effective than the direct methods because they can require far less computational effort and round-off error is reduced. This is particularly true when the matrix is **sparse**—that is, when it has a high percentage of zero entries.

Some additional material from linear algebra is needed to describe the convergence of the iterative methods. Principally, we need to have a measure of how close two vectors are to one another because the object of an iterative method is to determine an approximation that is within a certain tolerance of the exact solution.

In Section 7.2, the notion of a norm is used to show how various forms of distance between vectors can be described. We will also see how this concept can be extended to describe the norm of—and, consequently, the distance between—matrices. In Section 7.3, matrix eigenvalues and eigenvectors are described, and we consider the connection between these concepts and the convergence of an iterative method.

Section 7.4 describes the elementary Jacobi and Gauss-Seidel iterative methods. By analyzing the size of the largest eigenvalue of a matrix associated with an iterative method, we can determine conditions that predict the likelihood of convergence of the method. In Section 7.5 we introduce the SOR method. This is a commonly applied iterative technique because it reduces the approximation errors faster than the Jacobi and Gauss-Seidel methods.

In Section 7.6 we discuss some of the concerns that should be addressed when applying either an iterative or a direct technique for approximating the solution to a linear system.

The conjugate gradient method is presented in Section 7.7. This method, with preconditioning, is the technique most often used for sparse, positive-definite matrices.

7.2 Convergence of Vectors

The distance between the real numbers x and y is $|x - y|$. In Chapter 2 we saw that the stopping techniques for the iterative root-finding techniques used this measure to estimate the accuracy of approximate solutions and to determine when an approximation was sufficiently accurate. The iterative methods for solving systems of equations use similar logic, so the first step is to determine a way to measure the distance between n-dimensional vectors because this is the form that is taken by the solution to a system of equations.

Vector Norms

Let $\mathbb{R}^n$ denote the set of all n-dimensional column vectors with real number coefficients. It is a space-saving convenience to use the transpose notation presented in Section 6.4 when such a vector is represented in terms of its components. Generally, we write the vector

$$\mathbf{x} = \begin{bmatrix} x_1 \\ x_2 \\ \vdots \\ x_n \end{bmatrix} \quad \text{as} \quad \mathbf{x} = (x_1, x_2, \dots, x_n)^t.$$

Vector Norm on $\mathbb{R}^n$

A vector norm on $\mathbb{R}^n$ is a function, $\| \cdot \|$, from $\mathbb{R}^n$ into $\mathbb{R}$ with the following properties:

(i) $\|\mathbf{x}\| \geq 0$ for all $\mathbf{x} \in \mathbb{R}^n$,

(ii) $\|\mathbf{x}\| = 0$ if and only if $\mathbf{x} = (0, 0, \dots, 0)^t \equiv \mathbf{0}$,

(iii) $\|\alpha\mathbf{x}\| = |\alpha| \|\mathbf{x}\|$ for all $\alpha \in \mathbb{R}$ and $\mathbf{x} \in \mathbb{R}^n$,

(iv) $\|\mathbf{x} + \mathbf{y}\| \leq \|\mathbf{x}\| + \|\mathbf{y}\|$ for all $\mathbf{x}, \mathbf{y} \in \mathbb{R}^n$.

For our purposes, we need only two specific norms on $\mathbb{R}^n$. (A third is presented in Exercise 2.)

The l_2 and l_∞ norms for the vector $\mathbf{x} = (x_1, x_2, \dots, x_n)^t$ are defined by

$$\|\mathbf{x}\|_2 = \left\{ \sum_{i=1}^n x_i^2 \right\}^{1/2} \quad \text{and} \quad \|\mathbf{x}\|_\infty = \max_{1 \leq i \leq n} |x_i|.$$

The l_2 norm is called the **Euclidean norm** of the vector $\mathbf{x}$ because it represents the usual notion of distance from the origin when $\mathbf{x}$ is in $\mathbb{R}^1 \equiv \mathbb{R}$, $\mathbb{R}^2$, or $\mathbb{R}^3$. For example, the l_2 norm of the vector $\mathbf{x} = (x_1, x_2, x_3)^t$ gives the length of the straight line joining the points $(0, 0, 0)$ and (x_1, x_2, x_3); that is, the length of the shortest path between those two points. Figure 7.1 shows the boundary of those vectors in $\mathbb{R}^2$ and $\mathbb{R}^3$ that have l_2 norm less than 1. Figure 7.2 gives a similar illustration for the l_∞ norm.

Figure 7.1

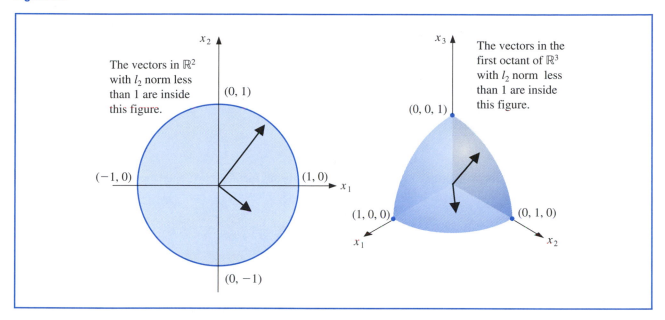

The vectors in $\mathbb{R}^2$ with l_2 norm less than 1 are inside this figure.

The vectors in the first octant of $\mathbb{R}^3$ with l_2 norm less than 1 are inside this figure.

Figure 7.2

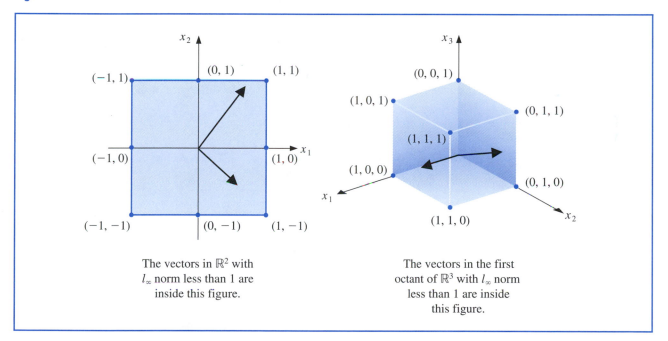

The vectors in $\mathbb{R}^2$ with l_∞ norm less than 1 are inside this figure.

The vectors in the first octant of $\mathbb{R}^3$ with l_∞ norm less than 1 are inside this figure.

Example 1   Determine the l_2 norm and the l_∞ norm of the vector $\mathbf{x} = (-1, 1, -2)^t$.

Solution   The vector $\mathbf{x} = (-1, 1, -2)^t$ in $\mathbb{R}^3$ has norms

$$\|\mathbf{x}\|_2 = \sqrt{(-1)^2 + (1)^2 + (-2)^2} = \sqrt{6}$$

and

$$\|\mathbf{x}\|_\infty = \max\{|-1|, |1|, |-2|\} = 2.$$

Notice that $\|\mathbf{x}\|_\infty < \|\mathbf{x}\|_2$ in this example. ▪

There are many forms of this inequality, hence many discoverers. Augustin Louis Cauchy (1789–1857) describes this inequality in 1821 in *Cours d'Analyse Algébrique*, the first rigorous calculus book. An integral form of the equality appears in the work of Viktor Yakovlevich Bunyakovsky (1804–1889) in 1859, and Hermann Amandus Schwarz (1843–1921) used a double integral form of this inequality in 1885. More details on the history can be found in [Stee].

Showing that $\|\mathbf{x}\|_\infty = \max_{1 \le i \le n} |x_i|$ satisfies the conditions necessary for a norm on $\mathbb{R}^n$ follows directly from the truth of similar statements concerning absolute values of real numbers. In the case of the l_2 norm, it is also easy to demonstrate the first three of the required properties, but the fourth,

$$\|\mathbf{x} + \mathbf{y}\|_2 \le \|\mathbf{x}\|_2 + \|\mathbf{y}\|_2,$$

is more difficult to show. To demonstrate this inequality we need the Cauchy-Buniakowsky-Schwarz inequality, which states that for any $\mathbf{x} = (x_1, x_2, \ldots, x_n)^t$ and $\mathbf{y} = (y_1, y_2, \ldots, y_n)^t$,

$$\sum_{i=1}^{n} |x_i y_i| \le \left\{ \sum_{i=1}^{n} x_i^2 \right\}^{1/2} \left\{ \sum_{i=1}^{n} y_i^2 \right\}^{1/2}. \tag{7.1}$$

With this it follows that $\|\mathbf{x} + \mathbf{y}\|_2 \le \|\mathbf{x}\|_2 + \|\mathbf{y}\|_2$ because

$$\|\mathbf{x} + \mathbf{y}\|_2^2 = \sum_{i=1}^{n} x_i^2 + 2 \sum_{i=1}^{n} x_i y_i + \sum_{i=1}^{n} y_i^2 \le \sum_{i=1}^{n} x_i^2 + 2 \sum_{i=1}^{n} |x_i y_i| + \sum_{i=1}^{n} y_i^2$$

$$\le \sum_{i=1}^{n} x_i^2 + 2 \left\{ \sum_{i=1}^{n} x_i^2 \right\}^{1/2} \left\{ \sum_{i=1}^{n} y_i^2 \right\}^{1/2} + \sum_{i=1}^{n} y_i^2 = (\|\mathbf{x}\|_2 + \|\mathbf{y}\|_2)^2.$$

Distance between Vectors in $\mathbb{R}^n$

The norm of a vector gives a measure for the distance between the vector and the origin, so the distance between two vectors is the norm of the difference of the vectors.

Distance between Vectors

If $\mathbf{x} = (x_1, x_2, \ldots, x_n)^t$ and $\mathbf{y} = (y_1, y_2, \ldots, y_n)^t$ are vectors in $\mathbb{R}^n$, the l_2 and l_∞ distances between $\mathbf{x}$ and $\mathbf{y}$ are defined by

$$\|\mathbf{x} - \mathbf{y}\|_2 = \left\{ \sum_{i=1}^{n} (x_i - y_i)^2 \right\}^{1/2} \quad \text{and} \quad \|\mathbf{x} - \mathbf{y}\|_\infty = \max_{1 \le i \le n} |x_i - y_i|.$$

Example 2 The linear system

$$3.3330 x_1 + 15920 x_2 - 10.333 x_3 = 15913,$$

$$2.2220 x_1 + 16.710 x_2 + 9.6120 x_3 = 28.544,$$

$$1.5611 x_1 + 5.1791 x_2 + 1.6852 x_3 = 8.4254$$

has the exact solution $\mathbf{x} = (x_1, x_2, x_3)^t = (1, 1, 1)^t$, and Gaussian elimination performed using five-digit rounding arithmetic and partial pivoting produces the approximate solution

$$\tilde{\mathbf{x}} = (\tilde{x}_1, \tilde{x}_2, \tilde{x}_3)^t = (1.2001, 0.99991, 0.92538)^t.$$

Determine the l_2 and l_∞ distances between the exact and approximate solutions.

Solution Measurements of $\mathbf{x} - \tilde{\mathbf{x}}$ are given by

$$\|\mathbf{x} - \tilde{\mathbf{x}}\|_2 = \left[(1 - 1.2001)^2 + (1 - 0.99991)^2 + (1 - 0.92538)^2\right]^{1/2}$$
$$= [(0.2001)^2 + (0.00009)^2 + (0.07462)^2]^{1/2} = 0.21356.$$

and

$$\|\mathbf{x} - \tilde{\mathbf{x}}\|_\infty = \max\{|1 - 1.2001|, |1 - 0.99991|, |1 - 0.92538|\}$$
$$= \max\{0.2001, 0.00009, 0.07462\} = 0.2001$$

Although the components $\tilde{x}_2$ and $\tilde{x}_3$ are good approximations to x_2 and x_3, the component $\tilde{x}_1$ is a poor approximation to x_1, and $|x_1 - \tilde{x}_1|$ dominates both norms. ∎

The distance concept in $\mathbb{R}^n$ is used to define the limit of a sequence of vectors. A sequence $\{\mathbf{x}^{(k)}\}_{k=1}^\infty$ of vectors in $\mathbb{R}^n$ is said to **converge** to $\mathbf{x}$ with respect to the norm $\|\cdot\|$ if, given any $\varepsilon > 0$, there exists an integer $N(\varepsilon)$ such that

$$\|\mathbf{x}^{(k)} - \mathbf{x}\| < \varepsilon \quad \text{for all } k \geq N(\varepsilon).$$

Example 3 Show that

$$\mathbf{x}^{(k)} = \left(x_1^{(k)}, x_2^{(k)}, x_3^{(k)}, x_4^{(k)}\right)^t = \left(1, 2 + \frac{1}{k}, \frac{3}{k^2}, e^{-k} \sin k\right)^t$$

converges to $\mathbf{x} = (1, 2, 0, 0)^t$ with respect to the l_∞ norm.

Solution Let $\varepsilon > 0$ be given. For each of the component functions,

$$\lim_{k \to \infty} 1 = 1, \quad \text{so an integer } N_1(\varepsilon) \text{ exists with} \quad \left|x_1^{(k)} - 1\right| < \varepsilon \quad \text{for all } k \geq N_1(\varepsilon),$$

$$\lim_{k \to \infty} (2 + 1/k) = 2, \quad \text{so an integer } N_2(\varepsilon) \text{ exists with} \quad \left|x_2^{(k)} - 2\right| < \varepsilon \quad \text{for all } k \geq N_2(\varepsilon),$$

$$\lim_{k \to \infty} 3/k^2 = 0, \quad \text{so an integer } N_3(\varepsilon) \text{ exists with} \quad \left|x_3^{(k)} - 0\right| < \varepsilon \quad \text{for all } k \geq N_3(\varepsilon),$$

$$\lim_{k \to \infty} e^{-k} \sin k = 0, \quad \text{so an integer } N_4(\varepsilon) \text{ exists with} \quad \left|x_4^{(k)} - 0\right| < \varepsilon \quad \text{for all } k \geq N_4(\varepsilon).$$

Let

$$N(\varepsilon) = \max\{N_1(\varepsilon), N_2(\varepsilon), N_3(\varepsilon), N_4(\varepsilon)\}.$$

Then when $k \geq N(\varepsilon)$, we have

$$||\mathbf{x}^{(k)} - \mathbf{x}||_\infty = \max\left\{\left|x_1^{(k)} - 1\right|, \left|x_2^{(k)} - 2\right|, \left|x_3^{(k)} - 0\right|, \left|x_4^{(k)} - 0\right|\right\} < \varepsilon,$$

so $\mathbf{x}^{(k)}$ converges to $\mathbf{x}$. ∎

In Example 3 we implicitly used the fact that a sequence of vectors $\{\mathbf{x}^{(k)}\}_{k=1}^\infty$ converges in the norm $\|\cdot\|_\infty$ to the vector $\mathbf{x}$ if and only if, for each $i = 1, 2, \ldots, n$, the sequence $\{x_i^{(k)}\}_{k=1}^\infty$ converges to x_i, the ith component of $\mathbf{x}$. This makes the determination of convergence for the norm $\|\cdot\|_\infty$ relatively easy.

To show directly that the sequence in Example 3 converges to $(1, 2, 0, 0)^t$ with respect to the l_2 norm is quite complicated. However, suppose that $\mathbf{x}$ is a vector in $\mathbb{R}^n$ and j is an index with the property that

$$\|\mathbf{x}\|_\infty = \max_{i=1,\ldots,n} |x_i| = |x_j|.$$

Then

$$\|\mathbf{x}\|_\infty^2 = |x_j|^2 = x_j^2 \le \sum_{i=1}^n x_i^2 = \|\mathbf{x}\|_2^2 \quad \text{and} \quad \|\mathbf{x}\|_2^2 = \sum_{i=1}^n x_i^2 \le \sum_{i=1}^n x_j^2 = n x_j^2 = n\|\mathbf{x}\|_\infty^2.$$

This gives the norm inequalities

$$\|\mathbf{x}\|_\infty \le \|\mathbf{x}\|_2 \le \sqrt{n}\|\mathbf{x}\|_\infty.$$

This implies that the sequence of vectors $\{\mathbf{x}^{(k)}\}$ also converges to $\mathbf{x}$ in $\mathbb{R}^n$ with respect to $\|\cdot\|_2$ if and only if $\lim_{k\to\infty} x_i^{(k)} = x_i$ for each $i = 1, 2, \ldots, n$, since this is when the sequence converges in the l_∞ norm.

In fact, it can be shown that all norms on $\mathbb{R}^n$ are equivalent with respect to convergence; that is,

- if $\|\cdot\|$ and $\|\cdot\|'$ are any two norms on $\mathbb{R}^n$ and $\{\mathbf{x}^{(k)}\}_{k=1}^\infty$ has the limit $\mathbf{x}$ with respect to $\|\cdot\|$, then $\{\mathbf{x}^{(k)}\}_{k=1}^\infty$ has the limit $\mathbf{x}$ with respect to $\|\cdot\|'$.

Since a vector sequence converges in the l_∞ norm precisely when each of its component sequences converges, we have the following.

Vector Sequence Convergence

The following statements are equivalent:

   (i)   The sequence of vectors $\{\mathbf{x}^{(k)}\}$ converges to $\mathbf{x}$ in some norm.

  (ii)   The sequence of vectors $\{\mathbf{x}^{(k)}\}$ converges to $\mathbf{x}$ in every norm.

 (iii)   For each of the component functions $x_i^{(k)}$ of $\mathbf{x}^{(k)}$, we have $\lim_{k\to\infty} x_i^{(k)} = x_i$.

Matrix Norms and Distances

In the subsequent sections, we will need methods for determining the distance between $n \times n$ matrices. This again requires the use of a norm.

Matrix Norm

A matrix norm on the set of all $n \times n$ matrices is a real-valued function, $\|\cdot\|$, defined on this set, satisfying for all $n \times n$ matrices A and B and all real numbers α:

   (i) $\|A\| \ge 0$,

  (ii) $\|A\| = 0$, if and only if A is O, the matrix with all zero entries,

 (iii) $\|\alpha A\| = |\alpha|\|A\|$,

 (iv) $\|A + B\| \le \|A\| + \|B\|$,

  (v) $\|AB\| \le \|A\|\|B\|$.

Every vector norm produces an associated natural matrix norm.

A **distance between $n \times n$ matrices** A and B with respect to this matrix norm is $\|A - B\|$. Although matrix norms can be obtained in various ways, the only norms we consider are those that are natural consequences of a vector norm.

Natural Matrix Norm

If $\| \cdot \|$ is a vector norm on $\mathbb{R}^n$, the natural matrix norm on the set of $n \times n$ matrices given by $\| \cdot \|$ is defined by

$$\|A\| = \max_{\|\mathbf{x}\|=1} \|A\mathbf{x}\|.$$

So, the l_2 and l_∞ matrix norms are, respectively,

$$\|A\|_2 = \max_{\|\mathbf{x}\|_2=1} \|A\mathbf{x}\|_2 \quad \text{(the } l_2 \text{ norm)} \quad \text{and} \quad \|A\|_\infty = \max_{\|\mathbf{x}\|_\infty=1} \|A\mathbf{x}\|_\infty \quad \text{(the } l_\infty \text{ norm)}.$$

When $n = 2$ these norms have the geometric representations shown in Figures 7.3 and 7.4.

Figure 7.3

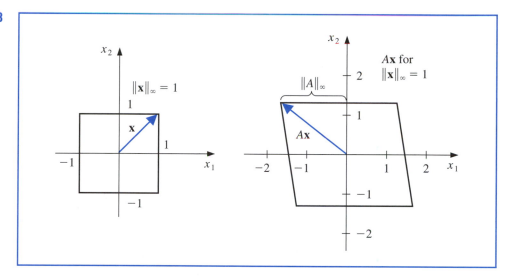

Figure 7.4

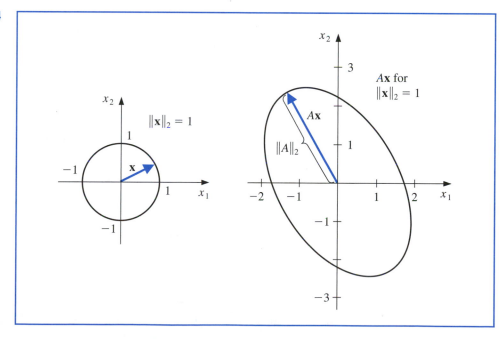

The l_∞ norm of a matrix has a representation with respect to the entries of the matrix that makes it particularly easy to compute.

---

l_∞ Matrix Norm Characterization

$$\|A\|_\infty = \max_{1 \le i \le n} \sum_{j=1}^{n} |a_{ij}|.$$

Example 4   Determine $\|A\|_\infty$ for the matrix

$$A = \begin{bmatrix} 1 & 2 & -1 \\ 0 & 3 & -1 \\ 5 & -1 & 1 \end{bmatrix}.$$

Solution   We have

$$\sum_{j=1}^{3} |a_{1j}| = |1| + |2| + |-1| = 4, \quad \sum_{j=1}^{3} |a_{2j}| = |0| + |3| + |-1| = 4,$$

and

$$\sum_{j=1}^{3} |a_{3j}| = |5| + |-1| + |1| = 7.$$

So $\|A\|_\infty = \max\{4, 4, 7\} = 7.$ ■

The l_2 norm of a matrix is not as easily determined, but in the next section we will discover an alternative method for finding this norm.

EXERCISE SET 7.2

1.   Find $\|\mathbf{x}\|_\infty$ and $\|\mathbf{x}\|_2$ for the following vectors.

   a. $\mathbf{x} = \left(3, -4, 0, \frac{3}{2}\right)^t$

   b. $\mathbf{x} = (2, 1, -3, 4)^t$

   c. $\mathbf{x} = (\sin k, \cos k, 2^k)^t$ for a fixed positive integer k

   d. $\mathbf{x} = (4/(k+1), 2/k^2, k^2 e^{-k})^t$ for a fixed positive integer k

2.   **a.**   Verify that $\| \cdot \|_1$ is a norm for $\mathbb{R}^n$ (called the l_1 norm), where

$$\|\mathbf{x}\|_1 = \sum_{i=1}^{n} |x_i|.$$

   b.   Find $\|\mathbf{x}\|_1$ for the vectors given in Exercise 1.

3.   Show that the following sequences are convergent, and find their limits.

   a. $\mathbf{x}^{(k)} = (1/k, e^{1-k}, -2/k^2)^t$

   b. $\mathbf{x}^{(k)} = (e^{-k} \cos k, k \sin(1/k), 3 + k^{-2})^t$

   c. $\mathbf{x}^{(k)} = \left(ke^{-k^2}, (\cos k)/k, \sqrt{k^2 + k} - k\right)^t$

   d. $\mathbf{x}^{(k)} = (e^{1/k}, (k^2 + 1)/(1 - k^2), (1/k^2)(1 + 3 + 5 + \cdots + (2k - 1)))^t$

4. Find $\| \cdot \|_\infty$ for the following matrices.

a. $\begin{bmatrix} 10 & 15 \\ 0 & 1 \end{bmatrix}$

b. $\begin{bmatrix} 10 & 0 \\ 15 & 1 \end{bmatrix}$

c. $\begin{bmatrix} 2 & -1 & 0 \\ -1 & 2 & -1 \\ 0 & -1 & 2 \end{bmatrix}$

d. $\begin{bmatrix} 4 & -1 & 7 \\ -1 & 4 & 0 \\ -7 & 0 & 4 \end{bmatrix}$

5. The following linear systems $A\mathbf{x} = \mathbf{b}$ have $\mathbf{x}$ as the actual solution and $\tilde{\mathbf{x}}$ as an approximate solution. Compute $\|\mathbf{x} - \tilde{\mathbf{x}}\|_\infty$ and $\|A\tilde{\mathbf{x}} - \mathbf{b}\|_\infty$.

a. $\dfrac{1}{2}x_1 + \dfrac{1}{3}x_2 = \dfrac{1}{63}$,
$\dfrac{1}{3}x_1 + \dfrac{1}{4}x_2 = \dfrac{1}{168}$,
$\mathbf{x} = \left(\dfrac{1}{7}, -\dfrac{1}{6} \right)^t$,
$\tilde{\mathbf{x}} = (0.142, -0.166)^t$.

b. $x_1 + 2x_2 + 3x_3 = 1$,
$2x_1 + 3x_2 + 4x_3 = -1$,
$3x_1 + 4x_2 + 6x_3 = 2$,
$\mathbf{x} = (0, -7, 5)^t$,
$\tilde{\mathbf{x}} = (-0.33, -7.9, 5.8)^t$.

c. $x_1 + 2x_2 + 3x_3 = 1$,
$2x_1 + 3x_2 + 4x_3 = -1$,
$3x_1 + 4x_2 + 6x_3 = 2$,
$\mathbf{x} = (0, -7, 5)^t$,
$\tilde{\mathbf{x}} = (-0.2, -7.5, 5.4)^t$.

d. $0.04x_1 + 0.01x_2 - 0.01x_3 = 0.06$,
$0.2x_1 + 0.5x_2 - 0.2x_3 = 0.3$,
$x_1 + 2x_2 + 4x_3 = 11$,
$\mathbf{x} = (1.827586, 0.6551724, 1.965517)^t$,
$\tilde{\mathbf{x}} = (1.8, 0.64, 1.9)^t$.

6. The l_1 matrix norm, defined by $\|A\|_1 = \max\limits_{\|\mathbf{x}\|_1 = 1} \|A\mathbf{x}\|_1$, can be computed using the formula

$$\|A\|_1 = \max_{1 \le j \le n} \sum_{i=1}^{n} |a_{ij}|,$$

where the l_1 vector norm is defined in Exercise 2. Find the l_1 norm of the matrices in Exercise 4.

7. Show by example that $\| \cdot \|_{\circledS}$, defined by $\|A\|_{\circledS} = \max\limits_{1 \le i, j \le n} |a_{ij}|$, does not define a matrix norm.

8. Show that $\| \cdot \|_{\circled1}$, defined by

$$\|A\|_{\circled1} = \sum_{i=1}^{n} \sum_{j=1}^{n} |a_{ij}|,$$

is a matrix norm. Find $\| \cdot \|_{\circled1}$ for the matrices in Exercise 4.

9. Show that if $\| \cdot \|$ is a vector norm on $\mathbb{R}^n$, then $\|A\| = \max_{\|\mathbf{x}\|=1} \|A\mathbf{x}\|$ is a matrix norm.

7.3 Eigenvalues and Eigenvectors

An $n \times m$ matrix can be considered as a function that uses matrix multiplication to take m-dimensional vectors into n-dimensional vectors. So an $n \times n$ matrix A takes the set of

n-dimensional vectors into itself. In this case certain nonzero vectors can have $\mathbf{x}$ and $A\mathbf{x}$ parallel, which means that a constant λ exists with $A\mathbf{x} = \lambda\mathbf{x}$, or that $(A - \lambda I)\mathbf{x} = \mathbf{0}$. There is a close connection between these numbers λ and the likelihood that an iterative method based on A will converge. We will consider the connection in this section.

For a square $n \times n$ matrix A, the **characteristic polynomial** of A is defined by

$$p(\lambda) = \det(A - \lambda I).$$

Because of the way the determinant of a matrix is defined, p is an nth-degree polynomial and, consequently, has at most n distinct zeros, some of which might be complex. These zeros of p are called the **eigenvalues** of the matrix A.

The result on page 256 in Chapter 6, then, implies that the following are equivalent:

- λ is an eigenvalue of A,

- $A - \lambda I$ does not have an inverse,

- there exists a vector $\mathbf{x} \neq \mathbf{0}$ with $A\mathbf{x} = \lambda\mathbf{x}$,

- $\det(A - \lambda I) = 0$.

If $\mathbf{x}$ is a nonzero vector with $A\mathbf{x} = \lambda\mathbf{x}$, then $\mathbf{x}$ is called an **eigenvector** of A corresponding to the eigenvalue λ. Note that if $\mathbf{x}$ is an eigenvector of A corresponding to the eigenvalue λ, then any nonzero scalar multiple $\alpha\mathbf{x}$ of $\mathbf{x}$ is also an eigenvector of A corresponding to λ because

> The prefix *eigen* comes from the German adjective meaning "to own" and is synonymous in English with the word *characteristic*. Each matrix has its own eigen- or characteristic equation, with corresponding eigen- or characteristic values and functions.

$$A(\alpha\mathbf{x}) = \alpha(A\mathbf{x}) = \alpha(\lambda\mathbf{x}) = \lambda(\alpha\mathbf{x}).$$

If $\mathbf{x}$ is an eigenvector associated with the eigenvalue λ, then $A\mathbf{x} = \lambda\mathbf{x}$, so the matrix A takes the vector $\mathbf{x}$ into a scalar multiple of itself. When λ is a real number and $\lambda > 1$, A has the effect of stretching $\mathbf{x}$ by a factor of λ. When $0 < \lambda < 1$, A shrinks $\mathbf{x}$ by a factor of λ. When $\lambda < 0$, the effects are similar, but the direction is reversed (see Figure 7.5).

Figure 7.5

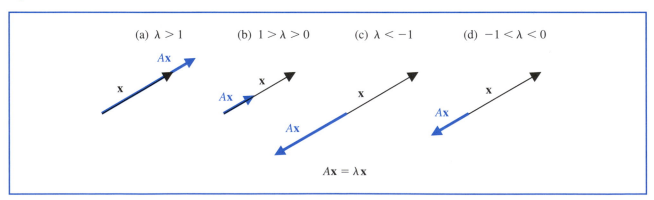

(a) $\lambda > 1$ (b) $1 > \lambda > 0$ (c) $\lambda < -1$ (d) $-1 < \lambda < 0$

$A\mathbf{x} = \lambda\mathbf{x}$

Example 1   Determine the eigenvalues and corresponding eigenvectors for the matrix

$$A = \begin{bmatrix} 2 & 0 & 0 \\ 1 & 1 & 2 \\ 1 & -1 & 4 \end{bmatrix}.$$

Solution The characteristic polynomial of A is

$$p(\lambda) = \det(A - \lambda I) = \det \begin{bmatrix} 2 - \lambda & 0 & 0 \\ 1 & 1 - \lambda & 2 \\ 1 & -1 & 4 - \lambda \end{bmatrix}$$

$$= -(\lambda^3 - 7\lambda^2 + 16\lambda - 12) = -(\lambda - 3)(\lambda - 2)^2,$$

so there are two eigenvalues of A: $\lambda_1 = 3$ and $\lambda_2 = 2$.

An eigenvector $\mathbf{x}_1 \neq \mathbf{0}$ corresponding to the eigenvalue $\lambda_1 = 3$ is a solution to the vector-matrix equation $(A - 3 \cdot I)\mathbf{x}_1 = \mathbf{0}$, so

$$\begin{bmatrix} 0 \\ 0 \\ 0 \end{bmatrix} = \begin{bmatrix} -1 & 0 & 0 \\ 1 & -2 & 2 \\ 1 & -1 & 1 \end{bmatrix} \cdot \begin{bmatrix} x_1 \\ x_2 \\ x_3 \end{bmatrix} = \begin{bmatrix} -x_1 \\ x_1 - 2x_2 + 2x_3 \\ x_1 - x_2 + x_3 \end{bmatrix},$$

which implies that $x_1 = 0$ and $x_2 = x_3$. Any nonzero value of x_3 produces an eigenvector for the eigenvalue $\lambda_1 = 3$. For example, when $x_3 = 1$ we have the eigenvector $\mathbf{x}_1 = (0, 1, 1)^t$. Any eigenvector of A corresponding to $\lambda = 3$ is a nonzero multiple of $\mathbf{x}_1$.

An eigenvector $\mathbf{x}_2 \neq \mathbf{0}$ of A associated with the eigenvalue $\lambda_2 = 2$ is a solution of the system $(A - 2I)\mathbf{x} = \mathbf{0}$, so

$$\begin{bmatrix} 0 \\ 0 \\ 0 \end{bmatrix} = \begin{bmatrix} 0 & 0 & 0 \\ 1 & -1 & 2 \\ 1 & -1 & 2 \end{bmatrix} \cdot \begin{bmatrix} x_1 \\ x_2 \\ x_3 \end{bmatrix} = \begin{bmatrix} 0 \\ x_1 - x_2 + 2x_3 \\ x_1 - x_2 + 2x_3 \end{bmatrix}.$$

In this case the eigenvector has only to satisfy the equation

$$x_1 - x_2 + 2x_3 = 0,$$

which can be done in various ways. For example, when $x_1 = 0$ we have $x_2 = 2x_3$, so one choice would be $\mathbf{x}_2 = (0, 2, 1)^t$. We could also choose $x_2 = 0$, which requires that $x_1 = -2x_3$. Hence $\mathbf{x}_3 = (-2, 0, 1)^t$ gives a second eigenvector for the eigenvalue $\lambda_2 = 2$, one that is not a multiple of $\mathbf{x}_2$.

The eigenvectors of A corresponding to the eigenvalue $\lambda_2 = 2$ generate an entire plane. This plane is described by all vectors of the form

$$\alpha \mathbf{x}_2 + \beta \mathbf{x}_3 = (-2\beta, 2\alpha, \alpha + \beta)^t,$$

for arbitrary constants α and β, provided that at least one of the constants is nonzero. ∎

The next example illustrates that even some very simple matrices can have no real eigenvalues.

Example 2 Show that there are no nonzero vectors $\mathbf{x}$ in $\mathbb{R}^2$ with $B\mathbf{x}$ parallel to $\mathbf{x}$ if

$$B = \begin{bmatrix} 0 & 1 \\ -1 & 0 \end{bmatrix}.$$

Solution The eigenvalues of B are the solutions to the characteristic polynomial

$$0 = \det(B - \lambda I) = \det \begin{bmatrix} -\lambda & 1 \\ -1 & -\lambda \end{bmatrix} = \lambda^2 + 1,$$

so the eigenvalues of B are the complex numbers $\lambda_1 = i$ and $\lambda_2 = -i$. A corresponding eigenvector $\mathbf{x}$ for λ_1 needs to satisfy

$$\begin{bmatrix} 0 \\ 0 \end{bmatrix} = \begin{bmatrix} -i & 1 \\ -1 & -i \end{bmatrix} \begin{bmatrix} x_1 \\ x_2 \end{bmatrix} = \begin{bmatrix} -ix_1 + x_2 \\ -x_1 - ix_2 \end{bmatrix},$$

that is, $0 = -ix_1 + x_2$, so $x_2 = ix_1$, and an eigenvector for $\lambda_1 = i$ is $(1, i)^t$. In a similar manner, an eigenvector for $\lambda_2 = -i$ is $(1, -i)^t$.

If $\mathbf{x}$ is an eigenvector of B, then exactly one of its components is real and the other is complex. As a consequence, there is no real constant λ and nonzero vector $\mathbf{x}$ in $\mathbb{R}^2$ with $B\mathbf{x} = \lambda\mathbf{x}$, and hence there is no nonzero vector $\mathbf{x}$ in $\mathbb{R}^2$ with $B\mathbf{x}$ parallel to $\mathbf{x}$. ■

MATLAB provides methods to directly compute the eigenvalues and eigenvectors of a matrix. We first define the matrix A by

`A = [1 0 2; 0 1 -1; -1 1 1]`

The characteristic polynomial is determined with

`p=poly(A)`

giving

$$p = 1.0000 \quad -3.0000 \quad 6.0000 \quad -4.0000$$

The numbers are the coefficients of the characteristic polynomial in descending order, so

$$p(\lambda) = \lambda^3 - 3\lambda^2 + 6\lambda - 4.$$

We can now compute the roots of the polynomial to obtain the eigenvalues with

`roots(p)`

The most direct way to obtain eigenvalues is with the `eig` command.

`eig(A)`

If we want the corresponding eigenvectors, we enter `eig` as

`[V, D] = eig(A)`

which produces the following matrix V and vector D. We have rounded the entries in V so that it will display on one line.

$$V = \begin{array}{lll} -0.70710678 & -0.70710678 & 0.70710678 \\ 0.35355339 + 0.00000000i & 0.35355339 - 0.00000000i & 0.70710678 \\ -0.00000000 - 0.61237244i & -0.00000000 + 0.61237244i & 0.00000000 \end{array}$$

$$D = \begin{array}{l} 0.999999999999999 + 1.732050807568876i \\ 0.999999999999999 - 1.732050807568876i \\ 1.000000000000000 \end{array}$$

The columns of V are eigenvectors of A corresponding to the eigenvalues in the rows of D.

The notions of eigenvalues and eigenvectors are introduced here for a specific computational convenience, but these concepts arise frequently in the study of physical systems. In fact, they are of sufficient interest that most of Chapter 9 is devoted to their approximation.

Spectral Radius

The **spectral radius** $\rho(A)$ of a matrix A is defined by

$$\rho(A) = \max |\lambda|, \quad \text{where } \lambda \text{ is an eigenvalue of } A.$$

(*Note*: For complex $\lambda = \alpha + \beta i$, we have $|\lambda| = (\alpha^2 + \beta^2)^{1/2}$.)

Example 3   Determine the spectral radius of the matrices

$$A = \begin{bmatrix} 2 & 0 & 0 \\ 1 & 1 & 2 \\ 1 & -1 & 4 \end{bmatrix} \quad \text{and} \quad B = \begin{bmatrix} 0 & 1 \\ -1 & 0 \end{bmatrix}.$$

Solution   In Example 1 we found that the eigenvalues of A were $\lambda_1 = 3$ and $\lambda_2 = 2$. So

$$\rho(A) = \max\{|3|, |2|\} = 3,$$

and in Example 2 we found that the eigenvalues of B were $\lambda_1 = i$ and $\lambda_2 = -i$. So

$$\rho(B) = \max\{\sqrt{1^2}, \sqrt{(-1)^2}\} = 1.$$ ∎

The spectral radius is closely related to the norm of a matrix.

---

l_2 Matrix Norm Characterization

If A is an $n \times n$ matrix, then

(i) $\|A\|_2 = [\rho(A^t A)]^{1/2}$;

(ii) $\rho(A) \leq \|A\|$ for any natural norm.

The first part of this result is the computational method for determining the l_2 norm of matrices that we mentioned at the end of the previous section.

Example 4   Determine the l_2 norm of

$$A = \begin{bmatrix} 1 & 1 & 0 \\ 1 & 2 & 1 \\ -1 & 1 & 2 \end{bmatrix}.$$

Solution   To apply part (i) of the l_2 Matrix Norm Characterization, we need to calculate $\rho(A^t A)$, so we need the eigenvalues of $A^t A$.

$$A^t A = \begin{bmatrix} 1 & 1 & -1 \\ 1 & 2 & 1 \\ 0 & 1 & 2 \end{bmatrix} \begin{bmatrix} 1 & 1 & 0 \\ 1 & 2 & 1 \\ -1 & 1 & 2 \end{bmatrix} = \begin{bmatrix} 3 & 2 & -1 \\ 2 & 6 & 4 \\ -1 & 4 & 5 \end{bmatrix}.$$

If

$$0 = \det(A^t A - \lambda I) = \det \begin{bmatrix} 3 - \lambda & 2 & -1 \\ 2 & 6 - \lambda & 4 \\ -1 & 4 & 5 - \lambda \end{bmatrix}$$

$$= -\lambda^3 + 14\lambda^2 - 42\lambda = -\lambda(\lambda^2 - 14\lambda + 42),$$

then $\lambda = 0$ or $\lambda = 7 \pm \sqrt{7}$. So

$$\|A\|_2 = \sqrt{\rho(A^t A)} = \sqrt{\max\{0, 7 - \sqrt{7}, 7 + \sqrt{7}\}} = \sqrt{7 + \sqrt{7}} \approx 3.106.$$ ∎

The operations in Example 4 can also be performed using MATLAB. First define

```
A = [1 1 0; 1 2 1; -1 1 2]
```

then compute its transpose and determine $A^t A$, and the eigenvalues of $A^t A$

```
u = eig(A'*A)
```

This gives the eigenvalues as

$$u = 0.000000000000003$$
$$4.354248688935409$$
$$9.645751311064592$$

The square root of the largest eigenvalue is the l_2 norm of A

```
sqrt(u(3))
```

which MATLAB gives as 3.105760987433610.

The l_2 norm of A can also be directly computed with

```
norm(A)
```

The l_∞ norm of A is found with `norm(A,Inf)`.

Convergent Matrices

In studying iterative matrix techniques, it is of particular importance to know when the powers of a matrix become small (that is, when all of the entries approach zero). We call an $n \times n$ matrix A **convergent** if, for each $i = 1, 2, \ldots, n$ and $j = 1, 2, \ldots, n$, we have

$$\lim_{k \to \infty} (A^k)_{ij} = 0.$$

Example 5 Show that

$$A = \begin{bmatrix} \frac{1}{2} & 0 \\ \frac{1}{4} & \frac{1}{2} \end{bmatrix}$$

is a convergent matrix.

Solution Computing the powers of A, we obtain:

$$A^2 = \begin{bmatrix} \frac{1}{4} & 0 \\ \frac{1}{4} & \frac{1}{4} \end{bmatrix}, \quad A^3 = \begin{bmatrix} \frac{1}{8} & 0 \\ \frac{3}{16} & \frac{1}{8} \end{bmatrix}, \quad A^4 = \begin{bmatrix} \frac{1}{16} & 0 \\ \frac{1}{8} & \frac{1}{16} \end{bmatrix},$$

and, in general,

$$A^k = \begin{bmatrix} (\frac{1}{2})^k & 0 \\ \frac{k}{2^{k+1}} & (\frac{1}{2})^k \end{bmatrix}.$$

So A is a convergent matrix because

$$\lim_{k \to \infty} \left(\frac{1}{2}\right)^k = 0 \quad \text{and} \quad \lim_{k \to \infty} \frac{k}{2^{k+1}} = 0. \qquad \blacksquare$$

The following important connection exists between the spectral radius of a matrix and the convergence of the matrix.

Convergent Matrix Equivalences

The following are equivalent statements:

(i) A is a convergent matrix.

(ii) $\lim_{n\to\infty} \|A^n\| = 0$, for some natural norm.

(iii) $\lim_{n\to\infty} \|A^n\| = 0$, for all natural norms.

(iv) $\rho(A) < 1$.

(v) $\lim_{n\to\infty} A^n \mathbf{x} = \mathbf{0}$, for every $\mathbf{x}$.

EXERCISE SET 7.3

1. Compute the eigenvalues and associated eigenvectors of the following matrices.

a. $\begin{bmatrix} 2 & -1 \\ -1 & 2 \end{bmatrix}$

b. $\begin{bmatrix} 0 & 1 \\ 1 & 1 \end{bmatrix}$

c. $\begin{bmatrix} 0 & \frac{1}{2} \\ \frac{1}{2} & 0 \end{bmatrix}$

d. $\begin{bmatrix} 1 & 1 \\ -2 & -2 \end{bmatrix}$

e. $\begin{bmatrix} 2 & 1 & 0 \\ 1 & 2 & 0 \\ 0 & 0 & 3 \end{bmatrix}$

f. $\begin{bmatrix} -1 & 2 & 0 \\ 0 & 3 & 4 \\ 0 & 0 & 7 \end{bmatrix}$

g. $\begin{bmatrix} 2 & 1 & 1 \\ 2 & 3 & 2 \\ 1 & 1 & 2 \end{bmatrix}$

h. $\begin{bmatrix} 3 & 2 & -1 \\ 1 & -2 & 3 \\ 2 & 0 & 4 \end{bmatrix}$

2. Find the spectral radius for each matrix in Exercise 1.

3. Show that

$$A_1 = \begin{bmatrix} 1 & 0 \\ \frac{1}{4} & \frac{1}{2} \end{bmatrix}$$

is not convergent, but

$$A_2 = \begin{bmatrix} \frac{1}{2} & 0 \\ 16 & \frac{1}{2} \end{bmatrix}$$

is convergent.

4. Which of the matrices in Exercise 1 are convergent?

5. Find the $\|\cdot\|_2$ norms of the matrices in Exercise 1.

6. Show that if λ is an eigenvalue of a matrix A and $\|\cdot\|$ is a vector norm, then an eigenvector $\mathbf{x}$ associated with λ exists with $\|\mathbf{x}\| = 1$.

7. Find 2×2 matrices A and B for which $\rho(A + B) > \rho(A) + \rho(B)$. (This shows that $\rho(A)$ cannot be a matrix norm.)

8.   Show that if A is symmetric, then $\|A\|_2 = \rho(A)$.

9.   Let λ be an eigenvalue of the $n \times n$ matrix A and $\mathbf{x} \neq \mathbf{0}$ be an associated eigenvector.

   a.   Show that λ is also an eigenvalue of A^t.

   b.   Show that for any integer $k \geq 1$, λ^k is an eigenvalue of A^k with eigenvector $\mathbf{x}$.

   c.   Show that if A^{-1} exists, then $1/\lambda$ is an eigenvalue of A^{-1} with eigenvector $\mathbf{x}$.

   d.   Let $\alpha \neq \lambda$ be given. Show that if $(A - \alpha I)^{-1}$ exists, then $1/(\lambda - \alpha)$ is an eigenvalue of $(A - \alpha I)^{-1}$ with eigenvector $\mathbf{x}$.

10.   In Exercise 8 of Section 6.4, it was assumed that the contribution a female beetle of a certain type made to the future years' beetle population could be expressed in terms of the matrix

$$A = \begin{bmatrix} 0 & 0 & 6 \\ \frac{1}{2} & 0 & 0 \\ 0 & \frac{1}{3} & 0 \end{bmatrix},$$

where the entry in the ith row and jth column represents the probabilistic contribution of a beetle of age j onto the next year's female population of age i.

   a.   Does the matrix A have any real eigenvalues? If so, determine them and any associated eigenvectors.

   b.   If a sample of this species was needed for laboratory test purposes that would have a constant proportion in each age group from year to year, what criteria could be imposed on the initial population to ensure that this requirement would be satisfied?

7.4 The Jacobi and Gauss-Seidel Methods

In this section we describe the elementary Jacobi and Gauss-Seidel iterative methods. These are classic methods that date to the late eighteenth century, but they find current application in problems where the matrix is large and has mostly zero entries in predictable locations. Applications of this type are common, for example, in the study of large integrated circuits and in the numerical solution of boundary-value problems and partial-differential equations.

General Iteration Methods

An iterative technique for solving the $n \times n$ linear system $A\mathbf{x} = \mathbf{b}$ starts with an initial approximation $\mathbf{x}^{(0)}$ to the solution $\mathbf{x}$ and generates a sequence of vectors $\{\mathbf{x}^{(k)}\}_{k=1}^{\infty}$ that converges to $\mathbf{x}$. These iterative techniques involve a process that converts the system $A\mathbf{x} = \mathbf{b}$ into an equivalent system of the form $\mathbf{x} = T\mathbf{x} + \mathbf{c}$ for some $n \times n$ matrix T and vector $\mathbf{c}$.

After the initial vector $\mathbf{x}^{(0)}$ is selected, the sequence of approximate solution vectors is generated by computing

$$\mathbf{x}^{(k)} = T\mathbf{x}^{(k-1)} + \mathbf{c}$$

for each $k = 1, 2, 3, \ldots$.

The following result provides an important connection between the eigenvalues of the matrix T and the expectation that the iterative method will converge.

Convergence and the Spectral Radius

The sequence

$$\mathbf{x}^{(k)} = T\mathbf{x}^{(k-1)} + \mathbf{c}$$

converges to the unique solution of $\mathbf{x} = T\mathbf{x} + \mathbf{c}$ for any $\mathbf{x}^{(0)}$ in $\mathbb{R}^n$ if and only if $\rho(T) < 1$.

Jacobi's Method

The **Jacobi** iterative method is produced by solving the ith equation in $A\mathbf{x} = \mathbf{b}$ for x_i to obtain, provided $a_{ii} \neq 0$,

$$x_i = \sum_{\substack{j=1 \\ j \neq i}}^{n} \left(-\frac{a_{ij}x_j}{a_{ii}} \right) + \frac{b_i}{a_{ii}}, \qquad \text{for } i = 1, 2, \ldots, n,$$

and generating each $x_i^{(k)}$ from components of $\mathbf{x}^{(k-1)}$, for $k \geq 1$, by

$$x_i^{(k)} = \frac{\sum_{\substack{j=1 \\ j \neq i}}^{n} \left(-a_{ij}x_j^{(k-1)} \right) + b_i}{a_{ii}}, \qquad \text{for } i = 1, 2, \ldots, n. \qquad (7.2)$$

Example 1 The linear system $A\mathbf{x} = \mathbf{b}$ given by

$$\begin{aligned}
E_1: && 10x_1 - && x_2 + && 2x_3 && && = 6, \\
E_2: && -x_1 + && 11x_2 - && x_3 + && 3x_4 && = 25, \\
E_3: && 2x_1 - && x_2 + && 10x_3 - && x_4 && = -11, \\
E_4: && && 3x_2 - && x_3 + && 8x_4 && = 15
\end{aligned}$$

Carl Gustav Jacob Jacobi (1804–1851) was initially recognized for his work in the area of number theory and elliptic functions, but his mathematical interests and abilities were very broad. He had a strong personality that was influential in establishing a research-oriented attitude that became the nucleus of a revival of mathematics at German universities in the nineteenth century.

has the unique solution $\mathbf{x} = (1, 2, -1, 1)^t$. Use Jacobi's iterative technique to find approximations $\mathbf{x}^{(k)}$ to $\mathbf{x}$ starting with $\mathbf{x}^{(0)} = (0, 0, 0, 0)^t$ until

$$\frac{\|\mathbf{x}^{(k)} - \mathbf{x}^{(k-1)}\|_\infty}{\|\mathbf{x}^{(k)}\|_\infty} < 10^{-3}.$$

Solution We first solve equation E_i for x_i, for each $i = 1, 2, 3, 4$, to obtain

$$\begin{aligned}
x_1 &= && \frac{1}{10}x_2 - \frac{1}{5}x_3 && && + \frac{3}{5}, \\
x_2 &= \frac{1}{11}x_1 && + \frac{1}{11}x_3 - \frac{3}{11}x_4 && && + \frac{25}{11}, \\
x_3 &= -\frac{1}{5}x_1 + \frac{1}{10}x_2 && + \frac{1}{10}x_4 && && - \frac{11}{10}, \\
x_4 &= && -\frac{3}{8}x_2 + \frac{1}{8}x_3 && && + \frac{15}{8}.
\end{aligned}$$

From the initial approximation $\mathbf{x}^{(0)} = (0, 0, 0, 0)^t$ we have $\mathbf{x}^{(1)}$ given by

$$\begin{aligned}
x_1^{(1)} &= && \frac{1}{10}x_2^{(0)} - \frac{1}{5}x_3^{(0)} && && + \frac{3}{5} &&= && 0.6000, \\
x_2^{(1)} &= \frac{1}{11}x_1^{(0)} && + \frac{1}{11}x_3^{(0)} - \frac{3}{11}x_4^{(0)} && && + \frac{25}{11} &&= && 2.2727, \\
x_3^{(1)} &= -\frac{1}{5}x_1^{(0)} + \frac{1}{10}x_2^{(0)} && + \frac{1}{10}x_4^{(0)} && && - \frac{11}{10} &&= && -1.1000, \\
x_4^{(1)} &= && -\frac{3}{8}x_2^{(0)} + \frac{1}{8}x_3^{(0)} && && + \frac{15}{8} &&= && 1.8750.
\end{aligned}$$

Additional iterates, $\mathbf{x}^{(k)} = (x_1^{(k)}, x_2^{(k)}, x_3^{(k)}, x_4^{(k)})^t$, are generated in a similar manner and are presented in Table 7.1.

Table 7.1

k	0	1	2	3	4	5	6	7	8	9	10
$x_1^{(k)}$	0.000	0.6000	1.0473	0.9326	1.0152	0.9890	1.0032	0.9981	1.0006	0.9997	1.0001
$x_2^{(k)}$	0.0000	2.2727	1.7159	2.053	1.9537	2.0114	1.9922	2.0023	1.9987	2.0004	1.9998
$x_3^{(k)}$	0.0000	−1.1000	−0.8052	−1.0493	−0.9681	−1.0103	−0.9945	−1.0020	−0.9990	−1.0004	−0.9998
$x_4^{(k)}$	0.0000	1.8750	0.8852	1.1309	0.9739	1.0214	0.9944	1.0036	0.9989	1.0006	0.9998

We stopped after 10 iterations because

$$\frac{\|\mathbf{x}^{(10)} - \mathbf{x}^{(9)}\|_\infty}{\|\mathbf{x}^{(10)}\|_\infty} = \frac{8.0 \times 10^{-4}}{1.9998} < 10^{-3}.$$

In fact, $\|\mathbf{x}^{(10)} - \mathbf{x}\|_\infty = 0.0002$. ■

Jacobi's method is written in the form $\mathbf{x}^{(k)} = T\mathbf{x}^{(k-1)} + \mathbf{c}$ by splitting A into its diagonal and off-diagonal parts. To see this, let

- D be the diagonal matrix whose diagonal entries are those of A,

- $-L$ be the strictly lower-triangular part of A, and

- $-U$ be the strictly upper-triangular part of A.

With this notation,

$$A = \begin{bmatrix} a_{11} & a_{12} & \cdots & a_{1n} \\ a_{21} & a_{22} & \cdots & a_{2n} \\ \vdots & \vdots & & \vdots \\ a_{n1} & a_{n2} & \cdots & a_{nn} \end{bmatrix}$$

is split into

$$A = \begin{bmatrix} a_{11} & 0 & \cdots & 0 \\ 0 & a_{22} & & \vdots \\ \vdots & & \ddots & 0 \\ 0 & \cdots & 0 & a_{nn} \end{bmatrix} - \begin{bmatrix} 0 & & \cdots & 0 \\ -a_{21} & & & \vdots \\ \vdots & & \ddots & \\ -a_{n1} & \cdots & -a_{n,n-1} & 0 \end{bmatrix} - \begin{bmatrix} 0 & -a_{12} & \cdots & -a_{1n} \\ \vdots & & \ddots & \\ & & & -a_{n-1,n} \\ 0 & \cdots & \cdots & 0 \end{bmatrix}$$

$$= D - L - U.$$

The equation $A\mathbf{x} = \mathbf{b}$ or $(D - L - U)\mathbf{x} = \mathbf{b}$ is then transformed into

$$D\mathbf{x} = (L + U)\mathbf{x} + \mathbf{b},$$

and, if D^{-1} exists—that is, if $a_{ii} \neq 0$ for each i—then

$$\mathbf{x} = D^{-1}(L + U)\mathbf{x} + D^{-1}\mathbf{b}.$$

This results in the matrix form of the Jacobi iterative technique:

$$\mathbf{x}^{(k)} = T_j\mathbf{x}^{(k-1)} + \mathbf{c}_j,$$

where $T_j = D^{-1}(L + U)$ and $\mathbf{c}_j = D^{-1}\mathbf{b}$. Since

$$\det D = a_{11} \cdot a_{22} \cdots a_{nn},$$

the diagonal matrix D is nonsingular precisely when $a_{ii} \neq 0$ for each $i = 1, 2, \ldots, n$.

The program JACITR71 implements the Jacobi method.

If $a_{ii} = 0$ for some i and the system is nonsingular, then JACITR71 performs a reordering of the equations so that no $a_{ii} = 0$. To speed convergence, the equations should be arranged so that a_{ii} is as large as possible.

The Gauss-Seidel Method

A likely improvement on the Jacobi method can be seen by reconsidering Eq. (7.2). In this equation, all the components of $\mathbf{x}^{(k-1)}$ are used to compute each of the $x_i^{(k)}$. But the components $x_1^{(k)}, \ldots, x_{i-1}^{(k)}$ of $\mathbf{x}^{(k)}$ have already been computed before we determine $x_i^{(k)}$, and they are expected to be better approximations to the actual solutions $x_1, \ldots, x_{i-1}$ than are $x_1^{(k-1)}, \ldots, x_{i-1}^{(k-1)}$. So we can use these most recently determined values to compute $x_i^{(k)}$. That is,

$$x_i^{(k)} = \frac{-\sum_{j=1}^{i-1}\left(a_{ij}x_j^{(k)}\right) - \sum_{j=i+1}^{n}\left(a_{ij}x_j^{(k-1)}\right) + b_i}{a_{ii}}, \tag{7.3}$$

for each $i = 1, 2, \ldots, n$. This modification is called the **Gauss-Seidel** iterative technique and is illustrated in the following example.

Example 2 Use the Gauss-Seidel iterative technique to find approximate solutions to

$$
\begin{aligned}
10x_1 - x_2 + 2x_3 &= 6, \\
-x_1 + 11x_2 - x_3 + 3x_4 &= 25, \\
2x_1 - x_2 + 10x_3 - x_4 &= -11, \\
3x_2 - x_3 + 8x_4 &= 15
\end{aligned}
$$

starting with $\mathbf{x} = (0,0,0,0)^t$ and iterating until

$$\frac{\|\mathbf{x}^{(k)} - \mathbf{x}^{(k-1)}\|_\infty}{\|\mathbf{x}^{(k)}\|_\infty} < 10^{-3}.$$

Phillip Ludwig Seidel (1821–1896) worked as an assistant to Jacobi solving problems on systems of linear equations that resulted from Gauss's work on least squares. These equations generally had off-diagonal elements that were much smaller than those on the diagonal, so the iterative methods were particularly effective. The iterative techniques now known as Jacobi and Gauss-Seidel were both known to Gauss before being applied in this situation, but Gauss's results were not often widely communicated.

Solution The solution $\mathbf{x} = (1, 2, -1, 1)^t$ was approximated by Jacobi's method in Example 1. For the Gauss-Seidel method we write the system, for each $k = 1, 2, \ldots$ as

$$
\begin{aligned}
x_1^{(k)} &= \frac{1}{10}x_2^{(k-1)} - \frac{1}{5}x_3^{(k-1)} + \frac{3}{5}, \\
x_2^{(k)} &= \frac{1}{11}x_1^{(k)} + \frac{1}{11}x_3^{(k-1)} - \frac{3}{11}x_4^{(k-1)} + \frac{25}{11}, \\
x_3^{(k)} &= -\frac{1}{5}x_1^{(k)} + \frac{1}{10}x_2^{(k)} + \frac{1}{10}x_4^{(k-1)} - \frac{11}{10}, \\
x_4^{(k)} &= -\frac{3}{8}x_2^{(k)} + \frac{1}{8}x_3^{(k)} + \frac{15}{8}.
\end{aligned}
$$

When $\mathbf{x}^{(0)} = (0,0,0,0)^t$, we have $\mathbf{x}^{(1)} = (0.6000, 2.3272, -0.9873, 0.8789)^t$. Subsequent iterations give the values in Table 7.2.

Because

$$\frac{\|\mathbf{x}^{(5)} - \mathbf{x}^{(4)}\|_\infty}{\|\mathbf{x}^{(5)}\|_\infty} = \frac{0.0008}{2.000} = 4 \times 10^{-4},$$

$\mathbf{x}^{(5)}$ is accepted as a reasonable approximation to the solution. Note that Jacobi's method in Example 1 required twice as many iterations for the same accuracy. ∎

Table 7.2

k	0	1	2	3	4	5
$x_1^{(k)}$	0.0000	0.6000	1.030	1.0065	1.0009	1.0001
$x_2^{(k)}$	0.0000	2.3272	2.037	2.0036	2.0003	2.0000
$x_3^{(k)}$	0.0000	−0.9873	−1.014	−1.0025	−1.0003	−1.0000
$x_4^{(k)}$	0.0000	0.8789	0.9844	0.9983	0.9999	1.0000

To write the Gauss-Seidel method in matrix form, multiply both sides of Eq. (7.3) by a_{ii} and collect all kth iterate terms, to give

$$a_{i1}x_1^{(k)} + a_{i2}x_2^{(k)} + \cdots + a_{ii}x_i^{(k)} = -a_{i,i+1}x_{i+1}^{(k-1)} - \cdots - a_{in}x_n^{(k-1)} + b_i,$$

for each $i = 1, 2, \ldots, n$. Writing all n equations gives

$$
\begin{aligned}
a_{11}x_1^{(k)} &= -a_{12}x_2^{(k-1)} - a_{13}x_3^{(k-1)} - \cdots - a_{1n}x_n^{(k-1)} + b_1, \\
a_{21}x_1^{(k)} + a_{22}x_2^{(k)} &= -a_{23}x_3^{(k-1)} - \cdots - a_{2n}x_n^{(k-1)} + b_2, \\
&\vdots \\
a_{n1}x_1^{(k)} + a_{n2}x_2^{(k)} + \cdots + a_{nn}x_n^{(k)} &= b_n.
\end{aligned}
$$

With the definitions of D, L, and U that we used previously, we have the Gauss-Seidel method represented by

$$(D - L)\mathbf{x}^{(k)} = U\mathbf{x}^{(k-1)} + \mathbf{b}.$$

If $(D - L)^{-1}$ exists, then

$$\mathbf{x}^{(k)} = T_g \mathbf{x}^{(k-1)} + \mathbf{c}_g, \qquad \text{for each } k = 1, 2, \ldots,$$

where $T_g = (D - L)^{-1}U$ and $c_g = (D - L)^{-1}\mathbf{b}$. Since $D - L$ is lower triangular,

$$\det(D - L) = a_{11} \cdot a_{22} \cdots a_{nn},$$

and $D - L$ is nonsingular precisely when $a_{ii} \neq 0$ for each $i = 1, 2, \ldots, n$.

The program GSEITR72 implements the Gauss-Seidel method.

The preceding discussion and the results of Examples 1 and 2 appear to imply that the Gauss-Seidel method is superior to the Jacobi method. This is almost always true, but there are linear systems for which the Jacobi method converges and the Gauss-Seidel method does not, as shown in Exercises 7 and 8.

If A is strictly diagonally dominant, then for any $\mathbf{b}$ and any choice of $\mathbf{x}^{(0)}$, the Jacobi and Gauss-Seidel methods will both converge to the unique solution of $A\mathbf{x} = \mathbf{b}$.

EXERCISE SET 7.4

1. Find the first 2 iterations of the Jacobi method for the following linear systems, using $\mathbf{x}^{(0)} = \mathbf{0}$:

   a. $3x_1 - x_2 + x_3 = 1,$
   $3x_1 + 6x_2 + 2x_3 = 0,$
   $3x_1 + 3x_2 + 7x_3 = 4.$

   b. $10x_1 - x_2 = 9,$
   $-x_1 + 10x_2 - 2x_3 = 7,$
   $-2x_2 + 10x_3 = 6.$

   c. $10x_1 + 5x_2 = 6,$
   $5x_1 + 10x_2 - 4x_3 = 25,$
   $-4x_2 + 8x_3 - x_4 = -11,$
   $-x_3 + 5x_4 = -11.$

   d. $4x_1 + x_2 - x_3 + x_4 = -2,$
   $x_1 + 4x_2 - x_3 - x_4 = -1,$
   $-x_1 - x_2 + 5x_3 + x_4 = 0,$
   $x_1 - x_2 + x_3 + 3x_4 = 1.$

e.
$$
\begin{aligned}
4x_1 + x_2 + x_3 \qquad\;\; + x_5 &= 6, \\
-x_1 - 3x_2 + x_3 + x_4 \qquad\;\; &= 6, \\
2x_1 + x_2 + 5x_3 - x_4 - x_5 &= 6, \\
-x_1 - x_2 - x_3 + 4x_4 \qquad\;\; &= 6, \\
2x_2 - x_3 + x_4 + 4x_5 &= 6.
\end{aligned}
$$

f.
$$
\begin{aligned}
4x_1 - x_2 \qquad\qquad\qquad\qquad\;\; &= 0, \\
-x_1 + 4x_2 - x_3 \qquad\qquad\qquad &= 5, \\
-x_2 + 4x_3 \qquad\qquad\;\; &= 0, \\
4x_4 - x_5 \qquad\; &= 6, \\
- x_4 + 4x_5 - x_6 &= -2, \\
- x_5 + 4x_6 &= 6.
\end{aligned}
$$

2.   Repeat Exercise 1 using the Gauss-Seidel method.

3.   Use the Jacobi method to solve the linear systems in Exercise 1, with $TOL = 10^{-3}$ in the l_∞ norm.

4.   Repeat Exercise 3 using the Gauss-Seidel method.

5.   The linear system

$$
\begin{aligned}
x_1 \qquad\quad - x_3 &= 0.2, \\
-\frac{1}{2}x_1 + x_2 - \frac{1}{4}x_3 &= -1.425, \\
x_1 - \frac{1}{2}x_2 + x_3 &= 2
\end{aligned}
$$

has the solution $(0.9, -0.8, 0.7)^t$.

a.   Is the coefficient matrix

$$
A = \begin{bmatrix} 1 & 0 & -1 \\ -\frac{1}{2} & 1 & -\frac{1}{4} \\ 1 & -\frac{1}{2} & 1 \end{bmatrix}
$$

strictly diagonally dominant?

b.   Compute the spectral radius of the Jacobi matrix T_j.

c.   Use the Jacobi iterative program JACITR71 to approximate the solution to the linear system with a tolerance of 10^{-2} and a maximum of 300 iterations.

d.   What happens in (c) when the system is changed to

$$
\begin{aligned}
x_1 \qquad\quad - 2x_3 &= 0.2, \\
-\frac{1}{2}x_1 + x_2 - \frac{1}{4}x_3 &= -1.425, \\
x_1 - \frac{1}{2}x_2 + x_3 &= 2.
\end{aligned}
$$

6.   Repeat Exercise 5 using the Gauss-Seidel program GSEITR72.

7.   The linear system

$$
\begin{aligned}
2x_1 - x_2 + x_3 &= -1, \\
2x_1 + 2x_2 + 2x_3 &= 4, \\
-x_1 - x_2 + 2x_3 &= -5
\end{aligned}
$$

has the solution $(1, 2, -1)^t$.

a.   Show that $\rho(T_j) = \frac{\sqrt{5}}{2} > 1$.

b.   Show that the Jacobi program JACITR71 with $x^{(0)} = 0$ fails to give a good approximation after 25 iterations.

c.   Show that $\rho(T_g) = \frac{1}{2}$.

d.   Use the Gauss-Seidel program GSEITR72 with $x^{(0)} = 0$ to approximate the solution to the linear system to within 10^{-5} in the l_∞ norm.

8. The linear system

$$x_1 + 2x_2 - 2x_3 = 7,$$
$$x_1 + x_2 + x_3 = 2,$$
$$2x_1 + 2x_2 + x_3 = 5$$

has the solution $(1, 2, -1)^t$.

a. Show that $\rho(T_j) = 0$.

b. Use the Jacobi program JACITR71 with $\mathbf{x}^{(0)} = \mathbf{0}$ to approximate the solution to the linear system to within 10^{-5} in the l_∞ norm.

c. Show that $\rho(T_g) = 2$.

d. Show that the Gauss-Seidel program GSEITR72 applied as in (b) fails to give a good approximation in 25 iterations.

9. Show that if A is strictly diagonally dominant, then $\|T_j\|_\infty < 1$.

7.5 The SOR Method

The SOR method is similar to the Jacobi and Gauss-Seidel methods, but it uses a scaling factor to more rapidly reduce the approximation error. In contrast to the classic methods discussed in the previous section, the SOR technique is a more recent innovation.

The SOR technique is one of a class of *relaxation* methods that compute approximations $\mathbf{x}^{(k)}$ by the formula

$$x_i^{(k)} = (1 - \omega)x_i^{(k-1)} + \frac{\omega}{a_{ii}}\left[b_i - \sum_{j=1}^{i-1} a_{ij}x_j^{(k)} - \sum_{j=i+1}^{n} a_{ij}x_j^{(k-1)}\right],$$

where ω is a scaling factor.

When $\omega = 1$, we have the Gauss-Seidel method. When $0 < \omega < 1$, the procedures are called **under-relaxation methods**.

When $1 < \omega$, the procedures are called **over-relaxation methods**, which are used to accelerate the convergence for systems that are convergent by the Gauss-Seidel technique. These methods are abbreviated **SOR** for **Successive Over-Relaxation** and are used for solving the linear systems that occur in the numerical solution of certain partial-differential equations.

To determine the matrix form of the SOR method, we rewrite the preceding equation as

$$a_{ii}x_i^{(k)} + \omega \sum_{j=1}^{i-1} a_{ij}x_j^{(k)} = (1 - \omega)a_{ii}x_i^{(k-1)} - \omega \sum_{j=i+1}^{n} a_{ij}x_j^{(k-1)} + \omega b_i,$$

so that in vector form we have

$$(D - \omega L)\mathbf{x}^{(k)} = [(1 - \omega)D + \omega U]\mathbf{x}^{(k-1)} + \omega\mathbf{b}.$$

If $(D - \omega L)^{-1}$ exists, then

The program SORITR73 implements the SOR method.

$$\mathbf{x}^{(k)} = T_\omega\mathbf{x}^{(k-1)} + \mathbf{c}_\omega,$$

where $T_\omega = (D - \omega L)^{-1}[(1 - \omega)D + \omega U]$ and $\mathbf{c}_\omega = \omega(D - \omega L)^{-1}\mathbf{b}$.

Example 1   The linear system $A\mathbf{x} = \mathbf{b}$ given by

$$
\begin{aligned}
4x_1 + 3x_2 \qquad &= 24, \\
3x_1 + 4x_2 - x_3 &= 30, \\
- x_2 + 4x_3 &= -24
\end{aligned}
$$

has the solution $(3, 4, -5)^t$. Compare the iterations from the Gauss-Seidel method and the SOR method with $\omega = 1.25$ using $\mathbf{x}^{(0)} = (1, 1, 1)^t$ for both methods.

Solution   For each $k = 1, 2, \ldots$, the equations for the Gauss-Seidel method are

$$
\begin{aligned}
x_1^{(k)} &= -0.75x_2^{(k-1)} + 6, \\
x_2^{(k)} &= -0.75x_1^{(k)} + 0.25x_3^{(k-1)} + 7.5, \\
x_3^{(k)} &= 0.25x_2^{(k)} - 6,
\end{aligned}
$$

and the equations for the SOR method with $\omega = 1.25$ are

$$
\begin{aligned}
x_1^{(k)} &= -0.25x_1^{(k-1)} - 0.9375x_2^{(k-1)} + 7.5, \\
x_2^{(k)} &= -0.9375x_1^{(k)} - 0.25x_2^{(k-1)} + 0.3125x_3^{(k-1)} + 9.375, \\
x_3^{(k)} &= 0.3125x_2^{(k)} - 0.25x_3^{(k-1)} - 7.5.
\end{aligned}
$$

The first 7 iterates for each method are listed in Tables 7.3 and 7.4. For the iterates to be accurate to seven decimal places, the Gauss-Seidel method requires 34 iterations, as opposed to 14 iterations for the SOR method with $\omega = 1.25$.   ∎

Table 7.3

k	0	1	2	3	4	5	6	7
$x_1^{(k)}$	1	5.250000	3.1406250	3.0878906	3.0549316	3.0343323	3.0214577	3.0134110
$x_2^{(k)}$	1	3.812500	3.8828125	3.9267578	3.9542236	3.9713898	3.9821186	3.9888241
$x_3^{(k)}$	1	-5.046875	-5.0292969	-5.0183105	-5.0114441	-5.0071526	-5.0044703	-5.0027940

Table 7.4

k	0	1	2	3	4	5	6	7
$x_1^{(k)}$	1	6.312500	2.6223145	3.1333027	2.9570512	3.0037211	2.9963276	3.0000498
$x_2^{(k)}$	1	3.5195313	3.9585266	4.0102646	4.0074838	4.0029250	4.0009262	4.0002586
$x_3^{(k)}$	1	-6.6501465	-4.6004238	-5.0966863	-4.9734897	-5.0057135	-4.9982822	-5.0003486

A reasonable question to ask is how the appropriate value of ω is chosen. Although no complete answer to this question is known for the general $n \times n$ linear system, the following result can be used in certain situations.

SOR Convergence

If A is a positive definite matrix and $0 < \omega < 2$, then the SOR method converges for any choice of initial approximate solution vector $\mathbf{x}^{(0)}$.
If, in addition, A is tridiagonal, then $\rho(T_g) = [\rho(T_j)]^2 < 1$, and the optimal choice of ω for the SOR method is

$$\omega = \frac{2}{1 + \sqrt{1 - [\rho(T_j)]^2}}.$$

With this choice of ω, we have $\rho(T_\omega) = \omega - 1$.

Example 2   Find the optimal choice of ω for the SOR method for the matrix

$$A = \begin{bmatrix} 4 & 3 & 0 \\ 3 & 4 & -1 \\ 0 & -1 & 4 \end{bmatrix}.$$

Solution   This matrix is tridiagonal, so we can apply the SOR Convergence result if we can also show that it is positive definite. The matrix is symmetric, and we will show that it is also positive definite by using part (v) of the result on page 268, as we did in the example that followed that result.

We have

$$\det(A) = 24, \quad \det\left(\begin{bmatrix} 4 & 3 \\ 3 & 4 \end{bmatrix}\right) = 7, \quad \text{and} \quad \det([4]) = 4,$$

which implies that A is positive definite.

Because

$$T_j = D^{-1}(L + U) = \begin{bmatrix} \frac{1}{4} & 0 & 0 \\ 0 & \frac{1}{4} & 0 \\ 0 & 0 & \frac{1}{4} \end{bmatrix} \begin{bmatrix} 0 & -3 & 0 \\ -3 & 0 & 1 \\ 0 & 1 & 0 \end{bmatrix} = \begin{bmatrix} 0 & -0.75 & 0 \\ -0.75 & 0 & 0.25 \\ 0 & 0.25 & 0 \end{bmatrix},$$

we have

$$T_j - \lambda I = \begin{bmatrix} -\lambda & -0.75 & 0 \\ -0.75 & -\lambda & 0.25 \\ 0 & 0.25 & -\lambda \end{bmatrix},$$

so

$$\det(T_j - \lambda I) = -\lambda(\lambda^2 - 0.625).$$

Thus

$$\rho(T_j) = \sqrt{0.625}$$

and

$$\omega = \frac{2}{1 + \sqrt{1 - [\rho(T_j)]^2}} = \frac{2}{1 + \sqrt{1 - 0.625}} \approx 1.24.$$

This explains the rapid convergence obtained in Example 1 when using $\omega = 1.25$.   ▪

EXERCISE SET 7.5

1. Find the first 2 iterations of the SOR method with $\omega = 1.1$ for the following linear systems, using $\mathbf{x}^{(0)} = \mathbf{0}$:

a. $\quad 3x_1 - \ x_2 + \ x_3 = 1,$
$\quad 3x_1 + 6x_2 + 2x_3 = 0,$
$\quad 3x_1 + 3x_2 + 7x_3 = 4.$

b. $\quad 10x_1 - \ x_2 \qquad\quad = 9,$
$\quad -x_1 + 10x_2 - \ 2x_3 = 7,$
$\qquad\quad -2x_2 + 10x_3 = 6.$

c. $\quad 10x_1 + \ 5x_2 \qquad\qquad\quad = 6,$
$\quad 5x_1 + 10x_2 - 4x_3 \qquad\quad = 25,$
$\qquad\quad -4x_2 + 8x_3 - \ x_4 = -11,$
$\qquad\qquad\quad - \ x_3 + 5x_4 = -11.$

d. $\quad 4x_1 + \ x_2 - \ x_3 + \ x_4 = -2,$
$\quad x_1 + 4x_2 - \ x_3 - \ x_4 = -1,$
$\quad -x_1 - \ x_2 + 5x_3 + \ x_4 = 0,$
$\quad x_1 - \ x_2 + \ x_3 + 3x_4 = 1.$

e. $\quad 4x_1 + \ x_2 + \ x_3 \qquad\ + \ x_5 = 6,$
$\quad -x_1 - 3x_2 + \ x_3 + \ x_4 \qquad = 6,$
$\quad 2x_1 + \ x_2 + 5x_3 - \ x_4 - \ x_5 = 6,$
$\quad -x_1 - \ x_2 - \ x_3 + 4x_4 \qquad = 6,$
$\qquad\quad 2x_2 - \ x_3 + \ x_4 + 4x_5 = 6.$

f. $\quad 4x_1 - \ x_2 \qquad\qquad\qquad\qquad = 0,$
$\quad -x_1 + 4x_2 - \ x_3 \qquad\qquad\quad = 5,$
$\qquad\quad -x_2 + 4x_3 \qquad\qquad\quad = 0,$
$\qquad\qquad\qquad\quad 4x_4 - \ x_5 \qquad\quad = 6,$
$\qquad\qquad\quad - \ x_4 + 4x_5 - \ x_6 = -2,$
$\qquad\qquad\qquad\qquad - \ x_5 + 4x_6 = 6.$

2. Use the SOR method with $\omega = 1.2$ to solve the linear systems in Exercise 1 with a tolerance $TOL = 10^{-3}$ in the l_∞ norm.

3. For those matrices in Exercise 1 that are both tridiagonal and positive definite, use the SOR method with the optimal choice of ω.

4. Suppose that an object can be at any one of $n + 1$ equally spaced points $x_0, x_1, \ldots, x_n$. When an object is at location x_i, it is equally likely to move to either x_{i-1} or x_{i+1} and cannot directly move to any other location. Consider the probabilities $\{P_i\}_{i=0}^n$ that an object starting at location x_i will reach the left endpoint x_0 before reaching the right endpoint x_n. Clearly, $P_0 = 1$ and $P_n = 0$. Since the object can move to x_i only from x_{i-1} or x_{i+1} and does so with probability $\frac{1}{2}$ for each of these locations,

$$P_i = \frac{1}{2} P_{i-1} + \frac{1}{2} P_{i+1}, \quad \text{for each } i = 1, 2, \ldots, n - 1.$$

a. Show that

$$
\begin{bmatrix}
1 & -\frac{1}{2} & 0 & \cdots & \cdots & \cdots & 0 \\
-\frac{1}{2} & 1 & -\frac{1}{2} & & & & \vdots \\
0 & -\frac{1}{2} & 1 & & & & \\
\vdots & & & \ddots & & & 0 \\
& & & & & & \\
& & & & -\frac{1}{2} & 1 & -\frac{1}{2} \\
0 & \cdots & \cdots & \cdots & 0 & -\frac{1}{2} & 1
\end{bmatrix}
\begin{bmatrix}
P_1 \\
P_2 \\
\vdots \\
\\
\\
P_{n-1}
\end{bmatrix}
=
\begin{bmatrix}
\frac{1}{2} \\
0 \\
\vdots \\
\\
\\
0
\end{bmatrix}.
$$

   b. Solve this system using $n = 10, 50$, and 100.

   c. Change the probabilities to α and $1 - \alpha$ for movement to the left and right, respectively, and derive the linear system similar to the one in (a).

   d. Repeat (b) using the system in (c) with $\alpha = \frac{1}{3}$.

5. The forces on the bridge truss shown here satisfy the equations in the following table:

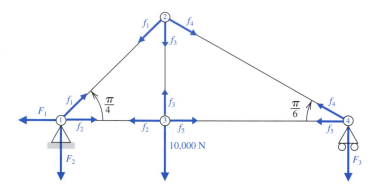

Joint	Horizontal Component	Vertical Component
①	$-F_1 + \frac{\sqrt{2}}{2} f_1 + f_2 = 0$	$\frac{\sqrt{2}}{2} f_1 - F_2 = 0$
②	$-\frac{\sqrt{2}}{2} f_1 + \frac{\sqrt{3}}{2} f_4 = 0$	$-\frac{\sqrt{2}}{2} f_1 - f_3 - \frac{1}{2} f_4 = 0$
③	$-f_2 + f_5 = 0$	$f_3 - 10{,}000 = 0$
④	$-\frac{\sqrt{3}}{2} f_4 - f_5 = 0$	$\frac{1}{2} f_4 - F_3 = 0$

This linear system can be placed in the matrix form

$$
\begin{bmatrix}
-1 & 0 & 0 & \frac{\sqrt{2}}{2} & 1 & 0 & 0 & 0 \\
0 & -1 & 0 & \frac{\sqrt{2}}{2} & 0 & 0 & 0 & 0 \\
0 & 0 & -1 & 0 & 0 & 0 & \frac{1}{2} & 0 \\
0 & 0 & 0 & -\frac{\sqrt{2}}{2} & 0 & -1 & \frac{1}{2} & 0 \\
0 & 0 & 0 & 0 & -1 & 0 & 0 & 1 \\
0 & 0 & 0 & 0 & 0 & 1 & 0 & 0 \\
0 & 0 & 0 & -\frac{\sqrt{2}}{2} & 0 & 0 & \frac{\sqrt{3}}{2} & 0 \\
0 & 0 & 0 & 0 & 0 & 0 & -\frac{\sqrt{3}}{2} & -1
\end{bmatrix}
\begin{bmatrix}
F_1 \\ F_2 \\ F_3 \\ f_1 \\ f_2 \\ f_3 \\ f_4 \\ f_5
\end{bmatrix}
=
\begin{bmatrix}
0 \\ 0 \\ 0 \\ 0 \\ 0 \\ 10{,}000 \\ 0 \\ 0
\end{bmatrix}.
$$

Approximate the solution of the resulting linear system to within 10^{-2} in the l_∞ norm using as initial approximation the vector all of whose entries are 1s and (i) the Gauss-Seidel method, (ii) the Jacobi method, and (iii) the SOR method with $\omega = 1.25$.

7.6 Error Bounds and Iterative Refinement

This section considers the errors in approximation that are likely to occur when solving linear systems by both direct and iterative methods. There is no universally superior technique for approximating the solution to linear systems, but some methods will give better results than others when the matrix satisfies certain conditions.

It seems intuitively reasonable that if $\tilde{\mathbf{x}}$ is an approximation to the solution $\mathbf{x}$ of $A\mathbf{x} = \mathbf{b}$ and the **residual vector**, defined by

$$\mathbf{r} = \mathbf{b} - A\tilde{\mathbf{x}},$$

has the property that if $\|\mathbf{r}\|$ is small, then $\|\mathbf{x} - \tilde{\mathbf{x}}\|$ should be small as well. This is often the case, but certain systems, which occur quite often in practice, fail to have this property.

Example 1 The linear system $A\mathbf{x} = \mathbf{b}$ given by

$$\begin{bmatrix} 1 & 2 \\ 1.0001 & 2 \end{bmatrix} \begin{bmatrix} x_1 \\ x_2 \end{bmatrix} = \begin{bmatrix} 3 \\ 3.0001 \end{bmatrix}$$

has the unique solution $\mathbf{x} = (1, 1)^t$. Determine the residual vector for the poor approximation $\tilde{\mathbf{x}} = (3, -0.0001)^t$.

Solution We have

$$\mathbf{r} = \mathbf{b} - A\tilde{\mathbf{x}} = \begin{bmatrix} 3 \\ 3.0001 \end{bmatrix} - \begin{bmatrix} 1 & 2 \\ 1.0001 & 2 \end{bmatrix} \begin{bmatrix} 3 \\ -0.0001 \end{bmatrix} = \begin{bmatrix} 0.0002 \\ 0 \end{bmatrix},$$

so $\|\mathbf{r}\|_\infty = 0.0002$. Although the norm of the residual vector is small, the approximation $\tilde{\mathbf{x}} = (3, -0.0001)^t$ is obviously quite poor; in fact, $\|\mathbf{x} - \tilde{\mathbf{x}}\|_\infty = 2$. ■

The difficulty in Example 1 is explained quite simply by noting that the solution to the system represents the intersection of the lines

$$l_1: x_1 + 2x_2 = 3 \text{ and } l_2: 1.0001x_1 + 2x_2 = 3.0001.$$

The point $(3, -0.0001)$ lies on l_2, and the lines are nearly the same. This means that $(3, -0.0001)$ also lies close to the line l_1, even though it differs significantly from the solution of the system, which is the intersection point $(1, 1)$. (See Figure 7.6.)

Figure 7.6

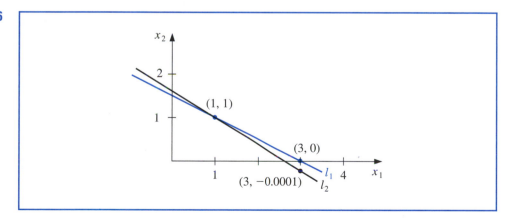

Example 1 was clearly constructed to show the difficulties that might—and, in fact, do—arise. Had the lines not been nearly coincident, we would expect a small residual vector to imply an accurate approximation. In the general situation, we cannot rely on the geometry of the system to give an indication of when problems might occur. We can, however, obtain this information by considering the norms of the matrix and its inverse.

Residual Vector Error Bounds

If $\tilde{\mathbf{x}}$ is an approximation to the solution of $A\mathbf{x} = \mathbf{b}$ and A is a nonsingular matrix, then for any natural norm,

$$\|\mathbf{x} - \tilde{\mathbf{x}}\| \leq \|\mathbf{b} - A\tilde{\mathbf{x}}\| \cdot \|A^{-1}\|$$

and

$$\frac{\|\mathbf{x} - \tilde{\mathbf{x}}\|}{\|\mathbf{x}\|} \leq \|A\| \cdot \|A^{-1}\| \frac{\|\mathbf{b} - A\tilde{\mathbf{x}}\|}{\|\mathbf{b}\|}, \quad \text{provided } \mathbf{x} \neq \mathbf{0} \text{ and } \mathbf{b} \neq \mathbf{0}.$$

This result implies that $\|A^{-1}\|$ and $\|A\| \cdot \|A^{-1}\|$ provide an indication of the connection between the residual vector and the accuracy of the approximation. In general, the relative error $\|\mathbf{x} - \tilde{\mathbf{x}}\|/\|\mathbf{x}\|$ is of most interest. Any convenient norm can be used for this approximation; the only requirement is that it be used consistently throughout.

Condition Numbers

The **condition number**, $K(A)$, of the nonsingular matrix A relative to a norm $\| \cdot \|$ is

$$K(A) = \|A\| \cdot \|A^{-1}\|.$$

Note that for any nonsingular matrix A and natural norm $\| \cdot \|$,

$$1 = \|I\| = \|A \cdot A^{-1}\| \leq \|A\| \cdot \|A^{-1}\| = K(A).$$

With this notation, we can reexpress the inequalities in the previous result as

$$\|\mathbf{x} - \tilde{\mathbf{x}}\| \leq K(A)\frac{\|\mathbf{b} - A\tilde{\mathbf{x}}\|}{\|A\|} \quad \text{and} \quad \frac{\|\mathbf{x} - \tilde{\mathbf{x}}\|}{\|\mathbf{x}\|} \leq K(A)\frac{\|\mathbf{b} - A\tilde{\mathbf{x}}\|}{\|\mathbf{b}\|}.$$

A matrix A is well-behaved (called **well-conditioned**) if $K(A)$ is close to 1, and A is not well-behaved (called **ill-conditioned**) when $K(A)$ is significantly greater than 1. Conditioning in this instance refers to the relative security that a small residual vector implies a correspondingly accurate approximate solution.

Example 2   Determine the condition number for the matrix

$$A = \begin{bmatrix} 1 & 2 \\ 1.0001 & 2 \end{bmatrix}.$$

Solution   We saw in Example 1 that the very poor approximation $(3, -0.0001)^t$ to the exact solution $(1, 1)^t$ had a residual vector with small norm, so we should expect the condition number of A to be large. We have $\|A\|_\infty = \max\{|1| + |2|, |1.001| + |2|\} = 3.0001$, which would not be considered large. However,

$$A^{-1} = \begin{bmatrix} -10000 & 10000 \\ 5000.5 & -5000 \end{bmatrix}, \quad \text{so} \quad \|A^{-1}\|_\infty = 20000,$$

and for the infinity norm,

$$K_\infty(A) = (20000)(3.0001) = 60002.$$

The size of the condition number for this example should certainly keep us from making hasty accuracy decisions based on the residual of an approximation. ▪

In MATLAB, the condition number $K_\infty(A)$ for the matrix in Example 2 can be computed using the command cond. To obtain the l_∞ condition number, use the command

cond(A,Inf)

MATLAB responds with

$$ans = 6.000199999999003e + 004$$

The default for cond is the l_2 condition number, and either of the commands cond(A) or cond(A,2) gives

$$K_2(A) = 5.000100002987370e + 004.$$

Iterative Refinement

The residual of an approximation can also be used to improve the accuracy of the approximation. Suppose that $\tilde{\mathbf{x}}$ is an approximation to the solution of the linear system $A\mathbf{x} = \mathbf{b}$ and that $\mathbf{r} = \mathbf{b} - \tilde{A}\mathbf{x}$ is the residual vector associated with $\tilde{\mathbf{x}}$. Consider $\tilde{\mathbf{y}}$, the approximate solution to the system $A\mathbf{y} = \mathbf{r}$. Then

$$\tilde{\mathbf{y}} \approx A^{-1}\mathbf{r} = A^{-1}(\mathbf{b} - A\tilde{\mathbf{x}}) = A^{-1}\mathbf{b} - A^{-1}A\tilde{\mathbf{x}} = \mathbf{x} - \tilde{\mathbf{x}}.$$

So

$$\mathbf{x} \approx \tilde{\mathbf{x}} + \tilde{\mathbf{y}}.$$

The program ITREF74 implements the Iterative Refinement method.

This new approximation $\tilde{\mathbf{x}} + \tilde{\mathbf{y}}$ is often much closer to the solution of $A\mathbf{x} = \mathbf{b}$ than is $\tilde{\mathbf{x}}$, and $\tilde{\mathbf{y}}$ is easy to determine because it involves the same matrix, A, as the original system. This technique is called **iterative refinement**, or *iterative improvement*, and is shown in the following Illustration. To increase accuracy, the residual vector is computed using double-digit arithmetic.

Illustration   The linear system given by

$$\begin{bmatrix} 3.3330 & 15920 & -10.333 \\ 2.2220 & 16.710 & 9.6120 \\ 1.5611 & 5.1791 & 1.6852 \end{bmatrix} \begin{bmatrix} x_1 \\ x_2 \\ x_3 \end{bmatrix} = \begin{bmatrix} 15913 \\ 28.544 \\ 8.4254 \end{bmatrix}$$

has the exact solution $\mathbf{x} = (1, 1, 1)^t$.

Using Gaussian elimination and five-digit rounding arithmetic leads successively to the augmented matrices

$$\begin{bmatrix} 3.3330 & 15920 & -10.333 & : & 15913 \\ 0 & -10596 & 16.501 & : & 10580 \\ 0 & -7451.4 & 6.5250 & : & -7444.9 \end{bmatrix}$$

and

$$\begin{bmatrix} 3.3330 & 15920 & -10.333 & : & 15913 \\ 0 & -10596 & 16.501 & : & -10580 \\ 0 & 0 & -5.0790 & : & -4.7000 \end{bmatrix}.$$

The approximate solution to this system is

$$\tilde{\mathbf{x}} = (1.2001, 0.99991, 0.92538)^t.$$

The residual vector corresponding to $\tilde{\mathbf{x}}$ is computed in double precision to be

$$\mathbf{r} = \mathbf{b} - A\tilde{\mathbf{x}}$$

$$= \begin{bmatrix} 15913 \\ 28.544 \\ 8.4254 \end{bmatrix} - \begin{bmatrix} 3.3330 & 15920 & -10.333 \\ 2.2220 & 16.710 & 9.6120 \\ 1.5611 & 5.1791 & 1.6852 \end{bmatrix} \begin{bmatrix} 1.2001 \\ 0.99991 \\ 0.92538 \end{bmatrix}$$

$$= \begin{bmatrix} 15913 \\ 28.544 \\ 8.4254 \end{bmatrix} - \begin{bmatrix} 15913.00518 \\ 28.26987086 \\ 8.611560367 \end{bmatrix} = \begin{bmatrix} -0.00518 \\ 0.27412914 \\ -0.186160367 \end{bmatrix},$$

so

$$\|\mathbf{r}\|_\infty = \|\mathbf{b} - A\tilde{\mathbf{x}}\|_\infty = 0.27413.$$

To use iterative refinement to improve this approximation, we now solve the system $A\mathbf{y} = \mathbf{r}$ for $\tilde{\mathbf{y}}$. Using five-digit arithmetic and Gaussian elimination, the approximate solution $\tilde{\mathbf{y}}$ to the equation $A\mathbf{y} = \mathbf{r}$ is

$$\tilde{\mathbf{y}} = (-0.20008, 8.9987 \times 10^{-5}, 0.074607)^t$$

and we have the improved approximation to the system $A\mathbf{x} = \mathbf{b}$ given by

$$\tilde{\mathbf{x}} + \tilde{\mathbf{y}} = (1.2001, 0.99991, 0.92538)^t + (-0.20008, 8.9987 \times 10^{-5}, 0.074607)^t$$

$$= (1.0000, 1.0000, 0.99999)^t.$$

This approximation has the residual vector

$$\|\tilde{\mathbf{r}}\|_\infty = \|\mathbf{b} - A(\tilde{\mathbf{x}} + \tilde{\mathbf{y}})\|_\infty = 0.0001.$$

If we were continuing the iteration processes, we would, of course, use $\tilde{\mathbf{x}} + \tilde{\mathbf{y}}$ as our starting values rather than $\tilde{\mathbf{x}}$. □

EXERCISE SET 7.6

1. Compute the l_∞ condition numbers of the following matrices.

   a. $\begin{bmatrix} \frac{1}{2} & \frac{1}{3} \\ \frac{1}{3} & \frac{1}{4} \end{bmatrix}$

   b. $\begin{bmatrix} 3.9 & 1.6 \\ 6.8 & 2.9 \end{bmatrix}$

   c. $\begin{bmatrix} 1 & 2 \\ 1.0001 & 2 \end{bmatrix}$

   d. $\begin{bmatrix} 1.003 & 58.09 \\ 5.550 & 321.8 \end{bmatrix}$

   e. $\begin{bmatrix} 1 & -1 & -1 \\ 0 & 1 & -1 \\ 0 & 0 & -1 \end{bmatrix}$

   f. $\begin{bmatrix} 0.04 & 0.01 & -0.01 \\ 0.2 & 0.5 & -0.2 \\ 1 & 2 & 4 \end{bmatrix}$

2. The following linear systems $A\mathbf{x} = \mathbf{b}$ have $\mathbf{x}$ as the actual solution and $\tilde{\mathbf{x}}$ as an approximate solution. Using the results of Exercise 1, compute $\|\mathbf{x} - \tilde{\mathbf{x}}\|_\infty$ and

$$K_\infty(A) \frac{\|\mathbf{b} - A\tilde{\mathbf{x}}\|_\infty}{\|A\|_\infty}.$$

a. $\dfrac{1}{2}x_1 + \dfrac{1}{3}x_2 = \dfrac{1}{63}$,

   $\dfrac{1}{3}x_1 + \dfrac{1}{4}x_2 = \dfrac{1}{168}$,

   $\mathbf{x} = \left(\dfrac{1}{7}, -\dfrac{1}{6}\right)^t$, $\tilde{\mathbf{x}} = (0.142, -0.166)^t$.

b. $3.9x_1 + 1.6x_2 = 5.5$,
   $6.8x_1 + 2.9x_2 = 9.7$,
   $\mathbf{x} = (1, 1)^t$, $\tilde{\mathbf{x}} = (0.98, 1.1)^t$.

c. $x_1 + 2x_2 = 3$,
   $1.0001x_1 + 2x_2 = 3.0001$,
   $\mathbf{x} = (1, 1)^t$, $\tilde{\mathbf{x}} = (0.96, 1.02)^t$.

d. $1.003x_1 + 58.09x_2 = 68.12$,
   $5.550x_1 + 321.8x_2 = 377.3$,
   $\mathbf{x} = (10, 1)^t$, $\tilde{\mathbf{x}} = (-10, 1)^t$.

e. $x_1 - x_2 - x_3 = 2\pi$,
   $\quad\quad x_2 - x_3 = 0$,
   $\quad\quad\quad -x_3 = \pi$,
   $\mathbf{x} = (0, -\pi, -\pi)^t$, $\tilde{\mathbf{x}} = (-0.1, -3.15, -3.14)^t$.

f. $0.04x_1 + 0.01x_2 - 0.01x_3 = 0.06$,
   $0.2x_1 + \quad 0.5x_2 - \quad 0.2x_3 = 0.3$,
   $\quad x_1 + \quad\quad 2x_2 + \quad\quad 4x_3 = 11$,
   $\mathbf{x} = (1.827586, 0.6551724, 1.965517)^t$, $\tilde{\mathbf{x}} = (1.8, 0.64, 1.9)^t$.

3.   The linear system

$$
\begin{bmatrix} 1 & 2 \\ 1.0001 & 2 \end{bmatrix} \begin{bmatrix} x_1 \\ x_2 \end{bmatrix} = \begin{bmatrix} 3 \\ 3.0001 \end{bmatrix}
$$

has the solution $(1, 1)^t$. Change A slightly to

$$
\begin{bmatrix} 1 & 2 \\ 0.9999 & 2 \end{bmatrix}
$$

and consider the linear system

$$
\begin{bmatrix} 1 & 2 \\ 0.9999 & 2 \end{bmatrix} \begin{bmatrix} x_1 \\ x_2 \end{bmatrix} = \begin{bmatrix} 3 \\ 3.0001 \end{bmatrix}.
$$

Compute the new solution using five-digit rounding arithmetic, and compare the change in A to the change in $\mathbf{x}$.

4.   The linear system $A\mathbf{x} = \mathbf{b}$ given by

$$
\begin{bmatrix} 1 & 2 \\ 1.00001 & 2 \end{bmatrix} \begin{bmatrix} x_1 \\ x_2 \end{bmatrix} = \begin{bmatrix} 3 \\ 3.00001 \end{bmatrix}
$$

has the solution $(1, 1)^t$. Use seven-digit rounding arithmetic to find the solution of the perturbed system

$$
\begin{bmatrix} 1 & 2 \\ 1.000011 & 2 \end{bmatrix} \begin{bmatrix} x_1 \\ x_2 \end{bmatrix} = \begin{bmatrix} 3.00001 \\ 3.00003 \end{bmatrix},
$$

and compare the change in A and $\mathbf{b}$ to the change in $\mathbf{x}$.

5.   (i) Use Gaussian elimination and three-digit rounding arithmetic to approximate the solutions to the following linear systems. (ii) Then use one iteration of iterative refinement to improve the approximation, and compare the approximations to the actual solutions.

a. $0.03x_1 + 58.9x_2 = 59.2$
$5.31x_1 - 6.10x_2 = 47.0$
Actual solution $(10, 1)^t$.

b. $3.3330x_1 + 15920x_2 + 10.333x_3 = 7953$
$2.2220x_1 + 16.710x_2 + 9.6120x_3 = 0.965$
$-1.5611x_1 + 5.1792x_2 - 1.6855x_3 = 2.714$
Actual solution $(1, 0.5, -1)^t$.

c. $1.19x_1 + 2.11x_2 - 100x_3 + x_4 = 1.12$
$14.2x_1 - 0.122x_2 + 12.2x_3 - x_4 = 3.44$
$100x_2 - 99.9x_3 + x_4 = 2.15$
$15.3x_1 + 0.110x_2 - 13.1x_3 - x_4 = 4.16$
Actual solution $(0.17682530, 0.01269269, -0.02065405, -1.18260870)^t$.

d. $\pi x_1 - ex_2 + \sqrt{2}x_3 - \sqrt{3}x_4 = \sqrt{11}$
$\pi^2 x_1 + ex_2 - e^2 x_3 + \dfrac{3}{7}x_4 = 0$
$\sqrt{5}x_1 - \sqrt{6}x_2 + x_3 - \sqrt{2}x_4 = \pi$
$\pi^3 x_1 + e^2 x_2 - \sqrt{7}x_3 + \dfrac{1}{9}x_4 = \sqrt{2}$
Actual solution $(0.78839378, -3.12541367, 0.16759660, 4.55700252)^t$.

6.   Repeat Exercise 5 using four-digit rounding arithmetic.

7.   The $n \times n$ *Hilbert* matrix, $H^{(n)}$, defined by

$$H_{ij}^{(n)} = \frac{1}{i + j - 1}, \quad 1 \le i, j \le n$$

is an ill-conditioned matrix that arises when solving for the coefficients of least squares polynomials (see Section 8.3, page 331).

a.   Show that

$$[H^{(4)}]^{-1} = \begin{bmatrix} 16 & -120 & 240 & -140 \\ -120 & 1200 & -2700 & 1680 \\ 240 & -2700 & 6480 & -4200 \\ -140 & 1680 & -4200 & 2800 \end{bmatrix},$$

and compute $K_\infty(H^{(4)})$.

b.   Show that

$$[H^{(5)}]^{-1} = \begin{bmatrix} 25 & -300 & 1050 & -1400 & 630 \\ -300 & 4800 & -18900 & 26880 & -12600 \\ 1050 & -18900 & 79380 & -117600 & 56700 \\ -1400 & 26880 & -117600 & 179200 & -88200 \\ 630 & -12600 & 56700 & -88200 & 44100 \end{bmatrix}$$

and compute $K_\infty(H^{(5)})$.

c.   Solve the linear system

$$H^{(4)} \begin{bmatrix} x_1 \\ x_2 \\ x_3 \\ x_4 \end{bmatrix} = \begin{bmatrix} 1 \\ 0 \\ 0 \\ 1 \end{bmatrix}$$

using three-digit rounding arithmetic, and compare the actual error to the residual vector error bound.

8.   **a.**   Use four-digit rounding arithmetic to compute the inverse H^{-1} of the 3×3 Hilbert matrix H.

b.   Use four-digit rounding arithmetic to compute $\hat{H} = (H^{-1})^{-1}$.

c.   Determine $\|H - \hat{H}\|_\infty$.

7.7 The Conjugate Gradient Method

The conjugate gradient method of Hestenes and Stiefel [HS] was originally developed as a direct method designed to solve an $n \times n$ positive definite linear system. As a direct method it is generally inferior to Gaussian elimination with pivoting because both methods require n steps to determine a solution, and the steps of the conjugate gradient method are more computationally expensive than those in Gaussian elimination.

Magnus Hestenes (1906–1991) and Eduard Stiefel (1907–1998) published the original paper on the conjugate gradient method in 1952 while working at the Institute for Numerical Analysis on the campus of UCLA.

However, the conjugate gradient method is useful when employed as an iterative approximation method for solving large sparse systems with nonzero entries occurring in predictable patterns. These problems frequently arise in the solution of boundary-value problems, and too much computation is required for direct methods in these situations. When the matrix has been preconditioned to make the calculations more effective, good results are obtained in only about $\sqrt{n}$ steps. Employed in this way, the method is preferred over Gaussian elimination and the previously discussed iterative methods.

Throughout this section we assume that the matrix A is positive definite. We will use the *inner product* notation

$$\langle \mathbf{x}, \mathbf{y} \rangle = \mathbf{x}^t \mathbf{y}, \tag{7.4}$$

where $\mathbf{x}$ and $\mathbf{y}$ are n-dimensional vectors. We will also need some additional standard results from linear algebra. A review of this material is found in Section 9.2.

The next result follows easily from the properties of transposes (see Exercise 12).

Inner Product Properties

For any vectors $\mathbf{x}$, $\mathbf{y}$, and $\mathbf{z}$ and any real number α, we have

(i) $\langle \mathbf{x}, \mathbf{y} \rangle = \langle \mathbf{y}, \mathbf{x} \rangle$;

(ii) $\langle \alpha \mathbf{x}, \mathbf{y} \rangle = \langle \mathbf{x}, \alpha \mathbf{y} \rangle = \alpha \langle \mathbf{x}, \mathbf{y} \rangle$;

(iii) $\langle \mathbf{x} + \mathbf{z}, \mathbf{y} \rangle = \langle \mathbf{x}, \mathbf{y} \rangle + \langle \mathbf{z}, \mathbf{y} \rangle$;

(iv) $\langle \mathbf{x}, \mathbf{x} \rangle \geq 0$;

(v) $\langle \mathbf{x}, \mathbf{x} \rangle = 0$ if and only if $\mathbf{x} = \mathbf{0}$.

When A is positive definite, $\langle \mathbf{x}, A\mathbf{x} \rangle = \mathbf{x}^t A \mathbf{x} > 0$ unless $\mathbf{x} = \mathbf{0}$. Also, because A is symmetric, we have

$$\langle \mathbf{x}, A\mathbf{y} \rangle = \mathbf{x}^t A \mathbf{y} = \mathbf{x}^t A^t \mathbf{y} = (A\mathbf{x})^t \mathbf{y} = \langle A\mathbf{x}, \mathbf{y} \rangle. \tag{7.5}$$

The following result is a basic tool in the development of the conjugate gradient method.

Minimization Condition for Positive Definite Matrices

The vector $\mathbf{x}$ is a solution to the positive definite linear system $A\mathbf{x} = \mathbf{b}$ if and only if $\mathbf{x}$ minimizes

$$g(\mathbf{x}) = \langle \mathbf{x}, A\mathbf{x} \rangle - 2\langle \mathbf{x}, \mathbf{b} \rangle.$$

To show this result we fix the vectors $\mathbf{x}$ and $\mathbf{v}$ and consider the single-variable function

$$h(t) = g(\mathbf{x} + t\mathbf{v}).$$

It is not difficult to show that $h(t)$ is a quadratic polynomial with a positive t^2 coefficient, so its minimum occurs when $h'(t) = 0$. Solving for this value of t and substituting into $h(t)$ gives the result. (See Exercise 14.)

To begin the conjugate gradient method, we choose $\mathbf{x}^{(0)}$, an approximate solution to $A\mathbf{x} = \mathbf{b}$, and $\mathbf{v}^{(1)} \neq \mathbf{0}$, which gives a *search direction* in which to move away from $\mathbf{x}^{(0)}$ to improve the approximation. Let

$$\mathbf{r}^{(0)} = \mathbf{b} - A\mathbf{x}^{(0)}$$

be the residual vector associated with $\mathbf{x}^{(0)}$ and

$$t_1 = \frac{\langle \mathbf{v}^{(1)}, \mathbf{b} - A\mathbf{x}^{(0)} \rangle}{\langle \mathbf{v}^{(1)}, A\mathbf{v}^{(1)} \rangle} = \frac{\langle \mathbf{v}^{(1)}, \mathbf{r}^{(0)} \rangle}{\langle \mathbf{v}^{(1)}, A\mathbf{v}^{(1)} \rangle}.$$

If $\mathbf{r}^{(0)} \neq \mathbf{0}$ and if $t_1 \neq 0$, then

$$\mathbf{x}^{(1)} = \mathbf{x}^{(0)} + t_1 \mathbf{v}^{(1)}$$

gives a smaller value for g than $g(\mathbf{x}^{(0)})$. So it is presumably closer to the solution $\mathbf{x}$ than is $\mathbf{x}^{(0)}$. To continue for $k = 2, 3, \ldots$, choose a new search direction $\mathbf{v}^{(k)}$ and compute

$$t_k = \frac{\langle \mathbf{v}^{(k)}, \mathbf{b} - A\mathbf{x}^{(k-1)} \rangle}{\langle \mathbf{v}^{(k)}, A\mathbf{v}^{(k)} \rangle} = \frac{\langle \mathbf{v}^{(k)}, \mathbf{r}^{(k-1)} \rangle}{\langle \mathbf{v}^{(k)}, A\mathbf{v}^{(k)} \rangle} \quad \text{and let} \quad \mathbf{x}^{(k)} = \mathbf{x}^{(k-1)} + t_k \mathbf{v}^{(k)},$$

The objective is to select the next direction $\mathbf{v}^{(k+1)}$ so that the sequence of approximations $\{\mathbf{x}^{(k)}\}$ converges rapidly to $\mathbf{x}$.

To choose the search directions, we construct the set of nonzero direction vectors $\{\mathbf{v}^{(1)}, \ldots, \mathbf{v}^{(n)}\}$ that satisfy

$$\langle \mathbf{v}^{(i)}, A\mathbf{v}^{(j)} \rangle = 0, \quad \text{if} \quad i \neq j.$$

This is called an A-**orthogonality condition**, and the set of vectors $\{\mathbf{v}^{(1)}, \ldots, \mathbf{v}^{(n)}\}$ is said to be A-**orthogonal**. It is not difficult to show that a set of A-orthogonal vectors associated with the positive definite matrix A is linearly independent. (See Exercise 13(a).)

In Chapter 9 (see page 364) we will see that if $\{\mathbf{v}^{(1)}, \ldots, \mathbf{v}^{(n)}\}$ is a linearly independent set, then every vector in $\mathbb{R}^n$ can be written as a linear combination of vectors in $\{\mathbf{v}^{(1)}, \ldots, \mathbf{v}^{(n)}\}$.

The following result states that this choice of search directions gives convergence in at most n steps, so as a direct method it produces the exact solution, assuming that the arithmetic is exact.

A-Orthogonality Convergence

Let $\{\mathbf{v}^{(1)}, \ldots, \mathbf{v}^{(n)}\}$ be an A-orthogonal set of nonzero vectors associated with the positive definite matrix A, and let $\mathbf{x}^{(0)}$ be arbitrary. Define

$$t_k = \frac{\langle \mathbf{v}^{(k)}, \mathbf{b} - A\mathbf{x}^{(k-1)} \rangle}{\langle \mathbf{v}^{(k)}, A\mathbf{v}^{(k)} \rangle} = \frac{\langle \mathbf{v}^{(k)}, \mathbf{r}^{(k-1)} \rangle}{\langle \mathbf{v}^{(k)}, A\mathbf{v}^{(k)} \rangle} \quad \text{and} \quad \mathbf{x}^{(k)} = \mathbf{x}^{(k-1)} + t_k \mathbf{v}^{(k)},$$

for $k = 1, 2, \ldots, n$. Then, assuming exact arithmetic, $A\mathbf{x}^{(n)} = \mathbf{b}$.

Example 1 For the linear system

$$A\mathbf{x} = \begin{bmatrix} 4 & 3 & 0 \\ 3 & 4 & -1 \\ 0 & -1 & 4 \end{bmatrix} \begin{bmatrix} x_1 \\ x_2 \\ x_3 \end{bmatrix} = \begin{bmatrix} 24 \\ 30 \\ -24 \end{bmatrix},$$

let $\mathbf{v}^{(1)} = (1, 0, 0)^t$, $\mathbf{v}^{(2)} = (-3/4, 1, 0)^t$, and $\mathbf{v}^{(3)} = (-3/7, 4/7, 1)^t$.

(a) Show that $\{\mathbf{v}^{(1)}, \mathbf{v}^{(2)}, \mathbf{v}^{(3)}\}$ is an A-orthogonal set.

(b) Show that the procedure in the A-orthogonal convergence result gives the exact solution $\mathbf{x} = (3, 4, -5)^t$ after 3 iterations starting with $\mathbf{x}^{(0)} = (0, 0, 0)^t$.

Solution We established in Example 2 of Section 7.5 (see page 300) that A is positive definite.

(a) For the given vectors we have

$$\langle \mathbf{v}^{(1)}, A\mathbf{v}^{(2)} \rangle = \mathbf{v}^{(1)t} A\mathbf{v}^{(2)} = [1, 0, 0] \begin{bmatrix} 4 & 3 & 0 \\ 3 & 4 & -1 \\ 0 & -1 & 4 \end{bmatrix} \begin{bmatrix} -\frac{3}{4} \\ 1 \\ 0 \end{bmatrix} = [1, 0, 0] \begin{bmatrix} 0 \\ \frac{7}{4} \\ -1 \end{bmatrix} = 0,$$

$$\langle \mathbf{v}^{(1)}, A\mathbf{v}^{(3)} \rangle = [1, 0, 0] \begin{bmatrix} 4 & 3 & 0 \\ 3 & 4 & -1 \\ 0 & -1 & 4 \end{bmatrix} \begin{bmatrix} -\frac{3}{7} \\ \frac{4}{7} \\ 1 \end{bmatrix} = [1, 0, 0] \begin{bmatrix} 0 \\ 0 \\ \frac{24}{7} \end{bmatrix} = 0,$$

and

$$\langle \mathbf{v}^{(2)}, A\mathbf{v}^{(3)} \rangle = \left[-\frac{3}{4}, 1, 0\right] \begin{bmatrix} 4 & 3 & 0 \\ 3 & 4 & -1 \\ 0 & -1 & 4 \end{bmatrix} \begin{bmatrix} -\frac{3}{7} \\ \frac{4}{7} \\ 1 \end{bmatrix} = \left[-\frac{3}{4}, 1, 0\right] \begin{bmatrix} 0 \\ 0 \\ \frac{24}{7} \end{bmatrix} = 0.$$

So $\{\mathbf{v}^{(1)}, \mathbf{v}^{(2)}, \mathbf{v}^{(3)}\}$ is an A-orthogonal set.

(b) Applying the iterations with $\mathbf{x}^{(0)} = (0, 0, 0)^t$ and $\mathbf{b} = (24, 30, -24)^t$ gives

$$\mathbf{r}^{(0)} = \mathbf{b} - A\mathbf{x}^{(0)} = \mathbf{b} - \mathbf{0} = (24, 30, -24)^t,$$

so

$$\langle \mathbf{v}^{(1)}, \mathbf{r}^{(0)} \rangle = \mathbf{v}^{(1)t}\mathbf{r}^{(0)} = 24, \quad \langle \mathbf{v}^{(1)}, A\mathbf{v}^{(1)} \rangle = 4, \quad \text{and} \quad t_1 = \frac{24}{4} = 6.$$

Hence

$$\mathbf{x}^{(1)} = \mathbf{x}^{(0)} + t_1\mathbf{v}^{(1)} = (0, 0, 0)^t + 6(1, 0, 0)^t = (6, 0, 0)^t.$$

Continuing, we have

$$\mathbf{r}^{(1)} = \mathbf{b} - A\mathbf{x}^{(1)} = (0, 12, -24)^t; \quad t_2 = \frac{\langle \mathbf{v}^{(2)}, \mathbf{r}^{(1)} \rangle}{\langle \mathbf{v}^{(2)}, A\mathbf{v}^{(2)} \rangle} = \frac{12}{7/4} = \frac{48}{7};$$

$$\mathbf{x}^{(2)} = \mathbf{x}^{(1)} + t_2\mathbf{v}^{(2)} = (6, 0, 0)^t + \frac{48}{7}\left(-\frac{3}{4}, 1, 0\right)^t = \left(\frac{6}{7}, \frac{48}{7}, 0\right)^t;$$

$$\mathbf{r}^{(2)} = \mathbf{b} - A\mathbf{x}^{(2)} = \left(0, 0, -\frac{120}{7}\right); \quad t_3 = \frac{\langle \mathbf{v}^{(3)}, \mathbf{r}^{(2)} \rangle}{\langle \mathbf{v}^{(3)}, A\mathbf{v}^{(3)} \rangle} = \frac{-120/7}{24/7} = -5;$$

and

$$\mathbf{x}^{(3)} = \mathbf{x}^{(2)} + t_3\mathbf{v}^{(3)} = \left(\frac{6}{7}, \frac{48}{7}, 0\right)^t + (-5)\left(-\frac{3}{7}, \frac{4}{7}, 1\right)^t = (3, 4, -5)^t.$$

We applied the technique $n = 3$ times, so this must be (and is) the actual solution. ∎

We now need a way to find an appropriate set of A-orthogonal search directions $\{\mathbf{v}^{(1)}, \mathbf{v}^{(2)}, \ldots, \mathbf{v}^{(n)}\}$. The conjugate gradient method chooses the $\{\mathbf{v}^{(k)}\}$ iteratively, as outlined in the following result.

The Conjugate Gradient Method

Given a linear system $A\mathbf{x} = \mathbf{b}$, select an initial approximation $\mathbf{x}^{(0)}$ and let

$$\mathbf{r}^{(0)} = \mathbf{b} - A\mathbf{x}^{(0)} \quad \text{and} \quad \mathbf{v}^{(1)} = \mathbf{r}^{(0)}.$$

For each $k = 1, 2, \ldots, n - 1$ construct

$$t_k = \frac{\|\mathbf{r}^{(k-1)}\|_2^2}{\langle \mathbf{v}^{(k)}, A\mathbf{v}^{(k)} \rangle}, \qquad\qquad \mathbf{x}^{(k)} = \mathbf{x}^{(k-1)} + t_k \mathbf{v}^{(k)},$$

$$\mathbf{r}^{(k)} = \mathbf{r}^{(k-1)} - t_k A\mathbf{v}^{(k)}, \qquad \text{and} \quad \mathbf{v}^{(k+1)} = \mathbf{r}^{(k)} + \frac{\|\mathbf{r}^{(k)}\|_2^2}{\|\mathbf{r}^{(k-1)}\|_2^2} \mathbf{v}^{(k)}.$$

Then $\{\mathbf{v}^{(1)}, \mathbf{v}^{(2)}, \ldots, \mathbf{v}^{(n)}\}$ is an A-orthogonal set. Moreover, if

$$t_n = \frac{\langle \mathbf{r}^{(n-1)}, \mathbf{r}^{(n-1)} \rangle}{\langle \mathbf{v}^{(n)}, A\mathbf{v}^{(n)} \rangle} \quad \text{and} \quad \mathbf{x}^{(n)} = \mathbf{x}^{(n-1)} + t_n \mathbf{v}^{(n)},$$

then $\mathbf{x}^{(n)}$ is the exact solution to $A\mathbf{x} = \mathbf{b}$, assuming the use of exact arithmetic.

Example 2   The linear system $A\mathbf{x} = \mathbf{b}$ given by

$$
\begin{aligned}
4x_1 + 3x_2 \qquad\quad &= 24, \\
3x_1 + 4x_2 - \ x_3 &= 30, \\
- \ x_2 + 4x_3 &= -24
\end{aligned}
$$

has solution $(3, 4, -5)^t$. Use the conjugate gradient method with $\mathbf{x}^{(0)} = (0, 0, 0)^t$ to approximate the solution.

Solution   The problem was considered in Example 1 of Section 7.5 where the Gauss-Seidel and the SOR (with a nearly optimal value of $\omega = 1.25$) methods were used. The SOR method was superior and a solution to within 3.5×10^{-4} in the l_∞ norm was found in 7 iterations.

For the conjugate gradient method we start with

$$\mathbf{x}^{(0)} = (0, 0, 0)^t, \quad \mathbf{r}^{(0)} = \mathbf{b} - A\mathbf{x}^{(0)} = \mathbf{b} = (24, 30, -24)^t, \quad \text{and}$$

$$\mathbf{v}^{(1)} = \mathbf{r}^{(0)} = (24, 30, -24)^t.$$

For the first iteration we compute

$$t_1 = \frac{\|\mathbf{r}^{(0)}\|_2^2}{\langle \mathbf{v}^{(1)}, A\mathbf{v}^{(1)} \rangle} = 0.1469072165;$$

$$\mathbf{x}^{(1)} = \mathbf{x}^{(0)} + t_1 \mathbf{v}^{(1)} = (3.525773196, 4.407216495, -3.525773196)^t;$$

$$\mathbf{r}^{(1)} = \mathbf{r}^{(0)} - t_1 A\mathbf{v}^{(1)} = (-3.32474227, -1.73195876, -5.48969072)^t;$$

and

$$\mathbf{v}^{(2)} = \mathbf{r}^{(1)} + \frac{\|\mathbf{r}^{(1)}\|_2^2}{\|\mathbf{r}^{(0)}\|_2^2} \mathbf{v}^{(1)} = (-2.807896697, -1.085901793, -6.006536293)^t.$$

The second iteration gives

$$t_2 = \frac{\|\mathbf{r}^{(1)}\|_2^2}{\langle \mathbf{v}^{(2)}, A\mathbf{v}^{(2)} \rangle} = 0.2378157558;$$

$$\mathbf{x}^{(2)} = \mathbf{x}^{(1)} + t_2\mathbf{v}^{(2)} = (2.858011121, 4.148971939, -4.954222164)^t;$$

$$\mathbf{r}^{(2)} = \mathbf{r}^{(1)} - t_2 A\mathbf{v}^{(2)} = (0.121039698, -0.124143281, -0.034139402)^t;$$

and

$$\mathbf{v}^{(3)} = \mathbf{r}^{(2)} + \frac{\|\mathbf{r}^{(2)}\|_2^2}{\|\mathbf{r}^{(1)}\|_2^2} \mathbf{v}^{(2)} = (0.1190554504, -0.1249106480, -0.03838400086)^t.$$

To complete the process we compute

$$t_3 = \frac{\|\mathbf{r}^{(2)}\|_2^2}{\langle \mathbf{v}^{(3)}, A\mathbf{v}^{(3)} \rangle} = 1.192628008$$

and

$$\mathbf{x}^{(3)} = \mathbf{x}^{(2)} + t_3\mathbf{v}^{(3)} = (2.999999998, 4.000000002, -4.999999998)^t.$$

The residual for this approximation

$$\mathbf{r}^{(3)} = \mathbf{r}^{(2)} - t_3 A\mathbf{v}^{(3)} = (0.36 \times 10^{-8}, 0.39 \times 10^{-8}, -0.141 \times 10^{-8})^t$$

would be zero if exact arithmetic had been used. However, even with the round-off error, this approximation is accurate to within 2×10^{-9} in the l_∞ norm. ■

Preconditioned Conjugate Gradient Method

Preconditioning replaces a given system with one having an equivalent solution but with better convergence characteristics.

We will now extend the conjugate gradient method to include *preconditioning*. If the matrix A is ill-conditioned, the conjugate gradient method is highly susceptible to round-off error. So, although the exact answer should be obtained in n steps, this is not likely to be the case. As a direct method the conjugate gradient method is not superior to Gaussian elimination with pivoting. The primary use of the conjugate gradient method is as an iterative method applied to a better-conditioned system. In this case an acceptable approximate solution is often obtained in about $\sqrt{n}$ steps.

Instead of solving the system $A\mathbf{x} = \mathbf{b}$, we select a *preconditioning matrix* C^{-1} and solve the system $\tilde{A}\tilde{\mathbf{x}} = \tilde{\mathbf{b}}$, where

$$\tilde{A} = C^{-1}A(C^{-1})^t, \quad \tilde{\mathbf{b}} = C^{-1}\mathbf{b}, \quad \text{and} \quad \tilde{\mathbf{x}} = C^t\mathbf{x}.$$

The preconditioning matrix is chosen so that the condition number of $\tilde{A}$ is significantly less than the condition number of A, and as a consequence the solution $\tilde{\mathbf{x}}$ that is obtained is not as susceptible to round-off error. Once $\tilde{\mathbf{x}}$ is determined, $\mathbf{x}$ is given by

$$\mathbf{x} = (C^t)^{-1}\tilde{\mathbf{x}} = (C^{-1})^t\tilde{\mathbf{x}}.$$

The l_∞ condition number of the matrix in Example 2 is $\frac{32}{3}$, which is not particularly large, and the matrix is only 3×3, so only 3 iterations of the conjugate gradient method are needed to get the exact solution, as we did in that example. To better demonstrate the method, we will consider an ill-conditioned 5×5 matrix in the next Illustration.

Illustration   Consider the linear system $A\mathbf{x} = \mathbf{b}$, where A is the 5×5 matrix

$$A = \begin{bmatrix} 0.2 & 0.1 & 1 & 1 & 0 \\ 0.1 & 4 & -1 & 1 & -1 \\ 1 & -1 & 60 & 0 & -2 \\ 1 & 1 & 0 & 8 & 4 \\ 0 & -1 & -2 & 4 & 700 \end{bmatrix} \quad \text{and} \quad \mathbf{b} = \begin{bmatrix} 1 \\ 2 \\ 3 \\ 4 \\ 5 \end{bmatrix}.$$

The matrix A is positive definite, but if we use the MATLAB command

```
cond(A,Inf)
```

we find that the l_∞ condition number of A is

$$K_\infty(A) = 13961.7.$$

As a consequence, A is quite ill-conditioned. However, suppose we choose as the conditioning matrix the diagonal matrix whose diagonal entries are the square roots of the reciprocals of the diagonal entries in A. That is, we define

$$C^{-1} = \begin{bmatrix} \frac{1}{\sqrt{0.2}} & 0 & 0 & 0 & 0 \\ 0 & \frac{1}{\sqrt{4}} & 0 & 0 & 0 \\ 0 & 0 & \frac{1}{\sqrt{60}} & 0 & 0 \\ 0 & 0 & 0 & \frac{1}{\sqrt{8}} & 0 \\ 0 & 0 & 0 & 0 & \frac{1}{\sqrt{700}} \end{bmatrix} = \begin{bmatrix} \sqrt{5} & 0 & 0 & 0 & 0 \\ 0 & \frac{1}{2} & 0 & 0 & 0 \\ 0 & 0 & \frac{\sqrt{15}}{30} & 0 & 0 \\ 0 & 0 & 0 & \frac{\sqrt{2}}{4} & 0 \\ 0 & 0 & 0 & 0 & \frac{\sqrt{7}}{70} \end{bmatrix}.$$

Since the matrix C^{-1} is diagonal, it is symmetric and

$$\tilde{A} = C^{-1}A(C^{-1})^t = C^{-1}AC^{-1}.$$

The new matrix $\tilde{A}$ has 1s on the main diagonal.

The MATLAB command $\text{cond}(\tilde{A},\text{Inf})$ can be used to find that

$$K_\infty(\tilde{A}) = 16.1154,$$

so $\tilde{A}$ is much better conditioned than is A, and the system

$$\tilde{A}\tilde{\mathbf{x}} = \tilde{\mathbf{b}}, \quad \text{where} \quad \tilde{\mathbf{b}} = C^{-1}\mathbf{b}$$

is expected to be stable with respect to round-off error. Once the solution $\tilde{\mathbf{x}}$ to this system is determined the solution to $A\mathbf{x} = \mathbf{b}$ can be found as

$$\mathbf{x} = C\tilde{\mathbf{x}}$$

where C is the diagonal matrix whose diagonal elements are the reciprocals of the diagonal elements of C^{-1}.

The results from applying a number of methods we have in this chapter to this system are shown in Table 7.5. Notice that the Conjugate Gradient method used all 5 iterations, which in theory would give the exact answer. However, the Preconditioned Conjugate Gradient method obtained superior results in only 4 iterations.   □

Table 7.5

Method	Number of Iterations	$\mathbf{x}^{(k)}$	$\|\mathbf{x}^* - \mathbf{x}^{(k)}\|_\infty$
Jacobi	49	$(7.86277141, 0.42320802, -0.07348669,$ $-0.53975964, 0.01062847)^t$	0.00305834
Gauss-Seidel	15	$(7.83525748, 0.42257868, -0.07319124,$ $-0.53753055, 0.01060903)^t$	0.02445559
SOR ($\omega = 1.25$)	7	$(7.85152706, 0.42277371, -0.07348303,$ $-0.53978369, 0.01062286)^t$	0.00818607
Conjugate Gradient	5	$(7.85341523, 0.42298677, -0.07347963,$ $-0.53987920, 0.008628916)^t$	0.00629785
Preconditioned Conjugate Gradient	4	$(7.85968827, 0.42288329, -0.07359878,$ $-0.54063200, 0.01064344)^t$	0.00009312

> The program PCCGRD75 implements the Preconditioned Conjugate Gradient method.

The program PCCGRD75 implements the preconditioned conjugate gradient method in a more efficient manner than that we have described, one that does not require finding the solution to the preconditioned system directly.

The preconditioned conjugate gradient method is often used in the solution of large linear systems in which the matrix is sparse and positive definite. These systems must be solved to approximate solutions to boundary-value problems in ordinary-differential equations (see Sections 11.3, 11.4, and 11.5). The larger the system, the more impressive the conjugate gradient method becomes because it significantly reduces the number of iterations required. In these systems, the preconditioning matrix C is approximately equal to L in the Cholesky factorization LL^t of A. Generally, small entries in A are ignored and Cholesky's method is applied to obtain what is called an incomplete LL^t factorization of A.

MATLAB contains several iterative methods that are also based on Krylov subspaces. For example, the command x= pcg(A,b) executes the preconditioned conjugate gradient method to solve the linear system $A\mathbf{x} = \mathbf{b}$. Some optional input parameters for pcg are TOL, a tolerance for convergence, MAXIT, the maximum number of iterations, and M, a preconditioner.

We have only touched on the details of the conjugate gradient method in order to give an indication of how it works and its importance. A good source for additional information is [Kelley].

EXERCISE SET 7.7

1. The linear system

$$x_1 + \frac{1}{2}x_2 = \frac{5}{21},$$

$$\frac{1}{2}x_1 + \frac{1}{3}x_2 = \frac{11}{84}.$$

has solution $(x_1, x_2)^t = (1/6, 1/7)^t$.

a. Solve the linear system using Gaussian elimination with two-digit rounding arithmetic.

b. Solve the linear system using the conjugate gradient method ($C = C^{-1} = I$) with two-digit rounding arithmetic.

c. Which method gives the better answer?

d. Choose $C^{-1} = D^{-1/2}$. Does this choice improve the conjugate gradient method?

2. The linear system

$$0.1x_1 + 0.2x_2 = 0.3,$$
$$0.2x_1 + 113x_2 = 113.2$$

has solution $(x_1, x_2)^t = (1, 1)^t$. Repeat the directions for Exercise 1 on this linear system.

3. The linear system

$$x_1 + \frac{1}{2}x_2 + \frac{1}{3}x_3 = \frac{5}{6},$$

$$\frac{1}{2}x_1 + \frac{1}{3}x_2 + \frac{1}{4}x_3 = \frac{5}{12},$$

$$\frac{1}{3}x_1 + \frac{1}{4}x_2 + \frac{1}{5}x_3 = \frac{17}{60}$$

has solution $(1, -1, 1)^t$.

a. Solve the linear system using Gaussian elimination with three-digit rounding arithmetic.

b. Solve the linear system using the conjugate gradient method with three-digit rounding arithmetic.

c. Does pivoting improve the answer in (a)?

d. Repeat (b) using $C^{-1} = D^{-1/2}$. Does this improve the answer in (b)?

4. Repeat Exercise 3 using single-precision arithmetic on a computer.

5. Use the program PCCGRD75 to perform two steps of the conjugate gradient method with $C = C^{-1} = I$ on each of the following linear systems. Compare the results to those obtained in Exercises 1 and 2 of Section 7.4 and Exercise 1 of Section 7.5.

a.
$$3x_1 - x_2 + x_3 = 1,$$
$$-x_1 + 6x_2 + 2x_3 = 0,$$
$$x_1 + 2x_2 + 7x_3 = 4.$$

b.
$$10x_1 - x_2 = 9,$$
$$-x_1 + 10x_2 - 2x_3 = 7,$$
$$- 2x_2 + 10x_3 = 6.$$

c.
$$10x_1 + 5x_2 = 6,$$
$$5x_1 + 10x_2 - 4x_3 = 25,$$
$$- 4x_2 + 8x_3 - x_4 = -11,$$
$$- x_3 + 5x_4 = -11.$$

d.
$$4x_1 + x_2 - x_3 + x_4 = -2,$$
$$x_1 + 4x_2 - x_3 - x_4 = -1,$$
$$-x_1 - x_2 + 5x_3 + x_4 = 0,$$
$$x_1 - x_2 + x_3 + 3x_4 = 1.$$

e.
$$4x_1 + x_2 + x_3 + x_5 = 6,$$
$$x_1 + 3x_2 + x_3 + x_4 = 6,$$
$$x_1 + x_2 + 5x_3 - x_4 - x_5 = 6,$$
$$x_2 - x_3 + 4x_4 = 6,$$
$$x_1 - x_3 + + 4x_5 = 6.$$

f.
$$4x_1 - x_2 = 0,$$
$$-x_1 + 4x_2 - x_3 = 5,$$
$$-x_2 + 4x_3 = 0,$$
$$4x_4 - x_5 = 6,$$
$$- x_4 + 4x_5 - x_6 = -2,$$
$$- x_5 + 4x_6 = 6.$$

6. Repeat Exercise 5 using $C^{-1} = D^{-1/2}$.

7. Use the program PCCGRD75 with $TOL = 10^{-3}$ in the l_∞ norm and $C = I$ for the systems in Exercise 5. Compare the results to those obtained in Exercises 3 and 4 of Section 7.4 and Exercise 2 of Section 7.5.

8. Repeat Exercise 7 using $C^{-1} = D^{-1/2}$.

9. Use (i) the Jacobi Method, (ii) the Gauss-Seidel method, (iii) the SOR method with $\omega = 1.3$, and (iv) the conjugate gradient method and preconditioning with $C^{-1} = D^{-1/2}$ to find solutions to the linear system $A\mathbf{x} = \mathbf{b}$ to within 10^{-5} in the l_∞ norm.

a.

$$a_{i,j} = \begin{cases} 4, & \text{when } j = i \text{ and } i = 1, 2, \ldots, 16, \\ -1, & \text{when } \begin{cases} j = i+1 \text{ and } i = 1, 2, 3, 5, 6, 7, 9, 10, 11, 13, 14, 15, \\ j = i-1 \text{ and } i = 2, 3, 4, 6, 7, 8, 10, 11, 12, 14, 15, 16, \\ j = i+4 \text{ and } i = 1, 2, \ldots, 12, \\ j = i-4 \text{ and } i = 5, 6, \ldots, 16, \end{cases} \\ 0, & \text{otherwise} \end{cases}$$

and

$$\mathbf{b} = (1.902207, 1.051143, 1.175689, 3.480083, 0.819600, -0.264419,$$
$$-0.412789, 1.175689, 0.913337, -0.150209, -0.264419, 1.051143,$$
$$1.966694, 0.913337, 0.819600, 1.902207)^t$$

b.

$$a_{i,j} = \begin{cases} 4, & \text{when } j = i \text{ and } i = 1, 2, \ldots, 25, \\ -1, & \text{when } \begin{cases} j = i+1 \text{ and } i = \begin{cases} 1, 2, 3, 4, 6, 7, 8, 9, 11, 12, 13, 14, \\ 16, 17, 18, 19, 21, 22, 23, 24, \end{cases} \\ j = i-1 \text{ and } i = \begin{cases} 2, 3, 4, 5, 7, 8, 9, 10, 12, 13, 14, 15, \\ 17, 18, 19, 20, 22, 23, 24, 25, \end{cases} \\ j = i+5 \text{ and } i = 1, 2, \ldots, 20, \\ j = i-5 \text{ and } i = 6, 7, \ldots, 25, \end{cases} \\ 0, & \text{otherwise} \end{cases}$$

and

$$\mathbf{b} = (1, 0, -1, 0, 2, 1, 0, -1, 0, 2, 1, 0, -1, 0, 2, 1, 0, -1, 0, 2, 1, 0, -1, 0, 2)^t$$

c.

$$a_{i,j} = \begin{cases} 2i, & \text{when } j = i \text{ and } i = 1, 2, \ldots, 40, \\ -1, & \text{when } \begin{cases} j = i+1 \text{ and } i = 1, 2, \ldots, 39, \\ j = i-1 \text{ and } i = 2, 3, \ldots, 40, \end{cases} \\ 0, & \text{otherwise} \end{cases}$$

and $b_i = 1.5i - 6$, for each $i = 1, 2, \ldots, 40$

10. Solve the linear system in Exercise 4(b) of Section 7.5 using the conjugate gradient method with $C^{-1} = I$.

11. Let

$$A_1 = \begin{bmatrix} 4 & -1 & 0 & 0 \\ -1 & 4 & -1 & 0 \\ 0 & -1 & 4 & -1 \\ 0 & 0 & -1 & 4 \end{bmatrix}, \quad -I = \begin{bmatrix} -1 & 0 & 0 & 0 \\ 0 & -1 & 0 & 0 \\ 0 & 0 & -1 & 0 \\ 0 & 0 & 0 & -1 \end{bmatrix},$$

$$\text{and} \quad O = \begin{bmatrix} 0 & 0 & 0 & 0 \\ 0 & 0 & 0 & 0 \\ 0 & 0 & 0 & 0 \\ 0 & 0 & 0 & 0 \end{bmatrix}.$$

Form the 16×16 matrix A in partitioned form,

$$A = \begin{bmatrix} A_1 & -I & O & O \\ -I & A_1 & -I & O \\ O & -I & A_1 & -I \\ O & O & -I & A_1 \end{bmatrix}.$$

Let $\mathbf{b} = (1, 2, 3, 4, 5, 6, 7, 8, 9, 0, 1, 2, 3, 4, 5, 6)^t$.

a. Solve $A\mathbf{x} = \mathbf{b}$ using the conjugate gradient method with tolerance 0.05.

b. Solve $A\mathbf{x} = \mathbf{b}$ using the preconditioned conjugate gradient method with $C^{-1} = D^{-1/2}$ and tolerance 0.05.

c. Is there any tolerance for which the methods of (a) and (b) require a different number of iterations?

12. Use the transpose properties given in Section 6.4 to show the Inner Product Properties given on the opening page of the section.

13. a. Show that an A-orthogonal set of nonzero vectors associated with a positive definite matrix is linearly independent.

b. Show that if $\{\mathbf{v}^{(1)}, \mathbf{v}^{(2)}, \dots, \mathbf{v}^{(n)}\}$ is a set of A-orthogonal nonzero vectors in $\mathbb{R}$ and $\mathbf{z}^t \mathbf{v}^{(i)} = \mathbf{0}$, for each $i = 1, 2, \dots, n$, then $\mathbf{z} = \mathbf{0}$.

14. Demonstrate the result in the Minimization Condition for Positive Definite Systems on page 309 by showing the following.

a. Show that $h(t) = g(\mathbf{x} + t\mathbf{v}) = g(\mathbf{x}) + 2t \langle \mathbf{v}, A\mathbf{x} - \mathbf{b} \rangle + t^2 \langle \mathbf{v}, A\mathbf{v} \rangle$.

b. Show that $h'(\hat{t}) = 0$ if and only if $\hat{t} = \frac{\langle \mathbf{v}, \mathbf{b} - A\mathbf{x} \rangle}{\langle \mathbf{v}, A\mathbf{v} \rangle}$.

c. Show that $h(\hat{t}) = g(\mathbf{x}) - \frac{\langle \mathbf{v}, \mathbf{b} - A\mathbf{x} \rangle^2}{\langle \mathbf{v}, A\mathbf{v} \rangle}$.

d. Show that the result follows from (c).

7.8 Survey of Methods and Software

In this chapter we have studied iterative techniques to approximate the solution of linear systems. We began with the Jacobi method and the Gauss-Seidel method to introduce the iterative methods. Both methods require an arbitrary initial approximation $\mathbf{x}^{(0)}$ and generate a sequence of vectors $\mathbf{x}^{(i+1)}$ using an equation of the form

$$\mathbf{x}^{(i+1)} = T\mathbf{x}^{(i)} + \mathbf{c}.$$

The method will converge if and only if the spectral radius of the iteration matrix $\rho(T) < 1$, and the smaller the spectral radius, the faster the convergence. Analysis of the residual vectors of the Gauss-Seidel technique led to the SOR iterative method, which involves a parameter ω to speed convergence. The preconditioned conjugate gradient method was introduced in Section 7.7.

These iterative methods and modifications are used extensively in the solution of linear systems that arise in the numerical solution of boundary value problems and partial differential equations (see Chapters 11 and 12). These systems are often very large, on the order of 10,000 equations in 10,000 unknowns, and are sparse with their nonzero entries in predictable positions. The iterative methods are also useful for other large sparse systems and are easily adapted for efficient use on parallel computers.

Almost all commercial and public domain packages that contain iterative methods for the solution of a linear system of equations require a preconditioner to be used with the method. Faster convergence of iterative solvers is often achieved by using a preconditioner. A preconditioner produces an equivalent system of equations that hopefully exhibits better convergence characteristics than the original system. The IMSL Library has a preconditioned conjugate gradient method. The NAG Library has several subroutines for the iterative solution of linear systems. All of the subroutines are based on Krylov subspaces. Saad [Sa2] has a detailed description of Krylov subspace methods. The packages LINPACK and LAPACK contain only direct methods for the solution of linear systems; however, the packages do contain many subroutines that are used by the iterative solvers. The public domain packages IML++, ITPACK, and SLAP contain iterative methods.

The concepts of condition number and poorly conditioned matrices were introduced in Section 7.6. Many of the subroutines for solving a linear system or for factoring a matrix into an LU factorization include checks for ill-conditioned matrices and also give an estimate of the condition number. LAPACK, LINPACK, the IMSL Library, and the NAG Library have subroutines that improve on a solution to a linear system that is poorly conditioned. The subroutines test the condition number and then use iterative refinement to obtain the most accurate solution possible given the precision of the computer.

More information on the use of iterative methods for solving linear systems can be found in Varga [Var], Young [Y], Hageman and Young [HY], and Axelsson [Ax]. Iterative methods for large sparse systems are discussed in Barrett et al, [Barr], Hackbusch [Hac], Kelley [Kelley], and Saad [Sa2].

Aleksei Nikolaevich Krylov (1863–1945) worked in applied mathematics, primarily in the areas of boundary value problems, the acceleration of convergence of Fourier series, and various classical problems involving mechanical systems. During the early 1930s he was the director of the Physics-Mathematics Institute of the Soviet Academy of Sciences.

CHAPTER

8

Approximation Theory

8.1 Introduction

Approximation theory involves two types of problems. One arises when a function is given explicitly, but we wish to find a "simpler" type of function, such as a polynomial, for representation. The other problem concerns fitting functions to given data and finding the "best" function in a certain class that can be used to represent the data. We will begin the chapter with this problem.

8.2 Discrete Least Squares Approximation

Consider the problem of estimating the values of a function at non-tabulated points, given the experimental data in Table 8.1.

Table 8.1

x_i	y_i	x_i	y_i
1	1.3	6	8.8
2	3.5	7	10.1
3	4.2	8	12.5
4	5.0	9	13.0
5	7.0	10	15.6

The interpolation in Chapter 3 requires a function that assumes the value of y_i at x_i for each $i = 1, 2, \ldots, 10$. Figure 8.1 shows a graph of the values in Table 8.1. From this graph, it appears that the actual relationship between x and y is linear. However, it is likely that no line precisely fits the data, because of errors in the data. In this case, it is unreasonable to require that the approximating function agree exactly with the given data. In fact, such a function would introduce oscillations that should not be present. For example, the graph of the ninth-degree interpolating polynomial for the data is shown in Figure 8.2. This polynomial is clearly a poor predictor of information between a number of the data points.

Figure 8.1

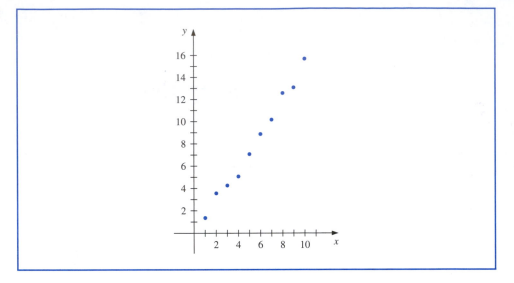

Figure 8.2

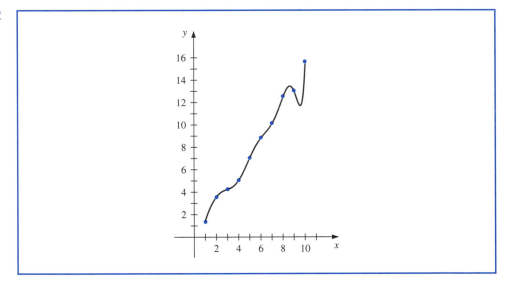

A better approach for a problem of this type would be to find the "best" (in some sense) approximating line, even if it did not agree precisely with the data at the points.

Let $a_1 x_i + a_0$ denote the ith value on the approximating line and y_i the ith given y-value. Then, assuming that there are no errors in the x-values in the data,

$$|y_i - (a_1 x_i + a_0)|$$

gives a measure for the error at the ith point when we use this approximating line. The problem of finding the equation of the best linear approximation in the absolute sense requires that values of a_0 and a_1 be found to minimize

$$E_\infty(a_0, a_1) = \max_{1 \leq i \leq 10} \{|y_i - (a_1 x_i + a_0)|\}.$$

This is commonly called a **minimax** problem and cannot be handled by elementary techniques. Another approach to determining the best linear approximation involves finding values of a_0 and a_1 to minimize

$$E_1(a_0, a_1) = \sum_{i=1}^{10} |y_i - (a_1 x_i + a_0)|.$$

This quantity is called the **absolute deviation**. To minimize a function of two variables, we need to set its partial derivatives to zero and simultaneously solve the resulting equations. In the case of the absolute deviation, we would need to find a_0 and a_1 with

$$0 = \frac{\partial}{\partial a_0} \sum_{i=1}^{10} |y_i - (a_1 x_i + a_0)| \quad \text{and} \quad 0 = \frac{\partial}{\partial a_1} \sum_{i=1}^{10} |y_i - (a_1 x_i + a_0)|.$$

The difficulty with this procedure is that the absolute-value function is not differentiable at zero; the derivative is 1 when the argument is positive and -1 when the argument is negative. So solutions to this pair of equations cannot necessarily be obtained.

Linear Least Squares

The **least squares** approach to this problem involves determining the best approximating line when the error involved is the sum of the squares of the differences between the y-values on the approximating line and the given y-values. Hence, constants a_0 and a_1 must be found that minimize the *total least squares error*:

$$E_2(a_0, a_1) = \sum_{i=1}^{10} |y_i - (a_1 x_i + a_0)|^2 = \sum_{i=1}^{10} (y_i - (a_1 x_i + a_0))^2.$$

The least squares method is the most convenient procedure for determining best linear approximations, and there are also important theoretical considerations that favor this method. The minimax approach generally assigns too much weight to a bit of data that is badly in error, whereas the absolute deviation method does not give sufficient weight to a point that is badly out of line. The least squares approach puts substantially more weight on a point that is out of line with the rest of the data but will not allow that point to dominate the approximation.

The general problem of fitting the best least squares line to a collection of data $\{(x_i, y_i)\}_{i=1}^{m}$ involves minimizing the total error

$$E_2(a_0, a_1) = \sum_{i=1}^{m} (y_i - (a_1 x_i + a_0))^2$$

with respect to the parameters a_0 and a_1. For a minimum to occur, we need

$$0 = \frac{\partial}{\partial a_0} \left(\sum_{i=1}^{m} (y_i - (a_1 x_i + a_0))^2 \right) = 2 \sum_{i=1}^{m} (y_i - a_1 x_i - a_0)(-1)$$

and

$$0 = \frac{\partial}{\partial a_1} \left(\sum_{i=1}^{m} (y_i - (a_1 x_i + a_0))^2 \right) = 2 \sum_{i=1}^{m} (y_i - a_1 x_i - a_0)(-x_i).$$

The word normal as used here implies perpendicular. The normal equations are obtained by finding perpendicular directions to a multidimensional surface.

These equations simplify to the **normal equations**

$$a_0 \sum_{i=1}^{m} x_i + a_1 \sum_{i=1}^{m} x_i^2 = \sum_{i=1}^{m} x_i y_i \quad \text{and} \quad a_0 \cdot m + a_1 \sum_{i=1}^{m} x_i = \sum_{i=1}^{m} y_i.$$

The solution to this system is as follows.

Linear Least Squares

The linear least squares solution for a given collection of data $\{(x_i, y_i)\}_{i=1}^{m}$ has the form $y = a_1 x + a_0$, where

$$a_0 = \frac{\left(\sum_{i=1}^{m} x_i^2\right)\left(\sum_{i=1}^{m} y_i\right) - \left(\sum_{i=1}^{m} x_i y_i\right)\left(\sum_{i=1}^{m} x_i\right)}{m\left(\sum_{i=1}^{m} x_i^2\right) - \left(\sum_{i=1}^{m} x_i\right)^2}$$

and

$$a_1 = \frac{m\left(\sum_{i=1}^{m} x_i y_i\right) - \left(\sum_{i=1}^{m} x_i\right)\left(\sum_{i=1}^{m} y_i\right)}{m\left(\sum_{i=1}^{m} x_i^2\right) - \left(\sum_{i=1}^{m} x_i\right)^2}.$$

Example 1 Find the least squares line approximating the data in Table 8.1.

Solution We first extend the table to include the terms x_i^2 and $x_i y_i$ that we will need for the solution, and sum the columns. This is shown in the first four columns of Table 8.2.

Table 8.2

x_i	y_i	x_i^2	$x_i y_i$	$P(x_i) = 1.538 x_i - 0.360$
1	1.3	1	1.3	1.18
2	3.5	4	7.0	2.72
3	4.2	9	12.6	4.25
4	5.0	16	20.0	5.79
5	7.0	25	35.0	7.33
6	8.8	36	52.8	8.87
7	10.1	49	70.7	10.41
8	12.5	64	100.0	11.94
9	13.0	81	117.0	13.48
10	15.6	100	156.0	15.02
55	81.0	385	572.4	$E_2 = \sum_{i=1}^{10}(y_i - P(x_i))^2 \approx 2.34$

For these data we have

$$a_0 = \frac{385(81) - 55(572.4)}{10(385) - (55)^2} = -0.360 \quad \text{and} \quad a_1 = \frac{10(572.4) - 55(81)}{10(385) - (55)^2} = 1.538,$$

so $P(x) = 1.538x - 0.360$. The graph of this line and the data points are shown in Figure 8.3. The approximate values given by the least squares technique at the data points are in Table 8.2. ▪

Figure 8.3

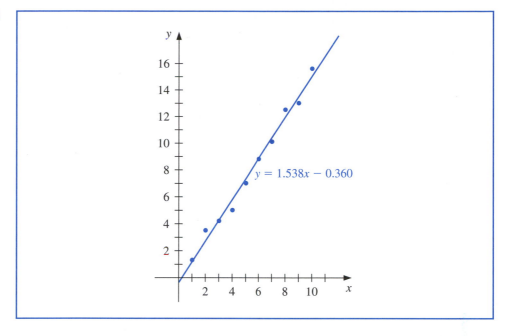

Polynomial Least Squares

The problem of approximating a set of data, $\{(x_i, y_i) \mid i = 1, 2, \ldots, m\}$, with a general algebraic polynomial

$$P_n(x) = a_n x^n + a_{n-1} x^{n-1} + \cdots + a_1 x + a_0$$

of degree $n < m - 1$ using least squares is handled in a similar manner. It requires choosing the constants $a_0, a_1, \ldots, a_n$ to minimize the *total least squares error*:

$$E_2 = \sum_{i=1}^{m} (y_i - P_n(x_i))^2.$$

For E_2 to be minimized, we need $\partial E_2 / \partial a_j = 0$ for each $j = 0, 1, \ldots, n$. This gives $n + 1$ **normal equations** in the $n + 1$ unknowns, a_j,

$$a_0 \sum_{i=1}^{m} x_i^0 + a_1 \sum_{i=1}^{m} x_i^1 + a_2 \sum_{i=1}^{m} x_i^2 + \cdots + a_n \sum_{i=1}^{m} x_i^n = \sum_{i=1}^{m} y_i x_i^0,$$

$$a_0 \sum_{i=1}^{m} x_i^1 + a_1 \sum_{i=1}^{m} x_i^2 + a_2 \sum_{i=1}^{m} x_i^3 + \cdots + a_n \sum_{i=1}^{m} x_i^{n+1} = \sum_{i=1}^{m} y_i x_i^1,$$

$$\vdots$$

$$a_0 \sum_{i=1}^{m} x_i^n + a_1 \sum_{i=1}^{m} x_i^{n+1} + a_2 \sum_{i=1}^{m} x_i^{n+2} + \cdots + a_n \sum_{i=1}^{m} x_i^{2n} = \sum_{i=1}^{m} y_i x_i^n.$$

The normal equations will have a unique solution provided that the x_i are distinct.

Example 2 Fit the data in Table 8.3 with the discrete least squares polynomial of degree at most 2.

Solution For this problem, $n = 2, m = 5$, and the three normal equations are

$$
\begin{aligned}
5a_0 + 2.5a_1 + 1.875a_2 &= 8.7680, \\
2.5a_0 + 1.875a_1 + 1.5625a_2 &= 5.4514, \\
1.875a_0 + 1.5625a_1 + 1.3828a_2 &= 4.4015.
\end{aligned}
$$

Table 8.3

i	x_i	y_i
1	0	1.0000
2	0.25	1.2840
3	0.50	1.6487
4	0.75	2.1170
5	1.00	2.7183

To solve this system using MATLAB, first define the matrix A and the vector **b** for the right side of the system. Since **b** is a column vector, its entries are separated by semicolons.

```
A = [5 2.5 1.875; 2.5 1.875 1.5625; 1.875 1.5625 1.3828]
b = [8.7680; 5.4514; 4.4015]
```

We can then use the `linsolve` command to obtain the coefficients of the discrete least squares polynomial of degree 2.

```
a = linsolve(A,b)
```

which produces

$$
\begin{aligned}
a = 1.005075518975769 \\
0.864675848193855 \\
0.843164151806143
\end{aligned}
$$

implying that the least squares polynomial of degree 2 fitting the data in Table 8.3 is approximately

$$P_2(x) = 1.0051 + 0.86468x + 0.84316x^2,$$

whose graph is shown in Figure 8.4.

Figure 8.4

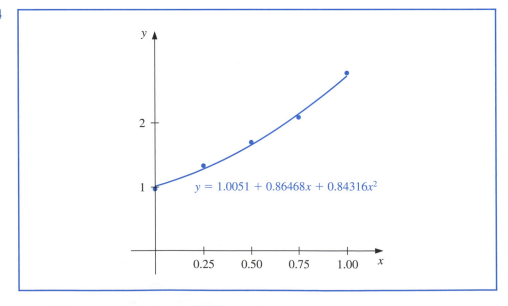

$y = 1.0051 + 0.86468x + 0.84316x^2$

The polynomial can also be computed in MATLAB using the `polyfit` command. First define the x values and y values by

```
x = [0 0.25 0.50 0.75 1.00];
y = [1.0000 1.2840 1.6487 2.1170 2.7183];
```

and then use the `polyfit` command:

```
p = polyfit(x,y,2)
```

At the given values of x_i we have the approximations shown in Table 8.4.

Table 8.4

i	1	2	3	4	5
x_i	0	0.25	0.50	0.75	1.00
y_i	1.0000	1.2840	1.6487	2.1170	2.7183
$P(x_i)$	1.0051	1.2740	1.6482	2.1279	2.7129
$y_i - P(x_i)$	−0.0051	0.0100	0.0004	−0.0109	0.0054

The total error,

$$E_2 = \sum_{i=1}^{5}(y_i - P(x_i))^2 = 2.74 \times 10^{-4},$$

is the least that can be obtained by using a polynomial of degree at most 2. ∎

EXERCISE SET 8.2

1. Compute the linear least squares polynomial for the data of Example 2.

2. Compute the least squares polynomial of degree 2 for the data of Example 1 and compare the total error E_2 for the two polynomials.

3. Find the least squares polynomials of degrees 1, 2, and 3 for the data in the following table. Compute the error E_2 in each case. Graph the data and the polynomials.

x_i	1.0	1.1	1.3	1.5	1.9	2.1
y_i	1.84	1.96	2.21	2.45	2.94	3.18

4. Find the least squares polynomials of degrees 1, 2, and 3 for the data in the following table. Compute the error E_2 in each case. Graph the data and the polynomials.

x_i	0	0.15	0.31	0.5	0.6	0.75
y_i	1.0	1.004	1.031	1.117	1.223	1.422

5. Given the following data

x_i	4.0	4.2	4.5	4.7	5.1	5.5	5.9	6.3	6.8	7.1
y_i	102.56	113.18	130.11	142.05	167.53	195.14	224.87	256.73	299.50	326.72

    a.   Construct the least squares polynomial of degree 1 and compute the error.

    b.   Construct the least squares polynomial of degree 2 and compute the error.

    c.   Construct the least squares polynomial of degree 3 and compute the error.

6.   Repeat Exercise 5 for the following data.

x_i	0.2	0.3	0.6	0.9	1.1	1.3	1.4	1.6
y_i	0.050446	0.098426	0.33277	0.72660	1.0972	1.5697	1.8487	2.5015

7.   Hooke's law states that when a force is applied to a spring constructed of uniform material, the length of the spring is a linear function of the force that is applied, as shown in the accompanying figure.

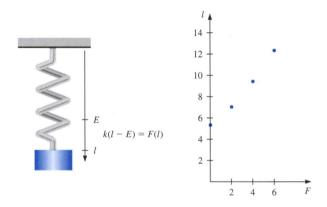

    a.   Suppose that $E = 5.3$ in. and that measurements are made of the length l in inches for applied weights $F(l)$ in pounds, as given in the following table. Find the least squares approximation for k.

$F(l)$	l
2	7.0
4	9.4
6	12.3

    b.   Additional measurements are made, giving the following additional data. Use these data to compute a new least squares approximation for k. Which of (a) or (b) best fits the total experimental data?

$F(l)$	l
3	8.3
5	11.3
8	14.4
10	15.9

8.   To determine a relationship between the number of fish and the number of species of fish in samples taken for a portion of the Great Barrier Reef, P. Sale and R. Dybdahl [SD] fit a linear least squares polynomial to the following collection of data, which were collected in samples over a 2-year period.

Let x be the number of fish in the sample and y be the number of species in the sample and determine the linear least squares polynomial for these data.

x	y	x	y	x	y
13	11	29	12	60	14
15	10	30	14	62	21
16	11	31	16	64	21
21	12	36	17	70	24
22	12	40	13	72	17
23	13	42	14	100	23
25	13	55	22	130	34

9. The following table lists the college grade-point averages of 20 mathematics and computer science majors, together with the scores that these students received on the mathematics portion of the ACT (American College Testing Program) test while in high school. Plot these data, and find the equation of the least squares line for this data. Do you think that the ACT scores are a reasonable predictor of college grade-point averages?

ACT Score	Grade-Point Average	ACT Score	Grade-Point Average
28	3.84	29	3.75
25	3.21	28	3.65
28	3.23	27	3.87
27	3.63	29	3.75
28	3.75	21	1.66
33	3.20	28	3.12
28	3.41	28	2.96
29	3.38	26	2.92
23	3.53	30	3.10
27	2.03	24	2.81

8.3 Continuous Least Squares Approximation

In this section we consider ways to construct a polynomial to approximate a given continuous function. Suppose $f \in C[a, b]$ and we want a polynomial of degree at most n,

$$P_n(x) = a_n x^n + a_{n-1} x^{n-1} + \cdots + a_1 x + a_0 = \sum_{k=0}^{n} a_k x^k,$$

to minimize the error

$$E(a_0, a_1, \ldots, a_n) = \int_a^b (f(x) - P_n(x))^2 \, dx = \int_a^b \left(f(x) - \sum_{k=0}^{n} a_k x^k \right)^2 dx.$$

A necessary condition for the numbers $a_0, a_1, \ldots, a_n$ to minimize the total error E is that at $(a_0, a_1, \ldots, a_n)$ we have

$$\frac{\partial E}{\partial a_j} = 0 \quad \text{for each } j = 0, 1, \ldots, n.$$

We can expand the integrand in this expression to

$$E = \int_a^b (f(x))^2 \, dx - 2 \sum_{k=0}^n a_k \int_a^b x^k f(x) \, dx + \int_a^b \left(\sum_{k=0}^n a_k x^k \right)^2 dx,$$

so

$$\frac{\partial E}{\partial a_j} = -2 \int_a^b x^j f(x) \, dx + 2 \sum_{k=0}^n a_k \int_a^b x^{j+k} \, dx.$$

for each $j = 0, 1, \ldots, n$. Setting these to zero and rearranging gives the $(n+1)$ linear **normal equations**

$$\sum_{k=0}^n a_k \int_a^b x^{j+k} \, dx = \int_a^b x^j f(x) \, dx, \quad \text{for each } j = 0, 1, \ldots, n.$$

These must be solved for the $n+1$ unknowns $a_0, a_1, \ldots, a_n$. The normal equations have a unique solution provided that $f \in C[a, b]$.

Example 1   Find the least squares approximating polynomial of degree 2 for the function $f(x) = \sin \pi x$ on the interval $[0, 1]$.

Solution   The normal equations for $P_2(x) = a_2 x^2 + a_1 x + a_0$ are

$$a_0 \int_0^1 1 \, dx + a_1 \int_0^1 x \, dx + a_2 \int_0^1 x^2 \, dx = \int_0^1 \sin \pi x \, dx,$$

$$a_0 \int_0^1 x \, dx + a_1 \int_0^1 x^2 \, dx + a_2 \int_0^1 x^3 \, dx = \int_0^1 x \sin \pi x \, dx,$$

$$a_0 \int_0^1 x^2 \, dx + a_1 \int_0^1 x^3 \, dx + a_2 \int_0^1 x^4 \, dx = \int_0^1 x^2 \sin \pi x \, dx.$$

Performing the integration yields

$$a_0 + \frac{1}{2} a_1 + \frac{1}{3} a_2 = \frac{2}{\pi}, \quad \frac{1}{2} a_0 + \frac{1}{3} a_1 + \frac{1}{4} a_2 = \frac{1}{\pi}, \quad \text{and} \quad \frac{1}{3} a_0 + \frac{1}{4} a_1 + \frac{1}{5} a_2 = \frac{\pi^2 - 4}{\pi^3}.$$

These three equations in three unknowns have the unique solution

$$a_0 = \frac{12\pi^2 - 120}{\pi^3} \approx -0.050465 \quad \text{and} \quad a_1 = -a_2 = \frac{720 - 60\pi^2}{\pi^3} \approx 4.12251.$$

Consequently, the least squares polynomial approximation of degree 2 for $f(x) = \sin \pi x$ on $[0, 1]$, shown in Figure 8.5, is

$$P_2(x) = -4.12251 x^2 + 4.12251 x - 0.050465.$$

Figure 8.5

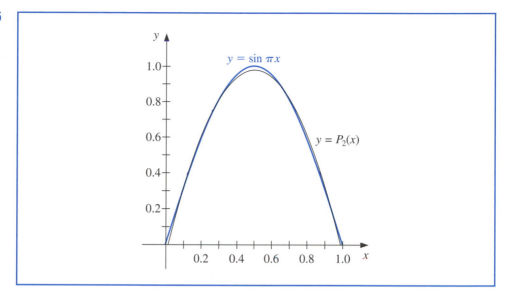

Example 1 illustrates the difficulty in obtaining a least squares polynomial approximation. An $(n + 1) \times (n + 1)$ linear system for the unknowns $a_0, \ldots, a_n$ must be solved, and the coefficients in the linear system are of the form

$$\int_a^b x^{j+k} \, dx = \frac{b^{j+k+1} - a^{j+k+1}}{j + k + 1}.$$

David Hilbert (1862–1943) was a dominant mathematician at the turn of the twentieth century. He is best remembered for giving a talk at the International Congress of Mathematicians in Paris in 1900 in which he posed 23 problems that he thought would be important for mathematicians in the next century to solve.

The matrix in this linear system is known as a *Hilbert matrix*, which is a classic example for demonstrating round-off error difficulties.

Another disadvantage to the technique used in Example 1 is similar to the situation that occurred when the Lagrange polynomials were first introduced in Chapter 3. We often don't know the degree of the approximating polynomial that is most appropriate, and the calculations to obtain the best nth-degree polynomial do not lessen the amount of work required to obtain the best polynomials of higher degree.

Both disadvantages are overcome by resorting to a technique that reduces the $n + 1$ equations in $n + 1$ unknowns to $n + 1$ equations, each of which contains only one unknown. This simplifies the problem to one that can be easily solved, but the technique requires some new concepts.

Linearly Independent Functions

The set of functions $\{\phi_0, \phi_1, \ldots, \phi_n\}$ is said to be **linearly independent** on $[a, b]$ if, whenever

$$c_0\phi_0(x) + c_1\phi_1(x) + \cdots + c_n\phi_n(x) = 0 \quad \text{for all } x \in [a, b]$$

we have $c_0 = c_1 = \cdots = c_n = 0$. Otherwise the set of functions is said to be **linearly dependent**.

Linearly independent sets of functions are basic to our discussion and, since the functions we are using for approximations are polynomials, the following result is fundamental.

Linearly Independent Sets of Polynomials

If $\phi_j(x)$ is a polynomial of degree j for each $j = 0, 1, \ldots, n$, then $\{\phi_0, \ldots, \phi_n\}$ is linearly independent on any interval $[a, b]$.

The situation illustrated in the following example demonstrates a fact that holds in a much more general setting. Let $\prod_n$ be the **set of all polynomials of degree at most n**.

- If $\{\phi_0(x), \phi_1(x), \ldots, \phi_n(x)\}$ is any collection of linearly independent polynomials in $\prod_n$, then every polynomial in $\prod_n$ can be written uniquely as a linear combination of $\{\phi_0(x), \phi_1(x), \ldots, \phi_n(x)\}$.

Example 2 Let

$$\phi_0(x) = 2, \quad \phi_1(x) = x - 3, \quad \text{and} \quad \phi_2(x) = x^2 + 2x + 7,$$

and

$$Q(x) = a_0 + a_1 x + a_2 x^2.$$

Show that **(a)** the set $\{\phi_0, \phi_1, \phi_2\}$ is linearly independent, and **(b)** there exist constants c_0, c_1, and c_2 such that $Q(x) = c_0 \phi_0(x) + c_1 \phi_1(x) + c_2 \phi_2(x)$.

Solution **(a)** We could use the result to establish the linear independence because $\phi_i(x)$ is of degree i for each $i = 0, 1$, and 2. But to emphasize the definition we will show linear independence directly.

Suppose that for all values of x we have

$$0 = c_0 \phi_0(x) + c_1 \phi_1(x) + c_2 \phi_2(x) = c_0(2) + c_1(x - 3) + c_2(x^2 + 2x + 7),$$

for some constants c_0, c_1, and c_2. Since $\phi_0(x), \phi_1(x)$, and $\phi_2(x)$ are polynomials of degrees 0, 1, and 2, respectively, we can take the second derivative of each side of the expression and conclude that

$$0 = 2c_2, \quad \text{which implies that} \quad c_2 = 0,$$

and

$$0 = c_0 \phi_0(x) + c_1 \phi_0(x) = c_0(2) + c_1(x - 3).$$

We can take the first derivative of this expression to conclude that

$$0 = c_1, \quad \text{which implies that} \quad c_1 = 0.$$

This leaves us with the expression

$$0 = c_0 \phi_0(x) = c_0(2), \quad \text{which implies that} \quad c_0 = 0.$$

These are the only constants for which

$$0 = c_0 \phi_0(x) + c_1 \phi_1(x) + c_2 \phi_2(x),$$

for all values of x, so the set $\{\phi_0, \phi_1, \phi_2\}$ is linearly independent.

(b) First note that

$$1 = \frac{1}{2}\phi_0(x), \quad x = \phi_1(x) + 3 = \phi_1(x) + \frac{3}{2}\phi_0(x),$$

and

$$x^2 = \phi_2(x) - 2x - 7 = \phi_2(x) - 2\left(\phi_1(x) + \frac{3}{2}\phi_0(x)\right) - 7\left(\frac{1}{2}\phi_0(x)\right)$$

$$= \phi_2(x) - 2\phi_1(x) - \frac{13}{2}\phi_0(x).$$

So

$$Q(x) = a_0\left(\frac{1}{2}\phi_0(x)\right) + a_1\left(\phi_1(x) + \frac{3}{2}\phi_0(x)\right) + a_2\left(\phi_2(x) - 2\phi_1(x) - \frac{13}{2}\phi_0(x)\right)$$

$$= \left(\frac{1}{2}a_0 + \frac{3}{2}a_1 - \frac{13}{2}a_2\right)\phi_0(x) + (a_1 - 2a_2)\phi_1(x) + a_2\phi_2(x). \qquad \blacksquare$$

Orthogonal Functions

To discuss general function approximation requires the introduction of the notions of weight functions and orthogonality. An integrable function w is called a **weight function** on the interval I if $w(x) \geq 0$ for all x in I, but w is not identically zero on any subinterval of I.

The purpose of a weight function is to assign varying degrees of importance to approximations on certain portions of the interval. For example, multiplying by the weight function

$$w(x) = \frac{1}{\sqrt{1 - x^2}}$$

places less emphasis near the center of the interval $(-1, 1)$ and more emphasis when $|x|$ is near 1 (see Figure 8.6). This particular weight function will be used in the next section.

Figure 8.6

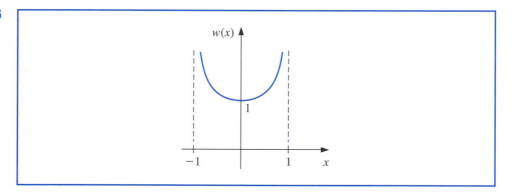

Suppose $\{\phi_0, \phi_1, \ldots, \phi_n\}$ is a set of linearly independent functions on $[a, b]$, w is a weight function for $[a, b]$, and, for $f \in C[a, b]$. A linear combination

$$P(x) = \sum_{k=0}^{n} a_k \phi_k(x)$$

is sought to minimize the error

$$E(a_0, \dots, a_n) = \int_a^b w(x) \left(f(x) - \sum_{k=0}^n a_k \phi_k(x) \right)^2 dx.$$

This problem reduces to the situation considered at the beginning of this section in the special case when $w(x) \equiv 1$ and $\phi_k(x) = x^k$ for each $k = 0, 1, \dots, n$.

The normal equations associated with this problem are derived from the fact that for each $j = 0, 1, \dots, n$,

$$0 = \frac{\partial E}{\partial a_j}(a_0, \dots, a_n) = 2 \int_a^b w(x) \left(f(x) - \sum_{k=0}^n a_k \phi_k(x) \right) \phi_j(x) \, dx.$$

The system of normal equations can be written

$$\int_a^b w(x) f(x) \phi_j(x) \, dx = \sum_{k=0}^n a_k \int_a^b w(x) \phi_k(x) \phi_j(x) \, dx, \quad \text{for each } j = 0, 1, \dots, n.$$

If the functions $\phi_0, \phi_1, \dots, \phi_n$ can be chosen so that

$$\int_a^b w(x) \phi_k(x) \phi_j(x) \, dx = \begin{cases} 0, & \text{when } j \neq k, \\ \alpha_k > 0, & \text{when } j = k, \end{cases} \tag{8.1}$$

for some positive numbers $\alpha_0, \alpha_1, \dots, \alpha_n$, then the normal equations reduce to

$$\int_a^b w(x) f(x) \phi_j(x) \, dx = a_j \int_a^b w(x) [\phi_j(x)]^2 \, dx = a_j \alpha_j$$

for each $j = 0, 1, \dots, n$, and are easily solved as

$$a_j = \frac{1}{\alpha_j} \int_a^b w(x) f(x) \phi_j(x) \, dx.$$

Hence the least squares approximation problem is greatly simplified when the functions $\phi_0, \phi_1, \dots, \phi_n$ are chosen to satisfy Eq. (8.1).

The word orthogonal means perpendicular. So in a sense, orthogonal functions are at right angles to one another.

The set of functions $\{\phi_0, \phi_1, \dots, \phi_n\}$ is said to be **orthogonal** for the interval $[a, b]$ with respect to the weight function w if for some positive numbers $\alpha_0, \alpha_1, \dots, \alpha_n$,

$$\int_a^b w(x) \phi_j(x) \phi_k(x) \, dx = \begin{cases} 0, & \text{when } j \neq k, \\ \alpha_k > 0, & \text{when } j = k. \end{cases}$$

If, in addition, $\alpha_k = 1$ for each $k = 0, 1, \dots, n$, the set is said to be **orthonormal**.

This definition, together with the remarks preceding it, implies the following.

Least Squares for Orthogonal Functions

If $\{\phi_0, \phi_1, \ldots, \phi_n\}$ is an orthogonal set of functions on an interval $[a, b]$ with respect to the weight function w, then the least squares approximation to f on $[a, b]$ with respect to w is

$$P(x) = \sum_{k=0}^{n} a_k \phi_k(x),$$

where

$$a_k = \frac{\int_a^b w(x)\phi_k(x)f(x)\,dx}{\int_a^b w(x)(\phi_k(x))^2\,dx} = \frac{1}{\alpha_k}\int_a^b w(x)\phi_k(x)f(x)\,dx.$$

The next result, which is based on the *Gram-Schmidt* process, describes a recursive procedure for constructing orthogonal polynomials on $[a, b]$ with respect to a weight function w.

Generating Sets of Orthogonal Polynomials

The set of polynomials $\{\phi_0(x), \phi_1(x), \ldots, \phi_n(x)\}$ defined in the following way is linearly independent and orthogonal on $[a, b]$ with respect to the weight function w.

$$\phi_0(x) \equiv 1, \qquad \phi_1(x) = x - B_1,$$

where

$$B_1 = \frac{\int_a^b xw(x)(\phi_0(x))^2\,dx}{\int_a^b w(x)(\phi_0(x))^2\,dx},$$

and when $k \geq 2$,

$$\phi_k(x) = (x - B_k)\phi_{k-1}(x) - C_k\phi_{k-2}(x),$$

where

$$B_k = \frac{\int_a^b xw(x)(\phi_{k-1}(x))^2\,dx}{\int_a^b w(x)(\phi_{k-1}(x))^2\,dx} \quad \text{and} \quad C_k = \frac{\int_a^b xw(x)\phi_{k-1}(x)\phi_{k-2}(x)\,dx}{\int_a^b w(x)(\phi_{k-2}(x))^2\,dx}.$$

Moreover, for any polynomial $Q_k(x)$ of degree $k < n$,

$$\int_a^b w(x)\phi_n(x)Q_k(x)\,dx = 0.$$

Illustration   The set of **Legendre polynomials**, $\{P_n(x)\}$, is orthogonal on $[-1, 1]$ with respect to the weight function $w(x) \equiv 1$. The classical definition of the Legendre polynomials requires that $P_n(1) = 1$ for each n, and a recursive relation is used to generate the polynomials when $n \geq 2$. This normalization will not be needed in our discussion, and the least squares approximating polynomials generated in either case are essentially the same.

Using the Gram-Schmidt process with $P_0(x) \equiv 1$ gives

$$B_1 = \frac{\int_{-1}^{1} x \, dx}{\int_{-1}^{1} dx} = 0 \quad \text{and} \quad P_1(x) = (x - B_1) P_0(x) = x.$$

Also,

$$B_2 = \frac{\int_{-1}^{1} x^3 \, dx}{\int_{-1}^{1} x^2 \, dx} = 0 \quad \text{and} \quad C_2 = \frac{\int_{-1}^{1} x^2 \, dx}{\int_{-1}^{1} 1 \, dx} = \frac{1}{3},$$

so

$$P_2(x) = (x - B_2) P_1(x) - C_2 P_0(x) = (x - 0)x - \frac{1}{3} \cdot 1 = x^2 - \frac{1}{3}.$$

Figure 8.7

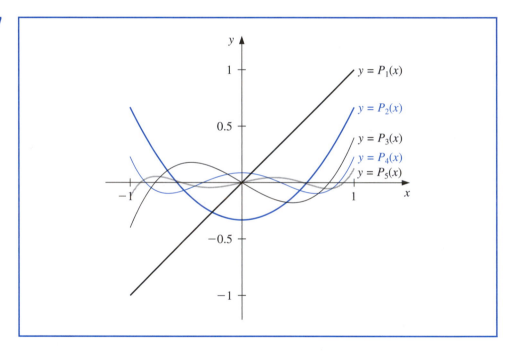

$y = P_1(x)$
$y = P_2(x)$
$y = P_3(x)$
$y = P_4(x)$
$y = P_5(x)$

Erhard Schmidt (1876–1959) received his doctorate under the supervision of David Hilbert in 1905 for a problem involving integral equations. Schmidt published a paper in 1907 in which he gave what is now called the Gram-Schmidt process for constructing an orthonormal basis for a set of functions. This generalized results of Jorgen Pedersen Gram (1850–1916) who considered this problem when studying least squares. Laplace, however, presented a similar process much earlier than either Gram or Schmidt.

The higher-degree Legendre polynomials shown in Figure 8.7 are derived in the same manner. However, the integration can be tedious. The MATLAB integration command quad can be used to compute the B_k and C_k. For example,

```
B3 = quad(inline('x.*(x.^2-1/3).^2','x'),-1,1)/
     quad(inline('(x.^2-1/3).^2','x'),-1,1)
```

produces

$$B3 = 3.903127820947816e - 017$$

approximately $B_3 = 0$, and

```
C3 = quad(inline('x.*(x.^2-1/3).*x','x'),-1,1)/
     quad(inline('x.^2','x'),-1,1)
```

produces

$$C3 = 0.266666666666667$$

that is, $C_3 = 4/15$. Thus

$$P_3(x) = x P_2(x) - \frac{4}{15} P_1(x) = x^3 - \frac{1}{3}x - \frac{4}{15}x = x^3 - \frac{3}{5}x.$$

The next two Legendre polynomials are

$$P_4(x) = x^4 - \frac{6}{7}x^2 + \frac{3}{35} \quad \text{and} \quad P_5(x) = x^5 - \frac{10}{9}x^3 + \frac{5}{21}x. \qquad \square$$

EXERCISE SET 8.3

1. Find the linear least squares polynomial approximation to $f(x)$ on the indicated interval if
   a. $f(x) = x^2 + 3x + 2$, $[0, 1]$;
   b. $f(x) = x^3$, $[0, 2]$;
   c. $f(x) = \frac{1}{x}$, $[1, 3]$;
   d. $f(x) = e^x$, $[0, 2]$;
   e. $f(x) = \frac{1}{2}\cos x + \frac{1}{3}\sin 2x$, $[0, 1]$;
   f. $f(x) = x \ln x$, $[1, 3]$.

2. Find the least squares polynomial approximation of degree 2 to the functions and intervals in Exercise 1.

3. Find the linear least squares polynomial approximation on the interval $[-1, 1]$ for the following functions.
   a. $f(x) = x^2 - 2x + 3$
   b. $f(x) = x^3$
   c. $f(x) = \frac{1}{x+2}$
   d. $f(x) = e^x$
   e. $f(x) = \frac{1}{2}\cos x + \frac{1}{3}\sin 2x$
   f. $f(x) = \ln(x + 2)$

4. Find the least squares polynomial approximation of degree 2 on the interval $[-1, 1]$ for the functions in Exercise 3.

5. Compute the error E for the approximations in Exercise 2.

6. Compute the error E for the approximations in Exercise 4.

7. Use the Gram-Schmidt process to construct $\phi_0(x), \phi_1(x), \phi_2(x)$, and $\phi_3(x)$ for the following intervals.
   a. $[0, 1]$
   b. $[0, 2]$
   c. $[1, 3]$

8. Repeat Exercise 1 using the results of Exercise 7.

9. Repeat Exercise 2 using the results of Exercise 7.

10. Use the Gram-Schmidt procedure to calculate L_1, L_2, and L_3, where $\{L_0(x), L_1(x), L_2(x), L_3(x)\}$ is an orthogonal set of polynomials on $(0, \infty)$ with respect to the weight functions $w(x) = e^{-x}$ and $L_0(x) \equiv 1$. The polynomials obtained from this procedure are called the **Laguerre polynomials**.

11. Use the Laguerre polynomials calculated in Exercise 10 to compute the least squares polynomials of degree 1, 2, and 3 on the interval $(0, \infty)$ with respect to the weight function $w(x) = e^{-x}$ for the following functions.
    a. $f(x) = x^2$
    b. $f(x) = e^{-x}$
    c. $f(x) = x^3$
    d. $f(x) = e^{-2x}$

12. Show that if $\{\phi_0, \phi_1, \dots, \phi_n\}$ is an orthogonal set of functions on $[a, b]$ with respect to the weight function w, then $\{\phi_0, \phi_1, \dots, \phi_n\}$ is a linearly independent set.

13. Suppose $\{\phi_0(x), \phi_1(x), \dots, \phi_n(x)\}$ is a linearly independent set in $\prod_n$. Show that for any element $Q(x)$ in $\prod_n$ there exist unique constants $c_0, c_1, \dots, c_n$ such that

$$Q(x) = \sum_{k=o}^{n} c_k \phi_k(x).$$

8.4   Chebyshev Polynomials

The **Chebyshev polynomials** $\{T_n(x)\}$ are orthogonal on $(-1, 1)$ with respect to the weight function

$$w(x) = (1 - x^2)^{-1/2} = \frac{1}{\sqrt{1 - x^2}}.$$

Pafnuty Lvovich Chebyshev (1821–1894) did exceptional mathematical work in many areas, including applied mathematics, number theory, approximation theory, and probability. In 1852 he traveled from St. Petersburg to visit mathematicians in France, England, and Germany. Lagrange and Legendre had studied individual sets of orthogonal polynomials, but Chebyshev was the first to see the important consequences of studying the theory in general. He developed the Chebyshev polynomials to study least squares approximation and probability, and then applied his results to interpolation, approximate quadrature, and other areas.

Although they can be derived by the method in the previous section, it is easier to give a definition and then show that they are the polynomials that satisfy the required orthogonality properties.

For $x \in [-1, 1]$, define

$$T_n(x) = \cos(n \arccos x) \quad \text{for each } n \geq 0.$$

It is not obvious from this definition that $T_n(x)$ is an nth degree polynomial in x, but we will now show that it is. First note that

$$T_0(x) = \cos 0 = 1 \quad \text{and} \quad T_1(x) = \cos(\arccos x) = x.$$

For $n \geq 1$ we introduce the substitution

$$\theta = \arccos x$$

to change the defining equation to

$$T_n(x) = T_n(\theta(x)) \equiv T_n(\theta) = \cos(n\theta), \quad \text{where } \theta \in [0, \pi].$$

A recurrence relation is derived by noting that

$$T_{n+1}(\theta) = \cos(n\theta + \theta) = \cos\theta \cos(n\theta) - \sin\theta \sin(n\theta)$$

and

$$T_{n-1}(\theta) = \cos(n\theta - \theta) = \cos\theta \cos(n\theta) + \sin\theta \sin(n\theta).$$

Adding these equations gives

$$T_{n+1}(\theta) + T_{n-1}(\theta) = 2\cos\theta \cos(n\theta) = 2\cos\theta \cdot T_n(\theta).$$

Returning to the variable x and solving for $T_{n+1}(x)$ we have, for each $n \geq 1$,

$$T_{n+1}(x) = 2xT_n(x) - T_{n-1}(x).$$

Since $T_0(x)$ and $T_1(x)$ are both polynomials in x, $T_{n+1}(x)$ will be a polynomial in x for each n.

Chebyshev Polynomials

$$T_0(x) = 1, \quad T_1(x) = x,$$

and, for $n \geq 1$, $T_{n+1}(x)$ is the polynomial of degree $n + 1$ given by

$$T_{n+1}(x) = 2xT_n(x) - T_{n-1}(x).$$

The recurrence relation implies that $T_n(x)$ is a polynomial of degree n with leading coefficient 2^{n-1}, when $n \geq 1$. The next three Chebyshev polynomials are

$$T_2(x) = 2xT_1(x) - T_0(x) = 2x^2 - 1, \quad T_3(x) = 2xT_2(x) - T_1(x) = 4x^3 - 3x,$$

and

$$T_4(x) = 2xT_3(x) - T_2(x) = 8x^4 - 8x^2 + 1.$$

The graphs of T_1, T_2, T_3, and T_4 are shown in Figure 8.8. Notice that each of the graphs is symmetric to either the origin or the y-axis, and that each assumes a maximum value of 1 and a minimum value of -1 on the interval $[-1, 1]$.

Figure 8.8

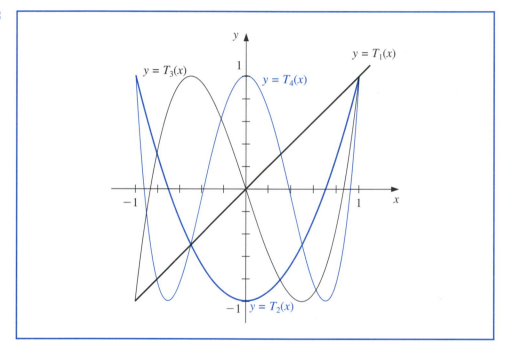

To show the orthogonality of the Chebyshev polynomials, consider

$$\int_{-1}^{1} w(x)T_n(x)T_m(x)dx = \int_{-1}^{1} \frac{T_n(x)T_m(x)}{\sqrt{1-x^2}}dx = \int_{-1}^{1} \frac{\cos(n \arccos x)\cos(m \arccos x)}{\sqrt{1-x^2}}dx.$$

Reintroducing the substitution $\theta = \arccos x$ gives

$$d\theta = -\frac{1}{\sqrt{1-x^2}}dx$$

and

$$\int_{-1}^{1} \frac{T_n(x)T_m(x)}{\sqrt{1-x^2}}dx = -\int_{\pi}^{0} \cos(n\theta)\cos(m\theta)d\theta = \int_{0}^{\pi} \cos(n\theta)\cos(m\theta)d\theta.$$

When $n \neq m$ we have

$$\cos(n\theta)\cos(m\theta) = \frac{1}{2}(\cos(n+m)\theta + \cos(n-m)\theta),$$

and

$$\int_{-1}^{1} \frac{T_n(x)T_m(x)}{\sqrt{1-x^2}}\,dx = \frac{1}{2}\int_{0}^{\pi}\cos((n+m)\theta)d\theta + \frac{1}{2}\int_{0}^{\pi}\cos((n-m)\theta)d\theta$$

$$= \left[\frac{1}{2(n+m)}\sin((n+m)\theta) + \frac{1}{2(n-m)}\sin(n-m)\theta\right]_{0}^{\pi} = 0.$$

By a similar technique, it can also be shown that

$$\int_{-1}^{1} \frac{(T_n(x))^2}{\sqrt{1-x^2}}\,dx = \frac{\pi}{2} \quad \text{for each } n \geq 1.$$

One of the important results about the Chebyshev polynomials concerns their zeros and extrema. These are easily verified by substitution into the polynomials and their derivatives.

Zeros and Extrema of Chebyshev Polynomials

The Chebyshev polynomial $T_n(x)$, of degree $n \geq 1$, has n simple zeros in $[-1, 1]$ at

$$\bar{x}_k = \cos\left(\frac{2k-1}{2n}\pi\right) \quad \text{for each } k = 1, 2, \ldots, n.$$

Moreover, T_n assumes its absolute extrema on $[-1, 1]$ at the $n+1$ points

$$\bar{x}'_k = \cos\left(\frac{k\pi}{n}\right) \quad \text{with} \quad T_n\left(\bar{x}'_k\right) = (-1)^k \quad \text{for each } k = 0, 1, \ldots, n.$$

The *monic Chebyshev polynomial* (polynomial with leading coefficient 1), $\tilde{T}_n(x)$, is derived from the Chebyshev polynomial, $T_n(x)$, by dividing by the leading coefficient, 2^{n-1}, when $n \geq 1$. So

$$\tilde{T}_0(x) = 1, \quad \text{and} \quad \tilde{T}_n(x) = 2^{1-n}T_n(x) \quad \text{for each } n \geq 1.$$

The monic Chebyshev polynomials satisfy the recurrence relation

$$\tilde{T}_2(x) = x\tilde{T}_1(x) - \frac{1}{2}\tilde{T}_0(x), \quad \tilde{T}_{n+1}(x) = x\tilde{T}_n(x) - \frac{1}{4}\tilde{T}_{n-1}(x),$$

for each $n \geq 2$. Because $\tilde{T}_n$ is a multiple of T_n, the zeros of $\tilde{T}_n$ also occur at

$$\bar{x}_k = \cos\left(\frac{2k-1}{2n}\pi\right) \quad \text{for each } k = 1, 2, \ldots, n,$$

and the extreme values of $\tilde{T}_n$ occur at

$$\bar{x}'_k = \cos\left(\frac{k\pi}{n}\right) \quad \text{with} \quad \tilde{T}_n(\bar{x}'_k) = \frac{(-1)^k}{2^{n-1}} \quad \text{for each } k = 0, 1, 2, \ldots, n.$$

Let $\widetilde{\Pi}_n$ denote the **set of all monic polynomials of degree n**. The following minimization property distinguishes the polynomials $\tilde{T}_n(x)$ from the other members of $\widetilde{\Pi}_n$.

Minimum Property of Monic Chebyshev Polynomials

The polynomial $\tilde{T}_n(x)$, when $n \geq 1$, has the property that

$$\frac{1}{2^{n-1}} = \max_{x \in [-1,1]} |\tilde{T}_n(x)| \leq \max_{x \in [-1,1]} |P_n(x)| \quad \text{for all } P_n \in \tilde{\Pi}_n.$$

Moreover, equality can occur only if $P_n = \tilde{T}_n$.

Minimizing Lagrange Interpolation Error

This result is used to tell us where to place interpolating nodes to minimize the error in Lagrange interpolation. The error form for the Lagrange polynomial applied to the interval $[-1, 1]$ states that if $x_0, \ldots, x_n$ are distinct numbers in the interval $[-1, 1]$ and if $f \in C^{n+1}[-1, 1]$, then, for each $x \in [-1, 1]$, a number $\xi(x)$ exists in $(-1, 1)$ with

$$f(x) - P(x) = \frac{f^{(n+1)}(\xi(x))}{(n+1)!} (x - x_0) \cdots (x - x_n),$$

where $P(x)$ is the Lagrange interpolating polynomial. Suppose that we want to minimize this error for all values of x in $[-1, 1]$. We have no control over $\xi(x)$, so we hope to minimize the error by shrewd placement of the nodes $x_0, \ldots, x_n$. This is equivalent to choosing $x_0, \ldots, x_n$ to minimize the quantity

$$|(x - x_0)(x - x_1) \cdots (x - x_n)|$$

throughout the interval $[-1, 1]$.

Since $(x - x_0)(x - x_1) \cdots (x - x_n)$ is a monic polynomial of degree $n+1$, the minimum is obtained when

$$(x - x_0)(x - x_1) \cdots (x - x_n) = \tilde{T}_{n+1}(x).$$

When x_k is chosen to be the $(k + 1)$st zero of $\tilde{T}_{n+1}$, that is, when x_k is

$$\overline{x}_{k+1} = \cos \frac{2k + 1}{2(n + 1)} \pi,$$

the maximum value of $|(x - x_0)(x - x_1) \cdots (x - x_n)|$ is minimized. Since

$$\max_{x \in [-1,1]} |\tilde{T}_{n+1}(x)| = \frac{1}{2^n}$$

this also implies that

$$\frac{1}{2^n} = \max_{x \in [-1,1]} |(x - \overline{x}_1)(x - \overline{x}_2) \cdots (x - \overline{x}_{n+1})| \leq \max_{x \in [-1,1]} |(x - x_0)(x - x_1) \cdots (x - x_n)|,$$

for any choice of $x_0, x_1, \ldots, x_n$ in the interval $[-1, 1]$.

Minimizing Lagrange Interpolation Error

If $P(x)$ is the interpolating polynomial of degree at most n with nodes at the roots of $T_{n+1}(x)$, then, for any $f \in C^{n+1}[-1, 1]$,

$$\max_{x \in [-1,1]} |f(x) - P(x)| \leq \frac{1}{2^n (n + 1)!} \max_{x \in [-1,1]} |f^{(n+1)}(x)|.$$

Minimizing Interpolation Error on Arbitrary Intervals

The technique for choosing points to minimize the interpolating error can be easily extended to a general closed interval $[a, b]$ by using the change of variable

$$\tilde{x} = \frac{1}{2} ((b - a)x + a + b)$$

to transform the numbers $\overline{x}_k$ in the interval $[-1, 1]$ into the corresponding numbers in the interval $[a, b]$, as shown in the next example.

Example 1   Let $f(x) = xe^x$ on $[0, 1.5]$. Compare the values given by the Lagrange polynomial with four equally spaced nodes with those given by the Lagrange polynomial with nodes given by zeros of the fourth Chebyshev polynomial.

Solution   The equally spaced nodes $x_0 = 0$, $x_1 = 0.5$, $x_2 = 1$, and $x_3 = 1.5$ give

$$L_0(x) = -1.3333x^3 + 4.0000x^2 - 3.6667x + 1,$$

$$L_1(x) = 4.0000x^3 - 10.000x^2 + 6.0000x,$$

$$L_2(x) = -4.0000x^3 + 8.0000x^2 - 3.0000x,$$

$$L_3(x) = 1.3333x^3 - 2.000x^2 + 0.66667x,$$

which produces the polynomial

$$P_3(x) = L_0(x)(0) + L_1(x)(0.5e^{0.5}) + L_2(x)e^1 + L_3(x)(1.5e^{1.5})$$
$$= 1.3875x^3 + 0.057570x^2 + 1.2730x.$$

For the second interpolating polynomial, we shift the zeros $\overline{x}_k = \cos((2k + 1)/8)\pi$, for $k = 0, 1, 2, 3$, of $\tilde{T}_4$ from $[-1, 1]$ to $[0, 1.5]$ using the linear transformation

$$\tilde{x}_k = \frac{1}{2} [(1.5 - 0)\overline{x}_k + (1.5 + 0)] = 0.75 + 0.75\overline{x}_k.$$

Because

$$\overline{x}_0 = \cos \frac{\pi}{8} = 0.92388, \quad \overline{x}_1 = \cos \frac{3\pi}{8} = 0.38268, \quad \overline{x}_2 = \cos \frac{5\pi}{8} = -0.38268, \quad \text{and}$$

$$\overline{x}_3 = \cos \frac{7\pi}{8} = -0.92388,$$

we have

$$\tilde{x}_0 = 1.44291, \quad \tilde{x}_1 = 1.03701, \quad \tilde{x}_2 = 0.46299, \quad \text{and} \quad \tilde{x}_3 = 0.05709.$$

The Lagrange coefficient polynomials for this set of nodes are

$$\tilde{L}_0(x) = 1.8142x^3 - 2.8249x^2 + 1.0264x - 0.049728,$$

$$\tilde{L}_1(x) = -4.3799x^3 + 8.5977x^2 - 3.4026x + 0.16705,$$

$$\tilde{L}_2(x) = 4.3799x^3 - 11.112x^2 + 7.1738x - 0.37415,$$

$$\tilde{L}_3(x) = -1.8142x^3 + 5.3390x^2 - 4.7976x + 1.2568.$$

The functional values required for these polynomials are given in the second column of Table 8.5. The interpolation polynomial of degree at most 3 is

$$\tilde{P}_3(x) = 1.3811x^3 + 0.044652x^2 + 1.3031x - 0.014352.$$

Table 8.5

x	$f(x) = xe^x$	$P_3(x)$	$\|xe^x - P_3(x)\|$	$\tilde{P}_3(x)$	$\|xe^x - \tilde{P}_3(x)\|$
0.15	0.1743	0.1969	0.0226	0.1868	0.0125
0.25	0.3210	0.3435	0.0225	0.3358	0.0148
0.35	0.4967	0.5121	0.0154	0.5064	0.0097
0.65	1.245	1.233	0.012	1.231	0.014
0.75	1.588	1.572	0.016	1.571	0.017
0.85	1.989	1.976	0.013	1.974	0.015
1.15	3.632	3.650	0.018	3.644	0.012
1.25	4.363	4.391	0.028	4.382	0.019
1.35	5.208	5.237	0.029	5.224	0.016

For comparison, Table 8.5 lists various values of x, together with the values of $f(x)$, $P_3(x)$, and $\tilde{P}_3(x)$. It can be seen from this table that, although the error using $P_3(x)$ is less than using $\tilde{P}_3(x)$ near the middle of the table, the maximum error involved with using $\tilde{P}_3(x)$, 0.0180, is considerably less than when using $P_3(x)$, which gives the error 0.0290. (See Figure 8.9.) ∎

Figure 8.9

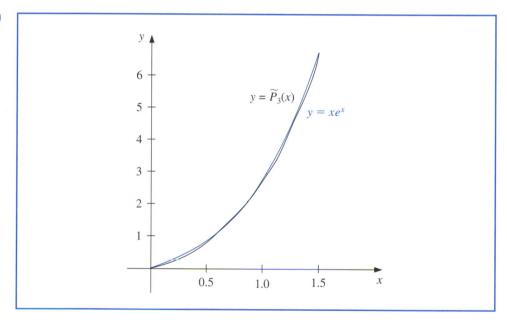

EXERCISE SET 8.4

1. Use the zeros of $\tilde{T}_3$ to construct an interpolating polynomial of degree 2 for the following functions on the interval $[-1, 1]$.

   a. $f(x) = e^x$ **b.** $f(x) = \sin x$

   c. $f(x) = \ln(x + 2)$ **d.** $f(x) = x^4$

2. Find a bound for the maximum error of the approximation in Exercise 1 on the interval $[-1, 1]$.

3. Use the zeros of $\tilde{T}_4$ to construct an interpolating polynomial of degree 3 for the functions in Exercise 1.

4. Repeat Exercise 2 for the approximations computed in Exercise 3.

5. Use the zeros of $\tilde{T}_3$ and transformations of the given interval to construct an interpolating polynomial of degree 2 for the following functions.

a. $f(x) = \dfrac{1}{x}, [1, 3]$

b. $f(x) = e^{-x}, [0, 2]$

c. $f(x) = \dfrac{1}{2}\cos x + \dfrac{1}{3}\sin 2x, [0, 1]$

d. $f(x) = x \ln x, [1, 3]$

6. Use the zeros of $\tilde{T}_4$ to construct an interpolating polynomial of degree 3 for the functions in Exercise 5.

7. Show that for any positive integers i and j with $i > j$ we have

$$T_i(x)T_j(x) = \frac{1}{2}[T_{i+j}(x) + T_{i-j}(x)].$$

8. Show that for each Chebyshev polynomial $T_n(x)$ we have

$$\int_{-1}^{1} \frac{[T_n(x)]^2}{\sqrt{1-x^2}}\, dx = \frac{\pi}{2}.$$

8.5 Rational Function Approximation

The class of algebraic polynomials has some distinct advantages for use in approximation. There is a sufficient number of polynomials to approximate any continuous function on a closed interval to within an arbitrary tolerance—polynomials are easily evaluated at arbitrary values—and the derivatives and integrals of polynomials exist and are easily determined. The disadvantage of using polynomials for approximation is their tendency for oscillation. This often causes error bounds in polynomial approximation to significantly exceed the average approximation error because error bounds are determined by the maximum approximation error. We now consider methods that spread the approximation error more evenly over the approximation interval.

A **rational function** r of degree N has the form

$$r(x) = \frac{p(x)}{q(x)},$$

where $p(x)$ and $q(x)$ are polynomials whose degrees sum to N.

Every polynomial is a rational function (simply let $q(x) \equiv 1$), so approximation by rational functions gives results with no greater error bounds than approximation by polynomials. However, rational functions whose numerator and denominator have the same or nearly the same degree often produce approximation results superior to polynomial methods for the same amount of computational effort. Rational functions have the added advantage of permitting efficient approximation of functions with infinite discontinuities near the interval of approximation. Polynomial approximation is generally unacceptable in this situation.

Padé Approximation

Suppose r is a rational function of degree $N = n + m$ of the form

$$r(x) = \frac{p(x)}{q(x)} = \frac{p_0 + p_1 x + \cdots + p_n x^n}{q_0 + q_1 x + \cdots + q_m x^m}$$

that is used to approximate a function f on a closed interval I containing zero. For r to be defined at zero requires that $q_0 \neq 0$. In fact, we can assume that $q_0 = 1$, for if this is not the case we simply replace $p(x)$ by $p(x)/q_0$ and $q(x)$ by $q(x)/q_0$. Consequently, there are $N + 1$ parameters $q_1, q_2, \ldots, q_m, p_0, p_1, \ldots, p_n$ available for the approximation of f by r.

The **Padé** approximation technique chooses the $N + 1$ parameters so that $f^{(k)}(0) = r^{(k)}(0)$ for each $k = 0, 1, \ldots, N$. Padé approximation is the extension of Taylor polynomial approximation to rational functions. In fact, when $n = N$ and $m = 0$, the Padé approximation is the Nth Taylor polynomial expanded about zero—that is, the Nth Maclaurin polynomial.

Consider the difference

$$f(x) - r(x) = f(x) - \frac{p(x)}{q(x)} = \frac{f(x)q(x) - p(x)}{q(x)} = \frac{f(x)\sum_{i=0}^{m} q_i x^i - \sum_{i=0}^{n} p_i x^i}{q(x)}$$

and suppose f has the Maclaurin series expansion $f(x) = \sum_{i=0}^{\infty} a_i x^i$. Then

$$f(x) - r(x) = \frac{\sum_{i=0}^{\infty} a_i x^i \sum_{i=0}^{m} q_i x^i - \sum_{i=0}^{n} p_i x^i}{q(x)}. \tag{8.2}$$

The object is to choose the constants $q_1, q_2, \ldots, q_m$ and $p_0, p_1, \ldots, p_n$ so that

$$f^{(k)}(0) - r^{(k)}(0) = 0 \quad \text{for each } k = 0, 1, \ldots, N.$$

This is equivalent to $f - r$ having a zero of multiplicity $N + 1$ at 0. As a consequence, we choose $q_1, q_2, \ldots, q_m$ and $p_0, p_1, \ldots, p_n$ so that the numerator on the right side of Eq. (8.2),

$$(a_0 + a_1 x + \cdots)(1 + q_1 x + \cdots + q_m x^m) - (p_0 + p_1 x + \cdots + p_n x^n),$$

has no terms of degree less than or equal to N.

To make better use of summation notation, we artificially define $p_{n+1} = p_{n+2} = \cdots = p_N = 0$ and $q_{m+1} = q_{m+2} = \cdots = q_N = 0$. The coefficient of x^k can then be written as

$$\left(\sum_{i=0}^{k} a_i q_{k-i} \right) - p_k.$$

The rational function for Padé approximation results from the solution of the $N + 1$ linear equations

$$\sum_{i=0}^{k} a_i q_{k-i} = p_k, \quad k = 0, 1, \ldots, N$$

in the $N + 1$ unknowns $q_1, q_2, \ldots, q_m, p_0, p_1, \ldots, p_n$.

> Henri Padé (1863–1953) gave a systematic study of what we call today Padé approximations in his doctoral thesis in 1892. He proved results on their general structure and also clearly set out the connection between Padé approximations and continued fractions. These ideas, however, had been studied by Daniel Bernoulli (1700–1782) and others as early as 1730. James Stirling (1692–1770) gave a similar method in *Methodus differentialis* published in the same year, and Euler used Padé-type approximations to find the sum of a series.

> The program PADEMD81 implements the Padé approximation method.

Example 1 The Maclaurin series expansion for e^{-x} is

$$\sum_{i=0}^{\infty} \frac{(-1)^i}{i!} x^i.$$

Find the Padé approximation to e^{-x} of degree 5 with $n = 3$ and $m = 2$.

Solution To find the Padé approximation we need to choose $p_0, p_1, p_2, p_3, q_1,$ and q_2 so that the coefficients of x^k for $k = 0, 1, \ldots, 5$ are 0 in the expression

$$\left(1 - x + \frac{x^2}{2} - \frac{x^3}{6} + \cdots \right)(1 + q_1 x + q_2 x^2) - (p_0 + p_1 x + p_2 x^2 + p_3 x^3).$$

Expanding and collecting terms produces

$$x^5: \quad -\frac{1}{120} + \frac{1}{24}q_1 - \frac{1}{6}q_2 = 0; \qquad x^2: \quad \frac{1}{2} - q_1 + q_2 = p_2;$$

$$x^4: \quad \frac{1}{24} - \frac{1}{6}q_1 + \frac{1}{2}q_2 = 0; \qquad x^1: \quad -1 + q_1 \qquad = p_1;$$

$$x^3: \quad -\frac{1}{6} + \frac{1}{2}q_1 - q_2 = p_3; \qquad x^0: \quad 1 \qquad = p_0.$$

This implies that $p_0 = 1$. To determine the other coefficients in MATLAB, define A and **b** by

```
A = [1 0 -1 0 0; -1 1 0 -1 0; 0.5 -1 0 0 1; -1/6 0.5 0 0 0;
    1/24 -1/6 0 0 0]
b = [1; -1/2; 1/6; -1/24; 1/120]
```

and then apply the command `linsolve`.

```
linsolve(A,b)
```

This produces

$$ans = 0.400000000000000$$
$$0.050000000000000$$
$$-0.600000000000000$$
$$0.150000000000000$$
$$0.016666666666667$$

As fractions these values are, respectively,

$$\frac{2}{5}, \quad \frac{1}{20}, \quad -\frac{3}{5}, \quad \frac{3}{20}, \quad \text{and} \quad -\frac{1}{60},$$

so the Padé approximation is

$$r(x) = \frac{1 - \frac{3}{5}x + \frac{3}{20}x^2 - \frac{1}{60}x^3}{1 + \frac{2}{5}x + \frac{1}{20}x^2}.$$

Table 8.6 lists values of $r(x)$ and $P_5(x)$, the fifth Maclaurin polynomial. The Padé approximation is clearly superior in this example. ▪

Table 8.6

x	e^{-x}	$P_5(x)$	$|e^{-x} - P_5(x)|$	$r(x)$	$|e^{-x} - r(x)|$
0.2	0.81873075	0.81873067	8.64×10^{-8}	0.81873075	7.55×10^{-9}
0.4	0.67032005	0.67031467	5.38×10^{-6}	0.67031963	4.11×10^{-7}
0.6	0.54881164	0.54875200	5.96×10^{-5}	0.54880763	4.00×10^{-6}
0.8	0.44932896	0.44900267	3.26×10^{-4}	0.44930966	1.93×10^{-5}
1.0	0.36787944	0.36666667	1.21×10^{-3}	0.36781609	6.33×10^{-5}

Continued Fraction Approximation

It is interesting to compare the number of arithmetic operations required for calculations of $P_5(x)$ and $r(x)$ in Example 1. Using nested multiplication, the fifth Taylor polynomial for $f(x) = e^{-x}$ can be expressed as

$$P_5(x) = 1 - x + \frac{1}{2}x^2 - \frac{1}{6}x^3 + \frac{1}{24}x^4 - \frac{1}{120}x^5$$

$$= \left(\left(\left(\left(-\frac{1}{120}x + \frac{1}{24}\right)x - \frac{1}{6}\right)x + \frac{1}{2}\right)x - 1\right)x + 1.$$

Assuming that the coefficients of $1, x, x^2, x^3, x^4$, and x^5 are represented as decimals, a single calculation of $P_5(x)$ in nested form requires five multiplications and five additions/subtractions.

Using nested multiplication, the Padé approximation in Example 1 is expressed as

$$r(x) = \frac{1 - \frac{3}{5}x + \frac{3}{20}x^2 - \frac{1}{60}x^3}{1 + \frac{2}{5}x + \frac{1}{20}x^2} = \frac{\left(\left(-\frac{1}{60}x + \frac{3}{20}\right)x - \frac{3}{5}\right)x + 1}{\left(\frac{1}{20}x + \frac{2}{5}\right)x + 1},$$

so a single calculation of $r(x)$ requires five multiplications, five additions/subtractions, and one division. Hence, computational effort appears to favor the polynomial approximation. However, by reexpressing $r(x)$ by continued division, we can write

$$r(x) = \frac{1 - \frac{3}{5}x + \frac{3}{20}x^2 - \frac{1}{60}x^3}{1 + \frac{2}{5}x + \frac{1}{20}x^2}$$

$$= \frac{-\frac{1}{3}x^3 + 3x^2 - 12x + 20}{x^2 + 8x + 20}$$

$$= -\frac{1}{3}x + \frac{17}{3} + \frac{\left(-\frac{152}{3}x - \frac{280}{3}\right)}{x^2 + 8x + 20}$$

$$= -\frac{1}{3}x + \frac{17}{3} + \frac{-\frac{152}{3}}{\left(\frac{x^2 + 8x + 20}{x + \frac{35}{19}}\right)}$$

or

$$r(x) = -\frac{1}{3}x + \frac{17}{3} - \frac{\frac{152}{3}}{\left(x + \frac{117}{19} + \frac{\frac{3125}{361}}{\left(x + \frac{35}{19}\right)}\right)}.$$

Written in this form, a single calculation of $r(x)$ requires one multiplication, five addition/subtractions, and two divisions. If the amount of computation required for division is approximately the same as for multiplication, the computational effort required for an evaluation of $P_5(x)$ significantly exceeds that required for an evaluation of $r(x)$.

Expressing a rational function approximation in this form is called **continued-fraction** approximation. This is a classical approximation technique of current interest because of its computational efficiency. It is, however, a specialized technique—one we will not discuss further.

Although the rational function approximation in Example 1 gave results superior to the polynomial approximation of the same degree, the approximation has a wide variation in accuracy; the approximation at 0.2 is accurate to within 8×10^{-9}, whereas, at 1.0 the approximation and the function agree only to within 7×10^{-5}. This accuracy variation is

Using continued fractions for rational approximation is a subject that has its roots in the works of Christopher Clavius (1537–1612). It was employed in the eighteenth and nineteenth centuries by, for example, Euler, Lagrange, and Hermite.

expected because the Padé approximation is based on a Taylor polynomial representation of e^{-x}, and the Taylor representation has a wide variation of accuracy in $[0.2, 1.0]$.

To obtain more uniformly accurate rational function approximations, we use the set of Chebyshev polynomials, a class that exhibits more uniform behavior. The general Chebyshev rational function approximation method proceeds in the same manner as Padé approximation, except that each x^k term in the Padé approximation is replaced by the kth-degree Chebyshev polynomial T_k. An introduction to this technique and more detailed references can be found in Burden and Faires [BF], pp. 533–537.

EXERCISE SET 8.5

1. Determine all Padé approximations for $f(x) = e^{2x}$ of degree 2. Compare the results at $x_i = 0.2i$, for $i = 1, 2, 3, 4, 5$, with the actual values $f(x_i)$.

2. Determine all Padé approximations for $f(x) = x \ln(x + 1)$ of degree 3. Compare the results at $x_i = 0.2i$, for $i = 1, 2, 3, 4, 5$, with the actual values $f(x_i)$.

3. Determine the Padé approximation of degree 5 with $n = 2$ and $m = 3$ for $f(x) = e^x$. Compare the results at $x_i = 0.2i$, for $i = 1, 2, 3, 4, 5$, with those from the fifth Maclaurin polynomial.

4. Repeat Exercise 3 using instead the Padé approximation of degree 5 with $n = 3$ and $m = 2$.

5. Determine the Padé approximation of degree 6 with $n = m = 3$ for $f(x) = \sin x$. Compare the results at $x_i = 0.1i$, for $i = 0, 1, \ldots, 5$, with the exact results and with the results of the sixth Maclaurin polynomial.

6. Determine the Padé approximations of degree 6 with (a) $n = 2, m = 4$ and (b) $n = 4, m = 2$ for $f(x) = \sin x$. Compare the results at each x_i to those obtained in Exercise 5.

7. Table 8.6 lists results of the Padé approximation of degree 5 with $n = 3$ and $m = 2$, the fifth Maclaurin polynomial, and the exact values of $f(x) = e^{-x}$ when $x_i = 0.2i$, for $i = 1, 2, 3, 4$, and 5. Compare these results with those produced from the other Padé approximations of degree 5.

   a. $n = 0, m = 5$ b. $n = 1, m = 4$
   c. $n = 2, m = 3$ d. $n = 4, m = 1$

8. Express the following rational functions in continued-fraction form.

   a. $\dfrac{x^2 + 3x + 2}{x^2 - x + 1}$ b. $\dfrac{4x^2 + 3x - 7}{2x^3 + x^2 - x + 5}$

   c. $\dfrac{2x^3 - 3x^2 + 4x - 5}{x^2 + 2x + 4}$ d. $\dfrac{2x^3 + x^2 - x + 3}{3x^3 + 2x^2 - x + 1}$

9. To accurately approximate $\sin x$ and $\cos x$ for inclusion in a mathematical library, we first restrict their domains. Given a real number x, divide by π to obtain the relation

$$|x| = M\pi + s, \quad \text{where } M \text{ is an integer and } |s| \leq \frac{\pi}{2}.$$

   a. Show that $\sin x = \text{sgn}(x)(-1)^M \sin s$.
   b. Construct a rational approximation to $\sin s$ using $n = m = 4$. Estimate the error when $0 \leq |s| \leq \frac{\pi}{2}$.
   c. Design an implementation of $\sin x$ using (a) and (b).
   d. Repeat (c) for $\cos x$ using the fact that $\cos x = \sin(x + \frac{\pi}{2})$.

10. To accurately approximate $f(x) = e^x$ for inclusion in a mathematical library, we first restrict the domain of f. Given a real number x, divide by $\ln \sqrt{10}$ to obtain the relation

$$x = M \ln \sqrt{10} + s,$$

where M is an integer and s is a real number satisfying $|s| \leq \frac{1}{2} \ln \sqrt{10}$.

a. Show that $e^x = e^s 10^{M/2}$.

b. Construct a rational function approximation for e^s using $n = m = 3$. Estimate the error when $0 \le |s| \le \frac{1}{2} \ln \sqrt{10}$.

c. Design an implementation of e^x using the results of (a) and (b) and the approximations

$$\frac{1}{\ln \sqrt{10}} = 0.8685889638 \quad \text{and} \quad \sqrt{10} = 3.162277660.$$

8.6 Trigonometric Polynomial Approximation

Trigonometric functions are used to approximate functions that have periodic behavior, that is, functions with the property that for some constant T, $f(x + T) = f(x)$ for all x. We can generally transform the problem so that $T = 2\pi$ and restrict the approximation to the interval $[-\pi, \pi]$.

Orthogonal Trigonometric Polynomials

During the late seventeenth and early eighteenth centuries, the Bernoulli family produced no less than eight important mathematicians and physicists. Some of the first Daniel Bernoulli's (1700–1782) most important work involved the pressure, density, and velocity of fluid flow, which produced what is known as the *Bernoulli principle.*

For each positive integer n, the set $\mathcal{T}_n$ of **trigonometric polynomials** of degree less than or equal to n is the set of all linear combinations of $\{\phi_0, \phi_1, \dots, \phi_{2n-1}\}$, where

$$\phi_0(x) = \frac{1}{2},$$

$$\phi_k(x) = \cos kx, \quad \text{for each } k = 1, 2, \dots, n,$$

and

$$\phi_{n+k}(x) = \sin kx, \quad \text{for each } k = 1, 2, \dots, n - 1.$$

(Some sources include an additional function in the set, $\phi_{2n}(x) = \sin nx$.)

The set $\{\phi_0, \phi_1, \dots, \phi_{2n-1}\}$ is orthogonal on $[-\pi, \pi]$ with respect to the weight function $w(x) \equiv 1$. For example, if $k \ne j$,

$$\int_{-\pi}^{\pi} \phi_{n+k}(x)\phi_j(x)\, dx = \int_{-\pi}^{\pi} \sin kx \cos jx\, dx.$$

The trigonometric identity

$$\sin kx \cos jx = \frac{1}{2} \sin(k + j)x + \frac{1}{2} \sin(k - j)x$$

can now be used to give

$$\int_{-\pi}^{\pi} \phi_{n+k}(x)\phi_j(x)\, dx = \frac{1}{2} \int_{-\pi}^{\pi} (\sin(k + j)x + \sin(k - j)x)\, dx$$

$$= \frac{1}{2} \left[\frac{-\cos(k + j)x}{k + j} - \frac{\cos(k - j)x}{k - j} \right]_{-\pi}^{\pi}.$$

But the cosine function is even, so

$$\cos(k + j)\pi = \cos(k + j)(-\pi) \quad \text{and} \quad \cos(k - j)\pi = \cos(k - j)(-\pi),$$

and

$$\int_{-\pi}^{\pi} \phi_{n+k}(x)\phi_j(x)\,dx = 0.$$

The result also holds when $k = j > 0$, for in this case we have

$$\int_{-\pi}^{\pi} \sin jx \cos jx\,dx = \left[\frac{1}{2j}(\sin jx)^2 \right]_{-\pi}^{\pi} = 0.$$

Showing orthogonality for the other possibilities from $\{\phi_0, \phi_1, \ldots, \phi_{2n-1}\}$ is similar and uses the appropriate trigonometric identities from the collection

$$\sin t_1 \sin t_2 = \frac{1}{2}[\cos(t_1 - t_2) - \cos(t_1 + t_2)],$$

$$\cos t_1 \cos t_2 = \frac{1}{2}[\cos(t_1 - t_2) + \cos(t_1 + t_2)], \qquad (8.3)$$

$$\sin t_1 \cos t_2 = \frac{1}{2}[\sin(t_1 - t_2) + \sin(t_1 + t_2)]$$

to convert the products into sums.

Orthogonal Trigonometric Polynomials

For a function $f \in C[-\pi, \pi]$, we want to find the *continuous least squares* approximation by functions of the form

$$S_n(x) = \frac{a_0}{2} + a_n \cos nx + \sum_{k=1}^{n-1}(a_k \cos kx + b_k \sin kx).$$

Since the set of functions $\{\phi_0, \phi_1, \ldots, \phi_{2n-1}\}$ is orthogonal on $[-\pi, \pi]$ with respect to $w(x) \equiv 1$, it follows from the result on page 335 and the equations in (8.3) that the appropriate selection of coefficients is

$$a_k = \frac{\int_{-\pi}^{\pi} f(x) \cos kx\,dx}{\int_{-\pi}^{\pi} (\cos kx)^2\,dx} = \frac{1}{\pi}\int_{-\pi}^{\pi} f(x) \cos kx\,dx, \quad \text{for each } k = 0, 1, 2, \ldots, n,$$

$$(8.4)$$

and

$$b_k = \frac{\int_{-\pi}^{\pi} f(x) \sin kx\,dx}{\int_{-\pi}^{\pi} (\sin kx)^2\,dx} = \frac{1}{\pi}\int_{-\pi}^{\pi} f(x) \sin kx\,dx, \quad \text{for each } k = 1, 2, \ldots, n-1.$$

$$(8.5)$$

Joseph Fourier (1768–1830) published his theory of trigonometric series in *Théorie analytique de la chaleur* to solve the problems of heat transfer and vibrations. He accompanied Napoleon Bonaparte on his Egyptian expedition in 1798, and was, for a time, governor of Lower Egypt.

The limit of $S_n(x)$ when $n \to \infty$ is called the **Fourier series** of f. Fourier series are used to describe the solution of various ordinary and partial-differential equations that occur in physical situations.

Example 1   Determine the trigonometric polynomial from $\mathcal{T}_n$ that approximates

$$f(x) = |x|, \quad \text{for } -\pi < x < \pi.$$

Solution   We first need to find the coefficients

$$a_0 = \frac{1}{\pi} \int_{-\pi}^{\pi} |x| \, dx = -\frac{1}{\pi} \int_{-\pi}^{0} x \, dx + \frac{1}{\pi} \int_{0}^{\pi} x \, dx = \frac{2}{\pi} \int_{0}^{\pi} x \, dx = \pi,$$

$$a_k = \frac{1}{\pi} \int_{-\pi}^{\pi} |x| \cos kx \, dx = \frac{2}{\pi} \int_{0}^{\pi} x \cos kx \, dx = \frac{2}{\pi k^2} \left[(-1)^k - 1 \right],$$

for each $k = 1, 2, \ldots, n$, and

$$b_k = \frac{1}{\pi} \int_{-\pi}^{\pi} |x| \sin kx \, dx = 0, \quad \text{for each } k = 1, 2, \ldots, n - 1.$$

That the b_k's are all 0 follows from the fact that $g(x) = |x| \sin kx$ is an odd function for each k, and the integral of a continuous odd function over an interval of the form $[-a, a]$ is 0. (See Exercises 11 and 12.) The trigonometric polynomial from $\mathcal{T}_n$ approximating f is, therefore,

$$S_n(x) = \frac{\pi}{2} + \frac{2}{\pi} \sum_{k=1}^{n} \frac{(-1)^k - 1}{k^2} \cos kx.$$

The first few trigonometric polynomials for $f(x) = |x|$ are shown in Figure 8.10.   ■

Figure 8.10

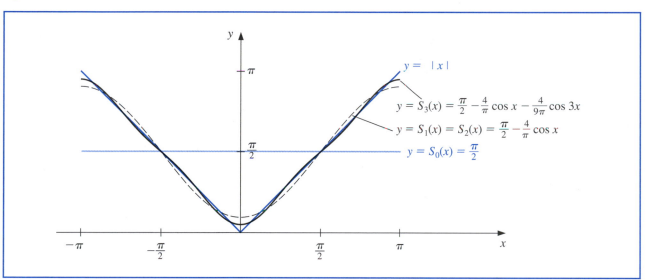

Discrete Trigonometric Approximation

There is a discrete analog to Fourier series that is useful for the least squares approximation and interpolation of large amounts of data when the data are given at equally spaced points. Suppose that a collection of $2m$ paired data points $\{(x_j, y_j)\}_{j=0}^{2m-1}$ is given, with the first elements in the pairs equally partitioning a closed interval. For convenience, we assume that the interval is $[-\pi, \pi]$ and that, as shown in Figure 8.11,

$$x_j = -\pi + \left(\frac{j}{m} \right) \pi \quad \text{for each } j = 0, 1, \ldots, 2m - 1.$$

Figure 8.11

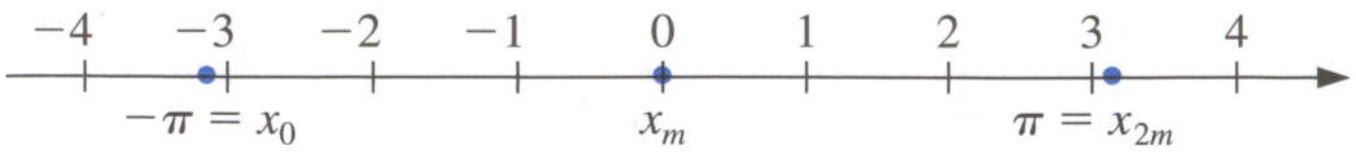

If this is not the case, a linear transformation can be used to change the data into this form.

The goal is to determine the trigonometric polynomial, $S_n(x)$, in $\mathcal{T}_n$ that minimizes

$$E(S_n) = \sum_{j=0}^{2m-1} (y_j - S_n(x_j))^2.$$

That is, we want to choose the constants $a_0, a_1, \ldots, a_n$ and $b_1, b_2, \ldots, b_{n-1}$ to minimize the total error

$$E(S_n) = \sum_{j=0}^{2m-1} \left[y_j - \left(\frac{a_0}{2} + a_n \cos nx_j + \sum_{k=1}^{n-1} (a_k \cos kx_j + b_k \sin kx_j) \right) \right]^2.$$

The determination of the constants is simplified by the fact that the set is orthogonal with respect to summation over the equally spaced points $\{x_j\}_{j=0}^{2m-1}$ in $[-\pi, \pi]$. By this we mean that, for each $k \neq l$,

$$\sum_{j=0}^{2m-1} \phi_k(x_j) \phi_l(x_j) = 0.$$

The orthogonality follows from the fact that if r and m are positive integers with $r < 2m$, we have (see Burden and Faires [BF], p. 543, for a verification)

$$\sum_{j=0}^{2m-1} \cos rx_j = 0 \quad \text{and} \quad \sum_{j=0}^{2m-1} \sin rx_j = 0.$$

To obtain the constants a_k for $k = 0, 1, \ldots, n$ and b_k for $k = 1, 2, \ldots, n-1$ in the summation

$$S_n(x) = \frac{a_0}{2} + a_n \cos nx + \sum_{k=1}^{n-1} (a_k \cos kx + b_k \sin kx),$$

we minimize the least squares sum

$$E(a_0, \ldots, a_n, b_1, \ldots, b_{n-1}) = \sum_{j=0}^{2m-1} (y_j - S_n(x_j))^2$$

by setting to zero the partial derivatives of E with respect to the a_k's and the b_k's. This implies that

$$a_k = \frac{1}{m} \sum_{j=0}^{2m-1} y_j \cos kx_j, \quad \text{for each } k = 0, 1, \ldots, n,$$

and

$$b_k = \frac{1}{m} \sum_{j=0}^{2m-1} y_j \sin kx_j, \quad \text{for each } k = 1, 2, \ldots, n-1.$$

Example 2   Find $S_2(x)$, the discrete least squares trigonometric polynomial of degree 2 for $f(x) = 2x^2 - 9$ when x is in $[-\pi, \pi]$.

Solution   We have $m = 2(2) - 1 = 3$, so the nodes are

$$x_j = \pi + \frac{j}{m}\pi \quad \text{and} \quad y_j = f(x_j) = 2x_j^2 - 9, \quad \text{for } j = 0, 1, 2, 3, 4, 5.$$

The trigonometric polynomial is

$$S_2(x) = \frac{1}{2}a_0 + a_2 \cos 2x + (a_1 \cos x + b_1 \sin x),$$

where

$$a_k = \frac{1}{3}\sum_{j=0}^{5} y_j \cos kx_j, \text{ for } k = 0, 1, 2, \quad \text{and} \quad b_1 = \frac{1}{3}\sum_{j=0}^{5} y_j \sin x_j.$$

The coefficients are

$$a_0 = \frac{1}{3}\left(f(-\pi) + f\left(-\frac{2\pi}{3}\right) + f\left(-\frac{\pi}{3}\right) + f(0) + f\left(\frac{\pi}{3}\right) + f\left(\frac{2\pi}{3}\right) \right)$$

$$= -4.10944566,$$

$$a_1 = \frac{1}{3}\left(f(-\pi)\cos(-\pi) + f\left(-\frac{2\pi}{3}\right)\cos\left(-\frac{2\pi}{3}\right) + f\left(-\frac{\pi}{3}\right)\cos\left(-\frac{\pi}{3}\right) \right.$$

$$\left. + f(0)\cos 0 + f\left(\frac{\pi}{3}\right)\cos\left(\frac{\pi}{3}\right) + f\left(\frac{2\pi}{3}\right)\cos\left(\frac{2\pi}{3}\right) \right) = -8.77298169,$$

$$a_2 = \frac{1}{3}\left(f(-\pi)\cos(-2\pi) + f\left(-\frac{2\pi}{3}\right)\cos\left(-\frac{4\pi}{3}\right) + f\left(-\frac{\pi}{3}\right)\cos\left(-\frac{2\pi}{3}\right) \right.$$

$$\left. + f(0)\cos 0 + f\left(\frac{\pi}{3}\right)\cos\left(\frac{2\pi}{3}\right) + f\left(\frac{2\pi}{3}\right)\cos\left(\frac{4\pi}{3}\right) \right) = 2.92432723,$$

and

$$b_1 = \frac{1}{3}\left(f(-\pi)\sin(-\pi) + f\left(-\frac{2\pi}{3}\right)\sin\left(-\frac{\pi}{3}\right) + f\left(-\frac{\pi}{3}\right)\left(-\frac{\pi}{3}\right) + f(0)\sin 0 \right.$$

$$\left. + f\left(\frac{\pi}{3}\right)\left(\frac{\pi}{3}\right) + f\left(\frac{2\pi}{3}\right)\left(\frac{2\pi}{3}\right) \right) = 0.$$

Thus

$$S_2(x) = \frac{1}{2}(-4.10944562) - 8.77298169 \cos x + 2.92432723 \cos 2x.$$

Figure 8.12 shows $f(x)$ and the discrete least squares trigonometric polynomial $S_2(x)$. ■

Figure 8.12

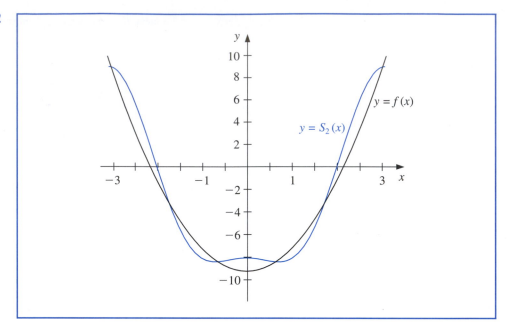

EXERCISE SET 8.6

1. Find the continuous least squares trigonometric polynomial $S_2(x)$ for $f(x) = x^2$ on $[-\pi, \pi]$.

2. Find the continuous least squares trigonometric polynomial $S_n(x)$ for $f(x) = x$ on $[-\pi, \pi]$.

3. Find the continuous least squares trigonometric polynomial $S_3(x)$ for $f(x) = e^x$ on $[-\pi, \pi]$.

4. Find the general continuous least squares trigonometric polynomial $S_n(x)$ for $f(x) = e^x$ on $[-\pi, \pi]$.

5. Find the general continuous least squares trigonometric polynomial $S_n(x)$ for

$$f(x) = \begin{cases} 0, & \text{if } -\pi < x \le 0, \\ 1, & \text{if } 0 < x < \pi. \end{cases}$$

6. Find the general continuous least squares trigonometric polynomial $S_n(x)$ for

$$f(x) = \begin{cases} -1, & \text{if } -\pi < x < 0, \\ 1, & \text{if } 0 \le x \le \pi. \end{cases}$$

7. Determine the discrete least squares trigonometric polynomial $S_n(x)$ on the interval $[-\pi, \pi]$ for the following functions, using the given values of m and n:

   a. $f(x) = \cos 2x$, $m = 4, n = 2$ b. $f(x) = \cos 3x$, $m = 4, n = 2$

   c. $f(x) = \sin \frac{1}{2}x + 2\cos \frac{1}{3}x$, $m = 6, n = 3$

   d. $f(x) = x^2 \cos x$, $m = 6, n = 3$

8. Compute the error $E(S_n)$ for each of the functions in Exercise 7.

9. Determine the discrete least squares trigonometric polynomial $S_3(x)$, using $m = 4$ for $f(x) = e^x \cos 2x$ on the interval $[-\pi, \pi]$. Compute the error $E(S_3)$.

10. Repeat Exercise 9 using $m = 8$. Compare the values of the approximating polynomials with the values of f at the points $\xi_j = -\pi + 0.2j\pi$, for $0 \le j \le 10$. Which approximation is better?

11. An *odd* function has the property that $f(-x) = -f(x)$ for all x in its domain. Show that for any continuous odd function f defined on the interval $[-a, a]$, we have $\int_{-a}^{a} f(x)\,dx = 0$.

12. An *even* function has the property that $f(-x) = f(x)$ for all x in its domain. Show that for any continuous even function f defined on the interval $[-a, a]$, we have $\int_{-a}^{a} f(x)\,dx = 2\int_{0}^{a} f(x)\,dx$.

13. Show that the functions $\phi_0(x) = 1/2, \phi_1(x) = \cos x, \ldots, \phi_n(x) = \cos nx, \phi_{n+1}(x) = \sin x, \ldots,$ $\phi_{2n-1}(x) = \sin(n-1)x$, are orthogonal on $[-\pi, \pi]$ with respect to $w(x) \equiv 1$.

14. In Example 1, the Fourier series was determined for $f(x) = |x|$. Use this series and the assumption that it represents f at zero to find the value of the convergent infinite series

$$\sum_{k=0}^{\infty} \frac{1}{(2k+1)^2}.$$

8.7 Fast Fourier Transforms

The *interpolatory* trigonometric polynomial on the $2m$ data points $\{(x_j, y_j)\}_{j=0}^{2m-1}$ is the least squares polynomial from $\mathcal{T}_m$ for this collection of points. The least squares trigonometric polynomial $S_n(x)$, when $n < m$, was found in Section 8.6 to be

$$S_n(x) = \frac{a_0}{2} + a_n \cos nx + \sum_{k=1}^{n-1}(a_k \cos kx + b_k \sin kx), \tag{8.6}$$

where

$$a_k = \frac{1}{m} \sum_{j=0}^{2m-1} y_j \cos kx_j \quad \text{for each } k = 0, 1, \ldots, n \tag{8.7}$$

and

$$b_k = \frac{1}{m} \sum_{j=0}^{2m-1} y_j \sin kx_j \quad \text{for each } k = 1, 2, \ldots, n-1. \tag{8.8}$$

We can use this form with $n = m$ for interpolation if we make a minor modification. For interpolation, we balance the system by replacing the term a_m with $a_m/2$. The interpolatory polynomial then has the form

$$S_m(x) = \frac{a_0 + a_m \cos mx}{2} + \sum_{k=1}^{m-1}(a_k \cos kx + b_k \sin kx),$$

where the coefficients a_k and b_k are given in Eqs. (8.4) and (8.5) on page 350.

The interpolation of large amounts of equally-spaced data by trigonometric polynomials can produce very accurate results. It is the appropriate approximation technique in areas involving digital filters, antenna field patterns, quantum mechanics, optics, and certain simulation problems. Until the middle of the 1960s, the method had not been extensively applied due to the number of arithmetic calculations required for the determination of the constants in the approximation. The interpolation of $2m$ data points requires approximately $(2m)^2$ multiplications and $(2m)^2$ additions by the direct calculation technique. The approximation of many thousands of data points is not unusual in areas requiring trigonometric interpolation, so the direct methods for evaluating the constants require multiplication and addition operations numbering in the millions. The computation time for this many calculations is prohibitive, and the round-off error would generally dominate the approximation.

In 1965 a paper by Cooley and Tukey [CT] described an alternative method of calculating the constants in the interpolating trigonometric polynomial. This method requires only $O(m \log_2 m)$ multiplications and $O(m \log_2 m)$ additions, provided m is chosen in an appropriate manner. For a problem with thousands of data points, this reduces the number of calculations from millions to thousands. The method had actually been discovered a number of years before the Cooley-Tukey paper appeared but had gone unnoticed by most researchers until that time. ([Brigh, pp. 8–9] contains a short, but interesting, historical summary of the method.)

The method described by Cooley and Tukey is generally known as the **Fast Fourier Transform (FFT) method** and led to a revolution in the use of interpolatory trigonometric polynomials. The method consists of organizing the problem so that the number of data points being used can be easily factored, particularly into powers of 2.

The relationship between the number, $2m$, of data points and the degree of the trigonometric polynomial used in the fast Fourier transform procedure allows some notational simplification. The equally-spaced nodes are given, for each $j = 0, 1, \ldots, 2m - 1$, by

$$x_j = -\pi + \left(\frac{j}{m}\right)\pi$$

and the coefficients, for each $k = 0, 1, \ldots, m$, by

$$a_k = \frac{1}{m}\sum_{j=0}^{2m-1} y_j \cos kx_j \quad \text{and} \quad b_k = \frac{1}{m}\sum_{j=0}^{2m-1} y_j \sin kx_j.$$

For notational convenience, b_0 and b_m have been added to the collection, but both are 0 and do not contribute to the sum.

Instead of directly evaluating the constants a_k and b_k, the fast Fourier transform procedure computes the complex coefficients c_k in the formula

$$\frac{1}{m}\sum_{k=0}^{2m-1} c_k e^{ikx},$$

where

$$c_k = \sum_{j=0}^{2m-1} y_j e^{\pi i j k/m}, \quad \text{for each } k = 0, 1, \ldots, 2m - 1. \tag{8.9}$$

Leonhard Euler first gave this formula in 1748 in *Introductio in analysin infinitorum*, which made the ideas of Johann Bernoulli more precise. This work based the calculus on the theory of elementary functions rather than curves.

Once the constants c_k have been determined, a_k and b_k can be recovered. To do this we need *Euler's formula*, which states that for all real numbers z we have

$$e^{iz} = \cos z + i \sin z,$$

where the complex number i satisfies $i^2 = -1$. Then, for each $k = 0, 1, \ldots, m$,

$$\frac{1}{m}c_k e^{-i\pi k} = \frac{1}{m}\sum_{j=0}^{2m-1} y_j e^{\pi i j k/m} e^{-i\pi k} = \frac{1}{m}\sum_{j=0}^{2m-1} y_j e^{ik(-\pi + (\pi j/m))}$$

$$= \frac{1}{m}\sum_{j=0}^{2m-1} y_j(\cos kx_j + i \sin kx_j),$$

so

$$\frac{1}{m}c_k e^{-i\pi k} = a_k + ib_k. \tag{8.10}$$

The operation reduction feature of the fast Fourier transform results from calculating the coefficients c_k in clusters. The following example gives a simple illustration of the technique.

Illustration   Consider the fast Fourier transform technique applied to $8 = 2^3$ data points $\{(x_j, y_j)\}_{j=0}^7$, where $x_j = -\pi + j\pi/4$, for each $j = 0, 1, \ldots, 7$. In this case $2m = 8$, so $m = 4 = 2^2$.
From Eq. (8.6) we have

$$S_4(x) = \frac{a_0 + a_4 \cos 4x}{2} + \sum_{k=1}^{3} (a_k \cos kx + b_k \sin kx),$$

where

$$a_k = \frac{1}{4} \sum_{j=0}^{7} y_j \cos kx_j \quad \text{and} \quad b_k = \frac{1}{4} \sum_{j=0}^{7} y_j \sin kx_j, \quad k = 0, 1, 2, 3, 4.$$

Define the Fourier transform as

$$\frac{1}{4} \sum_{j=0}^{7} c_k e^{ikx},$$

where

$$c_k = \sum_{j=0}^{7} y_j e^{ik\pi j/4}, \quad \text{for } k = 0, 1, \ldots, 7.$$

Then by Eq. (8.10), for $k = 0, 1, 2, 3, 4$, we have

$$\frac{1}{4} c_k e^{-ik\pi} = a_k + i b_k.$$

By direct calculation, the complex constants c_k are given by

$$c_0 = y_0 + y_1 + y_2 + y_3 + y_4 + y_5 + y_6 + y_7;$$

$$c_1 = y_0 + \left(\frac{i+1}{\sqrt{2}}\right) y_1 + i y_2 + \left(\frac{i-1}{\sqrt{2}}\right) y_3 - y_4 - \left(\frac{i+1}{\sqrt{2}}\right) y_5 - i y_6 - \left(\frac{i-1}{\sqrt{2}}\right) y_7;$$

$$c_2 = y_0 + i y_1 - y_2 - i y_3 + y_4 + i y_5 - y_6 - i y_7;$$

$$c_3 = y_0 + \left(\frac{i-1}{\sqrt{2}}\right) y_1 - i y_2 + \left(\frac{i+1}{\sqrt{2}}\right) y_3 - y_4 - \left(\frac{i-1}{\sqrt{2}}\right) y_5 + i y_6 - \left(\frac{i+1}{\sqrt{2}}\right) y_7;$$

$$c_4 = y_0 - y_1 + y_2 - y_3 + y_4 - y_5 + y_6 - y_7;$$

$$c_5 = y_0 - \left(\frac{i+1}{\sqrt{2}}\right) y_1 + i y_2 - \left(\frac{i-1}{\sqrt{2}}\right) y_3 - y_4 + \left(\frac{i+1}{\sqrt{2}}\right) y_5 - i y_6 + \left(\frac{i-1}{\sqrt{2}}\right) y_7;$$

$$c_6 = y_0 - i y_1 - y_2 + i y_3 + y_4 - i y_5 - y_6 + i y_7;$$

$$c_7 = y_0 - \left(\frac{i-1}{\sqrt{2}}\right) y_1 - i y_2 - \left(\frac{i+1}{\sqrt{2}}\right) y_3 - y_4 + \left(\frac{i-1}{\sqrt{2}}\right) y_5 + i y_6 + \left(\frac{i+1}{\sqrt{2}}\right) y_7.$$

Because of the small size of the collection of data points, many of the coefficients of the y_j in these equations are 1 or -1. This frequency will decrease in a larger application, so to count the computational operations accurately, multiplication by 1 or -1 will be included, even though it would not be necessary in this example. With this understanding, the direct computation of each equation requires 8 multiplications and 7 addition/subtractions, for a total of 64 multiplications and 56 additions/subtractions to determine $c_0, c_1, \ldots, c_7$.

To apply the fast Fourier transform procedure, we first define

$$d_0 = \frac{c_0 + c_4}{2} = y_0 + y_2 + y_4 + y_6; \qquad d_4 = \frac{c_2 + c_6}{2} = y_0 - y_2 + y_4 - y_6;$$

$$d_1 = \frac{c_0 - c_4}{2} = y_1 + y_3 + y_5 + y_7; \qquad d_5 = \frac{c_2 - c_6}{2} = i(y_1 - y_3 + y_5 - y_7);$$

$$d_2 = \frac{c_1 + c_5}{2} = y_0 + iy_2 - y_4 - iy_6; \qquad d_6 = \frac{c_3 + c_7}{2} = y_0 - iy_2 - y_4 + iy_6;$$

$$d_3 = \frac{c_1 - c_5}{2} \qquad\qquad\qquad\qquad d_7 = \frac{c_3 - c_7}{2}$$

$$= \left(\frac{i+1}{\sqrt{2}}\right)(y_1 + iy_3 - y_5 - iy_7); \qquad = \left(\frac{i-1}{\sqrt{2}}\right)(y_1 - iy_3 - y_5 + iy_7).$$

We then define

$$e_0 = \frac{d_0 + d_4}{2} = y_0 + y_4; \qquad e_4 = \frac{d_2 + d_6}{2} = y_0 - y_4;$$

$$e_1 = \frac{d_0 - d_4}{2} = y_2 + y_6; \qquad e_5 = \frac{d_2 - d_6}{2} = i(y_2 - y_6);$$

$$e_2 = \frac{id_1 + d_5}{2} = i(y_1 + y_5); \qquad e_6 = \frac{id_3 + d_7}{2} = \left(\frac{i-1}{\sqrt{2}}\right)(y_1 - y_5);$$

$$e_3 = \frac{id_1 - d_5}{2} = i(y_3 + y_7); \qquad e_7 = \frac{id_3 - d_7}{2} = i\left(\frac{i-1}{\sqrt{2}}\right)(y_3 - y_7).$$

Finally we define

$$f_0 = \frac{e_0 + e_4}{2} = y_0; \qquad f_4 = \frac{((i+1)/\sqrt{2})e_2 + e_6}{2} = \left(\frac{i-1}{\sqrt{2}}\right)y_1;$$

$$f_1 = \frac{e_0 - e_4}{2} = y_4; \qquad f_5 = \frac{((i+1)/\sqrt{2})e_2 - e_6}{2} = \left(\frac{i-1}{\sqrt{2}}\right)y_5;$$

$$f_2 = \frac{ie_1 + e_5}{2} = iy_2; \qquad f_6 = \frac{((i-1)/\sqrt{2})e_3 + e_7}{2} = \left(\frac{-i-1}{\sqrt{2}}\right)y_3;$$

$$f_3 = \frac{ie_1 - e_5}{2} = iy_6; \qquad f_7 = \frac{((i-1)/\sqrt{2})e_3 - e_7}{2} = \left(\frac{-i-1}{\sqrt{2}}\right)y_7.$$

The coefficients needed to compute the $c_0, \ldots, c_7, d_0, \ldots, d_7, e_0, \ldots, e_7$, and $f_0, \ldots, f_7$ are independent of the particular data points; they depend only on the fact that $m = 4$. For each m there is a unique set of constants $\{c_k\}_{k=0}^{2m-1}$, $\{d_k\}_{k=0}^{2m-1}$, $\{e_k\}_{k=0}^{2m-1}$, and $\{f_k\}_{k=0}^{2m-1}$. This portion of the work is not needed for a particular application, only the following calculations are required:

The f_k:

$$f_0 = y_0; \quad f_1 = y_4; \quad f_2 = iy_2; \quad f_3 = iy_6;$$

$$f_4 = \left(\frac{i-1}{\sqrt{2}}\right)y_1; \quad f_5 = \left(\frac{i-1}{\sqrt{2}}\right)y_5; \quad f_6 = -\left(\frac{i+1}{\sqrt{2}}\right)y_3; \quad f_7 = -\left(\frac{i+1}{\sqrt{2}}\right)y_7.$$

Once we have the f_k, we can find the e_k:

$$e_0 = f_0 + f_1; \quad e_1 = -i(f_2 + f_3); \quad e_2 = -\left(\frac{i-1}{\sqrt{2}}\right)(f_4 + f_5);$$

$$e_3 = -\left(\frac{i+1}{\sqrt{2}}\right)(f_6 + f_7); \quad e_4 = f_0 - f_1;$$

$$e_5 = f_2 - f_3; \quad e_6 = f_4 - f_5; \quad e_7 = f_6 - f_7.$$

Then, we can find the d_k:

$$d_0 = e_0 + e_1; \quad d_1 = -i(e_2 + e_3); \quad d_2 = e_4 + e_5; \quad d_3 = -i(e_6 + e_7);$$

$$d_4 = e_0 - e_1; \quad d_5 = e_2 - e_3; \quad d_6 = e_4 - e_5; \quad d_7 = e_6 - e_7.$$

Finally, we find the c_k:

$$c_0 = d_0 + d_1; \quad c_1 = d_2 + d_3; \quad c_2 = d_4 + d_5; \quad c_3 = d_6 + d_7;$$

$$c_4 = d_0 - d_1; \quad c_5 = d_2 - d_3; \quad c_6 = d_4 - d_5; \quad c_7 = d_6 - d_7.$$

Computing the constants $c_0, c_1, \ldots, c_7$ in this manner requires the number of operations shown in Table 8.7. Note again that multiplication by 1 or -1 has been included in the count, even though this does not require computational effort.

Table 8.7

Step	Multiplications/Divisions	Additions/Subtractions
(The f_k:)	8	0
(The e_k:)	8	8
(The d_k:)	8	8
(The c_k:)	0	8
Total	24	24

The lack of multiplications/divisions when finding the c_k reflects the fact that, for any m, the coefficients $\{c_k\}_{k=0}^{2m-1}$ are computed from $\{d_k\}_{k=0}^{2m-1}$ in the same manner:

$$c_k = d_{2k} + d_{2k+1} \quad \text{and} \quad c_{k+m} = d_{2k} - d_{2k+1}, \quad \text{for } k = 0, 1, \ldots, m-1,$$

so no complex multiplication is involved. In summary, the direct computation of the coefficients $c_0, c_1, \ldots, c_7$ requires 64 multiplications/divisions and 56 additions/subtractions. The fast Fourier transform technique reduces the computations to 24 multiplications/divisions and 24 additions/subtractions. □

The program FFTRNS82 implements the fast Fourier method.

Program FFTRNS82 performs the fast Fourier transform when $m = 2^p$ for some positive integer p. Modifications to the technique can be made when m takes other forms.

Example 1 Find the interpolating trigonometric polynomial of degree 2 on $[-\pi, \pi]$ for the data $\{(x_j, f(x_j))\}_{j=0}^{3}$, where $f(x) = 2x^2 - 9$,

$$a_k = \frac{1}{2}\sum_{j=0}^{3} f(x_j)\cos(kx_j) \quad \text{for } k = 0, 1, 2 \quad \text{and} \quad b_1 = \frac{1}{2}\sum_{j=0}^{3} f(x_j)\sin(x_j).$$

Solution We have

$$a_0 = \frac{1}{2}\left(f(-\pi) + f\left(-\frac{\pi}{2}\right) + f(0) + f\left(\frac{\pi}{2}\right) \right) = -3.19559339,$$

$$a_1 = \frac{1}{2}\left(f(-\pi)\cos(-\pi) + f\left(-\frac{\pi}{2}\right)\cos\left(-\frac{\pi}{2}\right) + f(0)\cos 0 + f\left(\frac{\pi}{2}\right)\cos\left(\frac{\pi}{2}\right) \right)$$

$$= -9.86960441,$$

$$a_2 = \frac{1}{2}\left(f(-\pi)\cos(-2\pi) + f\left(-\frac{\pi}{2}\right)\cos(-\pi) + f(0)\cos 0 + f\left(\frac{\pi}{2}\right)\cos(\pi) \right)$$

$$= 4.93480220,$$

and

$$b_1 = \frac{1}{2}\left(f(-\pi)\sin(-\pi) + f\left(-\frac{\pi}{2}\right)\sin\left(-\frac{\pi}{2}\right) + f(0)\sin 0 + f\left(\frac{\pi}{2}\right)\sin\left(\frac{\pi}{2}\right) \right) = 0.$$

So

$$S_2(x) = \frac{1}{2}(-3.19559339 + 4.93480220\cos 2x) - 9.86960441\cos x.$$

Figure 8.13 shows $f(x)$ and the interpolating trigonometric polynomial $S_2(x)$. ▪

Figure 8.13

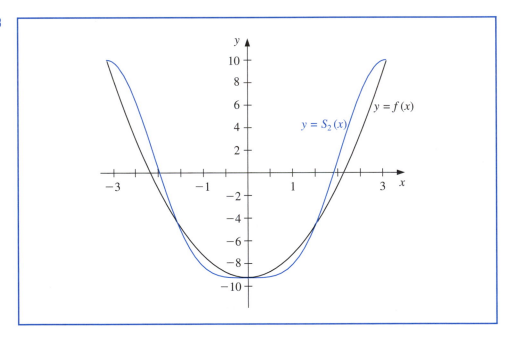

EXERCISE SET 8.7

1. Determine the trigonometric interpolating polynomial $S_2(x)$ of degree 2 on $[-\pi, \pi]$ for the following functions, and graph $y = f(x) - S_2(x)$.

   a. $f(x) = \pi(x - \pi)$

   b. $f(x) = x(\pi - x)$

     c. $f(x) = |x|$

     d. $f(x) = \begin{cases} -1, & \text{if } -\pi \le x \le 0, \\ 1, & \text{if } 0 < x \le \pi. \end{cases}$

2.   Determine the trigonometric interpolating polynomial of degree 4 for $f(x) = x(\pi - x)$ on the interval $[-\pi, \pi]$.

     a.   Use direct calculation.   **b.**   Use fast Fourier transforms.

3.   Use the fast Fourier transform method to compute the trigonometric interpolating polynomial of degree 4 on $[-\pi, \pi]$ for the following functions.

     a. $f(x) = \pi(x - \pi)$ **b.** $f(x) = |x|$

     c. $f(x) = \cos \pi x - 2 \sin \pi x$ **d.** $f(x) = x \cos x^2 + e^x \cos e^x$

4.   **a.**   Determine the trigonometric interpolating polynomial $S_4(x)$ of degree 4 for $f(x) = x^2 \sin x$ on the interval $[0, 1]$.

     b.   Compute $\int_0^1 S_4(x) \, dx$.

     c.   Compare the integral in part (b) to $\int_0^1 x^2 \sin x \, dx$.

5.   Use the approximations obtained in Exercise 3 to approximate the following integrals, and compare your results to the actual values.

     a. $\displaystyle\int_{-\pi}^{\pi} \pi(x - \pi) \, dx$ **b.** $\displaystyle\int_{-\pi}^{\pi} |x| \, dx$

     c. $\displaystyle\int_{-\pi}^{\pi} (\cos \pi x - 2 \sin \pi x) \, dx$ **d.** $\displaystyle\int_{-\pi}^{\pi} (x \cos x^2 + e^x \cos e^x) \, dx$

6.   Use FFTRNS82 to determine the trigonometric interpolating polynomial of degree 16 for $f(x) = x^2 \cos x$ on $[-\pi, \pi]$.

8.8   Survey of Methods and Software

In this chapter we have considered approximating data and functions with elementary functions. The elementary functions used were polynomials, rational functions, and trigonometric polynomials. We considered two types of approximations—discrete and continuous. Discrete approximations arise when approximating a finite set of data with an elementary function. Continuous approximations are used when the function to be approximated is known.

Discrete least squares techniques are recommended when the function is specified by giving a set of data that may not exactly represent the function. Least squares fit of data can take the form of a linear or other polynomial approximation or even an exponential form. These approximations are computed by solving sets of normal equations, as given in Section 8.2.

If the data are periodic, a trigonometric least squares fit might be appropriate. Because of the orthonormality of the trigonometric basis functions, the least squares trigonometric approximation does not require the solution of a linear system. For large amounts of periodic data, interpolation by trigonometric polynomials is also recommended. An efficient method of computing the trigonometric interpolating polynomial is given by the fast Fourier transform.

When the function to be approximated can be evaluated at any required argument, the approximations seek to minimize an integral instead of a sum. The continuous least squares polynomial approximations were considered in Section 8.3. Efficient computation of least squares polynomials led to orthonormal sets of polynomials, such as the Legendre and Chebyshev polynomials. Approximation by rational functions was studied in Section 8.5, where Padé approximation as a generalization of the Maclaurin polynomial was presented.

This method allows a more uniform method of approximation than polynomials. Continuous least squares approximation by trigonometric functions was discussed in Section 8.6, especially as it relates to Fourier series.

The IMSL Library and the NAG Libraries provide subroutines for least squares approximation of data. The approximation can be by polynomials, cubic splines, or a user's choice of basis functions. Chebyshev polynomials are also used in the constructions to minimize round-off error and enhance accuracy, and fast Fourier transformations are available.

The netlib Library contains the subroutine polfit.f under the package slatec to compute the polynomial least squares approximation to a discrete set of points. The subroutine pvalue.f can be used to evaluate the polynomial from polfit.f and any of its derivatives at a given point.

For further information on the general theory of approximation theory, see Powell [Po], Davis [Da], or Cheney [Ch]. A good reference for methods of least squares is Lawson and Hanson [LH], and information about Fourier transforms can be found in Van Loan [Van] and in Briggs and Hanson [BH].

CHAPTER

9

Approximating Eigenvalues

9.1 Introduction

Eigenvalues and eigenvectors were introduced in Chapter 7 in connection with the convergence of iterative methods for approximating the solution to a linear system. To determine the eigenvalues of an $n \times n$ matrix A, we construct the characteristic polynomial

$$p(\lambda) = \det(A - \lambda I)$$

and then determine its zeros. Finding the determinant of an $n \times n$ matrix is computationally expensive, and finding good approximations to the roots of $p(\lambda)$ is also difficult. In this chapter we explore other means for approximating the eigenvalues of a matrix, then in Section 9.6 we give an introduction to a technique for factoring a general $m \times n$ matrix into a form that has valuable applications in a number of areas.

9.2 Linear Algebra and Eigenvalues

In Chapter 7 we found that an iterative technique for solving a linear system will converge if all the eigenvalues associated with the problem have magnitude less than 1. The exact values of the eigenvalues in this case are not of primary importance—only the region of the complex plane in which they lie. An important result in this regard was first discovered by S. A. Geršgorin.

Geršgorin Circle Theorem

Let A be an $n \times n$ matrix and R_i denote the circle in the complex plane, $\mathcal{C}$, with center a_{ii} and radius $\sum_{j=1, j \neq i}^{n} |a_{ij}|$; that is,

$$R_i = \left\{ z \in \mathcal{C} \;\middle|\; |z - a_{ii}| \leq \sum_{j=1, j \neq i}^{n} |a_{ij}| \right\}.$$

The eigenvalues of A are contained within the union of these circles, $R = \bigcup_{i=1}^{n} R_i$. Moreover, the union of any k of the circles that do not intersect the remaining $(n - k)$ contains precisely k (counting multiplicities) of the eigenvalues.

Example 1   Determine the Geršgorin circles for the matrix

$$A = \begin{bmatrix} 4 & 1 & 1 \\ 0 & 2 & 1 \\ -2 & 0 & 9 \end{bmatrix},$$

and use these to find bounds for the spectral radius of A.

Solution   The circles in the Geršgorin Theorem are (see Figure 9.1)

$$R_1 = \{z \in \mathcal{C} \mid |z - 4| \leq 2\}, \quad R_2 = \{z \in \mathcal{C} \mid |z - 2| \leq 1\}, \quad \text{and} \quad R_3 = \{z \in \mathcal{C} \mid |z - 9| \leq 2\}.$$

Figure 9.1

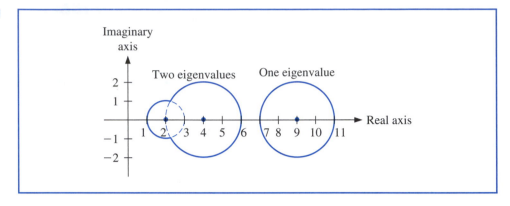

Because R_1 and R_2 are disjoint from R_3, there are precisely two eigenvalues within $R_1 \cup R_2$ and one within R_3. Moreover, $\rho(A) = \max_{1 \leq i \leq 3} |\lambda_i|$, so $7 \leq \rho(A) \leq 11$. ▪

Even when we need to find the eigenvalues, many techniques for their approximation are iterative. Determining regions in which they lie is the first step for finding the approximation, because it provides us with initial approximations.

Before considering additional results about eigenvalues and eigenvectors, we need some definitions and results from linear algebra. All the general results that will be needed in the remainder of this chapter are listed here for ease of reference. The proofs of many of the results that are not given are considered in the exercises, and all can be found in most standard texts on linear algebra (see, for example, [ND], [Poo], or [DG]).

The first definition parallels the definition for the linear independence of functions described in Chapter 8. In fact, much of what we will see in this section parallels material in that chapter.

Let $\{\mathbf{x}_1, \mathbf{x}_2, \mathbf{x}_3, \ldots, \mathbf{x}_k\}$ be a set of vectors. The set is **linearly independent** if the only solution to

$$\mathbf{0} = \alpha_1 \mathbf{x}_1 + \alpha_2 \mathbf{x}_2 + \alpha_3 \mathbf{x}_3 + \cdots + \alpha_k \mathbf{x}_k$$

is $\alpha_i = 0$, for each $i = 0, 1, \ldots, k$. Otherwise, the set of vectors is **linearly dependent**. Note that any set of vectors containing the zero vector is linearly dependent.

Semyon Aranovich Geršgorin (1901–1933) worked at the Petrograd Technological Institute until 1930, when he moved to the Leningrad Mechanical Engineering Institute. His 1931 paper *Über die Abgrenzung der Eigenwerte einer Matrix* ([Ger]) included what is now known as his Circle Theorem.

Unique Representation of Vectors

Suppose that $\{\mathbf{x}_1, \mathbf{x}_2, \mathbf{x}_3, \ldots, \mathbf{x}_n\}$ is a set of n linearly independent vectors in $\mathbb{R}^n$. Then for any vector $\mathbf{x} \in \mathbb{R}^n$, a unique collection of constants $\beta_1, \beta_2, \ldots, \beta_n$ exists with

$$\mathbf{x} = \beta_1 \mathbf{x}_1 + \beta_2 \mathbf{x}_2 + \beta_3 \mathbf{x}_3 + \cdots + \beta_n \mathbf{x}_n.$$

Any collection of n linearly independent vectors in $\mathbb{R}^n$ is called a **basis** for $\mathbb{R}^n$.

Example 2 **(a)** Show that $\mathbf{x}_1 = (1, 0, 0)^t$, $\mathbf{x}_2 = (-1, 1, 1)^t$, and $\mathbf{x}_3 = (0, 4, 2)^t$ is a basis for $\mathbb{R}^3$, and **(b)** given an arbitrary vector $\mathbf{x} \in \mathbb{R}^3$, find β_1, β_2, and β_3 with

$$\mathbf{x} = \beta_1 \mathbf{x}_1 + \beta_2 \mathbf{x}_2 + \beta_3 \mathbf{x}_3.$$

Solution **(a)** Let α_1, α_2, and α_3 be numbers with $\mathbf{0} = \alpha_1 \mathbf{x}_1 + \alpha_2 \mathbf{x}_2 + \alpha_3 \mathbf{x}_3$. Then

$$(0, 0, 0)^t = \alpha_1(1, 0, 0)^t + \alpha_2(-1, 1, 1)^t + \alpha_3(0, 4, 2)^t = (\alpha_1 - \alpha_2, \alpha_2 + 4\alpha_3, \alpha_2 + 2\alpha_3)^t,$$

so

$$\alpha_1 - \alpha_2 = 0, \quad \alpha_2 + 4\alpha_3 = 0, \quad \text{and} \quad \alpha_2 + 2\alpha_3 = 0.$$

The only solution to this system is $\alpha_1 = \alpha_2 = \alpha_3 = 0$, so this set $\{\mathbf{x}_1, \mathbf{x}_2, \mathbf{x}_3\}$ of 3 linearly independent vectors is a basis for $\mathbb{R}^3$.

(b) Let $\mathbf{x} = (x_1, x_2, x_3)^t$ be a vector in $\mathbb{R}^3$. Solving

$$\mathbf{x} = \beta_1 \mathbf{x}_1 + \beta_2 \mathbf{x}_2 + \beta_3 \mathbf{x}_3$$

$$= \beta_1(1, 0, 0)^t + \beta_2(-1, 1, 1)^t + \beta_3((0, 4, 2)^t = (\beta_1 - \beta_2, \beta_2 + 4\beta_3, \beta_2 + 2\beta_3)^t$$

is equivalent to solving for β_1, β_2, and β_3 in the system

$$\beta_1 - \beta_2 = x_1, \quad \beta_2 + 4\beta_3 = x_2, \quad \beta_2 + 2\beta_3 = x_3.$$

This system has the unique solution

$$\beta_1 = x_1 - x_2 + 2x_3, \quad \beta_2 = 2x_3 - x_2, \quad \text{and} \quad \beta_3 = \frac{1}{2}(x_2 - x_3). \qquad \blacksquare$$

The next result will be used in Section 9.3 to develop the Power method for approximating eigenvalues.

Linear Independence of Eigenvectors

If A is a matrix and $\lambda_1, \ldots, \lambda_k$ are distinct eigenvalues of A with associated eigenvectors $\mathbf{x}_1, \mathbf{x}_2, \ldots, \mathbf{x}_k$, then $\{\mathbf{x}_1, \mathbf{x}_2, \ldots, \mathbf{x}_k\}$ is a linearly independent set.

Example 3 Show that a basis can be formed for $\mathbb{R}^3$ using the eigenvectors of the 3×3 matrix

$$A = \begin{bmatrix} 2 & 0 & 0 \\ 1 & 1 & 2 \\ 1 & -1 & 4 \end{bmatrix}.$$

Solution In Example 1 of Section 7.3 (see page 286) we found that A has the characteristic polynomial

$$p(\lambda) = p(A - \lambda I) = (\lambda - 3)(\lambda - 2)^2.$$

Hence there are two distinct eigenvalues of A: $\lambda_1 = 3$ and $\lambda_2 = 2$. In that example we also found that $\lambda_1 = 3$ has the eigenvector $\mathbf{x}_1 = (0, 1, 1)^t$, and that there are two linearly independent eigenvectors $\mathbf{x}_2 = (0, 2, 1)^t$ and $\mathbf{x}_3 = (-2, 0, 1)^t$ corresponding to $\lambda_2 = 2$.

If the eigenvectors had corresponded to distinct eigenvalues, the previous result would ensure that

$$\{\mathbf{x}_1, \mathbf{x}_2, \mathbf{x}_3\} = \{(0, 1, 1)^t, (0, 2, 1)^t, (-2, 0, 1)^t\}$$

was linearly independent. However, they are not distinct, so we will use the definition as we did in Example 2(a). If

$$(0, 0, 0)^t = \alpha_1(0, 1, 1)^t + \alpha_2(0, 2, 1)^t + \alpha_3(-2, 0, 1)^t = (-2\alpha_3, \alpha_1 + 2\alpha_2, \alpha_1 + \alpha_2 + \alpha_3)^t,$$

then the first component tells us that $\alpha_3 = 0$. The second and third components imply that

$$0 = \alpha_1 + 2\alpha_2 \quad \text{and} \quad 0 = \alpha_1 + \alpha_2 \quad \text{so} \quad \alpha_2 = (\alpha_1 + 2\alpha_2) - (\alpha_1 + \alpha_2) = 0,$$

and $\alpha_1 = -\alpha_2 = 0$. Hence the set is linearly independent and forms a basis for $\mathbb{R}^3$. ▪

The eigenvectors of an $n \times n$ matrix do not always form a basis for $\mathbb{R}^n$, as the matrix in Exercise 5 illustrates. We will see that, when the number of linearly independent eigenvectors does not match the size of the matrix, there can be difficulties with the approximation methods for finding eigenvalues.

Orthogonal Vectors

In Section 8.3 we considered orthogonal and orthonormal sets of functions. Vectors with these properties are defined in a similar manner.

A set of vectors $\{\mathbf{v}_1, \mathbf{v}_2, \dots, \mathbf{v}_n\}$ is called **orthogonal** if $(\mathbf{v}_i)^t \mathbf{v}_j = 0$, for all $i \neq j$. If, in addition, $(\mathbf{v}_i)^t \mathbf{v}_i = 1$, for all $i = 1, 2, \dots, n$, then the set is called **orthonormal**.

Because $\mathbf{x}^t \mathbf{x} = \|\mathbf{x}\|_2^2$ for any $\mathbf{x}$ in $\mathbb{R}^n$, a set of orthogonal vectors $\{\mathbf{v}_1, \mathbf{v}_2, \dots, \mathbf{v}_n\}$ is orthonormal if and only if $\|\mathbf{v}_i\|_2 = 1$, for each $i = 1, 2, \dots, n$.

Linear Independence of Orthogonal Vectors

An orthogonal set of nonzero vectors is linearly independent.

Example 4 (a) Show that the vectors $\mathbf{v}_1 = (0, 4, 2)^t$, $\mathbf{v}_2 = (-5, -1, 2)^t$, and $\mathbf{v}_3 = (1, -1, 2)^t$ form an orthogonal set, and (b) use these to determine a set of orthonormal vectors.

Solution (a) We have $(\mathbf{v}_1)^t \mathbf{v}_2 = 0(-5) + 4(-1) + 2(2) = 0,$

$$(\mathbf{v}_1)^t \mathbf{v}_3 = 0(1) + 4(-1) + 2(2) = 0, \quad \text{and} \quad (\mathbf{v}_2)^t \mathbf{v}_3 = -5(1) - 1(-1) + 2(2) = 0,$$

so the vectors are orthogonal. They form a linearly independent set and a basis for $\mathbb{R}^3$. The l_2 norms of these vectors are

$$\|\mathbf{v}_1\|_2 = 2\sqrt{5}, \quad \|\mathbf{v}_2\|_2 = \sqrt{30}, \quad \text{and} \quad \|\mathbf{v}_3\|_2 = \sqrt{6}.$$

(b) The vectors

$$\mathbf{u}_1 = \frac{\mathbf{v}_1}{\|\mathbf{v}_1\|_2} = \left(\frac{0}{2\sqrt{5}}, \frac{4}{2\sqrt{5}}, \frac{2}{2\sqrt{5}}\right)^t = \left(0, \frac{2\sqrt{5}}{5}, \frac{\sqrt{5}}{5}\right)^t,$$

$$\mathbf{u}_2 = \frac{\mathbf{v}_2}{\|\mathbf{v}_2\|_2} = \left(\frac{-5}{\sqrt{30}}, \frac{-1}{\sqrt{30}}, \frac{2}{\sqrt{30}}\right)^t = \left(-\frac{\sqrt{30}}{6}, -\frac{\sqrt{30}}{30}, \frac{\sqrt{30}}{15}\right)^t,$$

$$\mathbf{u}_3 = \frac{\mathbf{v}_3}{\|\mathbf{v}_3\|_2} = \left(\frac{1}{\sqrt{6}}, \frac{-1}{\sqrt{6}}, \frac{2}{\sqrt{6}}\right)^t = \left(\frac{\sqrt{6}}{6}, -\frac{\sqrt{6}}{6}, \frac{\sqrt{6}}{3}\right)^t$$

form an orthonormal set because they inherit orthogonality from $\mathbf{v}_1, \mathbf{v}_2$, and $\mathbf{v}_3$, and additionally, $\|\mathbf{u}_1\|_2 = \|\mathbf{u}_2\|_2 = \|\mathbf{u}_3\|_2 = 1$. ∎

The **Gram-Schmidt** process for constructing a set of polynomials that are orthogonal with respect to a given weight function was described in Chapter 8 (see page 335). There is a parallel process, also known as Gram-Schmidt, that permits us to construct an orthogonal basis for $\mathbb{R}^n$ given a set of n linearly independent vectors in $\mathbb{R}^n$.

Gram-Schmidt Orthogonalization

Let $\{\mathbf{x}_1, \mathbf{x}_2, \ldots, \mathbf{x}_k\}$ be a set of k linearly independent vectors in $\mathbb{R}^n$. Then $\{\mathbf{v}_1, \mathbf{v}_2, \ldots, \mathbf{v}_k\}$, defined as follows, is orthogonal:

$$\mathbf{v}_1 = \mathbf{x}_1, \quad \mathbf{v}_2 = \mathbf{x}_2 - \left(\frac{\mathbf{v}_1^t \mathbf{x}_2}{\mathbf{v}_1^t \mathbf{v}_1}\right)\mathbf{v}_1, \quad \mathbf{v}_3 = \mathbf{x}_3 - \left(\frac{\mathbf{v}_1^t \mathbf{x}_3}{\mathbf{v}_1^t \mathbf{v}_1}\right)\mathbf{v}_1 - \left(\frac{\mathbf{v}_2^t \mathbf{x}_3}{\mathbf{v}_2^t \mathbf{v}_2}\right)\mathbf{v}_2,$$

and, in general, for $j = 2, \ldots, k$, $\qquad \mathbf{v}_j = \mathbf{x}_j - \sum_{i=1}^{j-1} \left(\frac{\mathbf{v}_i^t \mathbf{x}_j}{\mathbf{v}_i^t \mathbf{v}_i}\right)\mathbf{v}_i.$

To show this result, simply verify that for each $1 \leq i \leq k$ and $1 \leq j \leq k$, with $i \neq j$, we have $\mathbf{v}_i^t \mathbf{v}_j = 0$.

Note that when the original set of vectors $\{\mathbf{x}_1, \mathbf{x}_2, \ldots, \mathbf{x}_k\}$ forms a basis for $\mathbb{R}^n$, that is, when $k = n$, then the constructed vectors form an orthogonal basis for $\mathbb{R}^n$. From this we can form an orthonormal basis $\{\mathbf{u}_1, \mathbf{u}_2, \ldots, \mathbf{u}_n\}$ simply by defining, for each $i = 1, 2, \ldots, n$

$$\mathbf{u}_i = \frac{\mathbf{v}_i}{\|\mathbf{v}_i\|_2}.$$

The following example illustrates how an orthonormal basis for $\mathbb{R}^3$ can be constructed from three linearly independent vectors in $\mathbb{R}^3$.

Example 5   Use the Gram-Schmidt process to determine a set of orthonormal vectors from the linearly independent vectors

$$\mathbf{x}_1 = (0, 1, 1)^t, \quad \mathbf{x}_2 = (0, 2, 1)^t, \quad \text{and} \quad \mathbf{x}_3 = (-2, 0, 1)^t$$

that were eigenvectors for the matrix in Example 3.

Solution We have the orthogonal vectors $\mathbf{v}_1$, $\mathbf{v}_2$, and $\mathbf{v}_3$, given by

$$\mathbf{v}_1 = \mathbf{x}_1 = (0, 1, 1)^t$$

$$\mathbf{v}_2 = (0, 2, 1)^t - \left(\frac{((0, 1, 1)^t)^t (0, 2, 1)^t}{((0, 1, 1)^t)^t (0, 1, 1)^t} \right) (0, 1, 1)^t = (0, 2, 1)^t - \frac{3}{2}(0, 1, 1)^t$$

$$= \left(0, \frac{1}{2}, -\frac{1}{2} \right)^t,$$

and

$$\mathbf{v}_3 = (-2, 0, 1)^t - \left(\frac{((0, 1, 1)^t)^t (-2, 0, 1)^t}{((0, 1, 1)^t)^t (0, 1, 1)^t} \right) (0, 1, 1)^t$$

$$- \left(\frac{((0, \frac{1}{2}, -\frac{1}{2})^t)^t (-2, 0, 1)^t}{((0, \frac{1}{2}, -\frac{1}{2})^t)^t (0, \frac{1}{2}, -\frac{1}{2})^t} \right) (0, \frac{1}{2}, -\frac{1}{2})^t$$

$$= (-2, 0, 1)^t - \left(0, \frac{1}{2}, \frac{1}{2} \right)^t + \left(0, \frac{1}{2}, -\frac{1}{2} \right)^t = (-2, 0, 0)^t.$$

The l_2 norms of these vectors are

$$||\mathbf{v}_1||_2 = \sqrt{2}, \quad ||\mathbf{v}_2|||_2 = \frac{\sqrt{2}}{2}, \quad \text{and} \quad ||\mathbf{v}_3||_2 = 2,$$

so a set of orthonormal vectors corresponding to the orthogonal vectors are

$$\mathbf{u}_1 = \frac{\mathbf{v}_1}{||\mathbf{v}_1||_2} = \left(0, \frac{\sqrt{2}}{2}, \frac{\sqrt{2}}{2} \right)^t, \quad \mathbf{u}_2 = \frac{\mathbf{v}_2}{||\mathbf{v}_2||_2} = \left(0, \frac{\sqrt{2}}{2}, -\frac{\sqrt{2}}{2} \right)^t, \quad \text{and}$$

$$\mathbf{u}_3 = \frac{\mathbf{v}_3}{||\mathbf{v}_3||_2} = (-1, 0, 0)^t. \qquad ■$$

Orthogonal Matrices

We will now consider the connection between sets of vectors and the matrices formed using these vectors as their columns. We first consider some results about a class of special matrices. The terminology in the next definition follows from the fact that the columns of an orthogonal matrix form an orthogonal set of vectors.

> *It would probably be better to call orthogonal matrices* orthonormal *because the columns form not just an orthogonal set but an orthonormal set as well.*

A matrix Q is said to be **orthogonal** if its columns $\{\mathbf{q}_1^t, \mathbf{q}_2^t, \ldots, \mathbf{q}_n^t\}$ form an orthonormal set in $\mathbb{R}^n$.

The following are important properties of orthogonal matrices.

Orthogonal Matrix Properties

Suppose that Q is an orthogonal $n \times n$ matrix. Then

  (i) Q is invertible with $Q^{-1} = Q^t$.

  (ii)   For any $\mathbf{x}$ and $\mathbf{y}$ in $\mathbb{R}^n$, $(Q\mathbf{x})^t Q\mathbf{y} = \mathbf{x}^t \mathbf{y}$.

  (iii)   For any $\mathbf{x}$ in $\mathbb{R}^n$, $||Q\mathbf{x}||_2 = ||\mathbf{x}||_2$.

In addition, the converse of **(i)** also holds. That is,

- an invertible matrix Q with $Q^{-1} = Q^t$ is orthogonal.

As an example, the *permutation matrices* discussed in Section 6.5 have this property, so they are orthogonal matrices.

Property (iii) of the preceding result is often expressed by stating that

- orthogonal matrices are l_2-norm preserving.

As an immediate consequence of this property, every orthogonal matrix Q has $||Q||_2 = 1$.

Example 6   The matrix

$$Q = [\mathbf{u}_1, \mathbf{u}_2, \mathbf{u}_3] = \begin{bmatrix} 0 & 0 & -1 \\ \frac{\sqrt{2}}{2} & \frac{\sqrt{2}}{2} & 0 \\ \frac{\sqrt{2}}{2} & -\frac{\sqrt{2}}{2} & 0 \end{bmatrix}$$

was formed by using the orthonormal set of vectors found in Example 5 as its columns. Show that Q is an orthogonal matrix.

Solution   Note that

$$QQ^t = \begin{bmatrix} 0 & 0 & -1 \\ \frac{\sqrt{2}}{2} & \frac{\sqrt{2}}{2} & 0 \\ \frac{\sqrt{2}}{2} & -\frac{\sqrt{2}}{2} & 0 \end{bmatrix} \cdot \begin{bmatrix} 0 & \frac{\sqrt{2}}{2} & \frac{\sqrt{2}}{2} \\ 0 & \frac{\sqrt{2}}{2} & -\frac{\sqrt{2}}{2} \\ -1 & 0 & 0 \end{bmatrix} = \begin{bmatrix} 1 & 0 & 0 \\ 0 & 1 & 0 \\ 0 & 0 & 1 \end{bmatrix} = I.$$

This is sufficient to ensure that $Q^t = Q^{-1}$. Hence Q is an orthogonal matrix. ■

Similar Matrices

The next definition provides the basis for many of the techniques for determining the eigenvalues of a matrix.

Two matrices A and B are said to be **similar** if a nonsingular matrix S exists with $A = S^{-1}BS$.

An important feature of similar matrices is that they have the same eigenvalues.

Eigenvalues and Eigenvectors of Similar Matrices

Suppose A and B are similar matrices with $A = S^{-1}BS$ and λ is an eigenvalue of A with associated eigenvector $\mathbf{x}$. Then λ is an eigenvalue of B with associated eigenvector $S\mathbf{x}$.

A particularly important use of similarity occurs when an $n \times n$ matrix A is similar to a diagonal matrix; that is, when a diagonal matrix D and an invertible matrix S exist with

$$A = S^{-1}DS \quad \text{or equivalently} \quad D = SAS^{-1}.$$

In this case the matrix is said to be **diagonalizable**. The following result gives a characterization of diagonalizable matrices.

Matrices Similar to Diagonal Matrices

An $n \times n$ matrix A is similar to a diagonal matrix D if and only if A has n linearly independent eigenvectors. In this case, we define $D = S^{-1}AS$, where

- the columns of S consist of the eigenvectors, and

- the ith diagonal element of D is the eigenvalue of A that corresponds to the ith column of S.

In particular, because eigenvectors corresponding to distinct eigenvalues are orthogonal:

- An $n \times n$ matrix A that has n distinct eigenvalues is similar to a diagonal matrix.

The pair of matrices S and D is not unique. As the following example illustrates, any reordering of the columns of S and corresponding reordering of the diagonal elements of D will give a distinct pair.

Example 7 The characteristic polynomial of the matrix

$$A = \begin{bmatrix} 2 & 0 & 0 \\ 1 & 1 & 2 \\ 1 & -1 & 4 \end{bmatrix}$$

is

$$p(\lambda) = p(A - \lambda I) = (\lambda - 3)(\lambda - 2)^2,$$

as we saw is Example 1 of Section 7.3 and in Example 3 in this section. Eigenvectors are $\mathbf{x}_1 = (0, 1, 1)^t$ for the eigenvalue $\lambda_1 = 3$, and $\mathbf{x}_2 = (0, 2, 1)^t$ and $\mathbf{x}_3 = (-2, 0, 1)^t$ for $\lambda_2 = 2$. How many different diagonal matrices are similar to A?

Solution The number of different diagonal matrices similar to A is the number of ways that the eigenvalues can be arranged along the diagonal. Because 2 is a double zero of the characteristic polynomial of A, there are only 3 different diagonal matrices, which are

$$D_1 = \begin{bmatrix} 3 & 0 & 0 \\ 0 & 2 & 0 \\ 0 & 0 & 2 \end{bmatrix}, \quad D_2 = \begin{bmatrix} 2 & 0 & 0 \\ 0 & 3 & 0 \\ 0 & 0 & 2 \end{bmatrix}, \quad \text{and} \quad D_3 = \begin{bmatrix} 2 & 0 & 0 \\ 0 & 2 & 0 \\ 0 & 0 & 3 \end{bmatrix}.$$

If there had been 3 distinct eigenvalues, there would have been 6 different diagonal matrices. ■

The following result for symmetric matrices will be used in Section 9.6 when we construct the Singular Valued Decomposition of an arbitrary matrix. It tells us that symmetric matrices are characterized by the fact that they are diagonalizable and that the matrix used for the diagonalization is orthogonal.

Diagonalization of Symmetric Matrices

The $n \times n$ matrix A is symmetric if and only if there exists a diagonal matrix D and an orthogonal matrix Q with $A = QDQ^{-1} = QDQ^t$.

The following result follows from the diagonalization property of symmetric matrices. Its proof is outlined in Exercises 10 and 11.

Eigenvalues of Symmetric Matrices

Suppose that A is a symmetric $n \times n$ matrix. Then

(i) there exist n eigenvectors of A that form an orthonormal set, and

(ii) the eigenvalues of A are real numbers.

The final result in this section concerns the eigenvalues of positive definite matrices. Recall from Chapter 6 that a symmetric matrix A is called **positive definite** if for all nonzero vectors $\mathbf{x}$ we have $\mathbf{x}^t A \mathbf{x} > 0$. The following property makes positive definite matrices important in applications.

Eigenvalues of Positive Definite Matrices

A symmetric matrix A is positive definite if and only if all the eigenvalues of A are positive.

Illustration   The matrix

$$A = \begin{bmatrix} 2 & -1 & 0 \\ -1 & 2 & -1 \\ 0 & -1 & 2 \end{bmatrix}$$

can be shown to be positive definite by showing that, for any nonzero vector $\mathbf{x}$, we have

$$\mathbf{x}^t A \mathbf{x} = (x_1, x_2, x_3) \begin{bmatrix} 2 & -1 & 0 \\ -1 & 2 & -1 \\ 0 & -1 & 2 \end{bmatrix} (x_1, x_2, x_3)^t$$

$$= (x_1, x_2, x_3)(2x_1 - x_2, -x_1 + 2x_2 - x_3, -x_2 + 2x_3)^t$$

$$= x_1^2 + \left(x_1^2 - 2x_1x_2 + x_2^2 \right) + \left(x_2^2 - 2x_2x_3 + x_3^2 \right) + x_3^2$$

$$= x_1^2 + (x_1 - x_2)^2 + (x_2 - x_3)^2 + x_3^2.$$

So $\mathbf{x}^t A \mathbf{x} \geq 0$, and if it is 0 then all the terms in the last expression must be 0. This implies that $x_1 = x_2 = x_3 = 0$.

However, it is easier to use the result to establish that A is positive definite. The characteristic polynomial of A is

$$\rho(\lambda) = (2 - \lambda)^3 - 2(2 - \lambda) = (2 - \lambda)(\lambda^2 - 4\lambda + 2),$$

so the eigenvalues of the symmetric matrix A are $\lambda_1 = 2$, $\lambda_2 = 2 + \sqrt{2}$, and $\lambda_3 = 2 - \sqrt{2}$. These are all positive, so A is positive definite. □

EXERCISE SET 9.2

1. Use the Geršgorin Circle Theorem to determine bounds for **(i)** the eigenvalues and **(ii)** the spectral radius of the following matrices.

   a. $\begin{bmatrix} 1 & 0 & 0 \\ -1 & 0 & 1 \\ -1 & -1 & 2 \end{bmatrix}$
   b. $\begin{bmatrix} 4 & -1 & 0 \\ -1 & 4 & -1 \\ -1 & -1 & 4 \end{bmatrix}$

   c. $\begin{bmatrix} 3 & 2 & 1 \\ 2 & 3 & 0 \\ 1 & 0 & 3 \end{bmatrix}$
   d. $\begin{bmatrix} 4.75 & 2.25 & -0.25 \\ 2.25 & 4.75 & 1.25 \\ -0.25 & 1.25 & 4.75 \end{bmatrix}$

2. Find the eigenvalues and associated eigenvectors for the following 3×3 matrices. Is there a set of three linearly independent eigenvectors?

   a. $A = \begin{bmatrix} 2 & -3 & 6 \\ 0 & 3 & -4 \\ 0 & 2 & -3 \end{bmatrix}$
   b. $A = \begin{bmatrix} 1 & 0 & 0 \\ -1 & 0 & 1 \\ -1 & -1 & 2 \end{bmatrix}$

   c. $A = \begin{bmatrix} 2 & 0 & 1 \\ 0 & 2 & 0 \\ 1 & 0 & 2 \end{bmatrix}$
   d. $A = \begin{bmatrix} 2 & -1 & -1 \\ -1 & 2 & -1 \\ -1 & -1 & 2 \end{bmatrix}$

   e. $A = \begin{bmatrix} 1 & 1 & 1 \\ 1 & 1 & 0 \\ 1 & 0 & 1 \end{bmatrix}$
   f. $A = \begin{bmatrix} 2 & 1 & 1 \\ 1 & 2 & 1 \\ 1 & 1 & 2 \end{bmatrix}$

3. The matrices in Exercise 2(c), (d), (e), and (f) are symmetric.

   a. Are any positive definite?

   b. Consider the positive definite matrices in (a). Construct an orthogonal matrix Q for which $Q^t A Q = D$, a diagonal matrix, using the eigenvectors found in Exercise 2.

4. Show that the matrix given in Examples 3 and 7,

   $$A = \begin{bmatrix} 2 & 0 & 0 \\ 1 & 1 & 2 \\ 1 & -1 & 4 \end{bmatrix},$$

   is similar to the diagonal matrices

   $$D_1 = \begin{bmatrix} 3 & 0 & 0 \\ 0 & 2 & 0 \\ 0 & 0 & 2 \end{bmatrix}, \quad D_2 = \begin{bmatrix} 2 & 0 & 0 \\ 0 & 3 & 0 \\ 0 & 0 & 2 \end{bmatrix}, \quad \text{and} \quad D_3 = \begin{bmatrix} 2 & 0 & 0 \\ 0 & 2 & 0 \\ 0 & 0 & 3 \end{bmatrix}.$$

5. Show that no collection of eigenvectors of the 3×3 matrix

   $$A = \begin{bmatrix} 2 & 1 & 0 \\ 0 & 2 & 0 \\ 0 & 0 & 3 \end{bmatrix}$$

   can form a basis for $\mathbb{R}^3$.

6. Consider the sets of the following vectors. **(i)** Show that the set is linearly independent; **(ii)** use the Gram-Schmidt process to find a set of orthogonal vectors; **(iii)** determine a set of orthonormal vectors from the vectors in (ii).

   a. $\mathbf{x}_1 = (1, 1)^t, \ \mathbf{x}_2 = (-2, 1)^t$

   b. $\mathbf{x}_1 = (1, 1, 0)^t, \ \mathbf{x}_2 = (1, 0, 1)^t, \ \mathbf{x}_3 = (0, 1, 1)^t$

   c. $\mathbf{x}_1 = (1, 1, 1, 1)^t, \ \mathbf{x}_2 = (0, 2, 2, 2)^t, \ \mathbf{x}_3 = (1, 0, 0, 1)^t$

   d. $\mathbf{x}_1 = (2, 2, 3, 2, 3)^t, \ \mathbf{x}_2 = (2, -1, 0, -1, 0)^t, \ \mathbf{x}_3 = (0, 0, 1, 0, -1)^t, \ \mathbf{x}_4 = (1, 2, -1, 0, -1)^t,$
   $\mathbf{x}_5 = (0, 1, 0, -1, 0)^t$

7. Show that any four vectors in $\mathbb{R}^3$ are linearly dependent.

8.   Let Q be an orthogonal matrix. Show that $\|Q\|_2 = \|Q^t\|_2 = 1$.

9.   Let $\{\mathbf{v}_1, \ldots, \mathbf{v}_n\}$ be a set of nonzero orthonormal vectors in $\mathbb{R}^n$ and $\mathbf{x} \in \mathbb{R}^n$. Show that

$$\mathbf{x} = \sum_{k=1}^{n} c_k \mathbf{v}_k, \qquad \text{where } c_k = \mathbf{v}_k^t \mathbf{x}.$$

10.   Show that if A is a symmetric matrix, then there exist n eigenvectors of A that form an orthonormal set. [Hint: Consider the matrices Q and D for which the result about Diagonalization of Symmetric Matrices implies that $A = QDQ^t$. Show that, for each $1 \le i \le n$, the ith column of Q, $\mathbf{v}_i = (q_{1i}, q_{2i}, \ldots, q_{ni})^t$, is an eigenvector of A corresponding to the eigenvalue that is the ith diagonal element of D.]

11.   Use the result in Exercise 10 to show that the eigenvalues of a symmetric matrix are real numbers. [Hint: With the notation in the Hint to Exercise 10 we have $A\mathbf{v}_i = d_{ii}\mathbf{v}_i$, for each $1 \le i \le n$, and that $d_{ii} = \mathbf{v}_i^t A \mathbf{v}_i$.]

12.   In Exercise 11 of Section 6.6, a symmetric matrix

$$A = \begin{bmatrix} 1.59 & 1.69 & 2.13 \\ 1.69 & 1.31 & 1.72 \\ 2.13 & 1.72 & 1.85 \end{bmatrix}$$

was used to describe the average wing lengths of fruit flies that were offspring resulting from the mating of three mutants of the flies. The entry a_{ij} represents the average wing length of a fly that is the offspring of a male fly of type i and a female fly of type j.

a.   Find the eigenvalues and associated eigenvectors of this matrix.

b.   Use a result in this section to answer the question posed in (b) of Exercise 11, Section 6.6: Is this matrix positive definite?

9.3   The Power Method

The Power method is an iterative technique used to determine the *dominant* eigenvalue of a matrix; that is, the eigenvalue with the largest magnitude. By modifying the method it can also be used to determine other eigenvalues. One useful feature of the Power method is that it produces not only an eigenvalue, but an associated eigenvector. In fact, the Power method is often applied to find an eigenvector for an eigenvalue that is determined by some other means.

To apply the Power method, we first assume that the $n \times n$ matrix A has n eigenvalues, $\lambda_1, \lambda_2, \ldots, \lambda_n$, with an associated collection of linearly independent eigenvectors, $\{\mathbf{v}^{(1)}, \mathbf{v}^{(2)}, \mathbf{v}^{(3)}, \ldots, \mathbf{v}^{(n)}\}$. Moreover, we assume that A has precisely one eigenvalue, λ_1, that is largest in magnitude, and that

$$|\lambda_1| > |\lambda_2| \ge |\lambda_3| \ge \cdots \ge |\lambda_n| \ge 0.$$

If $\mathbf{x}$ is any vector in $\mathbb{R}^n$, the fact that $\{\mathbf{v}^{(1)}, \mathbf{v}^{(2)}, \mathbf{v}^{(3)}, \ldots, \mathbf{v}^{(n)}\}$ is linearly independent implies that constants $\beta_1, \beta_2, \ldots, \beta_n$ exist with

$$\mathbf{x} = \sum_{j=1}^{n} \beta_j \mathbf{v}^{(j)}.$$

Multiplying both sides of this equation successively by $A, A^2, \ldots, A^k$ gives

$$A\mathbf{x} = \sum_{j=1}^{n} \beta_j A\mathbf{v}^{(j)} = \sum_{j=1}^{n} \beta_j \lambda_j \mathbf{v}^{(j)},$$

$$A^2\mathbf{x} = \sum_{j=1}^{n} \beta_j \lambda_j A\mathbf{v}^{(j)} = \sum_{j=1}^{n} \beta_j \lambda_j^2 \mathbf{v}^{(j)},$$

$$\vdots$$

$$A^k\mathbf{x} = \sum_{j=1}^{n} \beta_j \lambda_j^k \mathbf{v}^{(j)}.$$

If λ_1^k is factored from each term on the right side of the last equation, then

$$A^k\mathbf{x} = \lambda_1^k \sum_{j=1}^{n} \beta_j \left(\frac{\lambda_j}{\lambda_1}\right)^k \mathbf{v}^{(j)}.$$

Since $|\lambda_1| > |\lambda_j|$ for all $j = 2, 3, \ldots, n$, we have $\lim_{k \to \infty} (\lambda_j/\lambda_1)^k = 0$, and

$$\lim_{k \to \infty} A^k\mathbf{x} = \lim_{k \to \infty} \lambda_1^k \beta_1 \mathbf{v}^{(1)}.$$

This sequence converges to zero if $|\lambda_1| < 1$ and diverges if $|\lambda_1| > 1$, provided, of course, that $\beta_1 \neq 0$. In neither situation will this limit permit us to determine λ_1. However, advantage can be made of this relationship by scaling the powers of $A^k\mathbf{x}$ in an appropriate manner to ensure that the limit is finite and nonzero.

The scaling begins by choosing an l_∞ unit vector $\mathbf{x}^{(0)}$ and a component $x_{p_0}^{(0)}$ of $\mathbf{x}^{(0)}$ with

$$x_{p_0}^{(0)} = \|\mathbf{x}^{(0)}\|_\infty = 1.$$

Let $\mathbf{y}^{(1)} = A\mathbf{x}^{(0)}$ and define $\mu^{(1)} = y_{p_0}^{(1)}$. Since

$$\mathbf{x}^{(0)} = \sum_{j=1}^{n} \beta_j \mathbf{v}^{(j)} \quad \text{and} \quad A\mathbf{x}^{(0)} = \sum_{j=1}^{n} \beta_j A\mathbf{v}^{(j)} = \sum_{j=1}^{n} \beta_j \lambda_j \mathbf{v}^{(j)},$$

we have

$$\mu^{(1)} = y_{p_0}^{(1)} = \frac{y_{p_0}^{(1)}}{x_{p_0}^{(0)}} = \frac{\beta_1 \lambda_1 v_{p_0}^{(1)} + \sum_{j=2}^{n} \beta_j \lambda_j v_{p_0}^{(j)}}{\beta_1 v_{p_0}^{(1)} + \sum_{j=2}^{n} \beta_j v_{p_0}^{(j)}} = \lambda_1 \left[\frac{\beta_1 v_{p_0}^{(1)} + \sum_{j=2}^{n} \beta_j (\lambda_j/\lambda_1) v_{p_0}^{(j)}}{\beta_1 v_{p_0}^{(1)} + \sum_{j=2}^{n} \beta_j v_{p_0}^{(j)}}\right].$$

Then let p_1 be the least integer such that

$$\left| y_{p_1}^{(1)} \right| = \|\mathbf{y}^{(1)}\|_\infty$$

and define the first approximation to the eigenvector as

$$\mathbf{x}^{(1)} = \frac{1}{y_{p_1}^{(1)}} \mathbf{y}^{(1)} = \frac{1}{y_{p_1}^{(1)}} A\mathbf{x}^{(0)}.$$

This continues with each iteration, calculating $\mathbf{x}^{(m-1)}$, $\mathbf{y}^{(m)}$, $\mu^{(m)}$, and $y_{p_m}^{(m)}$. The scalar sequence $\mu^{(m)}$ will converge to λ_1, and the vector sequence $\mathbf{x}^{(m-1)}$ will converge to an eigenvector of A associated with λ_1.

Accelerating Convergence

Choosing the least integer, p_m, for which $|y_{p_m}^{(m)}| = \|\mathbf{y}^{(m)}\|_\infty$ generally ensures that this index eventually becomes invariant. The rate at which $\{\mu^{(m)}\}_{m=1}^\infty$ converges to λ_1 is determined by the ratios $|\lambda_j/\lambda_1|^m$, for $j = 2, 3, \ldots, n$, and, in particular, by $|\lambda_2/\lambda_1|^m$; that is, the convergence is of order $O(|\lambda_2/\lambda_1|^m)$. Hence, there is a constant K such that, for large m,

$$|\mu^{(m)} - \lambda_1| \approx K \left| \frac{\lambda_2}{\lambda_1} \right|^m,$$

which implies that

$$\lim_{m \to \infty} \frac{\left| \mu^{(m+1)} - \lambda_1 \right|}{\left| \mu^{(m)} - \lambda_1 \right|} \approx \left| \frac{\lambda_2}{\lambda_1} \right| < 1.$$

The sequence $\{\mu^{(m)}\}$ converges linearly to λ_1, so Aitken's Δ^2 procedure can be used to speed the convergence.

In actuality, it is not necessary for the matrix to have n distinct eigenvalues for the Power method to converge. If the matrix has a unique dominant eigenvalue, λ_1, with multiplicity $r > 1$ and $\mathbf{v}^{(1)}, \mathbf{v}^{(2)}, \ldots, \mathbf{v}^{(r)}$ are linearly independent eigenvectors associated with λ_1, the procedure will still converge to λ_1. The sequence of vectors $\{\mathbf{x}^{(m)}\}_{m=0}^\infty$ depends on the choice of the initial vector $\mathbf{x}^{(0)}$ and converges to an eigenvector of λ_1 with l_∞ norm 1 that is a linear combination of $\mathbf{v}^{(1)}, \mathbf{v}^{(2)}, \ldots, \mathbf{v}^{(r)}$.

The Power method has the disadvantage that it is generally unknown at the outset whether the matrix has a single dominant eigenvalue and, when this is not the case, the method may not give convergence to an eigenvalue. Even when there is a single dominant eigenvalue, it is not known how $\mathbf{x}^{(0)}$ should be chosen to ensure that its representation will contain a nonzero contribution from the eigenvector associated with that eigenvalue.

> The program POWERM91 implements the Power method.

Example 1 Use the Power method program POWERM91 to approximate the dominant eigenvalue of the matrix

$$A = \begin{bmatrix} -4 & 14 & 0 \\ -5 & 13 & 0 \\ -1 & 0 & 2 \end{bmatrix},$$

and then apply Aitken's Δ^2 method to the approximations to accelerate the convergence.

Solution This matrix has eigenvalues $\lambda_1 = 6$, $\lambda_2 = 3$, and $\lambda_3 = 2$, so the Power method will converge. Let $\mathbf{x}^{(0)} = (1, 1, 1)^t$, then

$$\mathbf{y}^{(1)} = A\mathbf{x}^{(0)} = (10, 8, 1)^t,$$

so

$$\|\mathbf{y}^{(1)}\|_\infty = 10, \quad \mu^{(1)} = y_1^{(1)} = 10, \quad \text{and} \quad \mathbf{x}^{(1)} = \frac{\mathbf{y}^{(1)}}{10} = (1, 0.8, 0.1)^t.$$

Continuing in this manner leads to the values in Table 9.1, where $\hat{\mu}^{(m)}$ represents the sequence generated by Aitken's Δ^2 procedure. An approximation to the dominant eigenvalue, 6, at this stage is $\hat{\mu}^{(10)} = 6.000000$. The approximate l_∞-unit eigenvector for the eigenvalue 6 is $(\mathbf{x}^{(12)})^t = (1, 0.714316, -0.249895)^t$.

Table 9.1

m	$(\mathbf{x}^{(m)})^t$	$\mu^{(m)}$	$\hat{\mu}^{(m)}$
0	$(1, 1, 1)$		
1	$(1, 0.8, 0.1)$	10	6.266667
2	$(1, 0.75, -0.111)$	7.2	6.062473
3	$(1, 0.730769, -0.188803)$	6.5	6.015054
4	$(1, 0.722200, -0.220850)$	6.230769	6.004202
5	$(1, 0.718182, -0.235915)$	6.111000	6.000855
6	$(1, 0.716216, -0.243095)$	6.054546	6.000240
7	$(1, 0.715247, -0.246588)$	6.027027	6.000058
8	$(1, 0.714765, -0.248306)$	6.013453	6.000017
9	$(1, 0.714525, -0.249157)$	6.006711	6.000003
10	$(1, 0.714405, -0.249579)$	6.003352	6.000000
11	$(1, 0.714346, -0.249790)$	6.001675	
12	$(1, 0.714316, -0.249895)$	6.000837	

Although the approximation to the eigenvalue is correct to the places listed, the eigenvector approximation is less accurate to the true eigenvector, $(1, 5/7, -1/4)^t = (1, 0.714286, -0.25)^t$. ▪

MATLAB can be used to do the calculations in the Power method. For example, if we define the matrix A and the vector $\mathbf{x}^{(0)}$ with

```
A = [-4 14 0; -5 13 0; -1 0 2]
x0 = [1; 1; 1]
```

then

```
y1 = A*x0
```

gives $y^{(1)}$, so $\mu^{(1)} = y_{p_0}^{(1)} = 10$ and $p_1 = 1$. Then

```
x1 = 0.1*y1
y2 = A*x1
```

produces the vector $y2 \equiv \mathbf{y}^{(2)} = (7.2000, 5.4000, -0.8000)^t$. So the approximation to the eigenvalue is $\mu^{(2)} = y_{p_1}^{(2)} = y_1^{(2)} = 7.2$.

Symmetric Matrices

When A is symmetric, a variation in the choice of the vectors $\mathbf{x}^{(m)}$ and $\mathbf{y}^{(m)}$ and scalars $\mu^{(m)}$ can be made to significantly improve the rate of convergence of the sequence $\{\mu^{(m)}\}_{m=1}^{\infty}$ to the dominant eigenvalue λ_1. First, select $\mathbf{x}^{(0)}$ with $\|\mathbf{x}^{(0)}\|_2 = 1$. For each $m = 1, 2, \ldots$, define

$$\mu^{(m)} = \left(\mathbf{x}^{(m-1)}\right)^t A\mathbf{x}^{(m-1)} \quad \text{and} \quad \mathbf{x}^{(m)} = \frac{1}{\|A\mathbf{x}^{(m-1)}\|_2} A\mathbf{x}^{(m-1)}.$$

The program SYMPWR92 implements the Symmetric Power method.

The rate of convergence of the Power method is $O((\lambda_2/\lambda_1)^m)$, but, with this modification, the rate of convergence for symmetric matrices is $O((\lambda_2/\lambda_1)^{2m})$.

Example 2 Apply both the Power method and the Symmetric Power method to the matrix

$$A = \begin{bmatrix} 4 & -1 & 1 \\ -1 & 3 & -2 \\ 1 & -2 & 3 \end{bmatrix},$$

using Aitken's Δ^2 method to accelerate the convergence.

Solution This matrix has eigenvalues $\lambda_1 = 6$, $\lambda_2 = 3$, and $\lambda_3 = 1$. An eigenvector for the dominant eigenvalue 6 is $(1, -1, 1)^t$. Applying the Power method to this matrix with initial vector $(1, 0, 0)^t$ gives the values in Table 9.2.

Table 9.2

m	$(\mathbf{y}^{(m)})^t$	$\mu^{(m)}$	$\hat{\mu}^{(m)}$	$(\mathbf{x}^{(m)})^t$ with $\|\mathbf{x}^{(m)}\|_\infty = 1$
0				$(1, 0, 0)$
1	$(4, -1, 1)$	4		$(1, -0.25, 0.25)$
2	$(4.5, -2.25, 2.25)$	4.5	7	$(1, -0.5, 0.5)$
3	$(5, -3.5, 3.5)$	5	6.2	$(1, -0.7, 0.7)$
4	$(5.4, -4.5, 4.5)$	5.4	6.047617	$(1, -0.833\bar{3}, 0.833\bar{3})$
5	$(5.66\bar{6}, -5.166\bar{6}, 5.166\bar{6})$	$5.66\bar{6}$	6.011767	$(1, -0.911765, 0.911765)$
6	$(5.823529, -5.558824, 5.558824)$	5.823529	6.002931	$(1, -0.954545, 0.954545)$
7	$(5.909091, -5.772727, 5.772727)$	5.909091	6.000733	$(1, -0.976923, 0.976923)$
8	$(5.953846, -5.884615, 5.884615)$	5.953846	6.000184	$(1, -0.988372, 0.988372)$
9	$(5.976744, -5.941861, 5.941861)$	5.976744		$(1, -0.994163, 0.994163)$
10	$(5.988327, -5.970817, 5.970817)$	5.988327		$(1, -0.997076, 0.997076)$

We will now apply the Symmetric Power method to this matrix with the same initial vector $(1, 0, 0)^t$. The first steps are

$$\mathbf{x}^{(0)} = (1, 0, 0)^t, \quad A\mathbf{x}^{(0)} = (4, -1, 1)^t, \quad \mu^{(1)} = 4,$$

and

$$\mathbf{x}^{(1)} = \frac{1}{\|A\mathbf{x}^{(0)}\|_2} \cdot A\mathbf{x}^{(0)} = (0.942809, -0.235702, 0.235702)^t.$$

The remaining entries are shown in Table 9.3.

Table 9.3

m	$(\mathbf{y}^{(m)})^t$	$\mu^{(m)}$	$\hat{\mu}^{(m)}$	$(\mathbf{x}^{(m)})^t$ with $\|\mathbf{x}^{(m)}\|_2 = 1$
0	$(1, 0, 0)$			$(1, 0, 0)$
1	$(4, -1, 1)$	4	7	$(0.942809, -0.235702, 0.235702)$
2	$(4.242641, -2.121320, 2.121320)$	5	6.047619	$(0.816497, -0.408248, 0.408248)$
3	$(4.082483, -2.857738, 2.857738)$	5.666667	6.002932	$(0.710669, -0.497468, 0.497468)$
4	$(3.837613, -3.198011, 3.198011)$	5.909091	6.000183	$(0.646997, -0.539164, 0.539164)$
5	$(3.666314, -3.342816, 3.342816)$	5.976744	6.000012	$(0.612836, -0.558763, 0.558763)$
6	$(3.568871, -3.406650, 3.406650)$	5.994152	6.000000	$(0.595247, -0.568190, 0.568190)$
7	$(3.517370, -3.436200, 3.436200)$	5.998536	6.000000	$(0.586336, -0.572805, 0.572805)$
8	$(3.490952, -3.450359, 3.450359)$	5.999634		$(0.581852, -0.575086, 0.575086)$
9	$(3.477580, -3.457283, 3.457283)$	5.999908		$(0.579603, -0.576220, 0.576220)$
10	$(3.470854, -3.460706, 3.460706)$	5.999977		$(0.578477, -0.576786, 0.576786)$

The Symmetric Power method gives considerably faster convergence for this matrix than the Power method. Notice that the eigenvector approximations in the Power method converge to $(1, -1, 1)^t$, a vector with unit l_∞-norm. In the Symmetric Power method, the convergence is to the parallel vector $(\sqrt{3}/3, -\sqrt{3}/3, \sqrt{3}/3)^t$, which has unit l_2-norm. ∎

The following MATLAB commands can be used to compute the first rows of Table 9.3. Define A and $\mathbf{x}^{(0)} = (1, 0, 0)^t$ with

```
A = [4 -1 1; -1 3 -2; 1 -2 3]
x0 = [1; 0; 0]
```

Then

```
y1 = A*x0
n1 = norm(y1,2)
x1 = 1/n1*y1
```

produces the l_2 norm of the vector $\mathbf{y}^{(1)}$

$$n1 = 4.242640687119285$$

and $x1 \equiv \mathbf{x}^{(1)} = (0.942809041582063, -0.235702260395516, 0.235702260395516)^t$. The approximation $\mu^{(1)} = (\mathbf{y}^{(1)})^t \mathbf{x}^{(0)} = 4$ is found using

```
mu1 = y1'*x0
```

Then $\mathbf{y}^{(2)} = A\mathbf{x}^{(1)}$ is determined with

```
y2 = A*x1
```

This gives $\mathbf{y}^{(2)} = (4.242640687119286, -2.121320343559643, 2.121320343559643)^t$. To find $x^{(2)}$ we use the MATLAB commands

```
n2 = norm(y2,2)
x2 = 1/n2*y2
```

which gives

$$\|\mathbf{y}^{(2)}\|_2 = 5.196152422706632 \quad \text{and}$$
$$\mathbf{x}^{(2)} = (0.816496580927726, -0.408248290463863, 0.408248290463863)^t.$$

Then

```
mu2 = y2'*x1
```

produces the eigenvalue approximation

$$mu2 = 5.000000000000001$$

Inverse Power Method

The **Inverse Power Method** is a modification of the Power method that is used to determine the eigenvalue of A closest to a specified number q.

Suppose the matrix A has eigenvalues $\lambda_1, \ldots, \lambda_n$ with linearly independent eigenvectors $\mathbf{v}^{(1)}, \mathbf{v}^{(2)}, \ldots, \mathbf{v}^{(n)}$. The results in Exercise 9 of Section 7.3 imply that the eigenvalues of $(A - qI)^{-1}$, where $q \neq \lambda_i$ for each $i = 1, 2, \ldots, n$, are

$$\frac{1}{\lambda_1 - q}, \frac{1}{\lambda_2 - q}, \ldots, \frac{1}{\lambda_n - q}$$

with eigenvectors $\mathbf{v}^{(1)}, \mathbf{v}^{(2)}, \ldots, \mathbf{v}^{(n)}$. Applying the Power method to $(A - qI)^{-1}$ gives

$$\mathbf{y}^{(m)} = (A - qI)^{-1}\mathbf{x}^{(m-1)},$$

$$\mu^{(m)} = y_{p_{m-1}}^{(m)} = \frac{y_{p_{m-1}}^{(m)}}{x_{p_{m-1}}^{(m-1)}} = \frac{\sum_{j=1}^{n} \beta_j \dfrac{1}{(\lambda_j - q)^m} v_{p_{m-1}}^{(j)}}{\sum_{j=1}^{n} \beta_j \dfrac{1}{(\lambda_j - q)^{m-1}} v_{p_{m-1}}^{(j)}},$$

and

$$\mathbf{x}^{(m)} = \frac{\mathbf{y}^{(m)}}{y_{p_m}^{(m)}},$$

where, at each step, p_m represents the smallest integer for which $|y_{p_m}^{(m)}| = \|\mathbf{y}^{(m)}\|_{\infty}$. The sequence $\{\mu^{(m)}\}$ converges to $1/(\lambda_k - q)$, where

$$\frac{1}{|\lambda_k - q|} = \max_{1 \le i \le n} \frac{1}{|\lambda_i - q|},$$

and $\lambda_k \approx q + 1/\mu^{(m)}$ is the eigenvalue of A closest to q.

The choice of q determines the convergence, provided that $1/(\lambda_k - q)$ is a unique dominant eigenvalue of $(A - qI)^{-1}$ (although it may be a multiple eigenvalue). The closer q is to an eigenvalue λ_k, the faster the convergence because if λ represents the eigenvalue of A that is second closest to q, then the convergence is of order

$$O\left(\left|\frac{(\lambda - q)^{-1}}{(\lambda_k - q)^{-1}}\right|^m\right) = O\left(\left|\frac{\lambda_k - q}{\lambda - q}\right|^m\right).$$

The vector $\mathbf{y}^{(m)}$ is obtained from the equation

$$(A - qI)\mathbf{y}^{(m)} = \mathbf{x}^{(m-1)}.$$

Gaussian elimination with pivoting can be used to solve this linear system.

Although the Inverse Power method requires the solution of an $n \times n$ linear system at each step, the multipliers can be saved to reduce the computation. The selection of q can be based on the Geršgorin Theorem or on another means of approximating an eigenvalue.

One choice of q comes from the initial approximation $\mathbf{x}^{(0)}$ to the eigenvector:

$$q = \frac{\mathbf{x}^{(0)t} A \mathbf{x}^{(0)}}{\mathbf{x}^{(0)t} \mathbf{x}^{(0)}}. \tag{9.1}$$

This choice results from the observation that if $\mathbf{x}$ is an eigenvector of A with respect to the eigenvalue λ, then $A\mathbf{x} = \lambda\mathbf{x}$. So $\mathbf{x}^t A\mathbf{x} = \lambda\mathbf{x}^t\mathbf{x}$ and

$$\lambda = \frac{\mathbf{x}^t A\mathbf{x}}{\mathbf{x}^t\mathbf{x}} = \frac{\mathbf{x}^t A\mathbf{x}}{\|\mathbf{x}\|_2^2}.$$

If q is close to an eigenvalue, the convergence will be quite rapid. In fact, this method is often used to approximate an eigenvector when an approximate eigenvalue q is obtained by this or by some other technique.

The program INVPW93 implements the Inverse Power method.

The convergence of the Inverse Power method is linear, so Aitken's Δ^2 procedure can again be used to speed convergence. The following example illustrates the fast convergence of the Inverse Power method if q is close to an eigenvalue.

Example 3   Apply the Inverse Power method with $\mathbf{x}^{(0)} = (1, 1, 1)^t$ to the matrix

$$A = \begin{bmatrix} -4 & 14 & 0 \\ -5 & 13 & 0 \\ -1 & 0 & 2 \end{bmatrix} \quad \text{with} \quad q = \frac{\mathbf{x}^{(0)t} A \mathbf{x}^{(0)}}{\mathbf{x}^{(0)t} \mathbf{x}^{(0)}} = \frac{19}{3},$$

and use Aitken's $\triangle^2$ method to accelerate the convergence.

Solution   The Power method was applied to this matrix in Example 1 using the initial vector $\mathbf{x}^{(0)} = (1, 1, 1)^t$. It gave the approximate eigenvalue $\mu^{(12)} = 6.000837$ and eigenvector $(\mathbf{x}^{(12)})^t = (1, 0.714316, -0.249895)^t$.

For the Inverse Power method we consider

$$A - qI = \begin{bmatrix} -4 - \frac{19}{3} & 14 & 0 \\ -5 & 13 - \frac{19}{3} & 0 \\ -1 & 0 & 2 - \frac{19}{3} \end{bmatrix} = \begin{bmatrix} -\frac{31}{3} & 14 & 0 \\ -5 & \frac{20}{3} & 0 \\ -1 & 0 & -\frac{13}{3} \end{bmatrix}.$$

With $\mathbf{x}^{(0)} = (1, 1, 1)^t$, the method first finds $\mathbf{y}^{(1)}$ by solving $(A - qI)\mathbf{y}^{(1)} = \mathbf{x}^{(0)}$. This gives

$$\mathbf{y}^{(1)} = \left(-\frac{33}{5}, -\frac{24}{5}, \frac{84}{65} \right)^t = (-6.6, -4.8, 1.29\overline{2307692})^t.$$

So

$$\|\mathbf{y}^{(1)}\|_\infty = 6.6, \quad \mathbf{x}^{(1)} = \frac{1}{-6.6} \mathbf{y}^{(1)} = (1, 0.7272727, -0.1958042)^t,$$

and

$$\mu^{(1)} = -\frac{1}{6.6} + \frac{19}{3} = 6.1818182.$$

Subsequent results are listed in Table 9.4, and the right column lists the results of Aitken's $\triangle^2$ method applied to the $\mu^{(m)}$. These are clearly superior results to those obtained with the Power method.   ▪

Table 9.4

m	$\mathbf{x}^{(m)t}$	$\mu^{(m)}$	$\hat{\mu}^{(m)}$
0	$(1, 1, 1)$		
1	$(1, 0.7272727, -0.1958042)$	6.1818182	6.000098
2	$(1, 0.7155172, -0.2450520)$	6.0172414	6.000001
3	$(1, 0.7144082, -0.2495224)$	6.0017153	6.000000
4	$(1, 0.7142980, -0.2499534)$	6.0001714	6.000000
5	$(1, 0.7142869, -0.2499954)$	6.0000171	
6	$(1, 0.7142858, -0.2499996)$	6.0000017	

The following MATLAB commands use the Inverse Power method to generate the first two rows of Table 9.4.

```
A = [-4 14 0; -5 13 0; -1 0 2]
x0 = [1; 1; 1]
```

To compute the initial value of

$$q = (\mathbf{x}^{(0)})^t (A \mathbf{x}^{(0)}) / ((\mathbf{x}^{(0)})^t \mathbf{x}^{(0)})$$

we use

```
q = x0'*(A*x0)/(x0'*x0)
```

which gives

$$q = 6.333333333333333$$

The identity matrix I_3 is given by

```
I3 = eye(3)
```

and we obtain the matrix with

```
AQ = A - q*I3
```

To solve the linear system $(A - qI)\mathbf{y}^{(1)} = \mathbf{x}^{(0)}$ we use

```
y1 = linsolve(AQ,x0)
```

which gives

$$y1 = -6.599999999999953$$
$$-4.799999999999964$$
$$1.292307692307682$$

Then we find $\mathbf{x}^{(1)}$ and $\mu^{(1)}$

```
c = y1(1)
x1 = 1/c*y1
mu1 = q + 1/c
```

to obtain the first eigenvalue approximation $\mu^{(1)} = mu1 = 6.\overline{18}$.

Deflation Methods

Numerous techniques are available for obtaining approximations to the other eigenvalues of a matrix once an approximation to one of the eigenvalues has been computed. We will restrict our presentation to **deflation techniques**. These involve forming a new matrix B whose eigenvalues are the same as those of A, except that the known eigenvalue of A is replaced by the eigenvalue 0 in B. The following result justifies the procedure.

Eigenvalues and Eigenvectors of Deflated Matrices

Suppose $\lambda_1, \lambda_2, \ldots, \lambda_n$ are eigenvalues of A with associated eigenvectors $\mathbf{v}^{(1)}$, $\mathbf{v}^{(2)}, \ldots, \mathbf{v}^{(n)}$ and that λ_1 has multiplicity 1. If $\mathbf{x}$ is a vector with $\mathbf{x}^t\mathbf{v}^{(1)} = 1$, then

$$B = A - \lambda_1\mathbf{v}^{(1)}\mathbf{x}^t$$

has eigenvalues $0, \lambda_2, \lambda_3, \ldots, \lambda_n$ with associated eigenvectors $\mathbf{v}^{(1)}, \mathbf{w}^{(2)}, \mathbf{w}^{(3)}, \ldots, \mathbf{w}^{(n)}$, where $\mathbf{v}^{(i)}$ and $\mathbf{w}^{(i)}$ are related by the equation

$$\mathbf{v}^{(i)} = (\lambda_i - \lambda_1)\mathbf{w}^{(i)} + \lambda_1(\mathbf{x}^t\mathbf{w}^{(i)})\mathbf{v}^{(1)}, \tag{9.2}$$

for each $i = 2, 3, \ldots, n$.

Helmut Wielandt (1910–2001)
originally worked in permutation
groups, but during World War II
was engaged in research on
meteorology, cryptology, and
aerodynamics. This involved
vibration problems that required
the estimation of eigenvalues
associated with differential
equations and matrices.

Wielandt's deflation results from defining

$$\mathbf{x} = \frac{1}{\lambda_1 v_i^{(1)}} (a_{i1}, a_{i2}, \ldots , a_{in})^t,$$

where $v_i^{(1)}$ is a nonzero coordinate of the eigenvector $\mathbf{v}^{(1)}$ and the values $a_{i1}, a_{i2}, \ldots , a_{in}$ are the entries in the ith row of A. With this definition,

$$\mathbf{x}^t \mathbf{v}^{(1)} = \frac{1}{\lambda_1 v_i^{(1)}} (a_{i1}, a_{i2}, \ldots , a_{in}) \left(v_1^{(1)}, v_2^{(1)}, \ldots , v_n^{(1)} \right)^t = \frac{1}{\lambda_1 v_i^{(1)}} \sum_{j=1}^{n} a_{ij} v_j^{(1)},$$

where the sum is the ith coordinate of the product $A\mathbf{v}^{(1)}$. Since $A\mathbf{v}^{(1)} = \lambda_1 \mathbf{v}^{(1)}$, we have

$$\sum_{j=1}^{n} a_{ij} v_j^{(1)} = \lambda_1 v_i^{(1)}, \quad \text{so} \quad \mathbf{x}^t \mathbf{v}^{(1)} = \frac{1}{\lambda_1 v_i^{(1)}} \left(\lambda_1 v_i^{(1)} \right) = 1.$$

Hence $\mathbf{x}$ satisfies the hypotheses of the result concerning the eigenvalues of deflated matrices. Moreover, the ith row of $B = A - \lambda_1 \mathbf{v}^{(1)} \mathbf{x}^t$ consists entirely of zero entries.

Suppose that λ is an eigenvalue of B with eigenvector $\mathbf{w}$. Since $B\mathbf{w} = \lambda \mathbf{w}$ and the ith row of B has all zero entries, the ith component of $\mathbf{w}$ is also zero. Consequently, the ith column of the matrix B makes no contribution to the product $B\mathbf{w} = \lambda \mathbf{w}$, and the matrix B can be replaced by an $(n-1) \times (n-1)$ matrix B' obtained by deleting the ith row and column from B.

The matrix B' has eigenvalues $\lambda_2, \lambda_3, \ldots , \lambda_n$. If $|\lambda_2| > |\lambda_3|$, the Power method can be applied to the matrix B' to determine a dominant eigenvalue and an eigenvector, $\mathbf{w}^{(2)'}$ associated with λ_2, with respect to the matrix B'. To find the associated eigenvector $\mathbf{w}^{(2)}$ for the matrix B, insert a zero coordinate between the coordinates $w_{i-1}^{(2)'}$ and $w_i^{(2)'}$ of the $(n-1)$-dimensional vector $\mathbf{w}^{(2)'}$ and then calculate $\mathbf{v}^{(2)}$ by using Eq. (9.2).

The program WIEDEF94
implements the Wielandt
Deflation method.

Example 4   The matrix

$$A = \begin{bmatrix} 4 & -1 & 1 \\ -1 & 3 & -2 \\ 1 & -2 & 3 \end{bmatrix}$$

has the dominant eigenvalue $\lambda_1 = 6$ with associated unit eigenvector $\mathbf{v}^{(1)} = (1, -1, 1)^t$. Assume that this dominant eigenvalue is known and apply deflation to approximate another eigenvalue and eigenvector.

Solution   The procedure for obtaining a second eigenvalue λ_2 proceeds as follows:

$$\mathbf{x} = \frac{1}{6} \begin{bmatrix} 4 \\ -1 \\ 1 \end{bmatrix} = \left(\frac{2}{3}, -\frac{1}{6}, \frac{1}{6} \right)^t,$$

$$\mathbf{v}^{(1)} \mathbf{x}^t = \begin{bmatrix} 1 \\ -1 \\ 1 \end{bmatrix} \begin{bmatrix} \frac{2}{3}, & -\frac{1}{6}, & \frac{1}{6} \end{bmatrix} = \begin{bmatrix} \frac{2}{3} & -\frac{1}{6} & \frac{1}{6} \\ -\frac{2}{3} & \frac{1}{6} & -\frac{1}{6} \\ \frac{2}{3} & -\frac{1}{6} & \frac{1}{6} \end{bmatrix},$$

and

$$B = A - \lambda_1 \mathbf{v}^{(1)} \mathbf{x}^t = \begin{bmatrix} 4 & -1 & 1 \\ -1 & 3 & -2 \\ 1 & -2 & 3 \end{bmatrix} - 6 \begin{bmatrix} \frac{2}{3} & -\frac{1}{6} & \frac{1}{6} \\ -\frac{2}{3} & \frac{1}{6} & -\frac{1}{6} \\ \frac{2}{3} & -\frac{1}{6} & \frac{1}{6} \end{bmatrix} = \begin{bmatrix} 0 & 0 & 0 \\ 3 & 2 & -1 \\ -3 & -1 & 2 \end{bmatrix}.$$

Deleting the first row and column gives

$$B' = \begin{bmatrix} 2 & -1 \\ -1 & 2 \end{bmatrix},$$

which has eigenvalues $\lambda_2 = 3$ and $\lambda_3 = 1$. For $\lambda_2 = 3$, the eigenvector $\mathbf{w}^{(2)'}$ can be obtained by solving the linear system

$$(B' - 3I)\mathbf{w}^{(2)'} = \mathbf{0}, \quad \text{resulting in} \quad \mathbf{w}^{(2)'} = (1, -1)^t.$$

Adding a zero for the first component gives $\mathbf{w}^{(2)} = (0, 1, -1)^t$ and, from Eq. (9.2), we have the eigenvector $\mathbf{v}^{(2)}$ of A corresponding to $x_2 = 3$:

$$\mathbf{v}^{(2)} = (\lambda_2 - \lambda_1)\mathbf{w}^{(2)} + \lambda_1(\mathbf{x}^t\mathbf{w}^{(2)})\mathbf{v}^{(1)}$$

$$= (3 - 6)(0, 1, -1)^t + 6\left[\left(\frac{2}{3}, -\frac{1}{6}, \frac{1}{6}\right)(0, 1, -1)^t\right](1, -1, 1)^t = (-2, -1, 1)^t.$$

∎

Although deflation can be used to successively find approximations to all the eigenvalues and eigenvectors of a matrix, the process is susceptible to round-off error. Techniques based on similarity transformations are presented in the next two sections. These are generally preferable when approximations to all the eigenvalues are needed.

EXERCISE SET 9.3

1. Find the first 3 iterations obtained by the Power method applied to the following matrices.

   a. $\begin{bmatrix} 2 & 1 & 1 \\ 1 & 2 & 1 \\ 1 & 1 & 2 \end{bmatrix}$
   Use $\mathbf{x}^{(0)} = (1, -1, 2)^t$.

   b. $\begin{bmatrix} 1 & 1 & 1 \\ 1 & 1 & 0 \\ 1 & 0 & 1 \end{bmatrix}$
   Use $\mathbf{x}^{(0)} = (-1, 0, 1)^t$.

   c. $\begin{bmatrix} 1 & -1 & 0 \\ -2 & 4 & -2 \\ 0 & -1 & 2 \end{bmatrix}$
   Use $\mathbf{x}^{(0)} = (-1, 2, 1)^t$.

   d. $\begin{bmatrix} 4 & 1 & 1 & 1 \\ 1 & 3 & -1 & 1 \\ 1 & -1 & 2 & 0 \\ 1 & 1 & 0 & 2 \end{bmatrix}$
   Use $\mathbf{x}^{(0)} = (1, -2, 0, 3)^t$.

   e. $\begin{bmatrix} 5 & -2 & -\frac{1}{2} & \frac{3}{2} \\ -2 & 5 & \frac{3}{2} & -\frac{1}{2} \\ -\frac{1}{2} & \frac{3}{2} & 5 & -2 \\ \frac{3}{2} & -\frac{1}{2} & -2 & 5 \end{bmatrix}$
   Use $\mathbf{x}^{(0)} = (1, 1, 0, -3)^t$.

   f. $\begin{bmatrix} -4 & 0 & \frac{1}{2} & \frac{1}{2} \\ \frac{1}{2} & -2 & 0 & \frac{1}{2} \\ \frac{1}{2} & \frac{1}{2} & 0 & 0 \\ 0 & 1 & 1 & 4 \end{bmatrix}$
   Use $\mathbf{x}^{(0)} = (0, 0, 0, 1)^t$.

2. Repeat Exercise 1 using the Inverse Power method with q as given in Eq. (9.1).

3. Find the first 3 iterations obtained by the Symmetric Power method applied to the following matrices.

   a. $\begin{bmatrix} 2 & 1 & 1 \\ 1 & 2 & 1 \\ 1 & 1 & 2 \end{bmatrix}$
   Use $\mathbf{x}^{(0)} = (1, -1, 2)^t$.

   b. $\begin{bmatrix} 1 & 1 & 1 \\ 1 & 1 & 0 \\ 1 & 0 & 1 \end{bmatrix}$
   Use $\mathbf{x}^{(0)} = (-1, 0, 1)^t$.

c. $$\begin{bmatrix} 4.75 & 2.25 & -0.25 \\ 2.25 & 4.75 & 1.25 \\ -0.25 & 1.25 & 4.75 \end{bmatrix}$$
Use $\mathbf{x}^{(0)} = (0, 1, 0)^t$.

d. $$\begin{bmatrix} 4 & 1 & -1 & 0 \\ 1 & 3 & -1 & 0 \\ -1 & -1 & 5 & 2 \\ 0 & 0 & 2 & 4 \end{bmatrix}$$
Use $\mathbf{x}^{(0)} = (0, 1, 0, 0)^t$.

e. $$\begin{bmatrix} 4 & 1 & 1 & 1 \\ 1 & 3 & -1 & 1 \\ 1 & -1 & 2 & 0 \\ 1 & 1 & 0 & 2 \end{bmatrix}$$
Use $\mathbf{x}^{(0)} = (1, 0, 0, 0)^t$.

f. $$\begin{bmatrix} 5 & -2 & -\frac{1}{2} & \frac{3}{2} \\ -2 & 5 & \frac{3}{2} & -\frac{1}{2} \\ -\frac{1}{2} & \frac{3}{2} & 5 & -2 \\ \frac{3}{2} & -\frac{1}{2} & -2 & 5 \end{bmatrix}$$
Use $\mathbf{x}^{(0)} = (1, 1, 0, -3)^t$.

4. Use the Power method and Aitken's Δ^2 technique to approximate the dominant eigenvalue for the matrices in Exercise 1, iterating until $\|\mathbf{x}^{(m)} - \mathbf{x}^{(m-1)}\|_\infty < 10^{-4}$ or until the number of iterations exceeds 25.

5. Use the Power method and Wielandt deflation to approximate the second most dominant eigenvalues for the matrices in Exercise 1, iterating until $\|\mathbf{x}^{(m)} - \mathbf{x}^{(m-1)}\|_\infty < 10^{-4}$ or until the number of iterations exceeds 25.

6. Use the Symmetric Power method to approximate the dominant eigenvalue for the matrices given in Exercise 3, iterating until $\|\mathbf{x}^{(m)} - \mathbf{x}^{(m-1)}\|_2 < 10^{-4}$ or until the number of iterations exceeds 25.

7. Use the Inverse Power method on the matrices in Exercise 1, iterating until $\|\mathbf{x}^{(m)} - \mathbf{x}^{(m-1)}\|_\infty < 10^{-4}$ or until the number of iterations exceeds 25, and using the value of q given in Eq. (9.1).

8. Show that the ith row of $B = A - \lambda_1 \mathbf{v}^{(1)} \mathbf{x}^t$ is zero, where λ_1 is the largest value of A in absolute value, $\mathbf{v}^{(1)}$ is the associated eigenvector of A for λ_1, and $\mathbf{x} = 1/\lambda_1 v_i^{(1)} (a_{i1}, a_{i2}, \ldots, a_{in})^t$.

9. Following along the line of Exercise 8 in Section 6.4 and Exercise 10 in Section 7.3, suppose that a species of beetle has a life span of 4 years and that a female in the first year has a survival rate of $\frac{1}{2}$, in the second year a survival rate of $\frac{1}{4}$, and in the third year a survival rate of $\frac{1}{8}$. Suppose additionally that a female gives birth, on average, to two new females in the third year and to four new females in the fourth year. The matrix describing a single female's contribution in one year to the female population in the succeeding year is

$$A = \begin{bmatrix} 0 & 0 & 2 & 4 \\ \frac{1}{2} & 0 & 0 & 0 \\ 0 & \frac{1}{4} & 0 & 0 \\ 0 & 0 & \frac{1}{8} & 0 \end{bmatrix},$$

where again the entry in the ith row and jth column denotes the probabilistic contribution that a female of age j makes on the next year's female population of age i.

a. Use the Geršgorin Circle Theorem to determine a region in the complex plane containing all the eigenvalues of A.

b. Use the Power method to determine the dominant eigenvalue of the matrix and its associated eigenvector.

c. Use deflation to obtain a 3×3 matrix B with the same eigenvalues λ_2, λ_3, and λ_4 as A. Use the Inverse Power method with $q = -0.25$ to find λ_4, the only real eigenvalue of B.

d. Find the two additional (complex) eigenvalues of A.

e. What is your long-range prediction for the population of these beetles?

10. A linear dynamical system can be represented by the equations

$$\frac{d\mathbf{x}}{dt}(t) = A(t)\,\mathbf{x}(t) + B(t)\,\mathbf{u}(t), \qquad \mathbf{y}(t) = C(t)\,\mathbf{x}(t) + D(t)\,\mathbf{u}(t),$$

where A is an $n \times n$ variable matrix, B is an $n \times r$ variable matrix, C is an $m \times n$ variable matrix, D is an $m \times r$ variable matrix, $\mathbf{x}$ is an n-dimensional vector variable, $\mathbf{y}$ is an m-dimensional vector variable, and $\mathbf{u}$ is an r-dimensional vector variable. For the system to be stable, the matrix A must have all its eigenvalues with nonpositive real part for all t.

a. Is the system stable if

$$A(t) = \begin{bmatrix} -1 & 2 & 0 \\ -2.5 & -7 & 4 \\ 0 & 0 & -5 \end{bmatrix}?$$

b. Is the system stable if

$$A(t) = \begin{bmatrix} -1 & 1 & 0 & 0 \\ 0 & -2 & 1 & 0 \\ 0 & 0 & -5 & 1 \\ -1 & -1 & -2 & -3 \end{bmatrix}?$$

9.4 Householder's Method

Alston Householder (1904–1993) did research in mathematical biology before becoming the Director of the Oak Ridge National Laboratory in Tennessee in 1948. He began work on solving linear systems in the 1950s, which was when these methods were developed.

In the next section we use the QR method to reduce a symmetric tridiagonal matrix to a nearly diagonal matrix to which it is similar. The diagonal entries of the reduced matrix are approximations to the eigenvalues of both matrices. In this section, we consider the associated problem of reducing an arbitrary symmetric matrix to a similar tridiagonal matrix using a method devised by Alston Householder (see [Ho]). Although there is a clear connection between the problems we are solving in these two sections, Householder's method has wide application in areas other than eigenvalue approximation.

Householder's method is used to find a symmetric tridiagonal matrix B that is similar to a given symmetric matrix A. By the result on page 370, a symmetric matrix A is similar to a diagonal matrix D, and an orthogonal matrix Q exists with the property that

$$D = Q^{-1}AQ = Q^t AQ.$$

However, the matrix Q (and consequently D) is generally difficult to compute, and Householder's method offers a compromise.

Householder Transformations

Let $\mathbf{w}$ be in $\mathbb{R}^n$ with $\mathbf{w}^t \mathbf{w} = 1$. The $n \times n$ matrix

$$P = I - 2\mathbf{w}\mathbf{w}^t$$

is called a **Householder transformation**.

Householder transformations are used to selectively zero out blocks of entries in vectors or in columns of matrices in a manner that is extremely stable with respect to round-off error. An important property of Householder transformations follows. (See Exercise 3.)

Householder Transformations

Suppose that $P = I - 2\mathbf{w}\mathbf{w}^t$ is a Householder transformation, that is, $\mathbf{w}^t\mathbf{w} = 1$. Then P is symmetric and orthogonal, so $P^{-1} = P$.

Householder's method begins by determining a transformation $P^{(1)}$ with the property that $A^{(2)} = P^{(1)}AP^{(1)}$ has

$$a_{j1}^{(2)} = 0 \qquad \text{for each } j = 3, 4, \dots, n.$$

By symmetry, this also implies that $a_{1j}^{(2)} = 0$ for each $j = 3, 4, \dots, n.$

The vector $\mathbf{w} = (w_1, w_2, \ldots, w_n)^t$ is chosen so that $\mathbf{w}^t \mathbf{w} = 1$, the matrix

$$A^{(2)} = P^{(1)} A P^{(1)} = (I - 2\mathbf{w}\mathbf{w}^t) A (I - 2\mathbf{w}\mathbf{w}^t)$$

has the same entry as A in the first row and column, and $A^{(2)}$ has zeros for all the entries in the first column from rows 3 through n. That is,

$$a_{11}^{(2)} = a_{11} \quad \text{and} \quad a_{j1}^{(2)} = 0 \text{ for each } j = 3, 4, \ldots, n.$$

This imposes n conditions on the n unknowns $w_1, w_2, \ldots, w_n$. Setting $w_1 = 0$ ensures that $a_{11}^{(2)} = a_{11}$. Then we want

$$P^{(1)} = I - 2\mathbf{w}\mathbf{w}^t$$

to satisfy

$$P^{(1)}(a_{11}, a_{21}, a_{31}, \ldots, a_{n1})^t = (a_{11}, \alpha, 0, \ldots, 0)^t, \tag{9.3}$$

where α will be chosen later. To simplify notation, we define the vectors $\hat{\mathbf{w}}$ and $\hat{\mathbf{y}}$ in $\mathbb{R}^{n-1}$ by

$$\hat{\mathbf{w}} = (w_2, w_3, \ldots, w_n)^t, \quad \text{and} \quad \hat{\mathbf{y}} = (a_{21}, a_{31}, \ldots, a_{n1})^t,$$

and let $\hat{P}$ be the $(n-1) \times (n-1)$ Householder transformation

$$\hat{P} = I_{n-1} - 2\hat{\mathbf{w}}\hat{\mathbf{w}}^t.$$

Equation (9.3) can then be rewritten as

$$P^{(1)} \begin{bmatrix} a_{11} \\ a_{21} \\ a_{31} \\ \vdots \\ a_{n1} \end{bmatrix} = \left[\begin{array}{c:ccc} 1 & 0 & \cdots\cdots & 0 \\ \hdashline 0 & & & \\ \vdots & & \hat{P} & \\ \vdots & & & \\ 0 & & & \end{array} \right] \cdot \begin{bmatrix} a_{11} \\ \hline \mathbf{y} \end{bmatrix} = \begin{bmatrix} a_{11} \\ \hline \hat{P}\hat{\mathbf{y}} \end{bmatrix} = \begin{bmatrix} a_{11} \\ \hline \alpha \\ 0 \\ \vdots \\ 0 \end{bmatrix}$$

with

$$(\alpha, 0, \ldots, 0)^t = \hat{P}\hat{\mathbf{y}} = (I_{n-1} - 2\hat{\mathbf{w}}\hat{\mathbf{w}}^t)\hat{\mathbf{y}} = \hat{\mathbf{y}} - 2\hat{\mathbf{w}}(\hat{\mathbf{w}}^t\hat{\mathbf{y}}).$$

But $\hat{\mathbf{w}}^t\hat{\mathbf{y}}$ is a real number, so

$$(\alpha, 0, \ldots, 0)^t = \hat{P}\hat{\mathbf{y}} = \hat{\mathbf{y}} - 2(\hat{\mathbf{w}}^t\hat{\mathbf{y}})\hat{\mathbf{w}}. \tag{9.4}$$

Let $r = \hat{\mathbf{w}}^t\hat{\mathbf{y}}$. Then

$$(\alpha, 0, \ldots, 0)^t = (a_{21} - 2rw_2, a_{31} - 2rw_3, \ldots, a_{n1} - 2rw_n)^t.$$

Equating components gives

$$\alpha = a_{21} - 2rw_2 \quad \text{and} \quad 0 = a_{j1} - 2rw_j \quad \text{for each } j = 3, \ldots, n.$$

Thus

$$2rw_2 = a_{21} - \alpha \quad \text{and} \quad 2rw_j = a_{j1} \quad \text{for each } j = 3, \ldots, n. \tag{9.5}$$

Squaring both sides of each of the equations and summing gives

$$4r^2 \sum_{j=2}^n w_j^2 = (a_{21} - \alpha)^2 + \sum_{j=3}^n a_{j1}^2.$$

Since $\mathbf{w}^t \mathbf{w} = 1$ and $w_1 = 0$, we have $\sum_{j=2}^{n} w_j^2 = 1$ and

$$4r^2 = \sum_{j=2}^{n} a_{j1}^2 - 2\alpha a_{21} + \alpha^2. \tag{9.6}$$

Equation (9.4) and the fact that P is orthogonal imply that

$$\alpha^2 = (\alpha, 0, \ldots, 0)(\alpha, 0, \ldots, 0)^t = (\hat{P}\hat{\mathbf{y}})^t \hat{P}\hat{\mathbf{y}} = \hat{\mathbf{y}}^t \hat{P}^t \hat{P}\hat{\mathbf{y}} = \hat{\mathbf{y}}^t \hat{\mathbf{y}}.$$

Thus

$$\alpha^2 = \sum_{j=2}^{n} a_{j1}^2,$$

which, when substituted into Eq. (9.6) and simplified, gives

$$2r^2 = \sum_{j=2}^{n} a_{j1}^2 - \alpha a_{21}.$$

To ensure that $2r^2 = 0$ only if $a_{21} = a_{31} = \cdots = a_{n1} = 0$, we choose

$$\alpha = -(\text{sign } a_{21}) \left(\sum_{j=2}^{n} a_{j1}^2 \right)^{1/2}$$

which implies that

$$2r^2 = \sum_{j=2}^{n} a_{j1}^2 + |a_{21}| \left(\sum_{j=2}^{n} a_{j1}^2 \right)^{1/2}.$$

With this choice of α and $2r^2$, we solve the equations in (9.5) to obtain

$$w_2 = \frac{a_{21} - \alpha}{2r} \quad \text{and} \quad w_j = \frac{a_{j1}}{2r} \quad \text{for each } j = 3, \ldots, n.$$

To summarize the choice of $P^{(1)}$, we have

$$\alpha = -(\text{sign } a_{21}) \left(\sum_{j=2}^{n} a_{j1}^2 \right)^{1/2}, \qquad r = \left(\frac{1}{2}\alpha^2 - \frac{1}{2}a_{21}\alpha \right)^{1/2},$$

$$w_1 = 0, \quad w_2 = \frac{a_{21} - \alpha}{2r}, \quad w_j = \frac{a_{j1}}{2r} \quad \text{for each } j = 3, \ldots, n.$$

With this choice,

$$A^{(2)} = P^{(1)} A P^{(1)} = \begin{bmatrix} a_{11}^{(2)} & a_{12}^{(2)} & 0 & \cdots & 0 \\ a_{21}^{(2)} & a_{22}^{(2)} & a_{23}^{(2)} & \cdots & a_{2n}^{(2)} \\ 0 & a_{32}^{(2)} & a_{33}^{(2)} & \cdots & a_{3n}^{(2)} \\ \vdots & \vdots & \vdots & & \vdots \\ 0 & a_{n2}^{(2)} & a_{n3}^{(2)} & \cdots & a_{nn}^{(2)} \end{bmatrix}.$$

Having found $P^{(1)}$ and computed $A^{(2)}$, the process is repeated for $k = 2, 3, \ldots, n-2$ as follows:

$$\alpha = -\operatorname{sgn}\left(a_{k+1,k}^{(k)}\right)\left(\sum_{j=k+1}^{n}(a_{jk}^{(k)})^2\right)^{1/2}, \qquad r = \left(\frac{1}{2}\alpha^2 - \frac{1}{2}\alpha a_{k+1,k}^{(k)}\right)^{1/2},$$

$$w_1^{(k)} = w_2^{(k)} = \cdots = w_k^{(k)} = 0, \qquad w_{k+1}^{(k)} = \frac{a_{k+1,k}^{(k)} - \alpha}{2r}$$

$$w_j^{(k)} = \frac{a_{jk}^{(k)}}{2r}, \qquad \text{for each } j = k+2, k+3, \ldots, n,$$

$$P^{(k)} = I - 2\mathbf{w}^{(k)} \cdot (\mathbf{w}^{(k)})^t, \quad \text{and} \quad A^{(k+1)} = P^{(k)} A^{(k)} P^{(k)},$$

where

$$A^{(k+1)} = \begin{bmatrix} a_{11}^{(k+1)} & a_{12}^{(k+1)} & 0 & \cdots\cdots\cdots\cdots\cdots\cdots\cdots & 0 \\ a_{21}^{(k+1)} & & & & \vdots \\ 0 & & & & 0 \cdots\cdots 0 \\ \vdots & & a_{k+1,k}^{(k+1)} & a_{k+1,k+1}^{(k+1)} & a_{k+1,k+2}^{(k+1)} \cdots a_{k+1,n}^{(k+1)} \\ & & 0 & & \vdots \\ 0 \cdots\cdots 0 & & a_{n,k+1}^{(k+1)} & \cdots\cdots & a_{nn}^{(k+1)} \end{bmatrix}.$$

Continuing in this manner, the tridiagonal and symmetric matrix $A^{(n-1)}$ is formed, where

The program HSEHLD95 implements Householder's method.

$$A^{(n-1)} = P^{(n-2)} P^{(n-3)} \cdots P^{(1)} A P^{(1)} \cdots P^{(n-3)} P^{(n-2)}.$$

Example 1 Apply Householder transformations to the symmetric 4×4 matrix

$$A = \begin{bmatrix} 4 & 1 & -2 & 2 \\ 1 & 2 & 0 & 1 \\ -2 & 0 & 3 & -2 \\ 2 & 1 & -2 & -1 \end{bmatrix}$$

to produce a symmetric tridiagonal matrix that is similar to A.

Solution For the first application of a Householder transformation,

$$\alpha = -(1)\left(\sum_{j=2}^{4} a_{j1}^2\right)^{1/2} = -3, \; r = \left(\frac{1}{2}(-3)^2 - \frac{1}{2}(1)(-3)\right)^{1/2} = \sqrt{6},$$

$$\mathbf{w} = \left(0, \frac{\sqrt{6}}{3}, -\frac{\sqrt{6}}{6}, \frac{\sqrt{6}}{6}\right)^t,$$

$$
P^{(1)} = \begin{bmatrix} 1 & 0 & 0 & 0 \\ 0 & 1 & 0 & 0 \\ 0 & 0 & 1 & 0 \\ 0 & 0 & 0 & 1 \end{bmatrix} - 2\left(\frac{\sqrt{6}}{6}\right)^2 \begin{bmatrix} 0 \\ 2 \\ -1 \\ 1 \end{bmatrix} \cdot [0, 2, -1, 1]
$$

$$
= \begin{bmatrix} 1 & 0 & 0 & 0 \\ 0 & 1 & 0 & 0 \\ 0 & 0 & 1 & 0 \\ 0 & 0 & 0 & 1 \end{bmatrix} - \frac{1}{3} \begin{bmatrix} 0 & 0 & 0 & 0 \\ 0 & 4 & -2 & 2 \\ 0 & -2 & 1 & -1 \\ 0 & 2 & -1 & 1 \end{bmatrix} = \begin{bmatrix} 1 & 0 & 0 & 0 \\ 0 & -\frac{1}{3} & \frac{2}{3} & -\frac{2}{3} \\ 0 & \frac{2}{3} & \frac{2}{3} & \frac{1}{3} \\ 0 & -\frac{2}{3} & \frac{1}{3} & \frac{2}{3} \end{bmatrix},
$$

and

$$
A^{(2)} = P^{(1)} A P^{(1)} = \begin{bmatrix} 4 & -3 & 0 & 0 \\ -3 & \frac{10}{3} & 1 & \frac{4}{3} \\ 0 & 1 & \frac{5}{3} & -\frac{4}{3} \\ 0 & \frac{4}{3} & -\frac{4}{3} & -1 \end{bmatrix}.
$$

Continuing to the second iteration,

$$
\alpha = -\frac{5}{3}, \quad r = \frac{2\sqrt{5}}{3}, \quad \mathbf{w} = \left(0, 0, 2\sqrt{5}, \frac{\sqrt{5}}{5}\right)^t,
$$

and

$$
P^{(2)} = \begin{bmatrix} 1 & 0 & 0 & 0 \\ 0 & 1 & 0 & 0 \\ 0 & 0 & -\frac{3}{5} & -\frac{4}{5} \\ 0 & 0 & -\frac{4}{5} & \frac{3}{5} \end{bmatrix}.
$$

The symmetric tridiagonal matrix is

$$
A^{(3)} = P^{(2)} A^{(1)} P^{(2)} = \begin{bmatrix} 4 & -3 & 0 & 0 \\ -3 & \frac{10}{3} & -\frac{5}{3} & 0 \\ 0 & -\frac{5}{3} & -\frac{33}{25} & \frac{68}{75} \\ 0 & 0 & \frac{68}{75} & \frac{149}{75} \end{bmatrix}. \qquad \blacksquare
$$

In the next section, we will examine how the QR method can then be applied to $A^{(n-1)}$ to determine its eigenvalues, which are the same as those of the original matrix A.

EXERCISE SET 9.4

1. Use Householder's method to produce a symmetric tridiagonal matrix that is similar to the given matrix.

   a. $\begin{bmatrix} 12 & 10 & 4 \\ 10 & 8 & -5 \\ 4 & -5 & 3 \end{bmatrix}$

   b. $\begin{bmatrix} 2 & -1 & -1 \\ -1 & 2 & -1 \\ -1 & -1 & 2 \end{bmatrix}$

   c. $\begin{bmatrix} 1 & 1 & 1 \\ 1 & 1 & 0 \\ 1 & 0 & 1 \end{bmatrix}$

   d. $\begin{bmatrix} 4.75 & 2.25 & -0.25 \\ 2.25 & 4.75 & 1.25 \\ -0.25 & 1.25 & 4.75 \end{bmatrix}$

2. Use Householder's method to produce a symmetric tridiagonal matrix that is similar to the given matrix.

a. $\begin{bmatrix} 4 & -1 & -1 & 0 \\ -1 & 4 & 0 & -1 \\ -1 & 0 & 4 & -1 \\ 0 & -1 & -1 & 4 \end{bmatrix}$

b. $\begin{bmatrix} 5 & -2 & -0.5 & 1.5 \\ -2 & 5 & 1.5 & -0.5 \\ -0.5 & 1.5 & 5 & -2 \\ 1.5 & -0.5 & -2 & 5 \end{bmatrix}$

c. $\begin{bmatrix} 8 & 0.25 & 0.5 & 2 & -1 \\ 0.25 & -4 & 0 & 1 & 2 \\ 0.5 & 0 & 5 & 0.75 & -1 \\ 2 & 1 & 0.75 & 5 & -0.5 \\ -1 & 2 & -1 & -0.5 & 6 \end{bmatrix}$

d. $\begin{bmatrix} 2 & -1 & -1 & 0 & 0 \\ -1 & 3 & 0 & -2 & 0 \\ -1 & 0 & 4 & 2 & 1 \\ 0 & -2 & 2 & 8 & 3 \\ 0 & 0 & 1 & 3 & 9 \end{bmatrix}$

3. Suppose that $P = I - 2\mathbf{w}\mathbf{w}^t$ is a Householder transformation.

a. Show that P is symmetric.

b. Show that P is orthogonal.

9.5 The QR Method

The deflation methods discussed in Section 9.3 are not generally suitable for calculating all the eigenvalues of a matrix because of the growth of round-off error. In this section we consider the QR method, a matrix reduction technique used to simultaneously approximate all the eigenvalues of a symmetric matrix.

To apply the QR method, we begin with a symmetric matrix in tridiagonal form; that is, the only nonzero entries in the matrix lie either on the diagonal or on the subdiagonals directly above or below the diagonal. If this is not the form of the symmetric matrix, the first step is to apply Householder's method to compute a symmetric, tridiagonal matrix similar to the given matrix.

In the remainder of this section it will be assumed that the symmetric matrix for which these eigenvalues are to be calculated is tridiagonal. If we let A denote a matrix of this type, we can simplify the notation somewhat by labeling the entries of A as follows:

$$A = \begin{bmatrix} a_1 & b_2 & 0 & \cdots\cdots & 0 \\ b_2 & a_2 & b_3 & \ddots & \vdots \\ 0 & b_3 & a_3 & \ddots & 0 \\ \vdots & \ddots & \ddots & \ddots & b_n \\ 0 & \cdots\cdots & 0 & b_n & a_n \end{bmatrix}. \tag{9.7}$$

If $b_2 = 0$ or $b_n = 0$, then the 1×1 matrix $[a_1]$ or $[a_n]$ immediately produces an eigenvalue a_1 or a_n of A. The QR method takes advantage of this observation by successively decreasing the values of the entries below the main diagonal until $b_2 \approx 0$ or $b_n \approx 0$.

When $b_j = 0$ for some j, where $2 < j < n$, the problem can be reduced to considering, instead of A, the smaller matrices

$$\begin{bmatrix} a_1 & b_2 & 0 & \cdots & \cdots & 0 \\ b_2 & a_2 & b_3 & & & \vdots \\ 0 & b_3 & a_3 & & & 0 \\ \vdots & & & & b_{j-1} & \\ 0 & \cdots & 0 & b_{j-1} & a_{j-1} \end{bmatrix} \quad \text{and} \quad \begin{bmatrix} a_j & b_{j+1} & 0 & \cdots & \cdots & 0 \\ b_{j+1} & a_{j+1} & b_{j+2} & & & \vdots \\ 0 & b_{j+2} & a_{j+2} & & & 0 \\ \vdots & & & & b_n & \\ 0 & \cdots & 0 & b_n & a_n \end{bmatrix}. \quad (9.8)$$

If none of the b_j are zero, the QR method proceeds by forming a sequence of matrices $A = A^{(1)}, A^{(2)}, A^{(3)}, \ldots$, as follows:

- $A^{(1)} = A$ is factored as a product $A^{(1)} = Q^{(1)} R^{(1)}$, where $Q^{(1)}$ is orthogonal and $R^{(1)}$ is upper triangular.

- $A^{(2)}$ is defined as $A^{(2)} = R^{(1)} Q^{(1)}$.

In general, $A^{(i)}$ is factored as a product $A^{(i)} = Q^{(i)} R^{(i)}$ of an orthogonal matrix $Q^{(i)}$ and an upper triangular matrix $R^{(i)}$. Then $A^{(i+1)}$ is defined by the product of $R^{(i)}$ and $Q^{(i)}$ in the reverse direction $A^{(i+1)} = R^{(i)} Q^{(i)}$. Since $Q^{(i)}$ is orthogonal, $R^{(i)} = Q^{(i)^t} A^{(i)}$ and

$$A^{(i+1)} = R^{(i)} Q^{(i)} = (Q^{(i)^t} A^{(i)}) Q^{(i)} = Q^{(i)^t} A^{(i)} Q^{(i)}. \quad (9.9)$$

This ensures that $A^{(i+1)}$ is symmetric with the same eigenvalues as $A^{(i)}$. By the manner in which we define $R^{(i)}$ and $Q^{(i)}$, it also ensures that $A^{(i+1)}$ is tridiagonal. Continuing by induction, $A^{(i+1)}$ has the same eigenvalues as the original matrix A, and $A^{(i+1)}$ tends to a diagonal matrix with the eigenvalues of A along the diagonal.

Rotation Matrices

If A is the 2×2 rotation matrix
$$A = \begin{bmatrix} \cos\theta & -\sin\theta \\ \sin\theta & \cos\theta \end{bmatrix},$$
then $A\mathbf{x}$ is $\mathbf{x}$ rotated counterclockwise by the angle θ.

Rotation matrices are often called Givens transformations because they were introduced by James Wallace Givens (1910–1993) in the 1950s when he was at Argonne National Laboratories.

To describe the construction of the factoring matrices $Q^{(i)}$ and $R^{(i)}$, we need the notion of a *rotation matrix*.

A **rotation matrix** P differs from the identity matrix in at most four elements. These four elements are of the form

$$p_{ii} = p_{jj} = \cos\theta \quad \text{and} \quad p_{ij} = -p_{ji} = \sin\theta,$$

for some θ and some $i \neq j$.

For any rotation matrix P, the matrix AP differs from A only in the ith and jth columns and the matrix PA differs from A only in the ith and jth rows. For any $i \neq j$, the angle θ can be chosen so that the product PA has a zero entry for $(PA)_{ij}$. In addition, every rotation matrix P is orthogonal, because the definition implies that $PP^t = I$.

Example 1 Find a rotation matrix P with the property that PA has a zero entry in the second row and first column, for the tridiagonal matrix

$$A = \begin{bmatrix} 3 & 1 & 0 \\ 1 & 3 & 1 \\ 0 & 1 & 3 \end{bmatrix}.$$

Solution   The form of P we will use is

$$
P = \begin{bmatrix} \cos\theta & \sin\theta & 0 \\ -\sin\theta & \cos\theta & 0 \\ 0 & 0 & 1 \end{bmatrix} \quad \text{so} \quad PA = \begin{bmatrix} 3\cos\theta + \sin\theta & \cos\theta + 3\sin\theta & \sin\theta \\ -3\sin\theta + \cos\theta & -\sin\theta + 3\cos\theta & \cos\theta \\ 0 & 1 & 3 \end{bmatrix}.
$$

The angle θ is chosen so that $-3\sin\theta + \cos\theta = 0$; that is, so that $\tan\theta = \frac{1}{3}$. We want θ in the first quadrant, so

$$
\cos\theta = \frac{3\sqrt{10}}{10}, \quad \sin\theta = \frac{\sqrt{10}}{10}
$$

and

$$
PA = \begin{bmatrix} \frac{3\sqrt{10}}{10} & \frac{\sqrt{10}}{10} & 0 \\ -\frac{\sqrt{10}}{10} & \frac{3\sqrt{10}}{10} & 0 \\ 0 & 0 & 1 \end{bmatrix} \begin{bmatrix} 3 & 1 & 0 \\ 1 & 3 & 1 \\ 0 & 1 & 3 \end{bmatrix} = \begin{bmatrix} \sqrt{10} & \frac{3}{5}\sqrt{10} & \frac{1}{10}\sqrt{10} \\ 0 & \frac{4}{5}\sqrt{10} & \frac{3}{10}\sqrt{10} \\ 0 & 1 & 3 \end{bmatrix}.
$$

Note that the resulting matrix is neither symmetric nor tridiagonal.   ▪

The factorization of $A^{(1)}$ into $A^{(1)} = Q^{(1)}R^{(1)}$ uses a product of $n-1$ rotation matrices to construct

$$
R^{(1)} = P_n P_{n-1} \cdots P_2 A^{(1)}.
$$

We first choose the rotation matrix P_2 with

$$
p_{11} = p_{22} = \cos\theta_2 \quad \text{and} \quad p_{12} = -p_{21} = \sin\theta_2,
$$

where

$$
\sin\theta_2 = \frac{b_2}{\sqrt{b_2^2 + a_1^2}} \quad \text{and} \quad \cos\theta_2 = \frac{a_1}{\sqrt{b_2^2 + a_1^2}}.
$$

This choice gives

$$
(-\sin\theta_2)a_1 + (\cos\theta_2)b_2 = \frac{-b_2 a_1}{\sqrt{b_2^2 + a_1^2}} + \frac{a_1 b_2}{\sqrt{b_2^2 + a_1^2}} = 0.
$$

for the entry in the (2, 1) position; that is, in the second row and first column of the product $P_2 A^{(1)}$. So the matrix

$$
A_2^{(1)} = P_2 A^{(1)}
$$

has a zero in the (2, 1) position.

The multiplication $P_2 A^{(1)}$ affects both rows 1 and 2 of $A^{(1)}$, so the matrix $A_2^{(1)}$ does not necessarily retain zero entries in positions (1, 3), (1, 4), ..., and (1, n). However, $A^{(1)}$ is tridiagonal, so the (1, 4), ..., (1, n) entries of $A_2^{(1)}$ must also be 0. Only the (1, 3)-entry, the one in the first row and third column, can become nonzero in $A_2^{(1)}$.

Continuing, the matrix P_k is chosen so that the $(k, k-1)$ entry in $A_k^{(1)} = P_k A_{k-1}^{(1)}$ is zero. This results in the $(k-1, k+1)$ entry likely becoming nonzero. The matrix $A_k^{(1)}$ has the form

$$
A_k^{(1)} = \begin{bmatrix}
z_1 & q_1 & r_1 & 0 & \cdots & \cdots & \cdots & \cdots & 0 \\
0 & & & & & & & & \vdots \\
0 & & & & & & & & \vdots \\
\vdots & & 0 & z_{k-1} & q_{k-1} & r_{k-1} & & & \vdots \\
& & & 0 & x_k & y_k & 0 & & \\
& & & & b_{k+1} & a_{k+1} & b_{k+2} & & 0 \\
& & & & & & & & 0 \\
& & & & & & & & b_n \\
0 & \cdots & \cdots & \cdots & \cdots & \cdots & 0 & b_n & a_n
\end{bmatrix}
$$

and P_{k+1} has the form

$$
P_{k+1} = \left[\begin{array}{c|cc|c}
I_{k-1} & & O & O \\
\hline
& c_{k+1} & s_{k+1} & \\
O & & & O \\
& -s_{k+1} & c_{k+1} & \\
\hline
O & & O & I_{n-k-1}
\end{array}\right] \quad \leftarrow \text{row } k
\tag{9.10}
$$

$$
\uparrow \\
\text{column } k
$$

where 0 denotes the appropriately-dimensioned matrix with all zero entries.

The constants $c_{k+1} = \cos\theta_{k+1}$ and $s_{k+1} = \sin\theta_{k+1}$ in P_{k+1} are chosen so that the $(k+1, k)$ entry in $A_{k+1}^{(1)}$ is zero; that is, $-s_{k+1}x_k + c_{k+1}b_{k+1} = 0$. Since $c_{k+1}^2 + s_{k+1}^2 = 1$, the solution to this equation is

$$
s_{k+1} = \frac{b_{k+1}}{\sqrt{b_{k+1}^2 + x_k^2}} \quad \text{and} \quad c_{k+1} = \frac{x_k}{\sqrt{b_{k+1}^2 + x_k^2}},
$$

and $A_{k+1}^{(1)}$ has the form

$$
A_{k+1}^{(1)} = \begin{bmatrix}
z_1 & q_1 & r_1 & 0 & \cdots & \cdots & \cdots & \cdots & 0 \\
0 & & & & & & & & \vdots \\
0 & & & & & & & & \vdots \\
\vdots & & 0 & z_k & q_k & r_k & & & \vdots \\
& & & 0 & x_{k+1} & y_{k+1} & 0 & & \\
& & & & b_{k+2} & a_{k+2} & b_{k+3} & & 0 \\
& & & & & & & & 0 \\
& & & & & & & & b_n \\
0 & \cdots & \cdots & \cdots & \cdots & \cdots & 0 & b_n & a_n
\end{bmatrix}
\tag{9.11}
$$

Proceeding with this construction in the sequence $P_2, \ldots, P_n$ produces the upper triangular matrix

$$
R^{(1)} \equiv A_n^{(1)} =
\begin{bmatrix}
z_1 & q_1 & r_1 & 0 & \cdots\cdots\cdots & 0 \\
0 & & & & & \\
& & & & & 0 \\
\vdots & & & & & r_{n-2} \\
& & & & z_{n-1} & q_{n-1} \\
0 & \cdots\cdots\cdots\cdots & & & 0 & x_n
\end{bmatrix}. \qquad (9.12)
$$

Now the orthogonal matrix $Q^{(1)}$ is defined as $Q^{(1)} = P_2^t P_3^t \cdots P_n^t$. This gives

$$
A^{(2)} = R^{(1)} Q^{(1)} = R^{(1)} P_2^t P_3^t \cdots P_n^t = P_n \cdots P_2 A P_2^t P_3^t \cdots P_n^t.
$$

The matrix $A^{(2)}$ is tridiagonal, and, in general, the entries off the diagonal will be smaller in magnitude than the corresponding entries in $A^{(1)}$. The process is repeated to construct $A^{(3)}$, $A^{(4)}$, and so on.

Example 2 Apply one iteration of the QR method to the matrix that was given in Example 1:

$$
A =
\begin{bmatrix}
3 & 1 & 0 \\
1 & 3 & 1 \\
0 & 1 & 3
\end{bmatrix}.
$$

Solution Let $A^{(1)} = A$ be the given matrix and P_2 represent the rotation matrix determined in Example 1. We found, using the notation introduced in the QR method, that

$$
A_2^{(1)} = P_2 A^{(1)} =
\begin{bmatrix}
\frac{3\sqrt{10}}{10} & \frac{\sqrt{10}}{10} & 0 \\
-\frac{\sqrt{10}}{10} & \frac{3\sqrt{10}}{10} & 0 \\
0 & 0 & 1
\end{bmatrix}
\begin{bmatrix}
3 & 1 & 0 \\
1 & 3 & 1 \\
0 & 1 & 3
\end{bmatrix}
$$

$$
=
\begin{bmatrix}
\sqrt{10} & \frac{3}{5}\sqrt{10} & \frac{\sqrt{10}}{10} \\
0 & \frac{4\sqrt{10}}{5} & \frac{3\sqrt{10}}{10} \\
0 & 1 & 3
\end{bmatrix}
\equiv
\begin{bmatrix}
z_1 & q_1 & r_1 \\
0 & x_2 & y_2 \\
0 & b_3^{(1)} & a_3^{(1)}
\end{bmatrix}.
$$

Continuing, we have

$$
s_3 = \frac{b_3^{(1)}}{\sqrt{x_2^2 + (b_3^{(1)})^2}} = 0.36761 \quad \text{and} \quad c_3 = \frac{x_2}{\sqrt{x_2^2 + (b_3^{(1)})^2}} = 0.92998.
$$

so

$$
R^{(1)} \equiv A_3^{(1)} = P_3 A_2^{(1)} =
\begin{bmatrix}
1 & 0 & 0 \\
0 & 0.92998 & 0.36761 \\
0 & -0.36761 & 0.92998
\end{bmatrix}
\begin{bmatrix}
\sqrt{10} & \frac{3}{5}\sqrt{10} & \frac{\sqrt{10}}{10} \\
0 & \frac{4\sqrt{10}}{5} & \frac{3\sqrt{10}}{10} \\
0 & 1 & 3
\end{bmatrix}
$$

$$
=
\begin{bmatrix}
\sqrt{10} & \frac{3}{5}\sqrt{10} & \frac{\sqrt{10}}{10} \\
0 & 2.7203 & 1.9851 \\
0 & 0 & 2.4412
\end{bmatrix}
$$

and

$$Q^{(1)} = P_2^t P_3^t = \begin{bmatrix} \frac{3\sqrt{10}}{10} & -\frac{\sqrt{10}}{10} & 0 \\ \frac{\sqrt{10}}{10} & \frac{3\sqrt{10}}{10} & 0 \\ 0 & 0 & 1 \end{bmatrix} \begin{bmatrix} 1 & 0 & 0 \\ 0 & 0.92998 & -0.36761 \\ 0 & 0.36761 & 0.92998 \end{bmatrix}$$

$$= \begin{bmatrix} 0.94868 & -0.29409 & 0.11625 \\ 0.31623 & 0.88226 & -0.34874 \\ 0 & 0.36761 & 0.92998 \end{bmatrix}.$$

As a consequence,

$$A^{(2)} = R^{(1)}Q^{(1)} = \begin{bmatrix} \sqrt{10} & \frac{3}{5}\sqrt{10} & \frac{\sqrt{10}}{10} \\ 0 & 2.7203 & 1.9851 \\ 0 & 0 & 2.4412 \end{bmatrix} \begin{bmatrix} 0.94868 & -0.29409 & 0.11625 \\ 0.31623 & 0.88226 & -0.34874 \\ 0 & -0.36761 & 0.92998 \end{bmatrix}$$

$$= \begin{bmatrix} 3.6 & 0.86024 & 0 \\ 0.86024 & 3.12973 & 0.89740 \\ 0 & 0.89740 & 2.27027 \end{bmatrix}.$$

The off-diagonal elements of $A^{(2)}$ are smaller than those of $A^{(1)}$ by about 14%, so we have a reduction but it is not substantial. To decrease to below 0.001 we would need to perform 13 iterations of the QR method. Doing this gives

$$A^{(13)} = \begin{bmatrix} 4.4139 & 0.01941 & 0 \\ 0.01941 & 3.0003 & 0.00095 \\ 0 & 0.00095 & 1.5858 \end{bmatrix}.$$

This gives an approximate eigenvalue of $\left(A^{(13)}\right)_{33} = 1.5858$. The remaining eigenvalues are approximated by considering the reduced matrix

$$\begin{bmatrix} 4.4139 & 0.01941 \\ 0.01941 & 3.0003 \end{bmatrix},$$

which has approximate eigenvalues 4.4142 and 3.0000. ∎

Accelerating Convergence

If the eigenvalues of A have distinct moduli with $|\lambda_1| > |\lambda_2| > \cdots > |\lambda_n|$, then the rate of convergence of the entry $b_{j+1}^{(i+1)}$ to 0 in the matrix $A^{(i+1)}$ depends on the ratio $|\lambda_{j+1}/\lambda_j|$. The rate of convergence of $b_{j+1}^{(i+1)}$ to 0 determines the rate at which the entry $a_j^{(i+1)}$ converges to the jth eigenvalue λ_j. Thus, the rate of convergence can be slow if $|\lambda_{j+1}/\lambda_j|$ is not significantly less than 1.

To accelerate this convergence, a shifting technique is employed similar to that used with the Inverse Power method in Section 9.3. A constant σ is selected close to an eigenvalue of A. This modifies the factorization to choosing $Q^{(i)}$ and $R^{(i)}$ so that

$$A^{(i)} - \sigma I = Q^{(i)} R^{(i)}, \tag{9.13}$$

and, correspondingly, the matrix $A^{(i+1)}$ is defined to be

$$A^{(i+1)} = R^{(i)} Q^{(i)} + \sigma I. \tag{9.14}$$

With this modification, the rate of convergence of $b_{j+1}^{(i+1)}$ to 0 depends on the ratio $|(\lambda_{j+1} - \sigma)/(\lambda_j - \sigma)|$. This can result in a significant improvement over the original rate of convergence of $a_j^{(i+1)}$ to λ_j if σ is close to λ_{j+1} but not close to λ_j.

We change σ at each step so that when A has eigenvalues of distinct modulus, $b_n^{(i+1)}$ converges to 0 faster than $b_j^{(i+1)}$ for any integer j less than n. When $b_n^{(i+1)}$ is sufficiently small, we assume that $\lambda_n \approx a_n^{(i+1)}$, delete the nth row and column of the matrix, and proceed in the same manner to find an approximation to λ_{n-1}. The process is continued until an approximation has been determined for each eigenvalue.

The shifting technique chooses, at the ith step, the shifting constant σ_i, where σ_i is the eigenvalue closest to $a_n^{(i)}$ of the matrix

$$
E^{(i)} = \begin{bmatrix} a_{n-1}^{(i)} & b_n^{(i)} \\ b_n^{(i)} & a_n^{(i)} \end{bmatrix}.
$$

This shift translates the eigenvalues of A by a factor σ_i, and the convergence is usually cubic. The method accumulates these shifts until $b_n^{(i+1)} \approx 0$ and then adds the shifts to $a_n^{(i+1)}$ to approximate the eigenvalue λ_n.

If A has eigenvalues of the same modulus, $b_j^{(i+1)}$ may tend to 0 for some $j \neq n$ at a faster rate than $b_n^{(i+1)}$. In this case, the matrix-splitting technique described in Eq. (9.8) can be employed to reduce the problem to one involving a pair of matrices of reduced order.

> The program QRSYMT96 implements the QR method with shifting.

Example 3   Incorporate shifting into the QR method for the matrix

$$
A = \begin{bmatrix} 3 & 1 & 0 \\ 1 & 3 & 1 \\ 0 & 1 & 3 \end{bmatrix} = \begin{bmatrix} a_1^{(1)} & b_2^{(1)} & 0 \\ b_2^{(1)} & a_2^{(1)} & b_3^{(1)} \\ 0 & b_3^{(1)} & a_3^{(1)} \end{bmatrix}.
$$

Solution   To find the acceleration parameter for shifting requires finding the eigenvalues of

$$
\begin{bmatrix} a_2^{(1)} & b_3^{(1)} \\ b_3^{(1)} & a_3^{(1)} \end{bmatrix} = \begin{bmatrix} 3 & 1 \\ 1 & 3 \end{bmatrix},
$$

which are $\mu_1 = 4$ and $\mu_2 = 2$. The choice of eigenvalue closest to $a_3^{(1)} = 3$ is arbitrary, and we choose $\mu_2 = 2$ and shift by this amount. Then $\sigma_1 = 2$ and

$$
\begin{bmatrix} d_1 & b_2^{(1)} & 0 \\ b_2^{(1)} & d_2 & b_3^{(1)} \\ 0 & b_3^{(1)} & d_3 \end{bmatrix} = \begin{bmatrix} 1 & 1 & 0 \\ 1 & 1 & 1 \\ 0 & 1 & 1 \end{bmatrix}.
$$

Continuing the computation, using the notation of Eq. (9.11) on page 393, gives

$$
x_1 = 1, \quad y_1 = 1, \quad z_1 = \sqrt{2}, \quad c_2 = \frac{\sqrt{2}}{2}, \quad s_2 = \frac{\sqrt{2}}{2},
$$

$$
q_1 = \sqrt{2}, \quad x_2 = 0, \quad r_1 = \frac{\sqrt{2}}{2}, \quad \text{and} \quad y_2 = \frac{\sqrt{2}}{2},
$$

so

$$
A_2^{(1)} = \begin{bmatrix} z_1 & q_1 & r_1 \\ 0 & x_2 & y_2 \\ 0 & b_3 & a_3 \end{bmatrix} = \begin{bmatrix} \sqrt{2} & \sqrt{2} & \frac{\sqrt{2}}{2} \\ 0 & 0 & \frac{\sqrt{2}}{2} \\ 0 & 1 & 1 \end{bmatrix}.
$$

Further

$$
z_2 = 1, \quad c_3 = 0, \quad s_3 = 1, \quad q_2 = 1, \quad \text{and} \quad x_3 = -\frac{\sqrt{2}}{2},
$$

so, from Eq. (9.12) on page 394,

$$R^{(1)} = A_3^{(1)} = \begin{bmatrix} z_1 & q_1 & r_1 \\ 0 & z_2 & q_2 \\ 0 & 0 & x_3 \end{bmatrix} = \begin{bmatrix} \sqrt{2} & \sqrt{2} & \frac{\sqrt{2}}{2} \\ 0 & 1 & 1 \\ 0 & 0 & -\frac{\sqrt{2}}{2} \end{bmatrix}.$$

We also have

$$Q^{(1)} = P_2^t P_3^t = \begin{bmatrix} c_2 & -s_2 & 0 \\ s_2 & c_2 & 0 \\ 0 & 0 & 1 \end{bmatrix} \begin{bmatrix} 1 & 0 & 0 \\ 0 & c_3 & -s_3 \\ 0 & s_3 & c_3 \end{bmatrix}$$

$$= \begin{bmatrix} \frac{\sqrt{2}}{2} & -\frac{\sqrt{2}}{2} & 0 \\ \frac{\sqrt{2}}{2} & \frac{\sqrt{2}}{2} & 0 \\ 0 & 0 & 1 \end{bmatrix} \begin{bmatrix} 1 & 0 & 0 \\ 0 & 0 & -1 \\ 0 & 1 & 0 \end{bmatrix} = \begin{bmatrix} \frac{\sqrt{2}}{2} & 0 & \frac{\sqrt{2}}{2} \\ \frac{\sqrt{2}}{2} & 0 & -\frac{\sqrt{2}}{2} \\ 0 & 1 & 0 \end{bmatrix}.$$

So

$$A^{(2)} = R^{(1)} Q^{(1)} = \begin{bmatrix} \sqrt{2} & \sqrt{2} & \frac{\sqrt{2}}{2} \\ 0 & 1 & 1 \\ 0 & 0 & -\frac{\sqrt{2}}{2} \end{bmatrix} \begin{bmatrix} \frac{\sqrt{2}}{2} & 0 & \frac{\sqrt{2}}{2} \\ \frac{\sqrt{2}}{2} & 0 & -\frac{\sqrt{2}}{2} \\ 0 & 1 & 0 \end{bmatrix},$$

which implies that

$$A^{(2)} = \begin{bmatrix} 2 & \frac{\sqrt{2}}{2} & 0 \\ \frac{\sqrt{2}}{2} & 1 & -\frac{\sqrt{2}}{2} \\ 0 & -\frac{\sqrt{2}}{2} & 0 \end{bmatrix} = \begin{bmatrix} a_1^{(2)} & b_2^{(2)} & 0 \\ b_2^{(2)} & a_2^{(2)} & b_3^{(2)} \\ 0 & b_3^{(2)} & a_3^{(2)} \end{bmatrix}.$$

One iteration of the QR method is complete. Neither $b_2^{(2)} = \sqrt{2}/2$ nor $b_3^{(2)} = -\sqrt{2}/2$ is small, so another iteration of the QR method is performed. For this iteration, we calculate the eigenvalues $\frac{1}{2} \pm \frac{1}{2}\sqrt{3}$ of the matrix

$$\begin{bmatrix} a_2^{(2)} & b_3^{(2)} \\ b_3^{(2)} & a_3^{(2)} \end{bmatrix} = \begin{bmatrix} 1 & -\frac{\sqrt{2}}{2} \\ -\frac{\sqrt{2}}{2} & 0 \end{bmatrix},$$

and choose $\sigma_2 = \frac{1}{2} - \frac{1}{2}\sqrt{3}$, the closest eigenvalue to $a_3^{(2)} = 0$. Completing the calculations gives

$$A^{(3)} = \begin{bmatrix} 2.6720277 & 0.37597448 & 0 \\ 0.37597448 & 1.4736080 & 0.030396964 \\ 0 & 0.030396964 & -0.047559530 \end{bmatrix}.$$

If $b_3^{(3)} = 0.030396964$ is sufficiently small, then the approximation to the eigenvalue is

$$\lambda_3 \approx a_3^{(3)} + \sigma_1 + \sigma_2 = -0.047559530 + 2 + \frac{1}{2} - \frac{1}{2}\sqrt{3} \approx 1.5864151.$$

Deleting the third row and column gives

$$A^{(3)} = \begin{bmatrix} 2.6720277 & 0.37597448 \\ 0.37597448 & 1.4736080 \end{bmatrix},$$

James Hardy Wilkinson (1919–1986) is best known for his extensive work on numerical methods for solving systems of linear equations and eigenvalue problems. He also is known for his use of the numerical linear algebra technique of backward error analysis.

which has eigenvalues $\mu_1 = 2.7802140$ and $\mu_2 = 1.3654218$. Adding the shifts gives the approximations

$$\lambda_1 \approx 4.4141886 \quad \text{and} \quad \lambda_2 \approx 2.9993964.$$

The actual eigenvalues of the matrix A are 4.41420, 3.00000, and 1.58579, so the QR method gave four digits of accuracy in only two iterations. ▪

The QR factorization of a matrix is implemented in MATLAB using the function qr. A complete description of the QR method can be found in works of Wilkinson (see [Wil] or [WR]). There you will also find methods to employ if the matrix is not symmetric.

EXERCISE SET 9.5

1. Apply two iterations of the QR method without shifting to the following matrices.

a. $\begin{bmatrix} 2 & -1 & 0 \\ -1 & 2 & -1 \\ 0 & -1 & 2 \end{bmatrix}$ b. $\begin{bmatrix} 3 & 1 & 0 \\ 1 & 4 & 2 \\ 0 & 2 & 1 \end{bmatrix}$

c. $\begin{bmatrix} 4 & -1 & 0 \\ -1 & 3 & -1 \\ 0 & -1 & 2 \end{bmatrix}$ d. $\begin{bmatrix} 1 & 1 & 0 & 0 \\ 1 & 2 & -1 & 0 \\ 0 & -1 & 3 & 1 \\ 0 & 0 & 1 & 4 \end{bmatrix}$

e. $\begin{bmatrix} -2 & 1 & 0 & 0 \\ 1 & -3 & -1 & 0 \\ 0 & -1 & 1 & 1 \\ 0 & 0 & 1 & 3 \end{bmatrix}$ f. $\begin{bmatrix} 0.5 & 0.25 & 0 & 0 \\ 0.25 & 0.8 & 0.4 & 0 \\ 0 & 0.4 & 0.6 & 0.1 \\ 0 & 0 & 0.1 & 1 \end{bmatrix}$

2. Apply two iterations of the QR method without shifting to the following matrices.

a. $\begin{bmatrix} 2 & -1 & 0 \\ -1 & -1 & -2 \\ 0 & -2 & 3 \end{bmatrix}$ b. $\begin{bmatrix} 3 & 1 & 0 \\ 1 & 4 & 2 \\ 0 & 2 & 3 \end{bmatrix}$

c. $\begin{bmatrix} 4 & 2 & 0 & 0 & 0 \\ 2 & 4 & 2 & 0 & 0 \\ 0 & 2 & 4 & 2 & 0 \\ 0 & 0 & 2 & 4 & 2 \\ 0 & 0 & 0 & 2 & 4 \end{bmatrix}$ d. $\begin{bmatrix} 5 & -1 & 0 & 0 & 0 \\ -1 & 4.5 & 0.2 & 0 & 0 \\ 0 & 0.2 & 1 & -0.4 & 0 \\ 0 & 0 & -0.4 & 3 & 1 \\ 0 & 0 & 0 & 1 & 3 \end{bmatrix}$

3. Use the QR method given in QRSYMT96 with $TOL = 10^{-5}$ to determine all the eigenvalues for the matrices given in Exercise 1.

4. Use the Inverse Power method given in INVPWR93 with $TOL = 10^{-5}$ to determine the eigenvectors of the matrices in Exercise 1.

5. In Example 3 we used the acceleration technique for the QR method applied to the matrix

$$A = \begin{bmatrix} 3 & 1 & 0 \\ 1 & 3 & 1 \\ 0 & 1 & 3 \end{bmatrix}.$$

by choosing the shifting parameter $\sigma_1 = 2$, but we remarked that we could have chosen this shifting parameter as $\sigma_1 = 4$. Repeat the calculations of that example using this alternative choice.

6. **a.** Show that the rotation matrix $\begin{bmatrix} \cos\theta & -\sin\theta \\ \sin\theta & \cos\theta \end{bmatrix}$ applied to the vector $\mathbf{x} = (x_1, x_2)^t$ has the geometric effect of rotating $\mathbf{x}$ through the angle θ without changing its magnitude with respect to $\|\cdot\|_2$.

   b. Show that the magnitude of $\mathbf{x}$ with respect to $\|\cdot\|_\infty$ can be changed by a rotation matrix.

7. Let P be the rotation matrix with $p_{ii} = p_{jj} = \cos\theta$, and $p_{ij} = -p_{ji} = \sin\theta$ for $j < i$. Show that for any $n \times n$ matrix A:

$$(AP)_{pq} = \begin{cases} a_{pq}, & \text{if } q \neq i, j, \\ (\cos\theta)a_{pj} + (\sin\theta)a_{pi}, & \text{if } q = j, \\ (\cos\theta)a_{pi} - (\sin\theta)a_{pj}, & \text{if } q = i, \end{cases} \quad (PA)_{pq} = \begin{cases} a_{pq}, & \text{if } p \neq i, j, \\ (\cos\theta)a_{jq} - (\sin\theta)a_{iq}, & \text{if } p = j, \\ (\sin\theta)a_{jq} + (\cos\theta)a_{iq}, & \text{if } p = i. \end{cases}$$

8. **Jacobi's method** for a symmetric matrix A is described by

$$A_1 = A, \quad A_2 = P_1 A_1 P_1^t,$$

   and, in general,

$$A_{i+1} = P_i A_i P_i^t.$$

   The matrix A_{i+1} tends to a diagonal matrix, where P_i is a rotation matrix chosen to eliminate a large off-diagonal element in A_i. Suppose $a_{j,k}$ and $a_{k,j}$ are to be set to 0, where $j \neq k$. If $a_{jj} \neq a_{kk}$, then

$$(P_i)_{jj} = (P_i)_{kk} = \sqrt{\frac{1}{2}\left(1 + \frac{b}{\sqrt{c^2 + b^2}}\right)},$$

$$(P_i)_{kj} = \frac{c}{2(P_i)_{jj}\sqrt{c^2 + b^2}} = -(P_i)_{jk},$$

   where

$$c = 2a_{jk}\,\mathrm{sgn}(a_{jj} - a_{kk}) \quad \text{and} \quad b = |a_{jj} - a_{kk}|,$$

   or if $a_{jj} = a_{kk}$,

$$(P_i)_{jj} = (P_i)_{kk} = \frac{\sqrt{2}}{2}$$

   and

$$(P_i)_{kj} = -(P_i)_{jk} = \frac{\sqrt{2}}{2}.$$

   Jacobi's method is invoked by setting $a_{21} = 0$ and then setting $a_{31}, a_{32}, a_{41}, a_{42}, a_{43}, \ldots, a_{n,1}, \ldots,$ $a_{n,n-1}$ in turn to zero. This is repeated until a matrix A_k is computed with

$$\sum_{i=1}^{n} \sum_{\substack{j=1 \\ j \neq i}}^{n} |a_{ij}^{(k)}|$$

   sufficiently small. The eigenvalues of A can then be approximated by the diagonal entries of A_k. Repeat Exercise 3 using the Jacobi method.

9.6 Singular Value Decomposition

In Chapter 6 we saw how effective factorization can be when solving linear systems of the form $A\mathbf{x} = \mathbf{b}$ when the matrix A is used repeatedly for varying $\mathbf{b}$. In this section we will consider a technique for factoring a general $m \times n$ matrix. It has application in many areas, including least squares fitting of data, image compression, signal processing, and statistics.

The factorization of a general $m \times n$ matrix A we will consider is called the *Singular Value Decomposition*. This factorization takes the form

$$A = U S V^t,$$

where U is an $m \times m$ orthogonal matrix, V is an $n \times n$ orthogonal matrix, and S is an $m \times n$ matrix whose only nonzero elements lie along the main diagonal in a special way. We will assume throughout this section that $m \geq n$, and, in many important applications, m is very much larger than n.

Singular value decomposition has quite a long history, being first considered by mathematicians in the latter part of the nineteenth century. However, the important applications of the technique had to wait until computing power became available in the second half of the twentieth century, when algorithms could be developed for its efficient implementation. These were primarily the work of Gene Golub (1932–2007) in a series of papers in the 1960s and 1970s (see, in particular, [GK] and [GR]). A quite complete history of the technique can be found in a paper by G. W. Stewart, which is available on the Internet at the address given in [Stew2].

The following is one of the basic results in linear algebra.

Linear Independent Rows and Columns

The number of linearly independent rows of an $m \times n$ matrix A is the same as the number of linearly independent columns of A.

The next result gives some useful facts about the matrices AA^t and $A^t A$ used in constructing the singular value decomposition of A.

Matrices AA^t and $A^t A$

Let A be $m \times n$ matrix.

   (i)   The matrices AA^t and $A^t A$ are symmetric.

  (ii)   The eigenvalues of AA^t and $A^t A$ are real and nonnegative.

 (iii)   The nonzero eigenvalues of AA^t and $A^t A$ are the same.

The verification of these results is considered in Exercises 8 and 9.

Constructing a Singular Value Decomposition for an $m \times n$ Matrix A

A non-square matrix A, that is, a matrix with a different number of rows and columns, cannot have an eigenvalue because $A\mathbf{x}$ and $\mathbf{x}$ differ in size. However, there are numbers that play roles for non-square matrices that are similar to those played by eigenvalues for square matrices. One of the important features of the singular value decomposition of a general matrix is that it permits a generalization of eigenvalues and eigenvectors in this situation.

Recall that our objective is to determine a factorization of the $m \times n$ matrix A, where $m \geq n$, in the form

$$A = U S V^t,$$

where U is an $m \times m$ orthogonal matrix, V is $n \times n$ an orthogonal matrix, and S is an $m \times n$ diagonal matrix; that is, its only nonzero entries are $(S)_{ii} \equiv s_i \geq 0$, for $i = 1, \ldots , n$. (See Figure 9.2.)

Figure 9.2

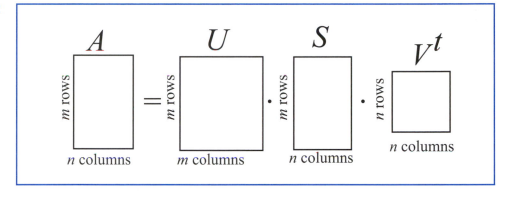

Constructing S in the Factorization $A = U S V^t$

We construct the matrix S by finding the eigenvalues of the symmetric $n \times n$ matrix $A^t A$. By result (ii) on page 400, these eigenvalues are all non-negative real numbers. We order them from largest to smallest and denote them

$$s_1^2 \geq s_2^2 \geq \cdots \geq s_k^2 > s_{k+1} = \cdots = s_n = 0.$$

That is, s_1^2 is the largest eigenvalue of $A^t A$ and s_k^2 is the smallest nonzero eigenvalue.

The positive square roots of the eigenvalues of $A^t A$ give the diagonal entries in S, which are called the **singular values** of A. Hence

$$S = \begin{bmatrix} s_1 & 0 & \cdots & 0 \\ 0 & s_2 & \ddots & \vdots \\ \vdots & \ddots & \ddots & 0 \\ 0 & \cdots & 0 & s_n \\ 0 & \cdots & \cdots & 0 \\ \vdots & & & \vdots \\ 0 & \cdots & \cdots & 0 \end{bmatrix},$$

where $s_i = 0$ when $k < i \leq n$.

Example 1   Determine the singular values of the 5×3 matrix

$$A = \begin{bmatrix} 1 & 0 & 1 \\ 0 & 1 & 0 \\ 0 & 1 & 1 \\ 0 & 1 & 0 \\ 1 & 1 & 0 \end{bmatrix}.$$

Solution   We have

$$A^t = \begin{bmatrix} 1 & 0 & 0 & 0 & 1 \\ 0 & 1 & 1 & 1 & 1 \\ 1 & 0 & 1 & 0 & 0 \end{bmatrix} \quad \text{so} \quad A^t A = \begin{bmatrix} 2 & 1 & 1 \\ 1 & 4 & 1 \\ 1 & 1 & 2 \end{bmatrix}.$$

The characteristic polynomial of $A^t A$ is

$$\rho(A^t A) = \lambda^3 - 8\lambda^2 + 17\lambda - 10 = (\lambda - 5)(\lambda - 2)(\lambda - 1),$$

so the eigenvalues of $A^t A$ are $\lambda_1 = s_1^2 = 5$, $\lambda_2 = s_2^2 = 2$, and $\lambda_3 = s_3^2 = 1$. As a consequence, the singular values of A are $s_1 = \sqrt{5}$, $s_2 = \sqrt{2}$, $s_3 = 1$, and in the singular value decomposition of A we have

$$S = \begin{bmatrix} \sqrt{5} & 0 & 0 \\ 0 & \sqrt{2} & 0 \\ 0 & 0 & 1 \\ 0 & 0 & 0 \\ 0 & 0 & 0 \end{bmatrix}. \qquad ■$$

In the special case when A is a symmetric $n \times n$ matrix, all the s_i^2 are eigenvalues of $A^2 = A^t A$, and these are the squares of the eigenvalues of A (see Exercise 9 of Section 7.2). So, in this case, the singular values are the absolute values of the eigenvalues of A. When A is also positive definite, the eigenvalues and singular values of A are the same.

Constructing V in the Factorization $A = U S V^t$

The $n \times n$ matrix $A^t A$ is symmetric, so $A^t A$ is diagonalizable by the result on page 370, and we have a factorization of the form

$$A^t A = V D V^t,$$

where D is an $n \times n$ diagonal matrix whose diagonal entries consist of the eigenvalues of $A^t A$, and V is an $n \times n$ orthogonal matrix whose ith column is an eigenvector with l_2-norm 1 corresponding to the eigenvalue that is the ith diagonal entry of D. The specific diagonal matrix depends on the order of the eigenvalues along the diagonal. We choose D so that these are written in decreasing order. This construction gives $D = S^t S$.

The columns, denoted $\mathbf{v}_1, \mathbf{v}_2, \ldots, \mathbf{v}_n$, of the orthogonal $n \times n$ matrix V are orthonormal eigenvectors corresponding to the eigenvalues on the diagonal of D. Multiple eigenvalues of $A^t A$ permit multiple choices of the corresponding eigenvectors, so although D is uniquely defined, the matrix V might not be. No problem though, we can choose any such V.

Constructing U in the Factorization $A = U S V^t$

To construct the orthogonal $m \times m$ matrix U, we first consider the nonzero values $s_1 \geq s_2 \geq \cdots \geq s_k > 0$ and the corresponding columns in V given by $\mathbf{v}_1, \mathbf{v}_2, \ldots, \mathbf{v}_k$. We define

$$\mathbf{u}_i = \frac{1}{s_i} A \mathbf{v}_i, \quad \text{for } i = 1, 2, \ldots, k. \tag{9.15}$$

We use these as the first k of the m columns of U. Because A is $m \times n$ and each $\mathbf{v}_i$ is $n \times 1$, the vector $\mathbf{u}_i$ is $m \times 1$, as required. In addition, for each $1 \leq i \leq k$ and $1 \leq j \leq k$, the fact that the vectors $\mathbf{v}_1, \mathbf{v}_2, \ldots, \mathbf{v}_n$ are eigenvectors of $A^t A$ that form an orthonormal set implies that

$$\mathbf{u}_i^t \mathbf{u}_j = \left(\frac{1}{s_i} A \mathbf{v}_i \right)^t \frac{1}{s_j} A \mathbf{v}_j = \frac{1}{s_i s_j} \left(\mathbf{v}_i^t A^t \right) A \mathbf{v}_j = \frac{1}{s_i s_j} \mathbf{v}_i^t \left(A^t A \mathbf{v}_j \right) = \frac{1}{s_i s_j} \mathbf{v}_i^t \left(s_j^2 \mathbf{v}_j \right)$$

$$= \frac{s_j}{s_i} \mathbf{v}_i^t \mathbf{v}_j = \begin{cases} 0 & \text{if } i \neq j, \\ 1 & \text{if } i = j. \end{cases}$$

So the first k columns of U form an orthonormal set of eigenvectors of AA^t. Since U is $m \times m$, we need $m - k$ additional columns of U. For this we first need to find $m - k$ vectors which when added to the vectors from the first k columns will be an orthonormal set. These additional vectors are chosen to be linearly independent and normalized eigenvectors corresponding to the eigenvalue 0 of the $m \times m$ matrix AA^t.

The matrix U will not be unique unless $k = m$, and then only if all the eigenvalues of $A^t A$ are unique. Non-uniqueness is of no concern; we only need one such orthogonal matrix U.

Verifying the Factorization $A = USV^t$

To verify that this process actually gives the factorization $A = USV^t$, first recall that the transpose of an orthogonal matrix is also the inverse of the matrix (see part (i) of the result on page 368). Hence to show that $A = USV^t$ we can show the equivalent statement $AV = US$.

The orthogonal vectors $\mathbf{v}_1, \mathbf{v}_2, \dots, \mathbf{v}_n$ form a basis for $\mathbb{R}^n$, $A\mathbf{v}_i = s_i\mathbf{u}_i$, for $i = 1, \dots, k$, and $A\mathbf{v}_i = \mathbf{0}$, for $i = k + 1, \dots, n$. Since S has all zero entries in all the rows below k, the product US will also have all zero entries below the kth row. So we only need to consider the first k rows in the product. And for this we have

$$AV = A[\mathbf{v}_1\ \mathbf{v}_2\ \cdots \mathbf{v}_k\ \mathbf{v}_{k+1}\ \cdots \mathbf{v}_n]$$

$$= [A\mathbf{v}_1\ A\mathbf{v}_2\ \cdots A\mathbf{v}_k\ A\mathbf{v}_{k+1}\ \cdots A\mathbf{v}_n]$$

$$= [s_1\mathbf{u}_1\ s_2\mathbf{u}_2\ \cdots s_k\mathbf{u}_k\ \mathbf{0}\ \cdots \mathbf{0}]$$

$$= [\mathbf{u}_1\ \mathbf{u}_2\ \cdots \mathbf{u}_k\ \mathbf{0}\ \cdots \mathbf{0}]
\begin{bmatrix}
s_1 & 0 & \cdots & 0 & 0 & \cdots & 0 \\
0 & \ddots & \ddots & \vdots & \vdots & & \vdots \\
\vdots & \ddots & \ddots & \vdots & \vdots & & \vdots \\
0 & \cdots & 0 & s_k & 0 & \cdots & 0 \\
0 & \cdots & \cdots & 0 & 0 & \cdots & 0 \\
\vdots & & & \vdots & \vdots & & \vdots \\
0 & \cdots & \cdots & 0 & 0 & \cdots & 0
\end{bmatrix} = US.$$

This completes the construction of the singular value decomposition of A.

Example 2   Determine the singular value decomposition of the 5×3 matrix

$$A = \begin{bmatrix} 1 & 0 & 1 \\ 0 & 1 & 0 \\ 0 & 1 & 1 \\ 0 & 1 & 0 \\ 1 & 1 & 0 \end{bmatrix}.$$

Solution   We found in Example 1 that A has the singular values $s_1 = \sqrt{5}$, $s_2 = \sqrt{2}$, and $s_3 = 1$, so

$$S = \begin{bmatrix} \sqrt{5} & 0 & 0 \\ 0 & \sqrt{2} & 0 \\ 0 & 0 & 1 \\ 0 & 0 & 0 \\ 0 & 0 & 0 \end{bmatrix}.$$

Eigenvectors of $A^t A$ corresponding to $s_1 = \sqrt{5}$, $s_2 = \sqrt{2}$, and $s_3 = 1$ are, respectively, $(1, 2, 1)^t$, $(1, -1, 1)^t$, and $(-1, 0, 1)^t$ (see Exercise 5). Normalizing these vectors and using the values for the columns of V gives

$$
V = \begin{bmatrix} \frac{\sqrt{6}}{6} & \frac{\sqrt{3}}{3} & -\frac{\sqrt{2}}{2} \\ \frac{\sqrt{6}}{3} & -\frac{\sqrt{3}}{3} & 0 \\ \frac{\sqrt{6}}{6} & \frac{\sqrt{3}}{3} & \frac{\sqrt{2}}{2} \end{bmatrix} \quad \text{and} \quad V^t = \begin{bmatrix} \frac{\sqrt{6}}{6} & \frac{\sqrt{6}}{3} & \frac{\sqrt{6}}{6} \\ \frac{\sqrt{3}}{3} & -\frac{\sqrt{3}}{3} & \frac{\sqrt{3}}{3} \\ -\frac{\sqrt{2}}{2} & 0 & \frac{\sqrt{2}}{2} \end{bmatrix}.
$$

The first 3 columns of U are therefore

$$
\mathbf{u}_1 = \frac{1}{\sqrt{5}} \cdot A \left(\frac{\sqrt{6}}{6}, \frac{\sqrt{6}}{3}, \frac{\sqrt{6}}{6} \right)^t = \left(\frac{\sqrt{30}}{15}, \frac{\sqrt{30}}{15}, \frac{\sqrt{30}}{10}, \frac{\sqrt{30}}{15}, \frac{\sqrt{30}}{10} \right)^t
$$

$$
\mathbf{u}_2 = \frac{1}{\sqrt{2}} \cdot A \left(\frac{\sqrt{3}}{3}, -\frac{\sqrt{3}}{3}, \frac{\sqrt{3}}{3} \right)^t = \left(\frac{\sqrt{6}}{3}, -\frac{\sqrt{6}}{6}, 0, -\frac{\sqrt{6}}{6}, 0 \right)^t
$$

$$
\mathbf{u}_3 = 1 \cdot A \left(-\frac{\sqrt{2}}{2}, 0, \frac{\sqrt{2}}{2} \right)^t = \left(0, 0, \frac{\sqrt{2}}{2}, 0, -\frac{\sqrt{2}}{2} \right)^t.
$$

To determine the two remaining columns of U, we first need vectors $\mathbf{x}_4$ and $\mathbf{x}_5$ that are orthogonal eigenvectors corresponding to the eigenvalue 0 of the 5×5 matrix $A A^t$. Two vectors that satisfy are

$$
\mathbf{x}_4 = (1, 1, -1, 1, -1)^t \quad \text{and} \quad \mathbf{x}_5 = (0, 1, 0, -1, 0)^t,
$$

which, when normalized, give

$$
\mathbf{u}_4 = \left(\frac{\sqrt{5}}{5}, \frac{\sqrt{5}}{5}, -\frac{\sqrt{5}}{5}, \frac{\sqrt{5}}{5}, -\frac{\sqrt{5}}{5} \right)^t \quad \text{and} \quad \mathbf{u}_5 = \left(0, \frac{\sqrt{2}}{2}, 0, -\frac{\sqrt{2}}{2}, 0 \right)^t.
$$

This produces the matrix U and the singular value decomposition as

$$
A = U S V^t = \begin{bmatrix} \frac{\sqrt{30}}{15} & \frac{\sqrt{6}}{3} & 0 & \frac{\sqrt{5}}{5} & 0 \\ \frac{\sqrt{30}}{15} & -\frac{\sqrt{6}}{6} & 0 & \frac{\sqrt{5}}{5} & \frac{\sqrt{2}}{2} \\ \frac{\sqrt{30}}{10} & 0 & \frac{\sqrt{2}}{2} & -\frac{\sqrt{5}}{5} & 0 \\ \frac{\sqrt{30}}{15} & -\frac{\sqrt{6}}{6} & 0 & \frac{\sqrt{5}}{5} & -\frac{\sqrt{2}}{2} \\ \frac{\sqrt{30}}{10} & 0 & -\frac{\sqrt{2}}{2} & -\frac{\sqrt{5}}{5} & 0 \end{bmatrix} \begin{bmatrix} \sqrt{5} & 0 & 0 \\ 0 & \sqrt{2} & 0 \\ 0 & 0 & 1 \\ 0 & 0 & 0 \\ 0 & 0 & 0 \end{bmatrix} \begin{bmatrix} \frac{\sqrt{6}}{6} & \frac{\sqrt{6}}{3} & \frac{\sqrt{6}}{6} \\ \frac{\sqrt{3}}{3} & -\frac{\sqrt{3}}{3} & \frac{\sqrt{3}}{3} \\ -\frac{\sqrt{2}}{2} & 0 & \frac{\sqrt{2}}{2} \end{bmatrix}. \quad \blacksquare
$$

MATLAB produces a singular value decomposition with the command svd. For the matrix in our example we first set

```
A = [1 0 1; 0 1 0; 0 1 1; 0 1 0; 1 1 0]
```

Then the command

```
[U, S, V] = svd(A)
```

produces the matrices U, S, and V. The entries in U have been rounded so that the matrix will fit on the page.

$$U = \begin{bmatrix} 0.36514837 & -0.81649658 & 0.00000000 & 0.41153421 & 0.17504169 \\ 0.36514837 & 0.40824829 & 0.00000000 & 0.13476900 & 0.82573441 \\ 0.54772256 & 0.00000000 & -0.70710679 & -0.41153421 & -0.17504169 \\ 0.36514837 & 0.40824829 & 0.00000000 & 0.68829942 & -0.47565103 \\ 0.54772256 & 0.00000000 & 0.70710679 & -0.41153421 & -0.17504169 \end{bmatrix},$$

$$S = \begin{bmatrix} 2.236067977499789 & 0 & 0 \\ 0 & 1.414213562373095 & 0 \\ 0 & 0 & 1.000000000000000 \\ 0 & 0 & 0 \\ 0 & 0 & 0 \end{bmatrix},$$

and

$$V = \begin{bmatrix} 0.408248290463863 & -0.577350269189626 & 0.707106781186547 \\ 0.816496580927726 & 0.577350269189626 & 0.000000000000000 \\ 0.408248290463863 & -0.577350269189625 & -0.707106781186548 \end{bmatrix}.$$

Least Squares Approximation

The singular value decomposition has application in many areas, one of which is an alternative means for finding the least squares polynomials for fitting data. Let A be an $m \times n$ matrix, with $m > n$, and $\mathbf{b}$ be a vector in $\mathbb{R}^m$. The least squares objective is to find a vector $\mathbf{x}$ in $\mathbb{R}^n$ that will minimize $||A\mathbf{x} - \mathbf{b}||_2$.

Suppose that the singular value decomposition of A is known, that is

$$A = U S V^t,$$

where U is an orthogonal $m \times m$ matrix, V is an orthogonal $n \times n$ matrix, and S is an $m \times n$ matrix that contains the nonzero singular values in decreasing order along the main diagonal in the first $k \leq n$ rows, and zero entries elsewhere. Because U and V are both orthogonal we have $U^{-1} = U^t$, $V^{-1} = V^t$, and by part (iii) of the Orthogonal Matrix Properties on page 368, U and V are both l_2 norm preserving. As a consequence,

$$||A\mathbf{x} - \mathbf{b}||_2 = ||U S V^t \mathbf{x} - U U^t \mathbf{b}||_2 = ||U \left(S V^t \mathbf{x} - U^t \mathbf{b} \right)||_2 = ||S V^t \mathbf{x} - U^t \mathbf{b}||_2.$$

Let $\mathbf{z} = V^t \mathbf{x}$ and $\mathbf{c} = U^t \mathbf{b}$. Then $S\mathbf{z} = (s_1 z_1, s_2 z_2, \ldots, s_k z_k, 0, \ldots, 0)^t$ and

$$||A\mathbf{x} - \mathbf{b}||_2 = ||S\mathbf{z} - \mathbf{c}||_2 = ||(s_1 z_1 - c_1, s_2 z_2 - c_2, \ldots, s_k z_k - c_k, -c_{k+1}, \ldots, -c_m)^t||_2$$

$$= \left\{ \sum_{i=1}^{k} (s_i z_i - c_i)^2 + \sum_{i=k+1}^{m} (c_i)^2 \right\}^{1/2}.$$

The norm is minimized when the vector $\mathbf{z}$ is chosen with

$$z_i = \begin{cases} \dfrac{c_i}{s_i}, & \text{when } i = 1, 2, \ldots, k, \\ \text{arbitrarily}, & \text{when } i = k+1, k+2, \ldots, m. \end{cases}$$

Because $\mathbf{c} = U^t \mathbf{b}$ and $\mathbf{x} = V\mathbf{z}$ are both easy to compute, the least squares solution is also easily found. In addition, the approximation error is

$$||A\mathbf{x} - \mathbf{b}||_2 = \left\{ \sum_{i=k+1}^{m} c_i^2 \right\}^{1/2}.$$

Example 3 Use the singular value decomposition technique to determine the least squares polynomial of degree two for the data given in Table 9.5.

Table 9.5

i	x_i	y_i
1	0	1.0000
2	0.25	1.2840
3	0.50	1.6487
4	0.75	2.1170
5	1.00	2.7183

Solution This problem was solved using normal equations in Example 2 of Section 8.2 (see page 326). Here we first need to determine the appropriate form for A, $\mathbf{x}$, and $\mathbf{b}$. In that example the problem was described as finding a_0, a_1, and a_2 with

$$P_2(x) = a_0 + a_1 x + a_2 x^2.$$

In order to express this in matrix form, we let

$$\mathbf{a} = \begin{bmatrix} a_0 \\ a_1 \\ a_2 \end{bmatrix}, \quad \mathbf{b} = \begin{bmatrix} y_0 \\ y_1 \\ y_2 \\ y_3 \\ y_4 \end{bmatrix} = \begin{bmatrix} 1.0000 \\ 1.2840 \\ 1.6487 \\ 2.1170 \\ 2.7183 \end{bmatrix}, \quad \text{and}$$

$$A = \begin{bmatrix} 1 & x_0 & x_0^2 \\ 1 & x_1 & x_1^2 \\ 1 & x_2 & x_2^2 \\ 1 & x_3 & x_3^2 \\ 1 & x_4 & x_4^2 \end{bmatrix} = \begin{bmatrix} 1 & 0 & 0 \\ 1 & 0.25 & 0.0625 \\ 1 & 0.5 & 0.25 \\ 1 & 0.75 & 0.5625 \\ 1 & 1 & 1 \end{bmatrix}.$$

The singular value decomposition of A has the form $A = U S V^t$, where

$$U = \begin{bmatrix} -0.2945 & -0.6327 & 0.6314 & -0.0143 & -0.3378 \\ -0.3466 & -0.4550 & -0.2104 & 0.2555 & 0.7505 \\ -0.4159 & -0.1942 & -0.5244 & -0.6809 & -0.2250 \\ -0.5025 & 0.1497 & -0.3107 & 0.6524 & -0.4505 \\ -0.6063 & 0.5767 & 0.4308 & -0.2127 & 0.2628 \end{bmatrix},$$

$$S = \begin{bmatrix} 2.7117 & 0 & 0 \\ 0 & 0.9371 & 0 \\ 0 & 0 & 0.1627 \\ 0 & 0 & 0 \\ 0 & 0 & 0 \end{bmatrix}, \quad \text{and} \quad V^t = \begin{bmatrix} -0.7987 & -0.4712 & -0.3742 \\ -0.5929 & 0.5102 & 0.6231 \\ 0.1027 & -0.7195 & 0.6869 \end{bmatrix}.$$

So

$$\mathbf{c} = U^t \begin{bmatrix} y_0 \\ y_1 \\ y_2 \\ y_3 \\ y_4 \end{bmatrix} = \begin{bmatrix} -0.2945 & -0.6327 & 0.6314 & -0.0143 & -0.3378 \\ -0.3466 & -0.4550 & -0.2104 & 0.2555 & 0.7505 \\ -0.4159 & -0.1942 & -0.5244 & -0.6809 & -0.2250 \\ -0.5025 & 0.1497 & -0.3107 & 0.6524 & -0.4505 \\ -0.6063 & 0.5767 & 0.4308 & -0.2127 & 0.2628 \end{bmatrix}^t \begin{bmatrix} 1 \\ 1.284 \\ 1.6487 \\ 2.117 \\ 2.7183 \end{bmatrix}$$

$$= \begin{bmatrix} -4.1372 \\ 0.3473 \\ 0.0099 \\ -0.0059 \\ 0.0155 \end{bmatrix},$$

and the components of $\mathbf{z}$ are

$$z_1 = \frac{c_1}{s_1} = \frac{-4.1372}{2.7117} = -1.526, \quad z_2 = \frac{c_2}{s_2} = \frac{0.3473}{0.9371} = 0.3706, \quad \text{and}$$

$$z_3 = \frac{c_3}{s_3} = \frac{0.0099}{0.1627} = 0.0609.$$

This gives the least squares coefficients in $P_2(x)$ as

$$\begin{bmatrix} a_0 \\ a_1 \\ a_2 \end{bmatrix} = \mathbf{a} = V\,\mathbf{z} = \begin{bmatrix} -0.7987 & -0.5929 & 0.1027 \\ -0.4712 & 0.5102 & -0.7195 \\ -0.3742 & 0.6231 & 0.6869 \end{bmatrix} \begin{bmatrix} -1.526 \\ 0.3706 \\ 0.0609 \end{bmatrix} = \begin{bmatrix} 1.005 \\ 0.8642 \\ 0.8437 \end{bmatrix},$$

which agrees with the results in Example 2 of Section 8.2. The least squares error using these values uses the last two components of $\mathbf{c}$, and is

$$E = ||A\mathbf{a} - \mathbf{b}||_2^2 = c_4^2 + c_5^2 = (-0.0059)^2 + (0.0155)^2 = 3.07 \times 10^{-4} \qquad \blacksquare$$

Other Applications

The reason for the importance of the singular value decomposition in many applications is that it permits us to obtain the most important features of an $m \times n$ matrix using a matrix that is often of significantly smaller size. Because the singular values are listed on the diagonal of S in decreasing order, retaining only the first k rows and columns of S produces an approximation to the matrix A. As an illustration, recall Figure 9.2 on page 401, which shows the singular value decomposition of an $m \times n$ matrix A.

Replace the $n \times n$ matrix S with the $k \times k$ matrix S_k that contains the most significant singular values. These would certainly be only those that are nonzero, but we might also delete some singular values that are relatively small.

Determine corresponding $k \times n$ and $m \times k$ matrices U_k and V_k^t, respectively, in accordance with the singular value decomposition procedure. This is shown shaded in Figure 9.3.

Figure 9.3

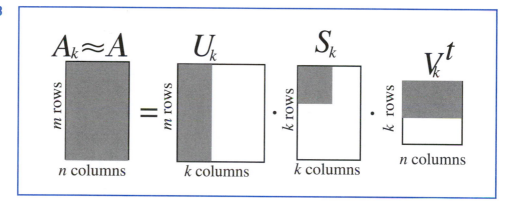

The new matrix $A_k = U_k\,S_k\,V_k^t$ is still of size $m \times n$ and would require $m \cdot n$ storage registers for its representation. However, in factored form, the storage requirement for the data is $m \cdot k$, for U_k, k for S_k, and $n \cdot k$ for V_k^t, for a total of $k(m + n + 1)$.

Suppose, for example, that $m = 2n$ and $k = n/3$. Then the original matrix A contains $mn = 2n^2$ items of data. The factorization producing A_k, however, contains only $mk = 2n^2/3$, for U_k, k for S_k, and $nk = n^2/3$ for V_k^t, items of data which occupy a total of $(n/3)(3n + 1)$ storage registers. This is a reduction of approximately 50% from the amount required to store the entire matrix A, and results in what is called *data compression*.

Illustration   In Example 2 we demonstrated that

$$
A = \begin{bmatrix} 1 & 0 & 1 \\ 0 & 1 & 0 \\ 0 & 1 & 1 \\ 0 & 1 & 0 \\ 1 & 1 & 0 \end{bmatrix} = USV^t = \begin{bmatrix} \frac{\sqrt{30}}{15} & \frac{\sqrt{6}}{3} & 0 & \frac{\sqrt{5}}{5} & 0 \\ \frac{\sqrt{30}}{15} & -\frac{\sqrt{6}}{6} & 0 & \frac{\sqrt{5}}{5} & \frac{\sqrt{2}}{2} \\ \frac{\sqrt{30}}{10} & 0 & \frac{\sqrt{2}}{2} & -\frac{\sqrt{5}}{5} & 0 \\ \frac{\sqrt{30}}{15} & -\frac{\sqrt{6}}{6} & 0 & \frac{\sqrt{5}}{5} & -\frac{\sqrt{2}}{2} \\ \frac{\sqrt{30}}{10} & 0 & -\frac{\sqrt{2}}{2} & -\frac{\sqrt{5}}{5} & 0 \end{bmatrix}
$$

$$
\times \begin{bmatrix} \sqrt{5} & 0 & 0 \\ 0 & \sqrt{2} & 0 \\ 0 & 0 & 1 \\ 0 & 0 & 0 \\ 0 & 0 & 0 \end{bmatrix} \begin{bmatrix} \frac{\sqrt{6}}{6} & \frac{\sqrt{6}}{3} & \frac{\sqrt{6}}{6} \\ \frac{\sqrt{3}}{3} & -\frac{\sqrt{3}}{3} & \frac{\sqrt{3}}{3} \\ -\frac{\sqrt{2}}{2} & 0 & \frac{\sqrt{2}}{2} \end{bmatrix}.
$$

Consider the reduced matrices associated with this factorization

$$
U_3 = \begin{bmatrix} \frac{\sqrt{30}}{15} & \frac{\sqrt{6}}{3} & 0 \\ \frac{\sqrt{30}}{15} & -\frac{\sqrt{6}}{6} & 0 \\ \frac{\sqrt{30}}{10} & 0 & \frac{\sqrt{2}}{2} \\ \frac{\sqrt{30}}{15} & -\frac{\sqrt{6}}{6} & 0 \\ \frac{\sqrt{30}}{10} & 0 & -\frac{\sqrt{2}}{2} \end{bmatrix}, \quad S_3 = \begin{bmatrix} \sqrt{5} & 0 & 0 \\ 0 & \sqrt{2} & 0 \\ 0 & 0 & 1 \end{bmatrix}, \quad \text{and}
$$

$$
V_3^t = \begin{bmatrix} \frac{\sqrt{6}}{6} & \frac{\sqrt{6}}{3} & \frac{\sqrt{6}}{6} \\ \frac{\sqrt{3}}{3} & -\frac{\sqrt{3}}{3} & \frac{\sqrt{3}}{3} \\ -\frac{\sqrt{2}}{2} & 0 & \frac{\sqrt{2}}{2} \end{bmatrix}.
$$

Then

$$
S_3 V_3^t = \begin{bmatrix} \frac{\sqrt{30}}{6} & \frac{\sqrt{30}}{3} & \frac{\sqrt{30}}{6} \\ \frac{\sqrt{6}}{3} & -\frac{\sqrt{6}}{3} & \frac{\sqrt{6}}{3} \\ -\frac{\sqrt{2}}{2} & 0 & \frac{\sqrt{2}}{2} \end{bmatrix} \quad \text{and} \quad A_3 = U_3 S_3 V_3^t = \begin{bmatrix} 1 & 0 & 1 \\ 0 & 1 & 0 \\ 0 & 1 & 1 \\ 0 & 1 & 0 \\ 1 & 1 & 0 \end{bmatrix} \equiv A. \quad \square
$$

Because the calculations in the Illustration were done using exact arithmetic, the matrix A_3 agreed precisely with the original matrix A. In general, finite-digit arithmetic would be used to perform the calculations, and absolute agreement would not be expected. The hope is that the data compression does not result in a matrix A_k that significantly differs from the original matrix A, and this depends on the relative magnitudes of the singular values of A.

Data compression has many applications. Consider, for example, the situation when A is a matrix consisting of pixels in a gray-scale photograph, perhaps taken from a great distance, such as a satellite photo of a portion of the Earth. The photograph likely includes *noise*, that is, data that does not truly represent the image, but rather represents the deterioration of the image by atmospheric particles, the quality of the lens, the reproduction process, etc. The noise data is incorporated in the data given in A, but hopefully this noise is much less significant than the true image. In this case, we would expect the larger singular values to represent the true image and the smaller singular values, those closest to zero, to be contributions of the noise. By performing a singular value decomposition that retains only those singular values above a certain threshold, we might be able to eliminate much of the noise and actually obtain an image that is not only smaller in size but is also a truer representation than the original photograph. (See [AP] for further details; in particular, their Figure 3.)

Additional important applications of the singular value decomposition include determining effective condition numbers for square matrices, determining the effective rank of a matrix, and removing signal noise. For more information on this important topic and a geometric interpretation of the factorization, see the survey paper by Kalman [Ka] and the references in that paper. For a more complete and extensive study of the theory, see Golub and Van Loan [GV].

EXERCISE SET 9.6

1. Determine the singular values of the following matrices.

   a. $A = \begin{bmatrix} 2 & 1 \\ 1 & 0 \end{bmatrix}$

   b. $A = \begin{bmatrix} 2 & 1 \\ 1 & 1 \\ 0 & 1 \end{bmatrix}$

   c. $A = \begin{bmatrix} 2 & 1 \\ -1 & 1 \\ 1 & 1 \\ 2 & -1 \end{bmatrix}$

   d. $A = \begin{bmatrix} 1 & 1 & 0 \\ -1 & 0 & 1 \\ 0 & 1 & -1 \\ 1 & 1 & -1 \end{bmatrix}$

2. Determine the singular values of the following matrices.

   a. $A = \begin{bmatrix} -1 & 1 \\ 1 & 1 \end{bmatrix}$

   b. $A = \begin{bmatrix} 1 & 1 & 0 \\ 1 & 0 & 1 \\ 0 & 1 & 1 \end{bmatrix}$

   c. $A = \begin{bmatrix} 1 & -1 \\ 1 & 1 \\ 0 & 1 \\ 1 & 0 \\ -1 & 1 \end{bmatrix}$

   d. $A = \begin{bmatrix} 0 & 1 & 1 \\ 0 & 1 & 0 \\ 1 & 1 & 0 \\ 0 & 1 & 0 \\ 1 & 0 & 1 \end{bmatrix}$

3. Determine a singular value decomposition for the matrices in Exercise 1.

4. Determine a singular value decomposition for the matrices in Exercise 2.

5. Let A be the matrix given in Example 1. Show that $(1, 2, 1)^t$, $(1, -1, 1)^t$, and $(-1, 0, 1)^t$ are eigenvectors of $A^t A$ corresponding to, respectively, the eigenvalues $\lambda_1 = 5$, $\lambda_2 = 2$, and $\lambda_3 = 1$.

6. Suppose that the $n \times n$ matrix A has the singular value decomposition $A = U S V^t$. Show that A^{-1} exists if and only if S^{-1} exists and find a singular value decomposition for A^{-1} when it exists.

7. Suppose that A is an $m \times n$ matrix. Show that the matrices $A^t A$ and $A A^t$ are both symmetric.

8. Suppose that A is an $m \times n$ matrix. Show that the eigenvalues of $A^t A$ are real and nonnegative by doing the following.

   a. Show that $0 \leq \mathbf{x}^t \mathbf{x}$ for any vector $\mathbf{x} \in \mathbb{R}^n$, with equality if and only if $\mathbf{x} = \mathbf{0}$.

   b. Show that for each eigenvalue λ of $A^t A$, there is an eigenvector $\mathbf{v}$ with $\mathbf{v}^t \mathbf{v} = 1$.

   c. Show that for the vector $\mathbf{v}$ in (b) we have $(A\mathbf{v})^t (A\mathbf{v}) = \lambda$.

   d. Use (c) to show that $\lambda \geq 0$.

9. Suppose that A is an $m \times n$ matrix.

   a. Show that if $\mathbf{v}$ is an eigenvector corresponding to the nonzero eigenvalue λ of $A^t A$, then $A\mathbf{v}$ is an eigenvector of $A A^t$ corresponding to λ.

   b. Show that if $\mathbf{v}$ is an eigenvector corresponding to the nonzero eigenvalue λ of $A A^t$, then $A^t \mathbf{v}$ is an eigenvector of $A^t A$ corresponding to λ.

   c. Show that the nonzero eigenvalues of $A^t A$ are the same as those of $A A^t$.

10. Show that if A is an $m \times n$ matrix and P is an $n \times n$ orthogonal matrix, then $P A$ has the same singular values as A. [Hint: Show that $(P A)^t (P A) = A^t A$.]

11. Suppose that A is an $n \times n$ invertible matrix with singular values $s_1, s_2, \ldots, s_n$.

    a. Use the results on page 289 to show that $||A||_2^2 = s_1^2$.

    b. Use the result in Exercise 9 of Section 7.3 to show that $||A^{-1}||_2^2 = 1/s_n^2$.

    c. Use (a) and (b) to show that $K_2(A) = \dfrac{s_1}{s_n}$.

12. Use the results in Exercise 11 to determine the condition numbers of the nonsingular square matrices in Exercises 1 and 2.

13. Show that the vectors $\mathbf{u}_1, \mathbf{u}_2, \ldots, \mathbf{u}_k$ defined in Eq. (9.15) are eigenvectors of AA^t corresponding, respectively, to $s_1^2, s_2^2, \ldots, s_k^2$.

14. Given the data

x_i	1.0	2.0	3.0	4.0	5.0
y_i	1.3	3.5	4.2	5.0	7.0

    a. Use the singular value decomposition technique to find the least squares polynomial of degree 1.

    b. Use the singular value decomposition technique to find the least squares polynomial of degree 2.

15. Given the data

x_i	1.0	1.1	1.3	1.5	1.9	2.1
y_i	1.84	1.96	2.21	2.45	2.94	3.18

    a. Use the singular value decomposition technique to find the least squares polynomial of degree 2.

    b. Use the singular value decomposition technique to find the least squares polynomial of degree 3.

9.7 Survey of Methods and Software

This chapter discussed the approximation of eigenvalues and eigenvectors. The Geršgorin circles give a crude approximation to the location of the eigenvalues of a matrix. The Power method can be used to find the dominant eigenvalue and an associated eigenvector for an arbitrary matrix A. If A is symmetric, the Symmetric Power method gives faster convergence to the dominant eigenvalue and an associated eigenvector. The Inverse Power method will approximate the eigenvalue closest to a given value and an associated eigenvector. This method is often used to refine an approximate eigenvalue and to compute an eigenvector once an eigenvalue has been found by some other technique.

Deflation methods, such as Wielandt deflation, obtain other eigenvalues once one of the eigenvalues is known. These methods are used only if a few eigenvalues are required because they are susceptible to round-off error. The Inverse Power method should be used to improve the accuracy of approximate eigenvalues obtained from a deflation technique.

Methods based on similarity transformations, such as Householder's method, are used to convert a symmetric matrix into a similar matrix that is tridiagonal. Techniques such as the QR method can then be applied to the tridiagonal matrix to obtain approximations to all the eigenvalues. The associated eigenvectors can be found by using an iterative method, such as the Inverse Power method, or by modifying the QR method to include the approximation of eigenvectors. We restricted our study to symmetric matrices and presented the QR method only to compute eigenvalues for the symmetric case.

The singular value decomposition is discussed in Section 9.6. It is used to factor an $m \times n$ matrix into the form $U S V^t$, where U is an orthogonal $m \times m$ matrix, V is an orthogonal $n \times n$ matrix, and S is an $m \times n$ matrix whose only nonzero entries are located along the main diagonal. This factorization has important applications that include image processing, data compression, and solving over-determined linear systems that arise in least squares approximations. The singular value decomposition requires the computation of eigenvalues and eigenvectors, so it is appropriate to have this technique conclude the chapter.

The subroutines in the IMSL and NAG Libraries are based on those contained in EISPACK and LAPACK, packages that were discussed in Section 1.5. In general, the subroutines transform a matrix into the appropriate form for the QR method or one of its modifications, such as the QL method. The subroutines approximate all the eigenvalues and can approximate an associated eigenvector for each eigenvalue. There are special routines that find all the eigenvalues within an interval or region or that find only the largest or smallest eigenvalue. Subroutines are also available to determine the accuracy of the eigenvalue approximation and the sensitivity of the process to round-off error.

The software package contained in netlib, ARPACK to solve large sparse eigenvalue problems, is based on the implicitly-restarted Arnoldi method by Sorensen [So]. The implicitly restarted Arnoldi method is a Krylov subspace method that finds a sequence of Krylov subspaces that converge to a subspace containing the eigenvalues.

The books by Wilkinson [Wil2] and by Wilkinson and Reinsch [WR] are classics in the study of eigenvalue problems. Stewart [Stew] is also a good source of information on the general problem, and Parlett [Par] considers the symmetric problem. A study of the nonsymmetric problem can be found in Saad [Sa1].

Systems of Nonlinear Equations

10.1 Introduction

A large part of the material in this book has involved the solution of systems of equations. Even so, to this point the methods have been appropriate only for systems of *linear* equations, equations in the variables $x_1, x_2, \ldots, x_n$ of the form

$$a_{i1}x_1 + a_{i2}x_2 + \cdots + a_{in}x_n = b_i$$

for $i = 1, 2, \ldots, n$. If you have wondered why we have not considered more general systems of equations, the reason is simple. It is much harder to approximate the solutions to a system of general, or *nonlinear*, equations.

Solving a system of nonlinear equations is a problem that is avoided when possible, customarily by approximating the nonlinear system by a system of linear equations. When this is unsatisfactory, the problem must be tackled directly. The most straightforward method of approach is to adapt the methods from Chapter 2 that approximate the solutions of a single nonlinear equation in one variable to apply when the single-variable problem is replaced by a vector problem that incorporates all the variables.

The principal tool in Chapter 2 was Newton's method, a technique that is generally quadratically convergent once a sufficiently accurate starting value is found. This is the first technique we modify to solve systems of nonlinear equations. Newton's method, as modified for systems of equations, is quite costly to apply, so, in Section 10.3, we describe how a modified Secant method can be used to obtain approximations more easily, although with a loss of the extremely rapid convergence that Newton's method provides.

Section 10.4 describes the method of Steepest Descent. This technique is only linearly convergent, but it does not require the accurate starting approximations needed for more rapidly-converging techniques. It is often used to find a good initial approximation for Newton's method or one of its modifications.

In Section 10.5, we give an introduction to continuation methods, which use a parameter to move from a problem with an easily determined solution to the solution of the original nonlinear problem.

A system of nonlinear equations has the form

$$f_1(x_1, x_2, \ldots, x_n) = 0,$$
$$f_2(x_1, x_2, \ldots, x_n) = 0,$$
$$\vdots$$
$$f_n(x_1, x_2, \ldots, x_n) = 0,$$

where each function f_i can be thought of as mapping a vector $\mathbf{x} = (x_1, x_2, \ldots, x_n)^t$ of the n-dimensional space $\mathbb{R}^n$ into the real line $\mathbb{R}$. A geometric representation of a nonlinear system when $n = 2$ is given in Figure 10.1.

Figure 10.1

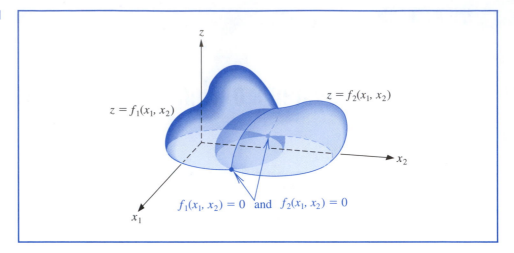

A general system of n nonlinear equations in n unknowns can be alternatively represented by defining a function $\mathbf{F}$, mapping $\mathbb{R}^n$ into $\mathbb{R}^n$, by

$$\mathbf{F}(x_1, x_2, \ldots, x_n) = (f_1(x_1, x_2, \ldots, x_n), f_2(x_1, x_2, \ldots, x_n), \ldots, f_n(x_1, x_2, \ldots, x_n))^t.$$

If vector notation is used to represent the variables $x_1, x_2, \ldots, x_n$, the nonlinear system assumes the form

$$\mathbf{F}(\mathbf{x}) = \mathbf{0}.$$

The functions $f_1, f_2, \ldots, f_n$ are the **coordinate functions** of $\mathbf{F}$.

Example 1   Place the 3×3 nonlinear system

$$3x_1 - \cos(x_2 x_3) - \frac{1}{2} = 0,$$

$$x_1^2 - 81(x_2 + 0.1)^2 + \sin x_3 + 1.06 = 0,$$

$$e^{-x_1 x_2} + 20x_3 + \frac{10\pi - 3}{3} = 0$$

in the form $\mathbf{F}(\mathbf{x}) = \mathbf{0}$.

Solution   Define the three coordinate functions f_1, f_2, and f_3 from $\mathbb{R}^3$ to $\mathbb{R}$ as

$$f_1(x_1, x_2, x_3) = 3x_1 - \cos(x_2 x_3) - \frac{1}{2},$$

$$f_2(x_1, x_2, x_3) = x_1^2 - 81(x_2 + 0.1)^2 + \sin x_3 + 1.06,$$

$$f_3(x_1, x_2, x_3) = e^{-x_1 x_2} + 20x_3 + \frac{10\pi - 3}{3}.$$

Then define $\mathbf{F}$ from $\mathbb{R}^3 \to \mathbb{R}^3$ by

$$\mathbf{F}(\mathbf{x}) = \mathbf{F}(x_1, x_2, x_3)$$
$$= (f_1(x_1, x_2, x_3), f_2(x_1, x_2, x_3), f_3(x_1, x_2, x_3))^t$$
$$= \left(3x_1 - \cos(x_2 x_3) - \frac{1}{2}, x_1^2 - 81(x_2 + 0.1)^2 + \sin x_3 + 1.06, e^{-x_1 x_2} + 20x_3 \right.$$
$$\left. + \frac{10\pi - 3}{3} \right)^t.$$

The original system is equivalent to $\mathbf{F}(\mathbf{x}) = \mathbf{0}$. ∎

Before discussing the solution of a system of nonlinear equations, we need some results concerning continuity and differentiability of functions from $\mathbb{R}^n$ into $\mathbb{R}$ and $\mathbb{R}^n$ into $\mathbb{R}^n$. These results parallel those given in Section 1.2 for a function from $\mathbb{R}$ into $\mathbb{R}$.

Let f be a function defined on a set $D \subset \mathbb{R}^n$ and mapping $\mathbb{R}^n$ into $\mathbb{R}$. The function f has the **limit** L at $\mathbf{x}_0$, written

$$\lim_{\mathbf{x} \to \mathbf{x}_0} f(\mathbf{x}) = L,$$

if, given any number $\varepsilon > 0$, a number $\delta > 0$ exists with the property that

$$|f(\mathbf{x}) - L| < \varepsilon \quad \text{whenever } \mathbf{x} \in D \quad \text{and} \quad 0 < \|\mathbf{x} - \mathbf{x}_0\| < \delta.$$

Any convenient norm can be used to satisfy the condition in this definition. The specific value of δ will depend on the norm chosen, but the existence and value of the limit L is independent of the norm.

The function f from $\mathbb{R}^n$ into $\mathbb{R}$ is **continuous** at $\mathbf{x}_0 \in D$ provided $\lim_{\mathbf{x} \to \mathbf{x}_0} f(\mathbf{x})$ exists and is $f(\mathbf{x}_0)$. In addition, f is **continuous on a set** D provided f is continuous at every point of D. This is expressed by writing $f \in C(D)$.

We define the limit and continuity concepts for functions from $\mathbb{R}^n$ into $\mathbb{R}^n$ by considering the coordinate functions from $\mathbb{R}^n$ into $\mathbb{R}$.

Let $\mathbf{F}$ be a function from $D \subset \mathbb{R}^n$ into $\mathbb{R}^n$ and suppose $\mathbf{F}$ has the representation

$$\mathbf{F}(\mathbf{x}) = (f_1(\mathbf{x}), f_2(\mathbf{x}), \dots, f_n(\mathbf{x}))^t,$$

where f_i is a function from $\mathbb{R}^n$ to $\mathbb{R}$ for each $i = 1, 2, \dots n$. We define the limit of $\mathbf{F}$ from $\mathbb{R}^n$ to $\mathbb{R}^n$ as

$$\lim_{\mathbf{x} \to \mathbf{x}_0} \mathbf{F}(\mathbf{x}) = \mathbf{L} = (L_1, L_2, \dots, L_n)^t$$

if and only if $\lim_{\mathbf{x} \to \mathbf{x}_0} f_i(\mathbf{x}) = L_i$ for each $i = 1, 2, \dots, n$.

The function $\mathbf{F}$ is **continuous** at $\mathbf{x}_0 \in D$ provided $\lim_{\mathbf{x} \to \mathbf{x}_0} \mathbf{F}(\mathbf{x})$ exists and is $\mathbf{F}(\mathbf{x}_0)$. In addition, $\mathbf{F}$ is **continuous on the set** D if $\mathbf{F}$ is continuous at each $\mathbf{x}$ in D.

These are the basic concepts we will need for the remainder of the chapter.

Continuity definitions for functions of n variables follow from those for a single variable by replacing, where necessary, absolute values by norms.

10.2 Newton's Method for Systems

Newton's method for approximating the solution p to the single nonlinear equation

$$f(x) = 0$$

requires an initial approximation p_0 to p and generates a sequence defined by

$$p_k = p_{k-1} - \frac{f(p_{k-1})}{f'(p_{k-1})}, \quad \text{for } k \geq 1.$$

To modify Newton's method to find the vector solution $\mathbf{p}$ to the vector equation

$$\mathbf{F}(\mathbf{x}) = \mathbf{0},$$

we first need to determine an initial approximation vector $\mathbf{p}^{(0)}$. We must then decide how to modify the single-variable Newton's method to a vector function method that will have the same convergence properties but not require division because this operation is undefined for vectors. We also need to replace the derivative of f in the single-variable version of Newton's method with something that is appropriate for the vector function $\mathbf{F}$.

The Jacobian Matrix

The derivative $f'(x)$ of the single-variable function $f(x)$ describes how the values of the function change relative to changes in the independent variable x. The vector function $\mathbf{F}$ has n different variables, $x_1, x_2, \ldots, x_n$, and n different component functions, $f_1, f_2, \ldots, f_n$, each of which can change as any one of the variables change. The appropriate derivative modification from the single-variable Newton's method to the vector form must involve all these n^2 possible changes, and the natural way to represent n^2 items is by an $n \times n$ matrix. Each change in a component function f_i at $\mathbf{x}$ with respect to the change in the variable x_j is described by the partial derivative

$$\frac{\partial f_i}{\partial x_j}(\mathbf{x}),$$

and the $n \times n$ matrix which replaces the derivative that occurs in the single-variable case is

$$J(\mathbf{x}) = \begin{bmatrix} \dfrac{\partial f_1}{\partial x_1}(\mathbf{x}) & \dfrac{\partial f_1}{\partial x_2}(\mathbf{x}) & \cdots & \dfrac{\partial f_1}{\partial x_n}(\mathbf{x}) \\[2ex] \dfrac{\partial f_2}{\partial x_1}(\mathbf{x}) & \dfrac{\partial f_2}{\partial x_2}(\mathbf{x}) & \cdots & \dfrac{\partial f_2}{\partial x_n}(\mathbf{x}) \\[1ex] \vdots & \vdots & & \vdots \\[1ex] \dfrac{\partial f_n}{\partial x_1}(\mathbf{x}) & \dfrac{\partial f_n}{\partial x_2}(\mathbf{x}) & \cdots & \dfrac{\partial f_n}{\partial x_n}(\mathbf{x}) \end{bmatrix}.$$

The matrix $J(\mathbf{x})$ is called the **Jacobian** matrix and has a number of applications in analysis. It might, in particular, be familiar due to its application in the multiple integration of a function of several variables over a region that requires a change of variables to be performed.

Newton's method for systems replaces division by the derivative in the single-variable case with multiplying by the inverse of the $n \times n$ Jacobian matrix in the vector situation. As a consequence, Newton's method for finding the solution $\mathbf{p}$ to the nonlinear system of equations represented by the vector equation $\mathbf{F}(\mathbf{x}) = \mathbf{0}$ has the form

$$\mathbf{p}^{(k)} = \mathbf{p}^{(k-1)} - [J(\mathbf{p}^{(k-1)})]^{-1}\mathbf{F}(\mathbf{p}^{(k-1)}), \quad \text{for } k \geq 1,$$

given the initial approximation $\mathbf{p}^{(0)}$ to the solution $\mathbf{p}$.

A weakness in Newton's method for systems arises from the necessity of inverting the matrix $J(\mathbf{p}^{(k-1)})$ at each iteration. In practice, explicit computation of the inverse of $J(\mathbf{p}^{(k-1)})$ is avoided by performing the operation in a two-step manner. First, a vector $\mathbf{y}^{(k-1)}$ is found that will satisfy

$$J(\mathbf{p}^{(k-1)})\mathbf{y}^{(k-1)} = -\mathbf{F}(\mathbf{p}^{(k-1)}).$$

After this has been accomplished, the new approximation, $\mathbf{p}^{(k)}$, is obtained by adding $\mathbf{y}^{(k-1)}$ to $\mathbf{p}^{(k-1)}$.

The program NWTSY101 implements Newton's method for systems of nonlinear equations.

Example 1 The nonlinear system

$$3x_1 - \cos(x_2 x_3) - \frac{1}{2} = 0,$$

$$x_1^2 - 81(x_2 + 0.1)^2 + \sin x_3 + 1.06 = 0,$$

$$e^{-x_1 x_2} + 20x_3 + \frac{10\pi - 3}{3} = 0$$

has the approximate solution $(0.5, 0, -0.52359877)^t$. Apply Newton's method to this problem with $\mathbf{p}^{(0)} = (0.1, 0.1, -0.1)^t$.

Solution Define

$$\mathbf{F}(x_1, x_2, x_3) = (f_1(x_1, x_2, x_3), f_2(x_1, x_2, x_3), f_3(x_1, x_2, x_3))^t,$$

where

$$f_1(x_1, x_2, x_3) = 3x_1 - \cos(x_2 x_3) - \frac{1}{2},$$

$$f_2(x_1, x_2, x_3) = x_1^2 - 81(x_2 + 0.1)^2 + \sin x_3 + 1.06,$$

and

$$f_3(x_1, x_2, x_3) = e^{-x_1 x_2} + 20x_3 + \frac{10\pi - 3}{3}.$$

The Jacobian matrix $J(\mathbf{x})$ for this system is

$$J(x_1, x_2, x_3) = \begin{bmatrix} 3 & x_3 \sin x_2 x_3 & x_2 \sin x_2 x_3 \\ 2x_1 & -162(x_2 + 0.1) & \cos x_3 \\ -x_2 e^{-x_1 x_2} & -x_1 e^{-x_1 x_2} & 20 \end{bmatrix}.$$

For $\mathbf{p}^{(0)} = (0.1, 0.1, -0.1)^t$ we have

$$\mathbf{F}(\mathbf{p}^{(0)}) = (-0.199995, -2.269833417, 8.462025346)^t$$

and

$$J(\mathbf{p}^{(0)}) = \begin{bmatrix} 3 & 9.999833334 \times 10^{-4} & 9.999833334 \times 10^{-4} \\ 0.2 & -32.4 & 0.9950041653 \\ -0.09900498337 & -0.09900498337 & 20 \end{bmatrix}.$$

Solving the linear system $J(\mathbf{p}^{(0)})\mathbf{y}^{(0)} = -\mathbf{F}(\mathbf{p}^{(0)})$ gives

$$\mathbf{y}^{(0)} = \begin{bmatrix} 0.3998696728 \\ -0.08053315147 \\ -0.4215204718 \end{bmatrix} \quad \text{and} \quad \mathbf{p}^{(1)} = \mathbf{p}^{(0)} + \mathbf{y}^{(0)} = \begin{bmatrix} 0.4998696728 \\ 0.01946684853 \\ -0.5215204718 \end{bmatrix}.$$

Continuing for $k = 2, 3, \ldots$, we have

$$
\begin{bmatrix} p_1^{(k)} \\ p_2^{(k)} \\ p_3^{(k)} \end{bmatrix} = \begin{bmatrix} p_1^{(k-1)} \\ p_2^{(k-1)} \\ p_3^{(k-1)} \end{bmatrix} + \begin{bmatrix} y_1^{(k-1)} \\ y_2^{(k-1)} \\ y_3^{(k-1)} \end{bmatrix},
$$

where

$$
\begin{bmatrix} y_1^{(k-1)} \\ y_2^{(k-1)} \\ y_3^{(k-1)} \end{bmatrix} = -\left(J\left(p_1^{(k-1)}, p_2^{(k-1)}, p_3^{(k-1)} \right) \right)^{-1} \mathbf{F}\left(p_1^{(k-1)}, p_2^{(k-1)}, p_3^{(k-1)} \right).
$$

At the kth step, the linear system $J(\mathbf{p}^{(k-1)})\mathbf{y}^{(k-1)} = -\mathbf{F}(\mathbf{p}^{(k-1)})$ must be solved, where

$$
J(\mathbf{p}^{(k-1)}) = \begin{bmatrix} 3 & p_3^{(k-1)} \sin p_2^{(k-1)} p_3^{(k-1)} & p_2^{(k-1)} \sin p_2^{(k-1)} p_3^{(k-1)} \\ 2p_1^{(k-1)} & -162\left(p_2^{(k-1)} + 0.1\right) & \cos p_3^{(k-1)} \\ -p_2^{(k-1)} e^{-p_1^{(k-1)} p_2^{(k-1)}} & -p_1^{(k-1)} e^{-p_1^{(k-1)} p_2^{(k-1)}} & 20 \end{bmatrix},
$$

$$
\mathbf{y}^{(k-1)} = \begin{bmatrix} y_1^{(k-1)} \\ y_2^{(k-1)} \\ y_3^{(k-1)} \end{bmatrix},
$$

and

$$
\mathbf{F}(\mathbf{p}^{(k-1)}) = \begin{bmatrix} 3p_1^{(k-1)} - \cos p_2^{(k-1)} p_3^{(k-1)} - \frac{1}{2} \\ \left(p_1^{(k-1)}\right)^2 - 81\left(p_2^{(k-1)} + 0.1\right)^2 + \sin p_3^{(k-1)} + 1.06 \\ e^{-p_1^{(k-1)} p_2^{(k-1)}} + 20 p_3^{(k-1)} + \frac{10\pi - 3}{3} \end{bmatrix}.
$$

The results using this iterative procedure are shown in Table 10.1.   ▪

Table 10.1

k	$p_1^{(k)}$	$p_2^{(k)}$	$p_3^{(k)}$	$\|\mathbf{p}^{(k)} - \mathbf{p}^{(k-1)}\|_\infty$
0	0.1000000000	0.1000000000	−0.1000000000	
1	0.4998696728	0.0194668485	−0.5215204718	0.4215204718
2	0.5000142403	0.0015885914	−0.5235569638	1.788×10^{-2}
3	0.5000000113	0.0000124448	−0.5235984500	1.576×10^{-3}
4	0.5000000000	8.516×10^{-10}	−0.5235987755	1.244×10^{-5}
5	0.5000000000	-1.375×10^{-11}	−0.5235987756	8.654×10^{-10}

The previous example illustrates that Newton's method can converge very rapidly once an approximation is obtained that is near the true solution. However, it is not always easy to determine starting values that will lead to a solution, and the method is computationally expensive. In the next section, we consider a method for overcoming the latter weakness. Good starting values can usually be found by the method discussed in Section 10.4.

EXERCISE SET 10.2

1. Give an example of a function $\mathbf{F} : \mathbb{R}^2 \to \mathbb{R}^2$ that is continuous at each point of $\mathbb{R}^2$ except at $(1, 0)$.

2. Give an example of a function $\mathbf{F} : \mathbb{R}^3 \to \mathbb{R}^3$ that is continuous at each point of $\mathbb{R}^3$ except at $(1, 2, 3)$.

3. Use Newton's method with $\mathbf{p}^{(0)} = \mathbf{0}$ to compute $\mathbf{p}^{(2)}$ for each of the following nonlinear systems.

   a. $$4x_1^2 - 20x_1 + \frac{1}{4}x_2^2 + 8 = 0,$$
   $$\frac{1}{2}x_1x_2^2 + 2x_1 - 5x_2 + 8 = 0.$$

   b. $$\sin(4\pi x_1 x_2) - 2x_2 - x_1 = 0,$$
   $$\left(\frac{4\pi - 1}{4\pi}\right)(e^{2x_1} - e) + 4ex_2^2 - 2ex_1 = 0.$$

   c. $$3x_1 - \cos(x_2 x_3) - \frac{1}{2} = 0,$$
   $$4x_1^2 - 625x_2^2 + 2x_2 - 1 = 0,$$
   $$e^{-x_1 x_2} + 20x_3 + \frac{10\pi - 3}{3} = 0.$$

   d. $$x_1^2 + x_2 \qquad\quad - 37 = 0,$$
   $$x_1 - x_2^2 \qquad\quad - 5 = 0,$$
   $$x_1 + x_2 + x_3 - \;\; 3 = 0.$$

4. Use Newton's method to find a solution to the following nonlinear systems in the given domain. Iterate until $\|\mathbf{p}^{(k)} - \mathbf{p}^{(k-1)}\|_\infty < 10^{-6}$.

   a. $$3x_1^2 - x_2^2 = 0,$$
   $$3x_1x_2^2 - x_1^3 - 1 = 0.$$
   Use $\mathbf{p}^{(0)} = (1, 1)^t$.

   b. $$\ln(x_1^2 + x_2^2) - \sin(x_1x_2) = \ln 2 + \ln \pi,$$
   $$e^{x_1 - x_2} + \cos(x_1x_2) = 0.$$
   Use $\mathbf{p}^{(0)} = (2, 2)^t$.

   c. $$x_1^3 + x_1^2 x_2 - x_1 x_3 + 6 = 0,$$
   $$e^{x_1} + e^{x_2} - x_3 = 0,$$
   $$x_2^2 - 2x_1 x_3 = 4.$$
   Use $\mathbf{p}^{(0)} = (-1, -2, 1)^t$.

   d. $$6x_1 - 2\cos(x_2x_3) - 1 = 0,$$
   $$9x_2 + \sqrt{x_1^2 + \sin x_3 + 1.06} + 0.9 = 0,$$
   $$60x_3 + 3e^{-x_1 x_2} + 10\pi - 3 = 0.$$
   Use $\mathbf{p}^{(0)} = (0, 0, 0)^t$.

5. The nonlinear system

   $$x_1^2 - x_2^2 + 2x_2 = 0,$$
   $$2x_1 + x_2^2 - 6 = 0.$$

   has four solutions. They are near $(-5, -4)^t$, $(2, -1)^t$, $(0.5, 2)^t$, and $(-2, 3)^t$. Use these points as initial approximations for Newton's method and iterate until $\|\mathbf{p}^{(k)} - \mathbf{p}^{(k-1)}\|_\infty < 10^{-6}$. Do the results justify using the stated points as initial approximations?

6. The nonlinear system

   $$2x_1 - 3x_2 + x_3 - 4 = 0,$$
   $$2x_1 + x_2 - x_3 + 4 = 0,$$
   $$x_1^2 + x_2^2 + x_3^2 - 4 = 0.$$

   has a solution near $(-0.5, -1.5, 1.5)^t$.

   a. Use this point as an initial approximation for Newton's method and iterate until $\|\mathbf{p}^{(k)} - \mathbf{p}^{(k-1)}\|_\infty < 10^{-6}$.

   b. Solve the first two equations for x_1 and x_3 in terms of x_2.

   c. Substitute the results of (b) into the third equation to obtain a quadratic equation in x_2.

   d. Solve the quadratic equation in (c) by the quadratic formula.

   e. Of the solutions in (a) and (d), which is closer to the initial approximation $(-0.5, -1.5, 1.5)^t$?

7.   The nonlinear system

$$3x_1 - \cos(x_2 x_3) - \frac{1}{2} = 0,$$

$$x_1^2 - 625x_2^2 - \frac{1}{4} = 0,$$

$$e^{-x_1 x_2} + 20x_3 + \frac{10\pi - 3}{3} = 0$$

has a singular Jacobian matrix at the solution. Apply Newton's method with $\mathbf{p}^{(0)} = (1, 1 - 1)^t$. Note that convergence may be slow or may not occur within a reasonable number of iterations.

8.   The nonlinear system

$$4x_1 - x_2 + x_3 = x_1 x_4,$$

$$-x_1 + 3x_2 - 2x_3 = x_2 x_4,$$

$$x_1 - 2x_2 + 3x_3 = x_3 x_4,$$

$$x_1^2 + x_2^2 + x_3^2 = 1$$

has six solutions.

a.   Show that if $(x_1, x_2, x_3, x_4)^t$ is a solution, then $(-x_1, -x_2, -x_3, x_4)^t$ is a solution.

b.   Use Newton's method three times to approximate all solutions. Iterate until $\left\| \mathbf{p}^{(k)} - \mathbf{p}^{(k-1)} \right\|_\infty < 10^{-5}$. Use the initial vectors $(1, 1, 1, 1)^t$, $(1, 0, 0, 0)^t$, and $(1, -1, 1, -1)^t$.

9.   Let A be an $n \times n$ matrix and $\mathbf{F}$ be the function from $\mathbb{R}^n$ to $\mathbb{R}^n$ defined by $\mathbf{F}(\mathbf{x}) = A\mathbf{x}$. What is the Jacobian matrix of $\mathbf{F}$?

10.   In Exercise 6 of Section 5.7, we considered the problem of predicting the population of two species that compete for the same food supply. In the problem, we made the assumption that the populations could be predicted by solving the system of equations

$$\frac{dx_1}{dt}(t) = x_1(t)\,(4 - 0.0003x_1(t) - 0.0004x_2(t))$$

and

$$\frac{dx_2}{dt}(t) = x_2(t)\,(2 - 0.0002x_1(t) - 0.0001x_2(t)).$$

In this exercise, we would like to consider the problem of determining equilibrium populations of the two species. The mathematical criteria that must be satisfied in order for the populations to be at equilibrium is that, simultaneously,

$$\frac{dx_1}{dt}(t) = 0 \quad \text{and} \quad \frac{dx_2}{dt}(t) = 0.$$

This occurs when the first species is extinct and the second species has a population of 20,000 or when the second species is extinct and the first species has a population of 13,333. Can an equilibrium occur in any other situation?

11.   The amount of pressure required to sink a large, heavy object in a soft homogeneous soil that lies above a hard base soil can be predicted by the amount of pressure required to sink smaller objects in the same soil. Specifically, the amount of pressure p required to sink a circular plate of radius r a distance d in the soft soil, where the hard base soil lies a distance $D > d$ below the surface, can be approximated by an equation of the form

$$p = k_1 e^{k_2 r} + k_3 r,$$

where k_1, k_2, and k_3 are constants, with $k_2 > 0$, depending on d and the consistency of the soil but not on the radius of the plate.

a.   Find the values of k_1, k_2, and k_3 if we assume that a plate of radius 1 in. requires a pressure of 10 lb/in.^{2} to sink 1 ft in a muddy field, a plate of radius 2 in. requires a pressure of 12 lb/in.^{2} to sink 1 ft, and a plate of radius 3 in. requires a pressure of 15 lb/in.^{2} to sink this distance (assuming that the mud is more than 1 ft deep).

b.   Use your calculations from (a) to predict the minimal size of circular plate that would be required to sustain a load of 500 lb on this field with sinkage of less than 1 ft.

10.3   Quasi-Newton Methods

A significant weakness of Newton's method for solving systems of nonlinear equations is the requirement that, at each iteration, a Jacobian matrix be computed and an $n \times n$ linear system solved that involves this matrix. To illustrate the magnitude of this weakness, consider the amount of computation associated with one iteration of Newton's method. The Jacobian matrix associated with a system of n nonlinear equations written in the form $\mathbf{F}(\mathbf{x}) = \mathbf{0}$ requires that the n^2 partial derivatives of the n component functions of $\mathbf{F}$ be determined and evaluated. In most situations, the exact evaluation of the partial derivatives is inconvenient, and in many applications it is impossible. This difficulty can generally be overcome by using finite-difference approximations to the partial derivatives. For example,

$$\frac{\partial f_j}{\partial x_k}(\mathbf{x}) \approx \frac{f_j(\mathbf{x} + h\mathbf{e}_k) - f_j(\mathbf{x})}{h},$$

where h is small in absolute value and $\mathbf{e}_k$ is the vector whose only nonzero entry is a 1 in the kth coordinate.

This approximation, however, still requires that at least n^2 scalar functional evaluations be performed to approximate the Jacobian matrix and does not decrease the amount of calculation, in general $O(n^3)$, required for solving the linear system involving this approximate Jacobian. The total computational effort for just one iteration of Newton's method is, consequently, at least $n^2 + n$ scalar functional evaluations (n^2 for the evaluation of the Jacobian matrix and n for the evaluation of $\mathbf{F}$) together with $O(n^3)$ arithmetic operations to solve the linear system. This amount of computational effort can be prohibitive except for relatively small values of n and easily-evaluated scalar functions.

In this section, we consider a generalization of the Secant method to systems of nonlinear equations; in particular, a technique known as **Broyden's method** (see [Broy]). The method requires only n scalar functional evaluations per iteration and also reduces the number of arithmetic calculations to $O(n^2)$. It belongs to a class of methods known as *least-change secant updates* that produce algorithms called *quasi-Newton*. These methods replace the Jacobian matrix in Newton's method with an approximation matrix that is updated at each iteration. The disadvantage to the method is that the quadratic convergence of Newton's method is lost. It is replaced by *superlinear* convergence, which implies that

$$\lim_{i \to \infty} \frac{\|\mathbf{p}^{(i+1)} - \mathbf{p}\|}{\|\mathbf{p}^{(i)} - \mathbf{p}\|} = 0,$$

where $\mathbf{p}$ denotes the solution to $\mathbf{F}(\mathbf{x}) = \mathbf{0}$, and $\mathbf{p}^{(i)}$ and $\mathbf{p}^{(i+1)}$ are consecutive approximations to $\mathbf{p}$. In most applications, the reduction to superlinear convergence is a more than acceptable trade-off for the decrease in the amount of computation.

An additional disadvantage of quasi-Newton methods is that, unlike Newton's method, they are not self-correcting. Newton's method, for example, will generally correct for round-off error with successive iterations, but unless special safeguards are incorporated, Broyden's method will not.

To describe Broyden's method, suppose that an initial approximation $\mathbf{p}^{(0)}$ is given to the solution $\mathbf{p}$ of $\mathbf{F}(\mathbf{x}) = \mathbf{0}$. We calculate the next approximation $\mathbf{p}^{(1)}$ in the same manner as Newton's method, or, if it is inconvenient to determine $J(\mathbf{p}^{(0)})$ exactly, we can use difference equations to approximate the partial derivatives. To compute $\mathbf{p}^{(2)}$, however, we depart from Newton's method and examine the Secant method for a single nonlinear equation. The Secant method differs from Newton's method because it uses

$$f'(p_1) \approx \frac{f(p_1) - f(p_0)}{p_1 - p_0}$$

instead of $f'(p_1)$. For nonlinear systems, $\mathbf{p}^{(1)} - \mathbf{p}^{(0)}$ is a vector, and the corresponding quotient is undefined. However, the method proceeds similarly in that we replace the matrix $J(\mathbf{p}^{(1)})$ in Newton's method by a matrix A_1 with the property that

$$A_1(\mathbf{p}^{(1)} - \mathbf{p}^{(0)}) = \mathbf{F}(\mathbf{p}^{(1)}) - \mathbf{F}(\mathbf{p}^{(0)}).$$

Any nonzero vector in $\mathbb{R}^n$ can be written as the sum of a multiple of $\mathbf{p}^{(1)} - \mathbf{p}^{(0)}$ and a multiple of a vector orthogonal to $\mathbf{p}^{(1)} - \mathbf{p}^{(0)}$. So, to uniquely define the matrix A_1, we need to specify how it acts on vectors orthogonal to $\mathbf{p}^{(1)} - \mathbf{p}^{(0)}$.

No information is available about the change in $\mathbf{F}$ in a direction orthogonal to $\mathbf{p}^{(1)} - \mathbf{p}^{(0)}$, so we simply require that no change occurs when defining A_1. That is,

$$A_1 \mathbf{z} = J(\mathbf{p}^{(0)})\mathbf{z} \quad \text{whenever} \quad (\mathbf{p}^{(1)} - \mathbf{p}^{(0)})^t \mathbf{z} = 0.$$

Thus any vector orthogonal to $\mathbf{p}^{(1)} - \mathbf{p}^{(0)}$ is unaffected by the update from $J(\mathbf{p}^{(0)})$, which was used to compute $\mathbf{p}^{(1)}$, to A_1, which is used in the determination of $\mathbf{p}^{(2)}$.

These conditions uniquely define A_1 (see Exercise 8) as

$$A_1 = J(\mathbf{p}^{(0)}) + \frac{[\mathbf{F}(\mathbf{p}^{(1)}) - \mathbf{F}(\mathbf{p}^{(0)}) - J(\mathbf{p}^{(0)})(\mathbf{p}^{(1)} - \mathbf{p}^{(0)})]}{\|\mathbf{p}^{(1)} - \mathbf{p}^{(0)}\|_2^2}(\mathbf{p}^{(1)} - \mathbf{p}^{(0)})^t. \qquad (10.1)$$

It is this matrix that is used in place of $J(\mathbf{p}^{(1)})$ to determine $\mathbf{p}^{(2)}$:

$$\mathbf{p}^{(2)} = \mathbf{p}^{(1)} - A_1^{-1}\mathbf{F}(\mathbf{p}^{(1)}).$$

Once $\mathbf{p}^{(2)}$ has been determined, the method can be repeated to determine $\mathbf{p}^{(3)}$, with A_1 used in place of $A_0 \equiv J(\mathbf{p}^{(0)})$ and with $\mathbf{p}^{(2)}$ and $\mathbf{p}^{(1)}$ in place of $\mathbf{p}^{(1)}$ and $\mathbf{p}^{(0)}$, respectively. To simplify the notation we introduce the variables

$$\mathbf{s}_i = \mathbf{p}^{(i)} - \mathbf{p}^{(i-1)} \quad \text{and} \quad \mathbf{y}_i = \mathbf{F}(\mathbf{p}^{(i)}) - \mathbf{F}(\mathbf{p}^{(i-1)}).$$

Then, once $\mathbf{p}^{(i)}$ has been determined, $\mathbf{p}^{(i+1)}$ can be computed by

$$A_i = A_{i-1} + \frac{[\mathbf{F}(\mathbf{p}^{(i)}) - \mathbf{F}(\mathbf{p}^{(i-1)})] - A_{i-1}(\mathbf{p}^{(i)} - \mathbf{p}^{(i-1)})}{\|\mathbf{p}^{(i)} - \mathbf{p}^{(i-1)}\|_2^2}(\mathbf{p}^{(i)} - \mathbf{p}^{(i-1)})^t$$

$$= A_{i-1} + \frac{\mathbf{y}_i - A_{i-1}\mathbf{s}_i}{\|\mathbf{s}_i\|_2^2}\mathbf{s}_i^t.$$

and

$$\mathbf{p}^{(i+1)} = \mathbf{p}^{(i)} - A_i^{-1}\mathbf{F}(\mathbf{p}^{(i)}).$$

If the method is performed as outlined, the number of scalar functional evaluations is reduced from $n^2 + n$ to n (those required for evaluating $\mathbf{F}(\mathbf{p}^{(i)})$), but the method still requires $O(n^3)$ calculations to solve the associated $n \times n$ linear system

$$A_i \mathbf{s}_{i+1} = -\mathbf{F}(\mathbf{p}^{(i)}).$$

Employing the method in this form would not be justified because of the reduction to superlinear convergence from the quadratic convergence of Newton's method. However, a significant improvement can be incorporated by employing a matrix-inversion formula.

Sherman-Morrison Formula

Sherman-Morrison Formula

If A is a nonsingular matrix and $\mathbf{x}$ and $\mathbf{y}$ are vectors with $\mathbf{y}^t A^{-1} \mathbf{x} \neq -1$, then $A + \mathbf{x}\mathbf{y}^t$ is nonsingular and

$$(A + \mathbf{x}\mathbf{y}^t)^{-1} = A^{-1} - \frac{A^{-1}\mathbf{x}\mathbf{y}^t A^{-1}}{1 + \mathbf{y}^t A^{-1}\mathbf{x}}.$$

This formula permits A_i^{-1} to be computed directly from A_{i-1}^{-1}, eliminating the need for a matrix inversion with each iteration. This computation involves only matrix-vector multiplication at each step and therefore requires only $O(n^2)$ arithmetic calculations.

By letting $A = A_{i-1}$, $\mathbf{x} = (\mathbf{y}_i - A_{i-1}\mathbf{s}_i)/\|\mathbf{s}_i\|_2^2$, and $\mathbf{y} = \mathbf{s}_i$, the Sherman-Morrison formula implies that

$$
\begin{aligned}
A_i^{-1} &= \left(A_{i-1} + \frac{\mathbf{y}_i - A_{i-1}\mathbf{s}_i}{\|\mathbf{s}_i\|_2^2}\mathbf{s}_i^t \right)^{-1} \\
&= A_{i-1}^{-1} - \frac{A_{i-1}^{-1} \left(\dfrac{\mathbf{y}_i - A_{i-1}\mathbf{s}_i}{\|\mathbf{s}_i\|_2^2} \right) \mathbf{s}_i^t A_{i-1}^{-1}}{1 + \mathbf{s}_i^t A_{i-1}^{-1} \left(\dfrac{\mathbf{y}_i - A_{i-1}\mathbf{s}_i}{\|\mathbf{s}_i\|_2^2} \right)} \\
&= A_{i-1}^{-1} - \frac{\left(A_{i-1}^{-1}\mathbf{y}_i - \mathbf{s}_i \right) \mathbf{s}_i^t A_{i-1}^{-1}}{\|\mathbf{s}_i\|_2^2 + \mathbf{s}_i^t A_{i-1}^{-1}\mathbf{y}_i - \|\mathbf{s}_i\|_2^2} \\
&= A_{i-1}^{-1} + \frac{\left(\mathbf{s}_i - A_{i-1}^{-1}\mathbf{y}_i \right) \mathbf{s}_i^t A_{i-1}^{-1}}{\mathbf{s}_i^t A_{i-1}^{-1}\mathbf{y}_i}.
\end{aligned}
$$

The program BROYM102 implements Broyden's method.

The calculation of A_i is bypassed, as is the necessity of solving the linear system.

Example 1 Use Broyden's method with $\mathbf{p}^{(0)} = (0.1, 0.1, -0.1)^t$ to approximate the solution to the nonlinear system

$$3x_1 - \cos(x_2 x_3) - \frac{1}{2} = 0,$$

$$x_1^2 - 81(x_2 + 0.1)^2 + \sin x_3 + 1.06 = 0,$$

$$e^{-x_1 x_2} + 20x_3 + \frac{10\pi - 3}{3} = 0.$$

Solution This system was solved by Newton's method in Example 1 of Section 10.2. The Jacobian matrix for this system is

$$J(x_1, x_2, x_3) = \begin{bmatrix} 3 & x_3 \sin x_2 x_3 & x_2 \sin x_2 x_3 \\ 2x_1 & -162(x_2 + 0.1) & \cos x_3 \\ -x_2 e^{-x_1 x_2} & -x_1 e^{-x_1 x_2} & 20 \end{bmatrix}.$$

Let $\mathbf{p}^{(0)} = (0.1, 0.1, -0.1)^t$ and

$$\mathbf{F}(x_1, x_2, x_3) = (f_1(x_1, x_2, x_3), f_2(x_1, x_2, x_3), f_3(x_1, x_2, x_3))^t,$$

where

$$f_1(x_1, x_2, x_3) = 3x_1 - \cos(x_2 x_3) - \frac{1}{2},$$

$$f_2(x_1, x_2, x_3) = x_1^2 - 81(x_2 + 0.1)^2 + \sin x_3 + 1.06,$$

and

$$f_3(x_1, x_2, x_3) = e^{-x_1 x_2} + 20x_3 + \frac{10\pi - 3}{3}.$$

For $\mathbf{p}^{(0)} = (0.1, 0.1, -0.1)^t$ we have

$$\mathbf{F}(\mathbf{p}^{(0)}) = \begin{bmatrix} -1.199950 \\ -2.269833 \\ 8.462025 \end{bmatrix}.$$

This implies that

$$A_0 = J(\mathbf{p}^{(0)})$$

$$= \begin{bmatrix} 3 & 9.999833 \times 10^{-4} & -9.999833 \times 10^{-4} \\ 0.2 & -32.4 & 0.9950042 \\ -9.900498 \times 10^{-2} & -9.900498 \times 10^{-2} & 20 \end{bmatrix}.$$

For this first iteration, we need to find the inverse of $(J(\mathbf{p}^{(0)}))$. However, for subsequent iterations, matrix inversion is not necessary. We have

$$A_0^{-1} = J\left(p_1^{(0)}, p_2^{(0)}, p_3^{(0)}\right)^{-1}$$

$$= \begin{bmatrix} 0.3333332 & 1.023852 \times 10^{-5} & 1.615701 \times 10^{-5} \\ 2.108607 \times 10^{-3} & -3.086883 \times 10^{-2} & 1.535836 \times 10^{-3} \\ 1.660520 \times 10^{-3} & -1.527577 \times 10^{-4} & 5.000768 \times 10^{-2} \end{bmatrix}.$$

So

$$\mathbf{p}^{(1)} = \mathbf{p}^{(0)} - A_0^{-1}\mathbf{F}(\mathbf{p}^{(0)}) = \begin{bmatrix} 0.4998697 \\ 1.946685 \times 10^{-2} \\ -0.5215205 \end{bmatrix},$$

$$\mathbf{F}(\mathbf{p}^{(1)}) = \begin{bmatrix} -3.394465 \times 10^{-4} \\ -0.3443879 \\ 3.188238 \times 10^{-2} \end{bmatrix},$$

$$\mathbf{y}_1 = \mathbf{F}(\mathbf{p}^{(1)}) - \mathbf{F}(\mathbf{p}^{(0)}) = \begin{bmatrix} 1.199611 \\ 1.925445 \\ -8.430143 \end{bmatrix},$$

$$\mathbf{s}_1 = \begin{bmatrix} 0.3998697 \\ -8.053315 \times 10^{-2} \\ -0.4215204 \end{bmatrix},$$

$$\mathbf{s}_1^t A_0^{-1} \mathbf{y}_1 = 0.3424604,$$

$$A_1^{-1} = A_0^{-1} + \frac{1}{0.3424604} \left[(\mathbf{s}_1 - A_0^{-1} \mathbf{y}_1) \mathbf{s}_1^t A_0^{-1} \right]$$

$$= \begin{bmatrix} 0.3333781 & 1.11050 \times 10^{-5} & 8.967344 \times 10^{-6} \\ -2.021270 \times 10^{-3} & -3.094849 \times 10^{-2} & 2.196906 \times 10^{-3} \\ 1.022214 \times 10^{-3} & -1.650709 \times 10^{-4} & 5.010986 \times 10^{-2} \end{bmatrix},$$

and

$$\mathbf{p}^{(2)} = \mathbf{p}^{(1)} - A_1^{-1} \mathbf{F}(\mathbf{p}^{(1)}) = \begin{bmatrix} 0.4999863 \\ 8.737833 \times 10^{-3} \\ -0.5231746 \end{bmatrix}.$$

Additional iterations are listed in Table 10.2. The 5th iteration of Broyden's method is slightly less accurate than was the 4th iteration of Newton's method given in the example at the end of the preceding section. ∎

Table 10.2

k	$p_1^{(k)}$	$p_2^{(k)}$	$p_3^{(k)}$	$\|\mathbf{p}^{(k)} - \mathbf{p}^{(k-1)}\|_2$
0	0.1000000	0.1000000	−0.1000000	
1	0.4998697	1.946685×10^{-2}	−0.5215205	5.93×10^{-1}
2	0.4999864	8.737839×10^{-3}	−0.5231746	1.0856×10^{-2}
3	0.5000066	8.672736×10^{-4}	−0.5235723	7.8806×10^{-3}
4	0.5000003	3.952827×10^{-5}	−0.5235977	8.2817×10^{-4}
5	0.5000000	1.934342×10^{-7}	−0.5235988	3.9351×10^{-5}

EXERCISE SET 10.3

1. Use Broyden's method with $\mathbf{p}^{(0)} = \mathbf{0}$ to compute $\mathbf{p}^{(2)}$ for each of the following nonlinear systems.

   a. $4x_1^2 - 20x_1 + \frac{1}{4}x_2^2 + 8 = 0,$

   $\frac{1}{2}x_1 x_2^2 + 2x_1 - 5x_2 + 8 = 0.$

   b. $\sin(4\pi x_1 x_2) - 2x_2 - x_1 = 0,$

   $\left(\frac{4\pi - 1}{4\pi} \right) (e^{2x_1} - e) + 4ex_2^2 - 2ex_1 = 0.$

   c. $3x_1 - \cos(x_2 x_3) - \frac{1}{2} = 0,$

   $4x_1^2 - 625x_2^2 + 2x_2 - 1 = 0,$

   $e^{-x_1 x_2} + 20x_3 + \frac{1}{3}(10\pi - 3) = 0.$

   d. $x_1^2 + x_2 - 37 = 0,$
   $x_1 - x_2^2 - 5 = 0,$
   $x_1 + x_2 + x_3 - 3 = 0.$

2. Use Broyden's method to approximate solutions to the nonlinear systems in Exercise 1 using the following initial approximations $\mathbf{p}^{(0)}$ until $\|\mathbf{p}^{(k)} - \mathbf{p}^{(k-1)}\|_\infty < 10^{-6}$.

   a. $(0, 0)^t$ b. $(0, 0)^t$ c. $(1, 1, 1)^t$ d. $(2, 1, -1)^t$

3.   Use Broyden's method to find a solution to the following nonlinear systems, iterating until $\|\mathbf{p}^{(k)} - \mathbf{p}^{(k-1)}\|_\infty < 10^{-6}$.

a. $3x_1^2 - x_2^2 = 0$
$3x_1 x_2^2 - x_1^3 - 1 = 0$
Use $\mathbf{p}^{(0)} = (1, 1)^t$.

b. $\ln(x_1^2 + x_2^2) - \sin(x_1 x_2) = \ln 2 + \ln \pi$
$e^{x_1 - x_2} + \cos(x_1 x_2) = 0$
Use $\mathbf{p}^{(0)} = (2, 2)^t$.

c. $x_1^3 + x_1^2 x_2 - x_1 x_3 + 6 = 0$
$e^{x_1} + e^{x_2} - x_3 = 0$
$x_2^2 - 2x_1 x_3 = 4$
Use $\mathbf{p}^{(0)} = (-1, -2, 1)^t$.

d. $6x_1 - 2\cos(x_2 x_3) - 1 = 0$
$9x_2 + \sqrt{x_1^2 + \sin x_3 + 1.06} + 0.9 = 0$
$60x_3 + 3e^{-x_1 x_2} + 10\pi - 3 = 0$
Use $\mathbf{p}^{(0)} = (0, 0, 0)^t$.

4.   The nonlinear system

$$3x_1 - \cos(x_2 x_3) - \frac{1}{2} = 0,$$

$$x_1^2 - 625x_2^2 - \frac{1}{4} = 0,$$

$$e^{-x_1 x_2} + 20x_3 + \frac{1}{3}(10\pi - 3) = 0$$

has a singular Jacobian matrix at the solution. Apply Broyden's method with $\mathbf{p}^{(0)} = (1, 1 - 1)^t$. Note that convergence may be slow or may not occur within a reasonable number of iterations.

5.   The nonlinear system

$$\begin{aligned}
4x_1 - \;\; x_2 + \;\; x_3 &= x_1 x_4, \\
-x_1 + 3x_2 - 2x_3 &= x_2 x_4, \\
x_1 - 2x_2 + 3x_3 &= x_3 x_4, \\
x_1^2 + \;\; x_2^2 + \;\; x_3^2 &= 1
\end{aligned}$$

has six solutions, and, as shown in Exercise 8 of Section 10.2, $(-x_1, -x_2, -x_3, x_4)$ is a solution whenever (x_1, x_2, x_3, x_4) is a solution. Use Broyden's method to approximate these solutions. Iterate until $\|\mathbf{p}^{(k)} - \mathbf{p}^{(k-1)}\|_\infty < 10^{-5}$. Use the initial vectors $(1, 1, 1, 1)^t$, $(1, 0, 0, 0)^t$, and $(1, -1, 1, -1)^t$.

6.   Show that if $\mathbf{0} \neq \mathbf{y} \in \mathbb{R}^n$ and $\mathbf{z} \in \mathbb{R}^n$, then $\mathbf{z} = \mathbf{z}_1 + \mathbf{z}_2$, where

$$\mathbf{z}_1 = \frac{\mathbf{y}^t \mathbf{z}}{\|\mathbf{y}\|_2^2} \mathbf{y}$$

is parallel to $\mathbf{y}$ and $\mathbf{z}_2 = \mathbf{z} - \mathbf{z}_1$ is orthogonal to $\mathbf{y}$.

7.   Show that if $\mathbf{z}$ is orthogonal to $\mathbf{p}^{(1)} - \mathbf{p}^{(0)}$, then for A_1 defined in Eq. (10.1) we have $A_1 = J(\mathbf{p}^{(0)})$.

8.   Let

$$A_1 = J(\mathbf{p}^{(0)}) + \frac{\left[\mathbf{F}(\mathbf{p}^{(1)}) - \mathbf{F}(\mathbf{p}^{(0)}) - J(\mathbf{p}^{(0)})(\mathbf{p}^{(1)} - \mathbf{p}^{(0)})\right](\mathbf{p}^{(1)} - \mathbf{p}^{(0)})^t}{\|\mathbf{p}^{(1)} - \mathbf{p}^{(0)}\|_2^2}.$$

a.   Show that $A_1(\mathbf{p}^{(1)} - \mathbf{p}^{(0)}) = \mathbf{F}(\mathbf{p}^{(1)}) - \mathbf{F}(\mathbf{p}^{(0)})$.

b.   Show that $A_1 \mathbf{z} = J(\mathbf{p}^{(0)})\mathbf{z}$ whenever $(\mathbf{p}^{(1)} - \mathbf{p}^{(0)})^t \mathbf{z} = 0$.

9. It can be shown that if A^{-1} exists and $\mathbf{x}, \mathbf{y} \in \mathbb{R}^n$, then $(A+\mathbf{xy}^t)^{-1}$ exists if and only if $\mathbf{y}^t A^{-1}\mathbf{x} \neq -1$. Use this result to verify the Sherman-Morrison formula: If A^{-1} exists and $\mathbf{y}^t A^{-1}\mathbf{x} \neq -1$, then $(A+\mathbf{xy}^t)^{-1}$ exists, and

$$(A + \mathbf{xy}^t)^{-1} = A^{-1} - \frac{A^{-1}\mathbf{xy}^t A^{-1}}{1 + \mathbf{y}^t A^{-1}\mathbf{x}}.$$

10.4 The Steepest Descent Method

The name for the Steepest Descent method follows from the three-dimensional application of pointing in the steepest downward direction.

The advantage of the Newton and quasi-Newton methods for solving systems of nonlinear equations is their speed of convergence once a sufficiently accurate approximation is known. A weakness of these methods is that an accurate initial approximation to the solution is needed to ensure convergence. The **method of Steepest Descent** will generally converge only linearly to the solution, but it is global in nature, that is, nearly any starting value will give convergence. As a consequence, it is often used to find sufficiently accurate starting approximations for the Newton-based techniques.

The method of Steepest Descent determines a local minimum for a multivariable function of the form $g: \mathbb{R}^n \rightarrow \mathbb{R}$. The method is valuable quite apart from providing starting values for solving nonlinear systems, but we will consider only this application.

The connection between the minimization of a function from $\mathbb{R}^n$ to $\mathbb{R}$ and the solution of a system of nonlinear equations is due to the fact that a system of the form

$$f_1(x_1, x_2, \ldots, x_n) = 0,$$
$$f_2(x_1, x_2, \ldots, x_n) = 0,$$
$$\vdots$$
$$f_n(x_1, x_2, \ldots, x_n) = 0,$$

has a solution at $\mathbf{x} = (x_1, x_2, \ldots, x_n)^t$ precisely when the function g from $\mathbb{R}^n$ to $\mathbb{R}$ defined by

$$g(x_1, x_2, \ldots, x_n) = \sum_{i=1}^{n} [f_i(x_1, x_2, \ldots, x_n)]^2$$

has the minimal value zero.

The method of Steepest Descent for finding a local minimum for an arbitrary function g from $\mathbb{R}^n$ into $\mathbb{R}$ can be intuitively described as follows:

- Evaluate g at an initial approximation $\mathbf{p}^{(0)} = (p_1^{(0)}, p_2^{(0)}, \ldots, p_n^{(0)})^t$.

- Determine a direction from $\mathbf{p}^{(0)}$ that results in a decrease in the value of g.

- Move an appropriate amount in this direction and call the new value $\mathbf{p}^{(1)}$.

- Repeat the steps with $\mathbf{p}^{(0)}$ replaced by $\mathbf{p}^{(1)}$.

The Gradient of a Function

Before describing how to choose the correct direction and the appropriate distance to move in this direction, we need to review some results from calculus. The Extreme Value Theorem implies that a differentiable single-variable function can have a relative minimum within

The root of *gradient* comes from the Latin word *gradi*, meaning "to walk". In this sense, the gradient of a surface is the rate at which it "walks uphill".

the interval only when the derivative is zero. To extend this result to multivariable functions, we need the following definition.

If $g : \mathbb{R}^n \to \mathbb{R}$, we define the **gradient** of g at $\mathbf{x} = (x_1, x_2, \ldots, x_n)^t$, $\nabla g(\mathbf{x})$, by

$$\nabla g(\mathbf{x}) = \left(\frac{\partial g}{\partial x_1}(\mathbf{x}), \frac{\partial g}{\partial x_2}(\mathbf{x}), \ldots, \frac{\partial g}{\partial x_n}(\mathbf{x}) \right)^t.$$

The gradient for a multivariable function is analogous to the derivative of a single variable function in the sense that a differentiable multivariable function can have a relative minimum at $\mathbf{x}$ only when the gradient at $\mathbf{x}$ is the zero vector.

The gradient has another important property connected with the minimization of multivariable functions. Suppose $\mathbf{v} = (v_1, v_2, \ldots, v_n)^t$ is a vector in $\mathbb{R}^n$ with $\|\mathbf{v}\|_2 = 1$. The **directional derivative** of g at $\mathbf{x}$ in the direction of $\mathbf{v}$ is defined by

$$D_{\mathbf{v}} g(\mathbf{x}) = \lim_{h \to 0} \frac{1}{h} [g(\mathbf{x} + h\mathbf{v}) - g(\mathbf{x})] = \mathbf{v}^t \cdot \nabla g(\mathbf{x}) = \sum_{i=1}^{n} v_i \frac{\partial g}{\partial x_i}(\mathbf{x}).$$

The directional derivative of g at $\mathbf{x}$ in the direction of $\mathbf{v}$ measures the change in the value of the function g relative to the change in the variable in the direction of $\mathbf{v}$.

A standard result from the calculus of multivariable functions states that the direction that produces the maximum increase for the directional derivative occurs when $\mathbf{v}$ is chosen in the direction of $\nabla g(\mathbf{x})$, provided that $\nabla g(\mathbf{x}) \neq \mathbf{0}$. So the maximum decrease will be in the direction of $-\nabla g(\mathbf{x})$.

• The direction of greatest decrease in the value of g at $\mathbf{x}$ is the direction given by $-\nabla g(\mathbf{x})$.

See Figure 10.2 for an illustration when g is a function of two variables.

Figure 10.2

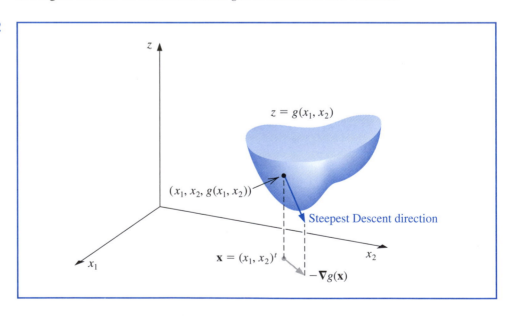

The objective is to reduce $g(\mathbf{x})$ to its minimal value of zero, so given the initial approximation $\mathbf{p}^{(0)}$, we choose

$$\mathbf{p}^{(1)} = \mathbf{p}^{(0)} - \alpha \nabla g(\mathbf{p}^{(0)}) \tag{10.2}$$

for some constant $\alpha > 0$.

The problem now reduces to choosing α so that $g(\mathbf{p}^{(1)})$ will be significantly less than $g(\mathbf{p}^{(0)})$. To determine an appropriate choice for the value α, we consider the single-variable function

$$h(\alpha) = g(\mathbf{p}^{(0)} - \alpha \nabla g(\mathbf{p}^{(0)})).$$

The value of α that minimizes h is the value needed for $\mathbf{p}^{(1)} = \mathbf{p}^{(0)} - \alpha \nabla g(\mathbf{p}^{(0)})$.

Finding a minimal value for h directly would require differentiating h and then solving a root-finding problem to determine the critical points of h. This procedure is generally too costly. Instead, we choose three numbers $\alpha_1 < \alpha_2 < \alpha_3$ that, we hope, are close to where the minimum value of $h(\alpha)$ occurs. Then we construct the quadratic polynomial $P(x)$ that interpolates h at α_1, α_2, and α_3. We define $\hat{\alpha}$ in $[\alpha_1, \alpha_3]$ so that $P(\hat{\alpha})$ is a minimum in $[\alpha_1, \alpha_3]$ and use $P(\hat{\alpha})$ to approximate the minimal value of $h(\alpha)$. Then $\hat{\alpha}$ is used to determine the new iterate for approximating the minimal value of g:

$$\mathbf{p}^{(1)} = \mathbf{p}^{(0)} - \hat{\alpha} \nabla g(\mathbf{p}^{(0)}).$$

Since $g(\mathbf{p}^{(0)})$ is available, we first choose $\alpha_1 = 0$ to minimize the computation. Next a number α_3 is found with $h(\alpha_3) < h(\alpha_1)$. (Since α_1 does not minimize h, such a number α_3 does exist.) Finally, α_2 is chosen to be $\alpha_3/2$.

The minimum value $\hat{\alpha}$ of $P(x)$ on $[\alpha_1, \alpha_3]$ occurs either at the only critical point of P or at the right endpoint α_3 because, by assumption, $P(\alpha_3) = h(\alpha_3) < h(\alpha_1) = P(\alpha_1)$. The critical point is easily determined because $P(x)$ is a quadratic polynomial.

The program STPDC103 implements the Steepest Descent method.

Program STPDC103 applies the method of Steepest Descent to approximate the minimal value of $g(\mathbf{x})$. To begin each iteration, the value 0 is assigned to α_1, and the value 1 is assigned to α_3. If $h(\alpha_3) \geq h(\alpha_1)$, then successive divisions of α_3 by 2 are performed and the value of α_3 is reassigned until $h(\alpha_3) < h(\alpha_1)$.

To employ the method to approximate the solution to the system

$$f_1(x_1, x_2, \ldots, x_n) = 0,$$
$$f_2(x_1, x_2, \ldots, x_n) = 0,$$
$$\vdots$$
$$f_n(x_1, x_2, \ldots, x_n) = 0,$$

we simply replace the function g with $\sum_{i=1}^{n} f_i^2$.

Example 1 Use the Steepest Descent method with $\mathbf{p}^{(0)} = (0, 0, 0)^t$ to find a reasonable starting approximation to the solution of the nonlinear system

$$f_1(x_1, x_2, x_3) = 3x_1 - \cos(x_2 x_3) - \frac{1}{2} = 0,$$

$$f_2(x_1, x_2, x_3) = x_1^2 - 81(x_2 + 0.1)^2 + \sin x_3 + 1.06 = 0,$$

$$f_3(x_1, x_2, x_3) = e^{-x_1 x_2} + 20x_3 + \frac{10\pi - 3}{3} = 0.$$

Solution   Let $g(x_1, x_2, x_3) = [f_1(x_1, x_2, x_3)]^2 + [f_2(x_1, x_2, x_3)]^2 + [f_3(x_1, x_2, x_3)]^2$. Then

$$\nabla g(x_1, x_2, x_3) \equiv \nabla g(\mathbf{x}) = \left(2f_1(\mathbf{x})\frac{\partial f_1}{\partial x_1}(\mathbf{x}) + 2f_2(\mathbf{x})\frac{\partial f_2}{\partial x_1}(\mathbf{x}) + 2f_3(\mathbf{x})\frac{\partial f_3}{\partial x_1}(\mathbf{x}), \right.$$

$$2f_1(\mathbf{x})\frac{\partial f_1}{\partial x_2}(\mathbf{x}) + 2f_2(\mathbf{x})\frac{\partial f_2}{\partial x_2}(\mathbf{x}) + 2f_3(\mathbf{x})\frac{\partial f_3}{\partial x_2}(\mathbf{x}),$$

$$\left. 2f_1(\mathbf{x})\frac{\partial f_1}{\partial x_3}(\mathbf{x}) + 2f_2(\mathbf{x})\frac{\partial f_2}{\partial x_3}(\mathbf{x}) + 2f_3(\mathbf{x})\frac{\partial f_3}{\partial x_3}(\mathbf{x}) \right)$$

$$= 2\mathbf{J}(\mathbf{x})^t \mathbf{F}(\mathbf{x}).$$

For $\mathbf{p}^{(0)} = (0, 0, 0)^t$, we have

$$g(\mathbf{p}^{(0)}) = f_1(0, 0, 0)^2 + f_2(0, 0, 0)^2 + f_3(0, 0, 0)^2$$

$$= \left(-\frac{3}{2}\right)^2 + (-81(0.01) + 1.06)^2 + \left(\frac{10\pi}{3}\right)^2 = 111.975,$$

and

$$z_0 = \|\nabla g(\mathbf{p}^{(0)})\|_2 = \|2\mathbf{J}(\mathbf{0})^t \mathbf{F}(\mathbf{0})\|_2 = 419.554.$$

Let

$$\mathbf{z} = \frac{1}{z_0}\nabla g(\mathbf{p}^{(0)}) = (-0.0214514, -0.0193062, 0.999583)^t.$$

With $\alpha_1 = 0$, we have $g_1 = g(\mathbf{p}^{(0)} - \alpha_1 \mathbf{z}) = g(\mathbf{p}^{(0)}) = 111.975$. We arbitrarily let $\alpha_3 = 1$ so that

$$g_3 = g(\mathbf{p}^{(0)} - \alpha_3 \mathbf{z}) = 93.5649.$$

Because $g_3 < g_1$, we accept α_3 and set $\alpha_2 = \alpha_3/2 = 0.5$. Evaluating g at $\mathbf{p}^{(0)} - \alpha_2 \mathbf{z}$ gives

$$g_2 = g(\mathbf{p}^{(0)} - \alpha_2 \mathbf{z}) = 2.53557.$$

We now find the quadratic polynomial that interpolates the data $(0, 111.975)$, $(1, 93.5649)$, and $(0.5, 2.53557)$. It is most convenient to use Newton's forward divided-difference interpolating polynomial for this purpose, which has the form

$$P(\alpha) = g_1 + h_1\alpha + h_3\alpha(\alpha - \alpha_2).$$

This interpolates

$$g(\mathbf{p}^{(0)} - \alpha\nabla g(\mathbf{p}^{(0)})) = g(\mathbf{p}^{(0)} - \alpha\mathbf{z})$$

at $\alpha_1 = 0$, $\alpha_2 = 0.5$, and $\alpha_3 = 1$ as follows:

$\alpha_1 = 0$, $g_1 = 111.975$,

$\alpha_2 = 0.5$, $g_2 = 2.53557$, $h_1 = \dfrac{g_2 - g_1}{\alpha_2 - \alpha_1} = -218.878$,

$\alpha_3 = 1$, $g_3 = 93.5649$, $h_2 = \dfrac{g_3 - g_2}{\alpha_3 - \alpha_2} = 182.059$, $h_3 = \dfrac{h_2 - h_1}{\alpha_3 - \alpha_1} = 400.937$.

This gives

$$P(\alpha) = 111.975 - 218.878\alpha + 400.937\alpha(\alpha - 0.5) = 400.937\alpha^2 - 419.346\alpha + 111.975$$

so

$$P'(\alpha) = 801.874\alpha - 419.346$$

and $P'(\alpha) = 0$ when $\alpha = \alpha_0 = 0.522959$. Since

$$g(\mathbf{p}^{(0)} - \alpha_0 \mathbf{z}) = 2.32762$$

is smaller than g_1 and g_3, we set

$$\mathbf{p}^{(1)} = \mathbf{p}^{(0)} - \alpha_0 \mathbf{z} = \mathbf{p}^{(0)} - 0.522959\mathbf{z} = (0.0112182, 0.0100964, -0.522741)^t$$

and

$$g(\mathbf{p}^{(1)}) = 2.32762.$$

Table 10.3 contains the remainder of the results. A true solution is $\mathbf{p} = (0.5, 0, -0.5235988)^t$, so $\mathbf{p}^{(2)}$ would likely be adequate as an initial approximation for Newton's method or Broyden's method. One of these quicker converging techniques would be appropriate at this stage because 70 iterations of the Steepest Descent method are required to find $\|\mathbf{p}^{(k)} - \mathbf{p}\|_\infty < 0.01$. ∎

Table 10.3

k	$p_1^{(k)}$	$p_2^{(k)}$	$p_3^{(k)}$	$g(p_1^{(k)}, p_2^{(k)}, p_3^{(k)})$
2	0.137860	−0.205453	−0.522059	1.27406
3	0.266959	0.00551102	−0.558494	1.06813
4	0.272734	−0.00811751	−0.522006	0.468309
5	0.308689	−0.0204026	−0.533112	0.381087
6	0.314308	−0.0147046	−0.520923	0.318837
7	0.324267	−0.00852549	−0.528431	0.287024

EXERCISE SET 10.4

1. Use the method of Steepest Descent to approximate a solution of the following nonlinear systems, iterating until $\|\mathbf{p}^{(k)} - \mathbf{p}^{(k-1)}\|_\infty < 0.05$.

   a. $4x_1^2 - 20x_1 + \frac{1}{4}x_2^2 + 8 = 0$

   $\frac{1}{2}x_1 x_2^2 + 2x_1 - 5x_2 + 8 = 0$

   b. $3x_1^2 - x_2^2 = 0$

   $3x_1 x_2^2 - x_1^3 - 1 = 0$

   c. $\ln(x_1^2 + x_2^2) - \sin(x_1 x_2) = \ln 2 + \ln \pi$

   $e^{x_1 - x_2} + \cos(x_1 x_2) = 0$

   d. $\sin(4\pi x_1 x_2) - 2x_2 - x_1 = 0$

   $\left(\frac{4\pi - 1}{4\pi}\right)(e^{2x_1} - e) + 4ex_2^2 - 2ex_1 = 0$

2. Use the results in Exercise 1 and Newton's method to approximate the solutions of the nonlinear systems in Exercise 1, iterating until $\|\mathbf{p}^{(k)} - \mathbf{p}^{(k-1)}\|_\infty < 10^{-6}$.

3. Use the method of Steepest Descent to approximate a solution of the following nonlinear systems, iterating until $\|\mathbf{p}^{(k)} - \mathbf{p}^{(k-1)}\|_\infty < 0.05$.

a. $15x_1 + x_2^2 - 4x_3 = 13$
    $x_1^2 + 10x_2 - x_3 = 11$
    $x_2^3 - 25x_3 = -22$

b. $10x_1 - 2x_2^2 + x_2 - 2x_3 - 5 = 0$
    $8x_2^2 + 4x_3^2 - 9 = 0$
    $8x_2x_3 + 4 = 0$

c. $x_1^3 + x_1^2x_2 - x_1x_3 + 6 = 0$
    $e^{x_1} + e^{x_2} - x_3 = 0$
    $x_2^2 - 2x_1x_3 = 4$

d. $x_1 + \cos(x_1x_2x_3) - 1 = 0$
    $(1 - x_1)^{1/4} + x_2 + 0.05x_3^2 - 0.15x_3 - 1 = 0$
    $-x_1^2 - 0.1x_2^2 + 0.01x_2 + x_3 - 1 = 0$

4. Use the results of Exercise 3 and Newton's method to approximate the solutions of the nonlinear systems in Exercise 3, iterating until $\|\mathbf{p}^{(k)} - \mathbf{p}^{(k-1)}\|_\infty < 10^{-6}$.

5. Use the method of Steepest Descent to approximate minima for the following functions, iterating until $\|\mathbf{p}^{(k)} - \mathbf{p}^{(k-1)}\|_\infty < 0.005$.

a. $g(x_1, x_2) = \cos(x_1 + x_2) + \sin x_1 + \cos x_2$

b. $g(x_1, x_2) = 100(x_1^2 - x_2)^2 + (1 - x_1)^2$

c. $g(x_1, x_2, x_3) = x_1^2 + 2x_2^2 + x_3^2 - 2x_1x_2 + 2x_1 - 2.5x_2 - x_3 + 2$

d. $g(x_1, x_2, x_3) = x_1^4 + 2x_2^4 + 3x_3^4 + 1.01$

6. a. Show that the quadratic polynomial that interpolates the function

$$h(\alpha) = g(\mathbf{p}^{(0)} - \alpha \nabla g(\mathbf{p}^{(0)}))$$

at $\alpha = 0$, α_2, and α_3 is

$$P(\alpha) = g(\mathbf{p}^{(0)}) + h_1\alpha + h_3\alpha(\alpha - \alpha_2)$$

where

$$h_1 = \frac{g(\mathbf{p}^{(0)} - \alpha_2\mathbf{z}) - g(\mathbf{p}^{(0)})}{\alpha_2},$$

$$h_2 = \frac{g(\mathbf{p}^{(0)} - \alpha_3\mathbf{z}) - g(\mathbf{p}^{(0)} - \alpha_2\mathbf{z})}{\alpha_3 - \alpha_2}, \quad \text{and} \quad h_3 = \frac{h_2 - h_1}{\alpha_3}$$

b. Show that the only critical point of P occurs at $\alpha_0 = 0.5(\alpha_2 - h_1/h_3)$.

10.5 Homotopy and Continuation Methods

Homotopy, or *continuation*, methods for nonlinear systems embed the problem to be solved within a collection of problems. Specifically, to solve a problem of the form

$$\mathbf{F}(\mathbf{x}) = \mathbf{0},$$

which has the unknown solution $\mathbf{p}$, we consider a family of problems described using a parameter λ that assumes values in $[0, 1]$. A problem with a known solution $\mathbf{x}(0)$ corresponds to $\lambda = 0$, and the problem with the unknown solution $\mathbf{x}(1) \equiv \mathbf{p}$ corresponds to $\lambda = 1$.

Suppose $\mathbf{x}(0)$ is an initial approximation to the solution $\mathbf{p}$ of $\mathbf{F}(\mathbf{x}) = \mathbf{0}$. Define

$$\mathbf{G} : [0, 1] \times \mathbb{R}^n \to \mathbb{R}^n$$

by

$$\mathbf{G}(\lambda, \mathbf{x}) = \lambda\mathbf{F}(\mathbf{x}) + (1 - \lambda)[\mathbf{F}(\mathbf{x}) - \mathbf{F}(\mathbf{x}(0))] = \mathbf{F}(\mathbf{x}) + (\lambda - 1)\mathbf{F}(\mathbf{x}(0)).$$

We will determine, for various values of λ, a solution to

$$\mathbf{G}(\lambda, \mathbf{x}) = \mathbf{0}.$$

When $\lambda = 0$, this equation assumes the form

$$\mathbf{0} = \mathbf{G}(0, \mathbf{x}) = \mathbf{F}(\mathbf{x}) - \mathbf{F}(\mathbf{x}(0)),$$

so $\mathbf{x}(0)$ is a solution. When $\lambda = 1$, the equation assumes the form

$$\mathbf{0} = \mathbf{G}(1, \mathbf{x}) = \mathbf{F}(\mathbf{x}),$$

and $\mathbf{x}(1) = \mathbf{p}$, the solution to the original problem.

A homotopy is a continuous deformation; a function that takes a real interval continuously into a set of functions.

The function $\mathbf{G}$, with the parameter λ, provides us with a family of functions that can lead from the known value $\mathbf{x}(0)$ to the solution $\mathbf{x}(1) = \mathbf{p}$. The function $\mathbf{G}$ is called a **homotopy** between the function $\mathbf{G}(0, \mathbf{x}) = \mathbf{F}(\mathbf{x}) - \mathbf{F}(\mathbf{x}(0))$ and the function $\mathbf{G}(1, \mathbf{x}) = \mathbf{F}(\mathbf{x})$.

Continuation Problem

The **continuation** problem is to:

Determine a way to proceed from the known solution $\mathbf{x}(0)$ of $\mathbf{G}(0, \mathbf{x}) = \mathbf{0}$ to the unknown solution $\mathbf{x}(1) = \mathbf{p}$ of $\mathbf{G}(1, \mathbf{x}) = \mathbf{0}$, which is the solution to $\mathbf{F}(\mathbf{x}) = \mathbf{0}$.

Assume that $\mathbf{x}(\lambda)$ is the unique solution to the equation

$$\mathbf{G}(\lambda, \mathbf{x}) = \mathbf{0}, \tag{10.3}$$

for each $\lambda \in [0, 1]$. The set $\{ \mathbf{x}(\lambda) \mid 0 \leq \lambda \leq 1 \}$ can be viewed as a curve in $\mathbb{R}^n$ from $\mathbf{x}(0)$ to $\mathbf{x}(1) = \mathbf{p}$ parameterized by λ. A continuation method finds a sequence of steps along this curve corresponding to $\{\mathbf{x}(\lambda_k)\}_{k=0}^m$, where $\lambda_0 = 0 < \lambda_1 < \cdots < \lambda_m = 1$.

If the functions $\lambda \to \mathbf{x}(\lambda)$ and $\mathbf{G}$ are differentiable, then taking the total derivative of $\mathbf{G}(\lambda, \mathbf{x})$ in Eq. (10.3) with respect to λ gives

$$\frac{\partial \mathbf{G}(\lambda, \mathbf{x}(\lambda))}{\partial \lambda} + \frac{\partial \mathbf{G}(\lambda, \mathbf{x}(\lambda))}{\partial \mathbf{x}} \mathbf{x}'(\lambda) = \mathbf{0}.$$

Solving for $\mathbf{x}'(\lambda)$ gives a system of differential equations

$$\mathbf{x}'(\lambda) = - \left[\frac{\partial \mathbf{G}(\lambda, \mathbf{x}(\lambda))}{\partial \mathbf{x}} \right]^{-1} \frac{\partial \mathbf{G}(\lambda, \mathbf{x}(\lambda))}{\partial \lambda} \tag{10.4}$$

with initial conditions given by $\mathbf{x}(0)$.

Since

$$\mathbf{G}(\lambda, \mathbf{x}(\lambda)) = \mathbf{F}(\mathbf{x}(\lambda)) + (\lambda - 1)\mathbf{F}(\mathbf{x}(0)),$$

we can determine both the partial derivative of $\mathbf{G}$ with respect to its first variable, λ,

$$\frac{\partial G}{\partial \lambda}(\lambda, \mathbf{x}(\lambda)) = \mathbf{F}(\mathbf{x}(0))$$

and, with respect to its second variable, **x**,

$$\frac{\partial \mathbf{G}}{\partial \mathbf{x}}(\lambda, \mathbf{x}(\lambda)) = \begin{bmatrix} \frac{\partial f_1}{\partial x_1}(\mathbf{x}(\lambda)) & \frac{\partial f_1}{\partial x_2}(\mathbf{x}(\lambda)) & \cdots & \frac{\partial f_1}{\partial x_n}(\mathbf{x}(\lambda)) \\ \frac{\partial f_2}{\partial x_1}(\mathbf{x}(\lambda)) & \frac{\partial f_2}{\partial x_2}(\mathbf{x}(\lambda)) & \cdots & \frac{\partial f_2}{\partial x_n}(\mathbf{x}(\lambda)) \\ \vdots & & & \vdots \\ \frac{\partial f_n}{\partial x_1}(\mathbf{x}(\lambda)) & \frac{\partial f_n}{\partial x_2}(\mathbf{x}(\lambda)) & \cdots & \frac{\partial f_n}{\partial x_n}(\mathbf{x}(\lambda)) \end{bmatrix} = J(\mathbf{x}(\lambda)),$$

the Jacobian matrix of **F** at **x**(λ).

Therefore the system of differential equations in Eq. (10.4) becomes

$$\mathbf{x}'(\lambda) = -\left(J(\mathbf{x}(\lambda))\right)^{-1} \mathbf{F}(\mathbf{x}(0)), \quad \text{for} \quad 0 \le \lambda \le 1,$$

with the initial conditions given by **x**(0).

The following result gives conditions under which the continuation method is feasible.

Homotopy Convergence

Let **F**(**x**) be continuously differentiable for $\mathbf{x} \in \mathbb{R}^n$. Suppose that the Jacobian matrix $J(\mathbf{x})$ is nonsingular for all $\mathbf{x} \in \mathbb{R}^n$ and that a constant M exists with $\| [J(\mathbf{x})]^{-1} \| \le M$, for all $\mathbf{x} \in \mathbb{R}^n$. Then, for any **x**(0) in $\mathbb{R}^n$, there exists a unique function **x**(λ), such that

$$\mathbf{G}(\lambda, \mathbf{x}(\lambda)) = \mathbf{0},$$

for all λ in [0, 1]. Moreover, **x**(λ) is continuously differentiable and

$$\mathbf{x}'(\lambda) = -\left(J(\mathbf{x}(\lambda))\right)^{-1} \mathbf{F}(\mathbf{x}(0)), \quad \text{for each } \lambda \in [0, 1].$$

The following Illustration shows the form of the system of differential equations associated with a nonlinear system of equations.

Illustration   Consider the nonlinear system

$$f_1(x_1, x_2, x_3) = 3x_1 - \cos(x_2 x_3) - 0.5 = 0,$$

$$f_2(x_1, x_2, x_3) = x_1^2 - 81(x_2 + 0.1)^2 + \sin x_3 + 1.06 = 0,$$

$$f_3(x_1, x_2, x_3) = e^{-x_1 x_2} + 20x_3 + \frac{10\pi - 3}{3} = 0.$$

The Jacobian matrix is

$$J(\mathbf{x}) = \begin{bmatrix} 3 & x_3 \sin x_2 x_3 & x_2 \sin x_2 x_3 \\ 2x_1 & -162(x_2 + 0.1) & \cos x_3 \\ -x_2 e^{-x_1 x_2} & -x_1 e^{-x_1 x_2} & 20 \end{bmatrix}.$$

Let $\mathbf{x}(0) = (0, 0, 0)^t$. Then

$$\mathbf{F}(\mathbf{x}(0)) = \begin{bmatrix} -1.5 \\ 0.25 \\ 10\pi/3 \end{bmatrix}.$$

The system of differential equations is

$$
\begin{bmatrix} x_1'(\lambda) \\ x_2'(\lambda) \\ x_3'(\lambda) \end{bmatrix} = - \begin{bmatrix} 3 & x_3 \sin x_2 x_3 & x_2 \sin x_2 x_3 \\ 2x_1 & -162(x_2 + 0.1) & \cos x_3 \\ -x_2 e^{-x_1 x_2} & -x_1 e^{-x_1 x_2} & 20 \end{bmatrix}^{-1} \begin{bmatrix} -1.5 \\ 0.25 \\ 10\pi/3 \end{bmatrix}. \qquad \square
$$

In general, the system of differential equations that we need to solve for our continuation problem has the form

$$
\frac{dx_1}{d\lambda}(\lambda) = \phi_1(\lambda, x_1, x_2, \dots, x_n),
$$

$$
\frac{dx_2}{d\lambda}(\lambda) = \phi_2(\lambda, x_1, x_2, \dots, x_n),
$$

$$
\vdots
$$

$$
\frac{dx_n}{d\lambda}(\lambda) = \phi_n(\lambda, x_1, x_2, \dots, x_n),
$$

where

$$
\begin{bmatrix} \phi_1(\lambda, x_1, \dots, x_n) \\ \phi_2(\lambda, x_1, \dots, x_n) \\ \vdots \\ \phi_n(\lambda, x_1, \dots, x_n) \end{bmatrix} = -J(x_1, \dots, x_n)^{-1} \begin{bmatrix} f_1(\mathbf{x}(0)) \\ f_2(\mathbf{x}(0)) \\ \vdots \\ f_n(\mathbf{x}(0)) \end{bmatrix}. \qquad (10.5)
$$

To use the Runge-Kutta method of order 4 to solve this system, we first choose an integer $N > 0$ and let $h = (1 - 0)/N$. Partition the interval $[0, 1]$ into N subintervals with the mesh points

$$
\lambda_j = jh, \quad \text{for each} \quad j = 0, 1, \dots, N.
$$

We use the notation w_{ij}, for each $j = 0, 1, \dots, N$ and $i = 1, \dots, n$, to denote an approximation to $x_i(\lambda_j)$; that is,

$$
w_{ij} \approx x_i(\lambda_j).
$$

For the initial conditions, set

$$
w_{1,0} = x_1(0), \quad w_{2,0} = x_2(0), \quad \dots, \quad w_{n,0} = x_n(0).
$$

Suppose $w_{1,j}, w_{2,j}, \dots, w_{n,j}$ have been computed. We obtain $w_{1,j+1}, w_{2,j+1}, \dots, w_{n,j+1}$ using the equations

$$
k_{1,i} = h\phi_i(\lambda_j, w_{1,j}, w_{2,j}, \dots, w_{n,j}), \quad \text{for each } i = 1, 2, \dots, n,
$$

$$
k_{2,i} = h\phi_i\left(\lambda_j + \frac{h}{2}, w_{1,j} + \frac{1}{2}k_{1,1}, w_{2,j} + \frac{1}{2}k_{1,2}, \dots, w_{n,j} + \frac{1}{2}k_{1,n}\right),
$$

$$
\text{for each } i = 1, 2, \dots, n,
$$

$$
k_{3,i} = h\phi_i\left(\lambda_j + \frac{h}{2}, w_{1,j} + \frac{1}{2}k_{2,1}, w_{2,j} + \frac{1}{2}k_{2,2}, \dots, w_{n,j} + \frac{1}{2}k_{2,n}\right),
$$

$$
\text{for each } i = 1, 2, \dots, n,
$$

$$
k_{4,i} = h\phi_i(\lambda_j + h, w_{1,j} + k_{3,1}, w_{2,j} + k_{3,2}, \dots, w_{n,j} + k_{3,n}), \quad \text{for each } i = 1, 2, \dots, n,
$$

and

$$w_{i,j+1} = w_{i,j} + \frac{1}{6}(k_{1,i} + 2k_{2,i} + 2k_{3,i} + k_{4,i}), \quad \text{for each } i = 1, 2, \ldots, n.$$

We use the vector notation

$$\mathbf{k}_1 = \begin{bmatrix} k_{1,1} \\ k_{1,2} \\ \vdots \\ k_{1,n} \end{bmatrix}, \quad \mathbf{k}_2 = \begin{bmatrix} k_{2,1} \\ k_{2,2} \\ \vdots \\ k_{2,n} \end{bmatrix}, \quad \mathbf{k}_3 = \begin{bmatrix} k_{3,1} \\ k_{3,2} \\ \vdots \\ k_{3,n} \end{bmatrix}, \quad \mathbf{k}_4 = \begin{bmatrix} k_{4,1} \\ k_{4,2} \\ \vdots \\ k_{4,n} \end{bmatrix}, \quad \text{and} \quad \mathbf{w}_j = \begin{bmatrix} w_{1,j} \\ w_{2,j} \\ \vdots \\ w_{n,j} \end{bmatrix}$$

to simplify the presentation. Eq. (10.5) implies that $\mathbf{x}(0) = \mathbf{x}(\lambda_0) = \mathbf{w}_0$, and for each $j = 0, 1, \ldots, N$,

$$\mathbf{k}_1 = h \begin{bmatrix} \phi_1(\lambda_j, \mathbf{w}_j) \\ \phi_2(\lambda_j, \mathbf{w}_j) \\ \vdots \\ \phi_n(\lambda_j, \mathbf{w}_j) \end{bmatrix} = h(-J(\mathbf{w}_j))^{-1}\mathbf{F}(\mathbf{x}(0)),$$

$$\mathbf{k}_2 = h\left(-J\left(\mathbf{w}_j + \frac{1}{2}\mathbf{k}_1\right)\right)^{-1}\mathbf{F}(\mathbf{x}(0)),$$

$$\mathbf{k}_3 = h\left(-J\left(\mathbf{w}_j + \frac{1}{2}\mathbf{k}_2\right)\right)^{-1}\mathbf{F}(\mathbf{x}(0)),$$

$$\mathbf{k}_4 = h(-J(\mathbf{w}_j + \mathbf{k}_3))^{-1}\mathbf{F}(\mathbf{x}(0)),$$

and

The program CONTM104 implements the Continuation method.

$$\mathbf{x}(\lambda_{j+1}) = \mathbf{x}(\lambda_j) + \frac{1}{6}(\mathbf{k}_1 + 2\mathbf{k}_2 + 2\mathbf{k}_3 + \mathbf{k}_4) = \mathbf{w}_j + \frac{1}{6}(\mathbf{k}_1 + 2\mathbf{k}_2 + 2\mathbf{k}_3 + \mathbf{k}_4).$$

Then $\mathbf{x}(\lambda_n) = \mathbf{x}(1)$ is our approximation to $\mathbf{p}$.

Example 1   Use the Continuation method with $\mathbf{x}(0) = (0, 0, 0)^t$ to approximate the solution to

$$f_1(x_1, x_2, x_3) = 3x_1 - \cos(x_2 x_3) - 0.5 = 0,$$

$$f_2(x_1, x_2, x_3) = x_1^2 - 81(x_2 + 0.1)^2 + \sin x_3 + 1.06 = 0,$$

$$f_3(x_1, x_2, x_3) = e^{-x_1 x_2} + 20x_3 + \frac{10\pi - 3}{3} = 0.$$

Solution   The Jacobian matrix is

$$J(\mathbf{x}) = \begin{bmatrix} 3 & x_3 \sin x_2 x_3 & x_2 \sin x_2 x_3 \\ 2x_1 & -162(x_2 + 0.1) & \cos x_3 \\ -x_2 e^{-x_1 x_2} & -x_1 e^{-x_1 x_2} & 20 \end{bmatrix}$$

and

$$\mathbf{F}(\mathbf{x}(0)) = (-1.5, 0.25, 10\pi/3)^t.$$

With $N = 4$ and $h = 0.25$, we have

$$\mathbf{k}_1 = h(-J(\mathbf{x}^{(0)}))^{-1}F(\mathbf{x}(0)) = 0.25 \begin{bmatrix} 3 & 0 & 0 \\ 0 & -16.2 & 1 \\ 0 & 0 & 20 \end{bmatrix}^{-1} \begin{bmatrix} -1.5 \\ 0.25 \\ 10\pi/3 \end{bmatrix}$$

$$= (0.125, -0.004222203325, -0.1308996939)^t;$$

$$\mathbf{k}_2 = h\left(-J(0.0625, -0.002111101663, -0.06544984695)\right)^{-1}(-1.5, 0.25, 10\pi/3)^t$$

$$= 0.25 \begin{bmatrix} 3 & -0.9043289149 \times 10^{-5} & -0.2916936196 \times 10^{-6} \\ 0.125 & -15.85800153 & 0.9978589232 \\ 0.002111380229 & -0.06250824706 & 20 \end{bmatrix}^{-1} \begin{bmatrix} -1.5 \\ 0.25 \\ 10\pi/3 \end{bmatrix}$$

$$= (0.1249999773, -0.003311761993, -0.1309232406)^t;$$

$$\mathbf{k}_3 = h\left(-J(0.06249998865, -0.001655880997, -0.0654616203)\right)^{-1}(-1.5, 0.25, 10\pi/3)^t$$

$$= (0.1249999844, -0.003296244825, -0.130920346)^t;$$

$$\mathbf{k}_4 = h\left(-J(0.1249999844, -0.003296244825, -0.130920346)\right)^{-1}(-1.5, 0.25, 10\pi/3)^t$$

$$= (0.1249998945, -0.00230206762, -0.1309346977)^t;$$

and

$$\mathbf{x}(\lambda_1) = \mathbf{w}_1 = \mathbf{w}_0 + \frac{1}{6}(\mathbf{k}_1 + 2\mathbf{k}_2 + 2\mathbf{k}_3 + \mathbf{k}_4)$$

$$= (0.1249999697, -0.00329004743, -0.1309202608)^t.$$

Continuing, we have

$$\mathbf{x}(\lambda_2) = \mathbf{w}_2 = (0.2499997679, -0.004507400128, -0.2618557619)^t,$$

$$\mathbf{x}(\lambda_3) = \mathbf{w}_3 = (0.3749996956, -0.003430352103, -0.3927634423)^t,$$

and

$$\mathbf{x}(\lambda_4) = \mathbf{x}(1) = \mathbf{w}_4 = (0.4999999954, 0.126782 \times 10^{-7}, -0.5235987758)^t.$$

These results are very accurate because the actual solution is $(0.5, 0, -0.52359877)^t$. ∎

Note that in the Runge-Kutta methods, the steps similar to

$$\mathbf{k}_i = h\left(-J(\mathbf{x}(\lambda_i) + \alpha_{i-1}\mathbf{k}_{i-1})\right)^{-1}\mathbf{F}(\mathbf{x}(0))$$

can be written as solving the linear system

$$J\left(\mathbf{x}(\lambda_i) + \alpha_{i-1}\mathbf{k}_{i-1}\right)\mathbf{k}_i = -h\mathbf{F}(\mathbf{x}(0))$$

for $\mathbf{k}_i$. This eliminates the need to determine the inverse of $-J(\mathbf{x}(\lambda_i) + \alpha_{i-1}\mathbf{k}_{i-1})$.

In the Runge-Kutta method of order 4, the calculation of each $\mathbf{w}_j$ requires solving four linear systems, one each when computing $\mathbf{k}_1$, $\mathbf{k}_2$, $\mathbf{k}_3$, and $\mathbf{k}_4$. So using N steps requires solving $4N$ linear systems. By comparison, Newton's method requires one matrix inversion per iteration. Therefore, the work involved for the Runge-Kutta method is roughly equivalent to $4N$ iterations of Newton's method.

An alternative in the continuation method is to use the modified Euler method or even Euler's method to decrease the number of linear systems that need to be solved. Another possibility is to use smaller values of N. The following Illustration shows these ideas.

Illustration   Table 10.4 summarizes a comparison of Euler's method, the Midpoint method, and the Runge-Kutta method of order 4 applied to the system

$$f_1(x_1, x_2, x_3) = 3x_1 - \cos(x_2 x_3) - 0.5 = 0,$$

$$f_2(x_1, x_2, x_3) = x_1^2 - 81(x_2 + 0.1)^2 + \sin x_3 + 1.06 = 0,$$

$$f_3(x_1, x_2, x_3) = e^{-x_1 x_2} + 20x_3 + \frac{10\pi - 3}{3} = 0,$$

with initial approximation $\mathbf{x}(0) = (0, 0, 0)^t$. The right-hand column in the table lists the number of linear systems that are required for the solution.

The most accurate approximation is obtained using the Runge-Kutta method with $N = 4$, which requires solving 16 linear systems. However, the Runge-Kutta method with $N = 1$ requires solving only 4 linear systems and has an error of

$$\|\mathbf{x}(1) - \mathbf{p}\|_\infty \le 1.2 \times 10^{-6},$$

which is very accurate for the computation involved.   □

Table 10.4

Method	N	$\mathbf{x}(1)$	Systems
Euler	1	$(0.5, -0.0168888133, -0.5235987755)^t$	1
Euler	4	$(0.499999379, -0.004309160698, -0.523679652)^t$	4
Midpoint	1	$(0.4999966628, -0.00040240435, -0.523815371)^t$	2
Midpoint	4	$(0.500000066, -0.00001760089, -0.5236127761)^t$	8
Runge-Kutta	1	$(0.4999989843, -0.1676151 \times 10^{-5}, -0.5235989561)^t$	4
Runge-Kutta	4	$(0.4999999954, 0.126782 \times 10^{-7}, -0.5235987758)^t$	16

The continuation method can be used as a stand-alone method not requiring a particularly good choice of $\mathbf{x}(0)$. However, the method can also be used to give an initial approximation for Newton's or Broyden's method. For example, the result obtained in the Illustration using Euler's method and $N = 1$ might be sufficient to start the more efficient Newton's or Broyden's methods and be better for this purpose than proceeding with the Continuation method, which requires more calculation.

EXERCISE SET 10.5

1. The nonlinear system

$$f_1(x_1, x_2) = x_1^2 - x_2^2 + 2x_2 = 0,$$

$$f_2(x_1, x_2) = 2x_1 + x_2^2 - 6 = 0$$

has two solutions, $(0.625204094, 2.179355825)^t$ and $(2.109511920, -1.334532188)^t$. Use the Continuation method and Euler's method with $N = 2$ to approximate the solutions when
   a. $\mathbf{x}(0) = (0, 0)^t$ b. $\mathbf{x}(0) = (1, 1)^t$ c. $\mathbf{x}(0) = (3, -2)^t$

2. Repeat Exercise 1 using the Runge-Kutta method of order 4 with $N = 1$.

3. Use the Continuation method and Euler's method with $N = 2$ on the following linear systems.

a. $4x_1^2 - 20x_1 + \frac{1}{4}x_2^2 + 8 = 0,$

$\frac{1}{2}x_1x_2^2 + 2x_1 - 5x_2 + 8 = 0.$

b. $\sin(4\pi x_1 x_2) - 2x_2 - x_1 = 0,$

$\left(\frac{4\pi - 1}{4\pi}\right)(e^{2x_1} - e) + 4ex_2^2 - 2ex_1 = 0.$

c. $3x_1 - \cos(x_2 x_3) - \frac{1}{2} = 0,$

$4x_1^2 - 625x_2^2 + 2x_2 - 1 = 0,$

$e^{-x_1 x_2} + 20x_3 + \frac{10\pi - 3}{3} = 0.$

d. $x_1^2 + x_2 - 37 = 0,$

$x_1 - x_2^2 - 5 = 0,$

$x_1 + x_2 + x_3 - 3 = 0.$

4. Use the Continuation method and the Runge-Kutta method of order 4 with $N = 1$ and $\mathbf{x}(0) = \mathbf{0}$ on the following linear systems.

a. $x_1(1 - x_1) + 4x_2 = 12,$

$(x_1 - 2)^2 + (2x_2 - 3)^2 = 25.$

b. $5x_1^2 - x_2^2 = 0,$

$x_2 - 0.25(\sin x_1 + \cos x_2) = 0.$

c. $15x_1 + x_2^2 - 4x_3 = 13,$

$x_1^2 + 10x_2 - x_3 = 11,$

$x_2^3 - 25x_3 = -22.$

d. $10x_1 - 2x_2^2 + x_2 - 2x_3 - 5 = 0,$

$8x_2^2 + 4x_3^2 - 9 = 0,$

$8x_2x_3 + 4 = 0.$

5. Use the Continuation method and the Runge-Kutta method of order 4 with $N = 1$ on Exercise 3 of Section 10.2. Compare the results with those obtained in that exercise.

6. Repeat Exercise 4 with $N = 2$.

7. Repeat Exercise 7 of Section 10.2 using the Continuation method and the Runge-Kutta method of order 4 with $N = 2$

8. Repeat Exercise 8 of Section 10.2 using the Continuation method and the Runge-Kutta method of order 4 with $N = 1$

10.6 Survey of Methods and Software

In this chapter we considered methods to approximate solutions to nonlinear systems

$$f_1(x_1, x_2, \ldots, x_n) = 0,$$

$$f_2(x_1, x_2, \ldots, x_n) = 0,$$

$$\vdots$$

$$f_n(x_1, x_2, \ldots, x_n) = 0.$$

Newton's method for systems requires a good initial approximation $(p_1^{(0)}, p_2^{(0)}, \ldots, p_n^{(0)})^t$ and generates a sequence

$$\mathbf{p}^{(k)} = \mathbf{p}^{(k-1)} - \left[J(\mathbf{p}^{(k-1)})\right]^{-1} \mathbf{F}(\mathbf{p}^{(k-1)}),$$

that converges rapidly to a solution $\mathbf{p}$ if $\mathbf{p}^{(0)}$ is sufficiently close to $\mathbf{p}$. However, Newton's method requires evaluating, or approximating, n^2 partial derivatives and solving an n by n linear system at each step, which requires $O(n^3)$ computations.

Broyden's method reduces the amount of computation at each step without significantly degrading the speed of convergence. This technique replaces the Jacobian matrix J with a matrix A_{k-1} whose inverse is directly determined at each step. This reduces the arithmetic computations from $O(n^3)$ to $O(n^2)$. Moreover, the only scalar function evaluations required are in evaluating the f_i, saving n^2 scalar function evaluations per step. Broyden's method also requires a good initial approximation, and it does not generally converge as rapidly as Newton's method.

The Steepest Descent method was presented as a way to obtain good initial approximations for Newton's and Broyden's methods. Although the Steepest Descent method does not give a rapidly-convergent sequence, it does not require a good initial approximation. The Steepest Descent method approximates a minimum of a multivariable function g. For our application, we choose

$$g(x_1, x_2, \dots, x_n) = \sum_{i=1}^{n} [f_i(x_1, x_2, \dots, x_n)]^2.$$

The minimum value of g is 0, which occurs when the functions f_i are simultaneously 0.

Homotopy and continuation methods are also used for nonlinear systems (see [AG]). In these methods, a given problem

$$\mathbf{F}(\mathbf{x}) = \mathbf{0}$$

is embedded in a one-parameter family of problems using a parameter λ that assumes values in $[0, 1]$. The original problem corresponds to $\lambda = 1$, and a problem with a known solution corresponds to $\lambda = 0$. For example, the set of problems

$$G(\lambda, \mathbf{x}) = \lambda \mathbf{F}(\mathbf{x}) + (1 - \lambda)(\mathbf{F}(\mathbf{x}) - \mathbf{F}(\mathbf{x}(0))) = \mathbf{0}, \quad \text{for } 0 \le \lambda \le 1,$$

with fixed $\mathbf{x}(0) \in \mathbb{R}^n$ forms a homotopy. When $\lambda = 0$, the solution is $\mathbf{x}(\lambda = 0) = \mathbf{x}(0)$. The solution, $\mathbf{p}$, to the original problem corresponds to $\mathbf{x}(\lambda = 1)$. A continuation method determines $\mathbf{x}(\lambda = 1) \approx \mathbf{p}$ by solving the sequence of problems corresponding to $\lambda_0 = 0 < \lambda_1 < \lambda_2 < \cdots < \lambda_m = 1$. The initial approximation to the solution of

$$\lambda_i \mathbf{F}(\mathbf{x}) + (1 - \lambda_i)(\mathbf{F}(\mathbf{x}) - \mathbf{F}(\mathbf{x_0})) = \mathbf{0}$$

is the solution, $\mathbf{x}(\lambda = \lambda_{i-1})$, to the problem

$$\lambda_{i-1} \mathbf{F}(\mathbf{x}) + (1 - \lambda_{i-1})(\mathbf{F}(\mathbf{x}) - \mathbf{F}(\mathbf{x_0})) = \mathbf{0}.$$

The package Hompack in the netlib Library solves a system of nonlinear equations by using various homotopy methods.

The methods in the IMSL and NAG Libraries are based on two subroutines contained in MINPACK, a public-domain package. Both methods use the Levenberg-Marquardt method, which is a weighted average of Newton's method and the Steepest Descent method. The weight is biased toward the Steepest Descent method until convergence is detected, at which time the weight is shifted toward the more rapidly convergent Newton's method. One subroutine uses a finite-difference approximation to the Jacobian, and the other requires a user-supplied subroutine to compute the Jacobian.

A comprehensive treatment of methods for solving nonlinear systems of equations can be found in Ortega and Rheinboldt [OR] and in Dennis and Schnabel [DenS]. Recent developments on iterative methods can be found in Argyros and Szidarovszky [AS], and information on the use of continuation methods is available in Allgower and Georg [AG].

CHAPTER

11

Boundary-Value Problems for Ordinary Differential Equations

11.1 Introduction

The differential equations in Chapter 5 are of first order and have one initial condition to satisfy. Later in that chapter we saw that the techniques could be extended to systems of equations and then to higher-order equations, but all the specified conditions were on the same endpoint. These are initial-value problems.

In this chapter we approximate the solution to **two-point boundary-value** problems, which are second-order differential equations with conditions imposed at two different points. The differential equations whose solutions we will approximate are of the form

$$y'' = f(x, y, y'), \qquad \text{for } a \leq x \leq b,$$

with the boundary conditions on the solution prescribed by

$$y(a) = \alpha \quad \text{and} \quad y(b) = \beta,$$

for some constants α and β. Such a problem has a unique solution provided that:

- The function f and its partial derivatives with respect to y and y' are continuous;

- The partial derivative of f with respect to y is positive; and

- The partial derivative of f with respect to y' is bounded.

These are all reasonable conditions for boundary-value problems representing physical problems.

11.2 The Linear Shooting Method

A boundary-value problem is **linear** when the function f has the form

$$f(x, y, y') = p(x)y' + q(x)y + r(x).$$

Linear problems occur frequently in applications and are much easier to solve than nonlinear problems. This is because adding any solution to the **nonhomogeneous** differential equation

$$y'' - p(x)y' - q(x)y = r(x)$$

A linear equation involves only linear powers of y and its derivatives.

to the complete solution of the **homogeneous** differential equation

$$y'' - p(x)y' - q(x)y = 0$$

gives all the solutions to the nonhomogeneous problem. The solutions of the homogeneous problem are easier to determine than are those of the nonhomogeneous problem. Moreover, to show that a linear problem has a unique solution, we need only show that p, q, and r are continuous and that the values of q are positive.

Linear Boundary-Value Problems

To approximate the unique solution to the linear boundary-value problem, first consider the two initial-value problems

$$y'' = p(x)y' + q(x)y + r(x), \quad \text{for } a \le x \le b, \quad \text{where } y(a) = \alpha \text{ and } y'(a) = 0, \quad (11.1)$$

and

$$y'' = p(x)y' + q(x)y, \quad \text{for } a \le x \le b, \quad \text{where } y(a) = 0 \text{ and } y'(a) = 1, \quad (11.2)$$

both of which are assumed to have unique solutions. Let $y_1(x)$ denote the solution to Eq. (11.1), $y_2(x)$ denote the solution to Eq. (11.2), and assume that $y(b) \ne 0$. (The situation when $y_2(b) = 0$ is considered in Exercise 8.) Then,

$$y(x) = y_1(x) + \frac{\beta - y_1(b)}{y_2(b)} y_2(x) \quad (11.3)$$

is the unique solution to the linear boundary-value problem

$$y'' = p(x)y' + q(x)y + r(x), \quad \text{for } a \le x \le b, \quad \text{with } y(a) = \alpha \text{ and } y(b) = \beta. \quad (11.4)$$

To verify this, first note that

$$y'' - p(x)y' - q(x)y = y_1'' - p(x)y_1' - q(x)y_1 + \frac{\beta - y_1(b)}{y_2(b)} [y_2'' - p(x)y_2' - q(x)y_2]$$

$$= r(x) + \frac{\beta - y_1(b)}{y_2(b)} \cdot 0 = r(x).$$

Moreover

$$y(a) = y_1(a) + \frac{\beta - y_1(b)}{y_2(b)} y_2(a) = y_1(a) + \frac{\beta - y_1(b)}{y_2(b)} \cdot 0 = \alpha$$

and

$$y(b) = y_1(b) + \frac{\beta - y_1(b)}{y_2(b)} y_2(b) = y_1(b) + \beta - y_1(b) = \beta.$$

Linear Shooting

This "shooting" method hits the target after one trial shot. In the next section we see that nonlinear problems require multiple shots.

The Linear Shooting method is based on the replacement of the boundary-value problem by the two initial-value problems, (11.1) and (11.2). Numerous methods are available from Chapter 5 for approximating the solutions $y_1(x)$ and $y_2(x)$, and once these approximations are available, the solution to the boundary-value problem is approximated using the weighted sum in Eq. (11.3). Graphically, the method has the appearance shown in Figure 11.1.

Figure 11.1

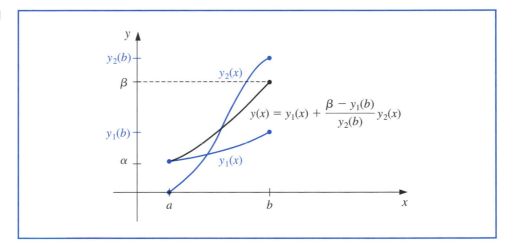

Program LINST111 incorporates the Runge-Kutta method of order 4 to find the approximations to $y_1(x)$ and $y_2(x)$, but any technique for approximating the solutions to initial-value problems can be substituted. The program has the additional feature of obtaining approximations for the derivative of the solution to the boundary-value problem in addition to the solution of the problem itself.

Example 1   Apply the Linear Shooting technique with $N = 10$ to the boundary-value problem

$$y'' = -\frac{2}{x}y' + \frac{2}{x^2}y + \frac{\sin(\ln x)}{x^2}, \quad \text{for } 1 \le x \le 2, \text{ with } y(1) = 1 \text{ and } y(2) = 2,$$

and compare the results to those of the exact solution

$$y = c_1 x + \frac{c_2}{x^2} - \frac{3}{10}\sin(\ln x) - \frac{1}{10}\cos(\ln x),$$

where

$$c_2 = \frac{1}{70}[8 - 12\sin(\ln 2) - 4\cos(\ln 2)] \approx -0.03920701320$$

and

$$c_1 = \frac{11}{10} - c_2 \approx 1.1392070132.$$

Solution   Applying the Linear Shooting technique to this problem requires approximating the solutions to the initial-value problems

$$y_1'' = -\frac{2}{x}y_1' + \frac{2}{x^2}y_1 + \frac{\sin(\ln x)}{x^2}, \quad \text{for } 1 \le x \le 2, \text{ with } y_1(1) = 1 \text{ and } y_1'(1) = 0,$$

and

$$y_2'' = -\frac{2}{x}y_2' + \frac{2}{x^2}y_2, \quad \text{for } 1 \le x \le 2, \text{ with } y_2(1) = 0 \text{ and } y_2'(1) = 1.$$

The first second-order differential equation is written as a system

$$u_1'(x) = u_2(x)$$

$$u_2'(x) = -\frac{2}{x}u_2(x) + \frac{2}{x^2}u_1(x) + \frac{\sin(\ln x)}{x^2}$$

with $u_1(1) = 1$ and $u_2(1) = 0$.

The second second-order differential equation is written as a system

$$v_1'(x) = v_2(x)$$

$$v_2'(x) = -\frac{2}{x}v_2(x) + \frac{2}{x^2}v_1(x)$$

with $v_1(1) = 0$ and $v_2(1) = 1$.

The results of the calculations, using the program LINST111 with $N = 10$ and $h = 0.1$, are given in Table 11.1. The value listed as $u_{1,i}$ approximates $y_1(x_i)$, the value $v_{1,i}$ approximates $y_2(x_i)$, and w_i approximates

$$y(x_i) = y_1(x_i) + \frac{2 - y_1(2)}{y_2(2)}y_2(x_i).$$

■

The accurate results in this example are due to the fact that the Runge-Kutta method of order 4 gives $O(h^4)$ approximations to the solutions of the initial-value problems.

Table 11.1

x_i	$u_{1,i} \approx y_1(x_i)$	$v_{1,i} \approx y_2(x_i)$	$w_i \approx y(x_i)$	$y(x_i)$	$|y(x_i) - w_i|$
1.0	1.00000000	0.00000000	1.00000000	1.00000000	
1.1	1.00896058	0.09117986	1.09262917	1.09262930	1.43×10^{-7}
1.2	1.03245472	0.16851175	1.18708471	1.18708484	1.34×10^{-7}
1.3	1.06674375	0.23608704	1.28338227	1.28338236	9.78×10^{-8}
1.4	1.10928795	0.29659067	1.38144589	1.38144595	6.02×10^{-8}
1.5	1.15830000	0.35184379	1.48115939	1.48115942	3.06×10^{-8}
1.6	1.21248372	0.40311695	1.58239245	1.58239246	1.08×10^{-8}
1.7	1.27087454	0.45131840	1.68501396	1.68501396	5.43×10^{-10}
1.8	1.33273851	0.49711137	1.78889854	1.78889853	5.05×10^{-9}
1.9	1.39750618	0.54098928	1.89392951	1.89392951	4.41×10^{-9}
2.0	1.46472815	0.58332538	2.00000000	2.00000000	

Reducing Round-Off Error

Unfortunately, there can be round-off error problems hidden in the Linear-Shooting method. If $y_1(x)$ rapidly increases as x goes from a to b, then $u_{1,N} \approx y_1(b)$ will be large. Should β be small in magnitude compared to $u_{1,N}$, the term $(\beta - u_{1,N})/v_{1,N}$ will be approximately $-u_{1,N}/v_{1,N}$. So the approximations

$$y(x_i) \approx w_i = u_{1,i} - \left(\frac{\beta - u_{1,N}}{v_{1,N}}\right)v_{1,i}, \approx u_{1,i} - \left(\frac{u_{1,N}}{v_{1,N}}\right)v_{1,i}$$

allow the possibility of a loss of accuracy due to cancelation. However, since $u_{1,i}$ is an approximation to $y_1(x_i)$, the behavior of y_1 can be easily monitored, and if $u_{1,i}$ increases rapidly from a to b, the shooting technique can be employed in the other direction—that is, solving instead the initial-value problems

$$y'' = p(x)y' + q(x)y + r(x), \quad \text{for } a \le x \le b, \quad \text{where} \quad y(b) = \beta \quad \text{and} \quad y'(b) = 0,$$

and

$$y'' = p(x)y' + q(x)y, \quad \text{for } a \le x \le b, \quad \text{where} \quad y(b) = 0 \quad \text{and} \quad y'(b) = 1.$$

If the reverse shooting technique still has round-off error problems, and if increased precision does not yield greater accuracy, other techniques must be employed. In general, however, if $u_{1,i}$ and $v_{1,i}$ are $O(h^n)$ approximations to $y_1(x_i)$ and $y_2(x_i)$, respectively, for each $i = 0, 1, \ldots, N$, then $w_{1,i}$ will be an $O(h^n)$ approximation to $y(x_i)$.

EXERCISE SET 11.2

1. The boundary-value problem

    $$y'' = 4(y - x), \qquad \text{for } 0 \le x \le 1 \quad \text{with } y(0) = 0 \text{ and } y(1) = 2$$

    has the solution $y(x) = e^2(e^4 - 1)^{-1}(e^{2x} - e^{-2x}) + x$. Use the Linear Shooting method to approximate the solution and compare the results to the exact solution.

    a.   With $h = \frac{1}{2}$
    b.   With $h = \frac{1}{4}$

2. The boundary-value problem

    $$y'' = y' + 2y + \cos x, \qquad \text{for } 0 \le x \le \frac{\pi}{2} \quad \text{with } y(0) = -0.3 \text{ and } y\left(\frac{\pi}{2}\right) = -0.1$$

    has the solution $y(x) = -\frac{1}{10}(\sin x + 3\cos x)$. Use the Linear Shooting method to approximate the solution and compare the results to the exact solution.

    a.   With $h = \frac{\pi}{4}$
    b.   With $h = \frac{\pi}{8}$

3. Use the Linear Shooting method to approximate the solution to the following boundary-value problems.

    a. $y'' = -3y' + 2y + 2x + 3$,   for $0 \le x \le 1$ with $y(0) = 2$ and $y(1) = 1$; use $h = 0.1$.
    b. $y'' = -\frac{4}{x}y' + \frac{2}{x^2}y - \frac{2\ln x}{x^2}$,   for $1 \le x \le 2$ with $y(1) = -\frac{1}{2}$ and $y(2) = \ln 2$; use $h = 0.05$.
    c. $y'' = -(x + 1)y' + 2y + (1 - x^2)e^{-x}$,   for $0 \le x \le 1$ with $y(0) = -1$ and $y(1) = 0$; use $h = 0.1$.
    d. $y'' = \frac{y'}{x} + \frac{3}{x^2}y + \frac{\ln x}{x} - 1$,   for $1 \le x \le 2$ with $y(1) = y(2) = 0$; use $h = 0.1$.

4. Although $q(x) < 0$ in the following boundary-value problems, unique solutions exist and are given. Use the Linear Shooting method to approximate the solutions to the following problems and compare the results to the exact solutions.

    a. $y'' + y = 0$,   for $0 \le x \le \frac{\pi}{4}$ with $y(0) = 1$ and $y\left(\frac{\pi}{4}\right) = 1$; use $h = \frac{\pi}{20}$; actual solution $y(x) = \cos x + (\sqrt{2} - 1)\sin x$.
    b. $y'' + 4y = \cos x$,   for $0 \le x \le \frac{\pi}{4}$ with $y(0) = 0$ and $y\left(\frac{\pi}{4}\right) = 0$; use $h = \frac{\pi}{20}$; actual solution $y(x) = -\frac{1}{3}\cos 2x - \frac{\sqrt{2}}{6}\sin 2x + \frac{1}{3}\cos x$.
    c. $y'' = -\frac{4}{x}y' - \frac{2}{x^2}y + \frac{2}{x^2}\ln x$,   for $1 \le x \le 2$ with $y(1) = \frac{1}{2}$ and $y(2) = \ln 2$; use $h = 0.05$; actual solution $y(x) = \frac{4}{x} - \frac{2}{x^2} + \ln x - \frac{3}{2}$.
    d. $y'' = 2y' - y + xe^x - x$,   for $0 \le x \le 2$ with $y(0) = 0$ and $y(2) = -4$; use $h = 0.2$; actual solution $y(x) = \frac{1}{6}x^3 e^x - \frac{5}{3}xe^x + 2e^x - x - 2$.

5. Use the Linear Shooting method to approximate the exact solution $y = e^{-10x}$ to the boundary-value problem

    $$y'' = 100y, \qquad \text{for } 0 \le x \le 1 \quad \text{with} \quad y(0) = 1 \quad \text{and} \quad y(1) = e^{-10}.$$

    Use **(a)** $h = 0.1$ and **(b)** $h = 0.05$.

6. Write the second-order initial-value problems (11.1) and (11.2) as first-order systems, and derive the equations necessary to solve the systems using the Runge-Kutta method of order 4 for systems.

7. Let u represent the electrostatic potential between two concentric metal spheres of radii R_1 and R_2 $(R_1 < R_2)$, such that the potential of the inner sphere is kept constant at V_1 volts and the potential of the outer sphere is 0 volts. The potential in the region between the two spheres is governed by Laplace's equation, which, in this particular application, reduces to

$$\frac{d^2u}{dr^2} + \frac{2}{r}\frac{du}{dr} = 0, \qquad \text{for } R_1 \le r \le R_2 \quad \text{with} \quad u(R_1) = V_1 \quad \text{and} \quad u(R_2) = 0.$$

Suppose $R_1 = 2$ cm, $R_2 = 4$ cm, and $V_1 = 110$ volts.

a. Approximate $u(3)$ using the Linear Shooting method.

b. Compare the results of (a) with the actual potential $u(3)$, where

$$u(r) = \frac{V_1 R_1}{r}\left(\frac{R_2 - r}{R_2 - R_1}\right).$$

8. Show that if y_2 is the solution to $y'' = p(x)y' + q(x)y$ and $y_2(a) = y_2(b) = 0$, then $y_2 \equiv 0$.

9. Consider the boundary-value problem

$$y'' + y = 0, \qquad \text{for } 0 \le x \le b \quad \text{with} \quad y(0) = 0 \quad \text{and} \quad y(b) = B.$$

Find choices for b and B so that the boundary-value problem has

a. No solution

b. Exactly one solution

c. Infinitely many solutions

10. Explain what happens when you attempt to apply the instructions in Exercise 9 to the boundary-value problem

$$y'' - y = 0, \qquad \text{for } 0 \le x \le b \quad \text{with} \quad y(0) = 0 \quad \text{and} \quad y(b) = B.$$

11.3   Linear Finite Difference Methods

The Shooting method discussed in Section 11.2 can have round-off error difficulties. The methods we present in this section have better round-off error stability, but they generally require more computation to obtain a specified accuracy.

Methods involving finite differences for solving boundary-value problems replace each of the derivatives in the differential equation with an appropriate difference-quotient approximation of the type considered in Section 4.9. The particular difference quotient is chosen to maintain a specified order of accuracy.

Discrete Approximation

The finite-difference method for the linear second-order boundary-value problem,

$$y'' = p(x)y' + q(x)y + r(x), \qquad \text{for } a \le x \le b, \quad \text{where} \quad y(a) = \alpha \quad \text{and} \quad y(b) = \beta,$$

requires that difference-quotient approximations be used for approximating both y' and y''. First, we select an integer $N > 0$ and divide the interval $[a, b]$ into $(N + 1)$ equal subintervals whose endpoints are the mesh points $x_i = a + ih$, for $i = 0, 1, \ldots, N + 1$, where $h = (b - a)/(N + 1)$.

At the interior mesh points, x_i, for $i = 1, 2, \ldots, N$, the differential equation to be approximated is

$$y''(x_i) = p(x_i)y'(x_i) + q(x_i)y(x_i) + r(x_i). \tag{11.5}$$

Expanding $y(x)$ in a third-degree Taylor polynomial about x_i evaluated at x_{i+1} and x_{i-1}, we have, assuming that $y \in C^4[x_{i-1}, x_{i+1}]$,

$$y(x_{i+1}) = y(x_i + h) = y(x_i) + hy'(x_i) + \frac{h^2}{2}y''(x_i) + \frac{h^3}{6}y'''(x_i) + \frac{h^4}{24}y^{(4)}(\xi_i^+),$$

for some ξ_i^+ in (x_i, x_{i+1}), and

$$y(x_{i-1}) = y(x_i - h) = y(x_i) - hy'(x_i) + \frac{h^2}{2}y''(x_i) - \frac{h^3}{6}y'''(x_i) + \frac{h^4}{24}y^{(4)}(\xi_i^-),$$

for some ξ_i^- in (x_{i-1}, x_i). If these equations are added, we have

$$y(x_{i+1}) + y(x_{i-1}) = 2y(x_i) + h^2 y''(x_i) + \frac{h^4}{24}\left[y^{(4)}\left(\xi_i^+\right) + y^{(4)}\left(\xi_i^-\right)\right],$$

and a simple algebraic manipulation gives

$$y''(x_i) = \frac{1}{h^2}\left[y(x_{i+1}) - 2y(x_i) + y(x_{i-1})\right] - \frac{h^2}{24}\left[y^{(4)}\left(\xi_i^+\right) + y^{(4)}\left(\xi_i^-\right)\right].$$

The Intermediate Value Theorem can be used to simplify this even further.

Centered-Difference Formula for $y''(x_i)$

$$y''(x_i) = \frac{1}{h^2}\left[y(x_{i+1}) - 2y(x_i) + y(x_{i-1})\right] - \frac{h^2}{12}y^{(4)}(\xi_i),$$

for some ξ_i in (x_{i-1}, x_{i+1}).

A centered-difference formula for $y'(x_i)$ is obtained in a similar manner.

Centered-Difference Formula for $y'(x_i)$

$$y'(x_i) = \frac{1}{2h}[y(x_{i+1}) - y(x_{i-1})] - \frac{h^2}{6}y'''(\eta_i),$$

for some η_i in (x_{i-1}, x_{i+1}).

The use of these centered-difference formulas in Eq. (11.5) results in the equation

$$\frac{y(x_{i+1}) - 2y(x_i) + y(x_{i-1})}{h^2} = p(x_i)\left[\frac{y(x_{i+1}) - y(x_{i-1})}{2h}\right]$$

$$+ q(x_i)y(x_i) + r(x_i) - \frac{h^2}{12}[2p(x_i)y'''(\eta_i) - y^{(4)}(\xi_i)].$$

A Finite-Difference method with truncation error of order $O(h^2)$ results from using this equation together with the boundary conditions $y(a) = \alpha$ and $y(b) = \beta$ to define

$$w_0 = \alpha, \qquad w_{N+1} = \beta,$$

and

$$\left(\frac{2w_i - w_{i+1} - w_{i-1}}{h^2}\right) + p(x_i)\left(\frac{w_{i+1} - w_{i-1}}{2h}\right) + q(x_i)w_i = -r(x_i),$$

for each $i = 1, 2, \ldots, N$.

In the form we will consider, the equation is rewritten as

$$-\left(1 + \frac{h}{2}p(x_i)\right)w_{i-1} + (2 + h^2 q(x_i))w_i - \left(1 - \frac{h}{2}p(x_i)\right)w_{i+1} = -h^2 r(x_i).$$

The resulting system of equations is expressed in the tridiagonal $N \times N$ matrix form $A\mathbf{w} = \mathbf{b}$, where

$$A = \begin{bmatrix} 2 + h^2 q(x_1) & -1 + \frac{h}{2}p(x_1) & 0 & \cdots\cdots\cdots\cdots\cdots\cdots & 0 \\ -1 - \frac{h}{2}p(x_2) & 2 + h^2 q(x_2) & -1 + \frac{h}{2}p(x_2) & & \vdots \\ 0 & \ddots & \ddots & \ddots & 0 \\ & & & & -1 + \frac{h}{2}p(x_{N-1}) \\ 0 & \cdots\cdots\cdots\cdots\cdots & 0 & -1 - \frac{h}{2}p(x_N) & 2 + h^2 q(x_N) \end{bmatrix},$$

$$\mathbf{w} = \begin{bmatrix} w_1 \\ w_2 \\ \vdots \\ w_{N-1} \\ w_N \end{bmatrix}, \quad \text{and} \quad \mathbf{b} = \begin{bmatrix} -h^2 r(x_1) + \left(1 + \frac{h}{2}p(x_1)\right)w_0 \\ -h^2 r(x_2) \\ \vdots \\ -h^2 r(x_{N-1}) \\ -h^2 r(x_N) + \left(1 - \frac{h}{2}p(x_N)\right)w_{N+1} \end{bmatrix}.$$

The program LINFD112 implements the Linear Finite-Difference method.

This system has a unique solution provided that p, q, and r are continuous on $[a, b]$, that $q(x) \geq 0$ on $[a, b]$, and that $h < 2/L$, where $L = \max_{a \leq x \leq b} |p(x)|$.

Example 1 Use the Finite-Difference method with $N = 9$ to approximate the solution to the linear boundary-value problem

$$y'' = -\frac{2}{x}y' + \frac{2}{x^2}y + \frac{\sin(\ln x)}{x^2}, \quad \text{for } 1 \leq x \leq 2, \text{ with } y(1) = 1 \text{ and } y(2) = 2.$$

Compare the results to those obtained using the Shooting method in Example 1 of Section 11.2.

Solution Since $N = 9$, we have

$$h = \frac{b - a}{n + 1} = \frac{2 - 1}{10} = 0.1,$$

which is the same spacing as in Example 1 of Section 11.2.

To use MATLAB to apply the Linear Finite-Difference method, we first define the endpoints of the interval, the boundary conditions, N, and h.

```
a = 1;b = 2;alpha = 1;beta = 2;N = 9;h = (b-a)/(N+1);
```

The functions $p(x), q(x)$, and $r(x)$ are defined with

```
p = inline('-2/t','t');q=inline('2/t^2','t');
r = inline('sin(log(t))/t^2','t');
```

The semicolons at the ends of the statement lines prevent MATLAB from echoing the input. We then initialize the vector of x values

```
x = [1.1 1.2 1.3 1.4 1.5 1.6 1.7 1.8 1.9];
```

We also place all 0 entries in the 9×9 array A and the right side **b** of the linear system

```
A = zeros(9,9);
b = zeros(9,1);
```

These will be overwritten as we generate the nonzero entries with

```
A(1,1) = 2+h*h*q(x(1));
A(1,2) = -1+h*p(x(1))/2;
b(1) = -h*h*r(x(1))+(1+h*p(x(1))/2)*alpha;
for i = 2:N-1
    A(i,i-1) = -1-h*p(x(i))/2;
    A(i,i) = 2+h*h*q(x(i));
    A(i,i+1) = -1+h*p(x(i))/2;
    b(i) = -h*h*r(x(i));
end;
A(N,N-1) = -1-h*p(x(N))/2;
A(N,N) = 2+h*h*q(x(N));
b(N) = -h*h*r(x(N))+(1-h*p(x(N))/2)*beta;
```

We now apply the `linsolve` command to solve the 9×9 linear system for **w**

```
w = linsolve(A,b')
```

giving

$$w = 1.092600520720135$$
$$1.187043128795073$$
$$1.283336870214396$$
$$1.381402046228608$$
$$1.481120262112219$$
$$1.582359895693457$$
$$1.684989018384561$$
$$1.788881746193746$$
$$1.893921099157133$$

The complete results are listed in Table 11.2.

These results are considerably less accurate than those obtained in Example 1 of Section 11.2. This is because the method used in that example involved a Runge-Kutta technique with local truncation error of order $O(h^4)$, whereas the difference method used here has local truncation error of order $O(h^2)$. ∎

Table 11.2

x_i	w_i	$y(x_i)$	$|w_i - y(x_i)|$
1.0	1.00000000	1.00000000	
1.1	1.09260052	1.09262930	2.88×10^{-5}
1.2	1.18704313	1.18708484	4.17×10^{-5}
1.3	1.28333687	1.28338236	4.55×10^{-5}
1.4	1.38140205	1.38144595	4.39×10^{-5}
1.5	1.48112026	1.48115942	3.92×10^{-5}
1.6	1.58235990	1.58239246	3.26×10^{-5}
1.7	1.68498902	1.68501396	2.49×10^{-5}
1.8	1.78888175	1.78889853	1.68×10^{-5}
1.9	1.89392110	1.89392951	8.41×10^{-6}
2.0	2.00000000	2.00000000	

To obtain a difference method with greater accuracy, we can proceed in a number of ways. Using fifth-order Taylor series for approximating $y''(x_i)$ and $y'(x_i)$ results in an error term involving h^4. However, this requires using multiples not only of $y(x_{i+1})$ and $y(x_{i-1})$, but also $y(x_{i+2})$ and $y(x_{i-2})$ in the approximation formulas for $y''(x_i)$ and $y'(x_i)$. This leads to difficulty at $i = 0$ and $i = N$. Moreover, the resulting system of equations is not in tridiagonal form, and the solution to the system requires many more calculations.

Employing Richardson's Extrapolation

Instead of obtaining a difference method with a higher-order error term as described in the previous paragraph, it is generally more satisfactory to consider a reduction in step size. This has the added advantage that Richardson's extrapolation technique can be used effectively because the error term is expressed in even powers of h with coefficients independent of h, provided y is sufficiently differentiable.

Example 2   Apply the program LINFD112 and Richardson's extrapolation to approximate the solution to the boundary-value problem

$$y'' = -\frac{2}{x}y' + \frac{2}{x^2}y + \frac{\sin(\ln x)}{x^2}, \quad \text{for } 1 \le x \le 2, \text{ with } y(1) = 1 \text{ and } y(2) = 2,$$

using $h = 0.1, 0.05$, and 0.025.

Solution   The results are listed in Table 11.3. The first extrapolation is

$$\text{Ext}_{1i} = \frac{1}{3}(4w_i(h = 0.05) - w_i(h = 0.1));$$

the second extrapolation is

$$\text{Ext}_{2i} = \frac{1}{3}(4w_i(h = 0.025) - w_i(h = 0.05));$$

and the final extrapolation is

$$\text{Ext}_{3i} = \frac{1}{15}(16\text{Ext}_{2i} - \text{Ext}_{1i}).$$

Table 11.3

x_i	$w_i(h = 0.05)$	$w_i(h = 0.025)$	Ext_{1i}	Ext_{2i}	Ext_{3i}
1.0	1.00000000	1.00000000	1.00000000	1.00000000	1.00000000
1.1	1.09262207	1.09262749	1.09262925	1.09262930	1.09262930
1.2	1.18707436	1.18708222	1.18708477	1.18708484	1.18708484
1.3	1.28337094	1.28337950	1.28338230	1.28338236	1.28338236
1.4	1.38143493	1.38144319	1.38144589	1.38144595	1.38144595
1.5	1.48114959	1.48115696	1.48115937	1.48115941	1.48115942
1.6	1.58238429	1.58239042	1.58239242	1.58239246	1.58239246
1.7	1.68500770	1.68501240	1.68501393	1.68501396	1.68501396
1.8	1.78889432	1.78889748	1.78889852	1.78889853	1.78889853
1.9	1.89392740	1.89392898	1.89392950	1.89392951	1.89392951
2.0	2.00000000	2.00000000	2.00000000	2.00000000	2.00000000

The values of $w_i(h = 0.1)$ are omitted from the table to save space, but they are listed in Table 11.2. The results for $w_i(h = 0.025)$ are accurate to approximately 3×10^{-6}, and the results of Ext_{3i} are correct to the decimal places listed. In fact, if sufficient digits had been used, the Ext_{3i} approximations would agree with the exact solution with maximum error of 6.3×10^{-11} at the mesh points. This is an impressive improvement. ■

EXERCISE SET 11.3

1. The boundary-value problem

$$y'' = 4(y - x), \qquad \text{for } 0 \le x \le 1 \quad \text{with} \quad y(0) = 0 \quad \text{and} \quad y(1) = 2$$

has the solution $y(x) = e^2(e^4 - 1)^{-1}(e^{2x} - e^{-2x}) + x$. Use the Linear Finite-Difference method to approximate the solution and compare the results to the exact solution.

   a. With $h = \frac{1}{2}$
   b. With $h = \frac{1}{4}$
   c. Use extrapolation to approximate $y(1/2)$

2. The boundary-value problem

$$y'' = y' + 2y + \cos x, \text{ for } 0 \le x \le \frac{\pi}{2} \text{ with } y(0) = -0.3 \text{ and } y\left(\frac{\pi}{2}\right) = -0.1$$

has the solution $y(x) = -\frac{1}{10}(\sin x + 3 \cos x)$. Use the Linear Finite-Difference method to approximate the solution and compare the results to the exact solution.

   a. With $h = \frac{\pi}{4}$
   b. With $h = \frac{\pi}{8}$
   c. Use extrapolation to approximate $y(\pi/4)$

3. Use the Linear Finite-Difference method to approximate the solution to the following boundary-value problems.

   a. $y'' = -3y' + 2y + 2x + 3$, for $0 \le x \le 1$ with $y(0) = 2$ and $y(1) = 1$; use $h = 0.1$.

   b. $y'' = -\frac{4}{x}y' + \frac{2}{x^2}y - \frac{2}{x^2}\ln x$, for $1 \le x \le 2$ with $y(1) = -\frac{1}{2}$ and $y(2) = \ln 2$; use $h = 0.05$.

   c. $y'' = -(x + 1)y' + 2y + (1 - x^2)e^{-x}$, for $0 \le x \le 1$ with $y(0) = -1$ and $y(1) = 0$; use $h = 0.1$.

   d. $y'' = \frac{y'}{x} + \frac{3}{x^2}y + \frac{\ln x}{x} - 1$, for $1 \le x \le 2$ for $y(1) = y(2) = 0$; use $h = 0.1$.

4. Although $q(x) < 0$ in the following boundary-value problems, unique solutions exist and are given. Use the Linear Finite-Difference method to approximate the solutions and compare the results to the exact solutions.

a. $y'' + y = 0$,   for $0 \le x \le \frac{\pi}{4}$ with $y(0) = 1$ and $y\left(\frac{\pi}{4}\right) = 1$; use $h = \frac{\pi}{20}$; actual solution $y(x) = \cos x + (\sqrt{2} - 1) \sin x$.

b. $y'' + 4y = \cos x$,   for $0 \le x \le \frac{\pi}{4}$ with $y(0) = 0$ and $y\left(\frac{\pi}{4}\right) = 0$; use $h = \frac{\pi}{20}$; actual solution $y(x) = -\frac{1}{3}\cos 2x - \frac{\sqrt{2}}{6}\sin 2x + \frac{1}{3}\cos x$.

c. $y'' = -\frac{4}{x}y' - \frac{2}{x^2}y + \frac{2\ln x}{x^2}$,   for $1 \le x \le 2$ with $y(1) = \frac{1}{2}$ and $y(2) = \ln 2$; use $h = 0.05$; actual solution $y(x) = \frac{4}{x} - \frac{2}{x^2} + \ln x - \frac{3}{2}$.

d. $y'' = 2y' - y + xe^x - x$,   for $0 \le x \le 2$ with $y(0) = 0$ and $y(2) = -4$; use $h = 0.2$; actual solution $y(x) = \frac{1}{6}x^3 e^x - \frac{5}{3}xe^x + 2e^x - x - 2$.

5.   Use the Linear Finite-Difference method to approximate the exact solution $y = e^{-10x}$ to the boundary-value problem

$$y'' = 100y, \qquad \text{for } 0 \le x \le 1 \quad \text{with} \quad y(0) = 1 \quad \text{and} \quad y(1) = e^{-10}.$$

Use **(a)** $h = 0.1$ and **(b)** $h = 0.05$.

6.   Repeat Exercise 3(a) and (b) using the extrapolation discussed in Example 2.

7.   The deflection of a uniformly loaded, long rectangular plate under an axial tension force is governed by a second-order differential equation. Let S represent the axial force and q, the intensity of the uniform load. The deflection w along the elemental length is given by

$$w''(x) = \frac{S}{D}w(x) - \frac{ql}{2D}x + \frac{q}{2D}x^2, \qquad \text{for } 0 \le x \le l \quad \text{with} \quad w(0) = w(l) = 0,$$

where l is the length of the plate and D is the flexural rigidity of the plate. Let $q = 200$ lb/in.^{2}, $S = 100$ lb/in., $D = 8.8 \times 10^7$ lb/in., and $l = 50$ in. Approximate the deflection at 1-inch intervals.

8.   The boundary-value problem governing the deflection of a beam with supported ends subject to uniform loading is

$$w''(x) = \frac{S}{EI}w(x) + \frac{q}{2EI}x(x - l), \qquad \text{for } 0 < x < l \quad \text{with} \quad w(0) = 0 \quad \text{and} \quad w(l) = 0.$$

Suppose the beam is a W10-type steel I-beam with the following characteristics: length $l = 120$ in., intensity of uniform load $q = 100$ lb/ft, modulus of elasticity $E = 3.0 \times 10^7$ lb/in.^{2}, stress at ends $S = 1000$ lb, and central moment of inertia $I = 625$ in.^{4}.

a.   Approximate the deflection $w(x)$ of the beam every 6 inches.

b.   The actual relationship is given by

$$w(x) = c_1 e^{ax} + c_2 e^{-ax} + b(x - l)x + c,$$

where $c_1 = 7.7042537 \times 10^4$, $c_2 = 7.9207462 \times 10^4$, $a = 2.3094010 \times 10^{-4}$, $b = -4.1666666 \times 10^{-3}$, and $c = -1.5625 \times 10^5$. Is the maximum error on the interval within 0.2 in.?

c.   State law requires that $\max_{0 < x < l} w(x) < 1/300$. Does this beam meet state code?

11.4 The Nonlinear Shooting Method

The shooting technique for the nonlinear second-order boundary-value problem

$$y'' = f(x, y, y'), \qquad \text{for } a \le x \le b, \quad \text{where} \quad y(a) = \alpha \quad \text{and} \quad y(b) = \beta, \qquad (11.6)$$

is similar to the Linear Shooting method, except that the solution to a nonlinear problem cannot be expressed as a linear combination of the solutions to two initial-value problems. Instead, we approximate the solution to the boundary-value problem by using the solutions to a sequence of initial-value problems involving a parameter t. These problems have the form

$$y'' = f(x, y, y'), \qquad \text{for } a \le x \le b, \quad \text{where} \quad y(a) = \alpha \quad \text{and} \quad y'(a) = t. \qquad (11.7)$$

We do this by choosing the parameters $t = t_k$ in a manner to ensure that

$$\lim_{k \to \infty} y(b, t_k) = y(b) = \beta,$$

where $y(x, t_k)$ denotes the solution to the initial-value problem (11.7) with $t = t_k$ and $y(x)$ denotes the solution to the boundary-value problem (11.6).

This technique is called a *shooting* method, by analogy to the procedure of firing objects at a stationary target.

We start with a parameter t_0 that determines the initial elevation at which the object is fired from the point (a, α) along the curve described by the solution to the initial-value problem (see Figure 11.2):

$$y'' = f(x, y, y'), \quad \text{for } a \leq x \leq b, \quad \text{where} \quad y(a) = \alpha \quad \text{and} \quad y'(a) = t_0.$$

Figure 11.2

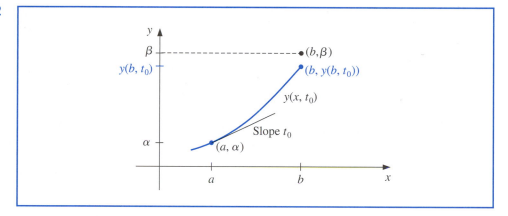

Shooting methods for nonlinear problems require iterations to approach the "target".

If $y(b, t_0)$ is not sufficiently close to β, we correct our approximation by choosing elevations t_1, t_2, and so on, until $y(b, t_k)$ is sufficiently close to "hitting" β. (See Figure 11.3.)

Figure 11.3

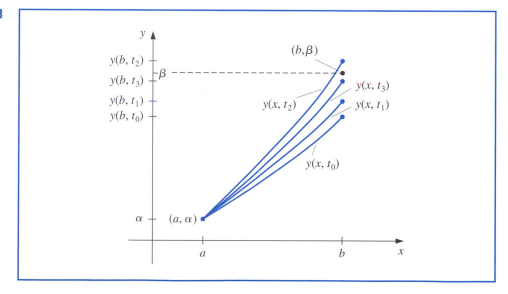

The problem is to determine the parameter t so that

$$y(b, t) - \beta = 0.$$

This is a nonlinear equation of the type considered in Chapter 2, and a number of methods are available.

To employ the *Secant method* to solve the problem, we choose initial approximations t_0 and t_1 to t, and then generate the remaining terms of the sequence by using the following procedure.

Secant Method Solution

Suppose that t_0 and t_1 are initial approximations to the parameter t that solves the nonlinear equation $y(b, t) - \beta = 0$. For each successive $k = 2, 3, \ldots$, solve the initial-value problem

$$y'' = f(x, y, y'), \qquad \text{for } a \leq x \leq b, \quad \text{where} \quad y(a) = \alpha,$$

with $y'(a) = t_{k-2}$ to find $y(b, t_{k-2})$ and with $y'(a) = t_{k-1}$ to find $y(b, t_{k-1})$. Define

$$t_k = t_{k-1} - \frac{(y(b, t_{k-1}) - \beta)(t_{k-1} - t_{k-2})}{y(b, t_{k-1}) - y(b, t_{k-2})}.$$

Then repeat the process with t_{k-1} replacing t_{k-2} and t_k replacing t_{k-1}.

Newton Iteration

To use the more powerful Newton's method to generate the sequence $\{t_k\}$, only one initial approximation, t_0, is needed. However, the iteration has the form

$$t_k = t_{k-1} - \frac{y(b, t_{k-1}) - \beta}{(dy/dt)(b, t_{k-1})},$$

and requires the knowledge of $(dy/dt)(b, t_{k-1})$. This presents a difficulty because an explicit representation for $y(b, t)$ is not known. We know only the values $y(b, t_0)$, $y(b, t_1), \ldots, y(b, t_{k-1})$.

To overcome this difficulty, first rewrite the initial-value problem, emphasizing that the solution depends on both x and t:

$$y''(x, t) = f(x, y(x, t), y'(x, t)), \quad \text{for } a \leq x \leq b, \text{ where } y(a, t) = \alpha \text{ and } y'(a, t) = t,$$

retaining the prime notation to indicate differentiation with respect to x. Since we are interested in determining $(dy/dt)(b, t)$ when $t = t_{k-1}$, we take the partial derivative with respect to t. This implies that

$$\frac{\partial y''}{\partial t}(x, t) = \frac{\partial f}{\partial t}(x, y(x, t), y'(x, t))$$

$$= \frac{\partial f}{\partial x}(x, y(x, t), y'(x, t))\frac{\partial x}{\partial t} + \frac{\partial f}{\partial y}(x, y(x, t), y'(x, t))\frac{\partial y}{\partial t}(x, t)$$

$$+ \frac{\partial f}{\partial y'}(x, y(x, t), y'(x, t))\frac{\partial y'}{\partial t}(x, t)$$

But x and t are independent, so $\frac{\partial x}{\partial t} = 0$, and the equation simplifies to

$$\frac{\partial y''}{\partial t}(x, t) = \frac{\partial f}{\partial y}(x, y(x, t), y'(x, t))\frac{\partial y}{\partial t}(x, t) + \frac{\partial f}{\partial y'}(x, y(x, t), y'(x, t))\frac{\partial y'}{\partial t}(x, t)$$

(11.8)

The initial conditions give

$$\frac{\partial y}{\partial t}(a, t) = 0 \quad \text{and} \quad \frac{\partial y'}{\partial t}(a, t) = 1.$$

If we simplify the notation by letting

$$z(x, t) = \frac{\partial y}{\partial t}(x, t)$$

and assume that the order of differentiation of x and t can be reversed, Eq. (11.8) becomes the linear initial-value problem

$$z''(x, t) = \frac{\partial f}{\partial y}(x, y, y')\, z(x, t) + \frac{\partial f}{\partial y'}(x, y, y')\, z'(x, t), \qquad \text{for } a \le x \le b,$$

where $z(a, t) = 0$ and $z'(a, t) = 1$. Newton's method therefore requires that two initial-value problems be solved for each iteration.

Newton's Method Solution

Suppose that t_0 is an initial approximation to the parameter t that solves the nonlinear equation $y(b, t) - \beta = 0$. For each successive $k = 1, 2, \ldots$, solve the initial-value problems

$$y'' = f(x, y, y'), \qquad \text{for } a \le x \le b, \quad \text{where} \quad y(a) = \alpha \quad \text{and} \quad y'(a) = t_{k-1}$$

and

$$z'' = f_y(x, y, y')z + f_{y'}(x, y, y')z', \quad \text{for } a \le x \le b, \quad \text{where}$$

$$z(a, t) = 0 \quad \text{and} \quad z'(a, t) = 1.$$

Then define

$$t_k = t_{k-1} - \frac{y(b, t_{k-1}) - \beta}{z(b, t_{k-1})}$$

and repeat the process with t_k replacing t_{k-1}.

The program NLINS113 implements the Nonlinear Shooting method.

In practice, none of these initial-value problems is likely to be solved exactly. The solutions are approximated by one of the methods discussed in Chapter 5. Program NLINS113 uses the Runge-Kutta method of order 4 to approximate both solutions required for Newton's method.

Example 1   Apply the Shooting method with Newton's method to the boundary-value problem

$$y'' = \frac{1}{8}(32 + 2x^3 - yy'), \quad \text{with } y(1) = 17 \text{ and } y(3) = \frac{43}{3}$$

for $1 \le x \le 3$. Use $N = 20$, $h = (3 - 1)/N = 0.2$, and let $x_i = a + ih = 1 + 0.2i$ for $i = 0, 1, \cdots, 20$. Let the maximum number of iterations be $M = 10$, and $TOL = 10^{-5}$. Compare the results with the exact solution $y(x) = x^2 + 16/x$.

Solution   We need to approximate solutions for $1 \le x \le 3$ to the initial-value problems

$$y'' = \frac{1}{8}(32 + 2x^3 - yy'), \quad \text{with } y(1) = 17 \text{ and } y'(1) = t_k,$$

and

$$z'' = \frac{\partial f}{\partial y}z + \frac{\partial f}{\partial y'}z' = -\frac{1}{8}(y'z + yz'), \quad \text{with } z(1) = 0 \text{ and } z'(1) = 1,$$

at each step in the iteration.

We write the first second-order equation as the system

$$u_1'(x) = u_2(x)$$

$$u_2'(x) = \frac{1}{8}(32 + 2x^3 - u_1(x)u_2(x))$$

with $u_1(1) = 17$ and $u_2(1) = t_k$.

The initial value of t_k is

$$t_0 = \frac{\beta - \alpha}{b - a} = \frac{43/3 - 17}{3 - 1} = -\frac{4}{3}.$$

For the next second-order equation we have

$$\frac{\partial f}{\partial y}(x, y, y') = -\frac{y'}{8} \quad \text{and} \quad \frac{\partial f}{\partial y'}(x, y, y') = -\frac{y}{8},$$

so our system of first-order differential equations becomes

$$v_1'(x) = v_2(x)$$

$$v_2'(x) = -\frac{y'}{8}v_1(x) - \frac{y}{8}v_2(x)$$

with $v_1(1) = 0$ and $u_2(1) = 1$.

We apply the Runge-Kutta method of order 4 to the first system obtaining

$$u_{1i} \approx u_1(x_i) \quad \text{for } i = 1, \ldots, 20$$

$$u_{2i} \approx u_2(x_i) = u_1'(x_i) \quad \text{for } i = 1, \ldots, 20.$$

To solve the second system using the Runge-Kutta method of order 4 requires the use of u_{1i} to approximate $y(x_i)$ and u_{2i} to approximate $y'(x_i)$ in the formulation of $k_{11}, k_{12}, k_{13}, k_{14}, k_{21}, k_{22}, k_{23}, k_{24}, v_{1i},$ and v_{2i}.

When

$$v_{1i} \approx v_1(x_i) \quad \text{for } i = 1, \ldots, 20$$

$$v_{2i} \approx v_2(x_i) \quad \text{for } i = 1, \ldots, 20$$

have been obtained, we test for convergence using

$$|y(b, t_0) - \beta| = \left| u_{1N} - \frac{43}{3} \right|.$$

If sufficient accuracy has not been obtained, we set

$$t_{k+1} = t_k - \frac{u_{1N} - \beta}{v_{1n}}$$

and repeat the Runge-Kutta process.

The stopping technique in NLINS113 requires

$$|w_{1,N}(t_k) - y(3)| \leq 10^{-5},$$

so we need 4 iterations and we find that $t_4 = -14.000203$.

The results obtained for this value of t are shown in Table 11.4. ∎

Table 11.4

x_i	$w_{1,i}$	$y(x_i)$	$|w_{1,i} - y(x_i)|$	x_i	$w_{1,i}$	$y(x_i)$	$|w_{1,i} - y(x_i)|$
1.1	15.755495	15.755455	4.06×10^{-5}	2.1	12.029066	12.029048	1.84×10^{-5}
1.2	14.773389	14.773333	5.60×10^{-5}	2.2	12.112741	12.112727	1.40×10^{-5}
1.3	13.997752	13.997692	5.94×10^{-5}	2.3	12.246532	12.246522	1.01×10^{-5}
1.4	13.388629	13.388571	5.71×10^{-5}	2.4	12.426673	12.426667	6.68×10^{-6}
1.5	12.916719	12.916667	5.23×10^{-5}	2.5	12.650004	12.650000	3.61×10^{-6}
1.6	12.560046	12.560000	4.64×10^{-5}	2.6	12.913847	12.913845	9.17×10^{-7}
1.7	12.301805	12.301765	4.02×10^{-5}	2.7	13.215924	13.215926	1.43×10^{-6}
1.8	12.128923	12.128889	3.14×10^{-5}	2.8	13.554282	13.554286	3.46×10^{-6}
1.9	12.031081	12.031053	2.84×10^{-5}	2.9	13.927236	13.927241	5.21×10^{-6}
2.0	12.000023	12.000000	2.32×10^{-5}	3.0	14.333327	14.333333	6.69×10^{-6}

Although Newton's method used with the shooting technique requires the solution of an additional initial-value problem, it will generally be faster than the Secant method. Both methods are only locally convergent because they require good initial approximations.

EXERCISE SET 11.4

1. Use the Nonlinear Shooting method with $h = 0.5$ to approximate the solution to the boundary-value problem

    $$y'' = -(y')^2 - y + \ln x, \qquad \text{for } 1 \leq x \leq 2 \quad \text{with} \quad y(1) = 0 \quad \text{and} \quad y(2) = \ln 2.$$

    Compare your results to the exact solution $y = \ln x$.

2. Use the Nonlinear Shooting method with $h = 0.25$ to approximate the solution to the boundary-value problem

    $$y'' = 2y^3, \qquad \text{for } -1 \leq x \leq 0 \quad \text{with } y(-1) = \frac{1}{2} \text{ and } y(0) = \frac{1}{3}.$$

    Compare your results to the exact solution $y(x) = 1/(x + 3)$.

3. Use the Nonlinear Shooting method to approximate the solution to the following boundary-value problems, iterating until $|w_{1,n} - \beta| \leq 10^{-4}$. The exact solution is given for comparison to your results.

    a. $y'' = y^3 - yy'$, for $1 \leq x \leq 2$ with $y(1) = \frac{1}{2}$ and $y(2) = \frac{1}{3}$; use $h = 0.1$ and compare the results to $y(x) = (x + 1)^{-1}$.

    b. $y'' = 2y^3 - 6y - 2x^3$, for $1 \leq x \leq 2$ with $y(1) = 2$ and $y(2) = \frac{5}{2}$; use $h = 0.1$ and compare the results to $y(x) = x + x^{-1}$.

c. $y'' = y' + 2(y - \ln x)^3 - x^{-1}$, for $2 \le x \le 3$ with $y(2) = \frac{1}{2} + \ln 2$ and $y(3) = \frac{1}{3} + \ln 3$; use $h = 0.1$ and compare the results to $y(x) = x^{-1} + \ln x$.

d. $y'' = \left[x^2(y')^2 - 9y^2 + 4x^6\right]/x^5$, for $1 \le x \le 2$ with $y(1) = 0$ and $y(2) = \ln 256$; use $h = 0.05$ and compare the results to $y(x) = x^3 \ln x$.

4. Use the Secant method with $t_0 = (\beta - \alpha)/(b - a)$ and $t_1 = t_0 + (\beta - y(b, t_0))/(b - a)$ to solve the problems in Exercises 3(a) and (c) and compare the number of iterations required with that of Newton's method.

5. The Van der Pol equation,

$$y'' - \mu(y^2 - 1)y' + y = 0, \qquad \text{for } \mu > 0,$$

governs the flow of current in a vacuum tube with three internal elements. Let $\mu = \frac{1}{2}$, $y(0) = 0$, and $y(2) = 1$. Approximate the solution $y(t)$ for $t = 0.2i$, where $1 \le i \le 9$.

11.5 Nonlinear Finite-Difference Methods

The difference method for the general nonlinear boundary-value problem

$$y'' = f(x, y, y'), \qquad \text{for } a \le x \le b, \quad \text{where} \quad y(a) = \alpha \quad \text{and} \quad y(b) = \beta,$$

is similar to the method applied to linear problems in Section 11.3. Here, however, the system of equations will not be linear, and an iterative process is used to approximate the solution.

As in the linear case, we divide $[a, b]$ into $(N + 1)$ equal subintervals whose endpoints are at $x_i = a + ih$ for $i = 0, 1, \ldots, N + 1$. Assuming that the exact solution has a bounded fourth derivative allows us to replace $y''(x_i)$ and $y'(x_i)$ in each of the equations by the appropriate centered-difference formula to obtain, for each $i = 1, 2, \ldots, N$,

$$\frac{y(x_{i+1}) - 2y(x_i) + y(x_{i-1})}{h^2} = f\left(x_i, y(x_i), \frac{y(x_{i+1}) - y(x_{i-1})}{2h} - \frac{h^2}{6}y'''(\eta_i)\right)$$

$$+ \frac{h^2}{12}y^{(4)}(\xi_i),$$

for some ξ_i and η_i in the interval (x_{i-i}, x_{i+1}).

The difference method results when the error terms are deleted and the boundary conditions are added. This produces the $N \times N$ nonlinear system

$$2w_1 - w_2 + h^2 f\left(x_1, w_1, \frac{w_2 - \alpha}{2h}\right) - \alpha = 0,$$

$$-w_1 + 2w_2 - w_3 + h^2 f\left(x_2, w_2, \frac{w_3 - w_1}{2h}\right) = 0,$$

$$\vdots$$

$$-w_{N-2} + 2w_{N-1} - w_N + h^2 f\left(x_{N-1}, w_{N-1}, \frac{w_N - w_{N-2}}{2h}\right) = 0,$$

$$-w_{N-1} + 2w_N + h^2 f\left(x_N, w_N, \frac{\beta - w_{N-1}}{2h}\right) - \beta = 0.$$

To approximate the solution to this system, we use Newton's method for nonlinear systems, as discussed in Section 10.2. A sequence of iterates $\{(w_1^{(k)}, w_2^{(k)}, \ldots, w_N^{(k)})^t\}$ is generated that converges to the solution of the system, provided that the initial approximation $(w_1^{(0)}, w_2^{(0)}, \ldots, w_N^{(0)})^t$ is sufficiently close to the true solution, $(w_1, w_2, \ldots, w_N)^t$.

Newton's Method for Iterations

Newton's method for nonlinear systems requires solving, at each iteration, an $N \times N$ linear system involving the Jacobian matrix. In our case, the Jacobian matrix is tridiagonal, and Crout factorization in Section 6.6 can be applied. The initial approximations $w_i^{(0)}$ to w_i for each $i = 1, 2, \ldots, N$, are obtained by passing a straight line through (a, α) and (b, β) and evaluating at x_i.

A good initial approximation might be required, so an upper bound for k should be specified and, if exceeded, a new initial approximation or a reduction in step size considered.

The program NLFDM114 implements the Nonlinear Finite-Difference method.

Example 1 Apply the program NLFDM114, with $h = 0.1$, to the nonlinear boundary-value problem

$$y'' = \frac{1}{8}(32 + 2x^3 - yy'), \quad \text{for } 1 \le x \le 3, \text{ with } y(1) = 17 \text{ and } y(3) = \frac{43}{3},$$

and compare the results to those obtained in Example 1 of Section 11.4.

Solution The stopping procedure in the program NLFDM114 is to iterate until values of successive iterates differ by less than 10^{-8}. This is accomplished with 4 iterations and gives the results in Table 11.5. They are less accurate than those obtained using the nonlinear shooting method, which gave results in the middle of the table accurate on the order of 10^{-5}. ∎

Table 11.5

x_i	w_i	$y(x_i)$	$|w_i - y(x_i)|$	x_i	w_i	$y(x_i)$	$|w_i - y(x_i)|$
1.0	17.000000	17.000000		2.0	11.997915	12.000000	2.085×10^{-3}
1.1	15.754503	15.755455	9.520×10^{-4}	2.1	12.027142	12.029048	1.905×10^{-3}
1.2	14.771740	14.773333	1.594×10^{-3}	2.2	12.111020	12.112727	1.707×10^{-3}
1.3	13.995677	13.997692	2.015×10^{-3}	2.3	12.245025	12.246522	1.497×10^{-3}
1.4	13.386297	13.388571	2.275×10^{-3}	2.4	12.425388	12.426667	1.278×10^{-3}
1.5	12.914252	12.916667	2.414×10^{-3}	2.5	12.648944	12.650000	1.056×10^{-3}
1.6	12.557538	12.560000	2.462×10^{-3}	2.6	12.913013	12.913846	8.335×10^{-4}
1.7	12.299326	12.301765	2.438×10^{-3}	2.7	13.215312	13.215926	6.142×10^{-4}
1.8	12.126529	12.128889	2.360×10^{-3}	2.8	13.553885	13.554286	4.006×10^{-4}
1.9	12.028814	12.031053	2.239×10^{-3}	2.9	13.927046	13.927241	1.953×10^{-4}
2.0	11.997915	12.000000	2.085×10^{-3}	3.0	14.333333	14.333333	

Employing Richardson's Extrapolation

Richardson's extrapolation can also be used for the Nonlinear Finite-Difference method. Table 11.6 lists the results when this method is applied to Example 1 using $h = 0.1, 0.05$, and 0.025, with 4 iterations in each case. The values of $w_i(h = 0.1)$ are omitted from the table because they are listed in Table 11.5.

The notation is the same as in Example 2 of Section 11.3, and the values of Ext_{3i} are all accurate to the places listed. In fact, if the data had been given to a sufficient accuracy and the tolerance increased, the actual maximum error for the values of Ext_{3i} would have been 3.68×10^{-10}.

Table 11.6

x_i	$w_i(h=0.05)$	$w_i(h=0.025)$	Ext_{1i}	Ext_{2i}	Ext_{3i}
1.0	17.00000000	17.00000000	17.00000000	17.00000000	17.00000000
1.1	15.75521721	15.75539525	15.75545543	15.75545460	15.75545455
1.2	14.77293601	14.77323407	14.77333479	14.77333342	14.77333333
1.3	13.99718996	13.99756690	13.99769413	13.99769242	13.99769231
1.4	13.38800424	13.38842973	13.38857346	13.38857156	13.38857143
1.5	12.91606471	12.91651628	12.91666881	12.91666680	12.91666667
1.6	12.55938618	12.55984665	12.56000217	12.56000014	12.56000000
1.7	12.30115670	12.30161280	12.30176684	12.30176484	12.30176471
1.8	12.12830042	12.12874287	12.12899094	12.12888902	12.12888889
1.9	12.03049438	12.03091316	12.03105457	12.03105275	12.03105263
2.0	11.99948020	11.99987013	12.00000179	12.00000011	12.00000000
2.1	12.02857252	12.02892892	12.02902924	12.02904772	12.02904762
2.2	12.11230149	12.11262089	12.11272872	12.11272736	12.11272727
2.3	12.24614846	12.24642848	12.24652299	12.24652182	12.24652174
2.4	12.42634789	12.42658702	12.42666773	12.42666673	12.42666667
2.5	12.64973666	12.64993420	12.65000086	12.65000005	12.65000000
2.6	12.91362828	12.91379422	12.91384683	12.91384620	12.91384615
2.7	13.21577275	13.21588765	13.21592641	13.21592596	13.21592593
2.8	13.55418579	13.55426075	13.55428603	13.55428573	13.55428571
2.9	13.92719268	13.92722921	13.92724153	13.92724139	13.92724138
3.0	14.33333333	14.33333333	14.33333333	14.33333333	14.33333333

EXERCISE SET 11.5

1. Use the Nonlinear Finite-Difference method with $h = 0.5$ to approximate the solution to the boundary-value problem

$$y'' = -(y')^2 - y + \ln x, \quad \text{for } 1 \le x \le 2 \quad \text{with} \quad y(1) = 0 \quad \text{and} \quad y(2) = \ln 2.$$

   Compare your results to the exact solution $y = \ln x$.

2. Use the Nonlinear Finite-Difference method with $h = 0.25$ to approximate the solution to the boundary-value problem

$$y'' = 2y^3, \quad \text{for } -1 \le x \le 0 \quad \text{with} \quad y(-1) = \frac{1}{2} \quad \text{and} \quad y(0) = \frac{1}{3}.$$

   Compare your results to the exact solution $y(x) = 1/(x+3)$.

3. Use the Nonlinear Finite-Difference method to approximate the solution to the following boundary-value problems, iterating until successive iterations differ by less than 10^{-4}. The exact solution is given for comparison to your results.

   a. $y'' = y^3 - yy'$, for $1 \le x \le 2$ with $y(1) = \frac{1}{2}$ and $y(2) = \frac{1}{3}$; use $h = 0.1$ and compare the results to $y(x) = (x+1)^{-1}$.

   b. $y'' = 2y^3 - 6y - 2x^3$, for $1 \le x \le 2$ with $y(1) = 2$ and $y(2) = \frac{5}{2}$; use $h = 0.1$ and compare the results to $y(x) = x + x^{-1}$.

   c. $y'' = y' + 2(y - \ln x)^3 - x^{-1}$, for $2 \le x \le 3$ with $y(2) = \frac{1}{2} + \ln 2$ and $y(3) = \frac{1}{3} + \ln 3$; use $h = 0.1$ and compare the results to $y(x) = x^{-1} + \ln x$.

   d. $y'' = (x^2(y')^2 - 9y^2 + 4x^6)/x^5$, for $1 \le x \le 2$ with $y(1) = 0$ and $y(2) = \ln 256$; use $h = 0.05$ and compare the results to $y(x) = x^3 \ln x$.

4. Repeat Exercise 3(a) and (b) using extrapolation.

5.   In Exercise 8 of Section 11.3, the deflection of a beam with supported ends subject to uniform loading
     was approximated. Using a more appropriate representation of curvature gives the differential equation

$$[1 + (w'(x))^2]^{-3/2} w''(x) = \frac{S}{EI} w(x) + \frac{q}{2EI} x(x - l), \qquad \text{for } 0 < x < l.$$

Approximate the deflection $w(x)$ of the beam every 6 in. and compare the results to those of Exercise
8 of Section 11.3.

11.6 Variational Techniques

John William Strutt Lord
Rayleigh (1842–1919), a
mathematical physicist who was
particularly interested in wave
propagation, received a Nobel
Prize in physics in 1904.

The Shooting method for approximating the solution to a boundary-value problem replaced
the given problem with initial-value problems. The finite-difference method replaced the
continuous operation of differentiation with the discrete operation of finite differences.
The Rayleigh-Ritz method is a variational technique that attacks the problem from a third
approach. The boundary-value problem is first reformulated as a problem of choosing,
from the set of all sufficiently differentiable functions satisfying the boundary conditions,
the function to minimize a certain integral. Then the set of feasible functions is reduced in
size and a function in this reduced set is found that minimizes the integral. This gives an
approximation to the solution to the original minimization problem and, as a consequence,
an approximation to the solution to the boundary-value problem.

Walter Ritz (1878–1909), a
theoretical physicist at Göttigen
University, published a paper on
a variational problem in 1909
[Ri]. He died of tuberculosis in
that same year.

To describe the Rayleigh-Ritz method, we consider approximating the solution to a
linear two-point boundary-value problem from beam-stress analysis. The boundary-value
problem is described by the differential equation

$$-\frac{d}{dx}\left(p(x)\frac{dy}{dx}\right) + q(x)y = f(x) \qquad \text{for } 0 \le x \le 1,$$

with the boundary conditions

$$y(0) = y(1) = 0.$$

This differential equation describes the deflection $y(x)$ of a beam of length 1 with variable
cross section given by $q(x)$. The deflection is due to the added stresses $p(x)$ and $f(x)$.

Variational Problems

As is the case with many boundary-value problems that describe physical phenomena,
the solution to the beam equation satisfies a *variational property*. The solution to the beam
equation is the function that minimizes a certain integral over all functions in the set $C_0^2[0, 1]$,
where we define

$$C_0^2[0, 1] = \left\{ u \in C^2[0, 1] \mid u(0) = u(1) = 0 \right\}.$$

Details concerning this connection can be found in *Spline Analysis* by Schultz [Schul],
pp. 88–89.

Variational Property for the Beam Equation

The function $y \in C_0^2[0, 1]$ is the unique solution to the boundary-value problem

$$-\frac{d}{dx}\left(p(x)\frac{dy}{dx}\right) + q(x)y = f(x), \quad \text{for } 0 \le x \le 1,$$

if and only if y is the unique function in $C_0^2[0, 1]$ that minimizes the integral

$$I[u] = \int_0^1 \left\{ p(x)\left[u'(x)\right]^2 + q(x)\left[u(x)\right]^2 - 2f(x)u(x) \right\} dx.$$

The Rayleigh-Ritz method approximates the solution y by minimizing the integral over a smaller set of functions. This subset consists of the linear combinations of certain basis functions $\phi_1, \phi_2, \ldots, \phi_n$. The basis functions are chosen to be linearly independent and to satisfy

$$\phi_i(0) = \phi_i(1) = 0, \quad \text{for each } i = 1, 2, \ldots, n.$$

An approximation $\phi(x) = \sum_{i=1}^n c_i\phi_i(x)$ to the solution $y(x)$ is obtained by finding constants $c_1, c_2, \ldots, c_n$ to minimize $I[\phi(x)] = I[\sum_{i=1}^n c_i\phi_i(x)]$.

From the variational property,

$$I[\phi(x)] = I\left[\sum_{i=1}^n c_i\phi_i(x)\right]$$

$$= \int_0^1 \left[p(x)\left[\sum_{i=1}^n c_i\phi_i'(x)\right]^2 + q(x)\left[\sum_{i=1}^n c_i\phi_i(x)\right]^2 - 2f(x)\sum_{i=1}^n c_i\phi_i(x) \right] dx.$$

For a minimum to occur, it is necessary to have

$$\frac{\partial I}{\partial c_j}[\phi(x)] = 0, \quad \text{for each } j = 1, 2, \ldots, n.$$

Differentiating with respect to the coefficients gives

$$\frac{\partial I}{\partial c_j}[\phi(x)] = \int_0^1 \left[2p(x)\sum_{i=1}^n c_i\phi_i'(x)\phi_j'(x) + 2q(x)\sum_{i=1}^n c_i\phi_i(x)\phi_j(x) - 2f(x)\phi_j(x) \right] dx,$$

so

$$0 = \sum_{i=1}^n \left[\int_0^1 \{p(x)\phi_i'(x)\phi_j'(x) + q(x)\phi_i(x)\phi_j(x)\}\, dx \right] c_i - \int_0^1 f(x)\phi_j(x)dx,$$

for each $j = 1, 2, \ldots, n$. These *normal equations* produce an $n \times n$ linear system $A\mathbf{c} = \mathbf{b}$ in the variables $c_1, c_2, \ldots, c_n$, where the symmetric matrix A is given by

$$a_{ij} = \int_0^1 [p(x)\phi_i'(x)\phi_j'(x) + q(x)\phi_i(x)\phi_j(x)]\, dx,$$

and the vector $\mathbf{b}$ has the coordinates

$$b_i = \int_0^1 f(x)\phi_i(x)dx.$$

Piecewise-Linear Basis

The most elementary choice of basis functions $\phi_1, \phi_2, \ldots, \phi_n$ involves piecewise linear polynomials. The first step is to form a partition of $[0, 1]$ by choosing points $x_0, x_1, \ldots, x_{n+1}$ with

$$0 = x_0 < x_1 < \cdots < x_n < x_{n+1} = 1.$$

Let $h_i = x_{i+1} - x_i$ for each $i = 0, 1, \ldots, n$, and define the basis functions $\phi_1(x), \phi_2(x)$, $\ldots, \phi_n(x)$ by

$$\phi_i(x) = \begin{cases} 0, & \text{for } 0 \leq x \leq x_{i-1}, \\ \dfrac{1}{h_{i-1}}(x - x_{i-1}), & \text{for } x_{i-1} < x \leq x_i, \\ \dfrac{1}{h_i}(x_{i+1} - x), & \text{for } x_i < x \leq x_{i+1}, \\ 0, & \text{for } x_{i+1} < x \leq 1, \end{cases} \tag{11.9}$$

for each $i = 1, 2, \ldots, n$. (See Figure 11.4.)

Figure 11.4

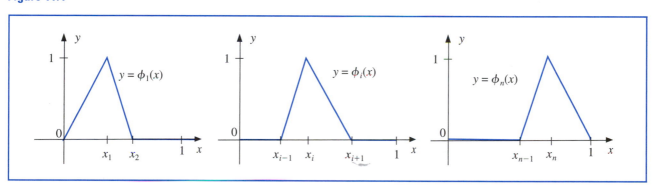

For each $j = 0, 1, \ldots, n$, the functions ϕ_i are piecewise linear. So their derivatives, ϕ_i', while not continuous, are constant on the open subinterval (x_j, x_{j+1}). In fact,

$$\phi_i'(x) = \begin{cases} 0, & \text{for } 0 < x < x_{i-1}, \\ \dfrac{1}{h_{i-1}}, & \text{for } x_{i-1} < x < x_i, \\ -\dfrac{1}{h_i}, & \text{for } x_i < x < x_{i+1}, \\ 0, & \text{for } x_{i+1} < x < 1, \end{cases}$$

for each $i = 1, 2, \ldots, n$. Because ϕ_i and ϕ_i' are nonzero only on (x_{i-1}, x_{i+1}),

$$\phi_i(x)\phi_j(x) \equiv 0 \quad \text{and} \quad \phi_i'(x)\phi_j'(x) \equiv 0, \quad \text{except when } j \text{ is } i-1, i, \text{ or } i+1.$$

As a consequence, the linear system reduces to an $n \times n$ tridiagonal linear system. The nonzero entries in A are

$$a_{ii} = \int_0^1 \{p(x)[\phi_i'(x)]^2 + q(x)[\phi_i(x)]^2\}\, dx$$

$$= \int_{x_{i-1}}^{x_i} \left(\frac{1}{h_{i-1}}\right)^2 p(x)dx + \int_{x_i}^{x_{i+1}} \left(\frac{-1}{h_i}\right)^2 p(x)dx$$

$$+ \int_{x_{i-1}}^{x_i} \left(\frac{1}{h_{i-1}}\right)^2 (x - x_{i-1})^2 q(x)dx + \int_{x_i}^{x_{i+1}} \left(\frac{1}{h_i}\right)^2 (x_{i+1} - x)^2 q(x)dx,$$

for each $i = 1, 2, \ldots, n$;

$$a_{i,i+1} = \int_0^1 \{p(x)\phi_i'(x)\phi_{i+1}'(x) + q(x)\phi_i(x)\phi_{i+1}(x)\}\, dx$$

$$= \int_{x_i}^{x_{i+1}} -\left(\frac{1}{h_i}\right)^2 p(x)dx + \int_{x_i}^{x_{i+1}} \left(\frac{1}{h_i}\right)^2 (x_{i+1} - x)(x - x_i)q(x)dx,$$

for each $i = 1, 2, \ldots, n - 1$; and

$$a_{i,i-1} = \int_0^1 \{p(x)\phi_i'(x)\phi_{i-1}'(x) + q(x)\phi_i(x)\phi_{i-1}(x)\}\, dx$$

$$= \int_{x_{i-1}}^{x_i} -\left(\frac{1}{h_{i-1}}\right)^2 p(x)dx + \int_{x_{i-1}}^{x_i} \left(\frac{1}{h_{i-1}}\right)^2 (x_i - x)(x - x_{i-1})q(x)dx,$$

for each $i = 2, \ldots, n$. The entries in **b** are

$$b_i = \int_0^1 f(x)\phi_i(x)dx = \int_{x_{i-1}}^{x_i} \frac{1}{h_{i-1}}(x - x_{i-1})f(x)dx + \int_{x_i}^{x_{i+1}} \frac{1}{h_i}(x_{i+1} - x)f(x)dx,$$

for each $i = 1, 2, \ldots, n$.

There are six types of integrals to be evaluated

$$Q_{1,i} = \left(\frac{1}{h_i}\right)^2 \int_{x_i}^{x_{i+1}} (x_{i+1} - x)(x - x_i)q(x)dx, \qquad \text{for each } i = 1, 2, \ldots, n - 1,$$

$$Q_{2,i} = \left(\frac{1}{h_{i-1}}\right)^2 \int_{x_{i-1}}^{x_i} (x - x_{i-1})^2 q(x)dx, \qquad \text{for each } i = 1, 2, \ldots, n,$$

$$Q_{3,i} = \left(\frac{1}{h_i}\right)^2 \int_{x_i}^{x_{i+1}} (x_{i+1} - x)^2 q(x)dx, \qquad \text{for each } i = 1, 2, \ldots, n,$$

$$Q_{4,i} = \left(\frac{1}{h_{i-1}}\right)^2 \int_{x_{i-1}}^{x_i} p(x)dx, \qquad \text{for each } i = 1, 2, \ldots, n + 1,$$

$$Q_{5,i} = \frac{1}{h_{i-1}} \int_{x_{i-1}}^{x_i} (x - x_{i-1})f(x)dx, \qquad \text{for each } i = 1, 2, \ldots, n,$$

and

$$Q_{6,i} = \frac{1}{h_i} \int_{x_i}^{x_{i+1}} (x_{i+1} - x) f(x) dx, \qquad \text{for each } i = 1, 2, \ldots, n.$$

Once the values of these integrals have been determined, we have

$$a_{i,i} = Q_{4,i} + Q_{4,i+1} + Q_{2,i} + Q_{3,i}, \qquad \text{for each } i = 1, 2, \ldots, n,$$

$$a_{i,i+1} = -Q_{4,i+1} + Q_{1,i}, \qquad \text{for each } i = 1, 2, \ldots, n-1,$$

$$a_{i,i-1} = -Q_{4,i} + Q_{1,i-1}, \qquad \text{for each } i = 2, 3, \ldots, n,$$

and

$$b_i = Q_{5,i} + Q_{6,i}, \qquad \text{for each } i = 1, 2, \ldots, n.$$

The entries in $\mathbf{c}$ are the unknown coefficients $c_1, c_2, \ldots, c_n$, from which the Rayleigh-Ritz approximation $\phi(x) = \sum_{i=1}^{n} c_i \phi_i(x)$ is constructed.

A practical difficulty with this method is the necessity of evaluating $6n$ integrals. The integrals can be evaluated either directly or by a quadrature formula such as Simpson's method.

Program PLRRG115 sets up the tridiagonal linear system and incorporates Crout factorization for tridiagonal systems to solve the system. The integrals $Q_{1,i}, \ldots, Q_{6,i}$ are approximated by first using piecewise linear interpolating polynomials to approximate each of the functions p, q, and f. Then these piecewise polynomials are integrated.

Because of the elementary nature of the next example, the integrals were found directly.

> The program PLRRG115 implements the linear Rayleigh-Ritz method.

Illustration   Consider the boundary-value problem

$$-y'' + \pi^2 y = 2\pi^2 \sin(\pi x), \quad \text{for } 0 \leq x \leq 1, \text{ with } y(0) = y(1) = 0.$$

Let $h_i = h = 0.1$, so that $x_i = 0.1i$, for each $i = 0, 1, \ldots, 9$. The integrals are

$$Q_{1,i} = 100 \int_{0.1i}^{0.1i+0.1} (0.1i + 0.1 - x)(x - 0.1i)\pi^2 \, dx = \frac{\pi^2}{60},$$

$$Q_{2,i} = 100 \int_{0.1i-0.1}^{0.1i} (x - 0.1i + 0.1)^2 \pi^2 \, dx = \frac{\pi^2}{30},$$

$$Q_{3,i} = 100 \int_{0.1i}^{0.1i+0.1} (0.1i + 0.1 - x)^2 \pi^2 \, dx = \frac{\pi^2}{30},$$

$$Q_{4,i} = 100 \int_{0.1i-0.1}^{0.1i} dx = 10,$$

$$Q_{5,i} = 10 \int_{0.1i-0.1}^{0.1i} (x - 0.1i + 0.1)2\pi^2 \sin \pi x \, dx$$

$$= -2\pi \cos 0.1\pi i + 20[\sin(0.1\pi i) - \sin((0.1i - 0.1)\pi)],$$

and

$$Q_{6,i} = 10 \int_{0.1i}^{0.1i+0.1} (0.1i + 0.1 - x)2\pi^2 \sin \pi x \, dx$$

$$= 2\pi \cos 0.1\pi i - 20[\sin((0.1i + 0.1)\pi) - \sin(0.1\pi i)].$$

The linear system $A\mathbf{c} = \mathbf{b}$ has

$$a_{i,i} = 20 + \frac{\pi^2}{15}, \qquad \text{for each } i = 1, 2, \ldots, 9,$$

$$a_{i,i+1} = -10 + \frac{\pi^2}{60}, \qquad \text{for each } i = 1, 2, \ldots, 8,$$

$$a_{i,i-1} = -10 + \frac{\pi^2}{60}, \qquad \text{for each } i = 2, 3, \ldots, 9,$$

and

$$b_i = 40 \sin(0.1\pi i)[1 - \cos 0.1\pi], \qquad \text{for each } i = 1, 2, \ldots, 9.$$

The solution to the tridiagonal linear system is

$$c_9 = 0.3102866742, \quad c_8 = 0.5902003271, \quad c_7 = 0.8123410598,$$

$$c_6 = 0.9549641893, \quad c_5 = 1.004108771, \quad c_4 = 0.9549641893,$$

$$c_3 = 0.8123410598, \quad c_2 = 0.5902003271, \quad c_1 = 0.3102866742.$$

The piecewise-linear approximation is

$$\phi(x) = \sum_{i=1}^{9} c_i \phi_i(x),$$

and the actual solution to the boundary-value problem is $y(x) = \sin \pi x$. Table 11.7 lists the error in the approximation at x_i, for each $i = 1, \ldots, 9$. ∎

Table 11.7

i	x_i	$\phi(x_i)$	$y(x_i)$	$|\phi(x_i) - y(x_i)|$
1	0.1	0.3102866742	0.3090169943	0.00127
2	0.2	0.5902003271	0.5877852522	0.00241
3	0.3	0.8123410598	0.8090169943	0.00332
4	0.4	0.9549641896	0.9510565162	0.00390
5	0.5	1.0041087710	1.0000000000	0.00411
6	0.6	0.9549641893	0.9510565162	0.00390
7	0.7	0.8123410598	0.8090169943	0.00332
8	0.8	0.5902003271	0.5877852522	0.00241
9	0.9	0.3102866742	0.3090169943	0.00127

The tridiagonal matrix A given by the piecewise linear basis functions is positive definite, so the linear system is stable with respect to round-off error and

$$|\phi(x) - y(x)| = O(h^2), \qquad \text{when } 0 \le x \le 1.$$

B-Spline Basis

The use of piecewise-linear basis functions results in an approximate solution that is continuous but not differentiable on [0, 1]. A more sophisticated set of basis functions is required to construct an approximation that has the two continuous derivatives stated in the Variational Principle on page 462. These basis functions are similar to the cubic interpolatory splines discussed in Section 3.5.

Recall that the cubic *interpolatory* spline S for a function f on the five nodes x_0, x_1, x_2, x_3, and x_4 is defined as follows:

(a) S is a cubic polynomial, denoted by S_j, on $[x_j, x_{j+1}]$, for $j = 0, 1, 2, 3$. (*This gives* 16 *selectable constants for S, 4 for each cubic*.)

(b) $S(x_j) = f(x_j)$, for $j = 0, 1, 2, 3, 4$ (5 *specified conditions*).

(c) $S_{j+1}(x_{j+1}) = S_j(x_{j+1})$, for $j = 0, 1, 2$ (3 *specified conditions*).

(d) $S'_{j+1}(x_{j+1}) = S'_j(x_{j+1})$, for $j = 0, 1, 2$ (3 *specified conditions*).

(e) $S''_{j+1}(x_{j+1}) = S''_j(x_{j+1})$, for $j = 0, 1, 2$ (3 *specified conditions*).

(f) One of the following boundary conditions is satisfied:

    (i) Free: $S''(x_0) = S''(x_4) = 0$ (2 *specified conditions*).

    (ii) Clamped: $S'(x_0) = f'(x_0)$ and $S'(x_4) = f'(x_4)$ (2 *specified conditions*).

Uniqueness of solution requires the number of constants in (a), 16, to be equal to the number of conditions in (b) through (f). There are 14 conditions specified in (b) through (e), so only one of the boundary conditions in (f) can be specified for the interpolatory cubic splines.

The cubic spline functions we will use for our basis functions are called **B-splines**, or *bell-shaped splines*. They differ from interpolatory splines in that both sets of boundary conditions in (f) must be satisfied. This requires the relaxation of two of the conditions in (b) through (e). The spline must have two continuous derivatives on $[x_0, x_4]$, so we cannot modify the conditions in (c) through (e). Therefore we must delete two of the requirements from the interpolation conditions.

There are 5 nodes, x_0, x_1, x_2, x_3, and x_4 and we must have interpolation at the endpoints x_0 and x_4. So we must relax the interpolation conditions at two of x_1, x_2, x_3. To maintain symmetry we relax the interpolation at x_2 and x_4. This implies that we need to modify condition (b) to

(b′) $S(x_j) = f(x_j)$ for $j = 0, 2, 4$.

The basic B-spline S shown in Figure 11.5 uses the equally spaced nodes $x_0 = -2, x_1 = -1$, $x_2 = 0, x_3 = 1$, and $x_4 = 2$. It satisfies the interpolatory conditions

(b′) $S(x_0) = 0, S(x_2) = 1, S(x_4) = 0$;

as well as both sets of boundary conditions

    (i) $S''(x_0) = S''(x_4) = 0$ and **(ii)** $S'(x_0) = S'(x_4) = 0.$

Figure 11.5

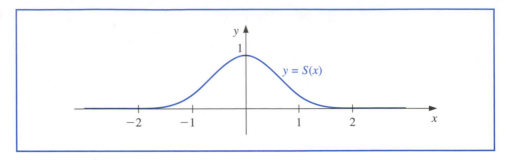

As a consequence, $S \in C^2(-\infty, \infty)$, and it is not difficult to verify that all of these conditions are satisfied for

$$
S(x) = \begin{cases}
0, & \text{for } x \leq -2, \\
\frac{1}{4}(2+x)^3, & \text{for } -2 \leq x \leq -1, \\
\frac{1}{4}\left[(2+x)^3 - 4(1+x)^3\right], & \text{for } -1 < x \leq 0, \\
\frac{1}{4}\left[(2-x)^3 - 4(1-x)^3\right], & \text{for } 0 < x \leq 1, \\
\frac{1}{4}(2-x)^3, & \text{for } 1 < x \leq 2, \\
0, & \text{for } 2 < x.
\end{cases}
\tag{11.10}
$$

To construct the basis functions ϕ_i in $C_0^2[0, 1]$, we first partition $[0, 1]$ by choosing a positive integer n and defining $h = 1/(n + 1)$. This produces the equally-spaced nodes $x_i = ih$, for each $i = 0, 1, \ldots, n + 1$. We then define the basis functions $\{\phi_i\}_{i=0}^{n+1}$ as

$$
\phi_i(x) = \begin{cases}
S\left(\frac{x}{h}\right) - 4S\left(\frac{x+h}{h}\right), & \text{for } i = 0, \\
S\left(\frac{x-h}{h}\right) - S\left(\frac{x+h}{h}\right), & \text{for } i = 1, \\
S\left(\frac{x-ih}{h}\right), & \text{for } 2 \leq i \leq n - 1, \\
S\left(\frac{x-nh}{h}\right) - S\left(\frac{x-(n+2)h}{h}\right), & \text{for } i = n, \\
S\left(\frac{x-(n+1)h}{h}\right) - 4S\left(\frac{x-(n+2)h}{h}\right), & \text{for } i = n + 1.
\end{cases}
$$

Then $\{\phi_i\}_{i=0}^{n+1}$ is a linearly independent set of cubic splines satisfying $\phi_i(0) = \phi_i(1) = 0$, for each $i = 0, 1, \ldots, n, n + 1$. The graphs of ϕ_i, for $2 \leq i \leq n - 1$, are shown in Figure 11.6 and the graphs of ϕ_0, ϕ_1, ϕ_n, and ϕ_{n+1} are in Figure 11.7.

Figure 11.6

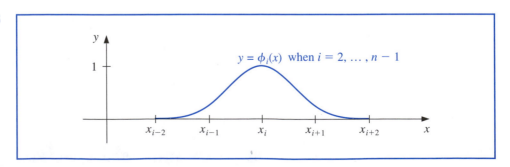

Figure 11.7

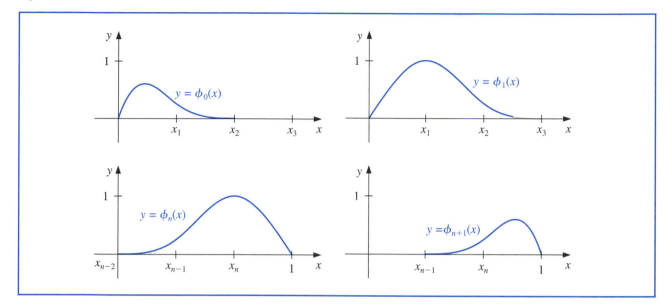

Since $\phi_i(x)$ and $\phi_i'(x)$ are nonzero only for $x_{i-2} \leq x \leq x_{i+2}$, the matrix in the Rayleigh-Ritz approximation is an $(n+1) \times (n+1)$ band matrix with bandwidth at most seven:

$$
A = \begin{bmatrix}
a_{00} & a_{01} & a_{02} & a_{03} & 0 & \cdots & & & & & 0 \\
a_{10} & a_{11} & a_{12} & a_{13} & a_{14} & & & & & & \\
a_{20} & a_{21} & a_{22} & a_{23} & a_{24} & a_{25} & & & & & \\
a_{30} & a_{31} & a_{32} & a_{33} & a_{34} & a_{35} & a_{36} & & & & \\
0 & & & & & & & & & & 0 \\
\vdots & & & & & & & & & & a_{n-2,n+1} \\
& & & & & & & & & & a_{n-1,n+1} \\
& & & & & & & & & & a_{n,n+1} \\
0 & \cdots & & & 0 & a_{n+1,n-2} & a_{n+1,n-1} & a_{n+1,n} & a_{n+1,n+1}
\end{bmatrix}, \quad (11.11)
$$

where

$$
a_{ij} = \int_0^1 \{ p(x)\phi_i'(x)\phi_j'(x) + q(x)\phi_i(x)\phi_j(x) \} \, dx,
$$

for each $i = 0, 1, \ldots, n+1$ and $j = 0, 1, \ldots, n+1$. The vector $\mathbf{b}$ has the entries

$$
b_i = \int_0^1 f(x)\phi_i(x)\,dx.
$$

The matrix A is positive definite, so the linear system $A\mathbf{c} = \mathbf{b}$ can be quickly solved by Cholesky's method or by Gaussian elimination, and there is stability with respect to round-off error.

The program CSRRG116 implements the Cubic B-Spline Rayleigh-Ritz method.

The integrations CSRRG116 are performed in a manner similar to the way program PLRRG115 does for the Piecewise Linear method. First, construct cubic spline interpolatory polynomials for p, q, and f using the methods presented in Section 3.5. Then approximate the integrands by products of cubic splines or derivatives of cubic splines. Since these integrands are piecewise polynomials, they can be integrated exactly on each subinterval and then summed. In general, this technique produces approximations $\phi(x)$ to $y(x)$ that satisfy

$$\left[\int_0^1 |y(x) - \phi(x)|^2 dx\right]^{1/2} = O(h^4), \qquad 0 \le x \le 1.$$

Illustration Consider the boundary-value problem

$$-y'' + \pi^2 y = 2\pi^2 \sin(\pi x), \quad \text{for } 0 \le x \le 1, \text{ with } y(0) = y(1) = 0.$$

In the Illustration on page 465 we let $h = 0.1$ and generated approximations using piecewise-linear basis functions.

Here we let $n = 3$, so that $h = 0.25$. The cubic spline basis functions are

$$\phi_0(x) = \begin{cases} 12x - 72x^2 + 112x^3, & \text{for } 0 \le x \le 0.25 \\ 2 - 12x + 24x^2 - 16x^3, & \text{for } 0.25 < x \le 0.5 \\ 0, & \text{otherwise} \end{cases}$$

$$\phi_1(x) = \begin{cases} 6x - 32x^3, & \text{for } 0 \le x \le 0.25 \\ -\frac{5}{4} + 21x - 60x^2 + 48x^3, & \text{for } 0.25 < x \le 0.5 \\ \frac{27}{4} - 27x + 36x^2 - 16x^3, & \text{for } 0.5 < x \le 0.75 \\ 0, & \text{otherwise} \end{cases}$$

$$\phi_2(x) = \begin{cases} 16x^3, & \text{for } 0 \le x \le 0.25 \\ 1 - 12x + 48x^2 - 48x^3, & \text{for } 0 < x \le 0.5 \\ -11 + 60x - 96x^2 + 48x^3, & \text{for } 0.5 < x \le 0.75 \\ 16 - 48x + 48x^2 - 16x^3, & \text{for } 0.75 < x \le 1 \\ 0, & \text{otherwise} \end{cases}$$

$$\phi_3(x) = \begin{cases} -\frac{1}{4} + 3x - 12x^2 + 16x^3, & \text{for } 0.25 < x \le 0.5 \\ \frac{31}{4} - 45x + 84x^2 - 48x^3, & \text{for } 0.5 < x \le 0.75 \\ -26 + 90x - 96x^2 + 32x^3, & \text{for } 0.75 < x \le 1 \\ 0, & \text{otherwise} \end{cases}$$

and

$$\phi_4(x) = \begin{cases} -2 + 12x - 24x^2 + 16x^3, & \text{for } 0.5 < x \le 0.75 \\ 52 - 204x + 264x^2 - 112x^3, & \text{for } 0.75 \le x \le 1 \\ 0, & \text{otherwise} \end{cases}$$

In this problem we have $p(x) = 1$ and $q(x) = \pi^2$, so the entries in the matrix A are generated as follows

$$a_{00} = \int_0^{0.5} \{1 \cdot [\phi_0'(x)]^2 + \pi^2 \cdot [\phi_0(x)]^2\} \, dx = 6.5463531,$$

$$a_{01} = a_{10} = \int_0^{0.5} \{1 \cdot \phi_0'(x)\phi_1'(x) + \pi^2 \cdot \phi_0(x)\phi_1(x)\} \, dx = 4.0764737,$$

$$a_{02} = a_{20} = \int_0^{0.5} \{1 \cdot \phi_0'(x)\phi_2'(x) + \pi^2 \cdot \phi_0(x)\phi_2(x)\} \, dx = -1.3722239,$$

$$a_{03} = a_{30} = \int_{0.25}^{0.5} \{1 \cdot \phi_0'(x)\phi_3'(x) + \pi^2 \cdot \phi_0(x)\phi_3(x)\} \, dx = -0.73898482,$$

$$a_{11} = \int_0^{0.75} \{1 \cdot [\phi_1'(x)]^2 + \pi^2 \cdot [\phi_1(x)]^2\} \, dx = 10.329086,$$

$$a_{12} = a_{21} = \int_0^{0.75} \{1 \cdot \phi_1'(x)\phi_2'(x) + \pi^2 \cdot \phi_1(x)\phi_2(x)\} \, dx = 0.26080684,$$

$$a_{13} = a_{31} = \int_{0.25}^{0.75} \{1 \cdot \phi_1'(x)\phi_3'(x) + \pi^2 \cdot \phi_1(x)\phi_3(x)\} \, dx = -1.6678178,$$

$$a_{14} = a_{41} = \int_{0.5}^{0.75} \{1 \cdot \phi_1'(x)\phi_4'(x) + \pi^2 \cdot \phi_1(x)\phi_4(x)\} \, dx = -0.73898482,$$

$$a_{22} = \int_0^{1} \{1 \cdot [\phi_2'(x)]^2 + \pi^2 \cdot [\phi_2(x)]^2\} \, dx = 8.6612683,$$

$$a_{23} = a_{32} = \int_{0.25}^{1} \{1 \cdot \phi_2'(x)\phi_3'(x) + \pi^2 \cdot \phi_2(x)\phi_3(x)\} \, dx = 0.26080684,$$

$$a_{24} = a_{42} = \int_{0.5}^{1} \{1 \cdot \phi_2'(x)\phi_4'(x) + \pi^2 \cdot \phi_2(x)\phi_4(x)\} \, dx = -1.3722239,$$

$$a_{33} = \int_{0.25}^{1} \{1 \cdot [\phi_3'(x)]^2 + \pi^2 \cdot [\phi_3(x)]^2\} \, dx = 10.329086,$$

$$a_{34} = a_{43} = \int_{0.5}^{1} \{1 \cdot \phi_3'(x)\phi_4'(x) + \pi^2 \cdot \phi_3(x)\phi_4(x)\} \, dx = 4.0764737,$$

$$a_{44} = \int_{0.5}^{1} \{1 \cdot [\phi_4'(x)]^2 + \pi^2 \cdot [\phi_4(x)]^2\} \, dx = 6.5463531,$$

and, because $f(x) = 2\pi^2 \sin \pi x$,

$$b_0 = \int_0^{0.5} (2\pi^2 \sin \pi x)\phi_0(x)dx = 1.0803542,$$

$$b_1 = \int_0^{0.75} (2\pi^2 \sin \pi x)\phi_1(x)dx = 4.7202512,$$

$$b_2 = \int_0^{1} (2\pi^2 \sin \pi x)\phi_2(x)dx = 6.6754433,$$

$$b_3 = \int_{0.25}^{1} (2\pi^2 \sin \pi x)\phi_3(x)dx = 4.7202512.$$

$$b_4 = \int_{0.5}^{1} (2\pi^2 \sin \pi x)\phi_4(x)dx = 1.08035418.$$

The solution to the system $A\mathbf{c} = \mathbf{b}$ is

$$c_0 = 0.00060266150, \quad c_1 = 0.52243908, \quad c_2 = 0.73945127,$$

$$c_3 = 0.52243908, \quad \text{and} \quad c_4 = 0.00060264906.$$

Table 11.8 lists the results obtained by applying the B-splines program CSRRG116 with $h = 0.25$. Notice that the result at $x_2 = 0.5$ is superior to the piecewise-linear result at $x_5 = 0.5$ in Table 11.7 on page 466, where we used the smaller step size $h = 0.1$. ☐

Table 11.8

i	x_i	$\phi(x_i)$	$y(x_i)$	$\|y(x_i) - \phi(x_i)\|$
0	0	0	0	0
1	0.25	0.70745256	0.70710678	0.00034578
2	0.5	1.0006708	1	0.0006708
3	0.75	0.70745256	0.70710678	0.00034578
4	1	0	0	0

EXERCISE SET 11.6

1. Use the Piecewise Linear method to approximate the solution to the boundary-value problem

$$y'' + \frac{\pi^2}{4}y = \frac{\pi^2}{16}\cos\frac{\pi}{4}x, \quad \text{for } 0 \le x \le 1 \quad \text{with } y(0) = y(1) = 0$$

using $x_0 = 0, x_1 = 0.3, x_2 = 0.7, x_3 = 1$ and compare the results to the exact solution $y(x) = -\frac{1}{3}\cos\frac{\pi}{2}x - \frac{\sqrt{2}}{6}\sin\frac{\pi}{2}x + \frac{1}{3}\cos\frac{\pi}{4}x$.

2. Use the Piecewise Linear method to approximate the solution to the boundary-value problem

$$-\frac{d}{dx}(xy') + 4y = 4x^2 - 8x + 1, \quad \text{for } 0 \le x \le 1 \quad \text{with } y(0) = y(1) = 0$$

using $x_0 = 0, x_1 = 0.4, x_2 = 0.8, x_3 = 1$ and compare the results to the exact solution $y(x) = x^2 - x$.

3. Use the Piecewise Linear method to approximate the solutions to the following boundary-value problems and compare the results to the exact solution:

   a. $-x^2 y'' - 2xy' + 2y = -4x^2$, for $0 \le x \le 1$ with $y(0) = y(1) = 0$; use $h = 0.1$; actual solution $y(x) = x^2 - x$.

   b. $-\frac{d}{dx}(e^x y') + e^x y = x + (2 - x)e^x$, for $0 \le x \le 1$ with $y(0) = y(1) = 0$; use $h = 0.1$; actual solution $y(x) = (x - 1)(e^{-x} - 1)$.

   c. $-\frac{d}{dx}(e^{-x} y') + e^{-x} y = (x - 1) - (x + 1)e^{-(x-1)}$, for $0 \le x \le 1$ with $y(0) = y(1) = 0$; use $h = 0.05$; actual solution $y(x) = x(e^x - e)$.

   d. $-(x + 1)y'' - y' + (x + 2)y = [2 - (x + 1)^2]e \ln 2 - 2e^x$, for $0 \le x \le 1$ with $y(0) = y(1) = 0$; use $h = 0.05$; actual solution $y(x) = e^x \ln(x + 1) - (e \ln 2)x$.

4. Use the Cubic Spline method with $n = 3$ to approximate the solution to each of the following boundary-value problems and compare the results to the exact solutions.

   a. $y'' + \frac{\pi^2}{4} y = \frac{\pi^2}{16} \cos \frac{\pi}{4} x$, for $0 \le x \le 1$ with $y(0) = 0$ and $y(1) = 0$

   b. $-\frac{d}{dx}(xy') + 4y = 4x^2 - 8x + 1$, for $0 \le x \le 1$ with $y(0) = 0$ and $y(1) = 0$

5. Repeat Exercise 3 using the Cubic B-Spline method.

6. Show that the boundary-value problem

$$-\frac{d}{dx}(p(x)y') + q(x)y = f(x), \qquad \text{for } 0 \le x \le 1 \quad \text{with } y(0) = \alpha \text{ and } y(1) = \beta,$$

can be transformed by the change of variable

$$z = y - \beta x - (1 - x)\alpha$$

into the form

$$-\frac{d}{dx}(p(x)z') + q(x)z = F(x), \quad 0 \le x \le 1, \qquad z(0) = 0, \qquad z(1) = 0.$$

7. Use Exercise 6 and the Piecewise Linear method with $n = 9$ to approximate the solution to the boundary-value problem

$$-y'' + y = x, \quad \text{for } 0 \le x \le 1 \quad \text{with} \quad y(0) = 1 \quad \text{and} \quad y(1) = 1 + e^{-1}.$$

8. Repeat Exercise 7 using the Cubic Spline method.

9. Show that the boundary-value problem

$$-\frac{d}{dx}(p(x)y') + q(x)y = f(x), \qquad \text{for } a \le x \le b \quad \text{with } y(a) = \alpha \text{ and } y(b) = \beta,$$

can be transformed into the form

$$-\frac{d}{dw}(p(w)z') + q(w)z = F(w), \qquad \text{for } 0 \le w \le 1 \quad \text{with } z(0) = 0 \text{ and } z(1) = 0,$$

by a method similar to that given in Exercise 6.

10. Show that the set of piecewise linear basis functions is linearly independent on $[0, 1]$.

11. Use the definition of positive definite to show that the matrix given by the piecewise linear basis functions is positive definite.

11.7 Survey of Methods and Software

In this chapter we discussed methods for approximating solutions to boundary-value problems. For the linear boundary-value problem

$$y'' = p(x)y' + q(x)y + r(x), \quad a \le x \le b, \quad y(a) = \alpha, \quad y(b) = \beta,$$

we considered both a linear shooting method and a finite-difference method to approximate the solution. The linear shooting method uses an initial-value technique to solve the problems

$$y'' = p(x)y' + q(x)y + r(x), \quad a \le x \le b, \quad y(a) = \alpha, \quad y'(a) = 0,$$

and

$$y'' = p(x)y' + q(x)y, \quad a \le x \le b, \quad y(a) = 0, \quad y'(a) = 1.$$

A weighted average of these solutions produces a solution to the linear boundary-value problem.

In the finite-difference method, we replaced y'' and y' with difference approximations and solved a linear system. Although the approximations may not be as accurate as the shooting method, there is less sensitivity to round-off error. Higher-order difference methods are available, or extrapolation can be used to improve accuracy.

For the nonlinear boundary problem

$$y'' = f(x, y, y'), \quad a \le x \le b, \quad y(a) = \alpha, \quad y(b) = \beta,$$

we also presented two methods. The nonlinear shooting method requires the solution of the initial-value problem

$$y'' = f(x, y, y'), \quad a \le x \le b, \quad y(a) = \alpha, \quad y'(a) = t,$$

for an initial choice of t. We improved the choice by using Newton's method to approximate the solution, t, to $y(b, t) = \beta$. This method required solving two initial-value problems at each iteration. The accuracy is dependent on the choice of method for solving the initial-value problems.

The finite-difference method for the nonlinear equation requires the replacement of y'' and y' by difference quotients, which results in a nonlinear system. This system is solved using Newton's method. Higher-order differences or extrapolation can be used to improve accuracy. Finite-difference methods tend to be less sensitive to round-off error than shooting methods.

The Rayleigh-Ritz method was illustrated by approximating the solution to the boundary-value problem

$$-\frac{d}{dx}\left(p(x)\frac{dy}{dx}\right) + q(x)y = f(x), \quad 0 \le x \le 1, \quad y(0) = y(1) = 0.$$

A piecewise-linear approximation or a cubic spline approximation can be obtained.

Most of the material concerning second-order boundary-value problems can be extended to problems with boundary conditions of the form

$$\alpha_1 y(a) + \beta_1 y'(a) = \alpha \quad \text{and} \quad \alpha_2 y(b) + \beta_2 y'(b) = \beta,$$

where $|\alpha_1| + |\beta_1| \neq 0$ and $|\alpha_2| + |\beta_2| \neq 0$, but some of the techniques become quite complicated. The reader who is interested in problems of this type is advised to consider a book specializing in boundary-value problems, such as [K,H].

MATLAB has a function for solving two point boundary value problems in ordinary differential equations. The function bvp4c is a finite difference code that implements the three-stage Lobatto IIIa formula. More information on this function can be found in the help option of MATLAB.

The IMSL and NAG Libraries contain methods for boundary-value problems. There are Finite-Difference methods and Shooting methods that are based on their adaptations of variable-step-size Runge-Kutta methods.

There are subroutines in the ODE package contained in the netlib Library to solve the linear and nonlinear two-point boundary-value problems, respectively. The routines are based on multiple shooting methods.

Further information on the general problems involved with the numerical solution to two-point boundary-value problems can be found in Keller [K,H] and in Bailey, Shampine, and Waltman [BSW]. Roberts and Shipman [RS] focuses on the shooting methods for the two-point boundary-value problem, and Pryce [Pr] restricts attention to Sturm-Liouville problems. The book by Ascher, Mattheij, and Russell [AMR] has a comprehensive presentation of multiple shooting and parallel shooting methods.

Numerical Methods for Partial-Differential Equations

12.1 Introduction

Physical problems that involve more than one variable are often expressed using equations involving partial derivatives. In this chapter, we present a brief introduction to some of the basic techniques available for approximating the solution to partial-differential equations involving two variables by showing how these techniques can be applied to certain standard physical problems.

12.2 Finite-Difference Methods for Elliptic Problems

The **elliptic** partial-differential equation we consider in this section is known as the **Poisson equation**:

$$\frac{\partial^2 u}{\partial x^2}(x, y) + \frac{\partial^2 u}{\partial y^2}(x, y) = f(x, y).$$

In this equation we assume that f describes the input to the problem on a plane region R with boundary S. Equations of this type arise in the study of various time-independent physical problems such as the steady-state distribution of heat in a plane region, the potential energy of a point in a plane acted on by gravitational forces in the plane, and two-dimensional steady-state problems involving incompressible fluids.

Additional constraints must be imposed to obtain a unique solution to the Poisson equation. For example, the study of the steady-state distribution of heat in a plane region requires that $f(x, y) \equiv 0$, resulting in a simplification to

$$\frac{\partial^2 u}{\partial x^2}(x, y) + \frac{\partial^2 u}{\partial y^2}(x, y) = 0,$$

which is called **Laplace's equation**. If the temperature within the region is determined by the temperature distribution on the boundary of the region, the constraints are called the **Dirichlet boundary conditions**, given by

$$u(x, y) = g(x, y)$$

for all (x, y) on S, the boundary of the region R. (See Figure 12.1.)

Figure 12.1

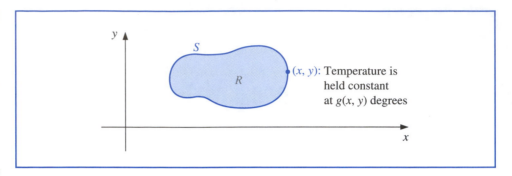

Siméon-Denis Poisson (1781–1840) was a student of Laplace and Legendre during the Napoleonic years in France. Later he assumed Fourier's professorship at the École Polytechnique where he worked on ordinary and partial differential equations, and later in life on probability theory.

The specific elliptic partial-differential equation we consider is

$$\nabla^2 u(x, y) \equiv \frac{\partial^2 u}{\partial x^2}(x, y) + \frac{\partial^2 u}{\partial y^2}(x, y) = f(x, y)$$

defined on a region

$$R = \{(x, y) \mid a < x < b, c < y < d\},$$

Pierre-Simon Laplace (1749–1827) worked in many mathematical areas, producing seminal papers in probability and mathematical physics. He published his major work on the theory of heat during the period 1817–1820.

with boundary conditions

$$u(x, y) = g(x, y) \qquad \text{for } (x, y) \in S,$$

where S denotes the boundary of R. If f and g are continuous on their domains, then this equation has a unique solution.

Selecting a Grid

Johann Peter Gustav Lejeune Dirichlet (1805–1859) made major contributions to the areas of number theory and the convergence of series. In fact, he could be considered the founder of Fourier series because, according to Riemann, he was the first to write a profound paper on this subject.

The method used here is similar to the Finite-Difference method for linear boundary-value problems, which was discussed in Section 11.3. The first step is to choose integers n and m and define step sizes h and k by $h = (b - a)/n$ and $k = (d - c)/m$. Partition the interval $[a, b]$ into n equal parts of width h and the interval $[c, d]$ into m equal parts of width k. We provide a grid on the rectangle R by drawing vertical and horizontal lines through the points with coordinates (x_i, y_j), where

$$x_i = a + ih \quad \text{and} \quad y_j = c + jk,$$

for each $i = 0, 1, \ldots, n$ and $j = 0, 1, \ldots, m$ (see Figure 12.2).

Figure 12.2

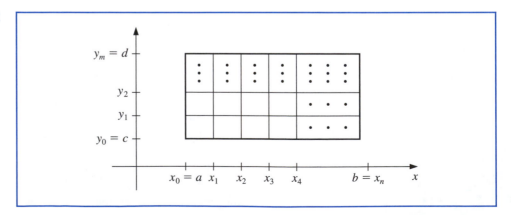

The lines $x = x_i$ and $y = y_j$ are **grid lines**, and their intersections are the **mesh points** of the grid. For each mesh point in the interior of the grid, we use the Taylor polynomial in the variable x about x_i to generate the centered-difference formula

$$\frac{\partial^2 u}{\partial x^2}(x_i, y_j) = \frac{u(x_{i+1}, y_j) - 2u(x_i, y_j) + u(x_{i-1}, y_j)}{h^2} - \frac{h^2}{12}\frac{\partial^4 u}{\partial x^4}(\xi_i, y_j),$$

for some ξ_i in (x_{i-1}, x_{i+1}), and the Taylor polynomial in the variable y about y_j to generate the centered-difference formula

$$\frac{\partial^2 u}{\partial y^2}(x_i, y_j) = \frac{u(x_i, y_{j+1}) - 2u(x_i, y_j) + u(x_i, y_{j-1})}{k^2} - \frac{k^2}{12}\frac{\partial^4 u}{\partial y^4}(x_i, \eta_j),$$

for some η_j in (y_{j-1}, y_{j+1}).

Using these formulas in the Poisson equation produces the following equations:

$$\frac{u(x_{i+1}, y_j) - 2u(x_i, y_j) + u(x_{i-1}, y_j)}{h^2} + \frac{u(x_i, y_{j+1}) - 2u(x_i, y_j) + u(x_i, y_{j-1})}{k^2}$$

$$= f(x_i, y_j) + \frac{h^2}{12}\frac{\partial^4 u}{\partial x^4}(\xi_i, y_j) + \frac{k^2}{12}\frac{\partial^4 u}{\partial y^4}(x_i, \eta_j),$$

for each $i = 1, 2, \ldots, n-1$ and $j = 1, 2, \ldots, m-1$. The boundary conditions give

$$u(x_i, y_0) = g(x_i, y_0) \quad \text{and} \quad u(x_i, y_m) = g(x_i, y_m),$$

for each $i = 1, 2, \ldots, n-1$, and for each $j = 0, 1, \ldots, m$,

$$u(x_0, y_j) = g(x_0, y_j) \quad \text{and} \quad u(x_n, y_j) = g(x_n, y_j).$$

Finite-Difference Method

In difference-equation form, this results in the *Finite-Difference* method for the Poisson equation, with error of order $O(h^2 + k^2)$.

Elliptic Finite-Difference Method

$$2\left[\left(\frac{h}{k}\right)^2 + 1\right] w_{ij} - (w_{i+1,j} + w_{i-1,j}) - \left(\frac{h}{k}\right)^2 (w_{i,j+1} + w_{i,j-1}) = -h^2 f(x_i, y_j),$$

for each $i = 1, 2, \ldots, n-1$ and $j = 1, 2, \ldots, m-1$, and

$$w_{0j} = g(x_0, y_j), \quad w_{nj} = g(x_n, y_j), \quad w_{i0} = g(x_i, y_0), \quad \text{and} \quad w_{im} = g(x_i, y_m),$$

for each $i = 1, 2, \ldots, n-1$ and $j = 0, 1, \ldots, m$, where w_{ij} approximates $u(x_i, y_j)$.

The typical equation involves approximations to $u(x_i, y_j)$ at the points

$$(x_{i-1}, y_j), \quad (x_i, y_j), \quad (x_{i+1}, y_j), \quad (x_i, y_{j-1}), \quad \text{and} \quad (x_i, y_{j+1}).$$

Reproducing the portion of the grid where these points are located (see Figure 12.3) shows that each equation involves approximations in a symmetric region about (x_i, y_j).

Figure 12.3

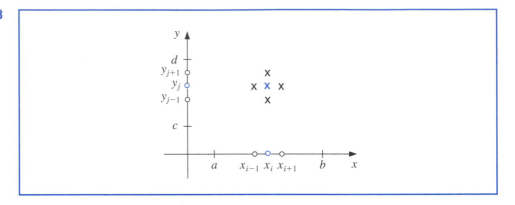

If we use the information from the boundary conditions in the system given by the Finite-Difference method (that is, at all points (x_i, y_j) that are adjacent to a boundary mesh point), we have an $(n - 1)(m - 1) \times (n - 1)(m - 1)$ linear system where the unknowns are the approximations w_{ij} to $u(x_i, y_j)$ at the interior mesh points.

The linear system involving these unknowns is expressed for matrix calculations more efficiently if the interior mesh points are relabeled by letting

$$P_l = (x_i, y_j) \quad \text{and} \quad w_l = w_{ij},$$

where $l = i + (m - 1 - j)(n - 1)$, for each $i = 1, 2, \ldots, n - 1$ and $j = 1, 2, \ldots, m - 1$. This, in effect, labels the mesh points consecutively from left to right and top to bottom. For example, with $n = 4$ and $m = 5$, the relabeling results in a grid whose points are shown in Figure 12.4. Labeling the points in this manner ensures that the system needed to determine the w_{ij} is a banded matrix with band width at most $2n - 1$.

Figure 12.4

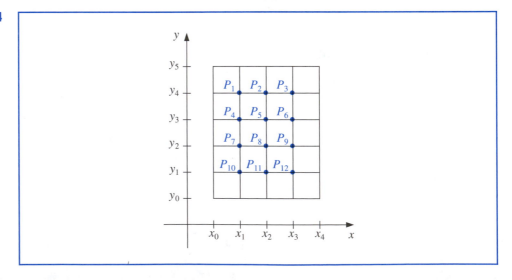

Example 1   Approximate the steady-state heat distribution in a thin square metal plate with side length 0.5 meters using $n = m = 4$. Two adjacent boundaries are held at $0°C$, and the heat on the other boundaries increases linearly from $0°C$ at one corner to $100°C$ where the sides meet.

Solution   Place the sides with the zero boundary conditions along the x- and y-axes. Then the problem is expressed as

$$\frac{\partial^2 u}{\partial x^2}(x, y) + \frac{\partial^2 u}{\partial y^2}(x, y) = 0,$$

for (x, y) in the set $R = \{(x, y) \mid 0 < x < 0.5, \ 0 < y < 0.5\}$. The boundary conditions are

$$u(0, y) = 0, \quad u(x, 0) = 0, \quad u(x, 0.5) = 200x, \quad \text{and } u(0.5, y) = 200y.$$

If $n = m = 4$, the problem has the grid given in Figure 12.5 and the difference equation is

$$4w_{i,j} - w_{i+1,j} - w_{i-1,j} - w_{i,j-1} - w_{i,j+1} = 0, \tag{12.1}$$

for each $i = 1, 2, 3$ and $j = 1, 2, 3$.

Figure 12.5

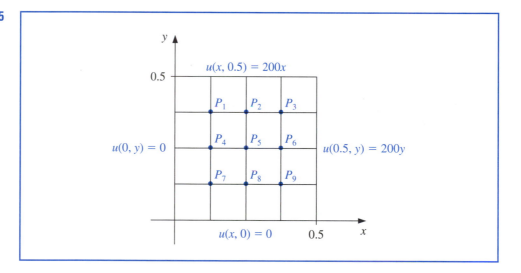

We can express this in terms of the relabeled interior grid points $w_i = u(P_i)$. We retain the double subscript notation on w to indicate the known points on the boundary. For example, at the point P_1 which gives the approximation to $u(0.125, 0.375)$, we have

$$4w_1 - w_2 - w_{0,3} - w_{1,4} - w_4 = 0 \quad \text{or} \quad 4w_1 - w_2 - w_4 = w_{0,3} + w_{1,4}.$$

In this equation, $w_{0,3} = u(0, 0.375) = 0$ and $w_{1,4} = u(0.125, 0.5) = 25$ are given by the boundary conditions, so

$$4w_1 - w_2 - w_4 = 0 + 25 = 25.$$

Expressing all the interior grid points in this manner gives the system of equations

$$P_1: \qquad 4w_1 - w_2 - w_4 = w_{0,3} + w_{1,4} = 0 + 25 = 25,$$

$$P_2: \qquad 4w_2 - w_3 - w_1 - w_5 = w_{2,4} = 50,$$

$$P_3: \qquad 4w_3 - w_2 - w_6 = w_{4,3} + w_{3,4} = 75 + 75 = 150,$$

$$P_4: \qquad 4w_4 - w_5 - w_1 - w_7 = w_{0,2} = 0,$$

$$P_5: \quad 4w_5 - w_6 - w_4 - w_2 - w_8 = 0,$$

$$P_6: \qquad 4w_6 - w_5 - w_3 - w_9 = w_{4,2} = 50,$$

$$P_7: \qquad 4w_7 - w_8 - w_4 = w_{0,1} + w_{1,0} = 0 + 0 = 0,$$

$$P_8: \qquad 4w_8 - w_9 - w_7 - w_5 = w_{2,0} = 0,$$

$$P_9: \qquad 4w_9 - w_8 - w_6 = w_{3,0} + w_{4,1} = 0 + 25 = 25.$$

Table 12.1

i	w_i
1	18.75
2	37.50
3	56.25
4	12.50
5	25.00
6	37.50
7	6.25
8	12.50
9	18.75

The program POIFD121 implements the Poisson Finite-Difference method.

So the linear system associated with this problem has the form

$$
\begin{bmatrix}
4 & -1 & 0 & -1 & 0 & 0 & 0 & 0 & 0 \\
-1 & 4 & -1 & 0 & -1 & 0 & 0 & 0 & 0 \\
0 & -1 & 4 & 0 & 0 & -1 & 0 & 0 & 0 \\
-1 & 0 & 0 & 4 & -1 & 0 & -1 & 0 & 0 \\
0 & -1 & 0 & -1 & 4 & -1 & 0 & -1 & 0 \\
0 & 0 & -1 & 0 & -1 & 4 & 0 & 0 & -1 \\
0 & 0 & 0 & -1 & 0 & 0 & 4 & -1 & 0 \\
0 & 0 & 0 & 0 & -1 & 0 & -1 & 4 & -1 \\
0 & 0 & 0 & 0 & 0 & -1 & 0 & -1 & 4
\end{bmatrix}
\begin{bmatrix}
w_1 \\ w_2 \\ w_3 \\ w_4 \\ w_5 \\ w_6 \\ w_7 \\ w_8 \\ w_9
\end{bmatrix}
=
\begin{bmatrix}
25 \\ 50 \\ 150 \\ 0 \\ 0 \\ 50 \\ 0 \\ 0 \\ 25
\end{bmatrix}.
$$

The values of $w_1, w_2, \ldots, w_9$ are found by applying the Gauss-Seidel method to this matrix, and are given in Table 12.1.

These answers are exact because the true solution, $u(x, y) = 400xy$, has $\dfrac{\partial^4 u}{\partial x^4} = \dfrac{\partial^4 u}{\partial y^4} \equiv 0$, and the truncation error is zero at each step.   ▪

Choice of Iterative Method

For simplicity, the Gauss-Seidel iterative procedure is used in the program POIFD121, but it is generally advisable to use a direct technique such as Gaussian elimination when the system is small, on the order of 100 or less, because the positive definiteness ensures stability with respect to round-off errors. Since the matrix is banded, methods in Section 6.6 could also be used. For large systems, an iterative method should be used. If the SOR method is used, the choice of the optimal ω in this situation comes from the fact that when A is decomposed into its diagonal D and upper- and lower-triangular parts U and L,

$$A = D - L - U,$$

and B is the matrix for the Jacobi method,

$$B = D^{-1}(L + U),$$

then the spectral radius of B is (see [Var] page 187)

$$\rho(B) = \frac{1}{2}\left[\cos\left(\frac{\pi}{m}\right) + \cos\left(\frac{\pi}{n}\right)\right].$$

By the SOR Convergent result on page 300, the value of ω is

$$\omega = \frac{2}{1 + \sqrt{1 - [\rho(B)]^2}} = \frac{4}{2 + \sqrt{4 - \left[\cos\left(\dfrac{\pi}{m}\right) + \cos\left(\dfrac{\pi}{n}\right)\right]^2}}.$$

Example 2 Use the Poisson Finite-Difference method with $n = 6$, $m = 5$, and a tolerance of 10^{-10} to approximate the solution to

$$\frac{\partial^2 u}{\partial x^2}(x, y) + \frac{\partial^2 u}{\partial y^2}(x, y) = xe^y, \quad 0 < x < 2, \quad 0 < y < 1,$$

with the boundary conditions

$$u(0, y) = 0, \quad u(2, y) = 2e^y, \quad 0 \le y \le 1,$$
$$u(x, 0) = x, \quad u(x, 1) = ex, \quad 0 \le x \le 2.$$

Compare the results with the exact solution $u(x, y) = xe^y$.

Solution Using the program POIFD121 with a maximum number of iterations set at $N = 100$ gives the results in Table 12.2. The stopping criterion for the Gauss-Seidel method in Step 17 requires that

$$\left| w_{ij}^{(l)} - w_{ij}^{(l-1)} \right| \le 10^{-10},$$

for each $i = 1, \ldots, 5$ and $j = 1, \ldots, 4$. The solution to the difference equation was accurately obtained, and the procedure stopped at $l = 61$. Even though consecutive values of the approximations agree to within 10^{-10}, the approximations $w_{i,j}$ to $u(x_i, y_j)$ do not have nearly this degree of accuracy. ∎

Table 12.2

i	j	x_i	y_j	$w_{i,j}^{(61)}$	$u(x_i, y_j)$	$\left| u(x_i, y_j) - w_{i,j}^{(61)} \right|$
1	1	0.3333	0.2000	0.40726	0.40713	1.30×10^{-4}
1	2	0.3333	0.4000	0.49748	0.49727	2.08×10^{-4}
1	3	0.3333	0.6000	0.60760	0.60737	2.23×10^{-4}
1	4	0.3333	0.8000	0.74201	0.74185	1.60×10^{-4}
2	1	0.6667	0.2000	0.81452	0.81427	2.55×10^{-4}
2	2	0.6667	0.4000	0.99496	0.99455	4.08×10^{-4}
2	3	0.6667	0.6000	1.2152	1.2147	4.37×10^{-4}
2	4	0.6667	0.8000	1.4840	1.4837	3.15×10^{-4}
3	1	1.0000	0.2000	1.2218	1.2214	3.64×10^{-4}
3	2	1.0000	0.4000	1.4924	1.4918	5.80×10^{-4}
3	3	1.0000	0.6000	1.8227	1.8221	6.24×10^{-4}
3	4	1.0000	0.8000	2.2260	2.2255	4.51×10^{-4}
4	1	1.3333	0.2000	1.6290	1.6285	4.27×10^{-4}
4	2	1.3333	0.4000	1.9898	1.9891	6.79×10^{-4}
4	3	1.3333	0.6000	2.4302	2.4295	7.35×10^{-4}
4	4	1.3333	0.8000	2.9679	2.9674	5.40×10^{-4}
5	1	1.6667	0.2000	2.0360	2.0357	3.71×10^{-4}
5	2	1.6667	0.4000	2.4870	2.4864	5.84×10^{-4}
5	3	1.6667	0.6000	3.0375	3.0369	6.41×10^{-4}
5	4	1.6667	0.8000	3.7097	3.7092	4.89×10^{-4}

EXERCISE SET 12.2

1. Use the Finite-Difference method to approximate the solution to the elliptic partial-differential equation

$$\frac{\partial^2 u}{\partial x^2} + \frac{\partial^2 u}{\partial y^2} = 4, \qquad 0 < x < 1, \qquad 0 < y < 2;$$
$$u(x, 0) = x^2, \qquad u(x, 2) = (x - 2)^2, \qquad 0 \le x \le 1;$$
$$u(0, y) = y^2, \qquad u(1, y) = (y - 1)^2, \qquad 0 \le y \le 2.$$

Use $h = k = \frac{1}{2}$ and compare the results to the exact solution $u(x, y) = (x - y)^2$.

2. Use the Finite-Difference method to approximate the solution to the elliptic partial-differential equation

$$\frac{\partial^2 u}{\partial x^2} + \frac{\partial^2 u}{\partial y^2} = 0, \qquad 1 < x < 2, \qquad 0 < y < 1;$$
$$u(x, 0) = 2 \ln x, \qquad u(x, 1) = \ln(x^2 + 1), \qquad 1 \le x \le 2;$$
$$u(1, y) = \ln(y^2 + 1), \qquad u(2, y) = \ln(y^2 + 4), \qquad 0 \le y \le 1.$$

Use $h = k = \frac{1}{3}$ and compare the results to the exact solution $u(x, y) = \ln(x^2 + y^2)$.

3. Use the Finite-Difference method to approximate the solutions to the following elliptic partial-differential equations:

a. $\dfrac{\partial^2 u}{\partial x^2} + \dfrac{\partial^2 u}{\partial y^2} = 0, \qquad 0 < x < 1, \qquad 0 < y < 1;$
    $u(x, 0) = 0, \qquad u(x, 1) = x, \qquad 0 \le x \le 1;$
    $u(0, y) = 0, \qquad u(1, y) = y, \qquad 0 \le y \le 1.$
    Use $h = k = 0.2$ and compare the results with the exact solution $u(x, y) = xy$.

b. $\dfrac{\partial^2 u}{\partial x^2} + \dfrac{\partial^2 u}{\partial y^2} = -(\cos(x + y) + \cos(x - y)), \quad 0 < x < \pi, \quad 0 < y < \dfrac{\pi}{2};$
    $u(0, y) = \cos y, \quad u(\pi, y) = -\cos y, \quad 0 \le y \le \dfrac{\pi}{2},$
    $u(x, 0) = \cos x, \quad u\left(x, \dfrac{\pi}{2}\right) = 0, \quad 0 \le x \le \pi.$
    Use $h = \pi/5$ and $k = \pi/10$ and compare the results with the exact solution $u(x, y) = \cos x \cos y$.

c. $\dfrac{\partial^2 u}{\partial x^2} + \dfrac{\partial^2 u}{\partial y^2} = (x^2 + y^2)e^{xy}, \qquad 0 < x < 2, \quad 0 < y < 1;$
    $u(0, y) = 1, \qquad u(2, y) = e^{2y}, \quad 0 \le y \le 1;$
    $u(x, 0) = 1, \qquad u(x, 1) = e^x, \quad 0 \le x \le 2.$
    Use $h = 0.2$ and $k = 0.1$, and compare the results with the exact solution $u(x, y) = e^{xy}$.

d. $\dfrac{\partial^2 u}{\partial x^2} + \dfrac{\partial^2 u}{\partial y^2} = \dfrac{x}{y} + \dfrac{y}{x}, \qquad 1 < x < 2, \quad 1 < y < 2;$
    $u(x, 1) = x \ln x, \qquad u(x, 2) = x \ln(4x^2), \quad 1 \le x \le 2;$
    $u(1, y) = y \ln y, \qquad u(2, y) = 2y \ln(2y), \quad 1 \le y \le 2.$
    Use $h = k = 0.1$ and compare the results with the exact solution $u(x, y) = xy \ln xy$.

4. Repeat Exercise 3(a) using extrapolation with $h_0 = 0.2$, $h_1 = h_0/2$, and $h_2 = h_0/4$.

5. A coaxial cable is made of a square inner conductor with side length 0.1 cm and a square outer conductor with side length 0.5 cm. The potential at a point in the cross section of the cable is described by Laplace's equation. Suppose that the inner conductor is kept at 0 volts and the outer conductor is kept at 110 volts. Find the potential between the two conductors by placing a grid with horizontal mesh spacing $h = 0.1$ cm and vertical mesh spacing $k = 0.1$ cm on the region

$$D = \{(x, y) \,|\, 0 \le x, y \le 0.5\}.$$

Approximate the solution to Laplace's equation at each grid point, and use the two sets of boundary conditions to derive a linear system to be solved by the Gauss-Seidel method.

6.   A 6-cm by 5-cm rectangular silver plate has heat being uniformly generated at each point at the rate $q = 1.5$ cal/cm$^3 \cdot$ s. Let x represent the distance along the edge of the plate of length 6 cm and y be the distance along the edge of the plate of length 5 cm. Suppose that the temperature u along the edges is kept at the following temperatures:

$$u(x, 0) = x(6 - x), \qquad u(x, 5) = 0, \qquad 0 \leq x \leq 6,$$
$$u(0, y) = y(5 - y), \qquad u(6, y) = 0, \qquad 0 \leq y \leq 5,$$

where the origin lies at a corner of the plate with coordinates $(0, 0)$ and the edges lie along the positive x- and y-axes. The steady-state temperature $u = u(x, y)$ satisfies Poisson's equation:

$$\frac{\partial^2 u}{\partial x^2}(x, y) + \frac{\partial^2 u}{\partial y^2}(x, y) = -\frac{q}{K}, \qquad 0 < x < 6, \qquad 0 < y < 5,$$

where K, the thermal conductivity, is 1.04 cal/cm $\cdot$ deg $\cdot$ s. Use the Finite-Difference method with $h = 0.4$ and $k = \frac{1}{3}$ to approximate the temperature $u(x, y)$.

12.3 Finite-Difference Methods for Parabolic Problems

In this section we consider the numerical solution to a problem involving a **parabolic** partial-differential equation of the form

$$\frac{\partial u}{\partial t}(x, t) - \alpha^2 \frac{\partial^2 u}{\partial x^2}(x, t) = 0,$$

where $\alpha \neq 0$. A physical problem that requires the solution to this equation is the flow of heat along a rod of length l (see Figure 12.6) that has a uniform temperature within each cross-sectional element. It requires the rod to be perfectly insulated on its lateral surface. The constant α is independent of the position in the rod and is determined by the heat-conductive properties of the material of which the rod is composed.

Figure 12.6

One of the typical sets of constraints for a heat-flow problem of this type is to specify the initial heat distribution in the rod,

$$u(x, 0) = f(x),$$

and to describe the behavior at the ends of the rod. For example, if the ends are held at constant temperatures U_1 and U_2, the boundary conditions have the form

$$u(0, t) = U_1 \quad \text{and} \quad u(l, t) = U_2,$$

and the heat distribution approaches the limiting temperature distribution

$$\lim_{t \to \infty} u(x, t) = U_1 + \frac{U_2 - U_1}{l} x.$$

If, instead, the rod is insulated so that no heat flows through the ends, the boundary conditions are

$$\frac{\partial u}{\partial x}(0, t) = 0 \quad \text{and} \quad \frac{\partial u}{\partial x}(l, t) = 0,$$

resulting in a constant temperature in the rod as the limiting case. The parabolic partial-differential equation is also of importance in the study of gas diffusion. In fact, it is known in some circles as the **diffusion equation**.

The specific parabolic partial-differential equation we will consider is

$$\frac{\partial u}{\partial t}(x, t) = \alpha^2 \frac{\partial^2 u}{\partial x^2}(x, t), \qquad \text{for } 0 < x < l \quad \text{and} \quad t > 0,$$

on the region

$$R = \{(x, t) \mid 0 < x < 1, 0 < t\}$$

subject to the conditions

$$u(0, t) = u(l, t) = 0 \qquad \text{for } t > 0, \quad \text{and} \quad u(x, 0) = f(x) \qquad \text{for } 0 \leq x \leq l.$$

The approach we use to approximate the solution to this problem involves finite differences similar to those in Section 12.2. First select an integer $m > 0$ and define $h = l/m$. Then select a time-step size k. The grid points for this situation are (x_i, t_j), where $x_i = ih$, for $i = 0, 1, \ldots, m$, and $t_j = jk$, for $j = 0, 1, \ldots$.

Forward-Difference Method

We obtain the Forward-Difference method by using the Taylor polynomial in t to form the difference quotient

$$\frac{\partial u}{\partial t}(x_i, t_j) = \frac{u(x_i, t_{j+1}) - u(x_i, t_j)}{k} - \frac{k}{2} \frac{\partial^2 u}{\partial t^2}(x_i, \mu_j),$$

for some μ_j in (t_j, t_{j+1}), and the Taylor polynomial in x to form the difference quotient

$$\frac{\partial^2 u}{\partial x^2}(x_i, t_j) = \frac{u(x_{i+1}, t_j) - 2u(x_i, t_j) + u(x_{i-1}, t_j)}{h^2} - \frac{h^2}{12} \frac{\partial^4 u}{\partial x^4}(\xi_i, t_j),$$

for some ξ_i in (x_{i-1}, x_{i+1}).

The parabolic partial-differential equation implies that at the interior grid point (x_i, t_j) we have

$$\frac{\partial u}{\partial t}(x_i, t_j) - \alpha^2 \frac{\partial^2 u}{\partial x^2}(x_i, t_j) = 0,$$

so the Forward-Difference method using the two difference quotients is

$$\frac{w_{i,j+1} - w_{ij}}{k} - \alpha^2 \frac{w_{i+1,j} - 2w_{ij} + w_{i-1,j}}{h^2} = 0,$$

where w_{ij} approximates $u(x_i, t_j)$. The error for this difference equation is

$$\tau_{ij} = \frac{k}{2} \frac{\partial^2 u}{\partial t^2}(x_i, \mu_j) - \alpha^2 \frac{h^2}{12} \frac{\partial^4 u}{\partial x^4}(\xi_i, t_j).$$

Solving the difference equation for $w_{i,j+1}$ gives

$$w_{i,j+1} = \left(1 - \frac{2\alpha^2 k}{h^2}\right) w_{ij} + \alpha^2 \frac{k}{h^2}(w_{i+1,j} + w_{i-1,j}),$$

for each $i = 1, 2, \ldots, m-1$ and $j = 1, 2, \ldots,$.

This equation is generally reexpressed using the positive constant $\lambda = \alpha^2 k / h^2$, so

$$w_{i,j+1} = (1 - 2\lambda)w_{ij} + \lambda(w_{i+1,j} + w_{i-1,j}).$$

Since the initial condition $u(x, 0) = f(x)$ implies that $w_{i0} = f(x_i)$, for each $i = 0, 1, \ldots, m$, these values can be used in the difference equation to find the value of w_{i1}, for each $i = 1, 2, \ldots, m-1$. The additional conditions $u(0, t) = 0$ and $u(l, t) = 0$ imply that $w_{01} = w_{m1} = 0$, so all the entries of the form w_{i1} can be determined. If the procedure is reapplied once all the approximations w_{i1}, are known, the values of $w_{i2}, w_{i3}, \ldots, w_{i,m-1}$ can be obtained in a similar manner.

The explicit nature of the Forward-Difference method implies that the $(m-1) \times (m-1)$ matrix associated with this system can be written in the tridiagonal form

$$A = \begin{bmatrix} (1-2\lambda) & \lambda & 0 & \cdots & \cdots & 0 \\ \lambda & (1-2\lambda) & \lambda & & & \vdots \\ 0 & & & & & 0 \\ \vdots & & & & & \lambda \\ 0 & \cdots & \cdots & 0 & \lambda & (1-2\lambda) \end{bmatrix}.$$

If we let

$$\mathbf{w}^{(0)} = (f(x_1), f(x_2), \ldots, f(x_{m-1}))^t$$

and

$$\mathbf{w}^{(j)} = (w_{1j}, w_{2j}, \ldots, w_{m-1,j})^t, \qquad \text{for each } j = 1, 2, \ldots,$$

then the approximate solution is given by

$$\mathbf{w}^{(j+1)} = A\mathbf{w}^{(j)}, \qquad \text{for each } j = 0, 1, 2, \ldots.$$

So $\mathbf{w}^{(j+1)}$ is obtained from $\mathbf{w}^{(j)}$ by a simple matrix multiplication, and is of order $O(k+h^2)$.

The Forward-Difference method involves, at a typical step, the mesh points

$$(x_{i-1}, t_j), \qquad (x_i, t_j), \qquad (x_i, t_{j+1}), \quad \text{and} \quad (x_{i+1}, t_j),$$

and uses approximations at the points marked with ×'s in Figure 12.7. The boundary and initial conditions give the values at the circled mesh points.

Figure 12.7

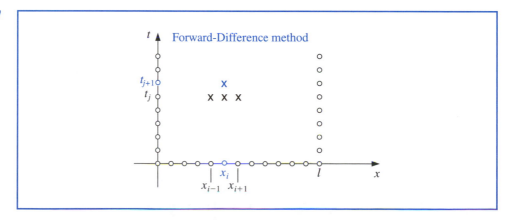

Example 1 Use steps sizes (a) $h = 0.1$ and $k = 0.0005$ and (b) $h = 0.1$ and $k = 0.01$ in the Forward-Difference method to approximate the solution to the heat equation

$$\frac{\partial u}{\partial t}(x, t) - \frac{\partial^2 u}{\partial x^2}(x, t) = 0, \quad 0 < x < 1, \quad 0 \leq t,$$

with boundary conditions

$$u(0, t) = u(1, t) = 0, \quad 0 < t,$$

and initial conditions

$$u(x, 0) = \sin(\pi x), \quad 0 \leq x \leq 1.$$

Compare the results at $t = 0.5$ to the exact solution

$$u(x, t) = e^{-\pi^2 t} \sin(\pi x).$$

Solution (a) The Forward-Difference method with $h = 0.1$, $k = 0.0005$ and $\lambda = (1)^2(0.0005/(0.1)^2) = 0.05$ gives the results in the third column of Table 12.3. As can be seen from the fourth column, these results are quite accurate.
(b) The Forward-Difference method with $h = 0.1$, $k = 0.01$ and $\lambda = (1)^2(0.01/(0.1)^2) = 1$ gives the results in the fifth column of Table 12.3. As can be seen from the sixth column, these results are worthless. ▪

Table 12.3

x_i	$u(x_i, 0.5)$	$w_{i,1000}$ $k = 0.0005$	$\|u(x_i, 0.5) - w_{i,1000}\|$	$w_{i,50}$ $k = 0.01$	$\|u(x_i, 0.5) - w_{i,50}\|$
0.0	0	0		0	
0.1	0.00222241	0.00228652	6.411×10^{-5}	8.19876×10^7	8.199×10^7
0.2	0.00422728	0.00434922	1.219×10^{-4}	-1.55719×10^8	1.557×10^8
0.3	0.00581836	0.00598619	1.678×10^{-4}	2.13833×10^8	2.138×10^8
0.4	0.00683989	0.00703719	1.973×10^{-4}	-2.50642×10^8	2.506×10^8
0.5	0.00719188	0.00739934	2.075×10^{-4}	2.62685×10^8	2.627×10^8
0.6	0.00683989	0.00703719	1.973×10^{-4}	-2.49015×10^8	2.490×10^8
0.7	0.00581836	0.00598619	1.678×10^{-4}	2.11200×10^8	2.112×10^8
0.8	0.00422728	0.00434922	1.219×10^{-4}	-1.53086×10^8	1.531×10^8
0.9	0.00222241	0.00228652	6.511×10^{-5}	8.03604×10^7	8.036×10^7
1.0	0	0		0	

To use MATLAB to determine the results in Example 1(a), we define l, α, m, h, and k with

```
l = 1;alpha = 1;m = 10;h = 1/m;k = 0.0005;
```

and then define $f(x)$. We initialize $(m - 1) \times (m - 1)$ matrix A and the vectors $\mathbf{x}$ and $\mathbf{u}$ by placing 0s in all their entries. The nonzero entries will be overwritten later.

```
f = inline('sin(pi*x)','x')
A = zeros(m-1,m-1);w = zeros(m-1,1);u = zeros(m-1,1);
```

The $(m - 1)$-dimensional vector $\mathbf{w}_0$ is defined by

```
for i = 1:m-1
    w(i) = f(i*h);
end;
```

and λ is

```
lambda = alpha^2*k/h^2
```

The nonzero entries of A are

```
A(1,1) = 1-2*lambda;
A(1,2) = lambda;
for i = 2:m-2
    A(i,i-1) = lambda;
    A(i,i+1) = lambda;
    A(i,i) = 1-2*lambda;
end;
A(m-1,m-2) = lambda;
A(m-1,m-1) = 1-2*lambda;
```

Since $k = 0.0005$, to obtain $\mathbf{w}^{(1000)}$ we raise A to the 1000th power and multiply it times $\mathbf{w}^{(0)}$

```
u = A^1000*w
```

which gives

$$u = 0.002286520786578$$
$$0.004349220987439$$
$$0.005986189135245$$
$$0.007037187382261$$
$$0.007399336697334$$
$$0.007037187382261$$
$$0.005986189135245$$
$$0.004349220987439$$
$$0.002286520786578$$

For the results in Example 1(**b**), MATLAB can be used with the same commands as in (**a**), with the replacement of the first command line with

```
l = 1;alpha = 1;m = 10;h = 1/m;k = 0.01;
```

and the last command line with

```
u = A^50*w
```

Stability Considerations

An error of order $O(k + h^2)$ is expected in Example 1. This is obtained with $h = 0.1$ and $k = 0.0005$, but it is certainly not obtained when $h = 0.1$ and $k = 0.01$. To explain the difficulty, we must look at the stability of the Forward-Difference method.

If the error $\mathbf{e}^{(0)} = (e_1^{(0)}, e_2^{(0)}, \ldots, e_{m-1}^{(0)})^t$ occurs in the initial data $\mathbf{w}^{(0)} = (f(x_1), f(x_2), \ldots, f(x_{m-1}))^t$, or in any particular step (the choice of the initial step is simply for convenience), then the error of $A\mathbf{e}^{(0)}$ propagates in $\mathbf{w}^{(1)}$ because

$$\mathbf{w}^{(1)} = A(\mathbf{w}^{(0)} + \mathbf{e}^{(0)}) = A\mathbf{w}^{(0)} + A\mathbf{e}^{(0)}.$$

This process continues. At the nth time step, the error in $\mathbf{w}^{(n)}$ due to $\mathbf{e}^{(0)}$ is $A^n \mathbf{e}^{(0)}$.

The Forward-Difference method is consequently conditionally stable; that is, stable precisely when these errors do not grow as n increases. But this is true if and only if

$$\|A^n \mathbf{e}^{(0)}\| \leq \|\mathbf{e}^{(0)}\|$$

for all n. To ensure that this is true, we need to have $\|A^n\| \leq 1$, a condition that requires that the spectral radius $\rho(A^n) = (\rho(A))^n \leq 1$. The Forward-Difference method is therefore stable only if $\rho(A) \leq 1$.

The eigenvalues of A are (see [BF] pages 735 and 857)

$$\mu_i = 1 - 4\lambda \left(\sin\left(\frac{i\pi}{2m}\right) \right)^2, \qquad \text{for each } i = 1, 2, \ldots, m-1,$$

so the condition for stability reduces to determining whether

$$\rho(A) = \max_{1 \leq i \leq m-1} \left| 1 - 4\lambda \left(\sin\left(\frac{i\pi}{2m}\right) \right)^2 \right| \leq 1,$$

which simplifies to

$$0 \leq \lambda \left(\sin\left(\frac{i\pi}{2m}\right) \right)^2 \leq \frac{1}{2}, \qquad \text{for each } i = 1, 2, \ldots, m-1.$$

Stability requires that this inequality condition hold as $h \to 0$ or, equivalently, as $m \to \infty$, so the fact that

$$\lim_{m \to \infty} \left[\sin\left(\frac{(m-1)\pi}{2m}\right) \right]^2 = 1$$

means that stability will occur only if $0 \leq \lambda \leq \frac{1}{2}$. Since $\lambda = \alpha^2(k/h^2)$, this inequality requires that h and k be chosen so that

$$\alpha^2 \frac{k}{h^2} \leq \frac{1}{2}.$$

In Example 1 we have $\alpha = 1$, so the condition is satisfied when $h = 0.1$ and $k = 0.0005$ because

$$\frac{0.0005}{(0.1)^2} = \frac{1}{20} < \frac{1}{2}.$$

But when k was increased to 0.01 with no corresponding increase in h, the ratio was

$$\frac{0.01}{(0.1)^2} = 1 > \frac{1}{2},$$

and stability problems became obvious.

Backward-Difference Method

To obtain a more stable method, we consider an *implicit-difference method* that results from using the backward-difference quotient for $(\partial u/\partial t)(x_i, t_j)$ in the form

$$\frac{\partial u}{\partial t}(x_i, t_j) = \frac{u(x_i, t_j) - u(x_i, t_{j-1})}{k} + \frac{k}{2}\frac{\partial^2 u}{\partial t^2}(x_i, \mu_j),$$

for some μ_j in (t_{j-1}, t_j). Substituting this equation and the centered-difference formula for $\partial^2 u/\partial x^2$ into the partial-differential equation gives

$$\frac{u(x_i, t_j) - u(x_i, t_{j-1})}{k} - \alpha^2 \frac{u(x_{i+1}, t_j) - 2u(x_i, t_j) + u(x_{i-1}, t_j)}{h^2}$$

$$= -\frac{k}{2}\frac{\partial^2 u}{\partial t^2}(x_i, \mu_j) - \alpha^2 \frac{h^2}{12}\frac{\partial^4 u}{\partial x^4}(\xi_i, t_j),$$

for some ξ_i in (x_{i-1}, x_{i+1}). The difference method this produces is called the *Backward-Difference method* for the heat equation.

Backward-Difference Method

$$\frac{w_{ij} - w_{i,j-1}}{k} - \alpha^2 \frac{w_{i+1,j} - 2w_{ij} + w_{i-1,j}}{h^2} = 0,$$

for each $i = 1, 2, \ldots, m - 1$, and $j = 1, 2, \ldots$.

The Backward-Difference method involves, at a typical step, the mesh points

$$(x_i, t_j), \qquad (x_i, t_{j-1}), \qquad (x_{i-1}, t_j), \quad \text{and} \quad (x_{i+1}, t_j),$$

and involves approximations at the points marked with $\times$'s in Figure 12.8.

Figure 12.8

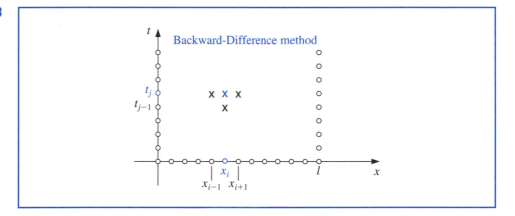

The boundary and initial conditions associated with the problem give information at the circled mesh points. There is no explicit procedure for finding the approximations in the Backward-Difference method. For example, the initial conditions give us values at (x_0, t_0), (x_0, t_1), (x_1, t_0), and (x_2, t_0), but this information will not directly give an approximation at (x_1, t_1) or at (x_2, t_1).

We again let λ denote the quantity $\alpha^2(k/h^2)$, and use the knowledge that $w_{i0} = f(x_i)$, for each $i = 1, 2, \ldots, m-1$, and $w_{mj} = w_{0j} = 0$, for each $j = 1, 2, \ldots,$. The equations for the Backward-Difference method are

$$(1 + 2\lambda)w_{1,j} - \lambda w_{2,j} = w_{1,j-1},$$

$$-\lambda w_{1,j} + (1 + 2\lambda)w_{2,j} - \lambda w_{3,j} = w_{2,j-1},$$

$$-\lambda w_{2,j} + (1 + 2\lambda)w_{3,j} - \lambda w_{4,j} = w_{3,j-1},$$

or, in general,

$$(1 + 2\lambda)w_{ij} - \lambda w_{i+1,j} - \lambda w_{i-1,j} = w_{i,j-1},$$

for each $i = 1, 2, \ldots, m-1$ and $j = 1, 2, \ldots$. The Difference method therefore has the matrix representation

$$
\begin{bmatrix}
(1+2\lambda) & -\lambda & 0 & \cdots & \cdots & 0 \\
-\lambda & \ddots & \ddots & & & \vdots \\
0 & \ddots & \ddots & \ddots & & 0 \\
\vdots & & \ddots & \ddots & \ddots & -\lambda \\
0 & \cdots & \cdots & 0 & -\lambda & (1+2\lambda)
\end{bmatrix}
\begin{bmatrix}
w_{1j} \\
w_{2j} \\
\vdots \\
\\
w_{m-1,j}
\end{bmatrix}
=
\begin{bmatrix}
w_{1,j-1} \\
w_{2,j-1} \\
\vdots \\
\\
w_{m-1,j-1}
\end{bmatrix},
$$

or $A\mathbf{w}^{(j)} = \mathbf{w}^{(j-1)}$.

Hence we must now solve a linear system to obtain $\mathbf{w}^{(j)}$ from $\mathbf{w}^{(j-1)}$. Since $\lambda > 0$, the matrix A is positive definite and strictly diagonally dominant, as well as being tridiagonal. To solve this system, we can use either Crout factorization for tridiagonal linear systems or an iterative technique such as the SOR or conjugate gradient methods.

Program HEBDM122 uses Crout factorization, which is acceptable unless m is large.

> The program HEBDM122 implements the Heat Equation Backward-Difference Method.

Example 2   Use the Backward-Difference method with $h = 0.1$ and $k = 0.01$ to approximate the solution to the heat equation

$$\frac{\partial u}{\partial t}(x, t) - \frac{\partial^2 u}{\partial x^2}(x, t) = 0, \quad 0 < x < 1, \quad 0 < t,$$

subject to the constraints

$$u(0, t) = u(1, t) = 0, \quad 0 < t, \quad u(x, 0) = \sin \pi x, \quad 0 \leq x \leq 1.$$

Solution   This problem was considered in Example 1 where we found that choosing $h = 0.1$ and $k = 0.0005$ gave quite accurate results. However, with the values $h = 0.1$ and $k = 0.01$ in that example the results were exceptionally poor. To demonstrate the unconditional stability of the Backward-Difference method, we will use $h = 0.1$ and $k = 0.01$ and again compare $w_{i,50}$ to $u(x_i, 0.5)$, where $i = 0, 1, \ldots, 10$.

To use MATLAB to obtain $\mathbf{w}^{(50)}$ from $\mathbf{w}^{(0)}$ we define $l, \alpha, m, h, k,$ and $f(x)$ with

```
l = 1;alpha = 1;m = 10;h = 1/m;k = 0.01;
f = inline('sin(pi*x)','x')
```

Then we initialize the matrix A and the vectors $\mathbf{w}$ and $\mathbf{u}$ by placing 0s in all their entries. The nonzero entries will be overwritten later.

```
A = zeros(m-1,m-1);w = zeros(m-1,1);u = zeros(m-1,1);
```

To determine λ we use

```
lambda = alpha^2*k/h^2
```

We generate the nonzero entries of A and overwrite the initial values of A with

```
A(1,1) = 1+2*lambda;
A(1,2) = -lambda;
for i = 2:m-2
    A(i,i-1) = -lambda;
    A(i,i+1) = -lambda;
    A(i,i) = 1+2*lambda;
end;
A(m-1,m-2) = -lambda;
A(m-1,m-1) = 1+2*lambda;
```

and update the vector **w** with

```
for i = 1:m-1
w(i) = f(i*h);
end
```

Then we solve the 50 linear systems in turn to obtain our approximation $\mathbf{w}^{(50)}$ to $u(x_i, 0.5)$ with

```
for i = 1:50
    u = linsolve(A,w);
    w = u;
end;
w
```

This produces

$$
\begin{aligned}
w = \ & 0.002898016645054 \\
& 0.005512355229220 \\
& 0.007587106076713 \\
& 0.008919178118941 \\
& 0.009378178863319 \\
& 0.008919178118941 \\
& 0.007587106076713 \\
& 0.005512355229220 \\
& 0.002898016645054
\end{aligned}
$$

The results listed in Table 12.4 have the same values of h and k as those in the fifth and sixth columns of Table 12.3. This implies that the Backward-Difference method is stable but the Forward-Difference method is not. ∎

Table 12.4

x_i	$w_{i,50}$	$u(x_i, 0.5)$	$|w_{i,50} - u(x_i, 0.5)|$
0.0	0	0	
0.1	0.00289802	0.00222241	6.756×10^{-4}
0.2	0.00551236	0.00422728	1.285×10^{-3}
0.3	0.00758711	0.00581836	1.769×10^{-3}
0.4	0.00891918	0.00683989	2.079×10^{-3}
0.5	0.00937818	0.00719188	2.186×10^{-3}
0.6	0.00891918	0.00683989	2.079×10^{-3}
0.7	0.00758711	0.00581836	1.769×10^{-3}
0.8	0.00551236	0.00422728	1.285×10^{-3}
0.9	0.00289802	0.00222241	6.756×10^{-4}
1.0	0	0	

Stability of the Backward-Difference Method

The reason the Backward-Difference method does not have the stability problems of the Forward-Difference method can be seen by analyzing the eigenvalues of the matrix A. For the Backward-Difference method the eigenvalues are (see [BF] page 737)

$$\mu_i = 1 + 4\lambda \left[\sin\left(\frac{i\pi}{2m}\right) \right]^2 \qquad \text{for each } i = 1, 2, \ldots, m-1;$$

and since $\lambda > 0$, we have $\mu_i > 1$ for all $i = 1, 2, \ldots, m-1$. This implies that A^{-1} exists because 0 is not an eigenvalue of A.

An error $\mathbf{e}^{(0)}$ in the initial data produces an error $(A^{-1})^n \mathbf{e}^{(0)}$ at the nth step. Since the eigenvalues of A^{-1} are the reciprocals of the eigenvalues of A, the spectral radius of A^{-1} is strictly less than 1 and the method is unconditionally stable; that is, stable independent of the choice of $\lambda = \alpha^2(k/h^2)$. The error for the method is of order $O(k + h^2)$, provided that the solution of the differential equation satisfies the usual differentiability conditions.

Crank-Nicolson Method

The weakness in the Backward-Difference method results from the fact that the error is $O(k + h^2)$, so it has a portion with order $O(k)$, which requires that time intervals be made much smaller than spatial intervals. It would be desirable to have a procedure with error of order $O(k^2 + h^2)$.

A method with error term $O(k^2 + h^2)$ is derived by averaging the Forward-Difference method at the jth step in t,

$$\frac{w_{i,j+1} - w_{ij}}{k} - \alpha^2 \frac{w_{i+1,j} - 2w_{ij} + w_{i-1,j}}{h^2} = 0,$$

which has error $(k/2)(\partial^2 u/\partial t^2)(x_i, \mu_j) + O(h^2)$, with the Backward-Difference method at the $(j+1)$st step in t,

$$\frac{w_{i,j+1} - w_{ij}}{k} - \alpha^2 \frac{w_{i+1,j+1} - 2w_{i,j+1} + w_{i-1,j+1}}{h^2} = 0,$$

which has error $-(k/2)(\partial^2 u/\partial t^2)(x_i, \hat{u}_j) + O(h^2)$. If we assume that

$$\frac{\partial^2 u}{\partial t^2}(x_i, \hat{\mu}_j) \approx \frac{\partial^2 u}{\partial t^2}(x_i, \mu_j),$$

Following work as a mathematical physicist during World War II, John Crank (1916–2006) did research in the numerical solution of partial differential equations; in particular, heat-conduction problems. The Crank-Nicolson method is based on work done with Phyllis Nicolson (1917–1968), a physicist at Leeds University. Their original paper on the method appeared in 1947 [CN].

then, in the averaged-difference method, the $O(k)$ portion is eliminated, and

$$\frac{w_{i,j+1} - w_{i,j}}{k} - \frac{\alpha^2}{2}\left[\frac{w_{i+1,j} - 2w_{i,j} + w_{i-1,j}}{h^2} + \frac{w_{i+1,j+1} - 2w_{i,j+1} + w_{i-1,j+1}}{h^2}\right] = 0,$$

has error of order $O(k^2 + h^2)$, provided, of course, that the usual differentiability conditions are satisfied. This is known as the **Crank-Nicolson method**.

If we let $\lambda = \alpha^2 \dfrac{k}{h^2}$, then these equations can be written in the form

$$-\frac{\lambda}{2}w_{i-1,j+1} + (1 + \lambda)w_{i,j+1} - \frac{\lambda}{2}w_{i+1,j+1} = \frac{\lambda}{2}w_{i-1,j} + (1 - \lambda)w_{i,j} + \frac{\lambda}{2}w_{i+1,j}.$$

Define

$$\mathbf{w}^{(j)} = (w_{1j}, w_{2j}, \ldots, w_{m-1,j})^t$$

for $j = 0, 1, 2, \ldots,$. Then these equations can be expressed in the form $A\mathbf{w}^{(j+1)} = B\mathbf{w}^{(j)}$, where the matrices A and B are

$$A = \begin{bmatrix} (1+\lambda) & -\frac{\lambda}{2} & 0 & \cdots & \cdots & 0 \\ -\frac{\lambda}{2} & & & & & \vdots \\ 0 & & & & & 0 \\ \vdots & & & & & -\frac{\lambda}{2} \\ 0 & \cdots & \cdots & 0 & -\frac{\lambda}{2} & (1+\lambda) \end{bmatrix}$$

and

$$B = \begin{bmatrix} (1-\lambda) & \frac{\lambda}{2} & 0 & \cdots & \cdots & 0 \\ \frac{\lambda}{2} & & & & & \vdots \\ 0 & & & & & 0 \\ \vdots & & & & & \frac{\lambda}{2} \\ 0 & \cdots & \cdots & 0 & \frac{\lambda}{2} & (1-\lambda) \end{bmatrix}.$$

The matrix A is positive definite, strictly diagonal dominant, and tridiagonal, so A is nonsingular. We could therefore express this matrix equation as $\mathbf{w}^{(j+1)} = A^{-1}B\mathbf{w}^{(j)}$. However, it is not computationally efficient to determine A^{-1}. Instead we use either Crout factorization for tridiagonal linear systems or the SOR method to obtain $\mathbf{w}^{(j+1)}$ from $\mathbf{w}^{(j)}$, for each $j = 0, 1, 2, \ldots$.

Figure 12.9 shows the interaction of the nodes for determining an approximation to $u(x_i, t_{j+1})$.

Figure 12.9

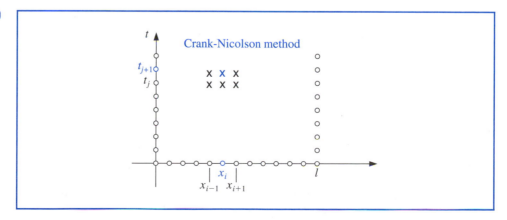

Program HECNM123 incorporates Crout factorization into the Crank-Nicolson technique.

Example 3 Use the Crank-Nicolson method with $h = 0.1$ and $k = 0.01$ to approximate the solution to the problem

> The program HECNM123 implements the Heat Equation Crank-Nicolson Method.

$$\frac{\partial u}{\partial t}(x, t) - \frac{\partial^2 u}{\partial x^2}(x, t) = 0, \quad 0 < x < 1 \quad 0 < t,$$

subject to the conditions

$$u(0, t) = u(1, t) = 0, \quad 0 < t,$$

and

$$u(x, 0) = \sin(\pi x), \quad 0 \le x \le 1.$$

Solution Choosing $h = 0.1$ and $k = 0.01$ gives $m = 10$, $N = 50$, and $\lambda = 1$ in program HECNM123. Recall that the Forward-Difference method gave dramatically poor results for this choice of h and k, but the Backward-Difference method gave results that were accurate to about 2×10^{-3} for entries in the middle of the table. The results in Table 12.5 indicate the increase in accuracy of the Crank-Nicolson method over the Backward-Difference method, the best of the two previously-discussed techniques. ▪

Table 12.5

x_i	$w_{i,50}$	$u(x_i, 0.5)$	$|w_{i,50} - u(x_i, 0.5)|$
0.0	0	0	
0.1	0.00230512	0.00222241	8.271×10^{-5}
0.2	0.00438461	0.00422728	1.573×10^{-4}
0.3	0.00603489	0.00581836	2.165×10^{-4}
0.4	0.00709444	0.00683989	2.546×10^{-4}
0.5	0.00745954	0.00719188	2.677×10^{-4}
0.6	0.00709444	0.00683989	2.546×10^{-4}
0.7	0.00603489	0.00581836	2.165×10^{-4}
0.8	0.00438461	0.00422728	1.573×10^{-4}
0.9	0.00230512	0.00222241	8.271×10^{-5}
1.0	0	0	

EXERCISE SET 12.3

1. Use the Backward-Difference method to approximate the solution to the following partial-differential equations.

   a. $\dfrac{\partial u}{\partial t} - \dfrac{\partial^2 u}{\partial x^2} = 0, \quad 0 < x < 2, \quad 0 < t;$

   $u(0, t) = u(2, t) = 0, \quad 0 < t,$

   $u(x, 0) = \sin \dfrac{\pi}{2} x, \quad 0 \le x \le 2.$

   Use $m = 4$, $T = 0.1$, and $N = 2$ and compare your answers to the exact solution $u(x, t) = e^{-(\pi^2/4)t} \sin(\pi x/2).$

b. $\dfrac{\partial u}{\partial t} - \dfrac{1}{16}\dfrac{\partial^2 u}{\partial x^2} = 0,$ $0 < x < 1,$ $0 < t;$

$u(0, t) = u(1, t) = 0,$ $0 < t,$

$u(x, 0) = 2 \sin 2\pi x,$ $0 \le x \le 1.$

Use $m = 3$, $T = 0.1$, and $N = 2$ and compare your answers to the exact solution $u(x, t) = 2e^{-(\pi^2/4)t} \sin 2\pi x$.

2.   Repeat Exercise 1 using the Crank-Nicolson method.

3.   Use the Forward-Difference method to approximate the solution to the following parabolic partial-differential equations.

a. $\dfrac{\partial u}{\partial t} - \dfrac{\partial^2 u}{\partial x^2} = 0,$ $0 < x < 2,$ $0 < t;$

$u(0, t) = u(2, t) = 0,$ $0 < t,$

$u(x, 0) = \sin 2\pi x,$ $0 \le x \le 2.$

(i) Use $h = 0.4$ and $k = 0.1$, and compare your answers at $t = 0.5$ to the exact solution $u(x, t) = e^{-4\pi^2 t} \sin 2\pi x$. (ii) Then use $h = 0.4$ and $k = 0.05$, and compare the answers.

b. $\dfrac{\partial u}{\partial t} - \dfrac{\partial^2 u}{\partial x^2} = 0,$ $0 < x < \pi,$ $0 < t;$

$u(0, t) = u(\pi, t) = 0,$ $0 < t,$

$u(x, 0) = \sin x,$ $0 \le x \le \pi.$

Use $h = \pi/10$ and $k = 0.05$ and compare your answers to the exact solution $u(x, t) = e^{-t} \sin x$ at $t = 0.5$.

c. $\dfrac{\partial u}{\partial t} - \dfrac{4}{\pi^2}\dfrac{\partial^2 u}{\partial x^2} = 0,$ $0 < x < 4,$ $0 < t;$

$u(0, t) = u(4, t) = 0,$ $0 < t,$

$u(x, 0) = \sin \dfrac{\pi}{4}x \left(1 + 2\cos\dfrac{\pi}{4}x\right),$ $0 \le x \le 4.$

Use $h = 0.2$ and $k = 0.04$. Compare your answers to the exact solution $u(x, t) = e^{-t} \sin(\pi/2)x + e^{-t/4} \sin(\pi/4)x$ at $t = 0.4$.

d. $\dfrac{\partial u}{\partial t} - \dfrac{1}{\pi^2}\dfrac{\partial^2 u}{\partial x^2} = 0,$ $0 < x < 1,$ $0 < 1;$

$u(0, t) = u(1, t) = 0,$ $0 < t,$

$u(x, 0) = \cos \pi \left(x - \dfrac{1}{2}\right),$ $0 \le x \le 1.$

Use $h = 0.1$ and $k = 0.04$. Compare your answers to the exact solution $u(x, t) = e^{-t} \cos \pi(x - \frac{1}{2})$ at $t = 0.4$.

4.   Repeat Exercise 3 using the Backward-Difference method.

5.   Repeat Exercise 3 using the Crank-Nicolson method.

6.   Modify the Backward-Difference method to accommodate the parabolic partial-differential equation

$$\dfrac{\partial u}{\partial t} - \dfrac{\partial^2 u}{\partial x^2} = F(x),\quad 0 < x < l,\quad 0 < t;$$

$$u(0, t) = u(l, t) = 0,\quad 0 < t,$$

$$u(x, 0) = f(x),\quad 0 \le x \le l.$$

7.   Use the results of Exercise 6 to approximate the solution to

$$\dfrac{\partial u}{\partial t} - \dfrac{\partial^2 u}{\partial x^2} = 2,\quad 0 < x < 1,\quad 0 < t;$$

$$u(0, t) = u(1, t) = 0,\quad 0 < t,$$

$$u(x, 0) = \sin \pi x + x(1 - x),$$

with $h = 0.1$ and $k = 0.01$. Compare your answer to the exact solution $u(x, t) = e^{-\pi^2 t} \sin \pi x + x(1 - x)$ at $t = 0.25$.

8. Modify the Backward-Difference method to accommodate the parabolic partial-differential equation

$$\frac{\partial u}{\partial t} - \alpha^2 \frac{\partial^2 u}{\partial x^2} = 0, \qquad 0 < x < l, \qquad 0 < t;$$

$$u(0, t) = \phi(t), \ u(l, t) = \Psi(t), \qquad 0 < t;$$

$$u(x, 0) = f(x), \qquad 0 \le x \le l,$$

where $f(0) = \phi(0)$ and $f(l) = \Psi(0)$.

9. The temperature $u(x, t)$ of a long, thin rod of constant cross section and homogeneous conducting material is governed by the one-dimensional heat equation. If heat is generated in the material, for example, by resistance to current or nuclear reaction, the heat equation becomes

$$\frac{\partial^2 u}{\partial x^2} + \frac{Kr}{\rho C} = K \frac{\partial u}{\partial t}, \qquad 0 < x < l, \qquad 0 < t,$$

where l is the length, ρ is the density, C is the specific heat, and K is the thermal diffusivity of the rod. The function $r = r(x, t, u)$ represents the heat generated per unit volume. Suppose that

$$l = 1.5 \text{ cm}, \qquad K = 1.04 \text{ cal/cm} \cdot \text{deg} \cdot \text{s},$$
$$\rho = 10.6 \text{ g/cm}^3, \qquad C = 0.056 \text{ cal/g} \cdot \text{deg},$$

and

$$r(x, t, u) = 5.0 \text{ cal/cm}^3 \cdot \text{s}.$$

If the ends of the rod are kept at $0°C$, then

$$u(0, t) = u(l, t) = 0, \qquad t > 0.$$

Suppose the initial temperature distribution is given by

$$u(x, 0) = \sin \frac{\pi x}{l}, \qquad 0 \le x \le l.$$

Use the results of Exercise 6 to approximate the temperature distribution with $h = 0.15$ and $k = 0.0225$.

10. Sagar and Payne [SP] analyze the stress-strain relationships and material properties of a cylinder subjected alternately to heating and cooling and consider the equation

$$\frac{\partial^2 T}{\partial r^2} + \frac{1}{r} \frac{\partial T}{\partial r} = \frac{1}{4K} \frac{\partial T}{\partial t}, \qquad \frac{1}{2} < r < 1, \qquad 0 < T,$$

where $T = T(r, t)$ is the temperature, r is the radial distance from the center of the cylinder, t is time, and K is a diffusivity coefficient.

a. Find approximations to $T(r, 10)$ for a cylinder with outside radius 1, given the initial and boundary conditions:

$$T(r, 0) = 200(r - 0.5), \qquad 0.5 \le r \le 1.$$

$$T\left(\frac{1}{2}, t\right) = t, \qquad 0 \le t \le 10,$$

$$T(1, t) = 100 + 40t, \qquad 0 \le t \le 10.$$

Use a modification of the Backward-Difference method with $K = 0.1$, $k = 0.5$, and $h = 0.1$.

b. Using the temperature distribution of (a), calculate the strain I by approximating the integral

$$I = \int_{0.5}^{1} \alpha T(r, t) r \, dr,$$

where $\alpha = 10.7$ and $t = 10$. Use the Composite Trapezoidal method with $n = 5$.

12.4 Finite-Difference Methods for Hyperbolic Problems

The problem studied in this section is the one-dimensional **wave equation**, which is an example of a **hyperbolic** partial-differential equation. Suppose an elastic string of length l is stretched between two supports at the same horizontal level (see Figure 12.10).

Figure 12.10

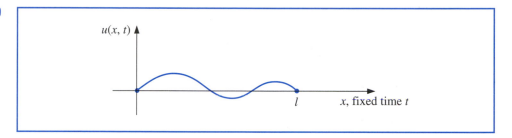

If the string is set to vibrate in a vertical plane, the vertical displacement $u(x, t)$ at a point x at time t satisfies the partial-differential equation

$$\frac{\partial^2 u}{\partial t^2}(x, t) = \alpha^2 \frac{\partial^2 u}{\partial x^2}(x, t), \quad \text{for } 0 < x < l \quad \text{and} \quad 0 < t, \text{ where } \alpha \neq 0,$$

provided that damping effects are neglected and the amplitude is not too large. To impose constraints on this problem, assume that the initial position and velocity of the string are given, respectively, by

$$u(x, 0) = f(x) \quad \text{and} \quad \frac{\partial u}{\partial t}(x, 0) = g(x), \qquad \text{for } 0 \leq x \leq l.$$

If the endpoints are fixed, we also have $u(0, t) = 0$ and $u(l, t) = 0$.

Other physical problems involving the hyperbolic partial-differential equation occur in the study of vibrating beams with one or both ends clamped, and in the transmission of electricity on a long line where there is some leakage of current to the ground.

The form of the wave equation that we will consider is given by the differential equation

$$\frac{\partial^2 u}{\partial t^2}(x, t) - \alpha^2 \frac{\partial^2 u}{\partial x^2}(x, t) = 0, \qquad \text{for } 0 < x < l \quad \text{and} \quad t > 0,$$

on the region $R = \{(x, t) \mid 0 < x < 1, 0 < t\}$, where α is a constant, subject to the boundary conditions

$$u(0, t) = u(l, t) = 0, \qquad \text{for } t > 0,$$

and the initial conditions

$$u(x, 0) = f(x), \quad \text{and} \quad \frac{\partial u}{\partial t}(x, 0) = g(x), \qquad \text{for } 0 \leq x \leq l.$$

To approximate the solution we first select an integer $m > 0$ and time-step size $k > 0$. With $h = l/m$, the mesh points (x_i, t_j) are defined by

$$x_i = ih \quad \text{and} \quad t_j = jk,$$

for each $i = 0, 1, \ldots, m$ and $j = 0, 1, \ldots$. At any interior mesh point (x_i, t_j), the wave equation becomes

$$\frac{\partial^2 u}{\partial t^2}(x_i, t_j) - \alpha^2 \frac{\partial^2 u}{\partial x^2}(x_i, t_j) = 0.$$

The Finite-Difference method is obtained using the centered-difference quotient for the second partial derivatives given by

$$\frac{\partial^2 u}{\partial t^2}(x_i, t_j) = \frac{u(x_i, t_{j+1}) - 2u(x_i, t_j) + u(x_i, t_{j-1})}{k^2} - \frac{k^2}{12}\frac{\partial^4 u}{\partial t^4}(x_i, \mu_j),$$

for some μ_j in (t_{j-1}, t_{j+1}) and

$$\frac{\partial^2 u}{\partial x^2}(x_i, t_j) = \frac{u(x_{i+1}, t_j) - 2u(x_i, t_j) + u(x_{i-1}, t_j)}{h^2} - \frac{h^2}{12}\frac{\partial^4 u}{\partial x^4}(\xi_i, t_j),$$

for some ξ_i in (x_{i-1}, x_{i+1}). Substituting these into the wave equation gives

$$\frac{u(x_i, t_{j+1}) - 2u(x_i, t_j) + u(x_i, t_{j-1})}{k^2} - \alpha^2 \frac{u(x_{i+1}, t_j) - 2u(x_i, t_j) + u(x_{i-1}, t_j)}{h^2}$$
$$= \frac{1}{12}\left[k^2 \frac{\partial^4 u}{\partial t^4}(x_i, \mu_j) - \alpha^2 h^2 \frac{\partial^4 u}{\partial x^4}(\xi_i, t_j) \right].$$

Neglecting the error term

$$\tau_{ij} = \frac{1}{12}\left[k^2 \frac{\partial^4 u}{\partial t^4}(x_i, \mu_j) - \alpha^2 h^2 \frac{\partial^4 u}{\partial x^4}(\xi_i, t_j) \right]$$

leads to the $O(h^2 + k^2)$ difference equation

$$\frac{w_{i,j+1} - 2w_{ij} + w_{i,j-1}}{k^2} - \alpha^2 \frac{w_{i+1,j} - 2w_{ij} + w_{i-1,j}}{h^2} = 0.$$

With $\lambda = \alpha k / h$, we can solve for $w_{i,j+1}$, the most advanced time-step approximation, to obtain

$$w_{i,j+1} = 2(1 - \lambda^2)w_{ij} + \lambda^2(w_{i+1,j} + w_{i-1,j}) - w_{i,j-1}.$$

This equation holds for each $i = 1, 2, \ldots, (m - 1)$ and $j = 1, 2, \ldots$. The boundary conditions give

$$w_{0j} = w_{mj} = 0, \qquad \text{for each } j = 1, 2, 3, \ldots,$$

and the initial condition implies that

$$w_{i0} = f(x_i), \qquad \text{for each } i = 1, 2, \ldots, m - 1.$$

Writing this set of equations in matrix form gives

$$\begin{bmatrix} w_{1,j+1} \\ w_{2,j+1} \\ \vdots \\ w_{m-1,j+1} \end{bmatrix} = \begin{bmatrix} 2(1-\lambda^2) & \lambda^2 & 0 & \cdots & \cdots & 0 \\ \lambda^2 & 2(1-\lambda^2) & \lambda^2 & & & \vdots \\ 0 & & & & & 0 \\ \vdots & & & & & \lambda^2 \\ 0 & \cdots & \cdots & 0 & \lambda^2 & 2(1-\lambda^2) \end{bmatrix} \begin{bmatrix} w_{1,j} \\ w_{2,j} \\ \vdots \\ w_{m-1,j} \end{bmatrix} - \begin{bmatrix} w_{1,j-1} \\ w_{2,j-1} \\ \vdots \\ w_{m-1,j-1} \end{bmatrix}.$$

(12.2)

To determine $w_{i,j+1}$ requires values from the jth and $(j-1)$st time steps. (See Figure 12.11.)

Figure 12.11

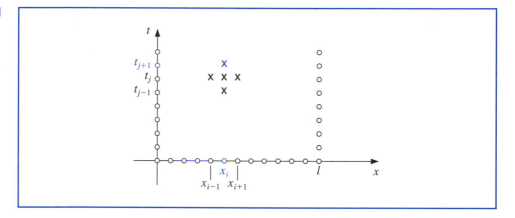

There is a starting problem because values for $j = 0$ are given in the initial conditions, but values for $j = 1$, which are needed to compute w_{i2}, must be obtained from the initial-velocity condition

$$\frac{\partial u}{\partial t}(x, 0) = g(x), \qquad \text{for } 0 \le x \le l.$$

One approach is to replace $\partial u / \partial t$ by a forward-difference approximation,

$$\frac{\partial u}{\partial t}(x_i, 0) = \frac{u(x_i, t_1) - u(x_i, 0)}{k} - \frac{k}{2}\frac{\partial^2 u}{\partial t^2}(x_i, \tilde{\mu}_i),$$

for some $\tilde{\mu}_i$ in $(0, t_1)$. Solving for $u(x_i, t_1)$ gives

$$u(x_i, t_1) = u(x_i, 0) + k\frac{\partial u}{\partial t}(x_i, 0) + \frac{k^2}{2}\frac{\partial^2 u}{\partial t^2}(x_i, \tilde{\mu}_i)$$

$$= u(x_i, 0) + kg(x_i) + \frac{k^2}{2}\frac{\partial^2 u}{\partial t^2}(x_i, \tilde{\mu}_i).$$

As a consequence,

$$w_{i1} = w_{i0} + kg(x_i), \qquad \text{for each } i = 1, \dots, m-1.$$

Improving the Initial Approximation

If we use the forward-difference approximation for $\partial u / \partial t$, the approximation to $u(x_i, t_1)$ has error of only $O(k)$, whereas the general difference equation for the hyperbolic equation is $O(h^2 + k^2)$. We now develop a method that gives a better approximation to $u(x_i, t_1)$.

Consider the equation

$$u(x_i, t_1) = u(x_i, 0) + k\frac{\partial u}{\partial t}(x_i, 0) + \frac{k^2}{2}\frac{\partial^2 u}{\partial t^2}(x_i, 0) + \frac{k^3}{6}\frac{\partial^3 u}{\partial t^3}(x_i, \hat{\mu}_i)$$

for some $\hat{\mu}_i$ in $(0, t_1)$, which comes from expanding $u(x_i, t_1)$ in a second Maclaurin polynomial in t.

If f'' exists, then

$$\frac{\partial^2 u}{\partial t^2}(x_i, 0) = \alpha^2 \frac{\partial^2 u}{\partial x^2}(x_i, 0) = \alpha^2 \frac{d^2 f}{dx^2}(x_i) = \alpha^2 f''(x_i)$$

and

$$u(x_i, t_1) = u(x_i, 0) + kg(x_i) + \frac{\alpha^2 k^2}{2} f''(x_i) + \frac{k^3}{6} \frac{\partial^3 u}{\partial t^3}(x_i, \hat{\mu}_i),$$

producing an approximation with error $O(k^3)$:

$$w_{i1} = w_{i0} + kg(x_i) + \frac{\alpha^2 k^2}{2} f''(x_i).$$

If $f''(x_i)$ is not readily available, we can use a centered-difference quotient to write

$$f''(x_i) = \frac{f(x_{i+1}) - 2f(x_i) + f(x_{i-1})}{h^2} - \frac{h^2}{12} f^{(4)}(\tilde{\xi}_i),$$

for some $\tilde{\xi}_i$ in (x_{i-1}, x_{i+1}). The approximation then becomes

$$\frac{u(x_i, t_1) - u(x_i, 0)}{k} = g(x_i) + \frac{k\alpha^2}{2h^2} [f(x_{i+1}) - 2f(x_i) + f(x_{i-1})] + O(k^2 + h^2 k),$$

or

$$u(x_i, t_1) = u(x_i, 0) + kg(x_i) + \frac{k^2 \alpha^2}{2h^2} [f(x_{i+1}) - 2f(x_i) + f(x_{i-1})] + O(k^3 + h^2 k^2).$$

Letting $\lambda = k\alpha/h$ and using the boundary conditions $u(x_i, 0) = f(x_i)$ gives

$$u(x_i, t_1) = u(x_i, 0) + kg(x_i) + \frac{\lambda^2}{2} [f(x_{i+1}) - 2f(x_i) + f(x_{i-1})] + O(k^3 + h^2 k^2)$$

$$= (1 - \lambda^2) f(x_i) + \frac{\lambda^2}{2} f(x_{i+1}) + \frac{\lambda^2}{2} f(x_{i-1}) + kg(x_i) + O(k^3 + h^2 k^2).$$

The difference equation for the wave equation

$$w_{i1} = (1 - \lambda^2) f(x_i) + \frac{\lambda^2}{2} f(x_{i+1}) + \frac{\lambda^2}{2} f(x_{i-1}) + kg(x_i) \qquad (12.3)$$

can be used to find w_{i1} for each $i = 1, 2, \ldots, m - 1$.

> The program WVFDM124 implements the Wave Equation Finite-Difference Method.

Program WVFDM124 uses Eq. (12.3) to find the w_{i1}. Then it uses the matrix equation (12.2) on page 498 to find the remaining approximations. It is assumed that there is an upper bound T for the value of t, to be used in the stopping technique, and that $k = T/N$, where N is also given.

Example 1   Approximate the solution to the hyperbolic problem

$$\frac{\partial^2 u}{\partial t^2}(x, t) - 4\frac{\partial^2 u}{\partial x^2}(x, t) = 0, \quad 0 < x < 1, \quad 0 < t,$$

with boundary conditions

$$u(0, t) = u(1, t) = 0, \quad \text{for} \quad 0 < t,$$

Table 12.6

x_i	$w_{i,20}$
0.0	0
0.1	1.618033989
0.2	1.902113033
0.3	0.618033989
0.4	−1.175570505
0.5	−2.000000000
0.6	−1.175570505
0.7	0.618033989
0.8	1.902113033
0.9	1.618033989
1.0	0

and initial conditions

$$u(x, 0) = \sin(\pi x), \quad 0 \le x \le 1, \quad \text{and} \quad \frac{\partial u}{\partial t}(x, 0) = 0, \quad 0 \le x \le 1,$$

using $h = 0.1$ and $k = 0.05$. Compare the results with the exact solution

$$u(x, t) = \sin \pi x \cos 2\pi t.$$

Solution   Choosing $h = 0.1$ and $k = 0.05$ gives $\lambda = 1$, $m = 10$, and $N = 20$. We will choose a maximum time $T = 1$ and apply the program WVFDM124. This produces the approximations $w_{i,N}$ to $u(0.1i, 1)$ for $i = 0, 1, \ldots, 10$. These results shown in Table 12.6 are correct to the places given.   ∎

The Explicit Finite-Difference method has stability problems, but there are implicit methods that are unconditionally stable.

EXERCISE SET 12.4

1. Use the Finite-Difference method with $m = 4$, $N = 4$, and $T = 1.0$ to approximate the solution to the wave equation

$$\frac{\partial^2 u}{\partial t^2} - \frac{\partial^2 u}{\partial x^2} = 0, \qquad 0 < x < 1, \qquad 0 < t;$$

$$u(0, t) = u(1, t) = 0, \qquad 0 < t,$$

$$u(x, 0) = \sin \pi x, \qquad 0 \le x \le 1,$$

$$\frac{\partial u}{\partial t}(x, 0) = 0, \qquad 0 \le x \le 1,$$

and compare your results at $t = 1.0$ to the exact solution $u(x, t) = \cos \pi t \sin \pi x$.

2. Use the Finite-Difference method with $m = 4$, $N = 4$, and $T = 0.5$ to approximate the solution to the wave equation

$$\frac{\partial^2 u}{\partial t^2} - \frac{1}{16\pi^2} \frac{\partial^2 u}{\partial x^2} = 0, \qquad 0 < x < 0.5, \qquad 0 < t;$$

$$u(0, t) = u(0.5, t) = 0, \qquad 0 < t,$$

$$u(x, 0) = 0, \qquad 0 \le x \le 0.5,$$

$$\frac{\partial u}{\partial t}(x, 0) = \sin 4\pi x, \qquad 0 \le x \le 0.5,$$

and compare your results at $t = 0.5$ to the exact solution $u(x, t) = \sin t \sin 4\pi x$.

3. Use the Finite-Difference method with

   a. $h = \pi/10$ and $k = 0.05$,

   b. $h = \pi/20$ and $k = 0.1$,

   c. $h = \pi/20$ and $k = 0.05$

   to approximate the solution to the wave equation

$$\frac{\partial^2 u}{\partial t^2} - \frac{\partial^2 u}{\partial x^2} = 0, \qquad 0 < x < \pi, \qquad 0 < t;$$

$$u(0, t) = u(\pi, t) = 0, \qquad 0 < t,$$

$$u(x, 0) = \sin x, \qquad 0 \le x \le \pi,$$

$$\frac{\partial u}{\partial t}(x, 0) = 0, \qquad 0 \le x \le \pi.$$

Compare your results at $t = 0.5$ to the exact solution $u(x, t) = \cos t \sin x$.

4. Use the Finite-Difference method with $h = k = 0.1$ to approximate the solution to the wave equation

$$\frac{\partial^2 u}{\partial t^2} - \frac{\partial^2 u}{\partial x^2} = 0, \qquad 0 < x < 1, \qquad 0 < t;$$

$$u(0, t) = u(1, t) = 0, \qquad 0 < t,$$

$$u(x, 0) = \sin 2\pi x, \qquad 0 \le x \le 1,$$

$$\frac{\partial u}{\partial t}(x, 0) = 2\pi \sin 2\pi x, \qquad 0 \le x \le 1,$$

and compare your results at $t = 0.3$ to the exact solution $u(x, t) = \sin 2\pi x(\cos 2\pi t + \sin 2\pi t)$.

5. Use the Finite-Difference method with $h = k = 0.1$ to approximate the solution to the wave equation

$$\frac{\partial^2 u}{\partial t^2} - \frac{\partial^2 u}{\partial x^2} = 0, \qquad 0 < x < 1, 0 < t;$$

$$u(0, t) = u(1, t) = 0, \qquad 0 < t,$$

$$u(x, 0) = \begin{cases} 1, & \text{if } 0 \le x \le \dfrac{1}{2}, \\ -1, & \text{if } \dfrac{1}{2} < x \le 1, \end{cases}$$

$$\frac{\partial u}{\partial t}(x, 0) = 0, \qquad 0 \le x \le 1.$$

6. The air pressure $p(x, t)$ in an organ pipe is governed by the wave equation

$$\frac{\partial^2 p}{\partial x^2} = \frac{1}{c^2} \frac{\partial^2 p}{\partial t^2}, \qquad 0 < x < l, \qquad 0 < t,$$

where l is the length of the pipe and c is a physical constant. If the pipe is open, the boundary conditions are given by

$$p(0, t) = p_0 \quad \text{and} \quad p(l, t) = p_0.$$

If the pipe is closed at the end where $x = l$, the boundary conditions are

$$p(0, t) = p_0 \quad \text{and} \quad \frac{\partial p}{\partial x}(l, t) = 0.$$

Assume that $c = 1, l = 1$, and the initial conditions are

$$p(x, 0) = p_0 \cos 2\pi x \quad \text{and} \quad \frac{\partial p}{\partial t}(x, 0) = 0, \qquad 0 \le x \le 1.$$

a. Use the Finite-Difference method to approximate the pressure for an open pipe with $p_0 = 0.9$ at $x = \frac{1}{2}$ for $t = 0.5$ and $t = 1$, using $h = k = 0.1$.

b. Modify the Finite-Difference method for the closed-pipe problem with $p_0 = 0.9$, and approximate $p(0.5, 0.5)$ and $p(0.5, 1)$ using $h = k = 0.1$.

7. In an electric transmission line of length l that carries alternating current of high frequency (called a *lossless* line), the voltage V and current i are described by

$$\frac{\partial^2 V}{\partial x^2} = LC \frac{\partial^2 V}{\partial t^2}, \qquad 0 < x < l, \qquad 0 < t,$$

$$\frac{\partial^2 i}{\partial x^2} = LC \frac{\partial^2 i}{\partial t^2}, \qquad 0 < x < l, \qquad 0 < t,$$

where L is the inductance per unit length and C is the capacitance per unit length. Suppose that the line is 200 ft long and the constants C and L are given by

$$C = 0.1 \text{ farads/ft} \quad \text{and} \quad L = 0.3 \text{ henries/ft}.$$

Suppose that the voltage and current also satisfy

$$V(0, t) = V(200, t) = 0, \qquad 0 < t,$$

$$V(x, 0) = 110 \sin \frac{\pi x}{200}, \qquad 0 \le x \le 200,$$

$$\frac{\partial V}{\partial t}(x, 0) = 0, \qquad 0 \le x \le 200,$$

$$i(0, t) = i(200, t) = 0, \qquad 0 < t,$$

$$i(x, 0) = 5.5 \cos \frac{\pi x}{200}, \qquad 0 \le x \le 200,$$

and

$$\frac{\partial i}{\partial t}(x, 0) = 0, \qquad 0 \le x \le 200.$$

Use the Finite-Difference method to approximate the voltage and current at $t = 0.2$ and $t = 0.5$ using $h = 10$ and $k = 0.1$.

12.5   Introduction to the Finite-Element Method

Finite element methods were developed by Alexander Hrennikoff (1896–1984) and Richard Courant (1888–1972) in the early 1940s. The first applications of the methods were in the 1950s in the aircraft industry. Use of the techniques followed a paper by Turner, Clough, Martin, and Topp [TCMT] that was published in 1956. Widespread application of the methods required large computer resources that were not available until the early 1970s.

The **Finite-Element method** for partial-differential equations is similar to the Rayleigh-Ritz method for approximating the solution to two-point boundary-value problems. It was originally developed for use in civil engineering but it is now used for approximating the solutions to partial-differential equations that arise in all areas of applied mathematics.

One advantage of the Finite-Element method over finite-difference methods is the relative ease with which the boundary conditions of the problem are handled. Many physical problems have boundary conditions involving derivatives and irregularly shaped boundaries. Boundary conditions of this type are difficult to handle using finite-difference techniques because each boundary condition involving a derivative must be approximated by a difference quotient at the grid points, and irregular shaping of the boundary makes placing the grid points difficult. The Finite-Element method includes the boundary conditions as integrals in a functional that is being minimized, so the construction procedure is independent of the particular boundary conditions of the problem.

In our discussion, the general form of the partial-differential equation will be

$$\frac{\partial}{\partial x}\left(p(x, y) \frac{\partial u}{\partial x} \right) + \frac{\partial}{\partial y}\left(q(x, y) \frac{\partial u}{\partial y} \right) + r(x, y)u(x, y) = f(x, y), \tag{12.4}$$

with (x, y) in $\mathcal{D}$, where $\mathcal{D}$ is a plane region with boundary $\mathcal{S}$.

Boundary conditions of the form

$$u(x, y) = g(x, y)$$

are imposed on a portion, $\mathcal{S}_1$, of the boundary. On the remainder of the boundary, $\mathcal{S}_2$, $u(x, y)$ is required to satisfy

$$p(x, y)\frac{\partial u}{\partial x}(x, y) \cos \theta_1 + q(x, y)\frac{\partial u}{\partial y}(x, y) \cos \theta_2 + g_1(x, y)u(x, y) = g_2(x, y),$$

where θ_1 and θ_2 are the direction angles of the outward normal to the boundary at the point (x, y), with $\theta_1 + \theta_2 = \pi/2$. (See Figure 12.12.)

Figure 12.12

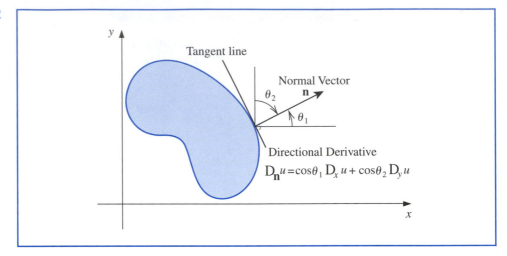

Physical problems in the areas of solid mechanics and elasticity have associated partial-differential equations of this type. The solution to such a problem is typically the minimization of a certain functional, involving integrals, over a class of functions determined by the problem.

Suppose $p, q, r,$ and f are all continuous in $\mathcal{D} \cup \mathcal{S}$, p and q have continuous first partial derivatives, and g_1 and g_2 are continuous on $\mathcal{S}_2$. Suppose, in addition, that $p(x, y) > 0$, $q(x, y) > 0, r(x, y) \leq 0,$ and $g_1(x, y) > 0$. Then a solution to our problem uniquely minimizes the functional

$$I[w] = \iint_{\mathcal{D}} \left\{ \frac{1}{2} \left[p(x, y) \left(\frac{\partial w}{\partial x} \right)^2 + q(x, y) \left(\frac{\partial w}{\partial y} \right)^2 - r(x, y)w^2 \right] + f(x, y)w \right\} dx\, dy$$

$$+ \int_{\mathcal{S}_2} \left\{ -g_2(x, y)w + \frac{1}{2}g_1(x, y)w^2 \right\} dS$$

over all twice continuously-differentiable functions $w \equiv w(x, y)$ satisfying $w(x, y) = g(x, y)$ on $\mathcal{S}_1$. The Finite-Element method approximates this solution by minimizing the functional I over a smaller class of functions, in a manner similar to the Rayleigh-Ritz method for the boundary-value problem in Chapter 11.

Defining the Elements

The first step is to divide the region into a finite number of sections, or elements, of a regular shape, either rectangles or triangles. (See Figure 12.13.)

Figure 12.13

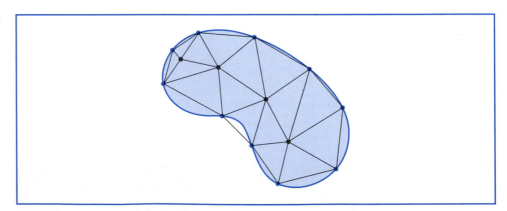

The set of functions used for approximation is generally a set of piecewise polynomials of fixed degree in x and y, and the approximation requires that the polynomials be pieced together in such a manner that the resulting function is continuous with an integrable or continuous first or second derivative on the entire region. Polynomials of linear type in x and y,

$$\phi(x, y) = a + bx + cy,$$

are commonly used with triangular elements, like those shown int Figure 12.13. Polynomials of bilinear type in x and y,

$$\phi(x, y) = a + bx + cy + dxy,$$

are used with rectangular elements.

For our discussion, suppose that the region $\mathcal{D}$ has been subdivided into triangular elements. The collection of triangles is denoted D, and the vertices of these triangles are called **nodes**. The method seeks an approximation of the form

$$\phi(x, y) = \sum_{i=1}^{m} \gamma_i \phi_i(x, y),$$

where $\phi_1, \phi_2, \dots, \phi_m$ are linearly independent piecewise-linear polynomials and γ_1, $\gamma_2, \dots, \gamma_m$ are constants. Some of these constants, say, $\gamma_{n+1}, \gamma_{n+2}, \dots, \gamma_m$, are used to ensure that the boundary condition

$$\phi(x, y) = g(x, y)$$

is satisfied on S_1. The remaining constants $\gamma_1, \gamma_2, \dots, \gamma_n$ are used to minimize the functional

$$I\left[\sum_{i=1}^{m} \gamma_i \phi_i(x, y)\right].$$

Since the functional is of the form

$$I[\phi(x, y)] = I\left[\sum_{i=1}^{m} \gamma_i \phi_i(x, y)\right]$$

$$= \iint_{\mathcal{D}} \left(\frac{1}{2}\left\{p(x, y)\left[\sum_{i=1}^{m} \gamma_i \frac{\partial \phi_i}{\partial x}(x, y)\right]^2 + q(x, y)\left[\sum_{i=1}^{m} \gamma_i \frac{\partial \phi_i}{\partial y}(x, y)\right]^2\right.\right.$$

$$\left.\left. - r(x, y)\left[\sum_{i=1}^{m} \gamma_i \phi_i(x, y)\right]^2\right\} + f(x, y)\sum_{i=1}^{m} \gamma_i \phi_i(x, y)\right) dy\, dx$$

$$+ \int_{S_2}\left\{-g_2(x, y)\sum_{i=1}^{m} \gamma_i \phi_i(x, y) + \frac{1}{2}g_1(x, y)\left[\sum_{i=1}^{m} \gamma_i \phi_i(x, y)\right]^2\right\} ds,$$

for a minimum to occur, considering I as a function of $\gamma_1, \gamma_2, \dots, \gamma_n$, it is necessary to have

$$\frac{\partial I}{\partial \gamma_j}[\phi(x, y)] = 0, \qquad \text{for each } j = 1, 2, \dots, n.$$

Performing the partial differentiation allows us to write this set of equations as a linear system,

$$A\mathbf{c} = \mathbf{b},$$

where $\mathbf{c} = (\gamma_1, \ldots, \gamma_n)^t$, and where $A = [\alpha_{ij}]$ and $\mathbf{b} = (\beta_1, \ldots, \beta_n)^t$ are defined by

$$\alpha_{ij} = \iint_{\mathcal{D}} \left[p(x, y) \frac{\partial \phi_i}{\partial x}(x, y) \frac{\partial \phi_j}{\partial x}(x, y) + q(x, y) \frac{\partial \phi_i}{\partial y}(x, y) \frac{\partial \phi_j}{\partial y}(x, y) \right. $$

$$\left. - r(x, y) \phi_i(x, y) \phi_j(x, y) \right] dx \, dy + \int_{\mathcal{S}_2} g_1(x, y) \phi_i(x, y) \phi_j(x, y) ds,$$

for each $i = 1, 2, \ldots, n$ and $j = 1, 2, \ldots, m$, and

$$\beta_i = -\iint_{\mathcal{D}} f(x, y) \phi_i(x, y) dx \, dy + \int_{\mathcal{S}_2} g_2(x, y) \phi_i(x, y) ds - \sum_{k=n+1}^{m} \alpha_{ik} \gamma_k,$$

for each $i = 1, \ldots, n$.

The particular choice of basis functions is important because the appropriate choice can often make the matrix A positive definite and banded. For our second-order problem, we assume that $\mathcal{D}$ is polygonal and that S is a contiguous set of straight lines. In this case, $\mathcal{D} = D$.

Triangulating the Region

To begin the procedure, we divide the region D into a collection of triangles $T_1, T_2, \ldots, T_M$, with the triangle T_i having three vertices, or nodes, denoted

$$V_j^{(i)} = \left(x_j^{(i)}, y_j^{(i)} \right), \qquad \text{for } j = 1, 2, 3.$$

To simplify the notation, we write $V_j^{(i)}$ simply as $V_j = (x_j, y_j)$ when working with the fixed triangle T_i. With each vertex V_j we associate a linear polynomial

$$N_j^{(i)}(x, y) \equiv N_j(x, y) = a_j + b_j x + c_j y, \quad \text{where} \quad N_j^{(i)}(x_k, y_k) = \begin{cases} 1, & \text{if } j = k, \\ 0, & \text{if } j \neq k. \end{cases}$$

For example, if $j = 2$ we have

$$a_j + b_j x_1 + c_j y_1 = 0, \quad a_j + b_j x_2 + c_j y_2 = 1, \quad \text{and} \quad a_j + b_j x_3 + c_j y_3 = 0.$$

This produces linear systems of the form

$$\begin{bmatrix} 1 & x_1 & y_1 \\ 1 & x_2 & y_2 \\ 1 & x_3 & y_3 \end{bmatrix} \begin{bmatrix} a_j \\ b_j \\ c_j \end{bmatrix} = \begin{bmatrix} 0 \\ 1 \\ 0 \end{bmatrix},$$

with the element 1 occurring in the jth row in the vector on the right.

Let $E_1, \ldots, E_n$ be a labeling of the nodes lying in $D \cup S$. With each node E_k, we associate a function ϕ_k that is linear on each triangle, has the value 1 at E_k, and is 0 at each of the other nodes. This choice makes ϕ_k identical to $N_j^{(i)}$ on triangle T_i when the node E_k is the vertex denoted $V_j^{(i)}$.

Illustration   Suppose that a finite-element problem contains the triangles T_1 and T_2 shown in Figure 12.14.

Figure 12.14

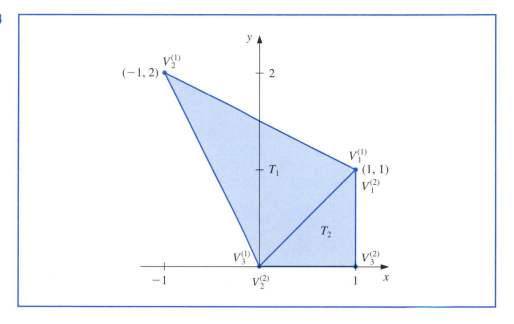

The linear function $N_1^{(1)}(x, y)$ that assumes the value 1 at $(1, 1)$ and the value 0 at both $(0, 0)$ and $(-1, 2)$ satisfies

$$a_1^{(1)} + b_1^{(1)}(1) + c_1^{(1)}(1) = 1,$$

$$a_1^{(1)} + b_1^{(1)}(-1) + c_1^{(1)}(2) = 0,$$

and

$$a_1^{(1)} + b_1^{(1)}(0) + c_1^{(1)}(0) = 0.$$

The solution to this system is $a_1^{(1)} = 0$, $b_1^{(1)} = \frac{2}{3}$, and $c_1^{(1)} = \frac{1}{3}$, so

$$N_1^{(1)}(x, y) = \frac{2}{3}x + \frac{1}{3}y.$$

In a similar manner, the linear function $N_1^{(2)}(x, y)$ that assumes the value 1 at $(1, 1)$ and the value 0 at both $(0, 0)$ and $(1, 0)$ satisfies

$$a_1^{(2)} + b_1^{(2)}(1) + c_1^{(2)}(1) = 1,$$

$$a_1^{(2)} + b_1^{(2)}(0) + c_1^{(2)}(0) = 0,$$

and

$$a_1^{(2)} + b_1^{(2)}(1) + c_1^{(2)}(0) = 0.$$

This implies that $a_1^{(2)} = 0$, $b_1^{(2)} = 0$, and $c_1^{(2)} = 1$. As a consequence, $N_1^{(2)}(x, y) = y$. Note that $N_1^{(1)}(x, y) = N_1^{(2)}(x, y)$ on the common boundary of T_1 and T_2 because $y = x$. □

Consider Figure 12.15, the upper left portion of the region shown in Figure 12.13. We will generate the entries in the matrix A that correspond to the nodes shown in this figure.

Figure 12.15

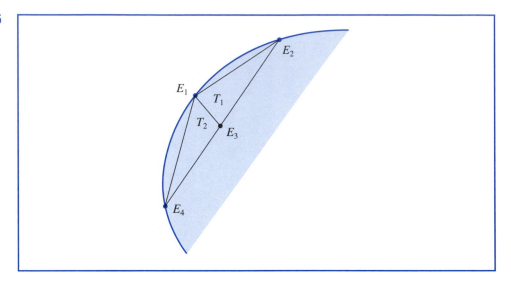

For simplicity, we assume that E_1 is not one of the nodes on S_2. The relationship between the nodes and the vertices of the triangles for this portion is

$$E_1 = V_3^{(1)} = V_1^{(2)}, \qquad E_4 = V_2^{(2)}, \qquad E_3 = V_2^{(1)} = V_3^{(2)}, \quad \text{and} \quad E_2 = V_1^{(1)}.$$

Since ϕ_1 and ϕ_3 are both nonzero on T_1 and T_2, the entries $\alpha_{1,3} = \alpha_{3,1}$ are computed by

$$\alpha_{1,3} = \iint_D \left[p(x, y)\frac{\partial\phi_1}{\partial x}\frac{\partial\phi_3}{\partial x} + q(x, y)\frac{\partial\phi_1}{\partial y}\frac{\partial\phi_3}{\partial y} - r(x, y)\phi_1\phi_3 \right] dx\, dy$$

$$= \iint_{T_1} \left[p(x, y)\frac{\partial\phi_1}{\partial x}\frac{\partial\phi_3}{\partial x} + q(x, y)\frac{\partial\phi_1}{\partial y}\frac{\partial\phi_3}{\partial y} - r(x, y)\phi_1\phi_3 \right] dx\, dy$$

$$+ \iint_{T_2} \left[p(x, y)\frac{\partial\phi_1}{\partial x}\frac{\partial\phi_3}{\partial x} + q(x, y)\frac{\partial\phi_1}{\partial y}\frac{\partial\phi_3}{\partial y} - r(x, y)\phi_1\phi_3 \right] dx\, dy.$$

On triangle T_1, we have

$$\phi_1(x, y) = N_3^{(1)}(x, y) = a_3^{(1)} + b_3^{(1)}x + c_3^{(1)}y$$

and

$$\phi_3(x, y) = N_2^{(1)}(x, y) = a_2^{(1)} + b_2^{(1)}x + c_2^{(1)}y,$$

so

$$\frac{\partial\phi_1}{\partial x}(x, y) = b_3^{(1)}, \quad \frac{\partial\phi_1}{\partial y}(x, y) = c_3^{(1)}, \quad \frac{\partial\phi_3}{\partial x}(x, y) = b_2^{(1)}, \quad \text{and} \quad \frac{\partial\phi_3}{\partial y}(x, y) = c_2^{(1)}.$$

Similarly, on T_2, we have

$$\phi_1(x, y) = N_1^{(2)}(x, y) = a_1^{(2)} + b_1^{(2)}x + c_1^{(2)}y$$

and

$$\phi_3(x, y) = N_3^{(2)}(x, y) = a_3^{(2)} + b_3^{(2)}x + c_3^{(2)}y,$$

so

$$\frac{\partial\phi_1}{\partial x}(x, y) = b_1^{(2)}, \qquad \frac{\partial\phi_1}{\partial y}(x, y) = c_1^{(2)}, \qquad \frac{\partial\phi_3}{\partial x}(x, y) = b_3^{(2)}, \quad \text{and} \quad \frac{\partial\phi_3}{\partial y}(x, y) = c_3^{(2)}.$$

Thus

$$\alpha_{13} = b_3^{(1)}b_2^{(1)} \iint_{T_1} p(x, y)dx\, dy + c_3^{(1)}c_2^{(1)} \iint_{T_1} q(x, y)dx\, dy$$

$$- \iint_{T_1} r(x, y)\left(a_3^{(1)} + b_3^{(1)}x + c_3^{(1)}y\right)\left(a_2^{(1)} + b_2^{(1)}x + c_2^{(1)}y\right)dx\, dy$$

$$+ b_1^{(2)}b_3^{(2)} \iint_{T_2} p(x, y)dx\, dy + c_1^{(2)}c_3^{(2)} \iint_{T_2} q(x, y)dx\, dy$$

$$- \iint_{T_2} r(x, y)\left(a_1^{(2)} + b_1^{(2)}x + c_1^{(2)}y\right)\left(a_3^{(2)} + b_3^{(2)}x + c_3^{(2)}y\right)dx\, dy.$$

All the double integrals over D reduce to double integrals over triangles. The usual procedure is to compute all possible integrals over the triangles and accumulate them into the correct entry α_{ij} in A.

Similarly, the double integrals of the form

$$\iint_D f(x, y)\phi_i(x, y)dx\, dy$$

are computed over triangles and then accumulated into the correct entry β_i of **b**. For example, to determine β_1 we need

$$-\iint_D f(x, y)\phi_1(x, y)dx\, dy = -\iint_{T_1} f(x, y)\left[a_3^{(1)} + b_3^{(1)}x + c_3^{(1)}y\right]dx\, dy$$

$$-\iint_{T_2} f(x, y)\left[a_1^{(2)} + b_1^{(2)}x + c_1^{(2)}y\right]dx\, dy.$$

Part of β_1 is contributed by ϕ_1 restricted to T_1 and the remainder by ϕ_1 restricted to T_2, since E_1 is a vertex of both T_1 and T_2. In addition, nodes that lie on S_2 have line integrals added to their entries in A and **b**.

Let F represent an arbitrary function from $\mathbb{R}^2$ to $\mathbb{R}$, and consider the evaluation of the double integral

$$\iint_T F(x, y)dy\, dx$$

over the triangle T with vertices (x_1, y_1), (x_2, y_2), and (x_3, y_3). First, we define

$$\Delta = \frac{1}{2}\det\begin{bmatrix} 1 & x_1 & y_1 \\ 1 & x_2 & y_2 \\ 1 & x_3 & y_3 \end{bmatrix}.$$

By a theorem in geometry, $|\delta|$ gives the area of the triangle T.

Let (x_4, y_4), (x_5, y_5), and (x_6, y_6) be the midpoints of the sides of the triangle T and let (x_7, y_7) be the centroid as shown in Figure 12.16.

Figure 12.16

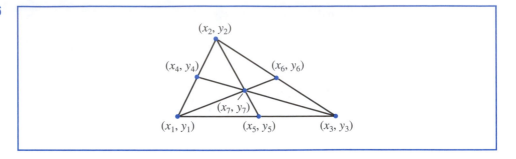

We have

$$x_4 = \frac{1}{2}(x_1 + x_2), \qquad y_4 = \frac{1}{2}(y_1 + y_2),$$

$$x_5 = \frac{1}{2}(x_1 + x_3), \qquad y_5 = \frac{1}{2}(y_1 + y_3),$$

$$x_6 = \frac{1}{2}(x_2 + x_3), \qquad y_6 = \frac{1}{2}(y_2 + y_3),$$

and, by a well-known theorem in geometry, the center is

$$x_7 = \frac{1}{3}(x_1 + x_2 + x_3), \qquad y_7 = \frac{1}{3}(y_1 + y_2 + y_3).$$

An $O(h^2)$ formula for the double integral is (see [ZM])

$$\iint_T F(x, y)dy\, dx = \frac{1}{2}|\Delta|\left\{ \frac{1}{20}(F(x_1, y_1) + F(x_2, y_2) + F(x_3, y_3)) \right.$$

$$\left. + \frac{2}{15}(F(x_4, y_4) + F(x_5, y_5) + F(x_6, y_6)) + \frac{9}{20}F(x_7, y_7) \right\}$$

We also need to compute the line integral

$$\int_L G(x, y)dS$$

where L is the line segment with endpoints (x_1, y_1) and (x_2, y_2). Let $x = x(t)$ and $y = y(t)$ be a parameterization of L with $(x_1, y_1) = (x(t_1), y(t_1))$ and $(x_2, y_2) = (x(t_2), y(t_2))$. Then

$$\int_L G(x, y)ds = \int_{t_1}^{t_2} G(x(t), y(t))\sqrt{[x'(t)]^2 + [y'(t)]^2}\, dt.$$

We can either evaluate the definite integral exactly or we use a method presented in Chapter 4.

Program FINEL125 performs the Finite Element method on a second-order elliptic differential equation in the form of Eq. (12.4) on page 503, where we have $p(x, y) > 0$, $q(x, y) > 0$, $r(x, y) \leq 0$, and $g_1(x, y) > 0$. In the program, all values of the matrix A and vector $\mathbf{b}$ are initially set to 0. After all the integrations have been performed, the 0 values are overwritten with the appropriate entries.

> The program FINEL125 implements the Finite Element method.

Illustration   The temperature, $u(x, y)$, in a two-dimensional region D satisfies Laplace's equation,

$$\frac{\partial^2 u}{\partial x^2}(x, y) + \frac{\partial^2 u}{\partial y^2}(x, y) = 0 \qquad \text{on } D.$$

Consider the region D shown in Figure 12.17 with boundary conditions given by

$$u(x, y) = 4, \qquad \text{for } (x, y) \text{ on } L_3 \text{ or } L_4,$$

$$\frac{\partial u}{\partial n}(x, y) = x, \qquad \text{for } (x, y) \text{ on } L_1,$$

$$\frac{\partial u}{\partial n}(x, y) = \frac{x + y}{\sqrt{2}}, \qquad \text{for } (x, y) \text{ on } L_2,$$

where $\partial u / \partial n$ denotes the directional derivative in the direction of the normal to the boundary of the region D at the point (x, y).

Figure 12.17

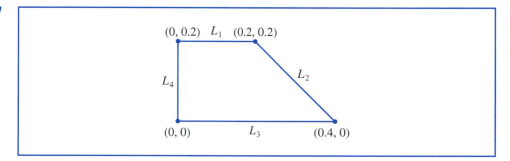

For this example, $S_1 = L_3 \cup L_4$ and $S_2 = L_1 \cup L_2$. Our functions are

$$p(x, y) = 1, \quad q(x, y) = 1, \quad r(x, y) = 0, \quad f(x, y) = 0$$

on D and its boundary. We also have

$$g(x, y) = 4, \qquad \text{on } S_1,$$

$$g_1(x, y) = 0, \qquad \text{on } S_2,$$

$$g_2(x, y) = x, \qquad \text{on } L_1 \quad \text{and} \quad g_2(x, y) = \frac{x + y}{\sqrt{2}} \quad \text{on } L_2.$$

We first subdivide D into triangles with the labeling shown in Figure 12.18.

Figure 12.18

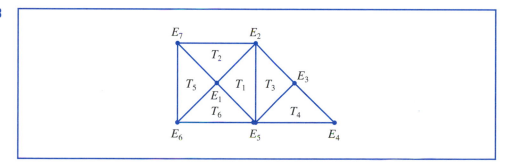

The nodes are given by $E_1 = (0.1, 0.1)$, $E_2 = (0.2, 0.2)$, $E_3 = (0.3, 0.1)$, $E_4 = (0.4, 0.0)$, $E_5 = (0.2, 0.0)$, $E_6 = (0.0, 0.0)$ and $E_7 = (0.0, 0.2)$. Since E_4, E_5, E_6, and E_7 are on S_1 and $g(x, y) = 4$, we have $\gamma_4 = \gamma_5 = \gamma_6 = \gamma_7 = 4$.

For each triangle T_i, when $1 \leq i \leq 6$, we assign the three vertices

$$V_j^{(i)} = \left(x_j^{(i)}, y_j^{(i)} \right) \qquad \text{for } j = 1, 2, 3$$

and calculate the area of the triangle. Because the triangles are all isosceles right triangles with hypotenuse 0.02, for each i we have

$$|\Delta_i| = \frac{1}{2} \left| \det \begin{bmatrix} 1 & x_1^{(i)} & y_1^{(i)} \\ 1 & x_2^{(i)} & y_2^{(i)} \\ 1 & x_3^{(i)} & y_3^{(i)} \end{bmatrix} \right| = \frac{1}{2}(\sqrt{0.02})^2 = 0.01.$$

This implies that the values of the functions

$$N_j^{(i)}(x, y) = a_j^{(i)} + b_j^{(i)} x + c_j^{(i)} y.$$

are as shown in Table 12.7.

Table 12.7

$i = 1$	$j = 1$	$v_1^{(1)} = (0.1, 0.1) = E_1$	$N_1^{(1)} = 2 - 10x$
	$j = 2$	$v_2^{(1)} = (0.2, 0.2) = E_2$	$N_2^{(1)} = -1 + 5x + 5y$
	$j = 3$	$v_3^{(1)} = (0.2, 0.0) = E_5$	$N_3^{(1)} = 5x - 5y$
$i = 2$	$j = 1$	$v_1^{(2)} = (0.1, 0.1) = E_1$	$N_1^{(2)} = 2 - 10y$
	$j = 2$	$v_2^{(2)} = (0.0, 0.2) = E_7$	$N_2^{(2)} = -5x + 5y$
	$j = 3$	$v_3^{(2)} = (0.2, 0.2) = E_2$	$N_3^{(2)} = -1 + 5x + 5y$
$i = 3$	$j = 1$	$v_1^{(3)} = (0.2, 0.2) = E_2$	$N_1^{(3)} = 1 - 5x + 5y$
	$j = 2$	$v_2^{(3)} = (0.3, 0.1) = E_3$	$N_2^{(3)} = -2 + 10x$
	$j = 3$	$v_3^{(3)} = (0.2, 0.0) = E_5$	$N_3^{(3)} = 2 - 5x - 5y$
$i = 4$	$j = 1$	$v_1^{(4)} = (0.3, 0.1) = E_3$	$N_1^{(4)} = 10y$
	$j = 2$	$v_2^{(4)} = (0.4, 0.0) = E_4$	$N_2^{(4)} = -1 + 5x - 5y$
	$j = 3$	$v_3^{(4)} = (0.2, 0.0) = E_5$	$N_3^{(4)} = 2 - 5x - 5y$
$i = 5$	$j = 1$	$v_1^{(5)} = (0.0, 0.2) = E_7$	$N_1^{(5)} = -5x + 5y$
	$j = 2$	$v_2^{(5)} = (0.1, 0.1) = E_1$	$N_2^{(5)} = 10x$
	$j = 3$	$v_3^{(5)} = (0.0, 0.0) = E_6$	$N_3^{(5)} = 1 - 5x - 5y$
$i = 6$	$j = 1$	$v_1^{(6)} = (0.0, 0.0) = E_6$	$N_1^{(6)} = 1 - 5x - 5y$
	$j = 2$	$v_2^{(6)} = (0.2, 0.0) = E_5$	$N_2^{(6)} = 5x - 5y$
	$j = 3$	$v_3^{(6)} = (0.1, 0.1) = E_1$	$N_3^{(6)} = 10y$

For each triangle T_i, when $1 \leq i \leq 6$, we approximate the double integrals

$$z_{j,k}^{(i)} = b_j^{(i)} b_k^{(i)} \iint_{T_i} p(x, y)dy\, dx + c_j^{(i)} c_k^{(i)} \iint_{T_i} q(x, y)dy\, dx$$

$$- \iint_{T_i} r(x, y) N_j^{(i)}(x, y) N_k^{(i)}(x, y)dy\, dx$$

$$= b_j^{(i)} b_k^{(i)} \frac{1}{2}|\Delta_i| + c_j^{(i)} c_k^{(i)} \frac{1}{2}|\Delta_i| - 0 = 0.01 \left(b_j^{(i)} b_k^{(i)} + c_j^{(i)} c_k^{(i)} \right)$$

corresponding to the vertices j and k, for each $j = 1, 2, 3$ and $k = 1, \ldots, j$. Further, for each vertex j and triangle T_i we have the double integral

$$H_j^{(i)} = -\iint_{T_i} f(x, y) N_j^{(i)}(x, y)dy\, dx = 0 \qquad \text{for } j = 1, 2, 3.$$

We let l_1 be the line from $(0.4, 0.0)$ to $(0.3, 0.1)$, l_2 be the line from $(0.3, 0.1)$ to $(0.2, 0.2)$, and l_3 be the line from $(0.2, 0.2)$ to $(0.0, 0.2)$. We use the parametrization

$$
\begin{aligned}
l_1: &\quad x = 0.4 - t, \quad y = t, &&\text{for} \quad 0 \le t \le 0.1, \\
l_2: &\quad x = 0.4 - t, \quad y = t, &&\text{for} \quad 0.1 \le t \le 0.2, \\
l_3: &\quad x = -t, \qquad\quad y = 0.2, &&\text{for} \quad -0.2 \le t \le 0.
\end{aligned}
$$

The line integrals that we calculate are over the edges of the triangle, which are on $\mathcal{S}_2 = L_1 \cup L_2$. Suppose that triangle T_i has an edge e_i on $\mathcal{S}_2$ from the vertex $V_j^{(i)}$ to $V_k^{(i)}$. The line integrals that we need are denoted by

$$
J_{k,j}^{(i)} = J_{j,k}^{(i)} = \int_{e_i} g_1(x, y) N_j^{(i)}(x, y) N_k^{(i)}(x, y) ds,
$$

$$
I_j^{(i)} = \int_{e_i} g_2(x, y) N_j^{(i)}(x, y) ds, \quad \text{and} \quad I_k^{(i)} = \int_{e_i} g_2(x, y) N_k^{(i)}(x, y) ds.
$$

Since $g_1(x, y) = 0$ on $\mathcal{S}_2$, we need to consider only the line integrals involving $g_2(x, y)$.

Triangle T_4 has an edge on L_4 with vertices $V_2^{(4)} = (0.4, 0.0) = E_4$ and $V_1^{(4)} = (0.3, 0.1) = E_3$. So

$$
\begin{aligned}
I_1^{(4)} &= \int_{l_1} g_2(x, y) N_1^{(4)}(x, y) ds = \int_{l_1} \frac{x + y}{\sqrt{2}} (10y) ds \\
&= \int_0^{0.1} \frac{0.4}{\sqrt{2}} (10t) \sqrt{(-1)^2 + 1} \, dt = \int_0^{0.1} 4t \, dt = 0.02,
\end{aligned}
$$

$$
\begin{aligned}
I_2^{(4)} &= \int_{l_1} g_2(x, y) N_2^{(4)}(x, y) ds = \int_{l_1} \frac{x + y}{\sqrt{2}} (-1 + 5x - 5y) ds \\
&= \int_0^{0.1} \frac{0.4}{\sqrt{2}} (-1 + 2 - 5t - 5t) \sqrt{2} \, dt = \int_0^{0.1} 0.4(1 - 10t) dt = 0.02.
\end{aligned}
$$

Triangle T_3 has an edge on L_2 with vertices $V_2^{(3)} = (0.3, 0.1) = E_3$ and $V_1^{(3)} = (0.2, 0.2) = E_2$. So

$$
\begin{aligned}
I_2^{(3)} &= \int_{l_2} g_2(x, y) N_2^{(3)}(x, y) ds = \int_{l_2} \frac{x + y}{\sqrt{2}} (-2 + 10x) ds \\
&= \int_{0.1}^{0.2} \frac{0.4}{\sqrt{2}} (-2 + 4 - 10t) \sqrt{2} \, dt = \int_{0.1}^{0.2} (0.8 - 4t) dt = 0.02,
\end{aligned}
$$

$$
\begin{aligned}
I_1^{(3)} &= \int_{l_2} g_2(x, y) N_1^{(3)}(x, y) ds = \int_{l_2} \frac{x + y}{\sqrt{2}} (1 - 5x + 5y) ds \\
&= \int_{0.1}^{0.2} \frac{0.4}{\sqrt{2}} (1 - 2 + 5t + 5t) \sqrt{2} \, dt = \int_{0.1}^{0.2} (-0.4 + 4t) dt = 0.02.
\end{aligned}
$$

Triangle T_2 has an edge on L_1 with vertices $V_3^{(2)} = (0.2, 0.2) = E_2$ and $V_2^{(2)} = (0.0, 0.2) = E_7$. So

$$I_3^{(2)} = \int_{l_3} g_2(x, y) N_3^{(2)}(x, y) ds = \int_{l_3} x(-1 + 5x + 5y) ds$$

$$= \int_{-0.2}^0 (-t)(-1 - 5t + 1)\sqrt{(-1)^2} \, dt = \int_{-0.2}^0 5t^2 \, dt = 0.01\overline{3},$$

$$I_2^{(2)} = \int_{l_3} g_2(x, y) N_2^{(2)}(x, y) ds = \int_{l_3} x(-5x + 5y) ds$$

$$= \int_{-0.2}^0 (-t)(5t + 1) dt = \int_{-0.2}^0 (-5t^2 - t) dt = 0.00\overline{6}.$$

Assembling all the elements gives

$$\alpha_{11} = z_{1,1}^{(1)} + z_{1,1}^{(2)} + z_{2,2}^{(5)} + z_{3,3}^{(6)}$$
$$= (-10)^2 0.01 + (-10)^2 0.01 + (10)^2 0.01 + (10)^2 0.01 = 4$$
$$\alpha_{12} = \alpha_{21} = z_{2,1}^{(1)} + z_{3,1}^{(2)} = (-50)0.01 + (-50)0.01 = -1$$
$$\alpha_{13} = \alpha_{31} = 0$$
$$\alpha_{22} = z_{2,2}^{(1)} + z_{3,3}^{(2)} + z_{1,1}^{(3)} = (50)0.01 + (50)0.01 + (50)0.01 = 1.5$$
$$\alpha_{23} = a_{32} = z_{2,1}^{(3)} = (-50)0.01 = -0.5$$
$$\alpha_{33} = z_{2,2}^{(3)} + z_{1,1}^{(4)} = (100)0.01 + (100)0.01 = 2$$
$$\beta_1 = -z_{3,1}^{(1)}\gamma_5 - z_{2,1}^{(2)}\gamma_7 - z_{2,1}^{(5)}\gamma_7 - z_{3,2}^{(5)}\gamma_6 - z_{3,1}^{(6)}\gamma_6 - z_{3,2}^{(6)}\gamma_5$$
$$= -4(-50)0.01 - 4(-50)0.01 - 4(-50)0.01$$
$$\quad - 4(-50)0.01 - 4(-50)0.01 - 4(-50)0.01 = 12$$
$$\beta_2 = -z_{3,2}^{(1)}\gamma_5 - z_{3,2}^{(2)}\gamma_7 - z_{3,1}^{(3)}\gamma_5 + I_1^{(3)} + I_3^{(2)}$$
$$= -4(25 - 25)0.01 - 4(-25 + 25)0.01 - 4(25 - 25)0.01$$
$$\quad + 0.02 + 0.01\overline{3} = 0.0\overline{3}$$
$$\beta_3 = -z_{3,2}^{(3)}\gamma_5 - z_{2,1}^{(4)}\gamma_4 - z_{3,1}^{(4)}\gamma_5 + I_1^{(4)} + I_2^{(3)}$$
$$= -4(-50)0.01 - 4(-50)0.01 - 4(-50)0.01 + 0.02 + 0.02 = 6.04$$

Thus

$$A = \begin{bmatrix} \alpha_{11} & \alpha_{12} & \alpha_{13} \\ \alpha_{21} & \alpha_{22} & \alpha_{23} \\ \alpha_{31} & \alpha_{32} & \alpha_{33} \end{bmatrix} = \begin{bmatrix} 4 & -1 & 0 \\ -1 & 1.5 & -0.5 \\ 0 & -0.5 & 2 \end{bmatrix} \quad \text{and} \quad \mathbf{b} = \begin{bmatrix} \beta_1 \\ \beta_2 \\ \beta_3 \end{bmatrix} = \begin{bmatrix} 12 \\ 0.0\overline{3} \\ 6.04 \end{bmatrix}$$

The linear system $A\mathbf{c} = \mathbf{b}$, where $\mathbf{c} = (\gamma_1, \gamma_2, \gamma_3)^t$ has solution

$$\gamma_1 = 4.00962963, \quad \gamma_2 = 4.03851852, \quad \text{and} \quad \gamma_3 = 4.02962963,$$

which gives the approximate solution $\phi(x, y)$ on the triangles

T_1: $\phi(x, y) = 4.00962963(2 - 10x) + 4.03851852(-1 + 5x + 5y) + 4(5x - 5y),$
T_2: $\phi(x, y) = 4.00962963(2 - 10y) + 4(-5x + 5y) + 4.03851852(-1 + 5x + 5y),$
T_3: $\phi(x, y) = 4.03851852(1 - 5x - 5y) + 4.02962963(-2 + 10x) + 4(2 - 5x - 5y),$
T_4: $\phi(x, y) = 4.02962963(10y) + 4(-1 + 5x - 5y) + 4(2 - 5x - 5y),$
T_5: $\phi(x, y) = 4(-5x + 5y) + 4.00962963(10x) + 4(1 - 5x - 5y),$
T_6: $\phi(x, y) = 4(1 - 5x - 5y) + 4(5x - 5y) + 4.00962963(10y).$

The actual solution to the boundary-value problem is $u(x, y) = xy + 4$. Table 12.8 compares the value of u to the value of ϕ at E_1, E_2, and E_3.   □

Table 12.8

x_i	y	$\phi(x, y)$	$u(x, y)$	$\|\phi(x, y) - u(x, y)\|$
0.1	0.1	4.00962963	4.01	0.00037037
0.2	0.2	4.03851852	4.04	0.00148148
0.3	0.1	4.02962963	4.03	0.00037037

Typically, the error for elliptic second-order problems with smooth coefficient functions is $O(h^2)$, where h is the maximum diameter of the circles that circumscribe the triangular elements. Piecewise bilinear basis functions on rectangular elements are also expected to give $O(h^2)$ results, where h is the maximum diagonal length of the rectangular elements. Other classes of basis functions can be used to give $O(h^4)$ results, but the construction is more complex. Efficient error theorems for finite-element methods are difficult to state and apply because the accuracy of the approximation depends on the continuity properties of the solution and the regularity of the boundary.

EXERCISE SET 12.5

1.   Use the Finite-Element method to approximate the solution to the following partial-differential equation (see the figure):

$$\frac{\partial}{\partial x}\left(y^2 \frac{\partial u}{\partial x}(x, y)\right) + \frac{\partial}{\partial y}\left(y^2 \frac{\partial u}{\partial y}(x, y)\right) - yu(x, y) = -x, \qquad (x, y) \in D,$$

$$u(x, 0.5) = 2x, \quad 0 \le x \le 0.5, \quad u(0, y) = 0, \qquad 0.5 \le y \le 1,$$

$$y^2 \frac{\partial u}{\partial x}(x, y)\cos\theta_1 + y^2 \frac{\partial u}{\partial y}(x, y)\cos\theta_2 = \frac{\sqrt{2}}{2}(y - x) \quad \text{for } (x, y) \in \mathcal{S}_2.$$

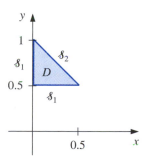

Let $M = 2$; T_1 have vertices $(0, 0.5)$, $(0.25, 0.75)$, $(0, 1)$; and T_2 have vertices $(0, 0.5)$, $(0.5, 0.5)$, and $(0.25, 0.75)$.

2. Repeat Exercise 1, using instead the triangles

$$
\begin{aligned}
T_1&: \quad (0, 0.75),\ (0, 1),\ (0.25, 0.75); \\
T_2&: \quad (0.25, 0.5),\ (0.25, 0.75),\ (0.5, 0.5); \\
T_3&: \quad (0, 0.5),\ (0, 0.75),\ (0.25, 0.75); \\
T_4&: \quad (0, 0.5),\ (0.25, 0.5),\ (0.25, 0.75).
\end{aligned}
$$

3. Use the Finite-Element method with the elements given in the accompanying figure to approximate the solution to the partial-differential equation

$$
\frac{\partial^2 u}{\partial x^2}(x, y) + \frac{\partial^2 u}{\partial y^2}(x, y) - 12.5\pi^2 u(x, y) = -25\pi^2 \sin \frac{5\pi}{2} x \sin \frac{5\pi}{2} y,
$$

for $0 < x < 0.4$ and $0 < y < 0.4$, subject to the Dirichlet boundary condition

$$
u(x, y) = 0.
$$

Compare the approximate solution to the exact solution

$$
u(x, y) = \sin \frac{5\pi}{2} x \sin \frac{5\pi}{2} y
$$

at the interior vertices and at the points $(0.125, 0.125)$, $(0.125, 0.25)$, $(0.25, 0.125)$, and $(0.25, 0.25)$.

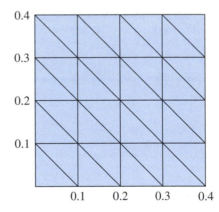

4. Repeat Exercise 3 with $f(x, y) = -25\pi^2 \cos \frac{5\pi}{2} x \cos \frac{5\pi}{2} y$, using the Neumann boundary condition

$$
\frac{\partial u}{\partial n}(x, y) = 0.
$$

The exact solution for this problem is

$$
u(x, y) = \cos \frac{5\pi}{2} x \cos \frac{5\pi}{2} y.
$$

5. The silver plate in the accompanying figure has heat being uniformly generated at each point at the rate $q = 1.5$cal/cm$^3 \cdot$s. The steady-state temperature $u(x, y)$ of the plate satisfies the Poisson equation

$$
\frac{\partial^2 u}{\partial x^2}(x, y) + \frac{\partial^2 u}{\partial y^2}(x, y) = \frac{-q}{k},
$$

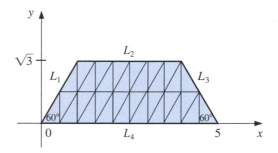

where k, the thermal conductivity, is 1.04 cal/cm·deg·s. Assume that the temperature is held at $15°C$ on L_2, that heat is lost on the slanted edges L_1 and L_3 according to the boundary condition $\partial u/\partial n = 4$, and that no heat is lost on L_4, that is, $\partial u/\partial n = 0$. Use the Finite-Element method and the triangles shown as elements to approximate the temperature of the plate at $(1, 0)$, $(4, 0)$, and $(\frac{5}{2}, \sqrt{3}/2)$.

12.6 Survey of Methods and Software

In this chapter, methods to approximate solutions to partial-differential equations were considered. We restricted our attention to Poisson's equation as an example of an elliptic partial-differential equation, the heat or diffusion equation as an example of a parabolic partial-differential equation, and the wave equation as an example of a hyperbolic partial-differential equation. Finite-difference approximations were discussed for these three examples.

Poisson's equation on a rectangle required the solution of a large sparse linear system, for which iterative techniques, such as the SOR or preconditioned conjugate gradient methods, are recommended. Three finite-difference methods were presented for the heat equation. The Forward-Difference method had stability problems, so the Backward-Difference method and the Crank-Nicolson methods were introduced. Although a tridiagonal linear system must be solved at each time step with these implicit methods, they are more stable than the explicit Forward-Difference method. The Finite-Difference method for the wave equation is explicit and can also have stability problems for certain choices of time and space discretizations.

In the last section of the chapter, we presented an introduction to the Finite-Element method for a specific elliptic partial-differential equation on a polygonal domain. Although our methods will work adequately for the problems and examples in the textbook, more powerful generalizations and modifications of these techniques are required for commercial applications.

We mention two subroutines from the IMSL Library. One is used to solve the partial-differential equation

$$\frac{\partial u}{\partial t} = F\left(x, t, u, \frac{\partial u}{\partial x}, \frac{\partial^2 u}{\partial x^2}\right)$$

with boundary conditions

$$\alpha(x, t)u(x, t) + \beta(x, t)\frac{\partial u}{\partial x}(x, t) = \gamma(x, t).$$

The method is based on collocation at Gaussian points on the x-axis for each value of t and uses cubic Hermite splines as basis functions. The other subroutine is used to solve

Poisson's equation on a rectangle. The method of solution is based on a choice of second- or fourth-order finite differences on a uniform mesh.

The NAG Library has a number of subroutines for partial-differential equations. One subroutine is used for Laplace's equation on an arbitrary domain in the xy-plane and another subroutine is used to solve a single parabolic partial-differential equation by the method of lines.

There are specialized packages for the Finite-Element method that are popular in engineering applications. The package FISHPACK in the netlib Library is used to solve separable elliptic partial-differential equations. General codes for partial-differential equations are difficult to write because of the problem of specifying domains other than common geometrical figures.

We have presented only a small sample of the many techniques used for approximating the solutions to the problems involving partial-differential equations. Further information on the general topic can be found in Lapidus and Pinder [LP], Twizell [Tw], and the book by Morton and Mayers [MM]. Software information can be found in Rice and Boisvert [RB] and in Bank [Ban].

Books that focus on finite-difference methods include Strikwerda [Stri], Thomas [Th], and Shashkov and Steinberg [ShS]. Axelsson and Barker [AB], Strange and Fix [SF], and Zienkiewicz and Morgan [ZM] are good sources for information on the finite-element method.

Bibliography

Text pages referring to the items are given in italics at the end of each reference.

[Ai]   Aitken, A. C. *On interpolation by iteration of proportional parts, without the use of differences.* Proc. Edinburgh Math. Soc. **3**(2) (1932), 56–76, QA1.E23   *52*

[AG]   Allgower, E. and K. Georg, *Numerical continuation methods: an introduction*, Springer-Verlag, New York, 1990, 388 pp. QA377.A56   *440*

[AP]   Andrews, H. C. and C. L. Patterson, *Outer product expansions and their uses in digital image processing*, American Mathematical Monthly **82**, No. 1 (1975), 1–13, QA1.A515   *409*

[AS]   Argyros, I. K. and F. Szidarovszky, *The theory and applications of iteration methods*, CRC Press, Boca Raton, FL, 1993, 355 pp. QA297.8.A74   *440*

[AMR]   Ascher, U. M., R. M. M. Mattheij, and R. D. Russell, *Numerical solution of boundary value problems for ordinary differential equations*, Prentice-Hall, Englewood Cliffs, NJ, 1988, 595 pp. QA379.A83   *474*

[Ax]   Axelsson, O., *Iterative solution methods*, Cambridge University Press, New York, 1994, 654 pp. QA297.8.A94   *319*

[AB]   Axelsson, O. and V. A. Barker, *Finite element solution of boundary value problems: theory and computation*, Academic Press, Orlando, FL, 1984, 432 pp. QA379.A9   *518*

[BSW]   Bailey, P. B., L. F. Shampine, and P. E. Waltman, *Nonlinear two-point boundary-value problems*, Academic Press, New York, 1968, 171 pp. QA372.B27   *474*

[Ban]   Bank, R. E., *PLTMG, A software package for solving elliptic partial differential equations: Users' Guide 7.0*, SIAM Publications, Philadelphia, PA, 1994, 128 pp. QA377.B26   *518*

[Barr] Barrett, R., et al., *Templates for the solution of linear systems: building blocks for iterative methods*, SIAM Publications, Philadelphia, PA, 1994, 112 pp. QA297.8.T45   *319*

[Ber]   Bernadelli, H., *Population waves*, Journal of the Burma Research Society **31** (1941), 1–18, DS527.B85   *258*

[BP]   Botha, J. F. and G. F. Pinder, *Fundamental concepts in the numerical solution of differential equations*, Wiley-Interscience, New York, 1983, 202 pp. QA374.B74   *228*

[BH]   Briggs, W. L. and V. E. Henson, *The DFT: an owner's manual for the discrete Fourier transform*, SIAM Publications, Philadelphia, PA, 1995, 434 pp. QA403.5.B75   *362*

[Brigh] Brigham, E. O., *The fast Fourier transform*, Prentice-Hall, Englewood Cliffs, NJ, 1974, 252 pp. QA403.B74   *356*

[Broy] Broyden, C. G., *A class of methods for solving nonlinear simultaneous equations*, Mathematics of Computation **19** (1965), 577–593, QA1.M4144   *421*

[BuR]   Bunch, J. R. and D. J. Rose (eds.), *Sparse matrix computations* (Proceedings of a conference held at Argonne National Laboratories, September 9–11, 1975), Academic Press, New York, 1976, 453 pp. QA188.S9   *276*

[BF]   Burden, R. L. and J. D. Faires *Numerical Analysis*, (Ninth edition), Brooks/Cole, Boston, MA, 2011, 872 pp. QA297.B84   *348, 352, 488, 492*

[Bur]   Burrage, K., 1995, *Parallel and sequential methods for ordinary differential equations*, Oxford University Press, New York, 446 pp. QA372.B883   *228*

[CF]   Chaitin-Chatelin, F. and V. Fraysse, *Lectures on finite precision computations*, SIAM Publications, Philadelphia, PA, 1996, 235 pp. QA297.C417   *31*

[CGGG]   Char, B. W., K. O. Geddes, W. M. Gentlemen, and G. H. Gonnet, *The design of Maple: a compact, portable, and powerful computer algebra system*, Computer Algebra. Lecture Notes in Computer Science No. 162, (J. A. Van Hulzen, ed.), Springer-Verlag, Berlin, 1983, 101–115 pp. QA155.7 E4 E85   *31*

[Ch]   Cheney, E. W., *Introduction to approximation theory*, McGraw-Hill, New York, 1966, 259 pp. QA221.C47   *362*

[CW]   Cody, W. J. and W. Waite, *Software manual for the elementary functions*, Prentice-Hall, Englewood Cliffs, NJ, 1980, 269 pp. QA331.C635   *31*

[CV]   Coleman, T. F. and C. Van Loan, *Handbook for matrix computations*, SIAM Publications, Philadelphia, PA, 1988, 264 pp. QA188.C65   *30, 276*

[CT]   Cooley, J. W. and J. W. Tukey, *An algorithm for the machine calculation of complex Fourier series*, Mathematics of Computation **19**, No. 90 (1965), 297–301, QA1.M4144   *356*

[CLRS]   Cormen, T. H., C. E. Leiserson, R. I. Rivest, and C. Stein, *Introduction to algorithms*, (Second edition), The MIT Press, Cambridge MA, 2001, 1180 pp. QA76.66.I5858   *27*

[Co]   Cowell, W. (ed.), *Sources and development of mathematical software*, Prentice-Hall, Englewood Cliffs, NJ, 1984, 404 pp. QA76.95.S68   *30*

[CN]   Crank, J. and P. Nicolson. *A practical method for numerical evaluation of solutions of partial differential equations of the heat-conduction type*, Proc. Cambridge Philos. Soc. **43** (1947), 0–67, Q41.C17   *493*

[Da]   Davis, P. J., *Interpolation and approximation*, Dover, New York, 1975, 393 pp. QA221.D33   *106, 362*

[DR]   Davis, P. J. and P. Rabinowitz, *Methods of numerical integration*, (Second edition), Academic Press, New York, 1984, 612 pp. QA299.3.D28   *172*

[Deb1]   De Boor, C., *On calculating with B-splines*, Journal of Approximation Theory **6** (1972), 50–62, QA221.J63   *467*

[Deb2]   De Boor, C., *A practical guide to splines*, Springer-Verlag, New York, 1978, 392 pp. QA1.A647 vol. 27   *105, 106*

[DG]   DeFranza, J. and D. Gagliardi, *Introduction to linear algebra*, McGraw-Hill, New York, 2009, 488 pp. QA184.2.D44   *364*

[DenS]   Dennis, J. E., Jr. and R. B. Schnabel, *Numerical methods for unconstrained optimization and nonlinear equations*, Prentice-Hall, Englewood Cliffs, NJ, 1983, 378 pp. QA402.5.D44   *440*

[Di]   Dierckx, P., *Curve and surface fitting with splines*, Oxford University Press, New York, 1993, 285 pp. QA297.6.D54   *106*

[DBMS]   Dongarra, J. J., J. R. Bunch, C. B. Moler, and G. W. Stewart, *LINPACK users guide*, SIAM Publications, Philadephia, PA, 1979, 367 pp. QA214.L56   *30*

[DRW]   Dongarra, J. J., T. Rowan, and R. Wade, *Software distributions using Xnetlib*, ACM Transactions on Mathematical Software **21**, No. 1 (1995), 79–88, QA76.6.A8   *30*

[DW]   Dongarra, J. and D. W. Walker, *Software libraries for linear algebra computation on high performance computers*, SIAM Review **37**, No. 2 (1995), 151–180, QA1.S2   *31*

[Do] Dormand, J. R., *Numerical methods for differential equations: a computational approach*, CRC Press, Boca Raton, FL, 1996, 368 pp. QA372.D67   *228*

[DoB] Dorn, G. L. and A. B. Burdick, *On the recombinational structure of complementation relationships in the m-dy complex of the* Drosophila melanogaster, Genetics **47** (1962), 503–518, QH431.G43   *275*

[E] Engels, H., *Numerical quadrature and cubature*, Academic Press, New York, 1980, 441 pp. QA299.3.E5   *172*

[Fe] Fehlberg, E., *Klassische Runge-Kutta Formeln vierter und niedrigerer Ordnung mit Schrittweiten-Kontrolle und ihre Anwendung auf Wärmeleitungsprobleme*, Computing **6** (1970), 61–71, QA76.C777   *206*

[FM] Forsythe, G. E. and C. B. Moler, *Computer solution of linear algebraic systems*, Prentice-Hall, Englewood Cliffs, NJ, 1967, 148 pp. QA297.F57   *276*

[Fr] Francis, J. G. F., *The QR transformation*, Computer Journal **4** (1961–2), Part I, 265–271; Part II, 332–345, QA76.C57   *395*

[Gar] Garbow, B. S., et al., *Matrix eigensystem routines: EISPACK guide extension*, Springer-Verlag, New York, 1977, 343 pp. QA193.M38   *30*

[Gea1] Gear, C. W., *Numerical initial-value problems in ordinary differential equations*, Prentice-Hall, Englewood Cliffs, NJ, 1971, 253 pp. QA372.G4   *228*

[Ger] Geršgorin, S. A. *Über die Abgrenzung der Eigenwerte einer Matrix*. Dokl. Akad. Nauk.(A), Otd. Fiz-Mat. Nauk. (1931), 749–754, QA1.A3493   *364*

[GL] George, A. and J. W. Liu, *Computer solution of large sparse positive definite systems*, Prentice-Hall, Englewood Cliffs, NJ, 1981, 324 pp. QA188.G46   *276*

[Go] Goldberg, D., *What every scientist should know about floating-point arithmetic*, ACM Computing Surveys **23**, No. 1 (1991), 5–48, QA76.5.A1   *31*

[GK] Golub, G.H. and W. Kahan, *Calculating the singular values and pseudo-inverse of a matrix*, SIAM J. Numer. Anal. **2**, Ser. B (1965) 205–224, QA297.A1S2   *400*

[GO] Golub, G. H. and J. M. Ortega, *Scientific computing: an introduction with parallel computing*, Academic Press, Boston, MA, 1993, 442 pp. QA76.58.G64   *31*

[GR] Golub, G. H. and C. Reinsch, *Singular value decomposition and least squares solutions*, Numerische Mathematik **14** (1970) 403–420, QA241.N9   *400*

[GV] Golub, G. H. and C. F. Van Loan, *Matrix computations*, (Third edition), Johns Hopkins University Press, Baltimore, MD, 1996, 694 pp. QA188.G65   *267, 276, 409*

[Hac] Hackbusch, W., *Iterative solution of large sparse systems of equations*, Springer-Verlag, New York, 1994, 429 pp. QA1.A647 vol. 95   *319*

[HY] Hageman, L. A. and D. M. Young, *Applied iterative methods*, Academic Press, New York, 1981, 386 pp. QA297.8.H34   *319*

[HNW1] Hairer, E., S. P. Nörsett, and G. Wanner, *Solving ordinary differential equations. Vol. 1: Nonstiff equations*, (Second revised edition), Springer-Verlag, Berlin, 1993, 519 pp. QA372.H16   *228*

[HNW2] Hairer, E., S. P. Nörsett, and G. Wanner, *Solving ordinary differential equations. Vol. 2: Stiff and differential-algebraic problems*, (Second revised edition), Springer, Berlin, 1996, 614 pp. QA372.H16   *228*

[He1] Henrici, P., *Discrete variable methods in ordinary differential equations*, John Wiley & Sons, New York, 1962, 407 pp. QA372.H48   *228*

[HS] Hestenes, M. R. and E. Stiefel, *Conjugate gradient methods in optimization*, Journal of Research of the National Bureau of Standards **49** (1952), 409–436, Q1.N34   *309*

[Heu] Heun, K., *Neue methode zur approximativen integration der differntialgleichungen einer unabhängigen veränderlichen*, Zeitschrift für Mathematik und Physik **45** (1900), 23–38, QA1.Z48 *186*

[Hi] Hildebrand, F. B., *Introduction to numerical analysis*, (Second edition), McGraw-Hill, New York, 1974, 669 pp. QA297.H54 *81*

[Ho] Householder, A. S., *The numerical treatment of a single nonlinear equation*, McGraw-Hill, New York, 1970, 216 pp. QA218.H68 *385*

[Joh] Johnston, R. L., *Numerical methods: a software approach*, John Wiley & Sons, New York, 1982, 276 pp. QA297.J64 *130*

[Ka] Kalman, D., *A singularly valuable decomposition: the SVD of a matrix*, The College Mathematics Journal **27** (1996), 2–23, QA11.A1 T9 *409*

[K,H] Keller, H. B., *Numerical methods for two-point boundary-value problems*, Blaisdell, Waltham, MA, 1968, 184 pp. QA372.K42 *474*

[K,J] Keller, J. B., *Probability of a shutout in racquetball*, SIAM Review **26**, No. 2 (1984), 267–268, QA1.S2 *50*

[Kelley] Kelley, C. T., *Iterative methods for linear and nonlinear equations*, SIAM Publications, Philadelphia, PA, 1995, 165 pp. QA297.8.K45 *315, 319*

[Ko] Köckler, N., *Numerical methods and scientific computing: using software libraries for problem solving*, Oxford University Press, New York, 1994, 328 pp. TA345.K653 *31*

[LP] Lapidus, L. and G. F. Pinder, *Numerical solution of partial differential equations in science and engineering*, John Wiley & Sons, New York, 1982, 677 pp. Q172.L36 *518*

[LH] Lawson, C. L. and R. J. Hanson, *Solving least squares problems*, SIAM Publications, Philadelphia, PA, 1995, 337 pp. QA275.L38 *362*

[Mo] Moler, C. B., *Demonstration of a matrix laboratory. Lecture notes in mathematics* (J. P. Hennart, ed.), Springer-Verlag, Berlin, 1982, 84–98 *31*

[MM] Morton, K. W. and D. F. Mayers, *Numerical solution of partial differential equations: an introduction*, Cambridge University Press, New York, 1994, 227 pp. QA377.M69 *518*

[N] Neville, E.H. *Iterative Interpolation*, J. Indian Math. Soc. **20** (1934), 87–120 *72*

[ND] Noble, B. and J. W. Daniel, *Applied linear algebra*, (Third edition), Prentice-Hall, Englewood Cliffs, NJ, 1988, 521 pp. QA184.N6 *364*

[OP] Ortega, J. M. and W. G. Poole, Jr., *An introduction to numerical methods for differential equations*, Pitman Publishing, Marshfield, MA, 1981, 329 pp. QA371.O65 *228*

[OR] Ortega, J. M. and W. C. Rheinboldt, *Iterative solution of nonlinear equations in several variables*, Academic Press, New York, 1970, 572 pp. QA297.8.O77 *440*

[Par] Parlett, B. N., *The symmetric eigenvalue problem*, Prentice-Hall, Englewood Cliffs, NJ, 1980, 348 pp. QA188.P37 *411*

[PF] Phillips, C. and T. L. Freeman, *Parallel numerical algorithms*, Prentice-Hall, New York, 1992, 315 pp. QA76.9.A43 F74 *31*

[Ph] Phillips, J., *The NAG Library: a beginner's guide*, Clarendon Press, Oxford, 1986, 245 pp. QA297.P35 *31*

[PDUK] Piessens, R., E. de Doncker-Kapenga, C. W. Überhuber, and D. K. Kahaner, *QUADPACK: a subroutine package for automatic integration*, Springer-Verlag, New York, 1983, 301 pp. QA299.3.Q36 *172*

[Pi] Pissanetzky, S., *Sparse matrix technology*, Academic Press, New York, 1984, 321 pp. QA188.P57 *276*

[Poo] Poole, D., *Linear algebra: a modern introduction*, (Third edition), Thomson Brooks/Cole, Belmont CA, 2011, 768 pp. QA184.2.P66 *364*

[Po] Powell, M. J. D., *Approximation theory and methods*, Cambridge University Press, Cambridge, 1981, 339 pp. QA221.P65 *106, 362*

[Pr] Pryce, J. D., *Numerical solution of Sturm-Liouville problems*, Oxford University Press, New York, 1993, 322 pp. QA379.P79 *474*

[RB] Rice, J. R. and R. F. Boisvert, *Solving elliptic problems using ELLPACK*, Springer-Verlag, New York, 1985, 497 pp. QA377.R53 *518*

[Ri] Ritz, W., *Über eine neue methode zur lösung gewisser variationsprobleme der mathematischen physik*, Journal für die reine und angewandte Mathematik, **135** (1909), pp. 1–61, QA1.J95 *461*

[RS] Roberts, S. and J. Shipman, *Two-point boundary value problems: shooting methods*, Elsevier, New York, 1972, 269 pp. QA372.R76 *474*

[RW] Rose, D. J. and R. A. Willoughby (eds.), *Sparse matrices and their applications* (Proceedings of a conference held at IBM Research, New York, September 9–10, 1971. 215 pp.), Plenum Press, New York, 1972, QA263.S94 *276*

[Sa1] Saad, Y., *Numerical methods for large eigenvalue problems*, Halsted Press, New York, 1992, 346 pp. QA188.S18 *411*

[Sa2] Saad, Y., *Iterative methods for sparse linear systems*, (Second edition), SIAM, Philadelphia, PA 2003, 528 pp. QA188.S17 *319*

[SP] Sagar, V. and D. J. Payne, *Incremental collapse of thick-walled circular cylinders under steady axial tension and torsion loads and cyclic transient heating*, Journal of the Mechanics and Physics of Solids **21**, No. 1 (1975), 39–54, TA350.J68 *496*

[SD] Sale, P. F. and R. Dybdahl, *Determinants of community structure for coral-reef fishes in experimental habitat*, Ecology **56** (1975), 1343–1355, QH540.E3 *328*

[Sche] Schendel, U., *Introduction to numerical methods for parallel computers*, (Translated by B.W. Conolly), Halsted Press, New York, 1984, 151 pp. QA297.S3813 *31*

[Scho] Schoenberg, I. J., *Contributions to the problem of approximation of equidistant data by analytic functions*, Quarterly of Applied Mathematics **4** (1946), Part A, 45–99; Part B, 112–141, QA1.A26 *106, 467*

[Schul] Schultz, M. H., *Spline analysis*, Prentice-Hall, Englewood Cliffs, NJ, 1973, 156 pp. QA211.S33 *106, 461*

[Schum] Schumaker, L. L., *Spline functions: basic theory*, Wiley-Interscience, New York, 1981, 553 pp. QA224.S33 *106*

[Sh] Shampine, L. F., *Numerical solution of ordinary differential equations*, Chapman & Hall, New York, 1994, 484 pp. QA372.S417 *228*

[ShS] Shashkov, M. and S. Steinberg, *Conservative finite-difference methods on general grids*, CRC Press, Boca Raton, FL, 1996, 359 pp. QA431.S484 *518*

[SJ] Sloan, I. H. and S. Joe, *Lattice methods for multiple integration*, Oxford University Press, New York, 1994, 239 pp. QA311.S56 *172*

[Sm,B] Smith, B. T., et al., *Matrix eigensystem routines: EISPACK guide*, (Second edition), Springer-Verlag, New York, 1976, 551 pp. QA193.M37 *30*

[So] Sorenson, D. C., *Implicitly restarted Arnoldi/Lanczos methods for large scale eigenvalue calculations*, *parallel numerical algorithms* (D. E. Keyes, A. Sameh, and V. Vankatakrishan, eds.), Kluwer Academic Publishers, Dordrecht, 1997, 119–166, QA76.9.A43 P35 *411*

[Stee] Steele, J. M., *The Cauchy-Schwarz master class*. Cambridge University Press, 2004, 306 pp. QA295.S78 *280*

[Stew]   Stewart, G. W., *Introduction to matrix computations*, Academic Press, New York, 1973, 441 pp. QA188.S7   *411, 276*

[Stew2]   Stewart, G. W., *On the early history of the singular value decomposition.* http://www.lib.umd.edu/drum/bitstream/1903/566/4/CS-TR-2855.pdf   *400*

[SF]   Strang, W. G. and G. J. Fix, *An analysis of the finite element method*, Prentice-Hall, Englewood Cliffs, NJ, 1973, 306 pp. TA335.S77   *518*

[Stri]   Strikwerda, J. C., *Finite difference schemes and partial differential equations*, (Second edition), SIAM Publications, Philadelphia, PA, 2004, 435 pp. QA374.S88   *518*

[Stro]   Stroud, A. H., *Approximate calculation of multiple integrals*, Prentice-Hall, Englewood Cliffs, NJ, 1971, 431 pp. QA311.S85   *172*

[StS]   Stroud, A. H. and D. Secrest, *Gaussian quadrature formulas*, Prentice-Hall, Englewood Cliffs, NJ, 1966, 374 pp. QA299.4.G4 S7   *135, 172*

[Th]   Thomas, J. W., *Numerical partial differential equations*, Springer-Verlag, New York, 1998, 445 pp. QA377.T495   *518*

[TCMT]   Turner, M. J., R. W. Clough, H. C. Martin, and L. J. Topp, *Stiffness and deflection of complex structures*, Journal of the Aeronautical Sciences **23** (1956), 805–824, TL501.I522   *503*

[Tw]   Twizell, E. H., *Computational methods for partial differential equations*, Ellis Horwood Ltd., Chichester, West Sussex, England, 1984, 276 pp. QA377.T95   *518*

[Van]   Van Loan, C. F., *Computational frameworks for the fast Fourier transform*, SIAM Publications, Philadelphia, PA, 1992, 273 pp. QA403.5.V35   *362*

[Var]   Varga, R. S., *Matrix iterative analysis*, (Second edition), Springer, New York, 2000, 358 pp. QA263.V3   *319, 480*

[Wil]   Wilkinson, J. H., *Rounding errors in algebraic processes*, Prentice-Hall, Englewood Cliffs, NJ, 1963, 161 pp. QA76.5.W53   *398*

[Wil2]   Wilkinson, J. H., *The algebraic eigenvalue problem*, Clarendon Press, Oxford, 1965, 662 pp. QA218.W5   *411*

[WR]   Wilkinson, J. H. and C. Reinsch (eds.), *Handbook for automatic computation. Vol. 2: Linear algebra*, Springer-Verlag, New York, 1971, 439 pp. QA251.W67   *30, 398, 411*

[Y]   Young, D. M., *Iterative solution of large linear systems*, Academic Press, New York, 1971, 570 pp. QA195.Y68   *319*

[ZM]   Zienkiewicz, O. C. and K. Morgan, *Finite elements and approximation*, John Wiley & Sons, New York, 1983, 328 pp. QA297.5.Z53   *510, 518*

Answers to Odd Exercises

1.2   Review of Calculus (Page 14)

1. For each part, $f \in C[a, b]$ on the given interval. Since $f(a)$ and $f(b)$ are of opposite sign, the Intermediate Value Theorem implies a number c exists with $f(c) = 0$.

3. For each part, $f \in C[a, b]$, f' exists on (a, b), and $f(a) = f(b) = 0$. Rolle's Theorem implies that a number c exists in (a, b) with $f'(c) = 0$. For (d), we can use $[a, b] = [-1, 0]$ or $[a, b] = [0, 2]$.

5. For $f(x) = x^3$ we have:

   a. $P_2(x) = 0$

   b. $R_2(0.5) = 0.125$; actual error $= 0.125$

   c. $P_2(x) = 1 + 3(x - 1) + 3(x - 1)^2$

   d. $R_2(0.5) = -0.125$; actual error $= -0.125$

7. Since

$$P_2(x) = 1 + x \quad \text{and} \quad R_2(x) = \frac{-2e^\xi(\sin\xi + \cos\xi)}{6}x^3$$

   for some number ξ between x and 0, we have the following:

   a. $P_2(0.5) = 1.5$ and $f(0.5) = 1.446889$. An error bound is 0.093222 and $|f(0.5) - P_2(0.5)| \le 0.0532$.

   b. $|f(x) - P_2(x)| \le 1.252$

   c. $\int_0^1 f(x)\, dx \approx 1.5$

   d. $|\int_0^1 f(x)\, dx - \int_0^1 P_2(x)\, dx| \le \int_0^1 |R_2(x)|\, dx \le 0.313$, and the actual error is 0.122.

9. The error is approximately 8.86×10^{-7}.

11. a. $P_3(x) = \frac{1}{3}x + \frac{1}{6}x^2 + \frac{23}{648}x^3$

    b. We have

$$f^{(4)}(x) = \frac{-119}{1296}e^{x/2}\sin\frac{x}{3} + \frac{5}{24}e^{x/2}\cos\frac{x}{3},$$

    so

$$|f^{(4)}(x)| \le |f^{(4)}(0.60473891)| \le 0.09787176 \quad \text{for } 0 \le x \le 1,$$

    and

$$|f(x) - P_3(x)| \le \frac{|f^{(4)}(\xi)|}{4!}|x|^4 \le \frac{0.09787176}{24}(1)^4 = 0.004077990.$$

13. A bound for the maximum error is 0.0026.

15. c. $\mathrm{erf}(1) \approx 0.8427008$

    d. $\mathrm{erf}(1) \approx 0.8427069$

1.3   Round-Off Error and Computer Arithmetic (Page 20)

1.

	Absolute Error	Relative Error
a.	0.001264	4.025×10^{-4}
b.	7.346×10^{-6}	2.338×10^{-6}
c.	2.818×10^{-4}	1.037×10^{-4}
d.	2.136×10^{-4}	1.510×10^{-4}
e.	2.647×10^{1}	1.202×10^{-3}
f.	1.454×10^{1}	1.050×10^{-2}
g.	420	1.042×10^{-2}
h.	3.343×10^{3}	9.213×10^{-3}

3.

	Approximation	Absolute Error	Relative Error
a.	134	0.079	5.90×10^{-4}
b.	133	0.499	3.77×10^{-3}
c.	2.00	0.327	0.195
d.	1.67	0.003	1.79×10^{-3}
e.	1.80	0.154	0.0788
f.	-15.1	0.0546	3.60×10^{-3}
g.	0.286	2.86×10^{-4}	10^{-3}
h.	0.00	0.0215	1.00

5.

	Approximation	Absolute Error	Relative Error
a.	133.9	0.021	1.568×10^{-4}
b.	132.5	0.001	7.55×10^{-6}
c.	1.700	0.027	0.01614
d.	1.673	0	0
e.	1.986	0.03246	0.01662
f.	-15.16	0.005377	3.548×10^{-4}
g.	0.2857	1.429×10^{-5}	5×10^{-5}
h.	-0.01700	0.0045	0.2092

7.

	Approximation	Absolute Error	Relative Error
a.	3.14557613	3.983×10^{-3}	1.268×10^{-3}
b.	3.14162103	2.838×10^{-5}	9.032×10^{-6}

9. b. The first formula gives -0.00658 and the second formula gives -0.0100. The true three-digit value is -0.0116, so the second formula is better.

11. a. $39.375 \leq$ volume ≤ 86.625

    b. $71.5 \leq$ surface area ≤ 119.5

13. a. $m = 17$

    b.

$$\binom{m}{k} = \frac{m!}{k!(m-k)!} = \frac{m(m-1)\cdots(m-k-1)(m-k)!}{k!(m-k)!} = \left(\frac{m}{k}\right)\left(\frac{m-1}{k-1}\right)\cdots\left(\frac{m-k-1}{1}\right)$$

    c. $m = 181707$

    d. 2,597,000; actual error 1960; relative error 7.541×10^{-4}

1.4 Errors in Scientific Computation (Page 28)

1.

	(i) x_1	(ii) Absolute Error	Relative Error		(i) x_2	(ii) Absolute Error	Relative Error
a.	92.26	1.542×10^{-2}	1.672×10^{-4}	**a.**	0.005419	6.273×10^{-7}	1.157×10^{-4}
b.	0.005421	1.264×10^{-6}	2.333×10^{-4}	**b.**	-92.26	4.580×10^{-3}	4.965×10^{-5}
c.	10.98	6.875×10^{-3}	6.257×10^{-4}	**c.**	0.001149	7.566×10^{-8}	6.584×10^{-5}
d.	-0.001149	7.566×10^{-8}	6.584×10^{-5}	**d.**	-10.98	6.875×10^{-3}	6.257×10^{-4}

3. a. -0.1000

   b. -0.1010

   c. Absolute error for (a) is 2.331×10^{-3} with relative error 2.387×10^{-2}.
   Absolute error for (b) is 3.331×10^{-3} with relative error 3.411×10^{-2}.

5. The fifth Maclaurin polynomials give the following approximations to $e^{-0.98}$:

   a. $\hat{P}_5(0.49) \approx 0.3743$

   b. The absolute and relative errors are, respectively, $1.011 \times 10{-3}$ and 2.694×10^{-3}

   c. $P_5(0.49)^{-1} \approx 0.3755$

   d. The absolute and relative errors are, respectively, 1.889×10^{-4} and 5.033×10^{-4}

	Approximation	Absolute Error	Relative Error
a. and **b.**	3.743	1.011×10^{-3}	2.694×10^{-3}
c. and **d.**	3.755	1.889×10^{-4}	5.033×10^{-4}

7. The approximate sums are 1.53 and 1.54, respectively. The actual value is 1.549. Significant round-off error occurs earlier with the first method.

9.

	(i) Approximation	(ii) Absolute Error	Relative Error
a.	2.715	3.282×10^{-3}	1.207×10^{-3}
b.	2.716	2.282×10^{-3}	8.394×10^{-4}
c.	2.716	2.282×10^{-3}	8.394×10^{-4}
d.	2.718	2.818×10^{-4}	1.037×10^{-4}

11. The rates of convergence are as follows:

   a. $O(h^2)$ **b.** $O(h)$ **c.** $O(h^2)$ **d.** $O(h)$

13. Since $\lim_{n \to \infty} x_n = \lim_{n \to \infty} x_{n+1} = x$ and $x_{n+1} = 1 + \frac{1}{x_n}$, we have $x = 1 + \frac{1}{x}$. This implies that $x = (1 + \sqrt{5})/2$. This number is called the *golden ratio*. It appears frequently in mathematics and the sciences.

2.2 The Bisection Method (Page 38)

1. $p_3 = 0.625$

3. The Bisection method gives the following:

   a. $p_7 = 0.5859$ **b.** $p_8 = 3.002$ **c.** $p_7 = 3.419$

5. a.

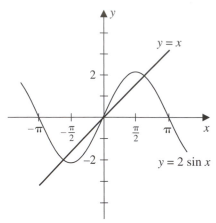

b. With $[1, 2]$, we have $p_7 = 1.8984$.

7. a. 2   **b.** -2 **c.** -1 **d.** 1

9. We have $\sqrt{3} \approx p_{14} = 1.7320$ using $[1, 2]$.

11. A bound is $n \geq 12$, and $p_{12} = 1.3787$.

13. Since $-1 < a < 0$ and $2 < b < 3$, we have $1 < a + b < 3$ or $1/2 < (a + b)/2 < 3/2$ in all cases. Further,

$$f(x) < 0, \quad \text{for } -1 < x < 0 \quad \text{and} \quad 1 < x < 2;$$

$$f(x) > 0, \quad \text{for } 0 < x < 1 \quad \text{and} \quad 2 < x < 3.$$

Thus $a_1 = a$, $f(a_1) < 0$, $b_1 = b$, and $f(b_1) > 0$.

a. Since $a + b < 2$, we have $p_1 = (a + b)/2$ and $1/2 < p_1 < 1$. Thus $f(p_1) > 0$. Hence, $a_2 = a_1 = a$ and $b_2 = p_1$. The only zero of f in $[a_2, b_2]$ is $p = 0$, so the convergence will be to 0.

b. Since $a + b > 2$, we have $p_1 = (a + b)/2$ and $1 < p_1 < 3/2$. Thus $f(p_1) < 0$. Hence, $a_2 = p_1$ and $b_2 = b_1 = b$. The only zero of f in $[a_2, b_2]$ is $p = 2$, so the convergence will be to 2.

c. Since $a + b = 2$, we have $p_1 = (a + b)/2 = 1$ and $f(p_1) = 0$. Thus a zero of f has been found on the first iteration. The convergence is to $p = 1$.

2.3   The Secant Method (Page 43)

1. a. $p_3 = 2.45454$ **b.** $p_3 = 2.44444$

3. Using the endpoints of the intervals as p_0 and p_1, we have the following:

   a. $p_{11} = 2.69065$ **b.** $p_7 = -2.87939$ **c.** $p_6 = 0.73909$ **d.** $p_5 = 0.96433$

5. Using the endpoints of the intervals as p_0 and p_1, we have the following:

   a. $p_{16} = 2.69060$ **b.** $p_6 = -2.87938$ **c.** $p_7 = 0.73908$ **d.** $p_6 = 0.96433$

7. For $p_0 = 0.1$ and $p_1 = 3$, we have $p_7 = 2.363171$.
   For $p_0 = 3$ and $p_1 = 4$, we have $p_7 = 3.817926$.
   For $p_0 = 5$ and $p_1 = 6$, we have $p_6 = 5.839252$.
   For $p_0 = 6$ and $p_1 = 7$, we have $p_9 = 6.603085$.

9. For $p_0 = 1$ and $p_1 = 2$, we have $p_5 = 1.73205068$, which compares to 14 iterations for the Bisection method.

11. For $p_0 = 0$ and $p_1 = 1$, the Secant method gives $p_7 = 0.589755$. The closest point on the graph is $(0.589755, 0.347811)$.

13. a. For $p_0 = -1$ and $p_1 = 0$, we have $p_{17} = -0.04065850$, and for $p_0 = 0$ and $p_1 = 1$, we have $p_9 = 0.9623984$.

   b. For $p_0 = -1$ and $p_1 = 0$, we have $p_5 = -0.04065929$, and for $p_0 = 0$ and $p_1 = 1$, we have $p_{12} = -0.04065929$. The Secant method fails to find the zero in $[0, 1]$.

15. With $p_0 = 6$, $p_1 = 7$, and tolerance 10^{-4}, the Secant method applied to $(x + \sqrt{x})(20 - x + \sqrt{20 - x}) - 155.55 = 0$ gives $p_5 = 6.512849$. The two numbers are approximately 6.512849 and 13.487151.

17. With $p_0 = -0.5$, $p_1 = -0.1$, and tolerance 10^{-5}, the Secant method gives $p_4 = -0.317062$.

2.4   *Newton's Method (Page 49)*

1. $p_2 = 2.60714$

3. a. For $p_0 = 2$, we have $p_5 = 2.69065$.   **b.** For $p_0 = -3$, we have $p_3 = -2.87939$.

   c. For $p_0 = 0$, we have $p_4 = 0.73909$.   **d.** For $p_0 = 0$, we have $p_3 = 0.96434$.

5. For $p_0 = 1.5$, we have $p_6 = 2.363171$.
For $p_0 = 3.5$, we have $p_5 = 3.817926$.
For $p_0 = 5.5$, we have $p_4 = 5.839252$.
For $p_0 = 7$, we have $p_5 = 6.603085$.

7. Newton's method gives the following:

   a. For $p_0 = 0.5$, we have $p_{13} = 0.567135$.   **b.** For $p_0 = -1.5$, we have $p_{23} = -1.414325$.

   c. For $p_0 = 0.5$, we have $p_{22} = 0.641166$.   **d.** For $p_0 = -0.5$, we have $p_{23} = -0.183274$.

9. With $p_0 = 1.5$, we have $p_3 = 1.73205081$ which compares to 14 iterations of the Bisection method and 5 iterations of the Secant method.

11. a. Since $f(x) = x^2 - 2$, we have $f'(x) = 2x$ and

$$p_{n+1} = p_n - \frac{f(p_n)}{f'(p_n)} = p_n - \frac{p_n^2 - 2}{2p_n} = \frac{1}{2}p_n + \frac{1}{p_n}.$$

   b. With $p_0 = 1$, we have $p_1 = 1.5000000$, $p_2 = 1.4166667$, $p_3 = 1.4142157$, $p_4 = 1.4142136$, and $p_5 = 1.4142136$. Since p_4 and p_5 agree to 7 decimal places, we are quite sure that these approximations are accurate to at least 10^{-7}.

13. Using tolerance 10^{-5}, Newton's method gives convergence to 4.4934095 provided p_0 is in [4.3, 4.5]. Newton's method does not converge to the smallest positive zero if $p_0 < 4.3$.

15. Using $p_0 = 0.75$, Newton's method gives $p_4 = 0.8423$.

17. The minimal interest rate is 6.67%.

19. a. $\frac{e}{3}, t = 3$ hours   **b.** 11 hours and 5 minutes   **c.** 21 hours and 14 minutes

2.5   *Error Analysis and Accelerating Convergence (Page 54)*

1. The results are listed in the following table.

	a.	**b.**	**c.**	**d.**
q_0	0.258684	0.907859	0.548101	0.731385
q_1	0.257613	0.909568	0.547915	0.736087
q_2	0.257536	0.909917	0.547847	0.737653
q_3	0.257531	0.909989	0.547823	0.738469
q_4	0.257530	0.910004	0.547814	0.738798
q_5	0.257530	0.910007	0.547810	0.738958

3. Newton's method gives $p_6 = -0.1828876$, and the improved value is $q_6 = -0.183387$.

5. a.   (i) Since $|p_{n+1} - 0| = \frac{1}{n+1} < \frac{1}{n} = |p_n - 0|$, the sequence $\left\{\frac{1}{n}\right\}$ converges linearly to 0.

     (ii) We need $\frac{1}{n} \leq 0.05$ or $n \geq 20$.

     (iii) Aitken's Δ^2 method gives $q_{10} = 0.04\overline{5}$.

   b.   (i) Since $|p_{n+1} - 0| = \frac{1}{(n+1)^2} < \frac{1}{n^2} = |p_n - 0|$, the sequence $\left\{\frac{1}{n^2}\right\}$ converges linearly to 0.

     (ii) We need $\frac{1}{n^2} \leq 0.05$ or $n \geq 5$.

     (iii) Aitken's Δ^2 method gives $q_2 = 0.0363$.

7. a. Since

$$\frac{|p_{n+1} - 0|}{|p_n - 0|^2} = \frac{10^{-2^{n+1}}}{(10^{-2^n})^2} = \frac{10^{-2^{n+1}}}{10^{-2 \cdot 2^n}} = \frac{10^{-2^{n+1}}}{10^{-2^{n+1}}} = 1,$$

the sequence is quadratically convergent.

b. Since

$$\frac{|p_{n+1} - 0|}{|p_n - 0|^2} = \frac{10^{-(n+1)^k}}{(10^{-n^k})^2} = \frac{10^{-(n+1)^k}}{10^{-2n^k}} = 10^{2n^k - (n+1)^k}$$

diverges, the sequence $p_n = 10^{-n^k}$ does not converge quadratically.

2.6   Müller's Method (Page 58)

1. **a.** For $p_0 = 1$, we have $p_{22} = 2.69065$.
   b. For $p_0 = 1$, we have $p_5 = 0.53209$; for $p_0 = -1$, we have $p_3 = -0.65270$; and for $p_0 = -3$, we have $p_3 = -2.87939$.
   c. For $p_0 = 1$, we have $p_5 = 1.32472$.
   d. For $p_0 = 1$, we have $p_4 = 1.12412$; and for $p_0 = 0$, we have $p_8 = -0.87605$.
   e. For $p_0 = 0$, we have $p_6 = -0.47006$; for $p_0 = -1$, we have $p_4 = -0.88533$; and for $p_0 = -3$, we have $p_4 = -2.64561$.
   f. For $p_0 = 0$, we have $p_{10} = 1.49819$.

3. The following table lists the initial approximation and the roots.

	p_0	p_1	p_2	Approximate Roots	Complex Conjugate Roots
a.	-1	0	1	$p_7 = -0.34532 - 1.31873i$	$-0.34532 + 1.31873i$
	0	1	2	$p_6 = 2.69065$	
b.	0	1	2	$p_6 = 0.53209$	
	1	2	3	$p_9 = -0.65270$	
	-2	-3	-2.5	$p_4 = -2.87939$	
c.	0	1	2	$p_5 = 1.32472$	
	-2	-1	0	$p_7 = -0.66236 - 0.56228i$	$-0.66236 + 0.56228i$
d.	0	1	2	$p_5 = 1.12412$	
	2	3	4	$p_{12} = -0.12403 + 1.74096i$	$-0.12403 - 1.74096i$
	-2	0	-1	$p_5 = -0.87605$	
e.	0	1	2	$p_{10} = -0.88533$	
	1	0	-0.5	$p_5 = -0.47006$	
	-1	-2	-3	$p_5 = -2.64561$	
f.	0	1	2	$p_6 = 1.49819$	
	-1	-2	-3	$p_{10} = -0.51363 - 1.09156i$	$-0.51363 + 1.09156i$
	1	0	-1	$p_8 = 0.26454 - 1.32837i$	$0.26454 + 1.32837i$

5. **a.** The roots are 1.244, 8.847, and -1.091. The critical points are 0 and 6.

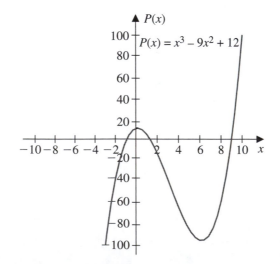

b. The roots are 0.5798, 1.521, 2.332, and −2.432, and the critical points are 1, 2.001, and −1.5.

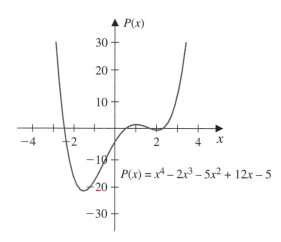

7. We define the polynomial using its coefficients in descending order with

p = [1 0 4 -4]

The call to the function `roots` is

r = roots(p)

which returns the three zeros of the polynomial

$$r = -0.423853799069784 + 2.130482604706657i$$
$$-0.423853799069784 - 2.130482604706657i$$
$$0.847707598139567$$

9. a. For $p_0 = 0.1$ and $p_1 = 1$, we have $p_{14} = 0.23233$.

b. For $p_0 = 0.55$, we have $p_6 = 0.23235$.

c. For $p_0 = 0.1$ and $p_1 = 1$, we have $p_8 = 0.23235$.

d. For $p_0 = 0.1$ and $p_1 = 1$, we have $p_{88} = 0.23035$.

e. For $p_0 = 0$, $p_1 = 0.25$, and $p_2 = 1$, we have $p_6 = 0.23235$.

11. The minimal material is approximately 573.64895 cm^2.

3.2 *Lagrange Polynomials (Page 73)*

1. a. (i) $P_1(x) = -0.29110731x + 1$; $P_1(0.45) = 0.86900171$; $|\cos 0.45 - P_1(0.45)| = 0.03144539$;

(ii) $P_2(x) = -0.43108687x^2 - 0.03245519x + 1$; $P_2(0.45) = 0.89810007$; $|\cos 0.45 - P_2(0.45)| = 0.0023470$

b. (i) $P_1(x) = 0.44151844x + 1$; $P_1(0.45) = 1.1986833$; $|\sqrt{1.45} - P_1(0.45)| = 0.00547616$;

(ii) $P_2(x) = -0.070228596x^2 + 0.483655598x + 1$; $P_2(0.45) = 1.20342373$; $|\sqrt{1.45} - P_2(0.45)| = 0.00073573$

c. (i) $P_1(x) = 0.78333938x$; $P_1(0.45) = 0.35250272$; $|\ln 1.45 - P_1(0.45)| = 0.01906083$;

(ii) $P_2(x) = -0.23389466x^2 + 0.92367618x$; $P_2(0.45) = 0.36829061$; $|\ln 1.45 - P_2(0.45)| = 0.00327294$

d. (i) $P_1(x) = 1.14022801x$; $P_1(0.45) = 0.51310260$; $|\tan 0.45 - P_1(0.45)| = 0.03004754$;

(ii) $P_2(x) = 0.86649261x^2 + 0.62033245x$; $P_2(0.45) = 0.45461436$; $|\tan 0.45 - P_2(0.45)| = 0.02844071$

3. a.

n	$x_0, x_1, ..., x_n$	$P_n(8.4)$
1	8.3, 8.6	17.87833
2	8.3, 8.6, 8.7	17.87716
3	8.3, 8.6, 8.7, 8.1	17.87714

b.

n	$x_0, x_1, ..., x_n$	$P_n(-\frac{1}{3})$
1	$-0.5, -0.25$	0.21504167
2	$-0.5, -0.25, 0.0$	0.16988889
3	$-0.5, -0.25, 0.0, -0.75$	0.17451852

c.

n	$x_0, x_1, ..., x_n$	$P_n(0.25)$
1	0.2, 0.3	-0.13869287
2	0.2, 0.3, 0.4	-0.13259734
3	0.2, 0.3, 0.4, 0.1	-0.13277477

d.

n	$x_0, x_1, ..., x_n$	$P_n(0.9)$
1	0.8, 1.0	0.44086280
2	0.8, 1.0, 0.7	0.43841352
3	0.8, 1.0, 0.7, 0.6	0.44198500

5. $\sqrt{3} \approx P_4\left(\frac{1}{2}\right) = 1.708\overline{3}$

7. a.

n	Actual Error	Error Bound
1	0.00118	0.00120
2	1.367×10^{-5}	1.452×10^{-5}

b.

n	Actual Error	Error Bound
1	4.0523×10^{-2}	4.5153×10^{-2}
2	4.6296×10^{-3}	4.6296×10^{-3}

c.

n	Actual Error	Error Bound
1	5.9210×10^{-3}	6.0971×10^{-3}
2	1.7455×10^{-4}	1.8128×10^{-4}

d.

n	Actual Error	Error Bound
1	2.7296×10^{-3}	1.4080×10^{-2}
2	5.1789×10^{-3}	9.2215×10^{-3}

9. $f(1.09) \approx P_3(1.09) = 0.2826$. The actual error is 4.3×10^{-5}, and an error bound is 7.4×10^{-6}. The discrepancy is due to the fact that the data are given to only four decimal places and only four-digit arithmetic is used.

11. $y = 4.25$

13. The largest possible step size is 0.004291932, so 0.004 would be a reasonable choice.

15. The difference between the actual value and the computed value is 2/3.

17. a.

x	erf(x)
0.0	0
0.2	0.2227
0.4	0.4284
0.6	0.6039
0.8	0.7421
1.0	0.8427

b. Linear interpolation with $x_0 = 0.2$ and $x_1 = 0.4$ gives erf$(\frac{1}{3}) \approx 0.3598$. Quadratic interpolation with $x_0 = 0.2$, $x_1 = 0.4$, and $x_2 = 0.6$ gives erf$(\frac{1}{3}) \approx 0.3632$. Since erf$(1/3) \approx 0.3626$, quadratic interpolation is more accurate.

3.3 Divided Differences (Page 81)

1. Newton's interpolatory divided-difference formula gives the following:

a. $P_1(x) = 16.9441 + 3.1041(x - 8.1)$; $P_1(8.4) = 17.87533$
$P_2(x) = P_1(x) + 0.06(x - 8.1)(x - 8.3)$; $P_2(8.4) = 17.87713$
$P_3(x) = P_2(x) + -0.00208333(x - 8.1)(x - 8.3)(x - 8.6)$; $P_3(8.4) = 17.87714$

b. $P_1(x) = -0.1769446 + 1.9069687(x - 0.6)$; $P_1(0.9) = 0.395146$
$P_2(x) = P_1(x) + 0.959224(x - 0.6)(x - 0.7)$; $P_2(0.9) = 0.4526995$
$P_3(x) = P_2(x) - 1.785741(x - 0.6)(x - 0.7)(x - 0.8)$; $P_3(0.9) = 0.4419850$

3. In the following equations we have $s = \frac{1}{h}(x - x_n)$.

a. $P_1(s) = 1.101 + 0.7660625s$; $f(-\frac{1}{3}) \approx P_1(-\frac{4}{3}) = 0.07958333$
$P_2(s) = P_1(s) + 0.406375s(s + 1)/2$; $f(-\frac{1}{3}) \approx P_2(-\frac{4}{3}) = 0.1698889$
$P_3(s) = P_2(s) + 0.09375s(s + 1)(s + 2)/6$; $f(-\frac{1}{3}) \approx P_3(-\frac{4}{3}) = 0.1745185$

b. $P_1(s) = 0.2484244 + 0.2418235s$; $f(0.25) \approx P_1(-1.5) = -0.1143108$
$P_2(s) = P_1(s) - 0.04876419s(s + 1)/2$; $f(0.25) \approx P_2(-1.5) = -0.1325973$
$P_3(s) = P_2(s) - 0.00283891s(s + 1)(s + 2)/6$; $f(0.25) \approx P_3(-1.5) = -0.1327748$

5. Using the formulas we have

    a. $f(0.05) \approx 1.05126$

    b. $f(0.65) \approx 1.91555$

7. $\Delta^3 f(x_0) = -6$ and $\Delta^4 f(x_0) = \Delta^5 f(x_0) = 0$, so the interpolating polynomial has degree 3.

9. $\Delta^2 P(10) = 1140$.

11. The approximation to $f(0.3)$ should be increased by 5.9375.

13. $f[x_0] = f(x_0) = 1$, $f[x_1] = f(x_1) = 3$, $f[x_0, x_1] = 5$.

3.4 Hermite Interpolation (Page 86)

1. The coefficients for the polynomials in divided-difference form are given in the following tables. For example, the polynomial in (a) is

$$H_3(x) = 17.56492 + 3.116256(x - 8.3) + 0.05948(x - 8.3)^2 - 0.00202222(x - 8.3)^2(x - 8.6).$$

a.	b.	c.	d.
17.56492	0.022363362	−0.02475	−0.62049958
3.116256	2.1691753	0.751	3.5850208
0.05948	0.01558225	2.751	−2.1989182
−0.00202222	−3.2177925	1	−0.490447
		0	0.037205
		0	0.040475
			−0.0025277777
			0.0029629628

3. a. We have $\sin 0.34 \approx H_5(0.34) = 0.33349$.

    b. The formula gives an error bound of 3.05×10^{-14}, but the actual error is 2.91×10^{-6}. The discrepancy is due to the fact that the data are given to only five decimal places.

    c. We have $\sin 0.34 \approx H_7(0.34) = 0.33350$. Although the error bound is now 5.4×10^{-20}, the accuracy of the given data dominates the calculations. This result is actually less accurate than the approximation in (b) because $\sin 0.34 = 0.333487$.

5. For 2(a) we have an error bound of 5.9×10^{-8}. The error bound for 2(c) is 0 because $f^{(n)}(x) \equiv 0$ for $n > 3$.

7. The Hermite polynomial generated from these data is

$$H_9(x) = 75x + 0.222222x^2(x - 3) - 0.0311111x^2(x - 3)^2$$

$$- 0.00644444x^2(x - 3)^2(x - 5) + 0.00226389x^2(x - 3)^2(x - 5)^2$$

$$- 0.000913194x^2(x - 3)^2(x - 5)^2(x - 8) + 0.000130527x^2(x - 3)^2(x - 5)^2(x - 8)^2$$

$$- 0.0000202236x^2(x - 3)^2(x - 5)^2(x - 8)^2(x - 13).$$

    a. The Hermite polynomial predicts a position of $H_9(10) = 743$ ft and a speed of $H_9'(10) = 48$ ft/s. Although the position approximation is reasonable, the low-speed prediction is suspect.

    b. To find the first time the speed exceeds 55 mi/h = $80.\overline{6}$ ft/s, we solve for the smallest value of x in the equation $80.\overline{6} = H_9'(x)$. This gives $x \approx 5.6488092$.

    c. The estimated maximum speed is $H_9'(12.37187) = 119.423$ ft/s ≈ 81.425 mi/h.

3.5 Spline Interpolation (Page 97)

1. $S(x) = x$ on $[0, 2]$

3. The equations of the respective free cubic splines are given by

$$S(x) = S_i(x) = a_i + b_i(x - x_i) + c_i(x - x_i)^2 + d_i(x - x_i)^3,$$

for x in $[x_i, x_{i+1}]$ and the coefficients in the following tables.

a.

i	a_i	b_i	c_i	d_i
0	17.564920	3.13410000	0.00000000	0.00000000

b.

i	a_i	b_i	c_i	d_i
0	0.22363362	2.17229175	0.00000000	0.00000000

c.

i	a_i	b_i	c_i	d_i
0	−0.02475000	1.03237500	0.00000000	6.50200000
1	0.33493750	2.25150000	4.87650000	−6.50200000

d.

i	a_i	b_i	c_i	d_i
0	−0.62049958	3.45508693	0.00000000	−8.9957933
1	0.28398668	3.18521313	−2.69873800	−0.94630333
2	0.00660095	2.61707643	−2.98262900	9.9420966

5. The equations of the respective clamped cubic splines are

$$s(x) = s_i(x) = a_i + b_i(x - x_i) + c_i(x - x_i)^2 + d_i(x - x_i)^3,$$

for x in $[x_i, x_{i+1}]$, where the coefficients are given in the following tables.

a.

i	a_i	b_i	c_i	d_i
0	17.564920	1.1162560	20.0600867	−44.4464666

b.

i	a_i	b_i	c_i	d_i
0	0.22363362	2.1691753	0.65914075	−3.2177925

c.

i	a_i	b_i	c_i	d_i
0	−0.02475000	0.75100000	2.5010000	1.0000000
1	0.33493750	2.18900000	3.2510000	1.0000000

d.

i	a_i	b_i	c_i	d_i
0	−0.62049958	3.5850208	−2.1498407	−0.49077413
1	−0.28398668	3.1403294	−2.2970730	−0.47458360
2	0.006600950	2.6666773	−2.4394481	−0.44980146

7. a. The equation of the spline is

$$S(x) = S_i(x) = a_i + b_i(x - x_i) + c_i(x - x_i)^2 + d_i(x - x_i)^3$$

on the interval $[x_i, x_{i+1}]$, where the coefficients are given in the following table.

x_i	a_i	b_i	c_i	d_i
0	1.0	−0.7573593	0.0	−6.627417
0.25	0.7071068	−2.0	−4.970563	6.627417
0.5	0.0	−3.242641	0.0	6.627417
0.75	−0.7071068	−2.0	4.970563	−6.627417

b. $\int_0^1 S(x)\, dx = 0.000000$

c. $S'(0.5) = -3.24264$, and $S''(0.5) = 0.0$

9. a. The equation of the spline is

$$s(x) = s_i(x) = a_i + b_i(x - x_i) + c_i(x - x_i)^2 + d_i(x - x_i)^3$$

on the interval $[x_i, x_{i+1}]$, where the coefficients are given in the following table.

x_i	a_i	b_i	c_i	d_i
0	1.0	−0.7573593	−5.193321	2.028118
0.25	0.7071068	−2.216388	−3.672233	4.896310
0.5	0.0	−3.134447	0.0	4.896310
0.75	−0.7071068	−2.216388	3.672233	2.028118

b. $\int_0^1 s(x)\,dx = 0.000000$

c. $s'(0.5) = -3.13445$, and $s''(0.5) = 0.0$

11. $a = 2$, $b = -1$, $c = -3$, $d = 1$

13. $B = \frac{1}{4}$, $D = \frac{1}{4}$, $b = -\frac{1}{2}$, $d = \frac{1}{4}$

15. Let $f(x) = a + bx + cx^2 + dx^3$. Clearly, f satisfies properties (a), (c), (d), (e) of the definition and f interpolates itself for any choice of $x_0, \ldots, x_n$. Since (ii) of (f) in the definition holds, f must be its own clamped cubic spline. However, $f''(x) = 2c + 6dx$ can be zero only at $x = -c/3d$. Thus, part (i) of (f) in the definition cannot hold at two values x_0 and x_n, and f cannot be a natural cubic spline.

17.

x_i	a_i	b_i	c_i	d_i
1960	179323	2397.312	0.00000	0.00588
1970	203302	2399.075	0.17627	−0.76838
1980	226542	2172.087	−22.87507	3.65764
1990	249633	2811.877	86.85402	−4.99517
2000	281442	3050.407	−63.00100	2.10003

$S(1950) = 155{,}344$, $S(1975) = 215{,}206$, and $S(2020) = 334{,}050$.

a. The actual population in 1950 was 151,326, so the approximations are probably not very accurate.

19. $S(x) = S_i(x) = a_i + b_i(x - x_i) + c_i(x - x_i)^2 + d_i(x - x_i)^3$ on $[x_i, x_{i+1}]$, where the coefficients of the spline on the subintervals are, for $i = 0, 1, 2$:

x_i	a_i	b_i	c_i	d_i
0	0	83.43	0	217.7
0.25	24.26	124.3	163.3	−374.4
0.5	59.68	135.7	−117.5	78.33

a. $s(0.75) = 87.487$, which yields a time of 1:27.49 compared to the actual time of 1:24.40 for the three quarters of a mile.

b. $s'(1.25) = 135.7 - 2(117.5)(1.25 - 0.5) + 3(78.33)(1.25 - 0.5)^2 = 91.64$ seconds per mile, so the predicted speed at the finish line is $3600/s'(1.25) = 39.28$ miles per hour.

3.6 Parametric Curves (Page 104)

1. Constructing parametric cubic Hermite approximations, we have

a. $x(t) = -10t^3 + 14t^2 + t$, $y(t) = -2t^3 + 3t^2 + t$

b. $x(t) = -10t^3 + 14.5t^2 + 0.5t$, $y(t) = -3t^3 + 4.5t^2 + 0.5t$

c. $x(t) = -10t^3 + 14t^2 + t$, $y(t) = -4t^3 + 5t^2 + t$

d. $x(t) = -10t^3 + 13t^2 + 2t$, $y(t) = 2t$

3. Constructing cubic Bezier polynomials, we have

a. $x(t) = -11.5t^3 + 15t^2 + 1.5t + 1$, $y(t) = -4.25t^3 + 4.5t^2 + 0.75t + 1$

b. $x(t) = -6.25t^3 + 10.5t^2 + 0.75t + 1$, $y(t) = -3.5t^3 + 3t^2 + 1.5t + 1$

c. For t between $(0, 0)$ and $(4, 6)$ we have

$$x(t) = -5t^3 + 7.5t^2 + 1.5t, \; y(t) = -13.5t^3 + 18t^2 + 1.5t,$$

and for t between $(4, 6)$ and $(6, 1)$ we have

$$x(t) = -5.5t^3 + 6t^2 + 1.5t + 4, \; y(t) = 4t^3 - 6t^2 - 3t + 6.$$

d. For t between $(0, 0)$ and $(2, 1)$ we have

$$x(t) = -5.5t^3 + 6t^2 + 1.5t, \; y(t) = -0.5t^3 + 1.5t,$$

for t between $(2, 1)$ and $(4, 0)$ we have

$$x(t) = -4t^3 + 3t^2 + 3t + 2, \; y(t) = -t^3 + 1,$$

and for t between $(4, 0)$ and $(6, -1)$ we have

$$x(t) = -8.5t^3 + 13.5t^2 - 3t + 4, \; y(t) = -3.25t^3 + 5.25t^2 - 3t.$$

4.2   Basic Quadrature Rules (Page 114)

1. The Midpoint rule gives the following approximations:

a. 0.1582031 **b.** −0.2666667 **c.** 0.1743309 **d.** 0.1516327

e. −0.6753247 **f.** −0.1768200 **g.** 0.1180292 **h.** 1.8039148

3. The Trapezoidal rule gives the following approximations:

a. 0.2656250 **b.** −0.2678571 **c.** 0.2280741 **d.** 0.1839397

e. −0.8666667 **f.** −0.1777643 **g.** 0.2180895 **h.** 4.1432597

5. Simpson's rule gives the following approximations:

a. 0.1940104 **b.** −0.2670635 **c.** 0.1922453 **d.** 0.1624017

e. −0.7391053 **f.** −0.1768216 **g.** 0.1513826 **h.** 2.583696

7. Formula (i) gives the following approximations:

a. 0.1938657 **b.** −0.2670631 **c.** 0.1922531 **d.** 0.1614099

e. −0.7364277 **f.** −0.1768207 **g.** 0.1515852 **h.** 2.585789

9. $f(1) = \frac{1}{2}$

11. $c_0 = \frac{1}{4}$, $c_1 = \frac{3}{4}$, and $x_1 = \frac{2}{3}$ gives exact results for all polynomial of degree less than or equal to 2.

13.

	(i) Midpoint rule	(ii) Trapezoidal rule	(iii) Simpson's rule
a.	4.83393	5.43476	5.03420
b.	-7.2×10^{-7}	1.6×10^{-6}	5.3×10^{-8}

4.3   Composite Quadrature Rules (Page 122)

1. The Composite Trapezoidal rule approximations are as follows:

a. 0.639900 **b.** 31.3653 **c.** 0.784241 **d.** −6.42872

e. −13.5760 **f.** 0.476977 **g.** 0.605498 **h.** 0.970926

3. The Composite Midpoint rule approximations are as follows:

a. 0.633096 **b.** 11.1568 **c.** 0.786700 **d.** −6.11274

e. −14.9985 **f.** 0.478751 **g.** 0.602961 **h.** 0.947868

5. Determining the values of n and h, we have the following results:

a. The Composite Trapezoidal rule requires $h < 0.000922295$ and $n \geq 2168$.

b. The Composite Simpson's rule requires $h < 0.037658$ and $n \geq 54$.

c. The Composite Midpoint rule requires $h < 0.00065216$ and $n \geq 3066$.

7. Determining the values of n and h, we have the following results:

a. The Composite Trapezoidal rule requires $h < 0.04382$ and $n \geq 46$. The approximation is 0.405471.

b. The Composite Simpson's rule requires $h < 0.44267$ and $n \geq 6$. The approximation is 0.405466.

c. The Composite Midpoint rule requires $h < 0.03098$ and $n \geq 64$. The approximation is 0.405460.

9. $\alpha = 1.5$

11. Approximating to 10^{-5}, we have the following results:

   a. 0.95449101, obtained using $n = 14$ in the Composite Simpson's rule.

   b. 0.99729312, obtained using $n = 20$ in the Composite Simpson's rule.

13. The length of the track is approximately 9858 ft.

15. Evaluating f, we have the following results:

   a. For $p_0 = 0.5$, we have $p_6 = 1.644854$ with $n = 20$.   **b.** For $p_0 = 0.5$, we have $p_6 = 1.645085$ with $n = 40$.

4.4   *Romberg Integration (Page 130)*

1. Romberg integration gives $R_{3,3}$ as follows:

   a. 0.1922593   **b.** 0.1606105   **c.** -0.1768200 **d.** 0.08875677

   e. 2.5879685   **f.** -0.7341567 **g.** 0.6362135   **h.** 0.6426970

3. Romberg integration gives the following values:

   a. 0.19225936 with $n = 4$ **b.** 0.16060279 with $n = 5$ **c.** -0.17682002 with $n = 4$ **d.** 0.088755284 with $n = 5$

   e. 2.5886286 with $n = 6$ **f.** -0.73396918 with $n = 6$ **g.** 0.63621335 with $n = 4$ **h.** 0.64269908 with $n = 5$

5. $R_{33} = 11.5246$

7. $f(2.5) \approx 0.43457$

9. $R_{31} = 5$

11. Let $N_2(h) = N\left(\frac{h}{3}\right) + \frac{1}{8}\left(N\left(\frac{h}{3}\right) - N(h)\right)$ and $N_3(h) = N_2\left(\frac{h}{3}\right) + \frac{1}{80}\left(N_2\left(\frac{h}{3}\right) - N_2(h)\right)$. Then $N_3(h)$ is an $O(h^6)$ approximation to M.

13. We have the following:

   a. L'Hôpital's Rule gives

$$\lim_{h \to 0} \frac{\ln(2+h) - \ln(2-h)}{h} = \lim_{h \to 0} \frac{D_h(\ln(2+h) - \ln(2-h))}{D_h(h)} = \lim_{h \to 0} \left(\frac{1}{2+h} + \frac{1}{2-h}\right) = 1,$$

so

$$\lim_{h \to 0} \left(\frac{2+h}{2-h}\right)^{1/h} = \lim_{h \to 0} e^{\frac{1}{h}[\ln(2+h) - \ln(2-h)]} = e^1 = e.$$

   b. $N(0.04) = 2.718644377221219$, $N(0.02) = 2.718372444800607$, $N(0.01) = 2.718304481241685$

   c. Let $N_2(h) = 2N\left(\frac{h}{2}\right) - N(h)$, $N_3(h) = N_2\left(\frac{h}{2}\right) + \frac{1}{3}[N_2\left(\frac{h}{2}\right) - N_2(h)]$. Then $N_2(0.04) = 2.718100512379995$, $N_2(0.02) = 2.718236517682763$, and $N_3(0.04) = 2.718281852783685$. $N_3(0.04)$ is an $O(h^3)$ approximation satisfying $|e - N_3(0.04)| \leq 0.5 \times 10^{-7}$.

   d.

$$N(-h) = \left(\frac{2-h}{2+h}\right)^{1/-h} = \left(\frac{2+h}{2-h}\right)^{1/h} = N(h)$$

   e. Let

$$e = N(h) + K_1 h + K_2 h^2 + K_3 h^3 + \cdots.$$

   Replacing h by $-h$ gives

$$e = N(-h) - K_1 h + K_2 h^2 - K_3 h^3 + \cdots,$$

   but $N(-h) = N(h)$, so

$$e = N(h) - K_1 h + K_2 h^2 - K_3 h^3 + \cdots.$$

   Thus

$$K_1 h + K_3 h^3 + \cdots = -K_1 h - K_3 h^3 \cdots,$$

and it follows that $K_1 = K_3 = K_5 = \cdots = 0$ and

$$e = N(h) + K_2 h^2 + K_4 h^4 + \cdots .$$

f. Let

$$N_2(h) = N\left(\frac{h}{2}\right) + \frac{1}{3}\left(N\left(\frac{h}{2}\right) - N(h)\right)$$

and

$$N_3(h) = N_2\left(\frac{h}{2}\right) + \frac{1}{15}\left(N_2\left(\frac{h}{2}\right) - N_2(h)\right).$$

Then

$$N_2(0.04) = 2.718281800660402, \ N_2(0.02) = 2.718281826722043$$

and

$$N_3(0.04) = 2.718281828459487.$$

$N_3(0.04)$ is an $O(h^6)$ approximation satisfying

$$|e - N_3(0.04)| \le 0.5 \times 10^{-12}.$$

4.5 Gaussian Quadrature (Page 137)

1. Gaussian quadrature gives the following:
   a. 0.1922687   **b.** 0.1594104   **c.** −0.1768190   **d.** 0.08926302
   e. 2.5913247   **f.** −0.7307230   **g.** 0.6361966   **h.** 0.6423172
3. Gaussian quadrature gives the following:
   a. 0.1922594   **b.** 0.1606028   **c.** −0.1768200   **d.** 0.08875529
   e. 2.5886327   **f.** −0.7339604   **g.** 0.6362133   **h.** 0.6426991
5. $a = 1, \ b = 1, \ c = \frac{1}{3}, \ d = -\frac{1}{3}$

4.6 Adaptive Quadrature (Page 143)

1. Simpson's rule gives the following:
   a. $S(1, 1.5) = 0.19224530$, $S(1, 1.25) = 0.039372434$, $S(1.25, 1.5) = 0.15288602$. The exact value is 0.19225935.
   b. $S(0, 1) = 0.16240168$, $S(0, 0.5) = 0.028861071$, $S(0.5, 1) = 0.13186140$, and the actual value is 0.16060279.
   c. $S(0, 0.35) = -0.17682156$, $S(0, 0.175) = -0.087724382$, $S(0.175, 0.35) = -0.089095736$. The exact value is −0.17682002.
   d. $S(0, \frac{\pi}{4}) = 0.087995669$, $S(0, \frac{\pi}{8}) = 0.0058315797$, $S(\frac{\pi}{8}, \frac{\pi}{4}) = 0.082877624$, and the actual value is 0.088755285.
   e. $S(0, \frac{\pi}{4}) = 2.5836964$, $S(0, \frac{\pi}{8}) = 0.33088926$, $S(\frac{\pi}{8}, \frac{\pi}{4}) = 2.2568121$. The exact value is 2.5886286.
   f. $S(1, 1.6) = -0.73910533$, $S(1, 1.3) = -0.26141244$, $S(1.3, 1.6) = -0.47305351$. The exact value is −0.73396917.
   g. $S(3, 3.5) = 0.63623873$, $S(3, 3.25) = 0.32567095$, $S(3.25, 3.5) = 0.31054412$. The exact value is 0.63621334.
   h. $S(0, \frac{\pi}{4}) = 0.64326905$, $S(0, \frac{\pi}{8}) = 0.37315002$, $S(\frac{\pi}{8}, \frac{\pi}{4}) = 0.26958270$. The exact value is 0.64269908.
3. Adaptive quadrature gives the following:
   a. 108.555281   **b.** −1724.966983   **c.** −15.306308   **d.** −18.945949

5. Adaptive quadrature gives the following:

$$\int_{0.1}^{2} \sin \frac{1}{x} \, dx = 1.1454 \quad \text{and} \quad \int_{0.1}^{2} \cos \frac{1}{x} \, dx = 0.67378.$$

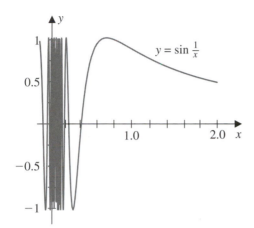

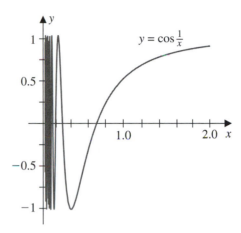

7. $\int_{0}^{2\pi} u(t) \, dt \approx 0.00001$

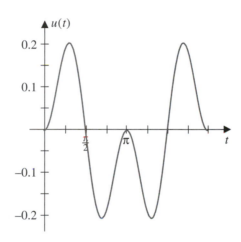

9. The approximate values of the Fresnel integrals $c(t)$ and $s(t)$ are given in the following table.

t	$c(t)$	$s(t)$
0.1	0.0999975	0.000523589
0.2	0.199921	0.00418759
0.3	0.299399	0.0141166
0.4	0.397475	0.0333568
0.5	0.492327	0.0647203
0.6	0.581061	0.0110498
0.7	0.659650	0.172129
0.8	0.722844	0.249325
0.9	0.764972	0.339747
1.0	0.779880	0.438245

4.7 Multiple Integrals (Page 155)

1. Composite Simpson's rule with $n = m = 4$ gives these values:
   a. 0.3115733   **b.** 0.2552526   **c.** 16.50864   **d.** 1.476684

3. Composite Simpson's rule first with $n = 4$ and $m = 8$, then with $n = 8$ and $m = 4$, and finally with $n = m = 6$ gives the following:
   a. 0.5119875, 0.5118533, 0.5118722. The exact result is 0.5118464.
   b. 1.718857, 1.718220, 1.718385. The exact result is 1.7182818.
   c. 1.001953, 1.000122, 1.000386. The exact result is 1.0000000.
   d. 0.7838542, 0.7833659, 0.7834362. The exact result is 0.7833333.
   e. -1.985611, -1.999182, -1.997353. The exact result is -2.0000000.
   f. 2.004596, 2.000879, 2.000980. The exact result is 2.0000000.
   g. 0.3084277, 0.3084562, 0.3084323. The exact result is 0.3084251.
   h. -22.61612, -19.85408, -20.14117. The exact result is -19.7392088.

5. Gaussian quadrature with $n = m = 2$ gives the following:
   a. 0.3115733   **b.** 0.2552446   **c.** 16.50863   **d.** 1.488875

7. Gaussian quadrature with $n = m = 3$, $n = 3$ and $m = 4$, $n = 4$ and $m = 3$, and $n = m = 4$ gives the following approximations. The exact results are listed in the solution to Exercise 3.
   a. 0.5118655, 0.5118445, 0.5118655, 0.5118445   **b.** 1.718163, 1.718302, 1.718139, 1.718277
   c. 1.000000, 1.000000, 1.0000000, 1.000000   **d.** 0.7833333, 0.7833333, 0.7833333, 0.7833333
   e. -1.991878, -2.000124, -1.991878, -2.000124 **f.** 2.001494, 2.000080, 2.001388, 1.999984
   g. 0.3084151, 0.3084145, 0.3084246, 0.3084245   **h.** -12.74790, -21.21539, -11.83624, -20.30373

9. Gaussian quadrature with $n = m = p = 2$ gives the first listed value. The second is the exact result.
   a. 5.204036, $e(e^{0.5} - 1)(e - 1)^2$ **b.** 0.08429784, $\dfrac{1}{12}$ **c.** 0.08641975, $\dfrac{1}{14}$
   d. 0.09722222, $\dfrac{1}{12}$ **e.** 7.103932, $2 + \dfrac{1}{2}\pi^2$ **f.** 1.428074, $\dfrac{1}{2}(e^2 + 1) - e$

11. Composite Simpson's rule with $n = m = 14$ gives 0.1479103 and Gaussian quadrature with $n = m = 4$ gives 0.1506823.

13. The area approximations are
   a. 1.0402528   **b.** 1.0402523

15. Gaussian quadrature with $n = m = p = 4$ gives 3.0521250. The exact result is 3.0521249.

4.8 Improper Integrals (Page 162)

1. Composite Simpson's rule gives the following:
   a. 0.5284163   **b.** 4.266654   **c.** 0.4329748   **d.** 0.8802210

3. Composite Simpson's rule gives the following:
   a. 0.4112649   **b.** 0.2440679   **c.** 0.05501681   **d.** 0.2903746

5. The escape velocity is approximately 6.9450 mi/s.

4.9 Numerical Differentiation (Page 170)

1. From the two-point formula we have the following approximations:
   a. $f'(0.5) \approx 0.8520$, $f'(0.6) \approx 0.8520$, $f'(0.7) \approx 0.7960$ **b.** $f'(0.0) \approx 3.7070$, $f'(0.2) \approx 3.1520$, $f'(0.4) \approx 3.1520$

3. For the endpoints of the tables we use the three-point endpoint formula. The other approximations come from the three-point midpoint formula.
   a. $f'(1.1) \approx 17.769705$, $f'(1.2) \approx 22.193635$, $f'(1.3) \approx 27.107350$, $f'(1.4) \approx 32.150850$
   b. $f'(8.1) \approx 3.092050$, $f'(8.3) \approx 3.116150$, $f'(8.5) \approx 3.139975$, $f'(8.7) \approx 3.163525$
   c. $f'(2.9) \approx 5.101375$, $f'(3.0) \approx 6.654785$, $f'(3.1) \approx 8.216330$, $f'(3.2) \approx 9.786010$
   d. $f'(2.0) \approx 0.13533150$, $f'(2.1) \approx -0.09989550$, $f'(2.2) \approx -0.3298960$, $f'(2.3) \approx -0.5546700$

5. The following results are using both the five-point endpoint formula and the five-point midpoint formula to approximate.

   a. The five-point endpoint formula gives $f'(2.1) \approx 3.899344$, $f'(2.2) \approx 2.876876$, $f'(2.5) \approx 1.544210$, and $f'(2.6) \approx 1.355496$. The five-point midpoint formula gives $f'(2.3) \approx 2.249704$ and $f'(2.4) \approx 1.837756$.

   b. The five-point endpoint formula gives $f'(-3.0) \approx -5.877358$, $f'(-2.8) \approx -5.468933$, $f'(-2.2) \approx -4.239911$, and $f'(-2.0) \approx -3.828853$. The five-point midpoint formula gives $f'(-2.6) \approx -5.059884$ and $f'(-2.4) \approx -4.650223$.

7. The approximation is -4.8×10^{-9}. $f''(0.5) = 0$. The error bound is 0.35874. The method is very accurate because the function is symmetric about $x = 0.5$.

9. We have

$$f'(3) \approx \frac{1}{12}[f(1) - 8f(2) + 8f(4) - f(5)] = 0.21062$$

with an error bound given by

$$\max_{1 \le x \le 5} \frac{|f^{(5)}(x)|h^4}{30} \le \frac{23}{30} = 0.7\overline{6}.$$

11. The optimal $h = 2\sqrt{\varepsilon/M}$, where $M = \max|f''(x)|$.

13. Since $e'(h) = -\varepsilon/h^2 + hM/3$, we have $e'(h) = 0$ if and only if $h = \sqrt[3]{3\varepsilon/M}$. Also, $e'(h) < 0$ if $h < \sqrt[3]{3\varepsilon/M}$ and $e'(h) > 0$ if $h > \sqrt[3]{3\varepsilon/M}$, so an absolute minimum for $e(h)$ occurs at $h = \sqrt[3]{3\varepsilon/M}$.

15. Using three-point formulas gives the following table.

Time	0	3	5	8	10	13
Speed	79	82.4	74.2	76.8	69.4	71.2

17. Expanding the function f in a fourth Taylor polynomial about x_0 evaluated at $x_0 + h$ and $x_0 - h$ gives

$$f(x_0 + h) = f(x_0) + f'(x_0)h + \frac{1}{2}f''(x_0)h^2 + \frac{1}{6}f'''(x_0)h^3 + \frac{1}{24}f^{(4)}(\xi_1)h^4$$

and

$$f(x_0 - h) = f(x_0) - f'(x_0)h + \frac{1}{2}f''(x_0)h^2 - \frac{1}{6}f'''(x_0)h^3 + \frac{1}{24}f^{(4)}(\xi_2)h^4.$$

Adding the equations gives

$$f(x_0 + h) + f(x_0 - h) = 2f(x_0) + h^2 f''(x_0) + \frac{h^4}{24}[f^{(4)}(\xi_1) + f^{(4)}(\xi_2)],$$

and

$$f''(x_0) = \frac{1}{h^2}[f(x_0 + h) - 2f(x_0) + f(x_0 - h)] - \frac{h^2}{24}[f^{(4)}(\xi_1) + f^{(4)}(\xi_2)].$$

If $f^{(4)}$ is continuous on $[x_0 - h, x_0 + h]$, the Intermediate Value Theorem implies that a number $\tilde{\xi}$ in $(x_0 - h, x_0 + h)$ exists with

$$f^{(4)}(\tilde{\xi}) = \frac{1}{2}\left[f^{(4)}(\xi_1) + f^{(4)}(\xi_2)\right].$$

So

$$f''(x_0) = \frac{1}{h^2}[f(x_0 + h) - 2f(x_0) + f(x_0 - h)] - \frac{h^2}{12}f^{(4)}(\xi).$$

5.2 Taylor Methods (Page 182)

1. Euler's method gives the approximations in the following tables.

a.

i	t_i	w_i	$y(t_i)$
1	0.500	0.0000000	0.2836165
2	1.000	1.1204223	3.2190993

b.

i	t_i	w_i	$y(t_i)$
1	2.500	2.0000000	1.8333333
2	3.000	2.6250000	2.5000000

c.

i	t_i	w_i	$y(t_i)$
1	1.250	2.7500000	2.7789294
2	1.500	3.5500000	3.6081977
3	1.750	4.3916667	4.4793276
4	2.000	5.2690476	5.3862944

d.

i	t_i	w_i	$y(t_i)$
1	0.250	1.2500000	1.3291498
2	0.500	1.6398053	1.7304898
3	0.750	2.0242547	2.0414720
4	1.000	2.2364573	2.1179795

3. Euler's method gives the approximations in the following tables.

a.

i	t_i	w_i	$y(t_i)$
2	1.200	1.0082645	1.0149523
5	1.500	1.0576682	1.0672624
7	1.700	1.1004322	1.1106551
10	2.000	1.1706516	1.1812322

b.

i	t_i	w_i	$y(t_i)$
2	1.400	0.4388889	0.4896817
5	2.000	1.4372511	1.6612818
7	2.400	2.4022696	2.8765514
10	3.000	4.5142774	5.8741000

c.

i	t_i	w_i	$y(t_i)$
2	0.400	−1.6080000	−1.6200510
5	1.000	−1.1992512	−1.2384058
7	1.400	−1.0797454	−1.1146484
10	2.000	−1.0181518	−1.0359724

d.

i	t_i	w_i	$y(t_i)$
2	0.2	0.1083333	0.1626265
5	0.5	0.2410417	0.2773617
7	0.7	0.4727604	0.5000658
10	1.0	0.9803451	1.0022460

5. Taylor's method of order 2 gives the approximations in the following tables.

a.

i	t_i	w_i	$y(t_i)$
1	0.50	0.12500000	0.28361652
2	1.00	2.02323897	3.21909932

b.

i	t_i	w_i	$y(t_i)$
1	2.50	1.75000000	1.83333333
2	3.00	2.42578125	2.50000000

c.

i	t_i	w_i	$y(t_i)$
1	1.25	2.78125000	2.77892944
2	1.50	3.61250000	3.60819766
3	1.75	4.48541667	4.47932763
4	2.00	5.39404762	5.38629436

d.

i	t_i	w_i	$y(t_i)$
1	0.25	1.34375000	1.32914981
2	0.50	1.77218707	1.73048976
3	0.75	2.11067606	2.04147203
4	1.00	2.20164395	2.11797955

7. Taylor's method of order 4 gives the approximations in the following tables.

a.

i	t_i	w_i	$y(t_i)$
2	1.2	1.0149771	1.0149523
4	1.4	1.0475619	1.0475339
6	1.6	1.0884607	1.0884327
8	1.8	1.1336811	1.1336536
10	2.0	1.1812594	1.1812322

b.

i	t_i	w_i	$y(t_i)$
2	1.4	0.4896141	0.4896817
4	1.8	1.1993085	1.1994386
6	2.2	2.2132495	2.2135018
8	2.6	3.6779557	3.6784753
10	3.0	5.8729143	5.8741000

c.

i	t_i	w_i	$y(t_i)$
2	0.4	−1.6201137	−1.6200510
4	0.8	−1.3359853	−1.3359632
6	1.2	−1.1663295	−1.1663454
8	1.6	−1.0783171	−1.0783314
10	2.0	−1.0359674	−1.0359724

d.

i	t_i	w_i	$y(t_i)$
2	0.2	0.1627236	0.1626265
4	0.4	0.2051833	0.2051118
6	0.6	0.3766352	0.3765957
8	0.8	0.6461246	0.6461052
10	1.0	1.0022549	1.0022460

9. a. Euler's method gives the approximations in the following table.

i	t_i	w_i	$y(t_i)$
1	1.05	-0.9500000	-0.9523810
2	1.10	-0.9045353	-0.9090909
11	1.55	-0.6263495	-0.6451613
12	1.60	-0.6049486	-0.6250000
19	1.95	-0.4850416	-0.5128205
20	2.00	-0.4712186	-0.5000000

b. Linear interpolation gives
(i) $y(1.052) \approx -0.9481814$, (ii) $y(1.555) \approx -0.6242094$, (iii) $y(1.978) \approx -0.4773007$.
The actual values are $y(1.052) = -0.9505703$, $y(1.555) = -0.6430868$, $y(1.978) = -0.5055612$.

c. Taylor's method of order 2 gives the approximations in the following table.

i	t_i	w_i	$y(t_i)$
1	1.05	-0.9525000	-0.9523810
2	1.10	-0.9093138	-0.9090909
11	1.55	-0.6459788	-0.6451613
12	1.60	-0.6258649	-0.6250000
19	1.95	-0.5139781	-0.5128205
20	2.00	-0.5011957	-0.5000000

d. Linear interpolation gives
(i) $y(1.052) \approx -0.9507726$, (ii) $y(1.555) \approx -0.6439674$, (iii) $y(1.978) \approx -0.5068199$.

e. Taylor's method of order 4 gives the approximations in the following table.

i	t_i	w_i	$y(t_i)$
1	1.05	-0.9523813	-0.9523810
2	1.10	-0.9090914	-0.9090909
11	1.55	-0.6451629	-0.6451613
12	1.60	-0.6250017	-0.6250000
19	1.95	-0.5128226	-0.5128205
20	2.00	-0.5000022	-0.5000000

f. Hermite interpolation gives
(i) $y(1.052) \approx -0.9505706$, (ii) $y(1.555) \approx -0.6430884$, (iii) $y(1.978) \approx -0.5055633$.

11. a. The approximate velocity using Taylor's methods of orders 2 and 4 are in the table.

i	t_i	Order 2	Order 4
2	0.2	5.86595	5.86433
5	0.5	2.82145	2.81789
7	0.7	0.84926	0.84455
10	1.0	-2.08606	-2.09015

b. The velocity is 0 in approximately 0.8 seconds.

5.3 Runge-Kutta Methods (Page 189)

1. a.

t	Modified Euler	$y(t)$
0.5	0.5602111	0.2836165
1.0	5.3014898	3.2190993

b.

t	Modified Euler	$y(t)$
2.5	1.8125000	1.8333333
3.0	2.4815531	2.5000000

c.

t	Modified Euler	$y(t)$
1.25	2.7750000	2.7789294
1.50	3.6008333	3.6081977
1.75	4.4688294	4.4793276
2.00	5.3728586	5.3862944

d.

t	Modified Euler	$y(t)$
0.25	1.3199027	1.3291498
0.50	1.7070300	1.7304898
0.75	2.0053560	2.0414720
1.00	2.0770789	2.1179795

3. a.

	Modified Euler	
t_i	w_i	$y(t_i)$
0.5	1.597265955	1.6
0.6	1.615015699	1.617647059
0.9	1.545108042	1.546961326
1.0	1.498430678	1.5

b.

	Modified Euler	
t_i	w_i	$y(t_i)$
1.1	−1.347996027	−1.347822707
1.2	−1.268565970	−1.268299404
1.9	−0.9395411781	−0.9392222368
2.0	−0.9105471247	−0.9102392264

c.

	Modified Euler	
t_i	w_i	$y(t_i)$
1.2	−1.72	−1.714285714
1.4	−1.561272503	−1.555555556
2.8	−1.219717333	−1.217391304
3.0	−1.202119310	−1.2

d.

	Modified Euler	
t_i	w_i	$y(t_i)$
0.5	1.289770701	1.289805276
0.6	1.380583709	1.380931216
0.9	1.631230851	1.632613182
1.0	1.700210296	1.701870053

5. a.

	Midpoint	
t_i	w_i	$y(t_i)$
1.2	1.0153257	1.0149523
1.5	1.0677427	1.0672624
1.7	1.1111478	1.1106551
2.0	1.1817275	1.1812322

b.

	Midpoint	
t_i	w_i	$y(t_i)$
1.4	0.4861770	0.4896817
2.0	1.6438889	1.6612818
2.4	2.8364357	2.8765514
3.0	5.7386475	5.8741000

c.

	Midpoint	
t_i	w_i	$y(t_i)$
0.4	−1.6192966	−1.6200510
1.0	−1.2402470	−1.2384058
1.4	−1.1175165	−1.1146484
2.0	−1.0382227	−1.0359724

d.

	Midpoint	
t_i	w_i	$y(t_i)$
0.2	0.1722396	0.1626265
0.5	0.2848046	0.2773617
0.7	0.5056268	0.5000658
1.0	1.0063347	1.0022460

7. a.

	Heun	
t_i	w_i	$y(t_i)$
0.50	0.2710885	0.2836165
1.00	3.1327255	3.2190993

b.

	Heun	
t_i	w_i	$y(t_i)$
2.50	1.8464828	1.8333333
3.00	2.5094123	2.5000000

c.

	Heun	
t_i	w_i	$y(t_i)$
1.25	2.7788462	2.7789294
1.50	3.6080529	3.6081977
1.75	4.4791319	4.4793276
2.00	5.3860533	5.3862944

d.

	Heun	
t_i	w_i	$y(t_i)$
0.25	1.3295717	1.3291498
0.50	1.7310350	1.7304898
0.75	2.0417476	2.0414720
1.00	2.1176975	2.1179795

9. a.

	Heun	
t_i	w_i	$y(t_i)$
0.5	1.599939902	1.6
0.6	1.617600330	1.617647059
0.9	1.546961530	1.546961326
1.0	1.500010266	1.5

b.

	Heun	
t_i	w_i	$y(t_i)$
1.1	−1.347797247	−1.347822707
1.2	−1.268261121	−1.268299404
1.9	−0.9391794862	−0.9392222368
2.0	−0.9101980983	−0.9102392264

c.

	Heun	
t_i	w_i	$y(t_i)$
1.2	−1.710175817	−1.714285714
1.4	−1.551807512	−1.555555556
2.8	−1.216030469	−1.217391304
3.0	−1.198763172	−1.2

d.

	Heun	
t_i	w_i	$y(t_i)$
0.5	1.289772720	1.289805276
0.6	1.380888251	1.380931216
0.9	1.632584856	1.632613182
1.0	1.701855837	1.701870053

11. a.

t_i	Runge-Kutta w_i	$y(t_i)$
1.2	1.0149520	1.0149523
1.5	1.0672620	1.0672624
1.7	1.1106547	1.1106551
2.0	1.1812319	1.1812322

b.

t_i	Runge-Kutta w_i	$y(t_i)$
1.4	0.4896842	0.4896817
2.0	1.6612651	1.6612818
2.4	2.8764941	2.8765514
3.0	5.8738386	5.8741000

c.

t_i	Runge-Kutta w_i	$y(t_i)$
0.4	−1.6200576	−1.6200510
1.0	−1.2384307	−1.2384058
1.4	−1.1146769	−1.1146484
2.0	−1.0359922	−1.0359724

d.

t_i	Runge-Kutta w_i	$y(t_i)$
0.2	0.1627655	0.1626265
0.5	0.2774767	0.2773617
0.7	0.5001579	0.5000658
1.0	1.0023207	1.0022460

13. a. $1.0221167 \approx y(1.25) = 1.0219569$, $1.1640347 \approx y(1.93) = 1.1643901$

  b. $1.9086500 \approx y(2.1) = 1.9249616$, $4.3105913 \approx y(2.75) = 4.3941697$

  c. $-1.1461434 \approx y(1.3) = -1.1382768$, $-1.0454854 \approx y(1.93) = -1.0412665$

  d. $0.3271470 \approx y(0.54) = 0.3140018$, $0.8967073 \approx y(0.94) = 0.8866318$

15. a. $1.0219569 = y(1.25) \approx 1.0219550$, $1.1643902 = y(1.93) \approx 1.1643898$

  b. $1.9249617 = y(2.10) \approx 1.9249217$, $4.3941697 = y(2.75) \approx 4.3939943$

  c. $-1.138268 = y(1.3) \approx -1.1383036$, $-1.0412666 = y(1.93) \approx -1.0412862$

  d. $0.31400184 = y(0.54) \approx 0.31410579$, $0.88663176 = y(0.94) \approx 0.88670653$

17. The equation for the Midpoint method has $w_0 = 1$ and

$$w_{i+1} = w_i + hf(t_i + 0.5h, w_i + 0.5hf(t_i, w_i))$$
$$= w_i + hf(t_i + 0.5h, w_i + 0.5h(-w_i + t_i + 1))$$
$$= w_i + h(-w_i - 0.5h(-w_i + t_i + 1) + t_i + 0.5h + 1)$$
$$= w_i(1 + h - 0.5h^2) + t_i(h - 0.5h^2) + h.$$

The equation for the Modified Euler method has $w_0 = 1$ and

$$w_{i+1} = w_i + 0.5h(f(t_i, w_i) + f(t_{i+1}, w_i + hf(t_i, w_i)))$$
$$= w_i + 0.5h(-w_i + t_i + 1 - w_i - h(-w_i + t_i + 1) + t_i + h + 1)$$
$$= w_i + 0.5h(-w_i + t_i + 1 - w_i - h(-w_i + t_i + 1) + t_i + h + 1)$$
$$= w_i(1 - h + 0.5h^2) + t_i(h - 0.5h^2) + h.$$

The Modified Euler method and the Midpoint method give the same results when applied to the initial-value problem $y' = f(t, y) = -y + t + 1'$, for $0 \le t \le 1$ and $y(0) = 1$ with any choice of h because the equation is linear in both y and t.

19. In 0.2 seconds we have approximately 2099 units of KOH.

5.4 *Predictor-Corrector Methods (Page 198)*

1. The Adams-Bashforth methods give the results in the following tables.

a.

t_i	2-step	3-step	4-step	5-step	$y(t_i)$
0.2	0.0268128	0.0268128	0.0268128	0.0268128	0.0268128
0.4	0.1200522	0.1507778	0.1507778	0.1507778	0.1507778
0.6	0.4153551	0.4613866	0.4960196	0.4960196	0.4960196
0.8	1.1462844	1.2512447	1.2961260	1.3308570	1.3308570
1.0	2.8241683	3.0360680	3.1461400	3.1854002	3.2190993

b.

t_i	2-step	3-step	4-step	5-step	$y(t_i)$
2.2	1.3666667	1.3666667	1.3666667	1.3666667	1.3666667
2.4	1.6750000	1.6857143	1.6857143	1.6857143	1.6857143
2.6	1.9632431	1.9794407	1.9750000	1.9750000	1.9750000
2.8	2.2323184	2.2488759	2.2423065	2.2444444	2.2444444
3.0	2.4884512	2.5051340	2.4980306	2.5011406	2.5000000

c.

t_i	2-step	3-step	4-step	5-step	$y(t_i)$
1.2	2.6187859	2.6187859	2.6187859	2.6187859	2.6187859
1.4	3.2734823	3.2710611	3.2710611	3.2710611	3.2710611
1.6	3.9567107	3.9514231	3.9520058	3.9520058	3.9520058
1.8	4.6647738	4.6569191	4.6582078	4.6580160	4.6580160
2.0	5.3949416	5.3848058	5.3866452	5.3862177	5.3862944

d.

t_i	2-step	3-step	4-step	5-step	$y(t_i)$
0.2	1.2529306	1.2529306	1.2529306	1.2529306	1.2529306
0.4	1.5986417	1.5712255	1.5712255	1.5712255	1.5712255
0.6	1.9386951	1.8827238	1.8750869	1.8750869	1.8750869
0.8	2.1766821	2.0844122	2.0698063	2.0789180	2.0789180
1.0	2.2369407	2.1115540	2.0998117	2.1180642	2.1179795

3. The Adams-Bashforth methods give the results in the following tables.

a.

t_i	2-step	3-step	4-step	5-step	$y(t_i)$
1.2	1.0161982	1.0149520	1.0149520	1.0149520	1.0149523
1.5	1.0697141	1.0664788	1.0675362	1.0671695	1.0672624
1.7	1.1133294	1.1097691	1.1109994	1.1105036	1.1106551
2.0	1.1840272	1.1803057	1.1815967	1.1810689	1.1812322

b.

t_i	2-step	3-step	4-step	5-step	$y(t_i)$
1.4	0.4867550	0.4896842	0.4896842	0.4896842	0.4896817
2.0	1.6377944	1.6584313	1.6603060	1.6613179	1.6612818
2.4	2.8163947	2.8667672	2.8735320	2.8762776	2.8765514
3.0	5.6491203	5.8268008	5.8589944	5.8706101	5.8741000

c.

t_i	2-step	3-step	4-step	5-step	$y(t_i)$
0.5	−1.5357010	−1.5381988	−1.5379372	−1.5378676	−1.5378828
1.0	−1.2374093	−1.2389605	−1.2383734	−1.2383693	−1.2384058
1.5	−1.0952910	−1.0950952	−1.0947925	−1.0948481	−1.0948517
2.0	−1.0366643	−1.0359996	−1.0359497	−1.0359760	−1.0359724

d.

t_i	2-step	3-step	4-step	5-step	$y(t_i)$
0.2	0.1739041	0.1627655	0.1627655	0.1627655	0.1626265
0.5	0.2846336	0.2732179	0.2780929	0.2769031	0.2773617
0.7	0.5042285	0.4972078	0.4998405	0.4988777	0.5000658
1.0	1.0037415	1.0020894	1.0064121	1.0073348	1.0022460

5. The Adams Fourth-Order Predictor-Corrector method gives the results in the following tables.

a.

t_i	w_i	$y(t_i)$
1.2	1.0149520	1.0149523
1.5	1.0672462	1.0672624
1.7	1.1106352	1.1106551
2.0	1.1812112	1.1812322

b.

t_i	w_i	$y(t_i)$
1.4	0.4896842	0.4896817
2.0	1.6612586	1.6612818
2.4	2.8765082	2.8765514
3.0	5.8739518	5.8741000

c.

t_i	w_i	$y(t_i)$
0.5	-1.5378788	-1.5378828
1.0	-1.2384134	-1.2384058
1.5	-1.0948609	-1.0948517
2.0	-1.0359757	-1.0359724

d.

t_i	w_i	$y(t_i)$
0.2	0.1627655	0.1626265
0.5	0.2769896	0.2773617
0.7	0.4998012	0.5000658
1.0	1.0021372	1.0022460

7. Milne-Simpson's Predictor-Corrector method gives the results in the following tables.

a.

i	t_i	w_i	$y(t_i)$
2	1.2	1.01495200	1.01495231
5	1.5	1.06725997	1.06726235
7	1.7	1.11065221	1.11065505
10	2.0	1.18122584	1.18123222

b.

i	t_i	w_i	$y(t_i)$
2	1.4	0.48968417	0.48968166
5	2.0	1.66126150	1.66128176
7	2.4	2.87648763	2.87655142
10	3.0	5.87375555	5.87409998

c.

i	t_i	w_i	$y(t_i)$
5	0.5	-1.53788255	-1.53788284
10	1.0	-1.23840789	-1.23840584
15	1.5	-1.09485532	-1.09485175
20	2.0	-1.03597247	-1.03597242

d.

i	t_i	w_i	$y(t_i)$
2	0.2	0.16276546	0.16262648
5	0.5	0.27741080	0.27736167
7	0.7	0.50008713	0.50006579
10	1.0	1.00215439	1.00224598

5.5 Extrapolation Methods (Page 203)

1. $y_{22} = 0.14846014$ approximates $y(0.1) = 0.14846010$.

3. The Extrapolation method gives the results in the following tables.

a.

i	t_i	w_i	h_i	k	y_i
1	1.05	1.10385729	0.05	2	1.10385738
2	1.10	1.21588614	0.05	2	1.21588635
3	1.15	1.33683891	0.05	2	1.33683925
4	1.20	1.46756907	0.05	2	1.46756957

b.

i	t_i	w_i	h_i	k	y_i
1	0.25	0.25228680	0.25	3	0.25228680
2	0.50	0.51588678	0.25	3	0.51588678
3	0.75	0.79594460	0.25	2	0.79594458
4	1.00	1.09181828	0.25	3	1.09181825

c.

i	t_i	w_i	h_i	k	y_i
1	1.50	-1.50000055	0.50	5	-1.50000000
2	2.00	-1.33333435	0.50	3	-1.33333333
3	2.50	-1.25000074	0.50	3	-1.25000000
4	3.00	-1.20000090	0.50	2	-1.20000000

d.

i	t_i	w_i	h_i	k	y_i
1	0.25	1.08708817	0.25	3	1.08708823
2	0.50	1.28980537	0.25	3	1.28980528
3	0.75	1.51349008	0.25	3	1.51348985
4	1.00	1.70187009	0.25	3	1.70187005

5. $P(5) \approx 56{,}751$.

5.6   Adaptive Techniques (Page 213)

1. a. $w_1 = 0.5620214 \approx y(t_1) = y(0.3426388) = 0.5619801$ **b.** $w_4 = 0.31055852 \approx y(t_4) = y(0.2) = 0.31055897$

3. The Runge-Kutta-Fehlberg method gives the results in the following tables.

a.

i	t_i	w_i	h_i	y_i
1	1.0500000	1.1038574	0.0500000	1.1038574
2	1.1000000	1.2158864	0.0500000	1.2158863
3	1.1500000	1.3368393	0.0500000	1.3368393
4	1.2000000	1.4675697	0.0500000	1.4675696

b.

i	t_i	w_i	h_i	y_i
1	0.2500000	0.2522868	0.2500000	0.2522868
2	0.5000000	0.5158867	0.2500000	0.5158868
3	0.7500000	0.7959445	0.2500000	0.7959446
4	1.0000000	1.0918182	0.2500000	1.0918183

c.

i	t_i	w_i	h_i	y_i
1	1.1382206	-1.7834313	0.1382206	-1.7834282
3	1.6364797	-1.4399709	0.3071709	-1.4399551
5	2.6364797	-1.2340532	0.5000000	-1.2340298
6	3.0000000	-1.2000195	0.3635203	-1.2000000

d.

i	t_i	w_i	h_i	y_i
1	0.2	1.0571819	0.2	1.0571810
2	0.4	1.2014801	0.2	1.2014860
3	0.6	1.3809214	0.2	1.3809312
4	0.8	1.5550243	0.2	1.5550314
5	1.0	1.7018705	0.2	1.7018701

5. The Adams Variable Step-Size Predictor-Corrector method gives the results in the following tables.

a.

i	t_i	w_i	h_i	y_i
1	1.05000000	1.10385717	0.05000000	1.10385738
2	1.10000000	1.21588587	0.05000000	1.21588635
3	1.15000000	1.33683848	0.05000000	1.33683925
4	1.20000000	1.46756885	0.05000000	1.46756957

b.

i	t_i	w_i	h_i	y_i
1	0.20000000	0.20120278	0.20000000	0.20120267
2	0.40000000	0.40861919	0.20000000	0.40861896
3	0.60000000	0.62585310	0.20000000	0.62585275
4	0.80000000	0.85397394	0.20000000	0.85396433
5	1.00000000	1.09183759	0.20000000	1.09181825

c.

i	t_i	w_i	h_i	y_i
5	1.16289739	-1.75426113	0.03257948	-1.75426455
10	1.32579477	-1.60547206	0.03257948	-1.60547731
15	1.57235777	-1.46625721	0.04931260	-1.46626230
20	1.92943707	-1.34978308	0.07694168	-1.34978805
25	2.47170180	-1.25358275	0.11633076	-1.25358804
30	3.00000000	-1.19999513	0.10299186	-1.20000000

d.

i	t_i	w_i	h_i	y_i
1	0.06250000	1.00583097	0.06250000	1.00583095
5	0.31250000	1.13099427	0.06250000	1.13098105
10	0.62500000	1.40361751	0.06250000	1.40360196
12	0.81250000	1.56515769	0.09375000	1.56514800
14	1.00000000	1.70186884	0.09375000	1.70187005

7. The current after 2 s is approximately $i(2) = 8.693$ amperes.

5.7 Methods for Systems of Equations (Page 221)

1. The Runge-Kutta for Systems method gives the results in the following tables.

a.

i	t_i	w_{1i}	u_{1i}	w_{2i}	u_{2i}
1	0.2	2.12036583	2.12500839	1.50699185	1.51158743
2	0.4	4.44122776	4.46511961	3.24224021	3.26598528
3	0.6	9.73913329	9.83235869	8.16341700	8.25629549
4	0.8	22.67655977	23.00263945	21.34352778	21.66887674
5	1.0	55.66118088	56.73748265	56.03050296	57.10536209

b.

i	t_i	w_{1i}	u_{1i}	w_{2i}	u_{2i}
1	0.5	0.95671390	0.95672798	-1.08381950	-1.08383310
2	1.0	1.30654440	1.30655930	-0.83295364	-0.83296776
3	1.5	1.34416716	1.34418117	-0.56980329	-0.56981634
4	2.0	1.14332436	1.14333672	-0.36936318	-0.36937457

c.

i	t_i	w_{1i}	u_{1i}	w_{2i}
1	0.5	0.70787076	0.70828683	-1.24988663
2	1.0	-0.33691753	-0.33650854	-3.01764179
3	1.5	-2.41332734	-2.41345688	-5.40523279
4	2.0	-5.89479008	-5.89590551	-8.70970537

i	t_i	u_{2i}	w_{3i}	u_{3i}
1	0.5	-1.25056425	0.39884862	0.39815702
2	1.0	-3.01945051	-0.29932294	-0.30116868
3	1.5	-5.40844686	-0.92346873	-0.92675778
4	2.0	-8.71450036	-1.32051165	-1.32544426

d.

i	t_i	w_{1i}	u_{1i}	w_{2i}
2	0.2	1.38165297	1.38165325	1.00800000
5	0.5	1.90753116	1.90753184	1.12500000
7	0.7	2.25503524	2.25503620	1.34300000
10	1.0	2.83211921	2.83212056	2.00000000

i	t_i	u_{2i}	w_{3i}	u_{3i}
2	0.2	1.00800000	-0.61833075	-0.61833075
5	0.5	0.12500000	-0.09090565	-0.09090566
7	0.7	1.34000000	0.26343971	0.26343970
10	1.0	2.00000000	0.88212058	0.88212056

3. First use the Runge-Kutta method of order four for systems to compute all starting values:

$$w_{1,0}, w_{2,0}, \ldots, w_{m,0}$$

$$w_{1,1}, w_{2,1}, \ldots, w_{m,1}$$

$$w_{1,2}, w_{2,2}, \ldots, w_{m,2}$$

$$w_{1,3}, w_{2,3}, \ldots, w_{m,3}.$$

Then for each $j = 3, 4, \ldots N - 1$, compute, for each $i = 1, \ldots, m$, the predictor values

$$w_{i,j+1,p} = w_{i,j} + \frac{h}{24}[55 f_i(t_i, w_{1,j}, \ldots, w_{m,j}) - 59 f_i(t_{j-1}, w_{1,j-1}, \ldots, w_{m,j-1})$$
$$+ 37 f_i(t_{j-2}, w_{1,j-2}, \ldots, w_{m,j-2}) - 9 f_i(t_{j-3}, w_{1,j-3}, \ldots, w_{m,j-3})],$$

and then the corrector values

$$w_{i,j+1} = w_{i,j} + \frac{h}{24}[9 f_i(t_{j+1}, w_{i,j+1,p}, \ldots, w_{m,j+1,p}) + 19 f_i(t_i, w_{1,j}, \ldots, w_{m,j})$$
$$- 5 f_i(t_{j-1}, w_{1,j-1}, \ldots, w_{m,j-1}) + f_i(t_{j-2}, w_{1,j-2}, \ldots, w_{m,j-2})].$$

5. The predicted number of prey, x_{1i}, and predators, x_{2i}, are given in the following table.

i	t_i	x_{1i}	x_{2i}
10	1.0	4393	1512
20	2.0	288	3175
30	3.0	32	2042
40	4.0	25	1258

A stable solution is $x_1 = 833.\overline{3}$ and $x_2 = 1500$.

5.8 Stiff Differential Equations (Page 227)

1. Euler's method gives the results in the following tables.

a.
i	t_i	w_i	$y(t_i)$
2	0.200	0.027182818	0.4493290
5	0.500	0.000027183	0.0301974
7	0.700	0.000000272	0.0049916
10	1.000	0.000000000	0.0003355

b.
i	t_i	w_i	$y(t_i)$
2	0.200	0.373333333	0.0461052
5	0.500	-0.933333333	0.2500151
7	0.700	0.146666667	0.4900003
10	1.000	1.333333333	1.0000000

c.
i	t_i	w_i	$y(t_i)$
2	0.500	16.47925	0.4794709
4	1.000	256.7930	0.8414710
6	1.500	4096.142	0.9974950
8	2.000	65523.12	0.9092974

d.
i	t_i	w_i	$y(t_i)$
2	0.200	6.128259	1.000000001
5	0.500	-378.2574	1.000000000
7	0.700	-6052.063	1.000000000
10	1.000	387332.0	1.000000000

3. The Adams Fourth-Order Predictor-Corrector method gives the results in the following tables.

a.
i	t_i	w_i	$y(t_i)$
2	0.200	0.4588119	0.4493290
5	0.500	-0.0112813	0.0301974
7	0.700	0.0013734	0.0049916
10	1.000	0.0023604	0.0003355

b.
i	t_i	w_i	$y(t_i)$
2	0.200	0.0792593	0.0461052
5	0.500	0.1554027	0.2500151
7	0.700	0.5507445	0.4900003
10	1.000	0.7278557	1.0000000

c.
i	t_i	w_i	$y(t_i)$
2	0.500	188.3082	0.4794709
4	1.000	38932.03	0.8414710
6	1.500	9073607	0.9974950
8	2.000	2115741299	0.9092974

d.
i	t_i	w_i	$y(t_i)$
2	0.200	-215.7459	1.000000000
5	0.500	-682637.0	1.000000000
7	0.700	-159172736	1.000000000
10	1.000	-566751172258	1.000000000

5. The following tables list the results of the Backward Euler method applied to the problems in Exercise 1.

a.

i	t_i	w_i	k	$y(t_i)$
2	0.20	0.75298666	2	0.44932896
5	0.50	0.10978082	2	0.03019738
7	0.70	0.03041020	2	0.00499159
10	1.00	0.00443362	2	0.00033546

b.

i	t_i	w_i	k	$y(t_i)$
2	0.20	0.08148148	2	0.04610521
5	0.50	0.25635117	2	0.25001513
7	0.70	0.49515013	2	0.49000028
10	1.00	1.00500556	2	1.00000000

c.

i	t_i	w_i	k	$y(t_i)$
2	0.50	0.50495522	2	0.47947094
4	1.00	0.83751817	2	0.84147099
6	1.50	0.99145076	2	0.99749499
8	2.00	0.90337560	2	0.90929743

d.

i	t_i	w_i	k	$y(t_i)$
2	0.20	1.00348713	3	1.00000001
5	0.50	1.00000262	2	1.00000000
7	0.70	1.00000002	1	1.00000000
10	1.00	1.00000000	1	1.00000000

6.2   Gaussian Elimination (Page 238)

1. a. Intersecting lines with solution $x_1 = x_2 = 1$.

  b. Intersecting lines with solution $x_1 = x_2 = 0$.

  c. One line, so there are an infinite number of solutions with $x_2 = \frac{3}{2} - \frac{1}{2}x_1$.

  d. Parallel lines, so there is no solution.

  e. One line, so there are an infinite number of solutions with $x_2 = -\frac{1}{2}x_1$.

  f. Three lines in the plane that do not intersect at a common point.

  g. Intersecting lines with solution $x_1 = \frac{2}{7}$ and $x_2 = -\frac{11}{7}$.

  h. Two planes in space that intersect in a line with $x_1 = -\frac{5}{4}x_2$ and $x_3 = \frac{3}{2}x_2 + 1$.

3. Gaussian elimination gives the following solutions:

  a. $x_1 = 1.1875, x_2 = 1.8125, x_3 = 0.875$ with one row interchange required.

  b. $x_1 = -1, x_2 = 0, x_3 = 1$ with no interchange required.

  c. $x_1 = 1.5, x_2 = 2, x_3 = -1.2, x_4 = 3$ with no interchange required.

  d. $x_1 = \frac{22}{9}, x_2 = -\frac{4}{9}, x_3 = \frac{4}{3}, x_4 = 1$ with one row interchange required.

  e. No unique solution.

  f. $x_1 = -1, x_2 = 2, x_3 = 0, x_4 = 1$ with one row interchange required.

5. a. When $\alpha = -1/3$, there is no solution.

  b. When $\alpha = 1/3$, there is an infinite number of solutions with $x_1 = x_2 + 1.5$, and x_2 is arbitrary.

  c. If $\alpha \neq \pm 1/3$, then the unique solution is

$$x_1 = \frac{3}{2(1+3\alpha)} \quad \text{and} \quad x_2 = \frac{-3}{2(1+3\alpha)}.$$

7. a. There is sufficient food to satisfy the average daily consumption.

  b. We could add 200 of species 1, or 150 of species 2, or 100 of species 3, or 100 of species 4.

  c. Assuming none of the increases indicated in part (b) was selected, species 2 could be increased by 650, or species 3 could be increased by 150, or species 4 could be increased by 150.

  d. Assuming none of the increases indicated in (b) or (c) was selected, species 3 could be increased by 150, or species 4 could be increased by 150.

6.3   Pivoting Strategies (Page 246)

1. a. None.   **b.** Interchange rows 2 and 3.   **c.** None.   **d.** Interchange rows 1 and 2.

3. a. Interchange rows 1 and 3, then interchange rows 2 and 3.   **b.** Interchange rows 2 and 3.

  c. Interchange rows 2 and 3.   **d.** Interchange rows 1 and 3, then interchange rows 2 and 3.

5. Gaussian elimination with three-digit chopping arithmetic gives the following results:

a. $x_1 = 30.0$, $x_2 = 0.990$ **b.** $x_1 = 1.00$, $x_2 = 9.98$

c. $x_1 = 0.00$, $x_2 = 10.0$, $x_3 = 0.142$ **d.** $x_1 = 12.0$, $x_2 = 0.492$, $x_3 = -9.78$

e. $x_1 = 0.206$, $x_2 = 0.0154$, $x_3 = -0.0156$, $x_4 = -0.716$ **f.** $x_1 = 0.828$, $x_2 = -3.32$, $x_3 = 0.153$, $x_4 = 4.91$

7. Gaussian elimination with partial pivoting and three-digit chopping arithmetic gives the following results:

a. $x_1 = 10.0$, $x_2 = 1.00$ **b.** $x_1 = 1.00$, $x_2 = 9.98$

c. $x_1 = -0.163$, $x_2 = 9.98$, $x_3 = 0.142$ **d.** $x_1 = 12.0$, $x_2 = 0.504$, $x_3 = -9.78$

e. $x_1 = 0.177$, $x_2 = -0.0072$, $x_3 = -0.0208$, $x_4 = -1.18$ **f.** $x_1 = 0.777$, $x_2 = -3.10$, $x_3 = 0.161$, $x_4 = 4.50$

9. Only (a), $\alpha = 6$

6.4 Linear Algebra and Matrix Inversion (Page 256)

1. a. $\begin{bmatrix} 1 & 0 & 0 \\ 1 & 2 & 0 \\ 9 & 5 & 1 \end{bmatrix}$ **b.** $\begin{bmatrix} 1 & -1 & 2 \\ 2 & -1 & 7 \\ -2 & 1 & -5 \end{bmatrix}$ **c.** $\begin{bmatrix} 1 & 0 & 0 \\ 2 & 1 & 0 \\ -7 & -2 & 1 \end{bmatrix}$ **d.** $\begin{bmatrix} 6 & -7 & 15 \\ 0 & -1 & 3 \\ 0 & 0 & 6 \end{bmatrix}$

3. a. Singular, $\det A = 0$ **b.** $\det A = -8$, $\det A^{-1} = -\dfrac{1}{8}$ **c.** Singular, $\det A = 0$

d. Singular, $\det A = 0$ **e.** $\det A = 28$, $\det A^{-1} = \dfrac{1}{28}$ **f.** $\det A = 3$, $\det A^{-1} = \dfrac{1}{3}$

5. a. Not true. One example is

$$A = \begin{bmatrix} 2 & 1 \\ 1 & 0 \end{bmatrix} \quad \text{and} \quad B = \begin{bmatrix} 1 & -1 \\ -1 & 2 \end{bmatrix}.$$

Then

$$AB = \begin{bmatrix} 1 & 0 \\ 1 & -1 \end{bmatrix}$$

is not symmetric.

b. True. Let A be a nonsingular symmetric matrix. From the properties of transposes and inverses we have $(A^{-1})^t = (A^t)^{-1}$. Thus $(A^{-1})^t = (A^t)^{-1} = A^{-1}$, and A^{-1} is symmetric.

c. Not true. Use the matrices A and B from (a).

7. a. The solution is $x_1 = 0$, $x_2 = 10$, and $x_3 = 26$.

b. We have $D_1 = -1$, $D_2 = 3$, $D_3 = 7$, and $D = 0$, and there are no solutions.

c. We have $D_1 = D_2 = D_3 = D = 0$, and there are infinitely many solutions.

9. a. For each $k = 1, 2, \ldots, m$, the number a_{ik} represents the total number of plants of type v_i eaten by herbivores in the species h_k. The number of herbivores of types h_k eaten by species c_j is b_{kj}. Thus the total number of plants of type v_i ending up in species c_j is $a_{i1}b_{1j} + a_{i2}b_{2j} + \cdots + a_{im}b_{mj} = (AB)_{ij}$.

b. We first assume $n = m = k$ so that the matrices will have inverses. Let $x_1, \ldots, x_n$ represent the vegetations of type $v_1, \ldots, v_n$, let $y_1, \ldots, y_n$ represent the number of herbivores of species $h_1, \ldots, h_n$, and let $z_1, \ldots, z_n$ represent the number of carnivores of species $c_1, \ldots, c_n$. If

$$\begin{bmatrix} x_1 \\ x_2 \\ \vdots \\ x_n \end{bmatrix} = A \begin{bmatrix} y_1 \\ y_2 \\ \vdots \\ y_n \end{bmatrix}, \quad \text{then} \quad \begin{bmatrix} y_1 \\ y_2 \\ \vdots \\ y_n \end{bmatrix} = A^{-1} \begin{bmatrix} x_1 \\ x_2 \\ \vdots \\ x_n \end{bmatrix}.$$

Thus $(A^{-1})_{i,j}$ represents the amount of type v_j plants eaten by one herbivore of species h_i. Similarly, if

$$\begin{bmatrix} y_1 \\ y_2 \\ \vdots \\ y_n \end{bmatrix} = B \begin{bmatrix} z_1 \\ z_2 \\ \vdots \\ z_n \end{bmatrix}, \quad \text{then} \quad \begin{bmatrix} z_1 \\ z_2 \\ \vdots \\ z_n \end{bmatrix} = B^{-1} \begin{bmatrix} y_1 \\ y_2 \\ \vdots \\ y_n \end{bmatrix}.$$

Thus $(B^{-1})_{i,j}$ represents the number of herbivores of species h_j eaten by a carnivore of species c_i. If $x = Ay$ and $y = Bz$, then $x = ABz$ and $z = (AB)^{-1}x$. But, $y = A^{-1}x$ and $z = B^{-1}y$, so $z = B^{-1}A^{-1}x$.

11. a. In component form:

$$(a_{11}x_1 - b_{11}y_1 + a_{12}x_2 - b_{12}y_2) + (b_{11}x_1 + a_{11}y_1 + b_{12}x_2 + a_{12}y_2)i = c_1 + id_1$$

$$(a_{21}x_1 - b_{21}y_1 + a_{22}x_2 - b_{22}y_2) + (b_{21}x_1 + a_{21}y_1 + b_{22}x_2 + a_{22}y_2)i = c_2 + id_2,$$

so

$$a_{11}x_1 + a_{12}x_2 - b_{11}y_1 - b_{12}y_2 = c_1$$
$$b_{11}x_1 + b_{12}x_2 + a_{11}y_1 + a_{12}y_2 = d_1$$
$$a_{21}x_1 + a_{22}x_2 - b_{21}y_1 - b_{22}y_2 = c_2$$
$$b_{21}x_1 + b_{22}x_2 + a_{21}y_1 + a_{22}y_2 = d_2$$

b. The system

$$\begin{bmatrix} 1 & 3 & 2 & -2 \\ -2 & 2 & 1 & 3 \\ 2 & 4 & -1 & -3 \\ 1 & 3 & 2 & 4 \end{bmatrix} \begin{bmatrix} x_1 \\ x_2 \\ y_1 \\ y_2 \end{bmatrix} = \begin{bmatrix} 5 \\ 2 \\ 4 \\ -1 \end{bmatrix}$$

has the solution $x_1 = -1.2$, $x_2 = 1$, $y_1 = 0.6$, and $y_2 = -1$.

6.5 Matrix Factorization (Page 265)

1. a. $x_1 = -3$, $x_2 = 3$, $x_3 = 1$

b. $x_1 = \dfrac{1}{2}$, $x_2 = -\dfrac{9}{2}$, $x_3 = \dfrac{7}{2}$

3. a. $P^t LU = \begin{bmatrix} 0 & 1 & 0 \\ 1 & 0 & 0 \\ 0 & 0 & 1 \end{bmatrix} \begin{bmatrix} 1 & 0 & 0 \\ 0 & 1 & 0 \\ 0 & -\frac{1}{2} & 1 \end{bmatrix} \begin{bmatrix} 1 & 1 & -1 \\ 0 & 2 & 3 \\ 0 & 0 & \frac{5}{2} \end{bmatrix}$

b. $P^t LU = \begin{bmatrix} 1 & 0 & 0 \\ 0 & 0 & 1 \\ 0 & 1 & 0 \end{bmatrix} \begin{bmatrix} 1 & 0 & 0 \\ 2 & 1 & 0 \\ 1 & 0 & 1 \end{bmatrix} \begin{bmatrix} 1 & 2 & -1 \\ 0 & -5 & 6 \\ 0 & 0 & 4 \end{bmatrix}$

c. $P^t LU = \begin{bmatrix} 1 & 0 & 0 & 0 \\ 0 & 0 & 0 & 1 \\ 0 & 1 & 0 & 0 \\ 0 & 0 & 1 & 0 \end{bmatrix} \begin{bmatrix} 1 & 0 & 0 & 0 \\ 2 & 1 & 0 & 0 \\ 1 & 0 & 1 & 0 \\ 3 & 0 & 0 & 1 \end{bmatrix} \begin{bmatrix} 1 & -2 & 3 & 0 \\ 0 & 5 & -2 & 1 \\ 0 & 0 & -1 & -2 \\ 0 & 0 & 0 & 3 \end{bmatrix}$

d. $P^t LU = \begin{bmatrix} 1 & 0 & 0 & 0 \\ 0 & 0 & 0 & 1 \\ 0 & 0 & 1 & 0 \\ 0 & 1 & 0 & 0 \end{bmatrix} \begin{bmatrix} 1 & 0 & 0 & 0 \\ 2 & 1 & 0 & 0 \\ 1 & 0 & 1 & 0 \\ 1 & 0 & 0 & 1 \end{bmatrix} \begin{bmatrix} 1 & -2 & 3 & 0 \\ 0 & 5 & -3 & -1 \\ 0 & 0 & -1 & -2 \\ 0 & 0 & 0 & 1 \end{bmatrix}$

5. a. $x_1 = 1$, $x_2 = 2$, $x_3 = -1$

b. $x_1 = 1$, $x_2 = 1$, $x_3 = 1$

c. $x_1 = 1.5$, $x_2 = 2$, $x_3 = -1.199998$, $x_4 = 3$

d. $x_1 = 2.939851$, $x_2 = 0.07067770$, $x_3 = 5.677735$, $x_4 = 4.379812$

6.6 Techniques for Special Matrices (Page 273)

1. (i) The symmetric matrices are in (a), (b), and (f).

(ii) The singular matrices are in (e) and (h).

(iii) The strictly diagonally dominant matrices are in (a), (b), (c), and (d).

(iv) The positive definite matrices are in (a) and (f).

3. Cholesky's LL^t factorization has the following for L.

a. $L = \begin{bmatrix} 1.414213 & 0 & 0 \\ -0.7071069 & 1.224743 & 0 \\ 0 & -0.8164972 & 1.154699 \end{bmatrix}$

b. $L = \begin{bmatrix} 2 & 0 & 0 & 0 \\ 0.5 & 1.658311 & 0 & 0 \\ 0.5 & -0.7537785 & 1.087113 & 0 \\ 0.5 & 0.4522671 & 0.08362442 & 1.240346 \end{bmatrix}$

c. $L = \begin{bmatrix} 2 & 0 & 0 & 0 \\ 0.5 & 1.658311 & 0 & 0 \\ -0.5 & -0.4522671 & 2.132006 & 0 \\ 0 & 0 & 0.9380833 & 1.766351 \end{bmatrix}$

d. $L = \begin{bmatrix} 2.449489 & 0 & 0 & 0 \\ 0.8164966 & 1.825741 & 0 & 0 \\ 0.4082483 & 0.3651483 & 1.923538 & 0 \\ -0.4082483 & 0.1825741 & -0.4678876 & 1.606574 \end{bmatrix}$

5. Crout factorization gives the following results:

a. $x_1 = 0.5, x_2 = 0.5, x_3 = 1$

b. $x_1 = -0.9999995, x_2 = 1.999999, x_3 = 1$

c. $x_1 = 1, x_2 = -1, x_3 = 0$

d. $x_1 = -0.09357798, x_2 = 1.587156, x_3 = -1.167431, x_4 = 0.5412844$

7. a. Not necessarily. Consider $\begin{bmatrix} 1 & 0 \\ 0 & 1 \end{bmatrix}$.

b. Yes, since $A = A^t$.

c. Yes, since $\mathbf{x}^t(A + B)\mathbf{x} = \mathbf{x}^t A\mathbf{x} + \mathbf{x}^t B\mathbf{x}$.

d. Yes, since $\mathbf{x}^t A^2\mathbf{x} = \mathbf{x}^t A^t A\mathbf{x} = (A\mathbf{x})^t(A\mathbf{x}) \geq 0$, and because A is nonsingular, equality holds only if $\mathbf{x} = \mathbf{0}$.

e. Not necessarily. Consider

$$A = \begin{bmatrix} 1 & 0 \\ 0 & 1 \end{bmatrix} \quad \text{and} \quad B = \begin{bmatrix} 2 & 0 \\ 0 & 2 \end{bmatrix}.$$

9. a. Since $\det A = 3\alpha - 2\beta$, A is singular if and only if $\alpha = 2\beta/3$.

b. $|\alpha| > 1, |\beta| < 1$

c. $\beta = 1$

d. $\alpha > \frac{2}{3}, \beta = 1$

11. a. Mating male i with female j produces offspring with the same wing characteristics as mating male j with female i.

b. No. Consider, for example, $\mathbf{x} = (1, 0, -1)^t$.

7.2 *Convergence of Vectors (Page 284)*

1. a. We have $\|\mathbf{x}\|_\infty = 4$ and $\|\mathbf{x}\|_2 = 5.220153$.

b. We have $\|\mathbf{x}\|_\infty = 4$ and $\|\mathbf{x}\|_2 = 5.477226$.

c. We have $\|\mathbf{x}\|_\infty = 2^k$ and $\|\mathbf{x}\|_2 = (1 + 4^k)^{1/2}$.

d. We have $\|\mathbf{x}\|_\infty = 4/(k + 1)$ and $\|\mathbf{x}\|_2 = (16/(k + 1)^2 + 4/k^4 + k^4 e^{-2k})^{1/2}$.

3. a. We have $\lim_{k\to\infty} \mathbf{x}^{(k)} = (0, 0, 0)^t$.   **b.** We have $\lim_{k\to\infty} \mathbf{x}^{(k)} = (0, 1, 3)^t$.

c. We have $\lim_{k\to\infty} \mathbf{x}^{(k)} = (0, 0, \frac{1}{2})^t$.   **d.** We have $\lim_{k\to\infty} \mathbf{x}^{(k)} = (1, -1, 1)^t$.

5. a. We have $\|\mathbf{x} - \hat{\mathbf{x}}\|_\infty = 8.57 \times 10^{-4}$ and $\|A\hat{\mathbf{x}} - \mathbf{b}\|_\infty = 2.06 \times 10^{-4}$.

b. We have $\|\mathbf{x} - \hat{\mathbf{x}}\|_\infty = 0.90$ and $\|A\hat{\mathbf{x}} - \mathbf{b}\|_\infty = 0.27$.

c. We have $\|\mathbf{x} - \hat{\mathbf{x}}\|_\infty = 0.5$ and $\|A\hat{\mathbf{x}} - \mathbf{b}\|_\infty = 0.3$.

d. We have $\|\mathbf{x} - \hat{\mathbf{x}}\|_\infty = 6.55 \times 10^{-2}$, and $\|A\hat{\mathbf{x}} - \mathbf{b}\|_\infty = 0.32$.

7. Let $A = \begin{bmatrix} 1 & 1 \\ 0 & 1 \end{bmatrix}$ and $B = \begin{bmatrix} 1 & 0 \\ 1 & 1 \end{bmatrix}$. Then $\|AB\|_\infty = 2$, but $\|A\|_\infty \cdot \|B\|_\infty = 1$.

9. It is not difficult to show that (*i*) holds. If $\|A\| = 0$, then $\|A\mathbf{x}\| = 0$ for all vectors $\mathbf{x}$ with $\|\mathbf{x}\| = 1$. Using $\mathbf{x} = (1, 0, \dots, 0)^t$, $\mathbf{x} = (0, 1, 0, \dots, 0)^t, \dots,$ and $\mathbf{x} = (0, \dots, 0, 1)^t$ successively implies that each column of A is zero. Thus $\|A\| = 0$ if and only if $A = 0$. Moreover,

$$\|\alpha A\| = \max_{\|\mathbf{x}\|=1} \|(\alpha A\mathbf{x})\| = |\alpha| \max_{\|\mathbf{x}\|=1} \|A\mathbf{x}\| = |\alpha| \cdot \|A\|,$$

$$\|A + B\| = \max_{\|\mathbf{x}\|=1} \|(A + B)\mathbf{x}\| \le \max_{\|\mathbf{x}\|=1}(\|A\mathbf{x}\| + \|B\mathbf{x}\|),$$

so

$$\|A + B\| \le \max_{\|\mathbf{x}\|=1} \|A\mathbf{x}\| + \max_{\|\mathbf{x}\|=1} \|B\mathbf{x}\| = \|A\| + \|B\|$$

and

$$\|AB\| = \max_{\|\mathbf{x}\|=1} \|(AB)\mathbf{x}\| = \max_{\|\mathbf{x}\|=1} \|A(B\mathbf{x})\|,$$

so

$$\|AB\| = \max_{\|\mathbf{x}\|=1} \|A(B\mathbf{x})\| \le \max_{\|\mathbf{x}\|=1} \|A\| \, \|B\mathbf{x}\| = \|A\| \max_{\|\mathbf{x}\|=1} \|B\mathbf{x}\| = \|A\| \, \|B\|.$$

7.3 Eigenvalues and Eigenvectors (Page 291)

1. a. The eigenvalue $\lambda_1 = 3$ has the eigenvector $\mathbf{x}_1 = (1, -1)^t$, and the eigenvalue $\lambda_2 = 1$ has the eigenvector $\mathbf{x}_2 = (1, 1)^t$.

b. The eigenvalue $\lambda_1 = \frac{1+\sqrt{5}}{2}$ has the eigenvector $\mathbf{x}_1 = (1, \frac{1+\sqrt{5}}{2})^t$, and the eigenvalue $\lambda_2 = \frac{1-\sqrt{5}}{2}$ has the eigenvector $\mathbf{x}_2 = \left(1, \frac{1-\sqrt{5}}{2}\right)^t$.

c. The eigenvalue $\lambda_1 = \frac{1}{2}$ has the eigenvector $\mathbf{x}_1 = (1, 1)^t$ and the eigenvalue $\lambda_2 = -\frac{1}{2}$ has the eigenvector $\mathbf{x}_2 = (1, -1)^t$.

d. The eigenvalue $\lambda_1 = 0$ has the eigenvector $\mathbf{x}_1 = (1, -1)^t$ and the eigenvalue $\lambda_2 = -1$ has the eigenvector $\mathbf{x}_2 = (1, -2)^t$.

e. The eigenvalue $\lambda_1 = \lambda_2 = 3$ has the eigenvectors $\mathbf{x}_1 = (0, 0, 1)^t$ and $\mathbf{x}_2 = (1, 1, 0)^t$, and the eigenvalue $\lambda_3 = 1$ has the eigenvector $\mathbf{x}_3 = (1, -1, 0)^t$.

f. The eigenvalue $\lambda_1 = 7$ has the eigenvector $\mathbf{x}_1 = (1, 4, 4)^t$, the eigenvalue $\lambda_2 = 3$ has the eigenvector $\mathbf{x}_2 = (1, 2, 0)^t$, and the eigenvalue $\lambda_3 = -1$ has the eigenvector $\mathbf{x}_3 = (1, 0, 0)^t$.

g. The eigenvalue $\lambda_1 = \lambda_2 = 1$ has the eigenvectors $\mathbf{x}_1 = (-1, 1, 0)^t$ and $\mathbf{x}_2 = (-1, 0, 1)^t$, and the eigenvalue $\lambda_3 = 5$ has the eigenvector $\mathbf{x}_3 = (1, 2, 1)^t$.

h. The eigenvalue $\lambda_1 = 3$ has the eigenvector $\mathbf{x}_1 = (-1, 1, 2)^t$, the eigenvalue $\lambda_2 = 4$ has the eigenvector $\mathbf{x}_2 = (0, 1, 2)^t$, and the eigenvalue $\lambda_3 = -2$ has the eigenvector $\mathbf{x}_3 = (-3, 8, 1)^t$.

3. Since

$$A_1^k = \begin{bmatrix} 1 & 0 \\ \frac{2^k-1}{2^{k+1}} & 2^{-k} \end{bmatrix}, \quad \text{we have } \lim_{k\to\infty} A_1^k = \begin{bmatrix} 1 & 0 \\ \frac{1}{2} & 0 \end{bmatrix}, \text{ so } A \text{ is not convergent.}$$

Also

$$A_2^k = \begin{bmatrix} 2^{-k} & 0 \\ \frac{16k}{2^{k-1}} & 2^{-k} \end{bmatrix}, \quad \text{so } \lim_{k\to\infty} A_2^k = \begin{bmatrix} 0 & 0 \\ 0 & 0 \end{bmatrix}, \text{ so } A \text{ is convergent.}$$

5. a. 3 **b.** 1.618034 **c.** 0.5 **d.** 3.162278

e. 3 **f.** 8.224257 **g.** 5.203527 **h.** 5.601152

7. Let $A = \begin{bmatrix} 1 & 1 \\ 0 & 1 \end{bmatrix}$ and $B = \begin{bmatrix} 1 & 0 \\ 1 & 1 \end{bmatrix}$. Then $\rho(A) = \rho(B) = 1$ and $\rho(A + B) = 3$.

9. a. Since

$$\det(A - \lambda I) = \det((A - \lambda I)^t) = \det(A^t - \lambda I^t) = \det(A^t - \lambda I),$$

λ is an eigenvalue of A if and only if λ is an eigenvalue of A^t.

b. If $A\mathbf{x} = \lambda\mathbf{x}$, then $A^2\mathbf{x} = \lambda A\mathbf{x} = \lambda^2\mathbf{x}$. By induction we have $A^n\mathbf{x} = \lambda^n\mathbf{x}$ for each positive integer n.

c. If $A\mathbf{x} = \lambda\mathbf{x}$ and A^{-1} exists, then $\mathbf{x} = \lambda A^{-1}\mathbf{x}$. Also, since A^{-1} exists, zero is not an eigenvalue of A, so $\lambda \neq 0$ and $\frac{1}{\lambda}\mathbf{x} = A^{-1}\mathbf{x}$. So $1/\lambda$ is an eigenvalue of A^{-1}.

d. Since $A\mathbf{x} = \lambda\mathbf{x}$, we have $(A - \alpha I)\mathbf{x} = (\lambda - \alpha)\mathbf{x}$, and since $(A - \alpha I)^{-1}$ exists and $\alpha \neq \lambda$, we have

$$\frac{1}{\lambda - \alpha}\mathbf{x} = (A - \alpha I)^{-1}\mathbf{x}.$$

7.4   The Jacobi and Gauss-Seidel Methods (Page 296)

1. Two iterations of Jacobi's method give the following results.

  a. $(0.1428571, -0.3571429, 0.4285714)^t$

  b. $(0.97, 0.91, 0.74)^t$

  c. $(-0.65, 1.65, -0.4, -2.475)^t$

  d. $(-0.5208333, -0.04166667, -0.2166667, 0.4166667)^t$

  e. $(1.325, -1.6, 1.6, 1.675, 2.425)^t$

  f. $(0.3125, 1.25, 0.3125, 1.375, 0.25, 1.375)^t$

3. Jacobi's Method gives the following results:

  a. $\mathbf{x}^{(10)} = (0.03507839, -0.2369262, 0.6578015)^t$

  b. $\mathbf{x}^{(6)} = (0.9957250, 0.9577750, 0.7914500)^t$

  c. $\mathbf{x}^{(22)} = (-0.7975853, 2.794795, -0.2588888, -2.251879)^t$

  d. $\mathbf{x}^{(14)} = (-0.7529267, 0.04078538, -0.2806091, 0.6911662)^t$

  e. $\mathbf{x}^{(12)} = (0.7870883, -1.003036, 1.866048, 1.912449, 1.985707)^t$

  f. $\mathbf{x}^{(9)} = (0.3570557, 1.428528, 0.357055, 1.571411, 0.2855225, 1.527141)^t$

5. a. A is not strictly diagonally dominant because of the last row.

  b.

$$T_j = \begin{bmatrix} 0 & 0 & 1 \\ 0.5 & 0 & 0.25 \\ -1 & 0.5 & 0 \end{bmatrix} \quad \text{and} \quad \rho(T_j) = 0.97210521 < 1.$$

Since T_j is convergent, the Jacobi method will converge.

  c. With $\mathbf{x}^{(0)} = (0, 0, 0)^t$, $\mathbf{x}^{(187)} = (0.90222655, -0.79595242, 0.69281316)^t$.

  d. In this case, $\rho(T_j) = 1.39331779371 > 1$. Since T_j is not convergent, the Jacobi method will not converge.

7. a.

$$T_j = \begin{bmatrix} 0 & \frac{1}{2} & -\frac{1}{2} \\ -1 & 0 & -1 \\ \frac{1}{2} & \frac{1}{2} & 0 \end{bmatrix} \quad \text{and} \quad \det(\lambda I - T_j) = \lambda^3 + \frac{5}{4}x.$$

Thus, the eigenvalues of T_j are 0 and $\pm\frac{\sqrt{5}}{2}i$, so $\rho(T_j) = \frac{\sqrt{5}}{2} > 1$.

  b. $\mathbf{x}^{(25)} = (-20.827873, 2.0000000, -22.827873)^t$

  c.

$$T_g = \begin{bmatrix} 0 & \frac{1}{2} & -\frac{1}{2} \\ 0 & -\frac{1}{2} & -\frac{1}{2} \\ 0 & 0 & -\frac{1}{2} \end{bmatrix} \quad \text{and} \quad \det(\lambda I - T_g) = \lambda\left(\lambda + \frac{1}{2}\right)^2.$$

Thus, the eigenvalues of T_g are 0, $-\frac{1}{2}$, and $-\frac{1}{2}$; and $\rho(T_g) = \frac{1}{2}$.

  d. $\mathbf{x}^{(23)} = (1.0000023, 1.9999975, -1.0000001)^t$ is within 10^{-5} in the l_∞ norm.

9. $T_j = (t_{ik})$ has entries given by

$$t_{ik} = \begin{cases} 0, & i = k \text{ for } 1 \le i \le n, \text{ and } 1 \le k \le n \\ -\dfrac{a_{ik}}{a_{ii}}, & i \ne k \text{ for } 1 \le i \le n, \text{ and } 1 \le k \le n. \end{cases}$$

Since A is strictly diagonally dominant,

$$\|T_j\|_\infty = \max_{1 \le i \le n} \sum_{\substack{k=1 \\ k \ne i}}^{n} \left| \frac{a_{ik}}{a_{ii}} \right| < 1.$$

7.5   The SOR Method (Page 301)

1. Two iterations of the SOR method give the following results:
   a. $(0.05410079, -0.2115435, 0.6477159)^t$
   b. $(0.9876790, 0.9784935, 0.7899328)^t$
   c. $(-0.71885, 2.818822, -0.2809726, -2.235422)^t$
   d. $(-0.6604902, 0.03700749, -0.2493513, 0.6561139)^t$
   e. $(1.079675, -1.260654, 2.042489, 1.995373, 2.049536)^t$
   f. $(0.378125, 1.445469, 0.359691, 1.458531, 0.307192, 1.572125)^t$

3. a. The tridiagonal matrices are in (b), (c), and (f).
   b. For $\omega = 1.012823$ we have

$$\mathbf{x}^{(4)} = (0.9957846, 0.9578935, 0.7915788)^t.$$

   c. For $\omega = 1.153499$ we have

$$\mathbf{x}^{(7)} = (-0.7977651, 2.795343, -0.2588021, -2.251760)^t.$$

   f. For $\omega = 1.108636$ we have

$$\mathbf{x}^{(6)} = (0.3571484, 1.428572, 0.3571425, 1.571427, 0.2857202, 1.5714277)^t.$$

5.(i). The solution vector using 30 iterations is

$$(0.00362, -6339.744638, -3660.253272, -8965.755808, 6339.744638, 10000,$$
$$-7320.508959, 6339.746729)^t.$$

(ii). The solution vector using 57 iterations is

$$(-0.002651, -6339.744637, -3660.255362, -8965.752851, 6339.748259, 10000,$$
$$-7320.506544, 6339.748258)^t.$$

(iii). The solution vector using 132 iterations is

$$(0.0045175, -6339.744528, -3660.253009, -8965.756179, 6339.743756, 10000,$$
$$-7320.509547, 6339.747544)^t.$$

7.6   Error Bounds and Iterative Refinement (Page 306)

1. The l_∞ condition numbers are as follows:
   a. 50   **b.** 241.37   **c.** 60,002   **d.** 339,866   **e.** 12   **f.** 198.17

3. The matrix is ill-conditioned since $K_\infty = 60002$. For the new system we have $\tilde{\mathbf{x}} = (-1.0000, 2.0000)^t$.

5. a. (i) $(-10.0, 1.01)^t$, (ii) $(10.0, 1.00)^t$

  b. (i) $(12.0, 0.499, -1.98)^t$, (ii) $(1.00, 0.500, -1.00)^t$

  c. (i) $(0.185, 0.0103, -0.0200, -1.12)^t$, (ii) $(0.177, 0.0127, -0.0207, -1.18)^t$

  d. (i) $(0.799, -3.12, 0.151, 4.56)^t$, (ii) $(0.758, -3.00, 0.159, 4.30)^t$

7. a. $K_\infty(H^{(4)}) = 28{,}375$

  b. $K_\infty(H^{(5)}) = 943{,}656$

  c. The actual solution is $\mathbf{x} = (-124, 1560, -3960, 2660)^t$.
  The approximate solution is $\tilde{\mathbf{x}} = (-124.2, 1563.8, -3971.8, 2668.8)^t$.
  The absolute error is $||\mathbf{x} - \tilde{\mathbf{x}}||_\infty = 11.8$.
  The relative error is

$$\frac{||\mathbf{x} - \tilde{\mathbf{x}}||_\infty}{||\mathbf{x}||_\infty} = 0.02980.$$

$$\frac{K_\infty(A)}{1 - K_\infty(A)\left(\frac{||\delta A||_\infty}{||A||_\infty}\right)} \left[\frac{||\delta b||_\infty}{||b||_\infty} + \frac{||\delta A||_\infty}{||A||_\infty}\right] = \frac{28375}{1 - 28375\left(\frac{6.\overline{6} \times 10^{-6}}{2.08\overline{3}}\right)}\left[0 + \frac{6.\overline{6} \times 10^{-6}}{2.08\overline{3}}\right] = 0.09987.$$

7.7 The Conjugate Gradient Method (Page 315)

1. a. $(0.18, 0.13)^t$

  b. $(0.19, 0.10)^t$

  c. Gaussian elimination gives the best answer because $\mathbf{v}^{(2)} = (0, 0)^t$ in the conjugate gradient method.

  d. $(0.13, 0.21)^t$. There is no improvement, although $\mathbf{v}^{(2)} \neq \mathbf{0}$.

3. a. $(1.00, -1.00, 1.00)^t$

  b. $(0.827, 0.0453, -0.0357)^t$

  c. The partial pivoting and scaled partial pivoting also give $(1.00, -1.00, 1.00)^t$.

  d. $(0.776, 0.238, -0.185)^t$;
  The residual from (3b) is $(-0.0004, -0.0038, 0.0037)^t$ and the residual from (3d) is $(0.0022, -0.0038, 0.0024)^t$.
  There does not appear to be much improvement, if any. Rounding error is more prevalent because of the increase in the number of matrix multiplications.

5. a. $\mathbf{x}^{(2)} = (0.1535933456, -0.1697932117, 0.5901172091)^t$, $||\mathbf{r}^{(2)}||_\infty = 0.221$.

  b. $\mathbf{x}^{(2)} = (0.9993129510, 0.9642734456, 0.7784266575)^t$, $||\mathbf{r}^{(2)}||_\infty = 0.144$.

  c. $\mathbf{x}^{(2)} = (-0.7290954114, 2.515782452, -0.6788904058, -2.331943982)^t$, $||\mathbf{r}^{(2)}||_\infty = 2.2$.

  d. $\mathbf{x}^{(2)} = (-0.7071108901, -0.0954748881, -0.3441074093, 0.5256091497)^t$, $||\mathbf{r}^{(2)}||_\infty = 0.39$.

  e. $\mathbf{x}^{(2)} = (0.5335968381, 0.9367588935, 1.339920949, 1.743083004, 1.743083004)^t$, $||\mathbf{r}^{(2)}||_\infty = 1.3$.

  f. $\mathbf{x}^{(2)} = (0.35714286, 1.42857143, 0.35714286, 1.57142857, 0.28571429, 1.57142857)^t$, $||\mathbf{r}^{(2)}||_\infty = 0$.

7. a. $\mathbf{x}^{(3)} = (0.06185567013, -0.1958762887, 0.6185567010)^t$, $||\mathbf{r}^{(3)}||_\infty = 0.4 \times 10^{-9}$.

  b. $\mathbf{x}^{(3)} = (0.9957894738, 0.9578947369, 0.7915789474)^t$, $||\mathbf{r}^{(3)}||_\infty = 0.1 \times 10^{-9}$.

  c. $\mathbf{x}^{(4)} = (-0.7976470579, 2.795294120, -0.2588235305, -2.251764706)^t$, $||\mathbf{r}^{(4)}||_\infty = 0.39 \times 10^{-7}$.

d. $\mathbf{x}^{(4)} = (-0.7534246575, 0.04109589039, -0.2808219179, 0.6917808219)^t$, $\|\mathbf{r}^{(4)}\|_\infty = 0.11 \times 10^{-9}$.

e. $\mathbf{x}^{(5)} = (0.4516129032, 0.7096774197, 1.677419355, 1.741935483, 1.806451613)^t$, $\|\mathbf{r}^{(5)}\|_\infty = 0.2 \times 10^{-9}$.

f. $\mathbf{x}^{(2)} = (0.35714286, 1.42857143, 0.35714286, 1.57142857, 0.28571429, 1.57142857)^t$, $\|\mathbf{r}^{(2)}\|_\infty = 0$.

9.

a.

	Jacobi 49 Iterations	Gauss-Seidel 28 Iterations	SOR ($\omega = 1.3$) 13 Iterations	Conjugate Gradient 9 Iterations
x_1	0.93406183	0.93406917	0.93407584	0.93407713
x_2	0.97473885	0.97475285	0.97476180	0.97476363
x_3	1.10688692	1.10690302	1.10691093	1.10691243
x_4	1.42346150	1.42347226	1.42347591	1.42347699
x_5	0.85931331	0.85932730	0.85933633	0.85933790
x_6	0.80688119	0.80690725	0.80691961	0.80692197
x_7	0.85367746	0.85370564	0.85371536	0.85372011
x_8	1.10688692	1.10690579	1.10691075	1.10691250
x_9	0.87672774	0.87674384	0.87675177	0.87675250
x_{10}	0.80424512	0.80427330	0.80428301	0.80428524
x_{11}	0.80688119	0.80691173	0.80691989	0.80692252
x_{12}	0.97473885	0.97475850	0.97476265	0.97476392
x_{13}	0.93003466	0.93004542	0.93004899	0.93004987
x_{14}	0.87672774	0.87674661	0.87675155	0.87675298
x_{15}	0.85931331	0.85933296	0.85933709	0.85933979
x_{16}	0.93406183	0.93407462	0.93407672	0.93407768

b.

	Jacobi 60 Iterations	Gauss-Seidel 35 Iterations	SOR ($\omega = 1.2$) 23 Iterations	Conjugate Gradient 11 Iterations
x_1	0.39668038	0.39668651	0.39668915	0.39669775
x_2	0.07175540	0.07176830	0.07177348	0.07178516
x_3	-0.23080396	-0.23078609	-0.23077981	-0.23076923
x_4	0.24549277	0.24550989	0.24551535	0.24552253
x_5	0.83405412	0.83406516	0.83406823	0.83407148
x_6	0.51497606	0.51498897	0.51499414	0.51500583
x_7	0.12116003	0.12118683	0.12119625	0.12121212
x_8	-0.24044414	-0.24040991	-0.24039898	-0.24038462
x_9	0.37873579	0.37876891	0.37877812	0.37878788
x_{10}	1.09073364	1.09075392	1.09075899	1.09076341
x_{11}	0.54207872	0.54209658	0.54210286	0.54211344
x_{12}	0.13838259	0.13841682	0.13842774	0.13844211
x_{13}	-0.23083868	-0.23079452	-0.23078224	-0.23076923
x_{14}	0.41919067	0.41923122	0.41924136	0.41925019
x_{15}	1.15015953	1.15018477	1.15019025	1.15019425
x_{16}	0.51497606	0.51499318	0.51499864	0.51500583
x_{17}	0.12116003	0.12119315	0.12120236	0.12121212
x_{18}	-0.24044414	-0.24040359	-0.24039345	-0.24038462
x_{19}	0.37873579	0.37877365	0.37878188	0.37878788
x_{20}	1.09073364	1.09075629	1.09076069	1.09076341
x_{21}	0.39668038	0.39669142	0.39669449	0.39669775
x_{22}	0.07175540	0.07177567	0.07178074	0.07178516
x_{23}	-0.23080396	-0.23077872	-0.23077323	-0.23076923
x_{24}	0.24549277	0.24551542	0.24551982	0.24552253
x_{25}	0.83405412	0.83406793	0.83407025	0.83407148

	Jacobi	Gauss-Seidel	SOR ($\omega = 1.1$)	Conjugate Gradient
c.	15	9	8	8
	Iterations	Iterations	Iterations	Iterations
x_1	-3.07611424	-3.07611739	-3.07611796	-3.07611794
x_2	-1.65223176	-1.65223563	-1.65223579	-1.65223582
x_3	-0.53282391	-0.53282528	-0.53282531	-0.53282528
x_4	-0.04471548	-0.04471608	-0.04471609	-0.04471604
x_5	0.17509673	0.17509661	0.17509661	0.17509661
x_6	0.29568226	0.29568223	0.29568223	0.29568218
x_7	0.37309012	0.37309011	0.37309011	0.37309011
x_8	0.42757934	0.42757934	0.42757934	0.42757927
x_9	0.46817927	0.46817927	0.46817927	0.46817927
x_{10}	0.49964748	0.49964748	0.49964748	0.49964748
x_{11}	0.52477026	0.52477026	0.52477026	0.52477027
x_{12}	0.54529835	0.54529835	0.54529835	0.54529836
x_{13}	0.56239007	0.56239007	0.56239007	0.56239009
x_{14}	0.57684345	0.57684345	0.57684345	0.57684347
x_{15}	0.58922662	0.58922662	0.58922662	0.58922664
x_{16}	0.59995522	0.59995522	0.59995522	0.59995523
x_{17}	0.60934045	0.60934045	0.60934045	0.60934045
x_{18}	0.61761997	0.61761997	0.61761997	0.61761998
x_{19}	0.62497846	0.62497846	0.62497846	0.62497847
x_{20}	0.63156161	0.63156161	0.63156161	0.63156161
x_{21}	0.63748588	0.63748588	0.63748588	0.63748588
x_{22}	0.64284553	0.64284553	0.64284553	0.64284553
x_{23}	0.64771764	0.64771764	0.64771764	0.64771764
x_{24}	0.65216585	0.65216585	0.65216585	0.65216585
x_{25}	0.65624320	0.65624320	0.65624320	0.65624320
x_{26}	0.65999423	0.65999423	0.65999423	0.65999422
x_{27}	0.66345660	0.66345660	0.66345660	0.66345660
x_{28}	0.66666242	0.66666242	0.66666242	0.66666242
x_{29}	0.66963919	0.66963919	0.66963919	0.66963919
x_{30}	0.67241061	0.67241061	0.67241061	0.67241060
x_{31}	0.67499722	0.67499722	0.67499722	0.67499721
x_{32}	0.67741692	0.67741692	0.67741691	0.67741691
x_{33}	0.67968535	0.67968535	0.67968535	0.67968535
x_{34}	0.68181628	0.68181628	0.68181628	0.68181628
x_{35}	0.68382184	0.68382184	0.68382184	0.68382184
x_{36}	0.68571278	0.68571278	0.68571278	0.68571278
x_{37}	0.68749864	0.68749864	0.68749864	0.68749864
x_{38}	0.68918652	0.68918652	0.68918652	0.68918652
x_{39}	0.69067718	0.69067718	0.69067718	0.69067717
x_{40}	0.68363346	0.68363346	0.68363346	0.68363349

11. a.

Solution	Residual
2.55613420	0.00668246
4.09171393	−0.00533953
4.60840390	−0.01739814
3.64309950	−0.03171624
5.13950533	0.01308093
7.19697808	−0.02081095
7.68140405	−0.04593118
5.93227784	0.01692180
5.81798997	0.04414047
5.85447806	0.03319707
5.94202521	−0.00099947
4.42152959	−0.00072826
3.32211695	0.02363822
4.49411604	0.00982052
4.80968966	0.00846967
3.81108707	−0.01312902

This converges in 6 iterations with tolerance 5.00×10^{-2} in the l_∞ norm and $\|\mathbf{r}^{(6)}\|_\infty = 0.046$.

b.

Solution	Residual
2.55613420	0.00668246
4.09171393	−0.00533953
4.60840390	−0.01739814
3.64309950	−0.03171624
5.13950533	0.01308093
7.19697808	−0.02081095
7.68140405	−0.04593118
5.93227784	0.01692180
5.81798996	0.04414047
5.85447805	0.03319706
5.94202521	−0.00099947
4.42152959	−0.00072826
3.32211694	0.02363822
4.49411603	0.00982052
4.80968966	0.00846967
3.81108707	−0.01312902

This converges in 6 iterations with tolerance 5.00×10^{-2} in the l_∞ norm and $\|\mathbf{r}^{(6)}\|_\infty = 0.046$.

c. All tolerances lead to the same convergence specifications.

13. a. Let $\{\mathbf{v}^{(1)}, \dots \mathbf{v}^{(n)}\}$ be a set of nonzero A-orthogonal vectors for the symmetric positive definite matrix A. Then $\langle \mathbf{v}^{(i)}, A\mathbf{v}^{(j)} \rangle = 0$, if $i \neq j$. Suppose

$$c_1 \mathbf{v}^{(1)} + c_2 \mathbf{v}^{(2)} + \dots + c_n \mathbf{v}^{(n)} = \mathbf{0},$$

where not all c_i are zero. Suppose k is the smallest integer for which $c_k \neq 0$. Then

$$c_k \mathbf{v}^{(k)} + c_{k+1} \mathbf{v}^{(k+1)} + \dots + c_n \mathbf{v}^{(n)} = \mathbf{0}.$$

We solve for $\mathbf{v}^{(k)}$ to obtain

$$\mathbf{v}^{(k)} = -\frac{c_{k+1}}{c_k} \mathbf{v}^{(k+1)} - \dots - \frac{c_n}{c_k} \mathbf{v}^{(n)}.$$

Multiplying by A gives

$$A\mathbf{v}^{(k)} = -\frac{c_{k+1}}{c_k} A\mathbf{v}^{(k+1)} - \dots - \frac{c_n}{c_k} A\mathbf{v}^{(n)},$$

so

$$\mathbf{v}^{(k)t} A \mathbf{v}^{(k)} = -\frac{c_{k+1}}{c_k} \mathbf{v}^{(k)t} A \mathbf{v}^{(k+1)} - \ldots - \frac{c_n}{c_k} \mathbf{v}^{(k)t} A \mathbf{v}^{(n)}$$

$$= -\frac{c_{k+1}}{c_k} \langle \mathbf{v}^{(k)}, A \mathbf{v}^{(k+1)} \rangle - \ldots - \frac{c_n}{c_k} \langle \mathbf{v}^{(k)}, A \mathbf{v}^{(n)} \rangle$$

$$= -\frac{c_{k+1}}{c_k} \cdot 0 - \ldots - \frac{c_n}{c_k} \cdot 0.$$

Since A is positive definite, $\mathbf{v}^{(k)} = \mathbf{0}$, which is a contradiction. Thus, all c_i must be zero, and $\{\mathbf{v}^{(1)}, \ldots, \mathbf{v}^{(n)}\}$ is linearly independent.

b. Let $\{\mathbf{v}^{(1)}, \ldots, \mathbf{v}^{(n)}\}$ be a set of nonzero A-orthogonal vectors for the symmetric positive definite matrix A, and let $\mathbf{z}$ be orthogonal to $\mathbf{v}^{(i)}$, for each $i = 1, \ldots, n$. From part (a), the set $\{\mathbf{v}^{(1)}, \ldots \mathbf{v}^{(n)}\}$ is linearly independent, so there is a collection of constants $\beta_1, \ldots, \beta_n$ with

$$\mathbf{z} = \sum_{i=1}^{n} \beta_i \mathbf{v}^{(i)}.$$

Hence,

$$\mathbf{z}^t \mathbf{z} = \sum_{i=1}^{n} \beta_i \mathbf{z}^t \mathbf{v}^{(i)} = \sum_{i=1}^{n} \beta_i \cdot 0 = 0,$$

and part (v) of the Inner Product Properties implies that $\mathbf{z} = \mathbf{0}$.

8.2 *Discrete Least Squares Approximation (Page 327)*

1. The linear least squares polynomial is $1.70784x + 0.89968$.

3. The least squares polynomials with their errors are:
$0.6208950 + 1.219621x$, with $E = 2.719 \times 10^{-5}$;
$0.5965807 + 1.253293x - 0.01085343x^2$, with $E = 1.801 \times 10^{-5}$;
$0.6290193 + 1.185010x + 0.03533252x^2 - 0.01004723x^3$, with $E = 1.741 \times 10^{-5}$.

5. a. The linear least squares polynomial is $72.0845x - 194.138$, with $E = 329$.

b. The least squares polynomial of degree 2 is $6.61821x^2 - 1.14352x + 1.23556$, with $E = 1.44 \times 10^{-3}$.

c. The least squares polynomial of degree 3 is $-0.0136742x^3 + 6.84557x^2 - 2.37919x + 3.42904$, with $E = 5.27 \times 10^{-4}$.

7. a. $k = 0.8996$, $E(k) = 0.295$

b. $k = 0.9052$, $E(k) = 0.128$
Thus (b) best fits the total experimental data.

9. Point average $= 0.101(\text{ACT score}) + 0.487$. It does somewhat predict because of the positive slope.

8.3 *Continuous Least Squares Approximation (Page 337)*

1. The linear least squares approximations are as follows:

a. $P_1(x) = 1.833333 + 4x$ **b.** $P_1(x) = -1.600003 + 3.600003x$ **c.** $P_1(x) = 1.140981 - 0.2958375x$

d. $P_1(x) = 0.1945267 + 3.000001x$ **e.** $P_1(x) = 0.6109245 + 0.09167105x$ **f.** $P_1(x) = -1.861455 + 1.666667x$

3. The linear least squares approximations on $[-1, 1]$ are as follows:

a. $P_1(x) = 3.333333 - 2x$ **b.** $P_1(x) = 0.6000025x$ **c.** $P_1(x) = 0.5493063 - 0.2958375x$

d. $P_1(x) = 1.175201 + 1.103639x$ **e.** $P_1(x) = 0.4207355 + 0.4353975x$ **f.** $P_1(x) = 0.6479184 + 0.5281226x$

5. The errors for the approximations in Exercise 2 are as follows:

    a. 0   **b.** 0.0457142   **c.** 0.000358354   **d.** 0.0106445   **e.** 0.0000134621

    f. 0.0000967795

7. The Gram-Schmidt process produces the following collections of polynomials:

    a. $\phi_0(x) = 1, \phi_1(x) = x - 0.5, \quad \phi_2(x) = x^2 - x + \frac{1}{6}, \quad$ and $\quad \phi_3(x) = x^3 - 1.5x^2 + 0.6x - 0.05$

    b. $\phi_0(x) = 1, \phi_1(x) = x - 1, \quad \phi_2(x) = x^2 - 2x + \frac{2}{3}, \quad$ and $\quad \phi_3(x) = x^3 - 3x^2 + \frac{12}{5}x - \frac{2}{5}$

    c. $\phi_0(x) = 1, \phi_1(x) = x - 2, \quad \phi_2(x) = x^2 - 4x + \frac{11}{3}, \quad$ and $\quad \phi_3(x) = x^3 - 6x^2 + 11.4x - 6.8$

9. The least squares polynomials of degree 2 are as follows:

    a. $P_2(x) = 3.833333\phi_0(x) + 4\phi_1(x) + 0.9999998\phi_2(x)$

    b. $P_2(x) = 2\phi_0(x) + 3.6\phi_1(x) + 3\phi_2(x)$

    c. $P_2(x) = 0.5493061\phi_0(x) - 0.2958369\phi_1(x) + 0.1588785\phi_2(x)$

    d. $P_2(x) = 3.194528\phi_0(x) + 3\phi_1(x) + 1.458960\phi_2(x)$

    e. $P_2(x) = 0.6567600\phi_0(x) + 0.09167105\phi_1(x) - 0.7375118\phi_2(x)$

    f. $P_2(x) = 1.471878\phi_0(x) + 1.666667\phi_1(x) + 0.2597705\phi_2(x)$

11. a. $2L_0(x) + 4L_1(x) + L_2(x)$ **b.** $\frac{1}{2}L_0(x) - \frac{1}{4}L_1(x) + \frac{1}{16}L_2(x) - \frac{1}{96}L_3(x)$

    c. $6L_0(x) + 18L_1(x) + 9L_2(x) + L_3(x)$ **d.** $\frac{1}{3}L_0(x) - \frac{2}{9}L_1(x) + \frac{2}{27}L_2(x) - \frac{4}{243}L_3(x)$

8.4 Chebyshev Polynomials (Page 343)

1. The interpolating polynomials of degree 2 are as follows:

    a. $P_2(x) = 2.377443 + 1.590534(x - 0.8660254) + 0.5320418(x - 0.8660254)x$

    b. $P_2(x) = 0.7617600 + 0.8796047(x - 0.8660254)$

    c. $P_2(x) = 1.052926 + 0.4154370(x - 0.8660254) - 0.1384262x(x - 0.8660254)$

    d. $P_2(x) = 0.5625 + 0.649519(x - 0.8660254) + 0.75x(x - 0.8660254)$

3. The interpolating polynomials of degree 3 are as follows:

    a. $P_3(x) = 2.519044 + 1.945377(x - 0.9238795) + 0.7047420(x - 0.9238795)(x - 0.3826834)$
            $+ 0.1751757(x - 0.9238795)(x - 0.3826834)(x + 0.3826834)$

    b. $P_3(x) = 0.7979459 + 0.7844380(x - 0.9238795) - 0.1464394(x - 0.9238795)(x - 0.3826834)$
            $- 0.1585049(x - 0.9238795)(x - 0.3826834)(x + 0.3826834)$

    c. $P_3(x) = 1.072911 + 0.3782067(x - 0.9238795) - 0.09799213(x - 0.9238795)(x - 0.3826834)$
            $+ 0.04909073(x - 0.9238795)(x - 0.3826834)(x + 0.3826834)$

    d. $P_3(x) = 0.7285533 + 1.306563(x - 0.9238795) + 0.9999999(x - 0.9238795)(x - 0.3826834)$

5. The zeros of $\tilde{T}_3$ produce the following interpolating polynomials of degree 2:

    a. $P_2(x) = 0.3489153 - 0.1744576(x - 2.866025) + 0.1538462(x - 2.866025)(x - 2)$

    b. $P_2(x) = 0.1547375 - 0.2461152(x - 1.866025) + 0.1957273(x - 1.866025)(x - 1)$

    c. $P_2(x) = 0.6166200 - 0.2370869(x - 0.9330127) - 0.7427732(x - 0.9330127)(x - 0.5)$

    d. $P_2(x) = 3.0177125 + 1.883800(x - 2.866025) + 0.2584625(x - 2.866025)(x - 2)$

7. If $i > j$, then

$$\frac{1}{2}(T_{i+j}(x) + T_{i-j}(x)) = \frac{1}{2}\left(\cos(i+j)\theta + \cos(i-j)\theta\right) = \cos i\theta \cos j\theta = T_i(x)T_j(x).$$

8.5 Rational Function Approximation (Page 348)

1. The Padé approximations of degree 2 for $f(x) = e^{2x}$ are
$n = 2, m = 0 : r_{2,0}(x) = 1 + 2x + 2x^2$,
$n = 1, m = 1 : r_{1,1}(x) = (1 + x)/(1 - x)$,
$n = 0, m = 2 : r_{0,2}(x) = (1 - 2x + 2x^2)^{-1}$.

i	x_i	$f(x_i)$	$r_{2,0}(x_i)$	$r_{1,1}(x_i)$	$r_{0,2}(x_i)$
1	0.2	1.4918	1.4800	1.5000	1.4706
2	0.4	2.2255	2.1200	2.3333	1.9231
3	0.6	3.3201	2.9200	4.0000	1.9231
4	0.8	4.9530	3.8800	9.0000	1.4706
5	1.0	7.3891	5.0000	Undefined	1.0000

3. $r_{2,3}(x) = (1 + \frac{2}{5}x + \frac{1}{20}x^2)/(1 - \frac{3}{5}x + \frac{3}{20}x^2 - \frac{1}{60}x^3)$

i	x_i	$f(x_i)$	$r_{2,3}(x_i)$
1	0.2	1.22140276	1.22140277
2	0.4	1.49182470	1.49182561
3	0.6	1.82211880	1.82213210
4	0.8	2.22554093	2.22563652
5	1.0	2.71828183	2.71875000

5. $r_{3,3}(x) = (x - \frac{7}{60}x^3)/(1 + \frac{1}{20}x^2)$. The coefficient of x^3 in the denominator is 0.

i	x_i	$f(x_i)$	6th Maclaurin Polynomial	$r_{3,3}(x_i)$
1	0.1	0.09983342	0.09966675	0.09938640
2	0.2	0.19866933	0.19733600	0.19709571
3	0.3	0.29552021	0.29102025	0.29246305
4	0.4	0.38941834	0.37875200	0.38483660
5	0.5	0.47942554	0.45859375	0.47357724

7. The Padé approximations of degree 5 are as follows:
a. $r_{0,5}(x) = (1 + x + \frac{1}{2}x^2 + \frac{1}{6}x^3 + \frac{1}{24}x^4 + \frac{1}{120}x^5)^{-1}$
b. $r_{1,4}(x) = (1 - \frac{1}{5}x)/(1 + \frac{4}{5}x + \frac{3}{10}x^2 + \frac{1}{15}x^3 + \frac{1}{120}x^4)$
c. $r_{2,3}(x) = (1 - \frac{2}{5}x + \frac{1}{20}x^2)/(1 + \frac{3}{5}x + \frac{3}{20}x^2 + \frac{1}{60}x^3)$
d. $r_{4,1}(x) = (1 - \frac{4}{5}x + \frac{3}{10}x^2 - \frac{1}{15}x^3 + \frac{1}{120}x^4)/(1 + \frac{1}{5}x)$

i	x_i	$f(x_i)$	$r_{0,5}(x_i)$	$r_{1,4}(x_i)$	$r_{2,3}(x_i)$	$r_{4,1}(x_i)$
1	0.2	0.81873075	0.81873081	0.81873074	0.81873076	0.81873077
2	0.4	0.67032005	0.67032276	0.67031942	0.67032040	0.67032099
3	0.6	0.54881164	0.54883296	0.54880635	0.54881490	0.54882143
4	0.8	0.44932896	0.44941181	0.44930678	0.44934366	0.44937931
5	1.0	0.36787944	0.36809816	0.36781609	0.36792453	0.36805556

9. a. Since

$$\sin|x| = \sin(M\pi + s) = \sin M\pi \cos s + \cos M\pi \sin s = (-1)^M \sin s,$$

we have

$$\sin x = \text{sign}(x) \sin |x| = \text{sign}(x)(-1)^M \sin s.$$

b. We have

$$\sin x \approx \left(s - \frac{31}{294}s^3 \right) \Big/ \left(1 + \frac{3}{49}s^2 + \frac{11}{5880}s^3 \right)$$

with $|\text{error}| \leq 2.84 \times 10^{-4}$.

c. Set $M = \text{round}(|x|/\pi)$; $s = |x| - M\pi$; $f_1 = \left(s - \frac{31}{294}s^3 \right) \Big/ \left(1 + \frac{3}{49}s^2 + \frac{11}{5880}s^4 \right)$. Then $f = (-1)^M f_1 \cdot x/|x|$ is the approximation.

d. Set $y = x + \frac{\pi}{2}$ and repeat (c) with y in place of x.

8.6 Trigonometric Polynomial Approximation (Page 354)

1. $S_2(x) = \dfrac{\pi^2}{3} - 4\cos x + \cos 2x$

3. $S_3(x) = 3.676078 - 3.676078\cos x + 1.470431\cos 2x - 0.7352156\cos 3x + 3.676078\sin x - 2.940862\sin 2x$

5. $S_n(x) = \dfrac{1}{2} + \dfrac{1}{\pi}\displaystyle\sum_{k=1}^{n-1} \dfrac{1 - (-1)^k}{k}\sin kx$

7. The trigonometric least squares polynomials are as follows:

a. $S_2(x) = \cos 2x$

b. $S_2(x) = 0$

c. $S_3(x) = 1.566453 + 0.5886815\cos x - 0.2700642\cos 2x + 0.2175679\cos 3x + 0.8341640\sin x - 0.3097866\sin 2x$

d. $S_3(x) = -2.046326 + 3.883872\cos x - 2.320482\cos 2x + 0.7310818\cos 3x$

9. The trigonometric least squares polynomial is

$S_3(x) = -0.4968929 + 0.2391965\cos x + 1.515393\cos 2x + 0.2391965\cos 3x - 1.150649\sin x$ with error $E(S_3) = 7.271197$.

11. Let $f(-x) = -f(x)$. The integral $\int_{-a}^{0} f(x)\,dx$ under the change of variable $t = -x$ transforms to

$$\int_{-a}^{0} f(x)\,dx = -\int_{a}^{0} f(-t)\,dt = \int_{0}^{a} f(-t)\,dt = -\int_{0}^{a} f(t)\,dt = -\int_{0}^{a} f(x)\,dx.$$

Thus,

$$\int_{-a}^{a} f(x)\,dx = \int_{-a}^{0} f(x)\,dx + \int_{0}^{a} f(x)\,dx = -\int_{0}^{a} f(x)\,dx + \int_{0}^{a} f(x)\,dx = 0.$$

13. Representative integrations that establish the orthogonality are:

$$\int_{-\pi}^{\pi} [\phi_0(x)]^2\,dx = \frac{1}{2}\int_{-\pi}^{\pi} dx = \pi,$$

$$\int_{-\pi}^{\pi} [\phi_k(x)]^2\,dx = \int_{-\pi}^{\pi} (\cos kx)^2\,dx = \int_{-\pi}^{\pi} \left[\frac{1}{2} + \frac{1}{2}\cos 2kx \right] dx = \pi + \left[\frac{1}{4k}\sin 2kx \right]_{-\pi}^{\pi} = \pi,$$

$$\int_{-\pi}^{\pi} \phi_k(x)\phi_0(x)\,dx = \left[\frac{1}{2}\int_{-\pi}^{\pi}\cos kx\,dx \right] = \frac{1}{2k}\sin kx \Big]_{-\pi}^{\pi} = 0,$$

and

$$\int_{-\pi}^{\pi} \phi_k(x)\phi_{n+j}(x)\,dx = \int_{-\pi}^{\pi}\cos kx \sin jx\,dx = \frac{1}{2}\int_{-\pi}^{\pi} [\sin(k+j)x - \sin(k-j)x]\,dx = 0.$$

8.7 *Fast Fourier Transforms (Page 360)*

1. The trigonometric interpolating polynomials are as follows:

   a. $S_2(x) = -12.33701 + 4.934802 \cos x - 2.467401 \cos 2x + 4.934802 \sin x$

   b. $S_2(x) = -6.168503 + 9.869604 \cos x - 3.701102 \cos 2x + 4.934802 \sin x$

   c. $S_2(x) = 1.570796 - 1.570796 \cos x$

   d. $S_2(x) = -0.5 - 0.5 \cos 2x + \sin x$

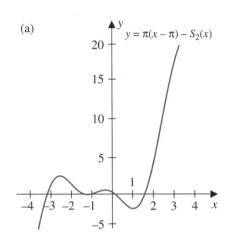

(a) $y = \pi(x - \pi) - S_2(x)$

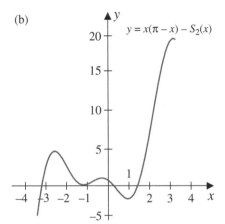

(b) $y = x(\pi - x) - S_2(x)$

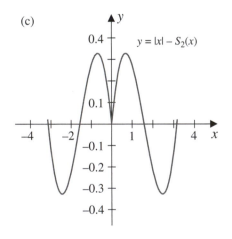

(c) $y = |x| - S_2(x)$

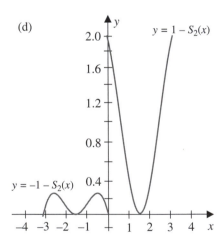

(d) $y = 1 - S_2(x)$ $y = -1 - S_2(x)$

3. The Fast Fourier Transform method gives the following trigonometric interpolating polynomials:

   a. $S_4(x) = -11.10331 + 2.467401 \cos x - 2.467401 \cos 2x + 2.467401 \cos 3x - 1.233701 \cos 4x + 5.956833 \sin x - 2.467401 \sin 2x + 1.022030 \sin 3x$

   b. $S_4(x) = 1.570796 - 1.340759 \cos x - 0.2300378 \cos 3x$

c. $S_4(x) = -0.1264264 + 0.2602724 \cos x - 0.3011140 \cos 2x + 1.121372 \cos 3x + 0.04589648 \cos 4x$
$-0.1022190 \sin x + 0.2754062 \sin 2x - 2.052955 \sin 3x$

d. $S_4(x) = -0.1526819 + 0.04754278 \cos x + 0.6862114 \cos 2x - 1.216913 \cos 3x + 1.176143 \cos 4x - 0.8179387 \sin x +$
$0.1802450 \sin 2x + 0.2753402 \sin 3x$

5.

	Approximation	Actual
a.	-69.76415	-62.01255
b.	9.869602	9.869604
c.	-0.7943605	-0.2739383
d.	-0.9593287	-0.9557781

9.2 Isolating Eigenvalues (Page 372)

1. a. The three eigenvalues are within $\{\lambda \mid |\lambda| \le 2\} \cup \{\lambda \mid |\lambda - 2| \le 2\}$, so $\rho(A) \le 4$.

 b. The three eigenvalues are within $R_1 = \{\lambda \mid |\lambda - 4| \le 2\}$, so $\rho(A) \le 6$.

 c. The three real eigenvalues satisfy $0 \le \lambda \le 6$, so $\rho(A) \le 6$.

 d. The three real eigenvalues satisfy $1.25 \le \lambda \le 8.25$, so $1.25 \le \rho(A) \le 8.25$.

3. a. The only matrices that are positive definite are 2(c) and 2(f).

 b. For 2(c) a possibility is

$$Q = \begin{bmatrix} 0 & \frac{\sqrt{2}}{2} & \frac{\sqrt{2}}{2} \\ 1 & 0 & 0 \\ 0 & \frac{\sqrt{2}}{2} & -\frac{\sqrt{2}}{2} \end{bmatrix} \quad \text{and} \quad D = \begin{bmatrix} 2 & 0 & 0 \\ 0 & 3 & 0 \\ 0 & 0 & 1 \end{bmatrix}$$

For 2(f) a possibility is

$$Q = \begin{bmatrix} \frac{\sqrt{2}}{2} & \frac{\sqrt{6}}{6} & \frac{\sqrt{3}}{3} \\ 0 & -\frac{\sqrt{6}}{3} & \frac{\sqrt{3}}{3} \\ -\frac{\sqrt{2}}{2} & \frac{\sqrt{6}}{6} & \frac{\sqrt{3}}{3} \end{bmatrix} \quad \text{and} \quad D = \begin{bmatrix} 1 & 0 & 0 \\ 0 & 1 & 0 \\ 0 & 0 & 4 \end{bmatrix}$$

5. The eigenvalues of A are $\lambda_1 = 3$ and $\lambda_2 = 2$. An eigenvector for $\lambda_1 = 3$ is $\mathbf{x}_1 = (0, 0, 1)^t$. For $\lambda_2 = 2$ an eigenvector $\mathbf{x}_2 = (x_1, x_2, x_3)^t$ needs to satisfy

$$\begin{bmatrix} 0 \\ 0 \\ 0 \end{bmatrix} = \begin{bmatrix} 0 & 1 & 0 \\ 0 & 0 & 0 \\ 0 & 0 & 1 \end{bmatrix} \begin{bmatrix} x_1 \\ x_2 \\ x_3 \end{bmatrix} = \begin{bmatrix} x_2 \\ 0 \\ x_3 \end{bmatrix}.$$

So $x_2 = 0$ and $x_3 = 0$. As a consequence, any eigenvector of $\lambda_2 = 2$ must be a multiple of $(0, 1, 0)^t$. Since there are only a total of two linearly independent eigenvectors of A, they cannot be a basis for $\mathbb{R}^3$.

7. Let $\mathbf{w} = (w_1, w_2, w_3)^t$, $\mathbf{x} = (x_1, x_2, x_3)^t$, $\mathbf{y} = (y_1, y_2, y_3)^t$, and $\mathbf{z} = (z_1, z_2, z_3)^t$ be in $\mathbb{R}^3$. First note that if $\{\mathbf{w}, \mathbf{x}, \mathbf{y}\}$ is linearly dependent, then $\{\mathbf{w}, \mathbf{x}, \mathbf{y}, \mathbf{z}\}$ also is linearly dependent.

Suppose that $\mathbf{w}$, $\mathbf{x}$, and $\mathbf{y}$ are linearly independent, and consider the linear system

$$\begin{bmatrix} w_1 & x_1 & y_1 \\ w_2 & x_2 & y_2 \\ w_3 & x_3 & y_3 \end{bmatrix} \begin{bmatrix} a \\ b \\ c \end{bmatrix} = \begin{bmatrix} 0 \\ 0 \\ 0 \end{bmatrix}.$$

Since $\mathbf{w}$, $\mathbf{x}$, and $\mathbf{y}$ are linearly independent, the only solution to this system is $a = b = c = 0$. This implies that there is also a unique solution to the linear system

$$\begin{bmatrix} w_1 & x_1 & y_1 \\ w_2 & x_2 & y_2 \\ w_3 & x_3 & y_3 \end{bmatrix} \begin{bmatrix} \hat{a} \\ \hat{b} \\ \hat{c} \end{bmatrix} = \begin{bmatrix} z_1 \\ z_2 \\ z_3 \end{bmatrix}.$$

Thus $\hat{a}\mathbf{w} + \hat{b}\mathbf{x} + \hat{c}\mathbf{y} - \mathbf{z} = \mathbf{0}$, and not all the coefficients are zero. Hence the set $\{\mathbf{w}, \mathbf{x}, \mathbf{y}, \mathbf{z}\}$ is linearly dependent.

9. Since $\{\mathbf{v}_i\}_{i=1}^n$ is linearly independent in $\mathbb{R}^n$, there exist numbers $c_1, \ldots, c_n$ with

$$\mathbf{x} = c_1\mathbf{v}_1 + \cdots + c_n\mathbf{v}_n.$$

Hence, for any k, with $1 \le k \le n$, $\mathbf{v}_k^t\mathbf{x} = c_1\mathbf{v}_k^t\mathbf{v}_1 + \cdots + c_n\mathbf{v}_k^t\mathbf{v}_n = c_k\mathbf{v}_k^t\mathbf{v}_k = c_k$.

11. Exercise 10 implies that $A = QDQ^t$ where D is a diagonal matrix with the eigenvalues of A as its diagonal entries and Q is an orthogonal matrix with the corresponding eigenvectors as its columns. Let $\mathbf{v}_i$ be the ith column of Q and d_{ii} be the ith diagonal entry of D, for each $i = 1, 2, \ldots, n$. Then $A\mathbf{v}_i = d_{ii}\mathbf{v}_i$, so

$$\mathbf{v}_i^t A\mathbf{v}_i = d_{ii}\mathbf{v}_i^t\mathbf{v}_i = d_{ii}||\mathbf{v}_i||_2^2 \quad \text{and} \quad d_{ii} = \frac{1}{||\mathbf{v}_i||_2^2}\mathbf{v}_i^t A\mathbf{v}_i.$$

Since all the components of $\mathbf{v}_i$ and all the entries of A are real numbers and $||\mathbf{v}_i||_2^2 \ne 0$, this implies that d_{ii} is also a real number, for each $i = 1, 2, \ldots, n$.

9.3 The Power Method (Page 383)

1. The approximate eigenvalues and approximate eigenvectors are as follows:

a. $\mu^{(3)} = 3.666667$, $\quad \mathbf{x}^{(3)} = (0.9772727, 0.9318182, 1)^t$

b. $\mu^{(3)} = 2.000000$, $\quad \mathbf{x}^{(3)} = (1, 1, 0.5)^t$

c. $\mu^{(3)} = 5.000000$, $\quad \mathbf{x}^{(3)} = (-0.2578947, 1, -0.2842105)^t$

d. $\mu^{(3)} = 5.038462$, $\quad \mathbf{x}^{(3)} = (1, 0.2213741, 0.3893130, 0.4045802)^t$

e. $\mu^{(3)} = 7.531073$, $\quad \mathbf{x}^{(3)} = (0.6886722, -0.6706677, -0.9219805, 1)^t$

f. $\mu^{(3)} = 4.106061$, $\quad \mathbf{x}^{(3)} = (0.1254613, 0.08487085, 0.00922509, 1)^t$

3. The approximate eigenvalues and approximate eigenvectors are as follows:

a. $\mu^{(3)} = 3.959538$, $\quad \mathbf{x}^{(3)} = (0.5816124, 0.5545606, 0.5951383)^t$

b. $\mu^{(3)} = 2.000000$, $\quad \mathbf{x}^{(3)} = (-0.6666667, -0.6666667, -0.3333333)^t$

c. $\mu^{(3)} = 7.189567$, $\quad \mathbf{x}^{(3)} = (0.5995308, 0.7367472, 0.3126762)^t$

d. $\mu^{(3)} = 6.037037$, $\quad \mathbf{x}^{(3)} = (0.5073714, 0.4878571, -0.6634857, -0.2536857)^t$

e. $\mu^{(3)} = 5.142562$, $\quad \mathbf{x}^{(3)} = (0.8373051, 0.3701770, 0.1939022, 0.3525495)^t$

f. $\mu^{(3)} = 8.593142$, $\quad \mathbf{x}^{(3)} = (-0.4134762, 0.4026664, 0.5535536, -0.6003962)^t$

5. The approximate eigenvalues and approximate eigenvectors are as follows:

a. $\lambda_1 \approx \mu^{(9)} = 3.999908$, $\quad \mathbf{x}^{(9)} = (0.9999943, 0.9999828, 1)^t$
$\lambda_2 \approx \mu^{(1)} = 1.000000$, $\quad \mathbf{x}^{(1)} = (-2.999908, 2.999908, 0)^t$

b. $\lambda_1 \approx \mu^{(13)} = 2.414214$, $\quad \mathbf{x}^{(13)} = (1, 0.7071429, 0.7070707)^t$
$\lambda_2 \approx \mu^{(1)} = 1.000000$, $\quad \mathbf{x}^{(1)} = (0, -1.414214, 1.414214)^t$

c. $\lambda_1 \approx \mu^{(9)} = 5.124749$, $\quad \mathbf{x}^{(9)} = (-0.2424476, 1, -0.3199733)^t$
$\lambda_2 \approx \mu^{(6)} = 1.636734$, $\quad \mathbf{x}^{(6)} = (1.783218, -1.135350, -3.124733)^t$

d. $\lambda_1 \approx \mu^{(24)} = 5.235861$, $\quad \mathbf{x}^{(24)} = (1, 0.6178361, 0.1181667, 0.4999220)^t$
$\lambda_2 \approx \mu^{(10)} = 3.618177$, $\quad \mathbf{x}^{(10)} = (0.7236390, -1.170593, 1.170675, -0.2763374)^t$

e. $\lambda_1 \approx \mu^{(17)} = 8.999667$, $\quad \mathbf{x}^{(17)} = (0.9999085, -0.9999078, -0.9999993, 1)^t$
$\lambda_2 \approx \mu^{(21)} = 5.000051$, $\quad \mathbf{x}^{(21)} = (1.999338, -1.999603, 1.999603, -2.000198)^t$

f. The method did not converge in 25 iterations. However, $\lambda_1 \approx \mu^{(363)} = 4.105309$, $\mathbf{x}^{(363)} = (0.06286299, 0.08702754, 0.01824680, 1)^t$, $\lambda_2 \approx \mu^{(15)} = -4.024308$, $\mathbf{x}^{(15)} = (-8.151965, 2.100699, 0.7519080, -0.3554941)^t$.

7. The approximate eigenvalues and approximate eigenvectors are as follows:

a. $\mu^{(9)} = 1.000015$, $\quad \mathbf{x}^{(9)} = (-0.1999939, 1, -0.7999909)^t$

b. $\mu^{(12)} = -0.4142136$, $\quad \mathbf{x}^{(12)} = (1, -0.7070918, -0.7071217)^t$

c. The method did not converge in 25 iterations. However, $\mu^{(42)} = 1.636636$, $\mathbf{x}^{(42)} = (-0.5706815, 0.3633636, 1)^t$.

d. $\mu^{(9)} = 1.381959$, $\mathbf{x}^{(9)} = (-0.3819400, -0.2361007, 0.2360191, 1)^t$

e. $\mu^{(6)} = 3.999997$, $\mathbf{x}^{(6)} = (0.9999939, 0.9999999, 0.9999940, 1)^t$

f. $\mu^{(3)} = 4.105293$, $\mathbf{x}^{(3)} = (0.06281419, 0.08704089, 0.01825213, 1)^t$

9. a. We have $|\lambda| \leq 6$ for all eigenvalues λ.

  b. The approximate eigenvalue is $\mu^{(133)} = 0.69766854$, with the approximate eigenvector $\mathbf{x}^{(133)} = (1, 0.7166727, 0.2568099, 0.04601217)^t$.

  c. Using the approximate eigenvalue and eigenvector computed in part (b) and Wielandt's deflation gives

$$B = \begin{bmatrix} 0 & -1.433345 & -2.866691 \\ 0.25 & -0.5136198 & -1.027240 \\ 0 & 0.03297566 & -0.1840487 \end{bmatrix},$$

  Applying the Inverse Power method with $q = -0.25$, $\mathbf{x}^{(0)} = (1, 1, 1)^t$, and tolerance 0.00001 gives the approximate eigenvalue -0.237313 and eigenvector $(1, -0.695123, 0.430345)^t$ after 5 iterations.

  d. The complex eigenvalues of A are $-0.230178 \pm 0.569659i$.

  e. The beetle population should approach zero because A is convergent.

9.4 Householder's Method (Page 389)

1. Householder's method produces the following tridiagonal matrices:

a. $\begin{bmatrix} 12.00000 & -10.77033 & 0.0 \\ -10.77033 & 3.862069 & 5.344828 \\ 0.0 & 5.344828 & 7.137931 \end{bmatrix}$

b. $\begin{bmatrix} 2.0000000 & 1.414214 & 0.0 \\ 1.414214 & 1.000000 & 0.0 \\ 0.0 & 0.0 & 3.0 \end{bmatrix}$

c. $\begin{bmatrix} 1.0000000 & -1.414214 & 0.0 \\ -1.414214 & 1.000000 & 0.0 \\ 0.0 & 0.0 & 1.000000 \end{bmatrix}$

d. $\begin{bmatrix} 4.750000 & -2.263846 & 0.0 \\ -2.263846 & 4.475610 & -1.219512 \\ 0.0 & -1.219512 & 5.024390 \end{bmatrix}$

3. a. Since $P = I - 2\mathbf{w}\mathbf{w}^t$, we have

$$P^t = (I - 2\mathbf{w}\mathbf{w}^t)^t = I^t - 2(\mathbf{w}\mathbf{w}^t)^t = I - 2(\mathbf{w}^t)^t\mathbf{w}^t = I - 2\mathbf{w}\mathbf{w}^t = P.$$

  b. Using (a) we have

$$P^t P = P^2 = (I - 2\mathbf{w}\mathbf{w}^t)^2 = I - 4\mathbf{w}\mathbf{w}^t + 4\mathbf{w}\mathbf{w}^t\mathbf{w}\mathbf{w}^t.$$

  But $\mathbf{w}^t\mathbf{w} = 1$, so

$$P^t P = I - 4\mathbf{w}\mathbf{w}^t + 4\mathbf{w}\mathbf{w}^t = I \quad \text{and} \quad P^t = P = P^{-1}.$$

9.5 The QR Method (Page 398)

1. Two iterations of the QR Algorithm without shifting produce the following matrices.

a. $A^{(3)} = \begin{bmatrix} 3.142857 & -0.559397 & 0.0 \\ -0.559397 & 2.248447 & -0.187848 \\ 0.0 & -0.187848 & 0.608696 \end{bmatrix}$

b. $A^{(3)} = \begin{bmatrix} 4.549020 & 1.206958 & 0.0 \\ 1.206958 & 3.519688 & 0.000725 \\ 0.0 & 0.000725 & -0.068708 \end{bmatrix}$

c. $A^{(3)} = \begin{bmatrix} 4.592920 & -0.472934 & 0.0 \\ -0.472934 & 3.108760 & -0.232083 \\ 0.0 & -0.232083 & 1.298319 \end{bmatrix}$

d. $A^{(3)} = \begin{bmatrix} 3.071429 & 0.855352 & 0.0 & 0.0 \\ 0.855352 & 3.314192 & -1.161046 & 0.0 \\ 0.0 & -1.1610446 & 3.331770 & 0.268898 \\ 0.0 & 0.0 & 0.268898 & 0.282609 \end{bmatrix}$

e. $A^{(3)} = \begin{bmatrix} -3.607843 & 0.612882 & 0.0 & 0.0 \\ 0.612882 & -1.395227 & -1.111027 & 0.0 \\ 0.0 & -1.111027 & 3.133919 & 0.346353 \\ 0.0 & 0.0 & 0.346353 & 0.869151 \end{bmatrix}$

f. $A^{(3)} = \begin{bmatrix} 1.013260 & 0.279065 & 0.0 & 0.0 \\ 0.279065 & 0.696255 & 0.107448 & 0.0 \\ 0.0 & 0.107448 & 0.843061 & 0.310832 \\ 0.0 & 0.0 & 0.310832 & 0.317424 \end{bmatrix}$

3. The matrices in Exercise 1 have the following eigenvalues, accurate to within 10^{-5}:

   a. 3.414214, 2.000000, 0.58578644   **b.** −0.06870782, 5.346462, 2.722246

   c. 1.267949, 4.732051, 3.000000   **d.** 4.745281, 3.177283, 1.822717, 0.2547188

   e. 3.438803, 0.8275517, −1.488068, −3.778287   **f.** 0.9948440, 1.189091, 0.5238224, 0.1922421

5. When the shifting parameter is $\gamma_2 = 2$ we have

$$A^{(2)} = \begin{bmatrix} -2 & \frac{1}{2}\sqrt{2} & 0 \\ \frac{1}{2}\sqrt{2} & -1 & \frac{1}{2}\sqrt{2} \\ 0 & \frac{1}{2}\sqrt{2} & 0 \end{bmatrix}, \quad A^{(3)} = \begin{bmatrix} -2.677203 & 0.375975 & 0 \\ 0.375975 & -1.473608 & 0.03039696 \\ 0 & 0.03039696 & 0.04755953 \end{bmatrix}$$

and

$$A^{(4)} = \begin{bmatrix} -2.799713 & 0.1993864 & 0 \\ 0.1993864 & -1.442885 & 4.5 \times 10^{-7} \\ 0 & 4.5 \times 10^{-7} & 2.2146 \times 10^{-5} \end{bmatrix}$$

   Adding the shifts gives the approximate eigenvalue 4.414214. The other approximate eigenvalues are 3.000000 and 1.585786.

7. Let $P = (p_{ij})$ be a rotation matrix with nonzero entries $p_{jj} = p_{ii} = \cos\theta$, $p_{ij} = -p_{ji} = \sin\theta$, and $p_{kk} = 1$, if $k \neq i$ and $k \neq j$. For any $n \times n$ matrix A,

$$(AP)_{rs} = \sum_{k=1}^{n} a_{rk} p_{ks}.$$

   If $s \neq i, j$, then $p_{ks} = 0$ unless $k = s$. Thus, $(AP)_{rs} = a_{rs}$.
   If $s = j$, then

$$(AP)_{rj} = a_{rj} p_{jj} + a_{ri} p_{ij} = a_{rj}\cos\theta + a_{ri}\sin\theta.$$

   If $s = i$, then

$$(AP)_{ri} = a_{rj} p_{ji} + a_{ri} p_{ii} = -a_{rj}\sin\theta + a_{ri}\cos\theta.$$

   Similarly, $(PA)_{rs} = \sum_{k=1}^{n} p_{rk} a_{ks}$. If $r \neq i, j$, then $p_{rk} = 0$ unless $r = k$. Thus $(PA)_{rs} = a_{rs}$.
   If $r = i$, then

$$(PA)_{is} = p_{ij} a_{js} + p_{ii} a_{is} = a_{js}\sin\theta + a_{is}\cos\theta.$$

   If $r = j$, then

$$(PA)_{js} = p_{jj} a_{js} + p_{ji} a_{is} = a_{js}\cos\theta - a_{is}\sin\theta.$$

9.6 Singular Value Decomposition (Page 409)

1. **a.** $s_1 = 1 + \sqrt{2}$, $s_2 = -1 + \sqrt{2}$ **b.** $s_1 = \sqrt{4 + \sqrt{10}}$, $s_2 = \sqrt{4 - \sqrt{10}}$

   c. $s_1 = 3.162278$, $s_2 = 2$ **d.** $s_1 = 2.645751$, $s_2 = 1$, $s_3 = 1$

3. a. $U = \begin{bmatrix} 0.923880 & 0.382683 \\ 0.382683 & -0.923880 \end{bmatrix}$, $S = \begin{bmatrix} 2.41421 & 0 \\ 0 & 0.414214 \end{bmatrix}$, $V^t = \begin{bmatrix} 0.923880 & 0.382683 \\ -0.382683 & 0.923880 \end{bmatrix}$

b. $U = \begin{bmatrix} 0.824736 & -0.391336 & 0.408248 \\ 0.521609 & 0.247502 & -0.816497 \\ 0.218482 & 0.886340 & 0.408248 \end{bmatrix}$, $S = \begin{bmatrix} 2.67624 & 0 \\ 0 & 0.915272 \\ 0 & 0 \end{bmatrix}$, $V^t = \begin{bmatrix} 0.811242 & 0.584710 \\ -0.584710 & 0.811242 \end{bmatrix}$

c. $U = \begin{bmatrix} -0.632456 & 0 & 0.258199 & -0.370901 & 0.629099 \\ 0 & -0.816497 & -0.430331 & -0.381832 & -0.048499 \\ 0.316228 & -0.408248 & 0.849731 & -0.075134 & -0.075134 \\ -0.316228 & -0.408248 & 0.010932 & 0.838799 & 0.172133 \\ 0.632456 & 0 & -0.161201 & 0.086066 & 0.752733 \end{bmatrix}$,

$S = \begin{bmatrix} 2.236070 & 0 \\ 0 & 1.732051 \\ 0 & 0 \\ 0 & 0 \\ 0 & 0 \end{bmatrix}$, $V^t = \begin{bmatrix} -0.707107 & 0.707107 \\ -0.707107 & -0.707107 \end{bmatrix}$

d. $U = \begin{bmatrix} -0.547723 & 0 & 0.707107 & -0.138916 & -0.425091 \\ -0.365148 & -0.408248 & 0 & -0.533212 & 0.644736 \\ -0.547723 & 0 & -0.707107 & -0.138916 & -0.425091 \\ -0.365148 & -0.408248 & 0 & 0.811044 & 0.205446 \\ -0.365148 & 0.816497 & 0 & -0.138916 & -0.425091 \end{bmatrix}$,

$S = \begin{bmatrix} 2.236070 & 0 & 0 \\ 0 & 1.414214 & 0 \\ 0 & 0 & 1 \\ 0 & 0 & 0 \\ 0 & 0 & 0 \end{bmatrix}$, $V^t = \begin{bmatrix} -0.408248 & -0.816497 & -0.408248 \\ 0.577350 & -0.577350 & 0.577350 \\ -0.707107 & 0 & 0.707107 \end{bmatrix}$

5. For the matrix A in Example 1 we have

$$A^t A = \begin{bmatrix} 1 & 0 & 0 & 0 & 1 \\ 0 & 1 & 1 & 1 & 1 \\ 1 & 0 & 1 & 0 & 0 \end{bmatrix} \begin{bmatrix} 1 & 0 & 1 \\ 0 & 1 & 0 \\ 0 & 1 & 1 \\ 0 & 1 & 0 \\ 1 & 1 & 0 \end{bmatrix} = \begin{bmatrix} 2 & 1 & 1 \\ 1 & 4 & 1 \\ 1 & 1 & 2 \end{bmatrix}$$

So $A^t A(1, 2, 1)^t = (5, 10, 5)^t = 5(1, 2, 1)^t$, $A^t A(1, -1, 1)^t = (2, -2, 2)^t = 2(1, -1, 1)^t$, and $A^t A(-1, 0, 1)^t = (-1, 0, 1)^t$.

7. Since

$$\left(A A^t\right)^t = \left(A^t\right)^t A^t = A A^t,$$

the matrix $A A^t$ is symmetric. That $A^t A$ is also symmetric can be shown interchanging the roles of A and A^t.

9. a. Suppose that $\lambda \mathbf{v} = A^t A \mathbf{v}$, where $\lambda \neq 0$ and $\mathbf{v} \neq \mathbf{0}$. Then

$$\lambda A \mathbf{v} = A(\lambda \mathbf{v}) = A(A^t A \mathbf{v}) = (A A^t) A \mathbf{v},$$

so $A \mathbf{v}$ is an eigenvector of $A A^t$ corresponding to λ provided that $A \mathbf{v} \neq \mathbf{0}$. But if $A \mathbf{v} = \mathbf{0}$, then

$$\mathbf{0} = A^t A \mathbf{v} = \lambda \mathbf{v},$$

which implies that either $\lambda = 0$ or $\mathbf{v} = \mathbf{0}$. This is a contradiction.

b. This follows from (a) with the roles of A and A^t interchanged.

c. This is a direct consequence of (a) and (b).

11. a. The result in part (i) on page 289 implies that $||A||_2 = [\rho(A^t A)]^{1/2}$. Since s_1^2 is the largest eigenvalue of AA^t and by Exercise 9(c) the nonzero eigenvalues of $A^t A$ and AA^t are the same, we have $||A||_2^2 = s_1^2$.

b. First recall from Chapter 6 that if A is an invertible matrix then $(A^{-1})^t = (A^t)^{-1}$. This implies that

$$||A^{-1}||_2^2 = \rho((A^{-1})^t A^{-1}) = \rho((A^t)^{-1} A^{-1}) = \rho((AA^t)^{-1}).$$

Exercise 9 of Section 7.3 implies that the eigenvalues of $(AA^t)^{-1}$ are the reciprocals of the eigenvalues of AA^t; that is,

$$\frac{1}{s_1^2}, \quad \frac{1}{s_2^2}, \dots, \frac{1}{s_n^2}.$$

The largest of these is $1/s_n^2$, so $||A^{-1}||_2^2 = 1/s_n^2$.

c. This follows directly from (a) and (b) because $K_2(A) = ||A|| \cdot ||A^{-1}|| = s_1 \cdot (1/s_n) = s_1/s_n$.

13. First note that $\mathbf{v}_i$ being an eigenvector of $A^t A$ with nonzero eigenvalue s_i^2 implies that

$$A^t \mathbf{u}_i = \frac{1}{s_i} A^t A \mathbf{v}_i = \frac{s_i^2}{s_i} \mathbf{v}_i = s_i \mathbf{v}_i \neq \mathbf{0}, \quad \text{so} \quad \mathbf{u}_i \neq \mathbf{0}.$$

Moreover

$$(AA^t)\mathbf{u}_i = (AA^t)\frac{1}{s_i} A\mathbf{v}_i = \frac{1}{s_i} A(A^t A\mathbf{v}_i) = \frac{1}{s_i} A(s_i^2 \mathbf{v}_i) = s_i A\mathbf{v}_i.$$

But

$$\mathbf{u}_i = \frac{1}{s_i} A\mathbf{v}_i \quad \text{so} \quad A\mathbf{v}_i = s_i \mathbf{u}_i,$$

which implies that

$$AA^t \mathbf{u}_i = s_i (s_i \mathbf{u}_i) = s_i^2 \mathbf{u}_i,$$

and that $\mathbf{u}_i$ is an eigenvector of AA^t corresponding to the eigenvalue s_i^2.

15. a. Use the tabulated values to construct

$$
\mathbf{b} = \begin{bmatrix} y_0 \\ y_1 \\ y_2 \\ y_3 \\ y_4 \\ y_5 \end{bmatrix} = \begin{bmatrix} 1.84 \\ 1.96 \\ 2.21 \\ 2.45 \\ 2.94 \\ 3.18 \end{bmatrix}
\quad \text{and} \quad
A = \begin{bmatrix} 1 & x_0 & x_0^2 \\ 1 & x_1 & x_1^2 \\ 1 & x_2 & x_2^2 \\ 1 & x_3 & x_3^2 \\ 1 & x_4 & x_4^2 \\ 1 & x_5 & x_5^2 \end{bmatrix} = \begin{bmatrix} 1 & 1.0 & 1.0 \\ 1 & 1.1 & 1.21 \\ 1 & 1.3 & 1.69 \\ 1 & 1.5 & 2.25 \\ 1 & 1.9 & 3.61 \\ 1 & 2.1 & 4.41 \end{bmatrix}.
$$

The matrix A has the singular value decomposition $A = U S V^t$, where

$$
U = \begin{bmatrix}
-0.203339 & -0.550828 & 0.554024 & 0.055615 & -0.177253 & -0.560167 \\
-0.231651 & -0.498430 & 0.185618 & 0.165198 & 0.510822 & 0.612553 \\
-0.294632 & -0.369258 & -0.337742 & -0.711511 & -0.353683 & 0.177288 \\
-0.366088 & -0.20758 & -0.576499 & 0.642950 & -0.264204 & -0.085730 \\
-0.534426 & 0.213281 & -0.200202 & -0.214678 & 0.628127 & -0.433808 \\
-0.631309 & 0.472467 & 0.414851 & 0.062426 & -0.343809 & 0.289864
\end{bmatrix},
$$

$$
S = \begin{bmatrix}
7.844127 & 0 & 0 \\
0 & 1.223790 & 0 \\
0 & 0 & 0.070094 \\
0 & 0 & 0 \\
0 & 0 & 0 \\
0 & 0 & 0
\end{bmatrix},
\quad \text{and} \quad
V^t = \begin{bmatrix}
-0.288298 & -0.475702 & -0.831018 \\
-0.768392 & -0.402924 & 0.497218 \\
0.571365 & -0.781895 & 0.249363
\end{bmatrix}.
$$

So

$$\mathbf{c} = U^t \mathbf{b} = \begin{bmatrix} -5.955009 \\ -1.185591 \\ -0.044985 \\ -0.003732 \\ -0.000493 \\ -0.001963 \end{bmatrix},$$

and the components of $\mathbf{z}$ are

$$z_1 = \frac{c_1}{s_1} = \frac{-5.955009}{7.844127} = -0.759168, \quad z_2 = \frac{c_2}{s_2} = \frac{-1.185591}{1.223790} = -0.968786,$$

and

$$z_3 = \frac{c_3}{s_3} = \frac{-0.044985}{0.070094} = -0.641784.$$

This gives the least squares coefficients in $P_2(x) = a_0 + a_1 x + a_2 x^2$ as

$$\begin{bmatrix} a_0 \\ a_1 \\ a_2 \end{bmatrix} = \mathbf{a} = V\mathbf{z} = \begin{bmatrix} 0.596581 \\ 1.253293 \\ -0.010853 \end{bmatrix}.$$

The least squares error using these values uses the last three components of $\mathbf{c}$, and is

$$E = \|A\mathbf{a} - \mathbf{b}\|_2^2 = c_4^2 + c_5^2 + c_6^2 = (-0.003732)^2 + (-0.000493)^2 + (-0.001963)^2 = 1.80 \times 10^{-5}.$$

b. Use the tabulated values to construct

$$\mathbf{b} = \begin{bmatrix} y_0 \\ y_1 \\ y_2 \\ y_3 \\ y_4 \\ y_5 \end{bmatrix} = \begin{bmatrix} 1.84 \\ 1.96 \\ 2.21 \\ 2.45 \\ 2.94 \\ 3.18 \end{bmatrix} \quad \text{and} \quad A = \begin{bmatrix} 1 & x_0 & x_0^2 & x_0^3 \\ 1 & x_1 & x_1^2 & x_1^3 \\ 1 & x_2 & x_2^2 & x_2^3 \\ 1 & x_3 & x_3^2 & x_3^3 \\ 1 & x_4 & x_4^2 & x_4^3 \\ 1 & x_5 & x_5^2 & x_5^3 \end{bmatrix} = \begin{bmatrix} 1 & 1.0 & 1.0 & 1.0 \\ 1 & 1.1 & 1.21 & 1.331 \\ 1 & 1.3 & 1.69 & 2.197 \\ 1 & 1.5 & 2.25 & 3.375 \\ 1 & 1.9 & 3.61 & 6.859 \\ 1 & 2.1 & 4.41 & 9.261 \end{bmatrix}.$$

The matrix A has the singular value decomposition $A = U S V^t$, where

$$U = \begin{bmatrix} -0.116086 & -0.514623 & 0.569113 & -0.437866 & -0.381082 & 0.246672 \\ -0.143614 & -0.503586 & 0.266325 & 0.184510 & 0.535306 & 0.578144 \\ -0.212441 & -0.448121 & -0.238475 & 0.484990 & 0.180600 & -0.655247 \\ -0.301963 & -0.339923 & -0.549619 & 0.038581 & -0.573591 & 0.400867 \\ -0.554303 & 0.074101 & -0.306350 & -0.636776 & 0.417792 & -0.115640 \\ -0.722727 & 0.399642 & 0.390359 & 0.363368 & -0.179026 & 0.038548 \end{bmatrix},$$

$$S = \begin{bmatrix} 14.506808 & 0 & 0 & 0 \\ 0 & 2.084909 & 0 & 0 \\ 0 & 0 & 0.198760 & 0 \\ 0 & 0 & 0 & 0.868328 \\ 0 & 0 & 0 & 0 \\ 0 & 0 & 0 & 0 \end{bmatrix},$$

and

$$V^t = \begin{bmatrix} -0.141391 & -0.246373 & -0.449207 & -0.847067 \\ -0.639122 & -0.566437 & -0.295547 & 0.428163 \\ 0.660862 & -0.174510 & -0.667840 & 0.294610 \\ -0.367142 & 0.766807 & -0.514640 & 0.111173 \end{bmatrix}.$$

So

$$\mathbf{c} = U^t \mathbf{b} = \begin{bmatrix} -5.632309 \\ -2.268376 \\ 0.036241 \\ 0.005717 \\ -0.000845 \\ -0.004086 \end{bmatrix},$$

and the components of $\mathbf{z}$ are

$$z_1 = \frac{c_1}{s_1} = \frac{-5.632309}{14.506808} = -0.388253, \quad z_2 = \frac{c_2}{s_2} = \frac{-2.268376}{2.084909} = -1.087998,$$

$$z_3 = \frac{c_3}{s_3} = \frac{0.036241}{0.198760} = 0.182336, \quad \text{and} \quad z_4 = \frac{c_4}{s_4} = \frac{0.005717}{0.868328} = 0.65843.$$

This gives the least squares coefficients in $P_2(x) = a_0 + a_1 x + a_2 x^2 + a_3 x^3$ as

$$\begin{bmatrix} a_0 \\ a_1 \\ a_2 \\ a_3 \end{bmatrix} = \mathbf{x} = V \mathbf{z} = \begin{bmatrix} 0.629019 \\ 1.185010 \\ 0.035333 \\ -0.010047 \end{bmatrix}.$$

The least squares error using these values uses the last two components of $\mathbf{c}$, and is

$$E = ||A\mathbf{x} - \mathbf{b}||_2^2 = c_5^2 + c_6^2 = (-0.000845)^2 + (-0.004086)^2 = 1.74 \times 10^{-5}.$$

The polynomials in (a) and (b) have about the same error because a_3 is close to 0.

10.2 Newton's Method for Systems (Page 419)

1. One example is $f(x_1, x_2) = \left(1, \dfrac{1}{|x_1 - 1| + |x_2|}\right)^t$.

3. **a.** $\mathbf{p}^{(2)} = (0.4958936, 1.983423)^t$ **b.** $\mathbf{p}^{(2)} = (-0.5131616, -0.01837622)^t$
   c. $\mathbf{p}^{(2)} = (0.5001667, 0.2508036, -0.5173874)^t$ **d.** $\mathbf{p}^{(2)} = (4.350877, 18.49123, -19.84211)^t$

5. **a.** With $\mathbf{p}^{(0)} = (-4, -5)^t$ we have $\mathbf{p}^{(3)} = (-4.86416785, -3.96589658)^t$.
   b. With $\mathbf{p}^{(0)} = (2, -1)^t$ we have $\mathbf{p}^{(4)} = ((2.10951192, -1.33453219)^t$.
   c. With $\mathbf{p}^{(0)} = (0.5, 2)^t$ we have $\mathbf{p}^{(4)} = (0.62520410, 2.17935582)^t$.
   d. With $\mathbf{p}^{(0)} = (-2, 3)^t$ we have $\mathbf{p}^{(4)} = (-1.87054817, 3.12107295)^t$.

7. With $\mathbf{p}^{(0)} = (1, 1 - 1)^t$ and $TOL = 10^{-6}$, we have $\mathbf{p}^{(20)} = (0.5, 9.5 \times 10^{-7}, -0.5235988)^t$.

9. When $\mathbf{F}(\mathbf{x}) = A\mathbf{x}$ the Jacobian matrix of $\mathbf{F}$ is simply A.

11. **a.** $k_1 = 8.77125, k_2 = 0.259690, k_3 = -1.37217$
    b. Solving the equation

$$\frac{500}{\pi r^2} = k_1 e^{k_2 r} + k_3 r$$

numerically gives $r = 3.18517$.

10.3 Quasi-Newton Methods (Page 425)

1. **a.** $\mathbf{p}^{(2)} = (0.4777920, 1.927557)^t$ **b.** $\mathbf{p}^{(2)} = (-0.3250070, -0.1386967)^t$
   c. $\mathbf{p}^{(2)} = (0.5002312, -1.080299, -0.52382394)^t$ **d.** $\mathbf{p}^{(2)} = (-67.00583, 38.31494, 31.69089)^t$

3. **a.** $\mathbf{p}^{(9)} = (0.5, 0.8660254)^t$ **b.** $\mathbf{p}^{(8)} = (1.772454, 1.772454)^t$
   c. $\mathbf{p}^{(9)} = (-1.456043, -1.664231, 0.4224934)^t$ **d.** $\mathbf{p}^{(5)} = (0.4981447, -0.1996059, -0.5288260)^t$

5. Using $\mathbf{p}^{(0)} = (1, 1, 1, 1)^t$ gives $\mathbf{p}^{(6)} = (0, 0.70710678, 0.70710678, 1)^t$.
   Using $\mathbf{p}^{(0)} = (1, 0, 0, 0)^t$ gives $\mathbf{p}^{(15)} = (0.81649659, 0.40824821, -0.40824837, 3)^t$.
   Using $\mathbf{p}^{(0)} = (1, -1, 1, -1)^t$ gives $\mathbf{p}^{(11)} = (0.57735034, -0.57735022, 0.57735024, 6)^t$.
   The other three solutions are $(0, -0.70710678, -0.70710678, 1)^t$, $(-0.81649659, -0.40824821, 0.40824837, 3)^t$,
   and $(-0.57735034, 0.57735022, -0.57735024, 6)^t$.

7. When $\mathbf{z}$ is orthogonal to $\mathbf{p}^{(1)} - \mathbf{p}^{(0)}$, $(\mathbf{p}^{(1)} - \mathbf{p}^{(0)})^t \mathbf{z} = \mathbf{0}$, so the second term on the right side of Eq. (10.1) is also $\mathbf{0}$. As a consequence, in this case $A_1 \mathbf{z}$ is simply $J(\mathbf{p}^{(0)})\mathbf{z}$.

9. We have

$$
\left(A^{-1} - \frac{A^{-1}\mathbf{xy}^t A^{-1}}{1 + \mathbf{y}^t A^{-1}\mathbf{x}} \right) (A + \mathbf{xy}^t) = A^{-1}A - \frac{A^{-1}\mathbf{xy}^t A^{-1} A}{1 + \mathbf{y}^t A^{-1}\mathbf{x}} + A^{-1}\mathbf{xy}^t - \frac{A^{-1}\mathbf{xy}^t A^{-1}\mathbf{xy}^t}{1 + \mathbf{y}^t A^{-1}\mathbf{x}}
$$

$$
= I - \frac{A^{-1}\mathbf{xy}^t}{1 + \mathbf{y}^t A^{-1}\mathbf{x}} + A^{-1}\mathbf{xy}^t - \frac{A^{-1}\mathbf{xy}^t A^{-1}\mathbf{xy}^t}{1 + \mathbf{y}^t A^{-1}\mathbf{x}}
$$

$$
= I - \frac{A^{-1}\mathbf{xy}^t - A^{-1}\mathbf{xy}^t - \mathbf{y}^t A^{-1}\mathbf{x}A^{-1}\mathbf{xy}^t + A^{-1}\mathbf{xy}^t A^{-1}\mathbf{xy}^t}{1 + \mathbf{y}^t A^{-1}\mathbf{x}}
$$

$$
= I + \frac{\mathbf{y}^t A^{-1}\mathbf{x}A^{-1}\mathbf{xy}^t - \mathbf{y}^t A^{-1}\mathbf{x}(A^{-1}\mathbf{xy}^t)}{1 + \mathbf{y}^t A^{-1}\mathbf{x}} = I.
$$

10.4 The Steepest Decent Method (Page 431)

1. **a.** With $\mathbf{p}^{(0)} = (0, 0)^t$, we have $\mathbf{p}^{(11)} = (0.4943541, 1.948040)^t$.
   b. With $\mathbf{p}^{(0)} = (1, 1)^t$, we have $\mathbf{p}^{(2)} = (0.4970073, 0.8644143)^t$.
   c. With $\mathbf{p}^{(0)} = (2, 2)^t$, we have $\mathbf{p}^{(1)} = (1.736083, 1.804428)^t$.
   d. With $\mathbf{p}^{(0)} = (0, 0)^t$, we have $\mathbf{p}^{(2)} = (-0.3610092, 0.05788368)^t$.

3. **a.** With $\mathbf{p}^{(0)} = (0, 0, 0)^t$, we have $\mathbf{p}^{(14)} = (1.043605, 1.064058, 0.9246118)^t$.
   b. With $\mathbf{p}^{(0)} = (0, 0, 0)^t$, we have $\mathbf{p}^{(9)} = (0.4932739, 0.9863888, -0.5175964)^t$.
   c. With $\mathbf{p}^{(0)} = (0, 0, 0)^t$, we have $\mathbf{p}^{(11)} = (-1.608296, -1.192750, 0.7205642)^t$.
   d. With $\mathbf{p}^{(0)} = (0, 0, 0)^t$, we have $\mathbf{p}^{(1)} = (0, 0.00989056, 0.9890556)^t$.

5. **a.** With $\mathbf{p}^{(0)} = (0, 0)^t$, we have $\mathbf{p}^{(8)} = (3.136548, 0)^t$ and $g(\mathbf{p}^{(8)}) = 0.005057848$.
   b. With $\mathbf{p}^{(0)} = (0, 0)^t$, we have $\mathbf{p}^{(13)} = (0.6157412, 0.3768953)^t$ and $g(\mathbf{p}^{(13)}) = 0.1481574$.
   c. With $\mathbf{p}^{(0)} = (0, 0, 0)^t$, we have $\mathbf{p}^{(5)} = (-0.6633785, 0.3145720, 0.5000740)^t$ and $g(\mathbf{p}^{(5)}) = 0.6921548$.
   d. With $\mathbf{p}^{(0)} = (1, 1, 1)^t$, we have $\mathbf{p}^{(4)} = (0.04022273, 0.01592477, 0.01594401)^t$ and $g(\mathbf{p}^{(4)}) = 1.010003$.

10.5 Homotopy and Continuation Methods (Page 438)

1. **a.** $(3, -2.25)^t$ **b.** $(0.42105263, 2.6184211)^t$ **c.** $(2.173110, -1.3627731)^t$

3. Using $\mathbf{x}(0) = \mathbf{0}$ in all parts gives:
   a. $(0.44006047, 1.8279835)^t$ **b.** $(-0.41342613, 0.096669468)^t$
   c. $(0.49858909, 0.24999091, -0.52067978)^t$ **d.** $(6.1935484, 18.532258, -21.725806)^t$

5. **a.** $(0.49950451, 0.86635691)^t$. This result is comparable since it required only 4 matrix inversions and this accuracy requires 5 iterations of Newton's method.
   b. $(1.7730066, 1.7703057)^t$. This result is comparable since it required only 4 matrix inversions and this accuracy requires 6 iterations of Newton's method.
   c. $(-1.4569217, -1.6645292, 0.42138616)^t$. This result is comparable since it required only 4 matrix inversions and this accuracy requires 5 iterations of Newton's method.
   d. $(0.49813364, -0.19957917, -0.52882773)^t$. This result is comparable since it required only 4 matrix inversions and this accuracy requires 5 iterations of Newton's method.

7. The Continuation method and the Runge-Kutta method of order 4 with $N = 2$ gives the approximation $(0.50024553, 0.078230039, -0.52156996)^t$.

11.2 The Linear Shooting Method (Page 445)

1. The Linear Shooting method gives the following results.

a.

i	x_i	w_{1i}	$y(x_i)$
1	0.5	0.82432432	0.82402714

b.

i	x_i	w_{1i}	$y(x_i)$
1	0.25	0.3937095	0.3936767
2	0.50	0.8240948	0.8240271
3	0.75	1.337160	1.337086

3. The Linear Shooting method gives the following results.

a.

i	x_i	w_{1i}	$y(x_i)$
3	0.3	0.7833204	0.7831923
6	0.6	0.6023521	0.6022801
9	0.9	0.8568906	0.8568760

b.

i	x_i	w_{1i}	$y(x_i)$
5	1.25	0.1676179	0.1676243
10	1.50	0.4581901	0.4581935
15	1.75	0.6077718	0.6077740

c.

i	x_i	w_{1i}	$y(x_i)$
3	0.3	−0.5185754	−0.5185728
6	0.6	−0.2195271	−0.2195247
9	0.9	−0.0406577	−0.0406570

d.

i	x_i	w_{1i}	$y(x_i)$
3	1.3	0.0655336	0.06553420
6	1.6	0.0774590	0.07745947
9	1.9	0.0305619	0.03056208

5. **a.** The Linear Shooting method with $h = 0.1$ gives the following results.

i	x_i	w_{1i}
3	0.3	0.05273437
5	0.5	0.00741571
8	0.8	0.00038976

b. The Linear Shooting method with $h = 0.05$ gives the following results.

i	x_i	w_{1i}
6	0.3	0.04990547
10	0.5	0.00673795
16	0.8	0.00033755

7. **a.** The approximate potential is $u(3) \approx 36.66702$ using $h = 0.1$ cm.

   b. The actual potential is $u(3) = 36.66667$.

9. **a.** There are no solutions if b is an integer multiple of π and $B \neq 0$.

   b. A unique solution exists whenever b is not an integer multiple of π.

   c. There are infinitely many solutions if b is an multiple integer of π and $B = 0$.

11.3 Linear Finite-Difference Methods (Page 451)

1. The Linear Finite-Difference method gives the following results.

a.

i	x_i	w_{1i}	$y(x_i)$
1	0.5	0.83333333	0.82402714

b.

i	x_i	w_{1i}	$y(x_i)$
1	0.25	0.39512472	0.39367669
2	0.50	0.82653061	0.82402714
3	0.75	1.33956916	1.33708613

c. $y(\frac{1}{2}) \approx \frac{1}{3}(4(0.82653061) - 0.83333333) = 0.82426304$

3. The Linear Finite-Difference method gives the following results.

a.

i	x_i	w_i	$y(x_i)$
2	0.2	1.018096	1.0221404
5	0.5	0.5942743	0.59713617
7	0.7	0.6514520	0.65290384

b.

i	x_i	w_i	$y(x_i)$
5	1.25	0.16797186	0.16762427
10	1.50	0.45842388	0.45819349
15	1.75	0.60787334	0.60777401

c.

i	x_i	w_{1i}	$y(x_i)$
3	0.3	−0.5183084	−0.5185728
6	0.6	−0.2192657	−0.2195247
9	0.9	−0.0405748	−0.04065697

d.

i	x_i	w_{1i}	$y(x_i)$
3	1.3	0.0654387	0.0655342
6	1.6	0.0773936	0.0774595
9	1.9	0.0305465	0.0305621

5. The Linear Finite-Difference method gives the following results.

a.

i	x_i	$w_i \, (h = 0.1)$
3	0.3	0.05572807
6	0.6	0.00310518
9	0.9	0.00016516

b.

i	x_i	$w_i \, (h = 0.05)$
6	0.3	0.05132396
12	0.6	0.00263406
18	0.9	0.00013340

7. The deflection at 10-inch intervals using a 1-inch stepsize is given in the table.

i	x_i	w_i
10	10.0	0.1098549
20	20.0	0.1761424
25	25.0	0.1849608
30	30.0	0.1761424
40	40.0	0.1098549

11.4   The Nonlinear Shooting Method (Page 457)

1. The Nonlinear Shooting method gives $w_1 = 0.405505 \approx \ln 1.5 = 0.405465$.

3. The Nonlinear Shooting method gives the following results.

a.

i	x_i	w_{1i}	$y(x_i)$
3	1.3	0.4347934	0.4347826
6	1.6	0.3846363	0.3846154
9	1.9	0.3448586	0.3448276

b.

i	x_i	w_{1i}	$y(x_i)$
3	1.3	2.069249	2.069231
6	1.6	2.225013	2.225000
9	1.9	2.426317	2.426316

c.

i	x_i	w_{1i}	$y(x_i)$
3	2.3	1.2676912	1.2676917
6	2.6	1.3401256	1.3401268
9	2.9	1.4095359	1.4095383

d.

i	x_i	w_{1i}	$y(x_i)$
5	1.25	0.4358290	0.4358272
10	1.50	1.3684496	1.3684447
15	1.75	2.9992010	2.9991909

5.

i	t_i	$w_{1i} \approx y(t_i)$	w_{2i}
3	0.6	0.71682963	0.92122169
5	1.0	1.00884285	0.53467944
8	1.6	1.13844628	−0.11915193

11.5   Nonlinear Finite-Difference Methods (Page 460)

1. The Nonlinear Finite-Difference method gives $w_1 = 0.4067967 \approx \ln 1.5 = 0.4054651$.

3. The Nonlinear Finite-Difference method gives the following results.

a.

i	x_i	w_{1i}	$y(x_i)$
3	1.3	0.4347972	0.4347826
6	1.6	0.3846286	0.3846154
9	1.9	0.3448316	0.3448276

b.

i	x_i	w_{1i}	$y(x_i)$
3	1.3	2.0694081	2.0692308
6	1.6	2.2250937	2.2250000
9	1.9	2.4263387	2.4263158

c.

i	x_i	w_{1i}	$y(x_i)$
3	2.3	1.2677078	1.2676917
6	2.6	1.3401418	1.3401268
9	2.9	1.4095432	1.4095383

d.

i	x_i	w_{1i}	$y(x_i)$
5	1.25	0.4345979	0.4358273
10	1.50	1.3662119	1.3684447
15	1.75	2.9969339	2.9991909

5.

i	x_i	w_i
5	30	0.01028080
10	60	0.01442767
15	90	0.01028080

Since the results are the same, adding the nonlinear term to the differential equation makes no difference.

11.6 Variational Techniques (Page 472)

1. The Piecewise Linear method gives $\phi(x) = -0.07713274\phi_1(x) - 0.07442678\phi_2(x)$. As a consequence, $\phi(x_1) = -0.07713274$ and $\phi(x_2) = -0.07442678$. The actual values are $y(x_1) = -0.07988545$ and $y(x_2) = -0.07712903$.

3. The Piecewise Linear method gives the following results.

a.

i	x_i	$\phi(x_i)$	$y(x_i)$
3	0.3	−0.212333	−0.21
6	0.6	−0.241333	−0.24
9	0.9	−0.090333	−0.09

b.

i	x_i	$\phi(x_i)$	$y(x_i)$
3	0.3	0.1815138	0.1814273
6	0.6	0.1805502	0.1804754
9	0.9	0.05936468	0.05934303

c.

i	x_i	$\phi(x_i)$	$y(x_i)$
5	0.25	−0.3585989	−0.3585641
10	0.50	−0.5348383	−0.5347803
15	0.75	−0.4510165	−0.4509614

d.

i	x_i	$\phi(x_i)$	$y(x_i)$
5	0.25	−0.1846134	−0.1845204
10	0.50	−0.2737099	−0.2735857
15	0.75	−0.2285169	−0.2284204

5. The Cubic B-Spline method gives the following results.

a.

i	x_i	$\phi(x_i)$	$y(x_i)$
3	0.3	−0.2100000	−0.21
6	0.6	−0.2400000	−0.24
9	0.9	−0.0900000	−0.09

b.

i	x_i	$\phi(x_i)$	$y(x_i)$
3	0.3	0.1814269	0.1814273
6	0.6	0.1804753	0.1804754
9	0.9	0.05934321	0.05934303

c.

i	x_i	$\phi(x_i)$	$y(x_i)$
5	0.25	−0.3585639	−0.3585641
10	0.50	−0.5347779	−0.5347803
15	0.75	−0.4509109	−0.4509614

d.

i	x_i	$\phi(x_i)$	$y(x_i)$
5	0.25	−0.1845191	−0.1845204
10	0.50	−0.2735833	−0.2735857
15	0.75	−0.2284186	−0.2284204

7. Exercise 6 and the Piecewise Linear method gives the following results.

i	x_i	$\phi(x_i)$	$y(x_i)$
3	0.3	1.0408182	1.0408182
6	0.6	1.1065307	1.1065306
9	0.9	1.3065697	1.3065697

9. A change in variable $w = (x - a)/(b - a)$ gives the boundary value problem

$$-\frac{d}{dw}(p((b-a)w+a)y') + (b-a)^2 q((b-a)w+a)y = (b-a)^2 f((b-a)w+a),$$

where $0 < w < 1$, $y(0) = \alpha$, and $y(1) = \beta$. Then Exercise 6 can be used.

11. Let $\mathbf{c} = (c_1, \dots, c_n)^t$ be any vector and let $\phi(x) = \sum_{j=1}^{n} c_j \phi_j(x)$. Then

$$\mathbf{c}^t A \mathbf{c} = \sum_{i=1}^{n} \sum_{j=1}^{n} a_{ij} c_i c_j = \sum_{i=1}^{n} \sum_{j=i-1}^{i+1} a_{ij} c_i c_j$$

$$= \sum_{i=1}^{n} \left[\int_0^1 \{ p(x) c_i \phi_i'(x) c_{i-1} \phi_{i-1}'(x) + q(x) c_i \phi_i(x) c_{i-1} \phi_{i-1}(x) \} \, dx \right.$$

$$+ \int_0^1 \{ p(x) c_i^2 [\phi_i'(x)]^2 + q(x) c_i^2 [\phi_i'(x)]^2 \} \, dx$$

$$\left. + \int_0^1 \{ p(x) c_i \phi_i'(x) c_{i+1} \phi_{i+1}'(x) + q(x) c_i \phi_i(x) c_{i+1} \phi_{i+1}(x) \} \, dx \right]$$

$$= \int_0^1 \{ p(x) [\phi'(x)]^2 + q(x) [\phi(x)]^2 \} \, dx.$$

So $\mathbf{c}^t A \mathbf{c} \geq 0$ with equality only if $\mathbf{c} = \mathbf{0}$. Since A is also symmetric, A is positive definite.

12.2 Elliptical Problems (Page 482)

1. The Poisson Equation Finite-Difference method gives the following results.

i	j	x_i	y_j	$w_{i,j}$	$u(x_i, y_j)$
1	1	0.5	0.5	0.0	0
1	2	0.5	1.0	0.25	0.25
1	3	0.5	1.5	1.0	1

3. The Poisson Equation Finite-Difference method gives the following results.

a. 30 iterations required:

i	j	x_i	y_j	$w_{i,j}$	$u(x_i, y_j)$
2	2	0.4	0.4	0.1599988	0.16
2	4	0.4	0.8	0.3199988	0.32
4	2	0.8	0.4	0.3199995	0.32
4	4	0.8	0.8	0.6399996	0.64

b. 29 iterations required:

i	j	x_i	y_j	$w_{i,j}$	$u(x_i, y_j)$
2	1	1.256637	0.3141593	0.2951855	0.2938926
2	3	1.256637	0.9424778	0.1830822	0.1816356
4	1	2.513274	0.3141593	−0.7721948	−0.7694209
4	3	2.513274	0.9424778	−0.4785169	−0.4755283

c. 126 iterations required:

i	j	x_i	y_j	$w_{i,j}$	$u(x_i, y_j)$
4	3	0.8	0.3	1.2714468	1.2712492
4	7	0.8	0.7	1.7509419	1.7506725
8	3	1.6	0.3	1.6167917	1.6160744
8	7	1.6	0.7	3.0659184	3.0648542

d. 127 iterations required:

i	j	x_i	y_j	$w_{i,j}$	$u(x_i, y_j)$
2	2	1.2	1.2	0.5251533	0.5250861
4	4	1.4	1.4	1.3190830	1.3189712
6	6	1.6	1.6	2.4065150	2.4064186
8	8	1.8	1.8	3.8088995	3.8088576

5. The approximate potential at some typical points is given in the following table.

i	j	x_i	y_j	$w_{i,j}$
1	4	0.1	0.4	88
2	1	0.2	0.1	66
4	2	0.4	0.2	66

12.3 Parabolic Problems (Page 494)

1. The Heat Equation Backward-Difference method gives the following results.

a.

i	j	x_i	t_j	$w_{i,j}$	$u(x_i, t_j)$
1	1	0.5	0.05	0.632952	0.652037
2	1	1.0	0.05	0.895129	0.883937
3	1	1.5	0.05	0.632952	0.625037
1	2	0.5	0.1	0.566574	0.552493
2	2	1.0	0.1	0.801256	0.781344
3	2	1.5	0.1	0.566574	0.552493

b.

i	j	x_i	t_j	$w_{i,j}$	$u(x_i, t_j)$
1	1	1/3	0.05	1.59728	1.53102
2	1	2/3	0.05	−1.59728	−1.53102
1	2	1/3	0.1	1.47300	1.35333
2	2	2/3	0.1	−1.47300	−1.35333

3. The Forward-Difference method gives the following results.

a. (i) For $h = 0.4$ and $k = 0.1$:

i	j	x_i	t_j	$w_{i,j}$	$u(x_i, t_j)$
2	5	0.8	0.5	3.035630	0
3	5	1.2	0.5	−3.035630	0
4	5	1.6	0.5	1.876122	0

(ii) For $h = 0.4$ and $k = 0.05$:

i	j	x_i	t_j	$w_{i,j}$	$u(x_i, t_j)$
2	10	0.8	0.5	0	0
3	10	1.2	0.5	0	0
4	10	1.6	0.5	0	0

b. For $h = \pi/10$ and $k = 0.05$:

i	j	x_i	t_j	$w_{i,j}$	$u(x_i, t_j)$
3	10	0.94247780	0.5	0.4864823	0.4906936
6	10	1.88495559	0.5	0.5718943	0.5768449
9	10	2.82743339	0.5	0.1858197	0.1874283

c. For $h = 0.2$ and $k = 0.04$:

i	j	x_i	t_j	$w_{i,j}$	$u(x_i, t_j)$
4	10	0.8	0.4	1.166149	1.169362
8	10	1.6	0.4	1.252413	1.254556
12	10	2.4	0.4	0.4681813	0.4665473
16	10	3.2	0.4	−0.1027637	−0.1056622

d. For $h = 0.1$ and $k = 0.04$:

i	j	x_i	t_j	$w_{i,j}$	$u(x_i, t_j)$
3	10	0.3	0.4	0.5397009	0.5423003
6	10	0.6	0.4	0.6344565	0.6375122
9	10	0.9	0.4	0.2061474	0.2071403

5. The Crank-Nicolson method gives the following results.

a. For $h = 0.4$ and $k = 0.1$:

i	j	x_i	t_j	$w_{i,j}$	$u(x_i, t_j)$
2	5	0.8	0.5	8.2×10^{-7}	0
3	5	1.2	0.5	-8.2×10^{-7}	0
4	5	1.6	0.5	5.1×10^{-7}	0

For $h = 0.4$ and $k = 0.05$:

i	j	x_i	t_j	$w_{i,j}$	$u(x_i, t_j)$
2	10	0.8	0.5	-2.6×10^{-6}	0
3	10	1.2	0.5	2.6×10^{-6}	0
4	10	1.6	0.5	-1.6×10^{-6}	0

b. For $h = \frac{\pi}{10}$ and $k = 0.05$:

i	j	x_i	t_j	$w_{i,j}$	$u(x_i, t_j)$
3	10	0.94247780	0.5	0.4926589	0.4906936
6	10	1.88495559	0.5	0.5791553	0.5768449
9	10	2.82743339	0.5	0.1881790	0.1874283

c. For $h = 0.2$ and $k = 0.04$:

i	j	x_i	t_j	$w_{i,j}$	$u(x_i, t_j)$
4	10	0.8	0.4	1.171532	1.169362
8	10	1.6	0.4	1.256005	1.254556
12	10	2.4	0.4	0.4654499	0.4665473
16	10	3.2	0.4	-0.1076139	-0.1056622

d. For $h = 0.1$ and $k = 0.04$:

i	j	x_i	t_j	$w_{i,j}$	$u(x_i, t_j)$
3	10	0.3	0.4	0.5440532	0.5423003
6	10	0.6	0.4	0.6395728	0.6375122
9	10	0.9	0.4	0.2078098	0.2071403

7. For the Modified Backward-Difference method, we have

i	j	x_i	y_j	$w_{i,j}$
3	25	0.3	0.25	0.2883460
5	25	0.5	0.25	0.3468410
8	25	0.8	0.25	0.2169217

9. For the Modified Backward-Difference method, we have

i	j	x_i	y_j	$w_{i,j}$
2	10	0.3	0.225	1.207730
5	10	0.75	0.225	1.836564
9	10	1.35	0.225	0.6928342

12.4 *Hyperbolic Problems (Page 501)*

1. The Wave Equation Finite-Difference method gives the following results.

i	j	x_i	t_j	$w_{i,j}$	$u(x_i, t_j)$
2	4	0.25	1.0	-0.7071068	-0.7071068
3	4	0.50	1.0	-1.0000000	-1.0000000
4	4	0.75	1.0	-0.7071068	-0.7071068

3. a. The Finite-Difference method with $h = \frac{\pi}{10}$ and $k = 0.05$ gives the following results.

i	j	x_i	t_j	$w_{i,j}$	$u(x_i, t_j)$
2	10	$\frac{\pi}{5}$	0.5	0.5163933	0.5158301
5	10	$\frac{\pi}{2}$	0.5	0.8785407	0.8775826
8	10	$\frac{4\pi}{5}$	0.5	0.5163933	0.5158301

b. The Finite-Difference method with $h = \frac{\pi}{20}$ and $k = 0.1$ gives the following results.

i	j	x_i	t_j	$w_{i,j}$
4	5	$\frac{\pi}{5}$	0.5	0.5159163
10	5	$\frac{\pi}{2}$	0.5	0.8777292
16	5	$\frac{4\pi}{5}$	0.5	0.5159163

c. The Finite-Difference method with $h = \frac{\pi}{20}$ and $k = 0.05$ gives the following results.

i	j	x_i	t_j	$w_{i,j}$
4	10	0.62831853	0.5	0.5159602
10	10	1.57079633	0.5	0.8778039
16	10	2.51327412	0.5	0.5159602

5. The Finite-Difference method gives the following results.

i	j	x_i	t_j	$w_{i,j}$
2	5	0.2	0.5	-1
5	5	0.5	0.5	0
8	5	0.8	0.5	1

7. Approximate voltages and currents are given in the following table.

i	j	x_i	t_j	Voltage	Current
5	2	50	0.2	77.769	3.88845
12	2	120	0.2	104.60	-1.69931
18	2	180	0.2	33.986	-5.22995
5	5	50	0.5	77.702	3.88510
12	5	120	0.5	104.51	-1.69785
18	5	180	0.5	33.957	-5.22453

12.5 Finite-Element Methods (Page 515)

1. With $E_1 = (0.25, 0.75)$, $E_2 = (0, 1)$, $E_3 = (0.5, 0.5)$, and $E_4 = (0, 0.5)$, the basis functions are

$$\phi_1(x, y) = \begin{cases} 4x & \text{on } T_1 \\ -2 + 4y & \text{on } T_2, \end{cases} \qquad \phi_2(x, y) = \begin{cases} -1 - 2x + 2y & \text{on } T_1 \\ 0 & \text{on } T_2, \end{cases}$$

$$\phi_3(x, y) = \begin{cases} 0 & \text{on } T_1 \\ 1 + 2x - 2y & \text{on } T_2, \end{cases} \quad \text{and} \quad \phi_4(x, y) = \begin{cases} 2 - 2x - 2y & \text{on } T_1 \\ 2 - 2x - 2y & \text{on } T_2. \end{cases}$$

and $\gamma_1 = 0.323825$, $\gamma_2 = 0$, $\gamma_3 = 1.0000$, and $\gamma_4 = 0$.

3. The Finite-Element method with $K = 8$, $N = 8$, $M = 32$, $n = 9$, $m = 25$, and $NL = 0$ and the element numbering in the figure gives the following results.

$$\gamma_1 = 0.511023 \qquad\qquad\qquad \gamma_2 = 0.720476$$

$$\gamma_3 = 0.507899 \qquad\qquad\qquad \gamma_4 = 0.720476$$

$$\gamma_5 = 1.01885 \qquad\qquad\qquad \gamma_6 = 0.720476$$

$$\gamma_7 = 0.507896 \qquad\qquad\qquad \gamma_8 = 0.720476$$

$$\gamma_9 = 0.511023 \qquad\qquad\qquad \gamma_i = 0, \quad 10 \le i \le 25$$

$$u(0.125, 0.125) \approx 0.614187, \qquad\qquad u(0.125, 0.25) \approx 0.690343,$$

$$u(0.25, 0.125) \approx 0.690343 \qquad\qquad \text{and} \quad u(0.25, 0.25) \approx 0.720476.$$

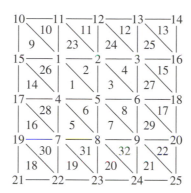

5. The Finite-Element method with $K = 0, N = 12, M = 32, n = 20, m = 27$, and $NL = 14$ and the element numbering in the figure gives the following results.

$\gamma_1 = 21.40335,$	$\gamma_8 = 24.19855,$	$\gamma_{15} = 20.23334,$	$\gamma_{22} = 15,$
$\gamma_2 = 19.87372,$	$\gamma_9 = 24.16799,$	$\gamma_{16} = 20.50056,$	$\gamma_{23} = 15,$
$\gamma_3 = 19.10019,$	$\gamma_{10} = 27.55237,$	$\gamma_{17} = 21.35070,$	$\gamma_{24} = 15,$
$\gamma_4 = 18.85895,$	$\gamma_{11} = 25.11508,$	$\gamma_{18} = 22.84663,$	$\gamma_{25} = 15,$
$\gamma_5 = 19.08533,$	$\gamma_{12} = 22.92824,$	$\gamma_{19} = 24.98178,$	$\gamma_{26} = 15,$
$\gamma_6 = 19.84115,$	$\gamma_{13} = 21.39741,$	$\gamma_{20} = 27.41907,$	$\gamma_{27} = 15,$
$\gamma_7 = 21.34694,$	$\gamma_{14} = 20.52179,$	$\gamma_{21} = 15.$	

We have

$$u(1, 0) \approx 22.92824, \quad u(4, 0) \approx 22.84663, \quad u\left(\frac{5}{2}, \frac{\sqrt{3}}{2}\right) \approx 18.85895.$$

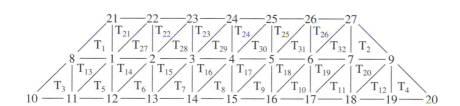

Index